Current and Future Developments in Food Science

(*Volume 3*)

Maria Luisa Marina (Ed.)

Green Extraction Techniques in Food Analysis

Edited by

Merichel Plaza & María Luisa Marina

Department of Analytical Chemistry
Physical Chemistry and Chemical Engineering
University of Alcalá, Alcalá de Henares, Madrid, Spain

Current and Future Developments in Food Science

Series Editor: Maria Luisa Marina

(Volume 3)

Green Extraction Techniques in Food Analysis

Editors: Merichel Plaza & María Luisa Marina

ISSN (Online): 2705-0815

ISSN (Print): 2705-0807

ISBN (Online): 978-981-5049-45-9

ISBN (Print): 978-981-5049-46-6

ISBN (Paperback): 978-981-5049-47-3

First published in 2023.

Bentham Science Publishers Pte. Ltd.
80 Robinson Road #02-00
Singapore 068898
Singapore
Email: subscriptions@benthamscience.net

BENTHAM SCIENCE

CONTENTS

PREFACE

In the last year, society has increased its interest in green technology due to higher awareness about the environment. Analytical chemistry plays a valuable role in the sustainable development of the planet. It is not just for pollutant monitoring in the environment, but also for the development of more sustainable processes. In this sense, the replacement of existing extraction techniques with more environmentally sustainable extraction techniques is an important step to the final greenness of the entire process. Also in food analysis, it is one of the main strategies to make it greener. One of the main aspects to be considered when processing to change from a current extraction method to a greener one is the choice of solvent. The degree of environmental impact changes according to the type of solvent used because it depends on the way natural resources are harvested, energy usage, and emissions to air and water from the production and the use of solvents, transportation, and disposal or recycling.

This book is composed of 14 chapters providing the reader with the last trends in the development and applications of green extraction techniques in food analysis. Chapter 1 presents the current state of the art of Green Analytical Chemistry and its main strategies for improving the sustainability of analytical methods, reducing their environmental impact, and offering solutions to the needs that arise from food analysis. Chapter 2 introduces the principles of green extraction techniques and makes an overview of the main green extraction techniques (such as microwave-assisted extraction, ultrasound-assisted extraction, subcritical water extraction, and supercritical fluid extraction). Chapter 3 provides an overview of the environmentally friendly solvents most commonly used in food analysis (including water, carbon dioxide, ethanol, ionic liquids, natural deep eutectic solvents, surfactants, and switchable solvents), in which the advantages and limitations of the use of these solvents in food analysis from the perspective of green analytical chemistry are described. In addition, some applications of the combination of environmentally friendly solvents with assisted extraction techniques and miniaturized techniques are discussed. In Chapters 4-6, recent applications of deep eutectic solvents, ionic liquids, and supramolecular solvents, respectively, in food analysis are described in depth. While Chapters 7-14 are focused on the applications of supercritical fluid extraction, gas-expanded liquids extraction, pressurized liquid extraction, microwave-assisted extraction, enzyme-assisted extraction, pulsed electric field, high-voltage electrical discharges, and high hydrostatic pressure in food analysis, respectively. Also, the main fundamentals/principles, instrumental setups, as well as, the advantages and drawbacks of the use of these green extraction techniques are broadly explained and detailed in these chapters.

This book is intended for a large audience comprising Ph.D. students and experts working on different topics related to food analysis with new technological and methodological approaches to make extraction techniques greener and more sustainable.

We would like to sincerely thank all authors who have collaborated in writing this book for their excellent contributions. We wish to express our scientific gratitude to them for sharing all their knowledge and experience on the development of more efficient green extraction processes.

Merichel Plaza & María Luisa Marina
Department of Analytical Chemistry
Physical Chemistry and Chemical Engineering
University of Alcalá, Alcalá de Henares, Madrid, Spain

List of Contributors

Ádina L. Santana	Grain Science and Industry Department, Kansas State University, 1301 N Mid Campus Drive, Manhattan, KS 66506, USA
Álvaro Santana-Mayor	Departamento de Química, Unidad Departamental de Química Analítica, Facultad de Ciencias, Universidad de La Laguna (ULL). Avenida Astrofísico Francisco Sánchez, s/nº. 38206 San Cristóbal de La Laguna (Tenerife), España
Adnane Remmal	Department of Biology, Faculty of Science Dhar El-Mahraz, University Sidi Mohammed Ben Abdellah, P.O. Box 1796, Fez30050, Morocco
Antonio V. Herrera-Herrera	Instituto Universitario de Bio-Orgánica Antonio González. Universidad de La Laguna (ULL), Avenida Astrofísico Francisco Sánchez, 2. 38206 San Cristóbal de La Laguna (Tenerife), España
Alfonso Jiménez	Department of Analytical Chemistry, Nutrition & Food Sciences, University of Alicante, Campus San Vicente, ES-03690, San Vicente del Raspeig (Alicante), Spain
Bárbara Socas-Rodríguez	Departamento de Química, Unidad Departamental de Química Analítica, Facultad de Ciencias, Universidad de La Laguna (ULL), Avenida Astrofísico Francisco Sánchez, s/no. 38206 San Cristóbal de La Laguna (Tenerife), España
Bienvenida Gilbert López	Analytical Chemistry Research Group (FQM-323), Department of Physical and Analytical Chemistry, University of Jaen, 23071 Jaen, Spain
Carlos Javier Pelegrín	Department of Analytical Chemistry, Nutrition & Food Sciences, University of Alicante, Campus San Vicente, ES-03690, San Vicente del Raspeig (Alicante), Spain
Diego Ballesteros-Vivas	Departamento de Nutrición y Bioquímica, Facultad de Ciencias, Pontificia Universidad Javeriana, Bogotá, Colombia
Elena Ibáñez	Foodomics Laboratory, Institute of Food Science Research (CIAL, CSIC), Nicolas Cabrera, 9, 28049 Madrid, Spain
Ester Hernández-Corroto	Universidad de Alcalá, Departamento de Química Analítica, Química Física e Ingeniería Química, Ctra. Madrid-Barcelona Km. 33.600, 28871 Alcalá de Henares (Madrid), Spain
Ergin Murat Altuner	Department of Biology, Faculty of Science, Kastamonu University, Kastamonu, Turkey
Eugène Vorobiev	Sorbonne Universités, Université de Technologie de Compiègne, Laboratoire Transformations Intégrées de la Matière Renouvelable (UTC/ESCOM, EA 4297 TIMR), Centre de Recherches de Royallieu, CS 60 319, 60 203 Compiègne Cedex, France
Fabián Parada Alfonso	Food Chemistry Research Group, Departamento de Química, Facultad de Ciencias, Universidad Nacional de Colombia, Bogotá, Colombia
Farid Chemat	Avignon University, INRAE, UMR 408, Green Extraction Team, F-84000 Avignon, France

Francesc A. Esteve-Turrillas Departamento de Química Analítica, Universidad de Valencia. Edificio Jeroni Muñoz, Calle Doctor Moliner 50. 46100 Burjassot (Valencia), España

Herminia Domínguez CINBIO, Centro de Investigaciones Biomédicas, Departamento de Enxeñería Química, Facultade de Ciencias, Universidade de Vigo, Campus Ourense, Edificio Politécnico, As Lagoas, 32004 Ourense, Spain

Jenny P. Ortega-Barbosa Food Chemistry Research Group, Departamento de Química, Facultad de Ciencias, Universidad Nacional de Colombia, Bogotá, Colombia

Jose A. Mendiola Foodomics Laboratory, Institute of Food Science Research (CIAL, CSIC), Nicolas Cabrera, 9, 28049 Madrid, Spain

Juliane Viganó Centro de Ciências da Natureza, Universidade Federal de São Carlos, Rod. Lauri Simões de Barros, km 12 – SP 189, Buri, SP 18290-000, Brazil

Juan S. Ortiz-Ramírez Food Chemistry Research Group, Departamento de Ingeniería Química y Ambiental, Facultad de Ingeniería, Universidad Nacional de Colombia, Bogotá, Colombia

Lidia Montero Applied Analytical Chemistry, University of Duisburg-Essen, Universitätsstrasse 5, 45141 Essen, Germany
Teaching and Research Center for Separation, University of Duisburg-Essen, Universitätsstrasse 5, 45141 Essen, Germany

Larry O. Chañi-Paucar School of Food Engineering, University of Campinas, Rua Monteiro Lobato, 80, Campinas, Brazil
Escuela Profesional de Ingeniería Agroindustrial, Universidad Nacional Autónoma Altoandina de Tarma (UNAAT), Jr. Huaraz 431, 12651, Junín, Perú

Luis I. Rodríguez-Varela Food Chemistry Research Group, Departamento de Ingeniería Química y Ambiental, Facultad de Ingeniería, Universidad Nacional de Colombia, Bogotá, Colombia

Lucía López Hortas CINBIO, Centro de Investigaciones Biomédicas, Departamento de Enxeñería Química, Facultade de Ciencias, Universidade de Vigo, Campus Ourense, Edificio Politécnico, As Lagoas, 32004 Ourense, Spain

M. Angela A. Meireles School of Food Engineering, University of Campinas, Rua Monteiro Lobato, 80, Campinas, Brazil

Malak Tabib Department of Biology, Faculty of Science, University Sidi Mohammed Ben Abdellah, Dhar El-Mahraz Campus, P.O. Box 1796, Fez 30050, Morocco
Avignon University, INRAE, UMR 408, Green Extraction Team, F-84000 Avignon, France

Miguel de la Guardia Departamento de Química Analítica, Universidad de Valencia. Edificio Jeroni Muñoz, Calle Doctor Moliner 50. 46100 Burjassot (Valencia), España

Miguel Ángel Rodríguez-Delgado Departamento de Química, Unidad Departamental de Química Analítica, Facultad de Ciencias, Universidad de La Laguna (ULL). Avenida Astrofísico Francisco Sánchez, s/nº. 38206 San Cristóbal de La Laguna (Tenerife), España

María Carmen Garrigós Department of Analytical Chemistry, Nutrition & Food Sciences, University of Alicante, Campus San Vicente, ES-03690, San Vicente del Raspeig (Alicante), Spain

Milena Álvarez Viñas CINBIO, Centro de Investigaciones Biomédicas, Departamento de Enxeñería Química, Facultade de Ciencias, Universidade de Vigo, Campus Ourense, Edificio Politécnico, As Lagoas, 32004 Ourense, Spain

María Dolores Torres CINBIO, Centro de Investigaciones Biomédicas, Departamento de Enxeñería Química, Facultade de Ciencias, Universidade de Vigo, Campus Ourense, Edificio Politécnico, As Lagoas, 32004 Ourense, Spain

María del Prado García Aparicio Universidad de Alcalá, Departamento de Química Analítica, Química Física e Ingeniería Química, Ctra. Madrid-Barcelona Km. 33.600, 28871 Alcalá de Henares (Madrid), Spain
Universidad de Alcalá, Instituto de Investigación Química "Andrés M. del Río", Ctra. Madrid-Barcelona Km. 33.600, 28871 Alcalá de Henares (Madrid), Spain

María Luisa Marina Universidad de Alcalá, Departamento de Química Analítica, Química Física e Ingeniería Química, Ctra. Madrid-Barcelona Km. 33.600, 28871 Alcalá de Henares (Madrid), Spain
Universidad de Alcalá, Instituto de Investigación Química "Andrés M. del Río", Ctra. Madrid-Barcelona Km. 33.600, 28871 Alcalá de Henares (Madrid), Spain

María Concepción García Universidad de Alcalá, Departamento de Química Analítica, Química Física e Ingeniería Química, Ctra. Madrid-Barcelona Km. 33.600, 28871 Alcalá de Henares (Madrid), Spain
Universidad de Alcalá, Instituto de Investigación Química "Andrés M. del Río", Ctra. Madrid-Barcelona Km. 33.600, 28871 Alcalá de Henares (Madrid), Spain

Njara Rakotomanomana Avignon University, INRAE, UMR 408, Green Extraction Team, F-84000 Avignon, France

Noelia Caballero-Casero Department of Analytical Chemistry, Institute of Chemistry for Energy and Environment, Edificio Anexo Marie Curie. Campus de Rabanales, University of Córdoba, 14071-Córdoba, Spain

Norelhouda Abderrezag Foodomics Laboratory, Institute of Food Science Research (CIAL, CSIC), Nicolas Cabrera, 9, 28049 Madrid, Spain
Laboratory of Environmental process engineering, Faculty of Process Engineering, University Salah Boubnider Constantine3, Nouvelle Ville Constantine, 25000 Constantine, Algeria

Natalie Álvarez Alarcon Food Chemistry Research Group, Departamento de Química, Facultad de Ciencias, Universidad Nacional de Colombia, Bogotá, Colombia

Noelia Flórez Fernández CINBIO, Centro de Investigaciones Biomédicas, Departamento de Enxeñería Química, Facultade de Ciencias, Universidade de Vigo, Campus Ourense, Edificio Politécnico, As Lagoas, 32004 Ourense, Spain

Nadia Boussetta Sorbonne Universités, Université de Technologie de Compiègne, Laboratoire Transformations Intégrées de la Matière Renouvelable (UTC/ESCOM, EA 4297 TIMR), Centre de Recherches de Royallieu, CS 60 319, 60 203 Compiègne Cedex, France

Paula Rodríguez-Seoane CINBIO, Centro de Investigaciones Biomédicas, Departamento de Enxeñería Química, Facultade de Ciencias, Universidade de Vigo, Campus Ourense, Edificio Politécnico, As Lagoas, 32004 Ourense, Spain

Priscilla Rocío Bautista Analytical Chemistry Research Group (FQM-323), Department of Physical and Analytical Chemistry, University of Jaen, 23071 Jaen, Spain

Priscilla C. Veggi Department of Chemical Engineering, Institute of Environmental, Chemical and Pharmaceutical Sciences, Universidade Federal de São Paulo, R. São Nicolau 210, Diadema, Brazil

Ruth Rodríguez-Ramos Departamento de Química, Unidad Departamental de Química Analítica, Facultad de Ciencias, Universidad de La Laguna (ULL). Avenida Astrofísico Francisco Sánchez, s/nº. 38206 San Cristóbal de La Laguna (Tenerife), España

Salvador Garrigues Departamento de Química Analítica, Universidad de Valencia. Edificio Jeroni Muñoz, Calle Doctor Moliner 50. 46100 Burjassot (Valencia), España,

Sergio Armenta Departamento de Química Analítica, Universidad de Valencia. Edificio Jeroni Muñoz, Calle Doctor Moliner 50. 46100 Burjassot (Valencia), España,

Soledad Rubio Department of Analytical Chemistry, Institute of Chemistry for Energy and Environment, Edificio Anexo Marie Curie. Campus de Rabanales, University of Córdoba, 14071-Córdoba, Spain

Vanesa Sanz CINBIO, Centro de Investigaciones Biomédicas, Departamento de Enxeñería Química, Facultade de Ciencias, Universidade de Vigo, Campus Ourense, Edificio Politécnico, As Lagoas, 32004 Ourense, Spain

Zully J. Suárez Montenegro Foodomics Laboratory, Institute of Food Science Research (CIAL, CSIC), Nicolas Cabrera, 9, 28049 Madrid, Spain
Facultad de Ingeniería Agroindustrial Universidad de Nariño, C/18 Cra. 50 Ciudadela Universitaria Torobajo, 520001 Pasto, Colombia

CHAPTER 1

Green Analytical Chemistry

Miguel de la Guardia[1,*], **Sergio Armenta**[1], **Francesc A. Esteve-Turrillas**[1] and **Salvador Garrigues**[1]

[1] *Departamento de Química Analítica, Universidad de Valencia. Edificio Jeroni Muñoz, Calle Doctor Moliner 50. 46100 Burjassot (Valencia), España*

Abstract: Food analysis demands are mandatory from quality, safety, and authentication point of view, and there is an increase in analytical activity in both the control laboratory and research and development. This chapter presents the current state-of-the-art of Green Analytical Chemistry and its main strategies for improving the sustainability of analytical methods, reducing their environmental impact, and offering solutions to the needs that arise from food analysis. Direct analysis is presented as the ideal method that avoids the use of solvents or reagents and the generation of waste. Miniaturization, automation, and the use of sustainable solvents, in addition to reducing energy consumption, are the basic strategies that allow us to achieve the objectives of Green Analytical Chemistry. The reduction of single-use plastic laboratory material and their waste has also been considered an objective for analytical method greenness.

Keywords: Agro-solvents, assisted extraction, Automation, Bio-solvents, Chemical imaging, Direct analysis, Eco-scale, Energy consumption, Food authentication, Food quality, Food safety, Green features, Greenness, Microextraction, Miniaturization, Plastic waste, Solvent consumption.

INTRODUCTION

Green Analytical Chemistry is a challenging strategy focused on the modification of conventional analytical methods in order to avoid or reduce the deleterious effects on both, users and the environment [1]. This green trend is of particular importance in the food analysis area, where an extremely high number of samples must be daily analysed all around the world in order to assess the food quality and food safety of raw and manufactured products. Moreover, food analysis is a diverse task and it may involve the determination of physico-chemical parameters and macronutrients, to the determination of micronutrients, bioactive components, and residues of pesticides. It also involves the detection of food adultera-

* **Corresponding Author Miguel de la Guardia:** Departamento de Química Analítica, Universidad de Valencia, Edificio Jeroni Muñoz, Calle Doctor Moliner 50, 46100 Burjassot (Valencia), España; Tel: +34 963544838; E-mail: miguel.delaguardia@uv.es

Merichel Plaza & Maria Luisa Marina (Eds.)

tion, fraud, and the geographical origin of food products. Most of these studies are carried out by using reference methods and guidelines from the *Codex Alimentarius* Commission of the Joint FAO/WHO Food Standards Programme [2], the Association of Official Agricultural Chemists (AOAC) [3], and European Food Safety Authority (EFSA) [4], which typically use conventional analytical techniques that are characterized by long analysis time, laborious sample preparation, and high consumption of reagents and solvents that very often involve the generation of toxic wastes [5].

The concept of Green Analytical Chemistry was firstly proposed by de la Guardia and Ruzicka in 1995 with the novel idea of environmentally conscientious Analytical Chemistry through miniaturization, containment, and reagent replacement [6]. Later in 1999, de la Guardia proposed the "integrated environmentally friendly approach", considering the side effects of chemical measurements that can be reduced by using new strategies for sampling, sample treatment, and chemometrics [7]. In this frame, methodologies like in-field sampling, on-line analysis, microwave-assisted treatment, automation through flow analysis, decontamination or passivation processes, and surface analysis were proposed in order to achieve excellent analytical figures of merit, but also considering external factors such as environmental safety, health, and social problems. Since that time, the Green Analytical Chemistry concept has been expanded and widely discussed in several books authored by researchers, such as Anastas in 1999 [8], Koel and Kaljurand in 2010 [9], de la Guardia and Armenta in 2011 [10], de la Guardia and Garrigues in 2012 [1], and, more recently, Płotka-Wasylka and Namieśnik in 2019 [11].

Since its inception, Green Analytical Chemistry concept has been running in parallel to Green Chemistry, focusing on chemical analysis and processes, respectively. Thus, the 12 principles of Green Chemistry defined by Anastas in 1998 [12] were adapted to an analytical focus by Galuzska *et al.* in 2013 [13]. The 12 principles of Green Analytical Chemistry are shown in Fig. (**1**), and they summarize the diverse trends proposed to reduce the environmental impact of analytical procedures. From a Food Analysis perspective, the main principles to have into account are those related to the implementation of *in situ* measurements and direct analytical techniques in order to avoid or reduce sample treatment, the reduction of the number of samples to be analysed, the use of automated and miniaturized methodologies, and the preferential use of multianalyte methods, improving in all cases the level of information obtained from sample measurements.

S	Select direct analytical technique
I	Integrate direct analytical processes and operations
G	Generate as little waste as possible and treat it properly
N	Never waste energy
I	Implement automation and miniaturization of methods
F	Favor reagents obtained from renewable source
I	Increase safety for operator
C	Carry out in-situ measurements
A	Avoid derivatization
N	Note that the sample number and size should be minimal
C	Choose multianalyte or multiparameter method
E	Eliminate or replace toxic reagents

Fig. (1). The 12 principles of Green Analytical Chemistry (adapted from reference [13]).

During these years, several criteria have been proposed to evaluate the green character of an analytical method, such as: the National Environmental Methods Index (NEMI) [14], Green Assessment Profile [15], color scale adapted-NEMI [16], penalty points and Eco-scale [17], Green Motion tool [18], E-factor [19], Green Certificate [20], and Green Analytical Procedure Index (GAPI) [21]. However, today there is not still a common agreement among analytical chemists about what is the definitive criterion to evaluate the green character of an analytical methodology. Thus, a homogeneous guideline is still required to quantify the green features of analytical methods and evaluate their environmental impact. Nevertheless, after two decades since the inception of the Green Analytical Chemistry concept, it can be confirmed that the environmental concern has increased with a significant positive balance.

Vibrational spectroscopy is widely applied in both, food laboratories and production lines using techniques based on infrared, Raman, and Hyperspectral Image System, usually associated with chemometrics [22]. These techniques allow the direct determination of food parameters without sample treatment or at least a minimal preparation, providing fast analytical tools with reduced waste generation and risks to the operator. In the same way, techniques like near and middle infrared spectroscopy, electronic tongue and nose, hyperspectral imaging, biosensors, and integrated multiple sensors have been proposed for the establishment of food safety and food quality [23].

Separation techniques like gas (GC) and liquid (LC) chromatography are typically employed for multianalyte determination procedures. Traditionally, LC is considered less green than GC because of the solvent consumption level, although LC offers a high potential to be greened by using bio-solvents and agro-solvents, high efficacy chromatographic columns, and multidimensional separations [24]. Moreover, the use of capillary electrophoresis shows significant green advantages such as high separation efficacy with short analysis time and reduced sample and solvent consumption [5]. The use of separation techniques involves a sample treatment in the majority of cases, whether to extract, enrich or isolate analytes from the matrix, being imperative in the case of solid samples where an extraction step must be carried out before the analytical determination. High-performance approaches have been employed to replace conventional extraction procedures, such as: ultrasound-assisted (UAE), head-space (HS), microwave-assisted (MAE), pressurized liquid (PLE), and supercritical fluid (SFE) extraction, allowing an efficient analyte extraction from the matrix, but also reducing extraction time, potential contaminations, solvent consumption, and waste generation [25]. Additionally, liquid samples are classically treated by liquid-liquid (LLE) and solid-phase (SPE) extraction approaches before analytical determinations in a selective and sensitive enough way, being labour-intensive techniques that consume high volumes of organic solvents and generate a high amount of residues. Thus, several alternatives have been proposed from a green point of view, based on microextraction procedures, such as solid-phase microextraction (SPME), stir bar sorptive extraction (SBSE), single-drop microextraction (SDME), dispersive liquid-liquid microextraction (DLLME), and hollow fiber liquid-phase microextraction (HF-LPME) [5], being nowadays very important in this field the use of smart materials [26].

In the last decade, the huge improvement of analytical instrumentation has contributed to the development of Foodomic methodologies, based on the massive determination of genes, proteins, and metabolites present in food in order to understand and predict the complexity of the Foodome [27]. Foodomics is mainly focussed on the use of untargeted approaches with high-performance analytical instruments, such as: high-resolution mass spectrometry, two-dimensional chromatography, and nuclear magnetic resonance, coupled to chemometric tools for the massive treatment of data. These approaches involve the use of high-performance instruments, but they should be complemented by a previous screening using fast methodologies such as direct analysis or image treatment approaches [28].

Moreover, the widespread use of smartphones connected to the Internet and storage of large databases in clouds may allow, in a near future, to integrate analytical tools for food analysis as apps [22]. These totally portable approaches would allow performing a quality control by food scanning directly at markets and homes [29].

FOOD ANALYSIS DEMANDS

Food quality control and food safety are important tasks which involve serious efforts from international organisms and laboratories in order to guarantee a high standard of imported, exported, and consumed foods, all around the world. The Food and Agriculture Organization (FAO) of the United Nations in the frame of the World Health Organization (WHO) involves an international framework for food control systems including World Trade Organizations, as Sanitary and Phytosanitary (SPS) and Technical Barriers to Trade (TBT) agreements and *Codex Alimentarius* [2]. So, this international structure evidences the importance of food and food safety and thus, the single way to assure the appropriate application of international norms and regulations must be based on a deep analytical chemistry control which, at the same time must be accurate and fast due to the urgent need of data obtention before the consume of many perishable products.

In the present ecological paradigm of Analytical Chemistry [16], the main stress has been focused on environmental and health aspects. However, we cannot forbid that an appropriate food intake is the basis of a healthy life and that contamination and contaminants are amplified through the food chain. So, at the same time, food analysis is a key subject in human health control and it provides valuable data about the state of ecosystems. Because of that, there is nowadays an increasing number of journals devoted to food analysis and related aspects, as it can be seen in Table **1**.

Table 1. Representative journals publishing articles especially devoted to food analysis and related aspects and their metrics.

Source title	Impact f	CiteScore	Cit [1]	Doc [2]	SNIP [3]	SJR [4]	Publisher
Trends in Food Science and Technology	11.077	14.2	11727	823	3.802	2.841	Elsevier
Comprehensive Reviews in Food Science and Food Safety	9.912	15.1	4273	283	4.268	2.760	Wiley-Blackwell

(Table 1) cont.....

Source title	Impact f	CiteScore	Cit [1]	Doc [2]	SNIP [3]	SJR [4]	Publisher
Annual review of food science and technology	8.960	16.6	1615	97	3.042	2.327	Annual Reviews Inc.
Critical Reviews in Food Science and Nutrition	7.862	13.2	12253	926	2.571	1.804	Taylor & Francis
Food Chemistry	6.306	10.7	81471	7623	2.370	1.775	Elsevier
Global Food Security	6.034	8.8	1710	195	2.621	2.309	Elsevier
Food and Energy SecurityOpen Access	5.242	6.9	540	78	2.212	1.287	Wiley-Blackwell
Antioxidants	5.014	4.9	3706	993	1.610	1.100	MDPI [5]
Food Research International	4.972	6.2	15780	2561	1.661	1.440	Elsevier
Food Quality and Preference	4.842	7.9	5715	719	1.837	1.296	Elsevier
Journal of Food and Drug AnalysisOpen Access	4.727	7.4	3353	451	1.877	1.043	Elsevier
Food and Chemical Toxicology	4.679	6.7	12860	1933	1.353	0.902	Elsevier
Current Opinion in Food Science	4.577	7.6	2773	364	1.682	1.466	Elsevier
Journal of Food Engineering	4.499	7.5	10533	1408	1.833	1.338	Elsevier
Food Control	4.258	8.4	19065	2258	1.733	1.430	Elsevier
Journal of Agricultural and Food Chemistry	4.192	6.1	61403	5313	1,388	1.086	American Chemical Society
Food Policy	4.189	7.0	2906	414	2.410	2.189	Elsevier
International Journal of Food Microbiology	4.187	7.4	9223	1254	1.545	1.364	Elsevier
Food and Function	4.171	5.6	12792	2302	1.201	1.035	Royal Society of Chemistry
Food Microbiology	4.156	7.1	5999	840	1.557	1.318	Elsevier
Food Reviews International	4.113	7.1	808	114	1.882	1.109	Taylor & Francis
Foods Open Access	4.092	1.3	1118	872	1.243	0.661	MDPI [5]
LWT - Food Science and Technology	4.006	6.4	23411	3661	1.642	1.313	Elsevier
Food and Bioproducts Processing	3.726	6.1	3059	503	1.448	1.027	Institution of Chemical Engineers
Journal of Food Composition and Analysis	3.721	5.5	3543	645	1.545	0.895	Elsevier
Journal of Functional Foods	3.701	5.9	12997	2197	1.235	0.998	Elsevier
Meat Science	3.644	6.7	7144	1070	1.781	1.417	Elsevier
Toxins	3.531	5.1	15096	2078	1.269	1.034	MDPI [5]

(Table 1) cont.....

Source title	Impact f	CiteScore	Cit [1]	Doc [2]	SNIP [3]	SJR [4]	Publisher
International Journal of Food Sciences and Nutrition	3.483	5.0	2056	411	0.812	0.671	Taylor & Francis
Food and Bioprocess Technology	3.356	5.6	4123	734	1.125	1.078	Springer Nature
Journal of Dairy Science	3.333	5.4	20918	3873	1.712	1.440	Elsevier
Food Engineering Reviews	3.294	8.7	1329	78	1.719	1.162	Springer Nature
Food Structure	3.064	3.7	317	86	1.422	0.858	Elsevier
International Journal of Food Science and Technology	2.773	4.1	5165	1252	1.054	0.798	Wiley-Blackwell
Food Analytical Methods	2.667	4.7	6638	1398	0.875	0.673	Springer Nature
Journal of the Science of Food and Agriculture	2.614	4.6	12202	2673	1.118	0.718	Wiley-Blackwell
Food Additives and Contaminants: Part B Surveillance	2.602	4.4	732	166	1.262	0.865	Taylor & Francis
Journal of Food Science	2.479	3.7	5597	1498	0.973	0.738	Wiley-Blackwell
European Food Research and Technology	2.366	3.8	3198	852	0.976	0.654	Springer Nature
Food Additives and Contaminants - Part A Chemistry. Analysis. Control. Exposure and Risk Assessment	2.340	3.8	2862	760	0.938	0.624	Taylor & Francis
Food Technology and Biotechnology Open Access	2.115	3.3	731	223	1.076	0.575	University of Zagreb
British Food Journal	2.102	3.1	2741	881	1.086	0.579	Emerald
Food Security	2.095	3.7	1328	357	1.190	0.903	Springer Nature
European Journal of Lipid Science and Technology	2.056	3.8	2789	738	0.863	0.669	Wiley-Blackwell
Journal of Food Science and Technology	1.946	3.3	6612	2022	1.082	0.666	Springer Nature
International Journal of Food Properties Open Access	1.808	3.3	3866	1165	0.969	0.562	Taylor & Francis
Journal of Food Quality Open Access	1.763	2.4	1291	528	1.012	0.513	Hindawi
Food Science and Technology International	1.654	2.7	687	255	0.820	0.554	SAGE
CYTA - Journal of Food Open Access	1.653	2.8	1148	417	0.987	0.519	Taylor & Francis

(Table 1) cont.....

Source title	Impact f	CiteScore	Cit [1]	Doc [2]	SNIP [3]	SJR [4]	Publisher
Journal of Food Measurement and Characterization	1.648	2.1	2047	964	0.772	0.492	Springer Nature
Journal of Food Protection	1.581	2.8	2896	1021	0.835	0.587	International Association for Food Protection
Agriculture and Food Security Open Access	1.537	2.5	495	198	0.905	0.414	Springer Nature
Food Quality and Safety Open Access	3,102	3.1	266	85	1.660	0.879	Oxford University Press
Agricultural and Food Science Open Access	0,731	1.7	156	92	0.654	0.285	M T T Agrifood Research Finland

Note: ([1]) Cit: citations 2016-19. ([2]) Doc: documents 2016-19. ([3]) SNIP: Source Normalized Impact per Paper. ([4]) SJR: SCImago Journal Rank. ([5]) MDPI: Multidisciplinary Digital Publishing Institute.

Most of these journals are included in the areas of Food Science in addition to those of Industrial and Manufacturing Engineering or Safety, Risk, Reliability, and Quality. Additionally, food analysis has moved from the bench to the real in-field application in which many new analytical methodologies have been developed concerning mineral elements [30], emerging contaminants [31], and green analytical procedures [22, 28, 32].

On considering the objectives of food analysis it must be noticed that they have moved from the initial interest on the main nutrients, like proteins, lipids, and carbohydrates to the increasing importance of oligocompounds, both, inorganic elements and organic molecules, and with an increasing level of complexity because the presence of lipids in foods strongly depends on their unsaturation level and plant origin [33], the origin of proteins can determine the quality of processed foods [34] and the different carbohydrates have different effects on human health [35]. Additionally, the determination of the total content of mineral elements does not provide a complete picture of their action, because depending on their oxidation state (Cr(III) or Cr(VI), Fe(II) or Se(IV)), inorganic form (AsO_4^{3-}, AsO_2^-, Hg^{2+}) or organic character (arsenobetaine, methyl mercury, phenyl mercury, tributyltin, …), foods could be safe or unsafe, related element available or no available as function of their oxidation state or chemical form and the associated toxicity to As or Hg decreased or increased as a function of their methylation grade [36]. Thus, for example, arsenic toxicity decreased on moving from inorganic to high methylated species, and on the contrary mercury toxicity increases on increasing methylation level. That means that Omic Sciences [37] and speciation [38, 39] are hot topics of today's research in Analytical Chemistry. So, to carry a deep characterization of foods is nowadays a complex activity which requires new developments to assure complete information from regulatory agencies responsible for supervising compliance with regulations to consumers.

Food authentication is required to avoid frauds, as those concerning the replacement of high-value items by less useful ones or the mixing of different categories of food components to obtain a cheaper one by non-correct preparation or production. A special topic related to food authentication is the existence of protected designation of origin (POD), protected geographical indication (PGI) and traditional specialty guaranteed (TSG). As instruments created by the European Union, try to highlight the special characteristics of food production in a specific area or by using special elements. These labels have created the need of new tools for authentication of specially labeled foods which in fact offer an added value for both, producers and consumers, and requires sophisticated methods of analysis to verify the discriminant composition of these special food products [40], based on the determination of element markers or particular compounds.

However, the main concern of the food analysis community is the guarantee of food safety in front of the presence of contaminants in human consume products and/or the degradation of food components due to the time or the preservation conditions. To do it, sensitive and selective enough methods are required to analyze classical toxic products as many transition metals, oxidants, biogenic amines, and also the so-called emerging contaminants [41]. So, there is an increasing demand for methods for ions, neutral molecules, or solids, like microplastics [42] or nanoparticles [43], which could damage our health in the short, middle, or long term.

On the other hand, food analysis does not concern at all with a homogeneous kind of samples, being possible to distinguish between natural or processed foods, liquid or solid samples, animal or vegetable origin products and it offers additional challenges due to the biological variability of samples and the need to drastically modify methods employed in one field to be applied successfully to another kind of samples. Besides looking for the enhancement of the main analytical features of methods regarding an improved sensitivity or selectivity, it is required to pay attention to applied properties in order to provide a fast analytical response for safety and economic reasons and, of course, to take care of methods greenness. All the aforementioned aspects have been summarized in Fig. (**2**).

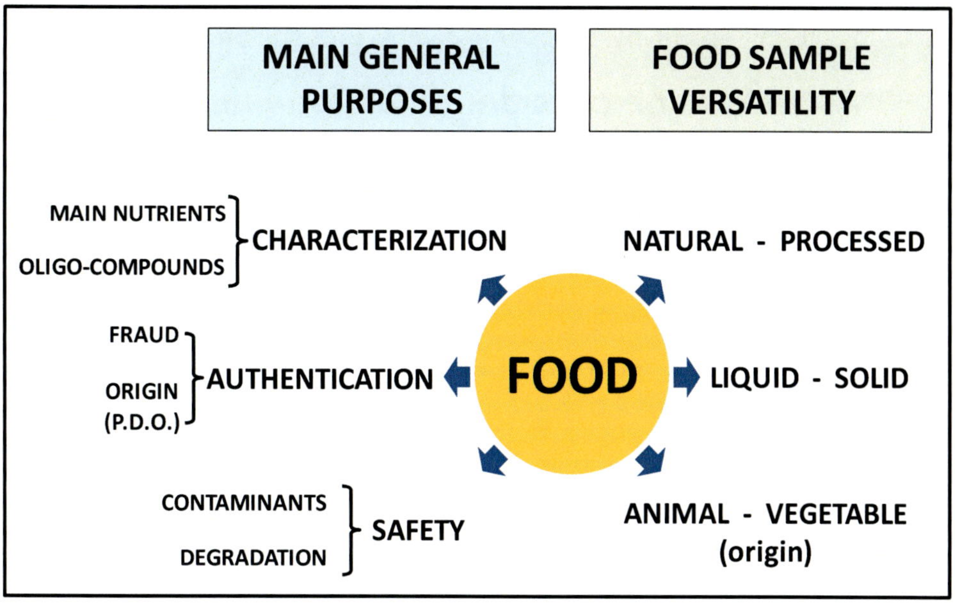

Fig. (2). The complexity of the main purposes of food analysis as a challenging task.

A simplistic approach could consider food samples as a more or less complex matrix which contains the target analytes. The physical state of samples and the complexity of matrix, together with the concentration level of selected analytes and the potential interferences, could orientate the appropriate selection of the best analytical method to be used. Considering the control of the food production chain, the authentication, and safety of food products it can be noticed that production area and technology, harvest conditions, food processing techniques, preservation, packaging of foods, food cooking, and its consumption are some of the main steps in which analytical efforts are required, as it has been summarized in Fig. (**3**).. Adulteration origin and trademark identification are hot topics with increasing economic importance in highly developed societies who carefully control their nutrition and try to distinguish their products from massive production practices thus moving to the production of organic foods, free from the use of pesticide, inorganic amendments, and other past and traditional practices.

In short, it can be concluded that nowadays many analytical laboratories all around the world try to evaluate many parameters in long series of food samples based on health, economic, environmental, and also political reasons, and to do it extra efforts, instruments, reagents, and consumables are required. So, in the end, a lot of wastes are generated and that creates the need to look at the possibilities offered by free or, at least, low deleterious environmental side effect green methods to determine as many as possible components and properties of foods.

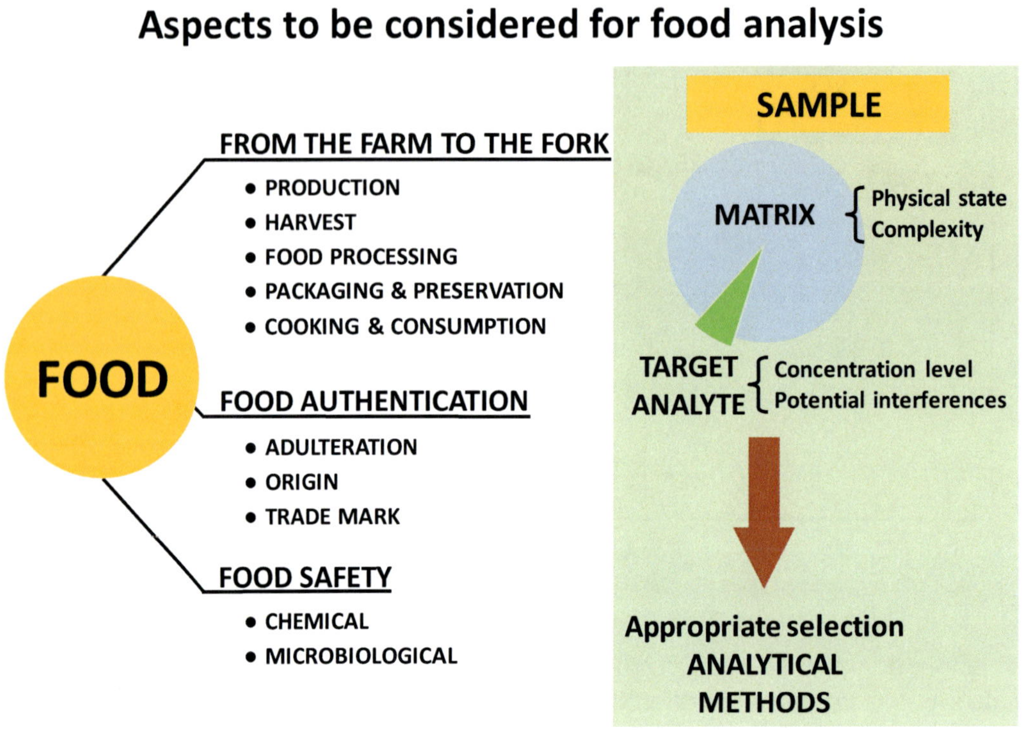

Fig. (3). The main aspects to be taken into account in food analysis.

GREEN ANALYTICAL SOLUTIONS

At it has been introduced in the first section of this chapter, the so called Green Analytical Chemistry has the main purpose of sustainability from both, environmental and economic viewpoints and thus the dream of a Green Analytical Chemistry in the field of food analysis is the obtention of a complete information about samples trough remote sensing or, at least, direct observation without any physical or chemical treatment. So, based on the drone technology coupled to hyperspectral camera images, it is possible nowadays to evaluate the production of fruits and vegetables [44], to evaluate fish production and to obtain in a fast way information about the size of specimens in fresh fish market [45]. Additionally, based on the development of image treatment algorithms, it is possible to evaluate fish freshness [46, 47], chicken meat freshness [48], or banana important indexes, as soluble solids, pH, titrable acidity, and firmness [49] without removing the banana skin or to determine the content of fat on processed meat products [50]. All the aforementioned parameters could be obtained without any treatment of samples and based on well-developed calibration models. Additionally, spectroscopy techniques as X-Ray fluorescence (XRF) [51], and infrared [52, 53] or Raman [54] spectroscopy could be employed to directly determine major and minor components in foods. However, the lack of sensitivity of the aforementioned techniques creates difficulties to determine trace compounds in foods. Nowadays atomic- spectroscopy-based methods, like inductively coupled plasma optical emission (ICP-OES) or inductively coupled plasma mass spectrometry (ICP-MS) are required for mineral element determinations [55] and chromatography [56] is the most useful technique for the determination of traces of organic compounds in foods. To do these determinations, a previous sample dissolution, matrix removal, and/or target analyte preconcentration are mandatory. So, the subject of this book covers a fundamental aspect for greening food analysis and extraction of active and toxic compounds from foods in a safe and environmentally friendly way.

As indicated in Fig. (**4**). There are many characteristics that we dream in a solvent to be used to extract active principles, analytes, or contaminants from foods. In short, it must be: i) non-toxic, ii) non-cumulative, iii) biodegradable, iv) with a high dissolving power, v) easy to be removed after extraction, vi) preferably from a natural origin, vii) renewable, and viii) easily available and cheap and, from a green analytical point of view, the aforementioned aspects must be considered in both, solvents and reagents to be employed in food analysis.

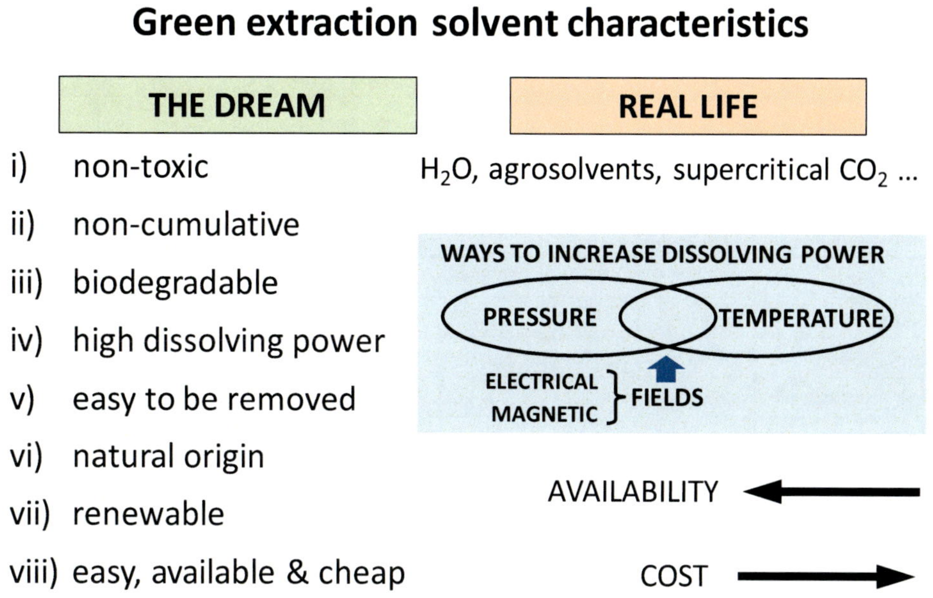

Fig. (4). The green extraction solvent ideal characteristics.

Obviously, the absence of toxicity and accumulative character is mandatory to assure solvent safety use from operators and to the environment. Being biodegradable is a requisite to avoid additional troubles to treat waste. Of course, a high dissolving power of the solvent in front of target analytes or active compounds is required to minimize the amount of solvent and reduce the extraction time and extraction conditions, such as temperature or pressure. After extraction, it could be nice for many analytical methods that the excess of solvent used could be removed easily and, if possible, at room temperature in order to favor analyte preconcentration and the possibility of solvent recovery. Natural origin solvents, specially agro-solvents, which are renewables, are interesting alternatives to petrol-based solvents. On the order hand, for sustainable economic reasons, the solvent to be employed in food extraction must be easily available and cheap to reduce total cost of extraction and the whole analytical processes.

Table 2. Main agro-solvents produced industrially or potentially exploitable and their physical-chemical properties (adapted from reference [57]).

Compound	CAS number	Molecular mass g/mol	Density (25 °C) g/mL	Boiling Point (750 mmHg)	Flash point °C	Viscosity (25 °C) Cp	Surface Tension (25 °C) mN/m	Evaporation rate (25 °C)
Water	7732-18-5	18.02	0.997	100	-	0.9	72	0.36
Ethanol	64-17-5	46.07	0.782	78.4	8.9	1.1	22.18	1.7
2-propanol	67-63-0	60.09	0.785	82.4	22.2	2.04	21	1.7
2-methyl-1-butanol	137-32-6	88.15	0.811	130	43.3	4.21	24.95	0.24
Butanol	71-36-3	74.12	0.806	117-118	35.0	2.62	23.35	0.44
1-octanol	111-87-5	130.23	0.824	196	81.1	7.41	27.1	0.007
1-decanol	112-30-1	158.28	0.826	231	82.2	11.9	28.7	0.001
Oleyl alcohol (65%)	143-28-2	268.48	0.846	207	>110	27.95	31.55	<0.005
Methyl acetate	79-20-9	74.08	0.934	57.5	-9.4	0.37	24.41	11.8
Ethyl acetate	141-78-6	88.11	0.898	76-78	-3.3	0.43	23.7	4.94
n-amyl acetate	628-63-7	130.18	0.872	142	23.9	0.86	25.07	0.4
Butyl acetate	123-86-4	116.16	0.878	124-126	22.2	0.68	24.81	1
Ethyl lactate	687-47-8	118.13	1.038	154	48.9	2.41	28.57	0.214
Butyl lactate	138-22-7	146.18	0.98	185-187	69.4	3.22	28	0.036
Diethyl succinate	123-25-1	174.19	1.043	217-218	90.6	2.48	32	0.009
DBE (Dupont) (*)	53-60-5	159	1.085	196-225	100.0	2.6	35.02	0.009
Decanedioic acid	109-43-3	314.46	0.932	345	>110	7.99	32.16	<0.005
Ethyl hexanoate	123-66-0	144.21	0.869	168	49.4	0.9	25.66	0.156
Ethyl octanoate	106-32-1	172.27	0.874	206-208	75.0	1.44	26.06	0.02

(Table 2) cont.....

Compound	CAS number	Molecular mass g/mol	Density (25 °C) g/mL	Boiling Point (750 mmHg)	Flash point °C	Viscosity (25 °C) Cp	Surface Tension (25 °C) mN/m	Evaporation rate (25 °C)
Ethyl decanoate	100-38-3	200.32	0.858	245	102.2	2.09	27.6	0.002
Methyl hexanoate	106-70-7	130.19	0.881	151	45.0	0.84	25.9	0.336
Methyl oleate	112-62-9	296.5	0.875	218	110.0	6.2	31.3	<0.005
Isopropyl palmitate	142-91-6	298.51	0.849	342	>110	6.38	28.8	<0.005
n-butyl stearate	123-95-5	340.59	0.857	350	160.0	8.26	33.2	<0.005
Glycerol carbonate	931-40-8	118.09	1.38	351	212.0	10	45.5	-
Isosorbide dimethyl ether	5306-85-4	174.19	1.17	315	103.0	6		
1,2-propanediol	57-55-6	76.09	1.032	187	107.2	43.22	36.52	0.01
1,3-propanediol	504-63-2	76.09	1.06	210-212	131	52.7	46.2	
Furfural	98-01-1	96.09	1.156	162	73.9	-	-	-
Furfuryl alcohol	98-00-0	98.1	1.131	170	65.0	4.76	38.45	0.04
Furan	110-00-9	68.08	0.933	32	-35.6	0.36	23.85	9.95
Tetrahydrofuran	109-99-9	72.1	0.882	67	-17.2	0.46	26.5	6.3
2-methyltetrahydrofuran	96-47-9	86.13	0.86	78-80	-11.0	4	-	
α-pinene	2437-95-8	136.24	0.854	155-156	32.2	1.32	27.6	0.41
α-terpineol	10482-56-1	154.25	0.929	217-218	89.4	36.5	31.6	0.005
(R)-(+)-limonene	5989-27-5	136.24	0.837	176	48.3	0.85	27.8	0.25
p-cymene	99-87-6	134.22	0.856	176-178	47.2	0.83	28.28	0.14
Ethyl methyl ketone	78-93-3	72.1	0.801	80	-3.3	0.43	24.18	4.03
Acetone	67-64-1	58.08	0.787	56	-17.8	0.31	22.86	6.06

(*) Mixture of refined dimethyl succinate, dimethyl glutarate, and dimethyl adipate

Water is, obviously, one of the solvents which has most of the aforementioned characteristics. However, it cannot dissolve all the target analytes and, because of that, on moving from dreams to the real-life, agro-solvents, like alcohols, and some terpenes, could offer interesting alternatives. In fact, Table **2**, taken from the book of Farid Chemat [57] provides a complete list of agro-solvents available in the market together with other agro-solvents that could be commercialized in a near future.

Another way to increase the dissolving power of water is the addition of amphiphilic compounds, like surfactants [58] suitable to create local polarity environments and different ordered media which favor the dissolution and extraction of nonpolar compounds. It can be done in an aqueous environment modified by the presence of micelles or microemulsified droplets of oil. It must be considered that the called surfactant media could be obtained from non-toxic and biodegradable tenside active molecules, which could drastically modify the solubility of the target molecules without using non-renewable and dangerous organic solvents, thus offering a green way to improve extraction steps. Additionally, CO_2 is freely available from the air and, under critical conditions of pressure and temperature, could move to the supercritical conditions at which it presents middle characteristics between gases and liquids with a high penetrability on solids and a polarity similar to cyclohexane. Supercritical CO_2 alone or combined with some alcohols could be really useful to increase the dissolving power of H_2O and agro-solvents but involves an increase in cost and technologies.

In addition to an appropriate selection of solvents, the use of pressure and temperature could enhance the extraction steps and nowadays the use of electric and magnetic fields could improve the efficiency and speed of many compounds extraction.

As indicated in Fig. (**5**), together with a proper selection of solvents to be used for food components extraction, matrix isolation or target analyte preconcentration, the adequate selection of the process concerns the main ways for greening extraction techniques.

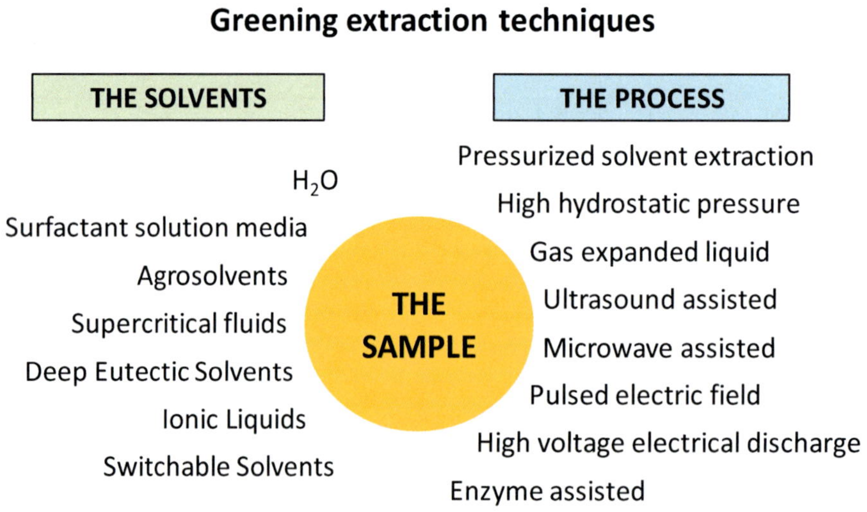

Fig. (5). Ways for greening extraction techniques.

The proper selection of as greenest as possible solvent is, of course, the best way to avoid the deleterious side effect of extraction steps, specifically regarding toxicity, flammability, non-renewable character of reagents, and troubles on managing solutions and wastes. Also, it must be taken into consideration the energy consumption and, because of that, the process could be an important matter as it will be evidenced in the next chapters in which pressurized solvent extraction, high hydrostatic pressure, gas expanded liquid extraction, ultrasound-assisted, microwave-assisted, pulsed electric field, high voltage electric discharges and enzyme assisted procedures will be discussed and all the aforementioned alternatives need to be evaluated from a green point of view.

As it can be summarized, the aforementioned techniques came from a clever combination of energy and pressure to favor solvent penetrability in order to improve analyte extraction in a short as possible period of time with a reduced energy consumption, saving also time and operator risks. It must be also indicated that sample treatment in closed systems enhances extraction efficiency and reduces analyte losses or contaminations together with the operator exposure to employed reagents.

In short, in spite of the matrix complexity or the analyte concentration level requirements involving the use of simple or combined extraction steps, it is possible to find analytical processes with green solutions for these steps [59] and there are no reasons to renounce to matrix isolation nor to the preconcentration of analyte before their quantitative determination for sustainable reasons.

CONCLUDING REMARKS

As it has been aforementioned in the previous sections, the tremendous development of Green Analytical Chemistry has been led by the social awareness on environmental problems. In this sense, analytical procedures are continuously developed based on their analytical features such as accuracy, precision, robustness, limits of detection and quantification, but also considering other parameters, such as reduction of reagents, solvents and energy consumption, operator risks, generated wastes, toxicity of both reagents and wastes, in summary, the so-called green features.

So, nowadays there is a tremendous impact of Green Analytical Chemistry in both, fundamental and application studies, trying to incorporate, together with the environmentally friendly aspects, the sustainability of analytical methods concerning cost and energy consumption.

FUTURE TRENDS

Considering the future trends of Green Analytical Chemistry, it can be noticed the importance of three basic tools: 1) direct analysis of untreated samples, 2) replacement of fossil-based solvents, and 3) cutting down on the use of single-use plastic. These research lines together with additional efforts in the acceptance and application of a simple tool to quantify the greenness of an analytical method, together with appropriate and teaching activities, will mark the future of Green Analytical Chemistry in the Foodomics area.

Direct analysis of samples based on the use of powerful techniques, such as infrared and Raman spectroscopy and chemical imaging is, obviously, the greenest way of analytical determinations. In this sense, the suitability of non-invasive infrared and Raman spectroscopy for on-line, in-line and at-line determination of food parameters and attributes has been deeply investigated. So, the development, validation, and routine application of advanced, robust chemometric tools for data treatment of spectra obtained directly from untreated food and beverage samples should be considered as a high priority research topic from a Green Analytical Chemistry point of view. New developments from both, basic software and new applications to solve real problems, are actually needed to improve the available methodologies and extend them to additional analytes and problems. The tremendous possibilities offered by software packages directly operating from the cloud, and/or the use of free apps operating in smart devices will be the key point to integrate advancements in chemometrics with the analytical measurements in the new era of Analytical Chemistry 4.0. In this sense, the development of direct analytical procedures based on the use of smartphones for data acquisition images has gained interest in the past decade.

Moreover, the research effort should be focused also on the development of effective and robust transfer calibration strategies together with the management of analytical big data to solve new and old problems in the Food area.

On the other hand, the continuous search for green alternatives for organic solvents is considered another important research trend in Green Analytical Chemistry. In this sense, the use of bio-solvents, obtained from renewable resources, can be considered an interesting alternative to fossil-derived solvents. Examples of bio-based solvents include ethanol, ethyl lactate that is obtained from corn and soybeans, and reacting ethanol and lactic acid and D-limonene among others. One of the main challenges of Green Analytical Chemistry is related to improving the purity of bio-solvents by addition of purification steps without compromising the green features of the procedure. Moreover, it should be considered that all bio-based solvents are not always greener than petroleum-based solvents, or at least, are not greener in all aspects.

At last but not least, analytical chemistry laboratories are facing the forgotten problem of plastic residues. Since the 1950s, the production of plastic has outpaced that of almost every other material. Plastic products are designed to be long-lasting and stable; however, half of the world's plastic production is intended to be used only once. Around 4900 million metric tons, representing 60% of all plastics ever produced, have been discarded and are accumulating in landfills or in the natural environment [60]. Plastic accumulation is causing the death of numerous seabirds and marine animals, after the ingestion of copious amounts of plastics and microplastics [61] with additional risks due to the fact that plastic materials could reach the food chain. One of the first measures implemented to reduce this plastic and microplastic contamination is the imposed charges by many governments to single-use plastic bags, bottles, and cooking utensils; but additional efforts to reduce this consumption in our daily work are still required.

Traditionally, it has been believed that organic solvents and inorganic acids provide the main bulk of any analytical laboratory waste, but recently, the scientific community has paid attention to plastic waste, a forgotten item that cannot be underestimated. According to Urbina *et al.*, biosciences labs generated in 2014, 5.5 million metric tons of plastic waste, accounting for about 2% of the plastic produced that year [62]. Disposable microcentrifuge tubes, pipette tips, and their boxes, Petri dishes, plastic test tubes, conical centrifuge tubes, plastic Pasteur pipettes, among others, become indispensable in chemical and biological laboratories. Unfortunately, all the aforementioned laboratory disposables are non-recyclable single-use plastic materials. However, a substantial proportion of plastic lab supplies, including pipette-tip boxes, can be reused after washing and sterilizing. In this sense, a company named Grenova Solutions has developed different devices to wash and sterilize contaminated pipette tips in large quantities for reuse, reducing plastic wastes, and increasing cost saving [63]. Moreover, different companies are committed to producing pipette tips boxes with recycled materials such as recycled fiber and recycled water bottles or renewable materials; such as compostable bioplastic with a degradation time of few months. The example of pipette tips can be extrapolated to other lab materials such as Petri dishes, polystyrene boxes for ice, gloves, and so on. In this sense, the companies Kimberly-Clark Professional® and TerraCycle® have partnered to create an interesting recycling program for disposable gloves so-called KIMTECH program [64]. The extension of the aforementioned changes is small for now. However, the implication of researchers, universities, and governments will, for sure, reduce laboratory plastic waste and it must be considered as an additional effort to be made for greening laboratory steps.

In summary, the development of the research lines commented through this chapter together with additional efforts in the acceptance and application of a simple tool to quantify the greenness of an analytical method, together with appropriate and teaching activities, will mark the future of Green Analytical Chemistry in the Foodomics area.

LIST OF ABBREVIATIONS

AOAC	Association of Official Agricultural Chemists
DLLME	Dispersive liquid-liquid microextraction
EFSA	European Food Safety Authority
FAO	Food and Agriculture Organization
GAPI	Green Analytical Procedure Index
GC	Gas chromatography
HF-LPME	Hollow fiber liquid-phase microextraction
HS	Head-space
ICP-OES	Inductively coupled plasma optical emission
ICP-MS	Inductively coupled plasma mass spectrometry
LC	Liquid chromatography
LLE	Liquid-liquid extraction
MAE	Microwave-assisted extraction
NEMI	National Environmental Methods Index
PLE	Pressurized liquid extraction
SBSE	Stir bar sorptive extraction
SDME	Single-drop microextraction
SFE	Supercritical fluid extraction
SPE	Solid-phase extraction
SPME	Solid-phase microextraction
PGI	Protected geographical indication
POD	Protected designation of origin
SPS	Sanitary and Phytosanitary
TBT	Technical Barriers to Trade
UAE	Ultrasound-assisted extraction
WHO	World Health Organization
XRF	X-Ray fluorescence
TSG	Traditional specialty guaranteed

CONSENT FOR PUBLICATION

Not applicable.

CONFLICT OF INTEREST

The authors declare no conflict of interest, financial or otherwise.

ACKNOWLEDGEMENT

Declared none.

REFERENCES

[1] de la Guardia M, Garrigues S, Eds. Handbook of Green Analytical Chemistry. John Wiley & Sons, Ltd 2012.
[http://dx.doi.org/10.1002/9781119940722]

[2] Codex Alimentarius Commission of the Joint FAO/WHO Food Standards Programme Website. http://www.fao.org/fao-who-codexalimentarius/en/

[3] Association of Official Agricultural Chemists. Official Methods of Analysis. 21st ed., Rockville, MD: AOAC 2019.

[4] European Food Safety Authority website. https://www.efsa.europa.eu/

[5] Demirhan B, Şatana Kara HE, Demirhan BE. Overview of Green Sample Preparation Techniques in Food Analysis. Intech 2017.
[http://dx.doi.org/10.5772/intechopen.68787]

[6] de la Guardia M, Ruzicka J. Guest editorial. Towards environmentally conscientious analytical chemistry through miniaturization, containment and reagent replacement. Analyst (Lond) 1995; 120(2): 17N-N.
[http://dx.doi.org/10.1039/an995200017n]

[7] de la Guardia M. An Integrated Approach of Analytical Chemistry. J Braz Chem Soc 1999; 10: 429-37.

[8] Anastas PT. Green Chemistry and the role of analytical methodology development. Crit Rev Anal Chem 1999; 29(3): 167-75.
[http://dx.doi.org/10.1080/10408349891199356]

[9] Koel M, Kaljurand M. Green Analytical Chemistry. Cambridge: RSC Publishing 2010.

[10] de la Guardia M, Armenta S. Comprehensive Analytical Chemistry, Vol 57. Green Analytical Chemistry: Theory and Practice, Elsevier: Amsterdam. 2011.

[11] Płotka-Wasylka J, Namieśnik J, Eds. Green Analytical Chemistry, Past, Present and Perspectives. Springer 2019.
[http://dx.doi.org/10.1007/978-981-13-9105-7]

[12] Anastas PT, Warner JC. Green Chemistry: Theory and Practice. New York: Oxford University Press 1998.

[13] Gałuszka A, Migaszewski Z, Namieśnik J. The 12 principles of green analytical chemistry and the SIGNIFICANCE mnemonic of green analytical practices. *TrAC-Trend.* Anal Chem 2013; 50: 78-84.

[14] National Environmental Methods Index (NEMI). 2002. https://www.nemi.gov/home/

[15] Raynie D, Driver JL. Green assessment of chemical methods. 13th Green Chemistry and Engineering Conference. Washington, DC. 2009.

[16] Garrigues S, de la Guardia M, Eds. Challenges in Green Analytical Chemistry. 2nd ed., Cambridge, UK: Royal Society of Chemistry 2020.
[http://dx.doi.org/10.1039/9781788016148]

[17] Gałuszka A, Konieczka P, Migaszewski ZM, Namieśnik J. Analytical Eco-Scale for assessing the greenness of analytical procedures. *TrAC-Trend.* Anal Chem 2012; 37: 61-72.

[18] Phan TVT, Gallardo C, Mane J. Green motion: a new and easy to use green chemistry metric from laboratories to industry. Green Chem 2015; 17(5): 2846-52.

[http://dx.doi.org/10.1039/C4GC02169J]

[19] Tobiszewski M, Marć M, Gałuszka A, Namieśnik J. Green chemistry metrics with special reference to green analytical chemistry. Molecules 2015; 20(6): 10928-46. [http://dx.doi.org/10.3390/molecules200610928] [PMID: 26076112]

[20] Armenta S, de la Guardia M, Namieśnik J, Green Microextraction. In: Valcárcel M, Cárdenas S, Lucena R (Eds.). Analytical microextraction techniques, Bentham Science Publishers: UAE 2016: pp. 3-27.

[21] Płotka-Wasylka J. A new tool for the evaluation of the analytical procedure: Green Analytical Procedure Index. Talanta 2018; 181: 204-9. [http://dx.doi.org/10.1016/j.talanta.2018.01.013] [PMID: 29426502]

[22] Pallone JAL, Caramês ETS, Alamar PD. Green analytical chemistry applied in food analysis: alternative techniques. Curr Opin Food Sci 2018; 22: 115-21. [http://dx.doi.org/10.1016/j.cofs.2018.01.009]

[23] Chen Q. Past and Present Research Systems of Green Chemistry. Org Chem Curr Res 2016; 5: 3.

[24] Płotka J, Tobiszewski M, Sulej AM, Kupska M, Górecki T, Namieśnik J. Green chromatography. J Chromatogr A 2013; 1307: 1-20. [http://dx.doi.org/10.1016/j.chroma.2013.07.099] [PMID: 23932374]

[25] Santana ÁL, Meireles MAA. Green analytical chemistry for food industries.Dr Inamuddin, R Boddula, A Asiri, Green Sustainable Process for Chemical and Environmental Engineering and Science, Analytical Techniques for Environmental and Industrial Analysis. Elsevier 2020; pp. 143-60.

[26] de la Guardia M, Esteve-Turrillas FA, Eds. Handbook of Smart Materials in Analytical Chemistry. Hoboken: John Wiley & Sons Ltd. 2019. [http://dx.doi.org/10.1002/9781119422587]

[27] Cifuentes A, Ed. Comprehensive Foodomics. Amsterdam: Elsevier 2020.

[28] Armenta S, Esteve-Turrillas FA, Garrigues S, de la Guardia M. Green Analytical Chemistry.Comprehensive Foodomics. Amsterdam: Elsevier 2020; pp. 483-93.

[29] Rateni G, Dario P, Cavallo F. Smartphone-Based Food Diagnostic Technologies: A Review. Sensors (Basel) 2017; 17(6): 1453. [http://dx.doi.org/10.3390/s17061453] [PMID: 28632188]

[30] de la Guardia M, Garrigues S, Eds. Handbook of Mineral Elements in Food. Chichester, UK: John Wiley & Sons Ltd. 2015. [http://dx.doi.org/10.1002/9781118654316]

[31] Farre M, Barcelo D. Analysis of emerging contaminants in food. *TrAC-Trend.* Anal Chem 2013; 43: 240-53.

[32] Castro-Puyana M, Mendiola JA, Ibañez E. Strategies for a cleaner new scientific discipline of green foodomics. *TrAC-Trend.* Anal Chem 2013; 52: 23-35.

[33] Valenzuela A, Morgado N. Trans fatty acid isomers in human health and in the food industry. Biol Res 1999; 32(4): 273-87. [http://dx.doi.org/10.4067/S0716-97601999000400007] [PMID: 10983247]

[34] Yada RY, Ed. Proteins in Food Processing.Woodhead Publishing Series in Food Science, Technology and Nutrition. Cambridge: Woodhead Publishing 2004.

[35] Coker JK, Moyne O, Rodionov DA, Zengler K. Carbohydrates great and small, from dietary fiber to sialic acids: How glycans influence the gut microbiome and affect human health. Gut Microbes 2021; 13(1): 1869502. [http://dx.doi.org/10.1080/19490976.2020.1869502] [PMID: 33615984]

[36] Duan P, Khan S, Ali N, *et al.* Biotransformation fate and sustainable mitigation of a potentially toxic

element of mercury from environmental matrices. Arab J Chem 2020; 13(9): 6949-65. [http://dx.doi.org/10.1016/j.arabjc.2020.06.041]

[37] Gallo M, Ferranti P. The evolution of analytical chemistry methods in foodomics. J Chromatogr A 2016; 1428: 3-15. [http://dx.doi.org/10.1016/j.chroma.2015.09.007] [PMID: 26363946]

[38] Kot A, Namiesnik J. The role of speciation in Analytical Chemistry. *TrAC-Trend.* Anal Chem 2000; 19(2-3): 69-79.

[39] Quiroz W. Speciation analysis in chemistry. ChemTexts 2021; 7(1): 7. [http://dx.doi.org/10.1007/s40828-020-00125-8]

[40] Agrimonti C, Marmiroli N. Food Genomics for the Characterization of PDO and PGI Virgin Olive Oils. Eur J Lipid Sci Technol 2019; 121(3): 1800132. [http://dx.doi.org/10.1002/ejlt.201800132]

[41] Kantiani L, Llorca M, Sanchís J, Farré M, Barceló D. Emerging food contaminants: a review. Anal Bioanal Chem 2010; 398(6): 2413-27. [http://dx.doi.org/10.1007/s00216-010-3944-9] [PMID: 20680618]

[42] Liu Q, Chen Z, Chen Y, Yang F, Yao W, Xie Y. Microplastics and Nanoplastics: Emerging Contaminants in Food. J Agric Food Chem 2021; 69(36): 10450-68. [http://dx.doi.org/10.1021/acs.jafc.1c04199] [PMID: 34473500]

[43] Blasco C, Pico Y. Determining nanomaterials in food. *TrAC-Trend.* Anal Chem 2011; 30(1): 84-99. [PMID: 21128611]

[44] Sinha JP. Aerial robot for smart farming and enhancing farmers' net benefit. Indian J Agric Sci 2020; 90(2): 258-67. [http://dx.doi.org/10.56093/ijas.v90i2.98997]

[45] Ubina N, Cheng SC, Chang CC, Chen HY. Evaluating fish feeding intensity in aquaculture with convolutional neural networks. Aquacult Eng 2021; 94: 102178. [http://dx.doi.org/10.1016/j.aquaeng.2021.102178]

[46] Dowlati M, Mohtasebi SS, de la Guardia M. Application of machine-vision techniques to fish-quality assessment. *TrAC-Trend.* Anal Chem 2012; 40: 168-79.

[47] Dowlati M, Mohtasebi SS, Omid M, Razavi SH, Jamzad M, de la Guardia M. Freshness assessment of gilthead sea bream (Sparus aurata) by machine vision based on gill and eye color changes. J Food Eng 2013; 119(2): 277-87. [http://dx.doi.org/10.1016/j.jfoodeng.2013.05.023]

[48] Taheri-Garavand A, Fatahi S, Shahbazi F, Guardia M. A nondestructive intelligent approach to real□time evaluation of chicken meat freshness based on computer vision technique. J Food Process Eng 2019; 42(4): e13039. [http://dx.doi.org/10.1111/jfpe.13039]

[49] Sanaeifar A, Bakhshipour A, de la Guardia M. Prediction of banana quality indices from color features using support vector regression. Talanta 2016; 148: 54-61. [http://dx.doi.org/10.1016/j.talanta.2015.10.073] [PMID: 26653423]

[50] Cruz-Fernández M, Luque-Cobija MJ, Cervera ML, Morales-Rubio A, de la Guardia M. Smartphone determination of fat in cured meat products. Microchem J 2017; 132: 8-14. [http://dx.doi.org/10.1016/j.microc.2016.12.020]

[51] Feng X, Zhang H, Yu P. X-ray fluorescence application in food, feed, and agricultural science: a critical review. Crit Rev Food Sci Nutr 2021; 61(14): 2340-50. [http://dx.doi.org/10.1080/10408398.2020.1776677] [PMID: 32543214]

[52] Yu H, Guo L, Kharbach M, Han W. Multi-Way Analysis Coupled with Near-Infrared Spectroscopy in Food Industry: Models and Applications. Foods 2021; 10(4): 802.

[http://dx.doi.org/10.3390/foods10040802] [PMID: 33917964]

[53] Karoui R, Downey G, Blecker C. Mid-infrared spectroscopy coupled with chemometrics: a tool for the analysis of intact food systems and the exploration of their molecular structure-quality relationships - a review. Chem Rev 2010; 110(10): 6144-68.
[http://dx.doi.org/10.1021/cr100090k] [PMID: 20804166]

[54] Yang TX, Zhao B, He LL. Raman instruments for food quality evaluation. Eval. Technol. Food Qual 2019; pp. 119-43.

[55] García Alonso JI, Marchante-Gayón JM, Moldovan M. New developments in food analysis by ICP-MS.Handbook of Mineral Elements in Food. Chichester, UK: John Wiley & Sons Ltd. 2015.

[56] Cacciola F, Donato P, Sciarrone D, Dugo P, Mondello L. Comprehensive Liquid Chromatography and Other Liquid-Based Comprehensive Techniques Coupled to Mass Spectrometry in Food Analysis. Anal Chem 2017; 89(1): 414-29.
[http://dx.doi.org/10.1021/acs.analchem.6b04370] [PMID: 28105838]

[57] Chemat F. Éco-extraction du vegetal Procédés innovants et solvents alternatifs. Paris: Dunond 2011.

[58] Moradi M, Yamini Y, Feizi N. Development and challenges of supramolecular solvents in liquid-based microextraction methods. *TrAC-Trend.* Anal Chem 2021; 138: 116231.

[59] Moreda-Piñeiro J, Moreda-Piñeiro A. Combined assisted extraction techniques as green sample pre-treatments in food analysis. *TrAC-Trend.* Anal Chem 2019; 118: 1-18.

[60] Geyer R, Jambeck JR, Law KL. Production, use, and fate of all plastics ever made. Sci Adv 2017; 3(7): e1700782.
[http://dx.doi.org/10.1126/sciadv.1700782] [PMID: 28776036]

[61] Young LC, Vanderlip C, Duffy DC, Afanasyev V, Shaffer SA. Bringing home the trash: do colony-based differences in foraging distribution lead to increased plastic ingestion in Laysan albatrosses? PLoS One 2009; 4(10): e7623.
[http://dx.doi.org/10.1371/journal.pone.0007623] [PMID: 19862322]

[62] Urbina MA, Watts AJR, Reardon EE. Labs should cut plastic waste too. Nature 2015; 528(7583): 479.
[http://dx.doi.org/10.1038/528479c] [PMID: 26701046]

[63] Grenova Inc. https://www.grenovasolutions.com

[64] Terracycle. The Kimtech™ Nitrile Glove Recycling Programme. https://www.terracycle.co.uk/en-GB/brigades/gloves

CHAPTER 2

Green Extraction Techniques

Malak Tabib[1,2], **Njara Rakotomanomana**[2], **Adnane Remmal**[1,*] and **Farid Chemat**[2,*]

[1] *Department of Biology, Faculty of Science Dhar El-Mahraz, University Sidi Mohammed Ben Abdellah, P.O. Box 1796, Fez30050, Morocco*

[2] *Avignon University, INRAE, UMR 408, Green Extraction Team, F-84000 Avignon, France*

Abstract: Green extraction of natural products was and will always remain an important research subject in various fields. It is based on developing techniques that meet the six principles of eco-extraction. This concept responds to the challenges of the 21st century, aiming to protect the environment, the operator, and the consumer by reducing hazardous solvent consumption and by favoring the use of more environmentally friendly methods. In this chapter, we review the principles of eco-extraction in detail, followed by an overview of four methods widely used in extraction, namely ultrasound-assisted extraction (UAE), microwave-assisted extraction (MAE), subcritical water extraction (SWE), and supercritical fluid extraction (SFE).

Keywords: Green extraction techniques, Microwave-assisted extraction (MAE), Principles, Subcritical water extraction (SWE), Supercritical fluid extraction (SFE), Ultrasound-assisted extraction (UAE).

INTRODUCTION

Sustainable development is one of the most commonly used terms in today's debates. It is considered a development that satisfies the needs of the present without compromising the ability of future generations to respond to their own needs. This demand affects all areas of society, including chemistry. It should reflect on how chemistry can contribute to greater sustainability in our society, now and in the future.

One of the contributions of chemistry to meet the challenge of greater sustainability in the development of our society is promoting sustainable chemist-

* **Corresponding Author Adnane Remmal:** Department of Biology, Faculty of Science Dhar El-Mahraz, University Sidi Mohammed Ben Abdellah, P.O. Box 1796, Fez 30050, Morocco; Tel: +212 661532398 E-mail: adnaneremmal@gmail.com

Corresponding Author Farid Chemat: Avignon University, INRAE, UMR 408, Green Extraction Team, F-84000 Avignon, France; Tel: +212 701292502; E-mail: chemat@univ-avignon.fr

Merichel Plaza & Maria Luisa Marina (Eds.)

ry in research and industrial production. Under the name of green chemistry (or in Europe also sustainable chemistry), many efforts have been made to make the chemistry of tomorrow less toxic and less dangerous. Green chemistry aims to make chemistry more energy efficient, reduce waste disposal, and/or produce innovative products using fewer natural resources. Alternative processes and reaction routes are designed, and new materials and products are developed, helping to ensure our current requirements, but taking greater account of the interests of future generations.

Extraction is considered a key step in food processing which consists of separating the desired compounds from the raw material and transferring these compounds into a solvent. It includes several methods, such as solvent extraction, ultrasound-assisted extraction, microwave-assisted extraction, Soxhlet extraction, supercritical fluid extraction, and subcritical water extraction. In the main, natural product extraction goes through the following phases: (1) the solvent penetrates the solid matrix; (2) the solute dissolves in the solvents; (3) the solute is diffused out of the solid matrix; (4) the extracted solutes are collected. The efficiency of the extraction is conditioned by various parameters, including particle size, the extraction solvent, the solvent-to-solid ratio, the extraction temperature, and duration.

These extraction methods allow for faster and more sustainable component separation, as fewer toxic solvents [1] and energy [2] are used. In addition, they simplify manipulation and sample preparation, give high purity and yield of the final extract, and eliminate post-treatment of wastewater [3]. Numerous classes of compounds such as vitamins, sugars, proteins, lipids, fibers, aromas, pigments, antioxidants, and other organic and mineral compounds have been extracted from various matrices, mainly insects [4], plant materials [5 - 8], and animal tissues [9].

This chapter provides an overview of existing knowledge on innovating methods of sample preparation of natural products. It gives the fundamental theoretical framework and a few details about the extraction using some of the most innovative, green, fast techniques such as ultrasound-assisted extraction, microwave-assisted extraction, subcritical water extraction, and supercritical fluid extraction by detailing their principles, instrumentations, and applications in food analysis.

GREEN EXTRACTION: DEFINITION AND PRINCIPLES

Despite the prejudicial opinions of some world leaders today, global environmental awareness continues to be on the increase. Terms such as green, biorefinery, and sustainability, are increasingly important in all facets of global development. This idea is closely associated with the principles of green extraction, which can be defined as a process of obtaining an extract using minimal hazardous / petroleum solvents by reducing energy consumption and waste as well as ensuring safe and high-quality extracts. It is a concept that seeks to meet the challenges of the 21st century by protecting the environment and the consumers, and at the same time, increasing competition between universities and industries to be more environmentally, economically, and innovative [2, 10, 11].

According to Chemat *et al.* (2012), The list of the "Six Principles of Green Extraction of Natural Products" can be consulted by industry and scientists as a direction to establish an innovative and green label, charter, and standard, and as a reflection to innovate not only in the process but in all aspects of solid-liquid extraction. The principles have been identified and described not as rules but more as innovative examples to follow, discovered by scientists, and successfully applied by industry [2].

Principle 1: Innovation by A Selection of Varieties and Use of Renewable Plant Resources

The rising claim of natural products and extracts to respond to the need of food, cosmetic and pharmaceutical industries is resulting in the over-exploitation of plant resources. Using agricultural by-products became trendy, providing additional income to primary producers and industries and improving the overall agricultural value chain. Nowadays, trends in botanical by-products are to increase their added value, from organic fertilizers and raw material pellets to functional ingredients [10, 12].

Plant/crop-based resources are defined as raw materials derived from the natural flora and the transformation processes in numerous industries (food, feed, fiber, *etc*.). An underlying hypothesis is that these resources are renewable over a short period, using annual crops, perennials, and short-rotation woody species. The reuse of these plant resources as raw materials for industrial production or as a source of energy is quite limited. This is due to the poor adaptation of the hydrocarbon processing system, which, unlike these materials, was developed in a more advanced way to use fossil fuels.

Renewable plant-based resources represent a strategic option to fulfill the growing need for industrial components, enabling economic, environmental, and societal benefits. The opportunity is advantageous. Nevertheless, it requires a foresight perspective, stakeholders' integration, investment in new approaches, and coordination of research to generate a safe future.

An example can be given with the plant breeding technique. Medicinal plants for instance play an important role in prevention and treatment. As stated by the

World Health Organization (WHO) they have been adopted by more than 80% of the world's population regularly such as in China or the countries of the African continent for several decades [13]. The relationship between medicines and the sustainability of medicinal plants is increasingly recognized and is receiving growing attention in international agreements and trade labeling schemes. Nonetheless, traditional companies have failed to care for the fair trade aspects of this sector [14].

Plant breeding represents a widely used alternative to the intensive use of these plants [15]. This technique aims to select new varieties with genetically transmitted sets of traits that are well suited to the objectives of producers and consumers. In this case, it involves the transmission of molecules with active principles. *Schizonepeta tenuifolia* for instance is a Chinese plant cultivated for its medicinal benefits. The breed was found to have high production, increased resistance to disease, and high concentrations of active compounds [13, 16].

Plant cultivation is another alternative for biodiversity conservation as the demand for medicinal plants is constantly growing [17]. A study was carried out on *Prunus africana* (Hook.f.) Kalkman, a sub-Saharan African plant, is mainly used for generations for its anti-prostate cancer effect [14, 18]. It is debarked from the roots to the branches, which causes its death. Within this context, new programs are being put in place to regenerate these plants and encourage their cultivation to put an end to their excessive exploitation.

Principle 2: Use of Alternative Solvents and Principally Water or Agro-solvents

Over the past few years, extraction processes have depended considerably on solvents, the vast majority of which are of petroleum origin, and suspected to be hazardous to human health and the environment. Solvent extraction has been used by industries for centuries in a variety of fields, ranging from chemical synthesis to waste treatment. A large part of these solvents is C (VOCs), with which the risk of fire and explosion is increased, resulting in environmental impacts promoting global warming. Similarly, simple exposure to those compounds can lead to some health issues. One of the most widely used solvents in extraction is *n*-hexane as the most effective for oil extraction. Despite its many advantages (easy removal due to its low boiling point, higher extraction yield compared to other solvents, low energy for extraction), exposure to this solvent can cause dizziness, and nausea and can heavily affect the nervous system. Furthermore, it has been classified as CMR 3 (Carcinogenic, mutagenic, and reprotoxic substances), which implies that it is suspected to be reprotoxic according to the European Directives and (REACH) regulations [19].

The selection of solvents is based on the following criteria: (i) workers' safety (hazardousness, carcinogenicity, mutagenicity), (ii) operational safety (flammability, explosiveness, volatility), (iii) environmental protection (persistence, contamination), (iv) sustainability of the process (recycling and reuse) [20]. Thus, it has become compulsory to replace them with greener alternatives despite their many advantages. Industries' interest is increasingly focused on green solvents (Fig. **1**).

The benefits of using these latter are numerous. They include low costs and extraction time, reduced risks due to overpressure, easy scale-up, enhanced, and extract purity. Among these solvents, we find bio-solvents which are defined as solvents produced from biomass sources such as energy crops (*e.g.*, corn), forest products (*e.g.*, wood), aquatic biomass (*e.g.*, microalgae), and waste materials (*e.g.*, urban wastes).

Water, for instance, is regarded as the greenest solvent since it is non-toxic for health and the environment. Besides, it is the least expensive, most abundant, non-inflammable solvent in nature. Water has the advantage to vary its physicochemical properties by changing the temperature. Hence, the use of subcritical water, based on increasing water temperature and pressure enough to keep it in the liquid state, expanded at an exponential rate over the past few years. Water molecule's small size (sphere diameter 2.75 □) helps substantially in the hydration of the solutes. As regards polarity, it is due to the partial positive charges of the two hydrogen atoms and the partial negative charge of the oxygen atom. This implies its interaction with both polar and non-polar molecules, thus extraction of a large range of molecules, such as sugars, proteins, organic acids, and inorganic substances [21].

Ethanol is another common bio-solvent. It has been in widespread use as a viable solvent in recent decades and is ranked as an environmentally desirable green product as it is obtained through the fermentation of renewable sources, including sugars, starches, and lignocellulose. Its relatively low price [22], low boiling point, and pure bioavailability make it suitable for use as a solvent on a wide scale.

Solvent-free alternatives are of great interest and became extensively applied. Their benefits are conspicuous: reduced prices and large volumes of solvent, easy scale-up, lowered extraction time and risks of overpressure and explosions, and higher purity of the extracts. Innovative techniques such as Microwave Hydrodiffusion and Gravity (MHG), Pulsed Electric Field (PEF), and Instant

Controlled Pressure Drop (French acronym: DIC, for "Détente Instantanée Contrôlée") have been developed and used increasingly during the past few years for the extraction of various products (aromas, oils, antioxidants, *etc.*) [2, 10, 12, 23].

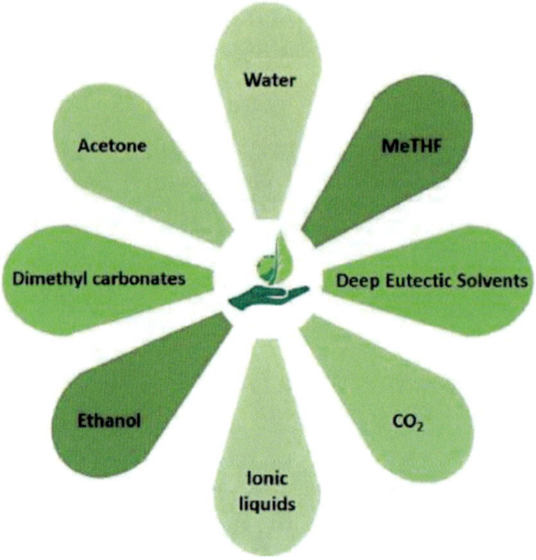

Fig. (1). Examples of green solvents.

Principle 3: Reduce Energy Consumption by Energy Recovery and Using Innovative Technologies

Energy consumption is one of the main concerns of industries, regardless of their field of activity as the costs can present significant expenses. It can be related to how much energy a production process consumes to be accomplished, such as heating, cooling, processing, and assembling, steam and cogeneration, lighting, heating, and cooling of buildings. In the words of Dr. Fatih Birol, IEA (International Energy Agency) Executive Director, *"The world urgently needs to put a laser-like focus on bringing down global emissions. This calls for a grand coalition encompassing governments, investors, companies, and everyone else who is committed to tackling climate change"* [24].

As reported by the World Energy Outlook 2019 [25], energy claim rises by 1% per year to 2040. Industries are increasingly focusing on low-carbon sources, mainly solar photovoltaics that contribute to more than half of this growth, and natural gas, which supplies another third. Oil demand stabilizes in the 2030s, and coal use declines. Despite that, some countries aspire to "net zero" put efforts to alter their energy consumption and supplies, but the impetus for clean energy technologies is insufficient to counterbalance the effects of extending population and economy. Hence, the world is a long way away from reaching sustainability goals. The challenge becomes in this case to prioritize and generalize the understanding of the use of energy sources to reduce energy consumption and promote recycling. Fortunately, most of the industry members are heading towards a decline in energy consumption by energy recovery and reduction of unit operations and favoring safe, robust and controlled processes.

The industrial sector can be classified into three groups: (i) energy-intensive manufacturing (food, pulp, and paper, basic chemicals, refining, iron and steel, non-ferrous metals, nonmetallic minerals), (ii) nonenergy-intensive manufacturing (other miscellaneous chemicals such as pharmaceuticals detergents, paint, coating and other industrials including electronic, transportation, machinery and electrical products), and (iii) nonmanufacturing (agriculture, mining, and construction). 26% of the most energy-intensive industries belong to chemistry, plastics manufacturing, and 16% in the food-processing sector. The field of plant extraction could be considered in both of these categories; there is a strong focus on the optimization of extraction processes for energy efficiency. In addition, several traditional extraction techniques require a lot of energy, for instance, distillation. This represents an opportunity to optimize energy consumption [25, 26].

Green chemistry relies to a growing extent on "low-energy" processes based on:

– Extraction at ambient temperature and pressure;
– Optimization of extraction time and solvent consumption;
– Development of innovative processes;
– Optimizing energy resources and promoting recycling.

For instance, Ultrasound and Microwave-assisted extractions have been employed widely as innovative techniques to enhance extraction yields while optimizing energy consumption. Their use in essential oil extraction, instead of hydrodistillation, enables them to increase and intensify extraction yield while decreasing both time and energy used. Ultrasound-Assisted Extraction (UAE) was employed by Altemimi *et al.* to optimize phenolic compounds extraction from peaches and pumpkins. It has proven its efficiency in terms of yield, extraction time, and energy consumption compared to conventional extraction [27].

In addition to the previously-cited techniques aiming to lower energy consumption, energy recovery represents an interesting and attractive option in the industrial field. As stated by Woolley *et al.*, it is based on the principle that *"the energy is never actually consumed, it is only converted from one form to another", and so there is a potential to capture this and utilize it as an energy supply"* [28].

Overall, the previously cited extraction technologies are energy-efficient given the fact that they allow reaching high yields in reduced time. Further innovative processes are in continuous development to meet the industry's need for energy savings while maintaining a constant quality of the extracts [2, 10].

Principle 4: Production of Co-Products Instead of Waste to Include the Bio and Agro-Refining Industry

The principle of a biorefinery can be compared to that of a petroleum refinery, where biomass is transformed into chemical products (*e.g.*, fuels). The difference lies in the raw material used and the technology employed. Among the several definitions of biorefinery, the most complete and exhaustive is "the optimized use of biomass for materials, chemicals, fuels and energy applications, where use relates to costs, economics, markets, yield, environment, impact, carbon balance, and social aspects. In other words, there needs to be optimized use of resources, maximized profitability, maximized benefits, and minimized wastes" [29].

The biorefinery concept was first derived from the oil refinery, in which various technologies were used to make multiple chemicals and fuels from petroleum [30]. As reported by Kamm & Kamm (2004) [31], biorefinery can be sorted into three phases: (i) phase I, where a biorefinery is not adaptable and uses one type of biomass, one process, and one specific product. An example of this is the dry-milling ethanol plant, where the corn is milled, saccharified, and fermented into ethanol [32]; (ii) in phase II, more products result from the process. For instance, wet milling also called wet media milling, is a process in which particles are dispersed in a liquid by shear, impact, crushing, or by attrition. This process is used for corn, where it is developed to effectively separate and purify its constituents (starch, oil, protein, and fiber) for use as food and feed ingredients, industrial products, or raw materials for processing into other value-added products. Other end-products are produced, such as syrup, ethanol, and lactic acid [33]. This type opens up many possibilities for linking industrial product lines to existing agricultural production units, thereby substantially improving the overall economic performance of the biorefinery process; (iii) phase III is not only capable of producing a wide range of value-added products, but also of using different types of raw materials and processing methods. Industries are moving towards this biorefinery phase, which will ensure a continuous supply, and therefore will improve the possibility of recovering industrial waste. This phase, therefore, focuses on the flexibility of the raw materials used to be able to adapt to changes in the demand and supply of final products.

Five dimensions must be taken into consideration to optimize the criteria of a biorefinery:

- *Economical*: create an autonomous biorefinery, without reinvestment to increase profitability. Indicators such as profit maximization or net present value allow the evaluation of this dimension. It is therefore important to include the diversification and sale of by-products;
- *Social*: offer more work opportunities, increase the safety of the products to be consumed, and promote responsible working conditions;
- *Environmental*: analyze the problems concerning the quality of air, soil, water, and the management of waste;
- *Technological*: take into consideration new techniques in use or under development, in addition to technological trends in the use and production of bio-based products.
- *Political*: support by sustainability initiatives (regulatory frameworks, certification systems, *etc.*). Generally, it concerns promotions or restrictions adopted by governments [34, 35].

The biorefinery concept covers a wide range of industries and can use a variety of biomass from forestry, agriculture, industry residues, or households including organic residues, wood, agricultural crops, *etc* [36]. It is expected to improve competitive and wealthy countries by sustainably providing bio-based products and energy. Numerous industries, namely the starch, sugar, paper, and pulp industries can be considered biorefineries [36]. Nowadays, a considerable number of biorefineries are developed and many research projects in this field are ongoing [10, 37]. For instance, microalgae have become a highly interesting raw material for renewable fuels. This industry fails to meet the economic needs of the consumers and the capital investment in the operations is overpowering. Interesting co-products result from extracting fractions of algae to enhance the economics of the biorefinery, such as pigments, lipids, carbohydrates, vitamins, antioxidants, and proteins that can be utilized in pharmaceutical, cosmetics, and food fields. Hence, the installation of a microalgae biorefinery is, therefore, conceivable to recover these high-value-added products and to ensure the sustainability of the process [38, 39] (Fig. **2**).

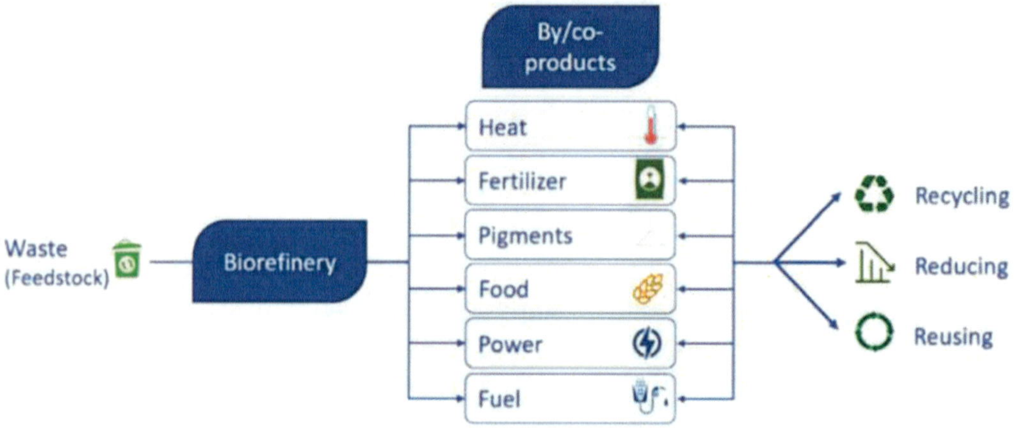

Fig. (2). Waste biorefinery for the valorization of value-added products.

Principle 5: Reduce Unit Operations and Favor Safe, Robust, and Controlled Processes

The challenge for industries in the 21st century focuses on the development of efficient processes in terms of extraction yields, cleanliness of the environment, optimization of space use, and operator safety. Conventional extraction techniques generally follow four steps [10]:

- Pre-treatment: cleaning, drying, and grinding into a powder to increase the surface contact area;
- First solid-liquid extraction with solvent;
- Second solid-liquid extraction by centrifugation or filtration;
- Elimination of the solvent to get the final pure extract.

Those processes are usually very long and high-energy consuming. This issue can be solved by intensification, a process that results in cleaner, safer, and more energy-effective techniques, while increasing the quality, purity, and yield of the products by reducing time, waste, and the quantity of solvents used, thus reducing the environmental footprint [40 - 42].

Intensification goes along with introducing greener and more sustainable techniques to progress substantially compared to other extraction methods. It focuses on replacing long processes, which are high-energy consuming (*e.g.* steam, cooling water), with shorter ones by reducing the number of unit operations, despite the multiple challenges (energy and raw material savings, safety control, reduction in waste, and ecological footprint, *etc.*) [2, 10] (Fig. **3**).

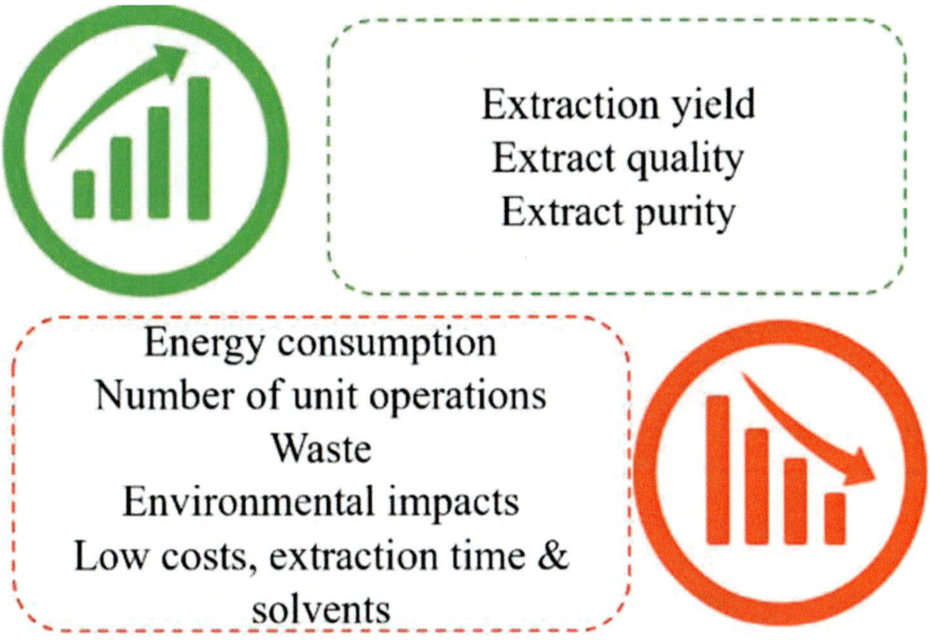

Fig. (3). Principle of intensification of the extraction process.

Solid-liquid extraction is a widely used technique in pharmaceutical, chemical, and biotechnological processes for the recovery of active compounds from plant materials. Processes such as steam distillation [43] or solvent extraction [44] can be high energy-demanding and time-consuming unit operations. Thus, the ford of high-performance techniques becomes requisite [45]. For instance, the application of ultrasound has drawn more attention in the past few years. This method is applied successfully to increase the yield of the extracted compounds. This can be explained by the cavitation phenomena, which produces physical effects (liquid turbulence and circulation currents) that cause a significant mass transfer rate [45]. Shirsath *et al.* applied ultrasonic-assisted extraction to obtain the curcumin from rhizomes of *Curcuma amada*. An interesting yield of 72% was recovered in 1 hour at 35 °C, which is higher compared to the conventional treatment where only 62% were recovered after 8 hours of extraction [46].

Further extraction applications on plants were applied with supercritical carbon dioxide CO_2. It is an alternative that allows the total removal of aqueous CO_2 from the extract after bringing back its pressure to the atmospheric level, thus getting back to the gaseous state. It also enables high yields in, shortened time, with no thermal and hydrolytic degradation of labile compounds and no toxic solvent residues [10, 47]. This extraction technique has been proven to be effective for the extraction of bioactive compounds from *Hibiscus sabdariffa*, where it has been demonstrated to be selective and suitable for organic acids, phenols, and other polar compounds [48].

Principle 6: Aim for A Non-Denatured and Biodegradable Extract Without Contaminants

After extraction, the resulting products have to satisfy several quality criteria. Counter to certain preconceived ideas, the term "natural" does not mean a product that is non-hazardous to the environment or humans. A natural substance is defined according to REACH legislation as "a naturally occurring substance as such, unprocessed or processed only by manual, mechanical or gravitational processes, by dissolution in water, by flotation, by extraction with water, or by steam distillation" [49].

In other terms, a natural substance cannot be chemically changed, thus its physicochemical properties should be appropriately conserved, and free from any additive or residue (pollutants, pesticides, heavy metals, toxins, *etc.*) [50]. The naturalness of a product is not always related to the solvent used for its extraction; in some cases, a product called natural is independent of the solvent used so long as its molecular structure remains intact. In other cases, the natural product is directly related to the type of solvent used, which must be natural or of natural origin. Be that as it may, the resulting extracts must be properly conserved to ensure their quality, preserve their active compounds, avoid denaturation, and ensure their antimicrobial quality for their safety.

According to the ISO 16128 standard, natural ingredients in the cosmetic field are exclusively obtained from plants, animals, microorganisms, or minerals using physical processes (such as grinding, drying, distillation), naturally occurring fermentation reactions, or other (permitted) preparation procedures (*e.g.*, solvent extraction, except non-natural solvents from petrochemicals) without the intention of chemically modifying the ingredient [51].

The resulting green extract or an eco-extract is determined using Life Cycle Analysis (LCA), an innovative framework developed in an environmental sustainability context [52]. As stated by Ilgin *et al.* [53], LCA is "a method used to evaluate the environmental impact of a product through its life cycle encompassing extraction and processing of the raw materials, manufacturing, distribution, use, recycling, and final disposal". This method aims at listing the various impacts of a process ranging from collecting raw materials to disposing of waste, taking into account the consumption of natural resources, energy, and environmental emissions. The latter can be evaluated by calculating the contribution of the flows to the environmental impacts (ecotoxicity, photochemical pollution, eutrophication, ozone layer destruction, *etc.*) [2, 54] (Fig. **4**).

Life-cycle analysis was first introduced in the 1960s when global concerns about the rapid exhaustion of finite raw materials and energy resources prompted an interest in finding ways to understand and predict future energy and resource supply and use. Today, the International Standards Organization released the ISO14040:2006 to define and discuss principles and framework guidelines of environmental management [52, 55].

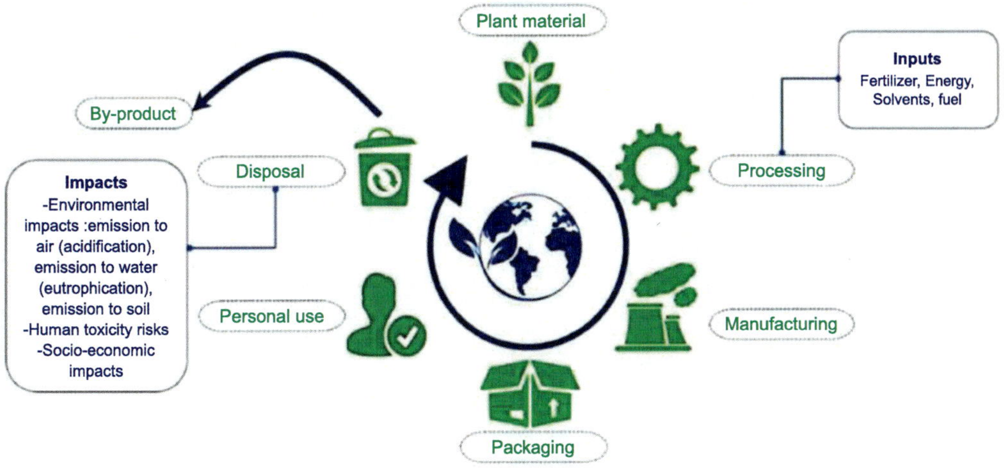

Fig. (4). Life Cycle Analysis (LCA) of natural products extraction

As many extraction techniques require the use of solvents other than water, several points of view are quite different from the definition of the "naturalness" of these resulting substances. Numerous certifications around the world have become essential to market certified "organic" products that have followed very stringent production conditions in various fields, such as the United States Department of Agriculture (USDA) Organic Certification, EU Organic Certification, or Ecocert Certification (Fig. **5**).

Overall, an "eco-extract" must be natural with high functionality properties (flavor, coloring, antioxidant…), low environmental impact and in compliance with specific legislations depending on the sector of application (food, pharmaceutical, cosmetic industry…).

Fig. (5). Examples of labels with organic/natural values.

GREEN EXTRACTION TECHNIQUES

Microwave Extraction

Microwaves are non-ionizing electromagnetic waves with a frequency ranging from 300 MHz to 300 GHz [56]. They serve many purposes, mainly for extraction thanks to their ability to convert partially electromagnetic energy to heat energy [57]. Microwave-assisted extraction (MAE) was first developed by Ganzler [58] for the extraction of biological compounds and their analysis. In the last two decades, many laboratories have studied the possibility of recovery of different classes of compounds (bioactive compounds, phytonutrients, functional food ingredients, pharma-active substances) [59] from plant or animal matrices [3]. Developments in microwave extraction have given two classes of technologies, including Microwave-Assisted Solvent Extraction (MASE) and Solvent-Free Microwave Extraction (SFME) [60].

Principle

Generally, the plant material is immersed in a non-absorbing microwave solvent and then exposed to microwaves. These latter penetrate easily into the matrix and cause an increase in temperature after interaction with polar molecules such as water [61]. MAE occurs as an efficient and rapid extraction method since it generates changes in the cell structure caused by electromagnetic waves [60, 62] compared to conventional methods. Various classes of compounds have been extracted efficiently, such as aromas, phenols, colors, and antioxidants, in addition to other primary and secondary metabolites from several matrices.

The extraction mechanism includes three successive steps: (i) desorption/breakdown of solutes from cells under high pressure and temperature conditions; (ii) diffusion of the solvent into the matrix; (iii) transfer of the solutes from the matrix to the extraction solvent then out of the matrix [63]. The process efficiency and its high extraction yield can be explained by two factors working simultaneously: better solubility and mass transfer and an enhanced disturbance of the surface equilibrium [62, 63] (Fig. **6**). Higher yields can be obtained by selecting solvents with a high tan δ value, and by increasing the temperature, which allows faster penetration of solvent into the cell walls. Tan δ or the dissipation factor is given by the following equation (1):

$$\tan \delta = \varepsilon'' / \varepsilon' \quad \textbf{(1)}$$

Where ε" is the dielectric loss which is the efficiency of converting microwaves into heating and ε' corresponds to the dielectric constant, *i.e.*, the measure of the ability to absorb microwave energy. Examples of solvents such as ethanol and methanol are less microwave absorbing than water due to their low dielectric constant, but the tan δ value will remain higher than water. Other solvents like chloroform or hexane do not absorb microwave energy, thus they produce no heat [57, 64].

In MAE, both heat and mass transfer occur from the inside to the outside, as opposed to conventional extraction, where the heat transfer occurs from the outside to the inside. Moreover, the heating of the solvent and the solid matrix is rapid, homogeneous, and effective [61].

Solvent-free microwave extraction (SFME) is a recent technique patented in 2004 and designed for essential oil recovery from aromatic plant material [65]. It is a combination between the distillation method and microwave heating. The extraction is achieved at atmospheric pressure. It is based on placing the biological material in a microwave reactor without any solvent, which is heated using microwave irradiation, leading to the evaporation of water situated in the cell that breaks out and releases exudates rich in active components, which in this case are oil molecules [64]. A cooling system cools continuously the extract, and the surplus water flows back to the extraction vessel to recover the in-situ water in the plant [62, 65].

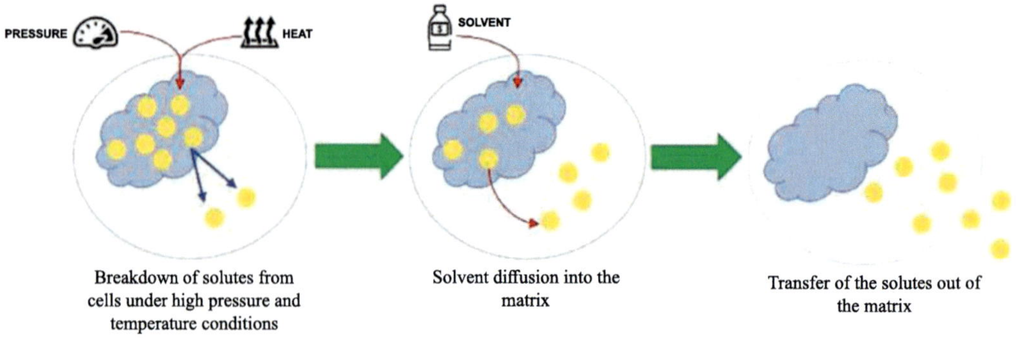

Fig. (6). Extraction mechanisms in MAE of natural products

Microwave Hydrodiffusion and Gravity (MHG) is a new process designed also for extracting essential oil with no solvent or water. Patented in 2008, this technique combines microwaves for the hydrodiffusion of the active compounds outside the biological material, and earth gravity to collect and separate the extracts [60, 66]. It consists in placing the plant material in the microwave reactor at atmospheric pressure. The in-situ water is then heated causing cell distension and the breaking out of the glands enclosing the essential oil. Then occurs the hydrodiffusion phenomenon, where the in-situ water and the freed essential oil are dropped by gravity out of the microwave reactor and are collected in a Pyrex vessel after passing through a cooling system. The extract is then transferred to a "Florentine flask", a traditional vessel used for water and oil separation. The essential oil floats at the top since it is lighter than the water. Unlike distillation, evaporation, MAE, and SFME, which use solvents and evaporation, MHG is a very advantageous technique since it is low energy-consuming.

Instrumentation

Microwave-assisted extraction systems are available in two forms: closed vessels brought about by controlled pressure and temperature, and open vessels with atmospheric pressure. Most closed systems are available as multi-mode microwaves, enabling random dispersion of the microwave radiations, which irradiate evenly the biological matrix and accelerate the mass transfer of active compounds (Fig. 7). In these methods, higher temperatures can be reached thanks to the increased pressure inside the extraction vessel, there is a low risk for losing volatile compounds and the contamination is decreased.

Other devices were developed such as focused microwave ovens, where only the part of the extraction vessel containing the sample is irradiated. In this case, the solvent is heated and refluxed through the sample to allow homogeneous and efficient heating. As for microwave-assisted extraction without solvent, two main techniques are patented: Solvent-Free Microwave Hydrodistillation (SMFE) and Microwave Hydrodiffusion and Gravity (MHG) (Fig. **8**).

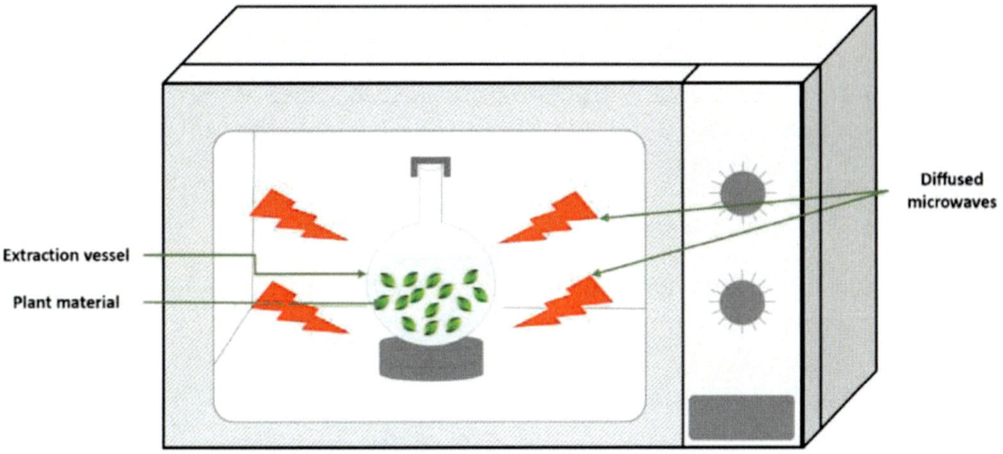

Fig. (7). Microwave-assisted solvent extraction.

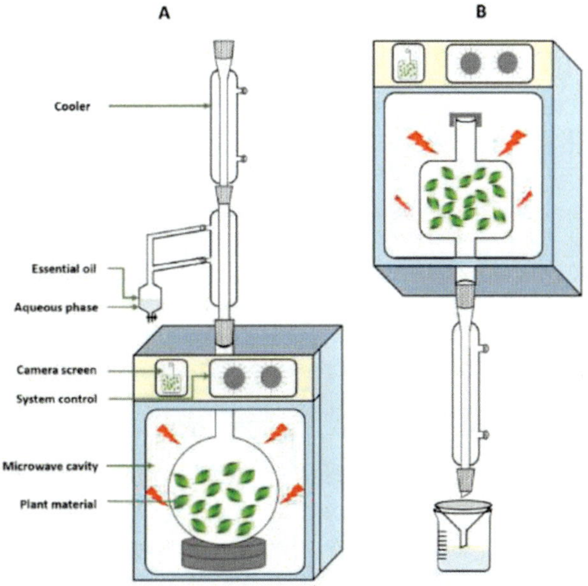

Fig. (8). Solvent-free microwave hydrodistillation (SFME) (A) and microwave hydrodiffusion and gravity (MHG) (B).

Open-vessel systems, on the other hand, help increase safety since the extraction is performed at atmospheric pressure and allows extraction over a large number of biological samples without a cooling system.

Microwave extraction is a potential green extracting technique designed to substitute the traditional solid-liquid extraction since it uses less solvent and achieves the extraction in less time with higher yields. This method is in continuous development; scientists are upgrading microwave extraction by deriving new techniques such as Ultrasound And Microwave-Assisted Extraction (UMAE), Vacuum Microwave HydroDistillation (VMHD), on-line Dynamic Microwave-Assisted Solvent Extraction (on-line DMASE), Microwave-Integrated Soxhlet Extraction (MIS), u-column Microwave-Assisted Solvent Extraction (U-column MASE) and more [60, 64]

Applications In Food Analysis

Several studies have been realized to compare MAE with conventional extractions (Table **1**), ranging from flavors and fragrances to oil and antioxidants. Three major parameters can be studied to enhance the efficiency and the yields, including irradiation time, power, and the solvent used.

Tran *et al.* [67] have described extracting essential oil from a Vietnamese variety of basil using Microwave-Assisted Hydro-Distillation (MAHD), which was chosen over a conventional method (hydrodistillation) for its reduced energy consumption, decreased time and cost, and low environmental footprint. This study was achieved with a Response Surface Methodology (RSM), which is widely used for optimization. In this instance, four factors were investigated: raw material size, water to leaves ratio, extraction time, and microwave power. The results demonstrated that an essential oil yield of 0.6% can be extracted from ground basil leaves under the optimal conditions of water to leaves ratio of 3, 2:1, an extraction time of 97 minutes, and a microwave power of 430 W. The main components of the extracted oil were Estragole (87.869%), α-Bergamotene (2.922%), τ-Cadinol (2.770%), and Linalool (1.347%).

Solvent-Free Microwave Extraction (SFME) is another advantageous technique widely employed for essential oil extraction from the fresh or premoistened dried matrix and is known for being efficient, low-energy, and time-consuming with very low environmental impact. Liu *et al.* [68] compared the efficiency of this method with conventional hydrodistillation to recover essential oil from *Cinnamomum camphora* leaves. This process was optimized with RSM coupled with a Box-Behnken design (BBD). SFME was proven more efficient with a yield of 3.51 ± 0.12% in 23 min *versus* 3.35 ± 0.10% in 240 min. Moreover, higher oxygenated compounds were extracted (83.93% *versus* 74.81%), with lower energy consumption (0.22 kWh *versus* 4 kWh) and a decreased environmental impact (177.87 g CO_2*versus* 3200 g CO_2).

Other studies were published for antioxidant extraction. Bouras *et al.* [69] have investigated the extraction of antioxidant compounds from *Quercus* bark using MAE. A comparison of the results of this extraction with another conventional one proves that microwave irradiation allows higher yields of antioxidants under optimized conditions (45 W, 60 min, 33% ethanol, and 0,38% methanol). Alongside that, extraction effectiveness tripled for total phenolic content and doubled for antioxidant recoveries compared to other conventional techniques with the same parameters but without microwaves. The resulting extracts were rich in some phenolic compounds, namely (+)-catechin, (−)-epicatechin, (−)-epigallocatechin, syringic acid, p-coumaric acid, sinapic acid, and naringenin. Antioxidants were also obtained using microwave irradiation from olive tree leaves. Şahin *et al.* [70] optimized conditions to get optimal total phenolic content (TPC) and oleuropein yields with a Solvent-Free Microwave-Assisted Extraction (SFMAE): 250 W, 2 min, and 5g of sample. The maximal yields of TPC and oleuropein were 2.480 ± 0.060 ppm and 0.060 ± 0.012 ppm respectively, and the extracts demonstrated anti-microbial activity against *S. aureus* and *S. epidermidis* with a minimum inhibitory concentration (MIC) value of 1.25 mg/mL.

Microwave-assisted extraction was used by Hu *et al.* [71] for oil extraction from Silkworm pupae; which resulted in a yield of 30.16% under optimal conditions of 360W, liquid to solid ratio of 7.5/1 mL/g, 29 minutes, and a solvent mixture of ethanol and hexane (1:1, v/v). Even though this yield is similar to the one obtained with the conventional Soxhlet extraction, the oil's oxidation stability and physicochemical properties were increased as well as its content in phenolic compounds. Interesting results were also observed in Yanik's report [72] on the microwave-assisted solvent extraction parameters of olive pomace oil. A maximum yield of 6.85 g/100 g dry matter was recovered with a closed vessel pressurized microwave under the following optimized conditions: 287 W, 10:1 solvent to matrix ratio, and 16 min. In addition to that, both the caffeic acid and tocopherols content were higher with a total of 985 mg/ kg oil and 278.07 mg/kg oil respectively, and a decreased amount of peroxides and polycyclic aromatic hydrocarbons content (17.8 meq O_2/kg oil and 0.44 µg benzo[α]pyrene /kg oil respectively).

Natural food color extraction with microwave irradiation was studied by Jafari *et al.* [73] in his paper. Saffron flowers represent a rich source of anthocyanins, which are widely used to substitute synthetic colorants in industries but are so vulnerable to drastic extraction parameters such as pH, temperature, light, solvents, *etc.* Microwave-assisted extraction occurs as a rapid and efficient technique since it destructs the cell wall and enhances quick solvent penetration. Optimization of the extraction process was studied (77.5 ml/g, 48 °C, 9.3 minutes), leading to a maximal yield of anthocyanins of 101 mg/g.

To conclude, MAE can be considered a boon for biological compounds' extraction thanks to its effective heating, fast energy transfer, reduced equipment size, low environmental impact, short extraction time, increased yields, and low solvent use.

Table 1. Microwave-Assisted Extraction.

Matrix	Analyte	Experimental Conditions	Detection and Analysis	References
Flavor and Fragrances				
Vietnamese basil leaves	Essential oil	MAHD **A,** atm **B**, 430 **C**, - **D**, 97 **E**, water-to-material ratio of 3.2:1 **F.** essential oil yield 0.6%. Major components: Estragole (87.869%), α-Bergamotene (2.922%), τ-Cadinol (2.770%), and Linalool (1.347%).	GC-MS	[67]
***Cinnamomum camphora* leaves**	Essential oil	SFME **A,** atm **B**, 580 **C**, - **D**, 23 **E**, - **F.** 3.51 ± 0.12% in 23 min for SFME *versus* 3.35 ± 0.10% in 240 min for conventional hydrodistillation	GC-MS	[68]
Antioxidant Extracts				
***Quercus* bark**	Polyphenols	MAE **A,** atm **B,** 45 **C,-D,** 60 **E,** 33% ethanol; 0.38% methanol **F.** extraction efficiencies increased by 3 times and 2 times respectively for total phenolic content and antioxidant recoveries compared to conventional methods that use the same parameters but without MAE.	LC-MS	[69]
Olive leaves	Polyphenols	SFMAE **A,** atm **B,** 250 **C,-D,** 2 **E,-F.** Best recoveries of TPC and oleuropein: 2.50 and 0.06 ppm of, respectively.	HPLC-DAD	[70]
Oil Extraction				
Silkworm pupae	Oil	MAE **A**, atm **B**, 360 **C**, - **D**, 29 **E**, ethanol - n-hexane (1:1, v/v) **F.** oil extracted by MAE: higher content of total phenolic, stronger antioxidant activities	GC-MS	[71]

(Table 1) cont.....

Matrix	Analyte	Experimental Conditions	Detection and Analysis	References
Olive pomace oil	Oil	MASE **A**, atm **B**, 287 **C**, - **D**, 16 **E**, 10:1 (solvent to-sample ratio) **F**. higher total phenolic content and amount of tocopherols (278.07 mg/kg oil) and lower peroxide value (17.8 meq O2/kg oil) and polycyclic aromatic hydrocarbons content (0.44 μg benzo[α]pyrene /kg oil) in microwave-assisted solvent extraction compared to the conventional industrial extraction method	GC-MS	[72]
Natural Food Colors Extraction				
Saffron (Crocus sativus) flower's tepal	Anthocyanins	MAE **A**, atm **B**, 360 **C**, 48 **D**, 9.3 **E**, solvent ratio to sample 77.5 ml/g **F**. rapid and efficient technique for saffron anthocyanins due to disruption of cell walls under microwave irradiation.	pH differential method	[73]
A: technique, **B**: pressure, **C**: power (W), **D**: temperature (°C), **E**: time (min), **F**: solvent.				

Ultrasound-Assisted Extraction

Extraction is a crucial step when it comes to the recovery of bioactive compounds from biological matrixes on the industrial scale. Greener techniques are increasingly being used to serve this purpose and to substitute conventional techniques that consume large amounts of solvent and result in relatively low yields with long extraction times. Ultrasound-assisted extraction (UAE) is a process for the extraction of a substance from any matrix to a suitable liquid phase (extraction medium), assisted by ultrasonic waves (> 20 kHz frequency) which propagate through the liquid media. It has proven its effectiveness in recovering compounds of interest and preserving their quality. The cavitation phenomenon effectively increases extraction yields while decreasing the amount of energy, solvent, and extraction time [74]. UAE has been considered a valuable tool in engineering processes and it was employed for the extraction of various active compounds, namely phenols, aromas, proteins, polysaccharides, and more [12, 75].

Principle

Ultrasonic waves have frequencies ranging from 20 kHz to 10 MHz, which are above the human hearing frequencies (20 Hz - 20 kHz) [60] (Fig. **9**). The main variables of an ultrasonic extraction are:

- Amplitude;
- Frequency: depends on the equipment utilized for the extraction;
- Power: represents the energy transmitted to the matrix through the medium. It is expressed as power (W), intensity (W/cm^2), or power density (W/cm^3 or W/mL). The ultrasound power can be calculated as follows (2):

$$P = m \times C_p \times \frac{dT}{dt} \quad \textbf{(2)}$$

Where *Cp* is the heat capacity at constant pressure (J/g K), *m* is the mass of the solvent (g), and *dT/dt* is the temperature variation according to time (K/min). As for the ultrasonic intensity, it can be expressed with the calculated power as shown in equation 3:

$$UI = \frac{4P}{2\pi D} \quad \textbf{(3)}$$

Two approaches to the application of ultrasound have been studied throughout the past years [75]:

- High-frequency low-intensity ultrasound: used for monitoring the quality of the products or processes (*e.g.*, evaluation of cheese maturation);
- Low-frequency high-power ultrasound: focuses on the improvement of the process (microbial inactivation, extraction, food cutting operation, emulsions, *etc.*).

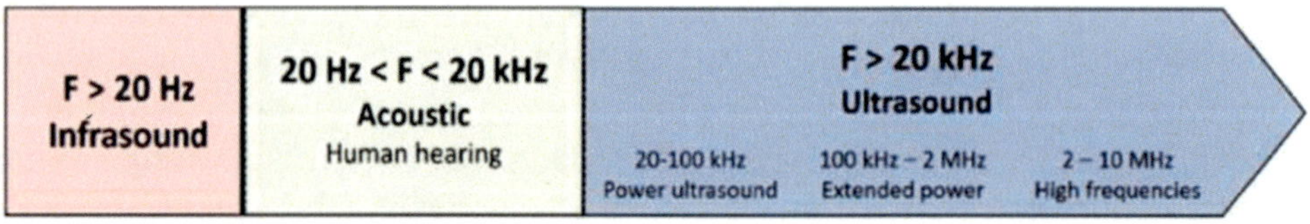

Fig. (9). Frequency ranges.

The extraction process can be explained as follows (Fig. **10**): ultrasonication occurs when the acoustic power input increases, inducing the production of cavitation bubbles, which are initiated in the nucleation sites (1). These bubbles go through a series of compression and rarefaction phases while the ultrasound waves pass through the solvent. After reaching a critical size, they implode during the compression cycle on cell surfaces as they generate micro-jetting (2) that engender numerous effects including erosion, peeling, and particle breakdown. This implosion can be explained by the conversion of sonic energy into mechanical energy due to the high pressure and temperature estimated to be up to 100 MPa and 5000 K respectively, which increases the rate of mass transfer of analytes to the solvent (3) [76, 77]. Furthermore, the bubbles provoke macro-turbulences leading to a micro-mixing [12, 60, 61, 78, 79]. At the molecular level, the nucleation sites are in fact voids created in the liquid once the molecules exceed the critical distance after being displaced and then replaced with their initial position due to the movements of the sound waves and the kinetic energy involved [60].

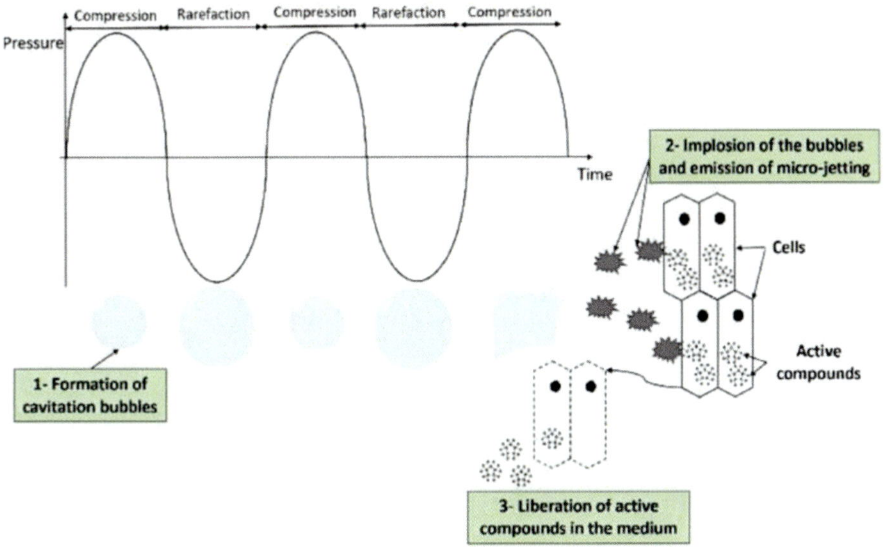

Fig. (10). Ultrasound-assisted extraction – mechanism.

Other applications of ultrasound have been studied based on different principles other than cavitation, such as uniform heat transfer (drying, freezing, cooking), increasing mass transfer (marinating), compression-rarefaction phenomenon (degassing), and vibration (filtrating and demoulding) [80].

Instrumentation

Three devices are commonly used in UAE: baths, reactors, and probes, all based on transducers and a source for ultrasonic power. Ultrasonic baths are the most available and widespread form of ultrasonic equipment and are available in the form of stainless-steel tanks with one or many transducers which convert electric energy into sound energy by mechanical vibrating at ultrasonic frequencies [76]. They can operate at 40 kHz and can be found equipped with temperature control. One advantageous aspect of this apparatus is that it can treat simultaneously two or more samples. Despite that, its low power, and low intensity delivered due to the attenuation by the water and the glass containers of the samples reduce its reproducibility [78]. The reactors are a form of the ultrasonic bath. They are very frequently used for extractions and are available in a cylindrical stainless-steel container, equipped with a double jacket for temperature control. They were developed by REUS, and they operate at 25 kHz.

High-power ultrasonic probes are the most widely used in extraction. They can deliver up to 100 times greater ultrasound intensity compared to ultrasonic baths. Probes with smaller surfaces are the most powerful (considering equation 2). They are generally connected to a transducer and they operate at 20 kHz while immersed in the medium, which results in direct delivery of the ultrasonic energy with a minimal loss. Probes are available in different lengths, tip shapes, and diameters. The ultrasonic intensity releases high energy and causes rapid heating of the medium. Thus, it is necessary to connect the system with a cooling double-jacket [76, 78] (Fig. **11**). Other devices were developed for larger volumes like continuous or recycled-flow sonoreactors, which consist of stainless steel reactors where the mixture is pumped continuously at atmospheric or high pressure. The system can be coupled with a double jacket for temperature control.

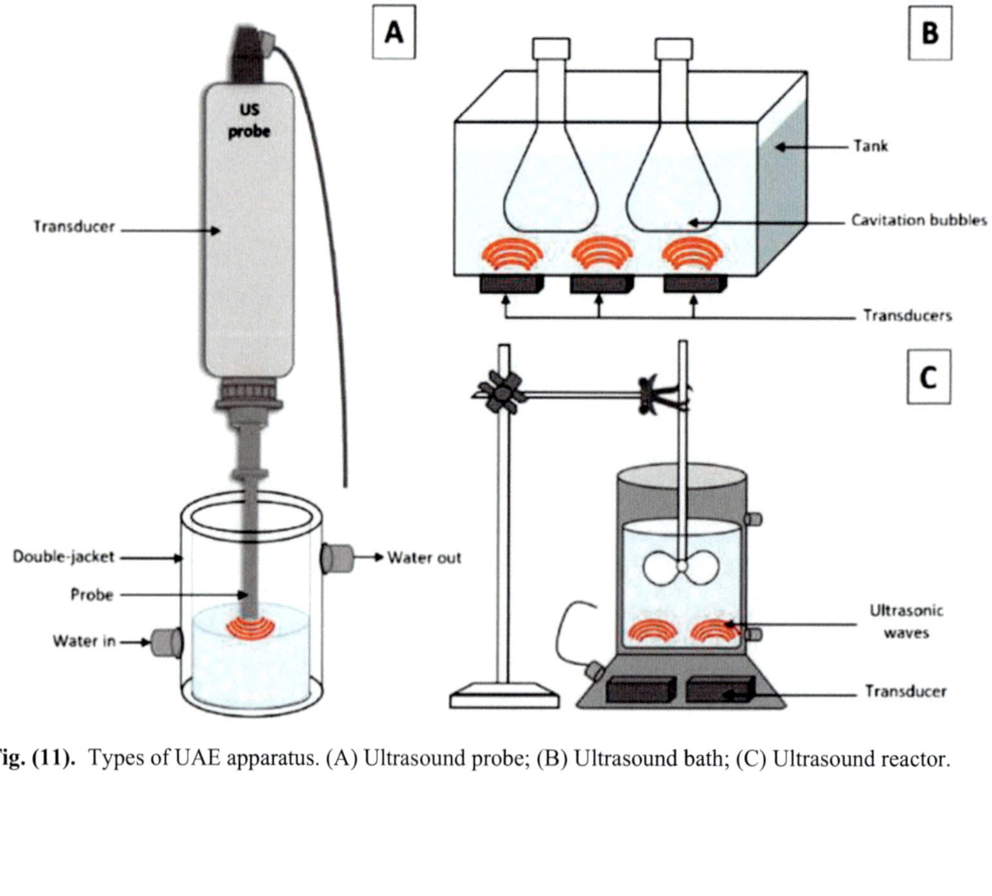

Fig. (11). Types of UAE apparatus. (A) Ultrasound probe; (B) Ultrasound bath; (C) Ultrasound reactor.

Up-scaling the UAE required upgraded equipment, namely multi-transducer ultrasound apparatus that provides higher power for large samples, and other combinations of the existing ultrasonic equipment with other techniques/solvents. For instance, Milani *et al.* [81] coupled ultrasound and Deep Eutectic Solvents (DES) for steviol glycosides extraction. DES represents an interesting alternative for a greener extraction and lower environmental impact. Ultrasounds were also coupled to microwaves for Harpagoside recovery from *Harpagophytum procumbens* [82] and were compared to conventional techniques. This method proved its efficiency since it was energy-effective, time, and solvent-saving. Other examples such as Sono-Soxhlets were used to improve oil extraction and to accelerate the conventional Soxhlet extraction along with reducing its solvent consumption [78].

The shapes and sizes of the ultrasonic equipment represent important parameters to optimize the extraction. It is important to flat bottom vessels with a minimal thickness to reduce the reflection and the attenuation of the ultrasonic waves [83]. The optimum size of the vessel and the position of the probe related to the transducer should be calculated to reduce energy loss.

Applications In Food Analysis

Ultrasound has been used on many matrices for the extraction of antioxidants, proteins, lipids, and other active compounds (Table **2**). This technique is increasingly used in an eco-extraction context, targeting the reduction of solvent, time, and energy consumption while increasing extraction yields, and thus replacing the conventional techniques.

Sicaire *et al.* [84] extracted oil from rapeseeds with an ultrasonic probe and compared it with a conventional maceration. The optimization was carried out by the RSM method, and the optimized parameters were defined as follows: 7.7 W/cm^2 for ultrasonic power intensity, 40 °C for processing temperature, and a solid/liquid ratio of 1/15 to obtain up to 97% of total oil. UAE has demonstrated its efficiency in terms of extraction, reduction of time, and solvent quantity as well as a better quality of the oil. In another study, ultrasound was explored for protein extraction from spirulina. Vernès *et al.* [85] optimized protein recovery using manothermosonication (MTS), a green and innovative technique that combines conventional ultrasound-assisted extraction and pressure and temperature control. It is indeed an effective method since it increased the yield to 229% compared to the conventional method without ultrasound, resulting in more than 28 g per 100 g of dry weight.

Ultrasound can also be exploited as a pre-treatment. Zhu *et al.* [86] extracted proteins and polyphenols from purple sweet potato under the following optimal conditions: 40 minutes, pH 2,5, 58% ethanol, and a supplementary hot extraction at 80° for up to 120 minutes. The resulting yield obtained was about 3.877 mg/g of polyphenols including 0.293 mg/g of anthocyanins, and 0.753 mg/g of proteins. The energy consumption calculated was advantageous and minimal (8406 J/mg).

Khadhraoui *et al.* [87] described the mechanisms responsible for the desired effect of ultrasounds on rosemary leaves. It highlights a succession of impacts, namely erosion; shear forces, sonoporation, fragmentation, capillary effect, and detexturation. UAE processing can also be interesting for combined utilization. In this context, another recent study of our team (Green laboratory-Avignon University in France) discussed by Tabib *et al.* [6] have described the separation of almond skin and the extraction of polyphenols using ultrasounds. The influencing parameters in this process include temperature, time, and ultrasonic intensity. Phenols extraction was 258% higher with UAE compared to the conventional method, and many effects were observed macroscopically and microscopically after observation by Scanning Electron Microscopy (SEM), such as erosion, fragmentation, and sonoporation.

Table 2. Ultrasound-Assisted Extraction.

Matrix	Analyte	Experimental Conditions	Detection and Analysis	References
Rapeseed oil	Fats	40 **A**, 15 **B**, 20 **C**, 7.7 W/cm^2**D**, Hexane **E**, Ultrasonic probe **F**	GC-FID	[84]
Spirulina	Proteins	24 **A**, 20 **B**, 20 **C**, 55 W/cm^2**D**, Sodium phosphate buffer 1:20 (g/g) **E**, Probe **F**	Lowry method	[85]
Purple Sweet Potato	Proteins, polyphenols	25 **A**, 40 **B**, 45 **C**, 178 W **D**, 1:20 (PSP/ Hydroalcoholic solution) **E**, Ultrasonic bath **F**	HPLC-DAD-ESI-MS^2	[86]
Rosemary leaves	Phenols	25 **A**, 60 **B**, 20 **C**, 41.7 W/cm^2 ± 0.4 **D**, demineralized water **E**, Probe **F**	UHPLC	[87]
Almonds	Polyphenols	20 **A**, 20 **B**, 20 **C**, 9.47 W/cm^2**D**, Distilled water **E**, Probe **F**	UPLC DAD/ ESI-MSn	[6]
Pomegranate wastes	Carotenoids	51,5 **A**, 30 **B**, 20 **C**, 130 W **D**, Sunflower oil (peel/solvent ratio 0,10) **E**, Probe **F**	Spectro -photo -metry 470 nm	[88]
Mulberries	Anthocyanins	48 **A**, 10 **B**, 24 W **C**, 200 **D**, Methanol 76% **E**, Probe **F**	UHPLC-QToF-MS	[89]

A: temperature (°C), **B**: time (min), **C**: frequency (kHz), **D**: power, **E**: solvent, **F**: type of ultrasound apparatus

A high yield of carotenoids recovered from pomegranate peel was observed by Goula *et al.* [88] with optimal conditions in UAE (51.5 °C, peel/solvent ratio of 0:10, 58.8% of amplitude level, and sunflower oil as a solvent). Similarly, optimal parameters were investigated for the determination of anthocyanin and total phenolic compounds in mulberries [89], including methanol composition, solvent ratio, temperature, amplitude, cycle, and pH. This method showed high reproducibility and repeatability and was proven to be easy and eco-friendly.

Subcritical Water Extraction

Water is very abundant and used in daily life, and its physicochemical variations make it a very interesting solvent. The subcritical state is reached when water is kept in a liquid state at a temperature between 100 and 374 °C under pressure between 1 and 221 bars [90]. Several studies have been conducted within this context to evaluate the feasibility of extracting bioactive compounds (polar and nonpolar) from natural matrices.

Principle

Subcritical water extraction (SWE), also called superheated water extraction or pressurized hot water extraction, is a green technique that uses only water as a solvent at its subcritical state [91, 92]. SWE depends on several parameters: temperature, pressure, solvent flow rate, and particle size. Other factors can also affect the extraction like co-solvents, time, and solvent/sample ratio. Other than the pressure and water temperature, SWE conditions could be compared to a conventional maceration.

The critical point can be defined as the state where water properties (thermal behavior, density, viscosity) are modified under two main parameters: temperature (374 °C) and pressure (221 bar). Under this point, water is kept in the liquid state by avoiding its vaporization along with varying its temperature between 100 and 374°C and its pressure between 1 and 221 bar [12, 93]. Changes in the chemical and physical properties of water lead to a variation in its polarity, measured by the dielectric constant ($\varepsilon = 80$ at 25 °C and atmospheric pressure) [92]. This parameter gets lower at the subcritical state, and water becomes similar to low-polar organic solvents allowing the extraction of medium polar compounds such as terpenoids, aromatics, or polyphenols. For instance, the dielectric constant of water at 250 °C and 50 bars equals that of ethanol ($\varepsilon = 27$) and becomes close to that of methanol ($\varepsilon = 33$) at 25 °C and 1 bar [12, 21, 91].

Subcritical water is a non-toxic, non-corrosive, and non-aggressive solvent. It enhances the extraction yield and the quality of the extracts in a shorter time. In addition to that, the viscosity is decreased and the diffusibility is increased. It is a better solvent for hydrophobic compounds compared to ambient water since its concentration in H_3O^+ and OH^- varies as the temperature increases [90]. Four major steps occur in SWE [94] (Fig. **12**):

- Step 1: solute desorption from the active sites in the matrix under high temperature and pressure;
- Step 2: diffusion of the solutes into the matrix;
- Step 3: diffusion of the solutes from the matrix into the solvent;
- Step 4: separation of the extracts from the solvent by chromatography.

Thus, several improvements can be considered, including the solubility and mass transfer from the matrix to the solvent as well as the destruction of the surface material promoted by the high pressure.

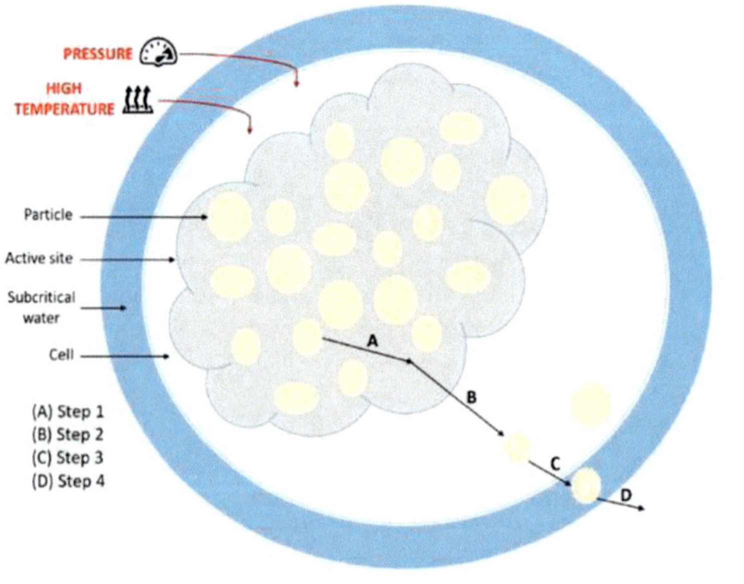

Fig. (12). Mechanism of active compounds extraction from a natural matrix using subcritical water as a solvent.

Instrumentation

SWE equipment includes two main methods of extraction: (i) static or batch mode, where we have two possible alternatives: in the first one, the plant material is introduced in a column with a neutral filler. The column is afterward put in an oven and filled with water using a pump. Under controlled temperature, pressure, and continuous stirring, the solvent, and the matrix stay in contact for a specific time. Then they are drained out with a valve along with the gas, cooled, and collected. The second way is a closed system under pressure and high temperature (Fig. **13**); (ii) dynamic or continuous mode is similar to percolation, where the subcritical water flows through the plant material placed previously in the extraction vessel with a controlled solvent flow rate. It consists of a pump, an extraction chamber, a collector, a heating system, and a pressure control system. A constant temperature is maintained throughout the extraction process thanks to a stainless tube coiled inside the heater [12, 94].

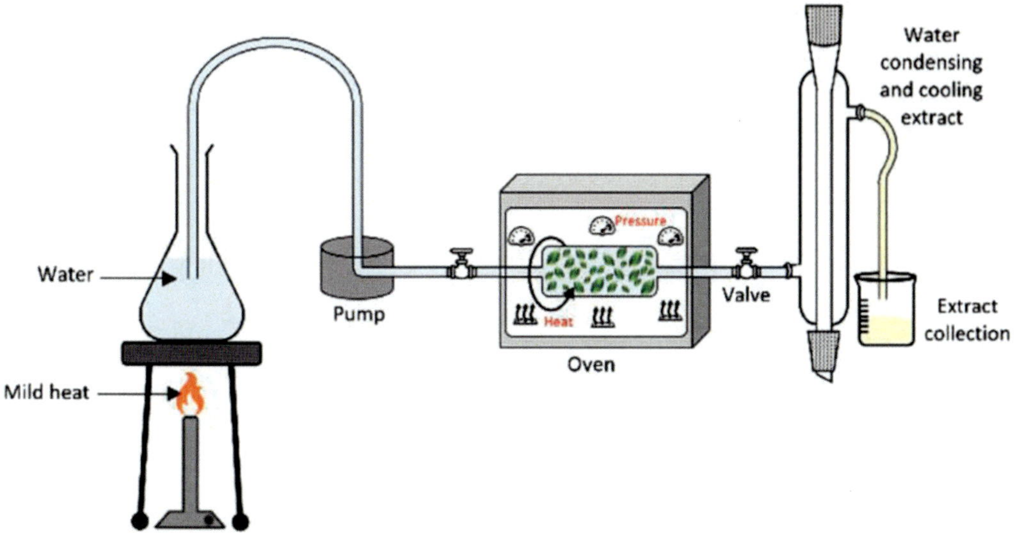

Fig. (13). Schematic representation of a subcritical water extraction unit; an example of a laboratory-scale batch setup.

SWE can be combined with other green techniques to improve its efficiency. For instance, microwave subcritical extraction is an interesting method since it decreases the pressure needed to reach the subcritical state and produces higher yields. A recent study carried out by Yang *et al.* [95] showed that this combination is a great and promising alternative for conventional extraction of steviol glycosides from *Stevia rebaudiana* leaves because it reduces significantly the extraction time from 45 min to only 1 minute with less solvent used to have the same yield.

Applications In Food Analysis

The feasibility of subcritical water extraction was explored on various matrices and several articles have been published in this regard. This technique is in full expansion and more and more trendy. Table **3** shows that SWE can be applied for the recovery of various compounds: carbohydrates [96, 97, 99, 101], phenolics

[96, 97, 100, 101], essential oil [98], sugars, and more. The interest in the industrialization of this process is increasing since it was initially used on a small scale.

Table 3. Subcritical Water Extraction.

Matrix		Analyte	Experimental Conditions	Detection and Analysis	References
Potato peel		Carbohydrates Phenolics antioxidants	40 **A**, 9 **B**, 190 **C**	HPLC-UV	[96]
Tobacco	Leaves	Carbohydrates phenolics	580 **A**, 25 **B**, 160 **C.** Solvent-solid ratio 23 mL/g	HPLC-VWD	[97]
	Scarp		580 **A**, 23 **B**, 150 **C.** Solvent-solid ratio 28 mL/g		
	Dust		580 **A**, 20 **B**, 160 **C.** Solvent-solid ratio 10 mL/g		
	Midrib		580 **A**, 25 **B**, 150 **C.** Solvent-solid ratio 30 mL/g		
Japanese mint		Essential oil	200 **A**, 5 **B**, 180 **C**	GC-FID	[98]
Chlorella pyrenoidosa		Proteins	55 **A**, 10 **B**, 270 **C.** 15% Microalgal Biomass	Lowry assay	[99]
		Carbohydrates	23,2 **A**, 5 **B**, 220 **C.** 10% Microalgal Biomass	Phenol-sulfuric acid method	
Kanuka		Phenols	40 **A**, 20 **B**, 170 **C.** solid-t--solvent ratio 15 g/L	HPLC-DAD	[100]
Ginseng roots		Sugars Phenols	60 **A**, 20*2 **B**, 200 **C.**	Phenol-sulfuric acid method; UFLC–MS/MS	[101]
		Proteins	60 **A**, 20*2 **B**, 180 **C**	Bradford assay	
A: Pressure (bar), **B**: time (min), **C**: Temperature (°C)					

Supercritical Fluid Extraction

Supercritical Fluid Extraction (SFE) is among the most widely-used techniques for the extraction of bioactive products from plants. It uses a fluid phase with properties between those of a gas and a liquid to influence the solubilization of the active compounds [102]. This method was first introduced to the large-scale setup in the 1970s for various applications, including recovering flavors and pungencies from spices, decaffeination of coffee and tea, and more [103]. The application of SFE for purification and extraction fields will keep on growing.

Principle

SFE is based on the use of supercritical fluids (SF) as a green alternative for organic solvents. The supercritical state is a homogeneous phase reached at high pressure and temperature above the critical point [104]. It combines the two characteristics of the liquid state and the gaseous state of the fluid. Indeed, the density of the supercritical fluid is liquid-like and higher than that of the gas state, the viscosity is gas-like, and the diffusivity is twice higher than that of the liquid state [105]. Variations in temperature and pressure lead to changes in the dielectric constant and the solubility of the SF, thus inducing a variable density and selectivity and allowing an enhanced mass transfer of different classes of active compounds [106, 107]. Other parameters that can influence the extraction process include the aspect of the solid matrix (particle size, porosity, sample conditions, pre-treatments...), the solvent and co-solvent (flow-rate, ratio), the solutes (polarity, volatility), the time, the extractor geometry, *etc* [105, 107, 108]. One of the main advantages of this technique is the easy removal of the solvent, which can be achieved by either decreasing the pressure or increasing the temperature after leaving the extraction column. In addition to that, it can be considered a promising technology thanks to its high selectivity, short extraction time, and non-hazardousness to the health and the environment to produce high purity extracts [109].

Supercritical carbon dioxide (scCO_2) is by far the most commonly used in the SFE. Compared to other solvents such as water, hexane, or methanol, CO_2 presents low temperature and pressure parameters (T_C = 31°C and P_c = 7.38 MPa). It is therefore an excellent solvent for thermolabile compounds since it does not degrade them and preserves their chemical composition [108]. Here, the active compound or the mixture of solutes is solubilized in the solvent and then dilated in a container at atmospheric pressure. Here, the CO_2 loses its supercritical solvent properties and releases the extracted compounds.

ScCO_2 is used in more than 90% of SFE extraction for its numerous advantages: availability, reusability, non-explosiveness, high density enabling high selectivity, low toxicity, non-flammability, and low price [104, 107]. Moreover, the separation phase of the extract and the solvent is simpler because the carbon dioxide depressurization leads to its return to the gaseous state and thus allows the recovery of the solvent that can be recirculated in the storage tank.

SFE applications cover the deterpenation of essential oils [110], chemical separations and purifications, regeneration of adsorbents, drying at critical points, wastewater, solid waste treatment, *etc*. This system has been developed over the last few decades, leading to a scaled-up process to reproduce the same extractions achieved at the laboratory scale.

Instrumentation

Like other techniques, the SFE equipment is available at the laboratory, pilot, or industrial scale. The common elements between the three scales are as follows: a pump for solvent, a modifier pump for organic solvents or water, a cell to keep the sample, a pressure regulator, an extraction vessel, and a fractionation vessel. For instance, when CO_2 is used as a solvent, it is transferred as a liquid (below 5 °C) to a heating zone where it is heated under high pressure to reach the supercritical state. It then flows through the extraction vessel through the biological material. The solvent-extract mixture arrives at the separator where the pressure is reduced. This induces the variation of the solvent density, and thus, its dissolving power, inducing the deposition of the extracts and the return of the CO_2 to the gaseous state after cooling and decompression. It can either be recycled or released into the atmosphere (Fig. **14** and **15**) [102, 104, 111].

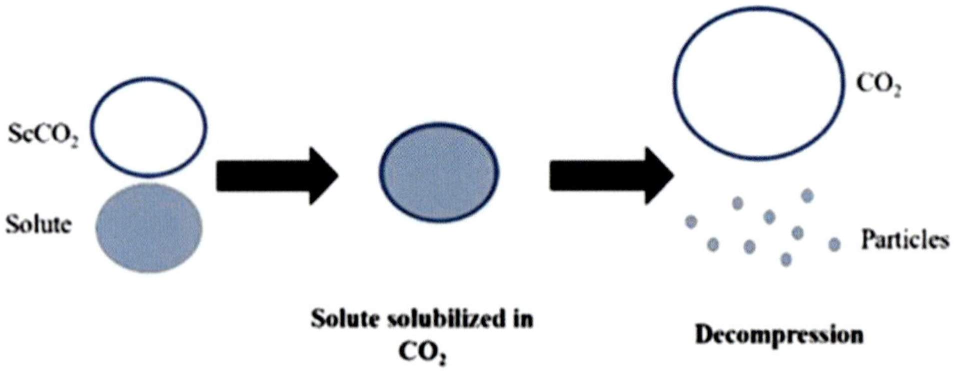

Fig. (14). Role of $scCO_2$ as a solvent in particle formation techniques [107].

A recent patent published by Zhang *et al.* described a new portable supercritical fluid extraction apparatus. Unlike industrial equipment, this device has the advantage of being lightweight and can be carried by a single person, with mechanically passive components. Various solvents can be used, including carbon dioxide, n-hexane, ethanol, methanol, and more [112].

Applications In Food Analysis

The application of SFE on an industrial scale is becoming more and more common to meet the demand for a more sustainable and eco-friendly process.

This technique uses reduced quantities of solvents, usually green, that can be recycled and enhances the extraction yields. Its selectivity allows it to extract several compounds of interest from various matrices, namely phenols, essential oils, and more [113 - 119] (Table **4**).

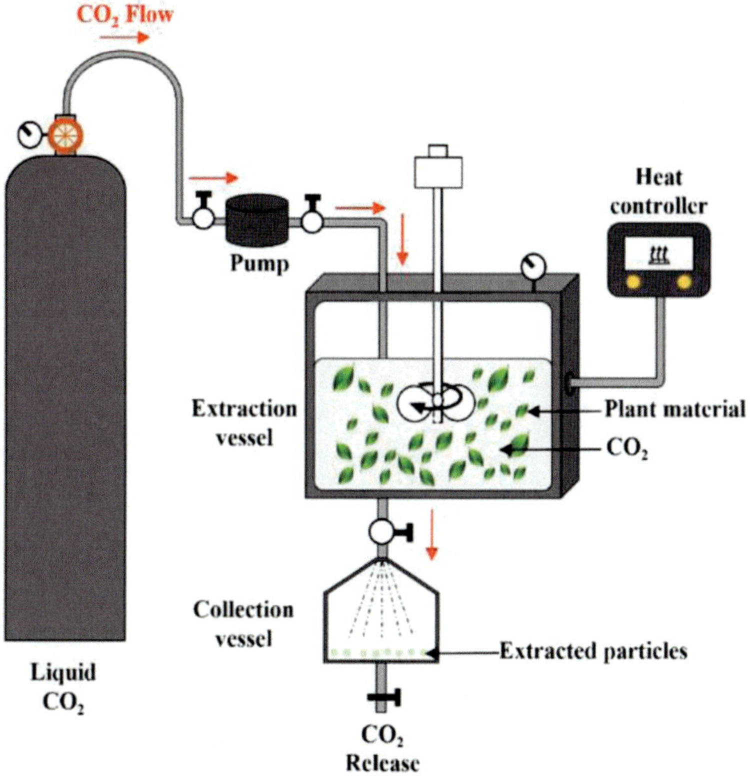

Fig. (15). Schematic diagram of SC-CO2 system.

COMPARISON AND FUTURE TRENDS

Table **5** compares the advantages and limitations of extraction techniques discussed in this chapter: ultrasound-assisted extraction (UAE), microwave-assisted extraction (MAE), subcritical water extraction (SWE), and supercritical fluid extraction (SFE). Here, many requirements in terms of quality control guidelines, reduction or substitution of solvent use, reduced extraction time, and low costs are being set out to avoid the hazardous aspect induced by these techniques, namely problems of water and environmental pollution that are linked directly to the intensive use of volatile organic solvents and higher energy inputs. In this context, a combination of green chemistry with more eco-friendly technologies becomes compulsory.

Table 4. Supercritical Fluid Extraction.

Matrix	Analyte	Experimental Conditions	Detection and Analysis	References
Cacao pod husk	Phenols	60 **A**, 299 **B**, 6 mL/min **C**, 150 **D**, CO_2 + 13,7% Ethanol **E.**	Folin-Ciocalteu method	[113]
Rosmarinus officinalis	Essential oil	40 **A**, 172.4 **B**, 126,2 ± 2 mL/min **C**, 180 **D**, CO_2 **E.**	GC-MS	[114]
Patchouli	Essential oil	40 **A**, 120 **B**,0,1667 g/s **C**, 240 **D**, CO_2 **E.** Particle size 0,3-0,6 mm	-	[115]
Clove (*Syzygium aromaticum*) leaves	Eugenol	40 **A**, 220 **B**, 1,99 g/min **C**, 80 **D**, CO_2 **E.**	GC-MS	[116]
Arbutus unedo	Phenols	55 **A**, 300 **B**, 10 g/min **C**, 90 **D**, CO_2 + 20% Ethanol **E.**	GC-MS, LC-MS/MS	[117]
Cannabis	Cannabinoids	60 **A**, 320 **B**, 150 g/min **C**, 600 **D**, CO_2 **E.**	UPLC-MS	[118]
Ripe bitter melon pericarp	β-carotene	70 **A**, ~390 **B**, 35 mL/min **C**, 190 **D**, CO_2 **E.**	SF chromatography - UPC^2 system	[119]
A: temperature (°C), **B**: pressure (bar), **C**: flow rate, **D**: time (min), **E**: solvent				

Table 5. Comparison of sample preparation techniques.

Name	**Ultrasound-Assisted Extraction (UAE)**	**Microwave-Assisted Extraction (MAE)**	**Subcritical Water Extraction (SWE)**	**Supercritical Fluid Extraction (SFE)**
Description of system	The sample is immersed in a solvent and submitted to ultrasound using a US probe or US bath.	The sample is placed in a vessel with and without solvent and submitted to microwave energy.	The sample is extracted under high pressure (1–22.1 MPa) and temperature (100 °C - 374 °C).	The sample is extracted using a supercritical fluid as a solvent (temperature and pressure are higher than the critical point).
Investment	Low	Moderate	High	Moderate
Ease of operation	Easy to use	Easy to use	Easy to use	Relativity easy to use
Solvent use	10 – 1000 mL per extraction	10 – 1000 mL per extraction	10 – 300mL per extraction	10 – 1000 mL per extraction
Sample size	1-100g	11-100g	1-100g	1-100g
Extraction time	10-60 min	3-30 min	5-30 min	5-30 min

(Table 5) cont.....

Name	Ultrasound-Assisted Extraction (UAE)	Microwave-Assisted Extraction (MAE)	Subcritical Water Extraction (SWE)	Supercritical Fluid Extraction (SFE)
Main disadvantages	Large solvent volume; filtration step required; non-selective	Extraction solvent must absorb microwave energy; filtration step required; low selectivity of solvents and molecules	Hard moisture removal; thermal degradation may occur at higher temperatures; expensive setup	High setup cost; technical knowledge of SF properties required (*e.g.*, phase behavior, cross-over region)
Main advantages	Easy to use; low energy input; rapid extraction; relatively low additional cost.	Rapid extraction; easy to handle; no solvent or moderate consumption; high stability and reproducibility, moderate cost.	Water as a solvent is green, cheap, and readily available; extraction of polar, moderately polar, low-polar, and non-polar compounds can be separated; high diffusion rate	A small amount of organic solvent or no solvent; preserves thermally labile compounds; selective extraction (small number of compounds extracted); is inexpensive to run.

CONCLUSION

Extraction is a limiting step with an important impact on the development, modernization, and valorization of biological material. As the industry develops, requirements for optimizing yields and extract quality are increasing. To serve this purpose, the existing techniques are developed to improve sample preparation and extraction parameters to efficiency while respecting the environmental aspect. An increasing number of reports are published in this regard, including combinations of existing techniques, or reducing hazardous solvent use.

LIST OF ABBREVIATIONS

UAE	Ultrasound-Assisted Extraction
MAE	Microwave-assisted Extraction
SWE	Subcritical Water Extraction
SFE	Supercritical Fluid Extraction
WHO	World Health Organization
VOCs	Volatile Organic Compounds
CMR	Carcinogenic, Mutagenic, and Reprotoxic substances

REACH	Registration, Evaluation, Authorization, and restriction of Chemicals
MHG	Microwave Hydrodiffusion and Gravity
PEF	Pulsed Electric Field
DIC	Détente Instantanée Contrôlée
ISO	International Organization for Standardization
LCA	Life Cycle Analysis
USDA	United States Department of Agriculture
MASE	Microwave-Assisted Solvent Extraction
SFME	Solvent-Free Microwave Extraction
UMAE	Ultrasound and Microwave-Assisted Extraction
VMHD	Vacuum Microwave HydroDistillation
DMASE	Dynamic Microwave-Assisted Solvent Extraction
MIS	Microwave-Integrated Soxhlet extraction
MAHD	Microwave-Assisted Hydro-Distillation
RSM	Response Surface Methodology
SFMAE	Solvent-Free Microwave-Assisted Extraction
DES	Deep Eutectic Solent
MTS	ManoThermoSonication
SEM	Scanning Electron Microscopy
GC-FID	Gas Chromatography-Flame Ionization Detection
HPLC-DAD-ESI-MS^2	High-Performance Liquid Chromatography coupled with Diode-Array Detection, ElectroSpray Ionization, and Mass Spectrometry
UHPLC	Ultra-High Performance Liquid Chromatography
UPLC-DAD-ESI-MSn	Ultra-high Performance Liquid-Chromatography coupled with Diode-Array Detection, ElectroSpray Ionization, and Mass Spectrometry
UHPLC-QToF-MS	Ultra-High Performance Liquid-Chromatography coupled with Quadrupole Time-Of-Flight and Mass Spectrometry
HPLC-VWD	High-Performance Liquid Chromatography coupled with Variable Wavelength Detector
HPLC-UV	High-Performance Liquid Chromatography coupled with UltraViolet
UFLC-MS/MS	UltraFast Liquid Chromatography coupled with Mass Spectrometry
scCO_2	supercritical CO_2
SF chromatography	Supercritical Fluid chromatography
UPC^2	UltraPerformance Convergence Chromatography
LC-MS/MS	Liquid-Chromatography coupled with Mass Spectrometry
GC-MS	Gas Chromatography coupled with Mass Spectrometry

SF Supercritical Fluid

CONSENT FOR PUBLICATION

Not applicable.

CONFLICT OF INTEREST

The authors declare no conflict of interest, financial or otherwise.

ACKNOWLEDGEMENT

Declared none.

REFERENCES

[1] Tiwari BK. Ultrasound: A clean, green extraction technology. Trends Analyt Chem 2015; 71: 100-9. [http://dx.doi.org/10.1016/j.trac.2015.04.013]

[2] Chemat F, Vian MA, Cravotto G. Green extraction of natural products: concept and principles. Int J Mol Sci 2012; 13(7): 8615-27. [http://dx.doi.org/10.3390/ijms13078615] [PMID: 22942724]

[3] Chemat F, Abert Vian M, Fabiano-Tixier AS, *et al.* A review of sustainable and intensified techniques for extraction of food and natural products. Green Chem 2020; 22(8): 2325-53. [http://dx.doi.org/10.1039/C9GC03878G]

[4] Ravi HK, Vian MA, Tao Y, *et al.* Alternative solvents for lipid extraction and their effect on protein quality in black soldier fly (Hermetia illucens) larvae. J Clean Prod 2019; 238: 117861. [http://dx.doi.org/10.1016/j.jclepro.2019.117861]

[5] Li Y, Fabiano-Tixier A-S, Chemat F. Essential oils: from conventional to green extraction. Essential oils as reagents in green chemistry. Springer 2014; pp. 9-20. [http://dx.doi.org/10.1007/978-3-319-08449-7_2]

[6] Tabib M, Tao Y, Ginies C, *et al.* A One-Pot Ultrasound-Assisted Almond Skin Separation/Polyphenols Extraction and its Effects on Structure, Polyphenols, Lipids, and Proteins Quality. Appl Sci (Basel) 2020; 10(10): 3628. [http://dx.doi.org/10.3390/app10103628]

[7] Hilali S, Fabiano-Tixier AS, Elmaataoui M, *et al.* Deodorization by solar steam distillation of rosemary leaves prior to solvent extraction of rosmarinic, carnosic, and ursolic acids. ACS Sustain Chem& Eng 2018; 6(8): 10969-79. [http://dx.doi.org/10.1021/acssuschemeng.8b02347]

[8] Esquivel-Hernández DA, Ibarra-Garza IP, Rodríguez-Rodríguez J, *et al.* Green extraction technologies for high-value metabolites from algae: a review. Biofuels Bioprod Biorefin 2017; 11(1): 215-31. [http://dx.doi.org/10.1002/bbb.1735]

[9] Ivanovs K, Blumberga D. Extraction of fish oil using green extraction methods: a short review. Energy Procedia 2017; 128: 477-83. [http://dx.doi.org/10.1016/j.egypro.2017.09.033]

[10] Chemat F, Abert-Vian M, Fabiano-Tixier AS, *et al.* Green extraction of natural products. Origins, current status, and future challenges. Trends Analyt Chem 2019; 118: 248-63. [http://dx.doi.org/10.1016/j.trac.2019.05.037]

[11] Herrero M, Ibañez E. Green extraction processes, biorefineries and sustainability: Recovery of high added-value products from natural sources. J Supercrit Fluids 2018; 134: 252-9. [http://dx.doi.org/10.1016/j.supflu.2017.12.002]

[12] Chemat F, Strube J. Green Extraction of Natural Products: Theory and Practice. Wiley 2015. [http://dx.doi.org/10.1002/9783527676828]

[13] Wang W, Xu J, Fang H, Li Z, Li M. Advances and challenges in medicinal plant breeding. Plant Sci 2020; 298: 110573. [http://dx.doi.org/10.1016/j.plantsci.2020.110573] [PMID: 32771174]

[14] Bodeker G, van 't Klooster C, Weisbord E. *Prunus africana* (Hook.f.) Kalkman: the overexploitation of a medicinal plant species and its legal context. J Altern Complement Med 2014; 20(11): 810-22. [http://dx.doi.org/10.1089/acm.2013.0459] [PMID: 25225776]

[15] Reynolds AG. Grapevine breeding programs for the wine industry. Elsevier 2015.

[16] Cao L, Jin Y, Wei J, Chu Q, Zhao R, Wang W. Comparison on agronomy and quality characters of selective strain of *Schizonepeta tenuifolia*. Zhongguo Zhongyao Zazhi 2009; 34(9): 1075-7. [PMID: 19685737]

[17] Wiersum KF, Dold AP, Husselman M, Cocks M. Cultivation of medicinal plants as a tool for biodiversity conservation and poverty alleviation in the Amatola region. South Africa: Frontis 2006; pp. 43-57. [http://dx.doi.org/10.1007/1-4020-5449-1_3]

[18] Komakech R, Kang J-H, Lee Y, Omujal F. A review of the potential of phytochemicals from *Prunus africana* (Hook f.) Kalkman stem bark for chemoprevention and chemotherapy of prostate cancer. Evid Based Complement Alternat Med. 2017; 2017: 3014019.

[19] Chemat F, Abert Vian M, Ravi HK, *et al.* Review of Alternative Solvents for Green Extraction of Food and Natural Products: Panorama, Principles, Applications and Prospects. Molecules 2019; 24(16): 3007. [http://dx.doi.org/10.3390/molecules24163007] [PMID: 31430982]

[20] Cvjetko Bubalo M, Vidović S, Radojčić Redovniković I, Jokić S. New perspective in extraction of plant biologically active compounds by green solvents. Food Bioprod Process 2018; 109: 52-73. [http://dx.doi.org/10.1016/j.fbp.2018.03.001]

[21] Castro-Puyana M, Marina ML, Plaza M. Water as green extraction solvent: Principles and reasons for its use. Curr Opin Green Sustain Chem 2017; 5: 31-6. [http://dx.doi.org/10.1016/j.cogsc.2017.03.009]

[22] Tekin K, Hao N, Karagoz S, Ragauskas AJ. Ethanol: A Promising Green Solvent for the Deconstruction of Lignocellulose. ChemSusChem 2018; 11(20): 3559-75. [http://dx.doi.org/10.1002/cssc.201801291] [PMID: 30152117]

[23] Chemat F, Fabiano-Tixier AS, Vian MA, Allaf T, Vorobiev E. Solvent-free extraction of food and natural products. Trends Analyt Chem 2015; 71: 157-68. [http://dx.doi.org/10.1016/j.trac.2015.02.021]

[24] Birol F. The world needs a grand coalition to tackle climate change. 2019.

[25] IEA, World Energy Outlook 2019. Paris, 2019.

[26] U.S. Energy Information Administration, Industrial sector energy consumption, p. 14, 2016.

[27] Altemimi A, Watson DG, Choudhary R, Dasari MR, Lightfoot DA. Ultrasound Assisted Extraction of Phenolic Compounds from Peaches and Pumpkins. PLoS One 2016; 11(2): e0148758. [http://dx.doi.org/10.1371/journal.pone.0148758] [PMID: 26885655]

[28] Woolley E, Luo Y, Simeone A. Industrial waste heat recovery: A systematic approach. Sustain Energy Technol Assess 2018; 29: 50-9.

[http://dx.doi.org/10.1016/j.seta.2018.07.001]

[29] Bridgwater A. Fast pyrolysis of biomass for the production of liquids. Biomass Combustion Science, Technology and Engineering. Elsevier 2013; pp. 130-71.
[http://dx.doi.org/10.1533/9780857097439.2.130]

[30] Vernès L, Li Y, Chemat F, Abert-Vian M. Biorefinery concept as a key for sustainable future to green chemistry-the case of microalgae, in *Plant Based.* Green Chemistry 20. Springer 2019; pp. 15-50.

[31] Kamm B, Kamm M. Principles of biorefineries. Appl Microbiol Biotechnol 2004; 64(2): 137-45.
[http://dx.doi.org/10.1007/s00253-003-1537-7] [PMID: 14749903]

[32] Kwiatkowski JR, McAloon AJ, Taylor F, Johnston DB. Modeling the process and costs of fuel ethanol production by the corn dry-grind process. Ind Crops Prod 2006; 23(3): 288-96.
[http://dx.doi.org/10.1016/j.indcrop.2005.08.004]

[33] Rausch KD, Hummel D, Johnson LA, May JB. Wet milling: The basis for corn biorefineries. Corn: Elsevier 2019; pp. 501-35.

[34] Bautista S, Narvaez P, Camargo M, Chery O, Morel L. Biodiesel-TBL+: A new hierarchical sustainability assessment framework of PC&I for biodiesel production – Part I. Ecol Indic 2016; 60: 84-107.
[http://dx.doi.org/10.1016/j.ecolind.2015.06.020]

[35] Espinoza Pérez AT, Camargo M, Narváez Rincón PC, Alfaro Marchant M. Key challenges and requirements for sustainable and industrialized biorefinery supply chain design and management: A bibliographic analysis. Renew Sustain Energy Rev 2017; 69: 350-9.
[http://dx.doi.org/10.1016/j.rser.2016.11.084]

[36] de Jong E, Jungmeier G. Biorefinery Concepts in Comparison to Petrochemical Refineries. Industrial Biorefineries & White Biotechnology. Elsevier 2015; pp. 3-33.
[http://dx.doi.org/10.1016/B978-0-444-63453-5.00001-X]

[37] Hingsamer M, Jungmeier G. Biorefineries. The Role of Bioenergy in the Bioeconomy. Elsevier 2019; pp. 179-222.
[http://dx.doi.org/10.1016/B978-0-12-813056-8.00005-4]

[38] Chew KW, Yap JY, Show PL, *et al.* Microalgae biorefinery: High value products perspectives. Bioresour Technol 2017; 229: 53-62.
[http://dx.doi.org/10.1016/j.biortech.2017.01.006] [PMID: 28107722]

[39] Venkata Mohan S, Nikhil GN, Chiranjeevi P, *et al.* Waste biorefinery models towards sustainable circular bioeconomy: Critical review and future perspectives. Bioresour Technol 2016; 215: 2-12.
[http://dx.doi.org/10.1016/j.biortech.2016.03.130] [PMID: 27068056]

[40] Reay D, Ramshaw C, Harvey A. Process Intensification: Engineering for efficiency, sustainability, and flexibility. Butterworth-Heinemann 2013.
[http://dx.doi.org/10.1016/B978-0-08-098304-2.00002-X]

[41] Garg N, Kontogeorgis GM, Woodley JM, Gani R. A Multi-stage and Multi-level Computer Aided Framework for Sustainable Process Intensification. Computer Aided Chemical Engineering. Elsevier 2018; Vol. 43: pp. 875-80.

[42] Baldea M. Multum in parvo: a process intensification retrospective and outlook. Computer aided chemical engineering. Elsevier 2014; Vol. 34: pp. 15-24.

[43] Cassel E, Vargas RMF, Martinez N, Lorenzo D, Dellacassa E. Steam distillation modeling for essential oil extraction process. Ind Crops Prod 2009; 29(1): 171-6.
[http://dx.doi.org/10.1016/j.indcrop.2008.04.017]

[44] Belova VV, Voshkin AA, Kholkin AI, Payrtman AK. Solvent extraction of some lanthanides from chloride and nitrate solutions by binary extractants. Hydrometallurgy 2009; 97(3-4): 198-203.
[http://dx.doi.org/10.1016/j.hydromet.2009.03.004]

[45] Shirsath SR, Sonawane SH, Gogate PR. Intensification of extraction of natural products using ultrasonic irradiations—A review of current status. Chem Eng Process 2012; 53: 10-23. [http://dx.doi.org/10.1016/j.cep.2012.01.003]

[46] Shirsath SR, Sable SS, Gaikwad SG, Sonawane SH, Saini DR, Gogate PR. Intensification of extraction of curcumin from Curcuma amada using ultrasound assisted approach: Effect of different operating parameters. Ultrason Sonochem 2017; 38: 437-45. [http://dx.doi.org/10.1016/j.ultsonch.2017.03.040] [PMID: 28633845]

[47] Araus K, Uquiche E, del Valle JM. Matrix effects in supercritical CO_2 extraction of essential oils from plant material. J Food Eng 2009; 92(4): 438-47. [http://dx.doi.org/10.1016/j.jfoodeng.2008.12.016]

[48] Pimentel-Moral S, Borrás-Linares I, Lozano-Sánchez J, Arráez-Román D, Martínez-Férez A, Segura-Carretero A. Supercritical CO_2 extraction of bioactive compounds from Hibiscus sabdariffa. J Supercrit Fluids 2019; 147: 213-21. [http://dx.doi.org/10.1016/j.supflu.2018.11.005]

[49] Générale de la Concurrence Direction. de la Consommation et de la Répression de Fraudes (DGCCRF), Emploi des termes « naturel, 100 % nature et de toute autre expression équivalente sur l'étiquetage des denrées alimentaires. 2009.

[50] Sanchez-Siles LM, Michel F, Román S, *et al.* The Food Naturalness Index (FNI): An integrative tool to measure the degree of food naturalness. Trends Food Sci Technol 2019; 91: 681-90. [http://dx.doi.org/10.1016/j.tifs.2019.07.015]

[51] Guissart E, Barragan-Montero V. Matières premières cosmétiques, p. 243.

[52] You S, Wang X. On the Carbon Abatement Potential and Economic Viability of Biochar Production Systems: Cost-Benefit and Life Cycle Assessment. Biochar from Biomass and Waste. Elsevier 2019; pp. 385-408.

[53] Ilgin MA, Gupta SM. Environmentally conscious manufacturing and product recovery (ECMPRO): A review of the state of the art. J Environ Manage 2010; 91(3): 563-91. [http://dx.doi.org/10.1016/j.jenvman.2009.09.037] [PMID: 19853369]

[54] Itskos G, *et al.* Energy and the Environment. Environment and Development. Elsevier 2016; pp. 363-452. [http://dx.doi.org/10.1016/B978-0-444-62733-9.00006-X]

[55] I. O. for Standardization, Environmental Management: Life Cycle Assessment; Principles and Framework, n° 2006.. ISO 2006.

[56] Letellier M, Budzinski H. Microwave assisted extraction of organic compounds. Analusis 1999; 27(3): 259-70. [http://dx.doi.org/10.1051/analusis:1999116]

[57] Mandal V, Mohan Y, Hemalatha S. Microwave assisted extraction-an innovative and promising extraction tool for medicinal plant research. Pharmacogn Rev 2007; 1(1): 7-18.

[58] Ganzler K, Salgó A, Valkó K. Microwave extraction. J Chromatogr A 1986; 371: 299-306. [http://dx.doi.org/10.1016/S0021-9673(01)94714-4] [PMID: 3558551]

[59] Both S, Strube J, Cravatto G. Mass transfer enhancement for solid-liquid extractions, Green Extr. Nat. Prod., 2015, 101-44.

[60] Pingret D, Fabiano-Tixier A S, Chemat F. Accelerated Methods for Sample Preparation in Food. In: Pawliszyn J (Ed.) Comprehensive Sampling and Sample Preparation: Analytical Techniques for Scientists. Academic Press 2012; Vol. 4: pp. 441-55. [http://dx.doi.org/10.1016/B978-0-12-381373-2.00143-5]

[61] Wang L, Weller CL. Recent advances in extraction of nutraceuticals from plants. Trends Food Sci Technol 2006; 17(6): 300-12.

[http://dx.doi.org/10.1016/j.tifs.2005.12.004]

[62] Chemat F, Cravotto G, Eds. Microwave-assisted Extraction for Bioactive Compounds. Boston, MA: Springer US 2013.
[http://dx.doi.org/10.1007/978-1-4614-4830-3]

[63] Alupului A, Calinescu I, Lavric V. Microwave extraction of active principles from medicinal plants. UPB Sci Bull Ser B 2012; 74(2): 129-42.

[64] Tatke P, Jaiswal Y. An overview of microwave assisted extraction and its applications in herbal drug research. Res J Med Plant 2011; 5(1): 21-31.
[http://dx.doi.org/10.3923/rjmp.2011.21.31]

[65] Chemat F, Lucchesi M, Smadia J. Solvent-free microwave extraction of volatile natural substances. Google Patents 2004.

[66] Vian MA, Fernandez X, Visinoni F, Chemat F. Microwave hydrodiffusion and gravity, a new technique for extraction of essential oils. J Chromatogr A 2008; 1190(1-2): 14-7.
[http://dx.doi.org/10.1016/j.chroma.2008.02.086] [PMID: 18343393]

[67] Tran T, Nguyen H, Nguyen D, *et al.* Optimization of Microwave-Assisted Extraction of Essential Oil from Vietnamese Basil (Ocimum basilicum L.) Using Response Surface Methodology. Processes (Basel) 2018; 6(11): 206.
[http://dx.doi.org/10.3390/pr6110206]

[68] Liu Z, Deng B, Li S, Zou Z. Optimization of solvent-free microwave assisted extraction of essential oil from Cinnamomum camphora leaves. Ind Crops Prod 2018; 124: 353-62.
[http://dx.doi.org/10.1016/j.indcrop.2018.08.016]

[69] Bouras M, Chadni M, Barba FJ, Grimi N, Bals O, Vorobiev E. Optimization of microwave-assisted extraction of polyphenols from Quercus bark. Ind Crops Prod 2015; 77: 590-601.
[http://dx.doi.org/10.1016/j.indcrop.2015.09.018]

[70] Şahin S, Samli R, Tan ASB, *et al.* Solvent-Free Microwave-Assisted Extraction of Polyphenols from Olive Tree Leaves: Antioxidant and Antimicrobial Properties. Molecules 2017; 22(7): 1056.
[http://dx.doi.org/10.3390/molecules22071056] [PMID: 28672807]

[71] Hu B, Li C, Zhang Z, *et al.* Microwave-assisted extraction of silkworm pupal oil and evaluation of its fatty acid composition, physicochemical properties and antioxidant activities. Food Chem 2017; 231: 348-55.
[http://dx.doi.org/10.1016/j.foodchem.2017.03.152] [PMID: 28450017]

[72] Yanık DK. Alternative to traditional olive pomace oil extraction systems: Microwave-assisted solvent extraction of oil from wet olive pomace. Lebensm Wiss Technol 2017; 77: 45-51.
[http://dx.doi.org/10.1016/j.lwt.2016.11.020]

[73] Jafari SM, Mahdavee Khazaei K, Assadpour E. Production of a natural color through microwave-assisted extraction of saffron tepal's anthocyanins. Food Sci Nutr 2019; 7(4): 1438-45.
[http://dx.doi.org/10.1002/fsn3.978] [PMID: 31024717]

[74] Kumar K, Srivastav S, Sharanagat VS. Ultrasound assisted extraction (UAE) of bioactive compounds from fruit and vegetable processing by-products: A review. Ultrason Sonochem 2021; 70: 105325.
[http://dx.doi.org/10.1016/j.ultsonch.2020.105325] [PMID: 32920300]

[75] Esclapez MD, García-Pérez JV, Mulet A, Cárcel JA. Ultrasound-assisted extraction of natural products. Food Eng Rev 2011; 3(2): 108-20.
[http://dx.doi.org/10.1007/s12393-011-9036-6]

[76] Vernès L, Vian M, Chemat F. Ultrasound, and Microwave as Green Tools for Solid-Liquid Extraction. Liquid-Phase Extraction. Elsevier 2020; pp. 355-74.
[http://dx.doi.org/10.1016/B978-0-12-816911-7.00012-8]

[77] Wen C, Zhang J, Zhang H, *et al.* Advances in ultrasound assisted extraction of bioactive compounds

from cash crops – A review. Ultrason Sonochem 2018; 48: 538-49.
[http://dx.doi.org/10.1016/j.ultsonch.2018.07.018] [PMID: 30080583]

[78] Chemat F, Rombaut N, Sicaire AG, Meullemiestre A, Fabiano-Tixier AS, Abert-Vian M. Ultrasound assisted extraction of food and natural products. Mechanisms, techniques, combinations, protocols and applications. A review. Ultrason Sonochem 2017; 34: 540-60.
[http://dx.doi.org/10.1016/j.ultsonch.2016.06.035] [PMID: 27773280]

[79] Pacheco-Fernández I, Pino V. Extraction With Ionic Liquids-Organic Compounds. Liquid-Phase Extraction. Elsevier 2020; pp. 499-537.
[http://dx.doi.org/10.1016/B978-0-12-816911-7.00017-7]

[80] Chemat F, Zill-e-Huma , Khan MK. Applications of ultrasound in food technology: Processing, preservation and extraction. Ultrason Sonochem 2011; 18(4): 813-35.
[http://dx.doi.org/10.1016/j.ultsonch.2010.11.023] [PMID: 21216174]

[81] Milani G, Vian M, Cavalluzzi MM, *et al.* Ultrasound and deep eutectic solvents: An efficient combination to tune the mechanism of steviol glycosides extraction. Ultrason Sonochem 2020; 69: 105255.
[http://dx.doi.org/10.1016/j.ultsonch.2020.105255] [PMID: 32682311]

[82] Baghdikian B, Filly A, Fabiano-Tixier AS, *et al.* Extraction by solvent using microwave and ultrasound-assisted techniques followed by HPLC analysis of Harpagoside from Harpagophytum procumbens and comparison with conventional solvent extraction methods. C R Chim 2016; 19(6): 692-8.
[http://dx.doi.org/10.1016/j.crci.2016.02.020]

[83] Santos HM, Lodeiro C, Capelo-Martínez J-L. The power of ultrasound. Ultrasound in chemistry: analytical applications. Wiley Online Library 2009; pp. 1-16.

[84] Sicaire AG, Vian MA, Fine F, Carré P, Tostain S, Chemat F. Ultrasound induced green solvent extraction of oil from oleaginous seeds. Ultrason Sonochem 2016; 31: 319-29.
[http://dx.doi.org/10.1016/j.ultsonch.2016.01.011] [PMID: 26964955]

[85] Vernès L, Abert-Vian M, El Maâtaoui M, Tao Y, Bornard I, Chemat F. Application of ultrasound for green extraction of proteins from spirulina. Mechanism, optimization, modeling, and industrial prospects. Ultrason Sonochem 2019; 54: 48-60.
[http://dx.doi.org/10.1016/j.ultsonch.2019.02.016] [PMID: 30827903]

[86] Zhu Z, Guan Q, Koubaa M, *et al.* HPLC-DAD-ESI-MS2 analytical profile of extracts obtained from purple sweet potato after green ultrasound-assisted extraction. Food Chem 2017; 215: 391-400.
[http://dx.doi.org/10.1016/j.foodchem.2016.07.157] [PMID: 27542491]

[87] Khadhraoui B, Turk M, Fabiano-Tixier AS, *et al.* Histo-cytochemistry and scanning electron microscopy for studying spatial and temporal extraction of metabolites induced by ultrasound. Towards chain detexturation mechanism. Ultrason Sonochem 2018; 42: 482-92.
[http://dx.doi.org/10.1016/j.ultsonch.2017.11.029] [PMID: 29429695]

[88] Goula AM, Ververi M, Adamopoulou A, Kaderides K. Green ultrasound-assisted extraction of carotenoids from pomegranate wastes using vegetable oils. Ultrason Sonochem 2017; 34: 821-30.
[http://dx.doi.org/10.1016/j.ultsonch.2016.07.022] [PMID: 27773309]

[89] Espada-Bellido E, Ferreiro-González M, Carrera C, Palma M, Barroso CG, Barbero GF. Optimization of the ultrasound-assisted extraction of anthocyanins and total phenolic compounds in mulberry (Morus nigra) pulp. Food Chem 2017; 219: 23-32.
[http://dx.doi.org/10.1016/j.foodchem.2016.09.122] [PMID: 27765221]

[90] Chakraborty S, Shaik L, Gokhale JS. Subcritical Water: An Innovative Processing Technology. Innovative Food Processing Technologies 2021; pp. 552-66.
[http://dx.doi.org/10.1016/B978-0-08-100596-5.22966-1]

[91] Ko MJ, Nam HH, Chung MS. Subcritical water extraction of bioactive compounds from Orostachys

japonicus A. Berger (Crassulaceae). Sci Rep 2020; 10(1): 10890.
[http://dx.doi.org/10.1038/s41598-020-67508-2] [PMID: 32616865]

[92] Gbashi S, Adebo OA, Piater L, Madala NE, Njobeh PB. Subcritical water extraction of biological materials. Separ Purif Rev 2017; 46(1): 21-34.
[http://dx.doi.org/10.1080/15422119.2016.1170035]

[93] Smith RM. Superheated water: the ultimate green solvent for separation science. Anal Bioanal Chem 2006; 385(3): 419-21.
[http://dx.doi.org/10.1007/s00216-006-0437-y] [PMID: 16715273]

[94] Zhang J, Wen C, Zhang H, Duan Y, Ma H. Recent advances in the extraction of bioactive compounds with subcritical water: A review. Trends Food Sci Technol 2020; 95: 183-95.
[http://dx.doi.org/10.1016/j.tifs.2019.11.018]

[95] Yang Z, Uhler B, Lipkie T. Microwave-Assisted Subcritical Water Extraction of Steviol Glycosides From Stevia rebaudiana Leaves. Nat. Prod. Commun., 2019, 14(6), 1934578X19860003.
[http://dx.doi.org/10.1177/1934578X19860003]

[96] Alvarez VH, Cahyadi J, Xu D, Saldaña MDA. Optimization of phytochemicals production from potato peel using subcritical water: Experimental and dynamic modeling. J Supercrit Fluids 2014; 90: 8-17.
[http://dx.doi.org/10.1016/j.supflu.2014.02.013]

[97] Jokić S, Gagić T, Knez Ž, Banožić M, Škerget M. Separation of active compounds from tobacco waste using subcritical water extraction. J Supercrit Fluids 2019; 153: 104593.
[http://dx.doi.org/10.1016/j.supflu.2019.104593]

[98] Chiou TY, Nomura S, Konishi M, *et al.* Conversion and Hydrothermal Decomposition of Major Components of Mint Essential Oil by Small-Scale Subcritical Water Treatment. Molecules 2020; 25(8): 1953.
[http://dx.doi.org/10.3390/molecules25081953] [PMID: 32331471]

[99] Zainan NH, Thiruvenkadam S, Danquah MK, Harun R. Biochemical analysis and potential applications of aqueous and solid products generated from subcritical water extraction of microalgae Chlorella pyrenoidosa biomass. J Appl Phycol 2020; 32(1): 111-26.
[http://dx.doi.org/10.1007/s10811-019-01960-0]

[100] Essien S, Young B, Baroutian S. Subcritical water extraction for selective recovery of phenolic bioactives from kānuka leaves. J Supercrit Fluids 2020; 158: 104721.
[http://dx.doi.org/10.1016/j.supflu.2019.104721]

[101] Zhang Y, Zhang Y, Taha AA, *et al.* Subcritical water extraction of bioactive components from ginseng roots (Panax ginseng C.A. Mey). Ind Crops Prod 2018; 117: 118-27.
[http://dx.doi.org/10.1016/j.indcrop.2018.02.079]

[102] Gandhi K, Arora S, Kumar A. Industrial applications of supercritical fluid extraction: A review. Int J Chem Stud 2017; 5(3): 336-40.

[103] Manjare SD, Dhingra K. Supercritical fluids in separation and purification: A review. Mater Sci Energy Technol 2019; 2(3): 463-84.
[http://dx.doi.org/10.1016/j.mset.2019.04.005]

[104] Khaw KY, Parat MO, Shaw PN, Falconer JR. Solvent Supercritical Fluid Technologies to Extract Bioactive Compounds from Natural Sources: A Review. Molecules 2017; 22(7): 1186.
[http://dx.doi.org/10.3390/molecules22071186] [PMID: 28708073]

[105] Chemat F, Vian MA, Eds. Alternative Solvents for Natural Products Extraction. Berlin, Heidelberg: Springer Berlin Heidelberg 2014.
[http://dx.doi.org/10.1007/978-3-662-43628-8]

[106] Attard TM, Bukhanko N, Eriksson D, *et al.* Supercritical extraction of waxes and lipids from biomass: A valuable first step towards an integrated biorefinery. J Clean Prod 2018; 177: 684-98.
[http://dx.doi.org/10.1016/j.jclepro.2017.12.155]

[107] Chemat F, Vorobiev E, Eds. Green food processing techniques. 1st Ed. Cambridge: Elsevier, 2019.

[108] Molino A, Mehariya S, Di Sanzo G, *et al.* Recent developments in supercritical fluid extraction of bioactive compounds from microalgae_ Role of key parameters, technological achievements, and challenges. Journal of CO_2 Utilization, 2020, 36, 196-209.

[109] Uddin MS, Sarker MZI, Ferdosh S, *et al.* Phytosterols and their extraction from various plant matrices using supercritical carbon dioxide: a review. J Sci Food Agric 2015; 95(7): 1385-94. [http://dx.doi.org/10.1002/jsfa.6833] [PMID: 25048690]

[110] Danielski L, Ferreira SRS, Hense H, Martínez J, Brunner G. Deterpenation of Citrus Peel Oils with Supercritical Carbon Dioxide-A Review. Tree For Sci Biotechnol 2008; 2(1): 10-22.

[111] Sapkale GN, Patil SM, Surwase US, Bhatbhage PK. Supercritical fluid extraction. Int J Chem Sci 2010; 8(2): 729-43.

[112] Zhang X, Janzen EC, Liu J. Portable supercritical fluid extraction apparatus. Google Patents 2020.

[113] Valadez-Carmona L, Ortiz-Moreno A, Ceballos-Reyes G, Mendiola JA, Ibáñez E. Valorization of cacao pod husk through supercritical fluid extraction of phenolic compounds. J Supercrit Fluids 2018; 131: 99-105.
[http://dx.doi.org/10.1016/j.supflu.2017.09.011]

[114] Conde-Hernández LA, Espinosa-Victoria JR, Trejo A, Guerrero-Beltrán JÁ. CO 2 -supercritical extraction, hydrodistillation and steam distillation of essential oil of rosemary (Rosmarinus officinalis). J Food Eng 2017; 200: 81-6.
[http://dx.doi.org/10.1016/j.jfoodeng.2016.12.022]

[115] Soh SH, Agarwal S, Jain A, Lee LY, Chin SK, Jayaraman S. Mathematical modeling of mass transfer in supercritical fluid extraction of patchouli oil. Eng Rep 2019; 1(4): e12051.
[http://dx.doi.org/10.1002/eng2.12051]

[116] Frohlich PC, Santos KA, Palú F, Cardozo-Filho L, da Silva C, da Silva EA. Evaluation of the effects of temperature and pressure on the extraction of eugenol from clove (Syzygium aromaticum) leaves using supercritical CO_2. J Supercrit Fluids 2019; 143: 313-20.
[http://dx.doi.org/10.1016/j.supflu.2018.09.009]

[117] Alexandre A, Serra A T, Matias A A, Duarte C M M, Bronze M R. Supercritical fluid extraction of Arbutus unedo distillate residues-Impact of process conditions on antithe proliferative response of extracts. J. CO2 Util., 2020, 37, 29-38.

[118] Rochfort S, Isbel A, Ezernieks V, *et al.* Utilisation of Design of experiments Approach to optimize Supercritical fluid extraction of Medicinal cannabis. Sci Rep 2020; 10(1): 9124.
[http://dx.doi.org/10.1038/s41598-020-66119-1] [PMID: 32499550]

[119] Patel AS, Kar A, Dash S, Dash SK. Supercritical fluid extraction of β-carotene from ripe bitter melon pericarp. Sci Rep 2019; 9(1): 19266.
[http://dx.doi.org/10.1038/s41598-019-55481-4] [PMID: 31848376]

CHAPTER 3

Environmentally Friendly Solvents

Lidia Montero[1,2], **Priscilla Rocío Bautista**[3] and **Bienvenida Gilbert López**[3,*]

[1] *Applied Analytical Chemistry, University of Duisburg-Essen, Universitätsstrasse 5, 45141 Essen, Germany*

[2] *Teaching and Research Center for Separation, University of Duisburg-Essen, Universitätsstrasse 5, 45141 Essen, Germany*

[3] *Analytical Chemistry Research Group (FQM-323), Department of Physical and Analytical Chemistry, University of Jaen, 23071 Jaen, Spain*

Abstract: The present chapter aims to provide a brief overview of the environmentally friendly solvents most commonly used in food analysis, including water, carbon dioxide, ethanol, ionic liquids, (natural) deep eutectic solvents (NA)DES, surfactants, and switchable solvents. A general outlook of their properties, production sources, and classification is provided. The advantages and limitations of the use of these solvents in food analysis are evaluated from the point of view of Green Analytical Chemistry. Some recent applications have been selected to illustrate the potential of environmentally friendly solvents in combination with assisted extraction techniques and miniaturized techniques for the development of green extraction methods in food analysis.

Keywords: Analytical Chemistry, Environmentally friendly solvents, Food analysis, Green Chemistry, Green extraction, Green solvent, Sample preparation.

INTRODUCTION

Environmentally friendly solvents, commonly designated as "green" solvents, respond to the need of minimizing the impact of chemical processes in the environment. These solvents show better EHS (environmental, health, and safety) properties than the traditional hazardous ones, but other aspects are also taken into consideration, such as energy flows and costs related to their production, performance, and disposal. In this context, an environmentally friendly solvent may be defined as one solvent that fulfills some of the following twelve criteria [1]: (1) availability at a large scale, (2) competitive price, (3) recyclability, (4) purity grade, (5) synthesis, (6) negligible toxicity, (7) biodegradability, (8) perfor-

* **Corresponding Author Bienvenida Gilbert López:** Analytical Chemistry Research Group (FQM-323), Department of Physical and Analytical Chemistry, University of Jaen, 23071 Jaen, Spain; E-mail: adnaneremmal@gmail.com and E-mail: bgilbert@ujaen.es

Merichel Plaza & Maria Luisa Marina (Eds.)

mance, (9) thermal and chemical stability, (10) flammability, (11) transport and storage, (12) renewability. The use of these solvents is encompassed under the principles of Green Analytical Chemistry (GAC) [2], which aim, among other goals, to reduce the amounts of solvents and reagents used in analytical methods along with the waste generated, and the replacement of toxic solvents. In this regard, sample preparation, which consumes more than 80% of the analysis time, is a key step in the development of any analytical application [3]. It is obvious that the greenest approach is the direct analysis of samples, but even though fields such as ambient mass spectrometry have developed in recent years, direct analysis of food is still a complex issue [4]. Therefore, the current approaches to make methods for food analysis and foodomics greener are more focused on the miniaturization of analytical techniques and the use of green solvents [5, 6].

In terms of solvent consumption, the employment of environmentally friendly solvents instead of more hazardous ones is critical in studies aiming at the recovery of bioactive compounds from natural sources to produce food-related products. The most common conventional techniques used for the extraction of valuable compounds from food and other natural sources are solid-liquid extraction (SLE) and liquid-liquid extraction (LLE). These techniques are characterized by simplicity and the use of not expensive instrumentation, but one of their main drawbacks is the consumption of considerable amounts of organic solvents [7]. Among the strategies followed to seek greener alternatives to these methods, the use of the so-called assisted extraction techniques is a widely used approach. The application of ultrasounds, microwaves, or high pressures allows the reduction of extraction time and solvent consumption. Another approach to reducing the consumption of organic solvents is the use of miniaturized extraction techniques, such as liquid-phase micro extraction techniques (LPME), or miniaturized solid-phase extraction (μSPE) [8].

According to GAC principles, the reduction of the amount of solvents used is not sufficient, but additionally, toxic solvents should be replaced by greener ones, preferably obtained from renewable sources. Therefore, it is important to have some tools that allow the classification of environmentally friendly solvents into groups, helping the analyst in the selection of the most suitable solvent or solvent mixture for a particular food application. In the next section, different selection guides are presented, and the considerations to be taken into account are discussed. Following, a detailed description of different solvents and their applications in food analysis is presented.

CLASSIFICATION AND SELECTION OF ENVIRONMENTALLY FRIENDLY SOLVENTS

The selection of an environmentally friendly solvent for a particular application is a challenging task. Considering that the concept of green solvent is based on the fulfillment of a sort of criteria related to green chemistry [1], it is easy to figure out that there is not a straightforward definition and classification of green solvents. In addition, a one-to-one replacement of conventional-to-green solvents is difficult and has to be based on a two-side comparison: on one hand, the green solvent should own better EHS properties than the solvent to be replaced and, on the other hand, the physicochemical properties and analytical performance of the new solvent should be at least similar to the one to be replaced. To this aim, efforts have been taken to develop different computer-aided selection tools that put environmentally friendly solvents into groups.

Among the classification tools based on EHS characteristics of solvents, it is worth mentioning the solvent selection guide CHEM21, although it was designed for the pharmaceutical industry [9]. It establishes a scoring system based on environmental issues such as volatility, recyclability or bioaccumulation, health issues related to occupational hazards, and safety issues such as flammability. One of the strengths of this guide is that it is based on easy-to-access information contained in the solvents Safety Data Sheets. The outcome of the CHEM21 guide is a ranking that groups the solvents into four categories, namely highly hazardous, hazardous, problematic, and recommended solvents. Another approach is the combination of EHS information with life cycle assessment (LCA) methodology, which considers energy use during solvent production and waste treatment [10]. LCA methodology may be also applied to the entire chemical process (and not only to the solvents), which is the so-called "cradle-to-grave" approach [11]. Likewise, the "analytical Eco-Scale" is a tool designed for GAC that evaluates the analytical method as a whole, considering not only the hazards associated with the solvents but also the energy consumed by the instrumentation and the waste generated [12]. This scale starts from 100 points, from which penalty points are subtracted, thus being the greenest approach the highest scored one. It should be noted that in the last years the trend is moving towards the development of green metrics that assess the entire method of analysis, instead of focusing on a simple replacement of solvents and reagents for greener ones [13 - 15].

Regarding the tools available for the comparison of solvent properties, Hansen solubility parameters (HSP) [16, 17] or Kamlet-Taft solvatochromic parameters [18, 19] can be used to compare solvent properties. Both models consider molecular interactions such as hydrogen-bond donating ability (acidity), hydrogen-bond accepting ability (basicity), and polarity/polarizability. Also, both models have the limitation of needing literature data or time-consuming experimental calculation of the parameters describing intermolecular forces. The COSMO-RS (COnductor-like Screening MOdel for Real Solvent) tool also allows for solvents classification, including newly designed solvents [20]. This includes Density Functional Theory (DFT) calculations and groups of the solvents by principal component analysis (PCA).

Table 1. Advantages and disadvantages of environmentally friendly solvents [17, 20, 21].

Solvent family	Advantages	Disadvantages	Example of Solvents
Water	-Non-flammable, non-toxic, renewable	-	H_2O
Supercritical fluids	-Non-flammable, inert, non-toxic, recyclable	-High-pressure equipment and operational costs	$scCO_2$
Bio-based solvents	-From renewable feedstock -Efficient, and cost-effective	-Need a consistent feedstock supply -May not have a good EHS profile - More research on kinetic performance and thermodynamic studies is needed	Ethanol, ethyl lactate, ethyl acetate, α-pinene, d-limonene
Ionic liquids	-Non-volatile, thermally stable, non-flammable	- High cost - May not have good EHS profile	$[Bmim]BF_4$
Deep Eutectic Solvents (DES)	-Non-volatile, thermally stable, low-cost synthesis, wide polarity, bio-degradable	-	ChCl:urea
Natural Deep Eutectic solvents (NADES)	-Non-volatile, thermally stable, low-cost synthesis -Natural origin (primary metabolites)	-Viscosity - Take away natural resources that could be used for other purposes	Lactic acid–choline chloride
Fluorinated solvents	-Inert, low toxicity, non-flammable	-	C_8F_{18}
Liquid polymers	-Non-volatile, biocompatible	-	Poly(ethyleneglycols) poly(dimethylsiloxane)
Eco-friendly solvents	-Good EHS properties	-Not bio-based	Propylene carbonate, 3-methoxy-3-methyl-butan-1-ol, dimethyl sulfoxide

ChCl: choline chloride

Despite the difficulties to group environmentally friendly solvents and the variety of classifications provided by the different approaches, there is a consensus (with small differences) in the specific literature in ordering the solvents in the following seven families [21]: (i) water, (ii) supercritical fluids, (iii) bio-based solvents, (iv) ionic liquids and eutectic mixtures, (v) fluorinated solvents, (vi) liquid polymers, and (vii) eco-friendly solvents. The main advantages and drawbacks of each family are summarized in Table **1**.

In this chapter, the environmentally friendly solvents have been divided into four groups, namely (i) water, (ii) liquefied gases, (iii) bio-based solvents, and (iv) neoteric solvents (amphiphilic solvents, ionic liquids, eutectic mixtures, and switchable solvents), as summarized in Fig. (**1**). From each group, the most representative solvents used in food analysis will be discussed and some of their applications in food analyses will be presented. In addition, it should be kept in mind that solvent residues in extracts from natural products intended to be used as food ingredients have to comply with the regulation applied to food additives; in this context, it is common to use the solvents included in the GRAS (Generally Recognized As Safe) inventory of the USFDA (United States Food and Drug Administration) [22], independently of their greenness.

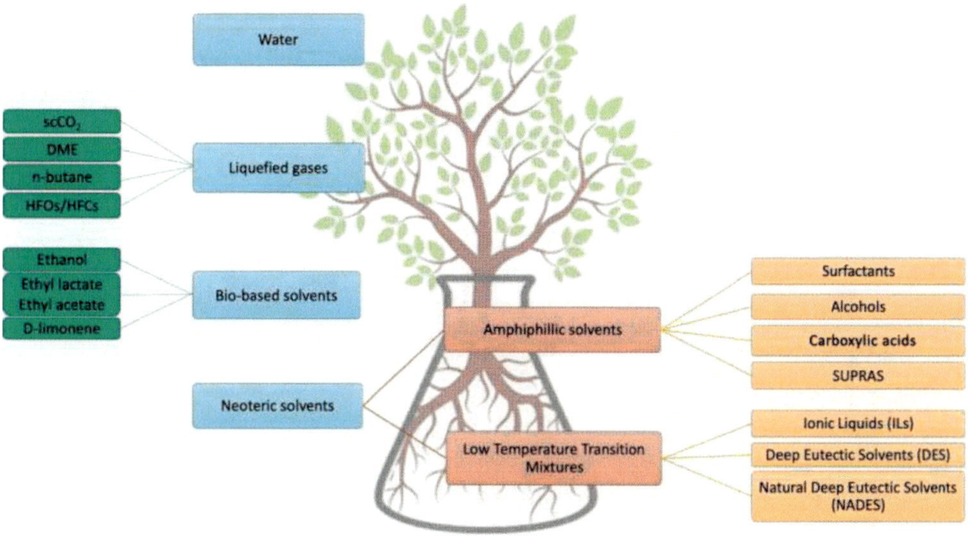

Fig. (1). Different families of environmentally friendly solvents.

Water constitutes its own category due to its unique properties as a solvent, which may be tuned according to pressure and temperature. Supercritical carbon dioxide ($scCO_2$) is the most representative solvent from the family of liquefied gases, which are gases that are compressed during the extraction process to work in the liquid state. Therefore, solvents included in these two groups are good candidates to be used in pressurized green extraction techniques, such as pressurized hot water extraction (PHWE) and supercritical fluid extraction (SFE), respectively.

Bio-based solvents are a heterogeneous group of solvents obtained from renewable feedstocks through biorefineries. Currently, plant biomass is the primary source of chemicals for the replacement of petroleum-derived solvents, including trees, crops, agricultural byproducts, or aquatic biomass such as algae [21]. Vegetable oils constitute the source of biodiesel and glycerol-related solvents (*e.g.* glycerol carbonate, glycerol-containing eutectic solvents). Sugar polymers contained in plant materials, such as cellulose, sucrose, or starch are hydrolyzed into simple sugars and then subjected to different processes that end in the production of solvents, *e.g.* ethanol, ethyl lactate, 2-methyl tetrahydrofuran (2-MeTHF). As regards agri-food waste, some bio-based solvents may be obtained from orange peel essential oil, including d-limonene or α-pinene [1, 23]. Nevertheless, the obtention of bio-based solvents from renewable sources does not imply a low EHS impact or economic viability. Some examples of bio-based solvents with good EHS profile and their physicochemical properties are included in Table **2**. These solvents have been used in food applications in combination with green extraction techniques such as pressurized liquid extraction (PLE).

Table 2. Physicochemical properties of some environmentally friendly solvents [21, 24, 25].

Solvent	Formula	Molecular Weight, g/mol	Dielectric constant, ε (20 °C)	Octanol/Water Partition Coefficient, logP	Boiling Point, °C	Viscosity, m [Pa.s] (25 °C)
Water	H_2O	18.01	80.40	n/a	100	0.89
Carbon dioxide	CO_2	44.01	n/a	0.83	-78.5	0.02
Ethanol	CH_3CH_2OH	46.07	22.40	-0.31	78	1.07
Ethyl lactate	$C_5H_{10}O_3$	118.13	15.70	0.18	154	n/a
Ethyl acetate	$CH_3COOC_2H_5$	88.11	6.08	0.73	77	0.42
d-Limonene	$C_{10}H_{16}$	136.24	2.30	4.57	176	0.90

n/a: not available

"Neoteric solvents" is the name given to a miscellaneous category that includes different families of new solvents. We have included amphiphilic solvents (covering surfactants, alcohols, and carboxylic acids) [26], ionic liquids, eutectic mixtures, and switchable solvents. Most of these types of solvents are binary mixtures, and their particular characteristics are further discussed in dedicated sections. In some specific literature, the term low transition temperature mixtures

(LTTMs) is used to encompass deep eutectic solvents (DES), natural deep eutectic solvents (NADES), and ionic liquids [27]. These new solvents have been mainly used in liquid-phase microextraction techniques and assisted extraction techniques, which consume a low volume of solvents. Also, they have been used in microwave-assisted extraction (MAE) and ultrasound-assisted extraction (UAE) techniques.

WATER

The use of water as a solvent involves several attractive advantages, including its cheap price and its pure grade readily available. In addition, water is non-toxic, does not produce hazardous, and it is simple to handle. Therefore, water presents environmental and economic characteristics that make this solvent one of the best alternatives for environmentally friendly extraction procedures [28 - 30]. Besides these benefits, water presents special physicochemical advantages, which make it a unique extraction solvent.

Water presents excellent capacities for solving polar and ionic compounds. This ability is related to the molecular structure of water, which is able to form dipoles due to the partial positive charge of the two hydrogen atoms and the negative oxygen atom charge of the water molecule. The two poles aid the formation of hydrogen bonds with other water molecules as well as with polar and ionic compounds, promoting their solvation [31]. The ability of water to form hydrogen bonds produces a strong intermolecular water interaction, resulting in high viscosity and surface tension [28]. Other physicochemical characteristics that differentiate water from other organic solvents are the high dielectric constant of water (ε= 80, at room temperature) and polarizability π^* [29].

On the other hand, water is a liquid solvent under atmospheric conditions, and the liquid state of water is maintained up to 100 ºC. However, temperature highly affects the physicochemical properties of water. For example, the solvation power of water changes when the temperature increases, and at high temperatures, water is able to dissolve compounds with a lower polarity that are not possible to dissolve at room temperatures. Moreover, at high temperatures, water also reduces its dielectric constant, viscosity, and surface tension [29, 32, 33]. Fig. (**2**) depicts the modification of the different physicochemical properties of water according to the temperature. All these modifications produced by the increment of the temperature are of great interest for the use of water as an extraction solvent because the extraction efficiency of water is improved by an increment in the solvent-matrix interaction that allows a better penetration in the matrix, as well as in the diffusivity and solubility of the compounds (mass transfer rate) [32, 34].

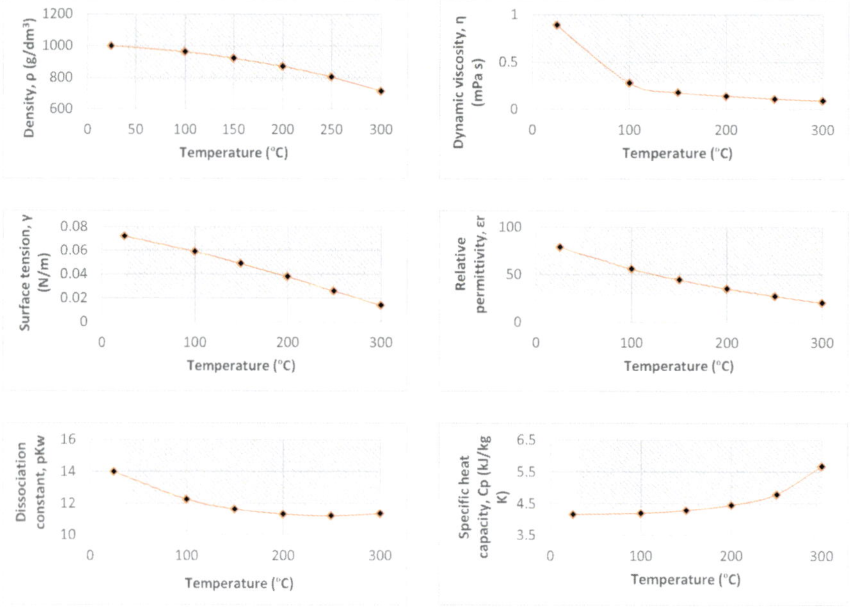

Fig. (2). Modifications of the physicochemical characteristics of water accordingly to the temperature.

Together with the temperature, pressure is another parameter that significantly affects the physicochemical characteristics of water. In fact, all the modifications produced by temperature are increased by the increment of the pressure. Pressure can be used to maintain the liquid state of water even at high temperatures (subcritical conditions) (374 °C at 22.1 MPa). Under these conditions, the dielectric constant of water is modified achieving similar values to the dielectric constant of organic solvents at room temperature. For example, at 5 MPa and 250 °C, the dielectric constant of water is $\varepsilon = 27$, while the ε of methanol and ethanol are 33 and 24 at 25 °C, respectively [33]. This effect implies that subcritical water presents similar solvation power for non-polar molecules than some organic solvents [35]. Therefore, the controlled modification of temperature and pressure over the physicochemical properties of water increases the polarity range of molecules that can be extracted using water.

Different extraction techniques have been carried out using water as extraction solvent, the most common techniques are SLE, UAE, MAE, or PLE. Each extraction technique applies specific temperatures and pressure conditions depending on the procedural characteristics of the technique, as well as on the compounds of interest. Then, water can be used at different states for the extraction, for instance, water at room temperature, hot water from 25 to 100 °C, subcritical water (100-374 °C), and supercritical water (temperature and pressure over the critical point) [29]. That way, in each state, the extraction properties of water will be different according to the selected temperature and pressure. The extractions performed using water at high pressures and temperatures (subcritical conditions) are called subcritical water extractions (SWE), also known as superheated water extraction (SHWE) or pressurized hot water extraction (PHWE). As explained before, water under subcritical conditions provides unique extraction properties, which are relevant to carry out green extraction of natural products reducing the environmental impact and solvent costs [36].

SWE has been employed for the recovery of polar valuable component groups of natural sources, mainly, polysaccharides and phenolic compounds. For instance, the effect of the SWE extraction parameters on the polysaccharides from the fungus *Lentinus edodes* (commonly known as shiitake) was evaluated by Zhang *et al.* [37]. During the extraction optimization, they observed that the molecular weight of the extracted polysaccharides increased accordingly with the temperature, achieving long chains of a maximum molecular weight of 2.13×10^5 g/mol at 130 ºC. However, temperatures over 130 ºC produced breakings on the chain of the polysaccharides, and hence, shorter molecules were detected. Besides, the temperature had an impact on the polysaccharide chain structure: temperatures higher than 150 ºC for 15 minutes produced the disruption of the typical triple-helical configuration of the chain. All the extracted polysaccharides were heteropolysaccharides consisting of units of glucose, mannose, and galactose at different molar ratios. Some of these polysaccharides showed *in vitro* inhibitory effect against the human cancer cell lines HepG2 and MCT-7. Other reported interesting polysaccharides extracted by SWE are pectin from sunflowers with low esterification degree and excellent thermal stability [38]. Fucoidans are special bioactive polysaccharides found exclusively in brown algae. Several SWE applications for the recovery of fucoidans have been reported. Saravana *et al.* achieved an extraction yield of fucoidans of 13.56% using SWE from the brown algae *Saccharina japonica.* The highest temperatures used in the SWE optimization helped to increase the solvent diffusion rate and mass transfer at the same time that the viscosity was reduced, promoting the dilution of the fucoidans in the extraction media. The solid-to-liquid ratio (S/L) had an important impact on the extraction yield of these polysaccharides (lower extraction yields were obtained at higher S/L) [39]. Alboofetileh *et al.* also used SWE for the extraction of fucoidans from the algae *Nizamuddinia zanardinii* obtaining extracts with a complex monosaccharide composition (fucose, mannose, galactose, xylose, and glucose) and with antioxidant, anticancer, and immunomodulatory activities [40].

The other main group of compounds that has shown successful behavior under SWE is the phenolic compounds. Many different natural matrices have been submitted to the SWE extraction for the recovery of phenolic compounds in the last years, such as ginger [41], coffee [42], chamomile [43], sorghum [44], *Morus nigra* L. fruit [45] or ginseng root [46]. An interesting observation made by Cheing *et al.* was that the temperature could be correlated with the type of phenolic compound that is extracted. They observed that each kind of phenolic compound presented an optimal extraction temperature, being this optimal temperature higher for the phenolic compounds with a lower number of hydroxyl groups, and lower for those with a larger number of hydroxyl groups [47].

Interestingly, the SWE extraction of different phenolic compounds has been commonly included in biorefinery or valorization processes. This trend is acquiring great importance and the use of food by-products is currently a global tendency to face environmental and sustainability issues. The use of waste produced in the food industry together with the application of green extraction techniques is the best combination to achieve that goal. That way, the SWE extractions of phenolic compounds from kiwifruit pomace [48], chestnut shells [49], pistachio hulls [50], onion skin [51], petals of saffron [52], mandarin and orange peel [53, 54], papaya seeds [55], cocoa shells [56], or peach palm [57] have been reported.

Other appreciated compounds from food by-products have been also extracted by SWE for their valorization. For example, the recovery of fructooligosaccharides (FOS) and β-ecdysone from the roots and aerial parts of ginseng was studied with the aim of using the parts of the plants that are considered agricultural waste. The root extract showed 8.8 g FOS per 100 g of SWE extract and 0.7 g of β-ecdysone per 100 g extract. The economical evaluation of the process showed good profitability for the production of these valuable compounds [58]. The pectin from by-products from the cocoa industry has been also recovered by SWE, with enhanced pectin recovery in comparison to conventional extraction methods like citric acid solution extraction [59], as well as the pectin from pomelo peels [60]. Although the raw materials were different in these two applications, in both of them the optimal extraction temperature of pectin was almost the same, 121 and 120 ºC, respectively. The main difference can be observed in the pressure applied for the pectin extraction in each matrix (10.34 and 1 MPa, respectively) probably due to the difference in the matrix structure. Other different compounds that have been recovered from food by-products are the oil and saponins from *Camellia oleifera* seeds [61]. This SWE application gave rise to a rich oil and saponin extract containing valuable bioactive compounds with great oxidative resistance.

Besides, SWE has been also employed for the extraction of peptides from tuna skin [62] and microalgae [63]. Table **3** summarizes the SWE extraction methods developed in the last years for the extraction of natural products.

Table 3. Applications of SWE for environmentally friendly extraction of natural ingredients.

Water (SWE)		
Food or Ingredient Under Study	**Extraction Method**	**Refs.**
Morus nigra L. fruit	L/S: 5:1 Time: 60 min Temperature: 60 °C and 120 °C Pressure: 0.1 MPa	[45]
Bioactive components from ginseng roots (*Panax ginseng* C.A. Mey)	L/S ratio: 30:1 Time: 20 min (x2) Temperature: 180-200 °C Pressure: 6.0 MPa	[46]
Flavanones from defatted orange peel	Sample and flow rate: 5.0 g sample, 10 mL/min Temperature: 150 °C Pressure: 10 MPa	[64]
Fucoidans from the brown alga *Nizamuddinia zanardinii*	L/S ratio: 21:1 Time: 29 min Temperature: 150 ºC Pressure: 0.75 MPa	[40]
Peptides from tuna skin collagen	L/S ratio: 200:1 Time: 5 min Temperature: 280 ºC Pressure: 8 MPa	[62]
Peptides from the microalgae *Spirulina platensis*	USW L/S ratio: 20:1 Time: 60 min Temperature: 120 °C Pressure and power: 10 MPa, 200 W	[63]
Phenolic compounds from sorghum bran	L/S ratio: 35:1 Time: 21 min Temperature: 144.5 °C	[44]
Polysaccharides from shiitake	Time: 15 min Temperature: 120-130 ºC Pressure: 5 MPa	[37]
Pectin from sunflower	L/S ratio: 7:1 Time: 20 min Temperature: 120 ºC Pressure: 0.8 MPa	[38]

(Table 3) cont.....

Water (SWE)		
Food or Ingredient Under Study	**Extraction Method**	**Refs.**
FOS and β-ecdysone from ginseng	FOS: Time: 5 min Temperature: 120 ºC β-ecdysone (roots): Time:15 min Temperature: 120 ºC Pressure: 12 Mpa β-ecdysone (aerial parts): Time: 5 min Temperature: 80 ºC Pressure: 2 Mpa	[58]
Pectin from cocoa by-products	Time: 30 min Temperature: 121 ºC Pressure: 10.34 MPa	[59]
Pectin from pomelo peel	Temperature: 120 ºC Pressure: 3 MPa	[60]
Fucoidans from *Saccharina japonica*	L/S ratio: 25:1 Time: 11.98 min Temperature: 127.01 ºC Pressure: 8 MPa	[39]
Fucoidans from *Nizamuddinia zanardinii*	S/L ratio: 0.04:1 Time: 29 min Temperature: 150 ºC Pressure: 0.75 MPa	[40]
Oil and saponins from *Camellia oleifera* seeds	Oil: L/S ratio:10.79:1[1] Time: 32.03 min Temperature: 133.59 ºC Pressure: 3 MPa Saponins: L/S ratio: 8.33:1 Time: 32.07 min Temperature: 121.11 ºC Pressure: 3 MPa	[61]

L/S ratio: liquid to sample ratio (*v/w*); SWE, subcritical water extraction; USW, ultrasound coupled with subcritical water.

LIQUEFIED GASES

The extraction of non-polar compounds usually is performed using *n*-hexane as an extraction solvent because of its low polarity, besides its boiling point is relatively low, which is convenient to evaporate the solvent from the final extract. However, this organic solvent is toxic, flammable, and generates environmental pollution.

Besides, *n*-hexane is produced from petroleum, and its use should be reduced due to the big environmental impact associated with it [65, 66]. A green alternative for the extraction of non-polar compounds is the use of liquefied gases, which are gases that are cooled or compressed during the extraction process to work under liquid state. This characteristic is really interesting because once the extraction is finished the solvent return to the gas phase and it is automatically separated from the valuable extract without using further evaporation steps [67]. Therefore, using liquefied gases, extraction with low-energy consumption and extracts with low residual traces of solvents are accomplished. The most common liquefied gases used for green solvent extractions are supercritical CO_2 ($scCO_2$), dimethyl ether (DME), *n*-butane, and hydrofluoroolefins (HFOs) and hydrofluorocarbons (HFCs). All these solvents have some common advantages compared to other green solvents like their low cost, low viscosity, and low boiling points [66].

Supercritical CO_2 ($scCO_2$)

$ScCO_2$ is a liquefied gas with a larger number of applications as a solvent for the recovery and extraction of non-polar compounds. The wide application of $scCO_2$ as an extraction solvent is due to multiple reasons, for example, it is recognized as GRAS for the food industry, it is not expensive and it is easily available. Its use is safe since it is not flammable and non-toxic [30, 68, 69].

To reach the supercritical conditions of any solvent, including CO_2, both, the temperature and the pressure of the solvent should be higher than its critical point. In this sense, during the extraction process, one of the main advantages of using $scCO_2$ is that CO_2 reaches the supercritical condition at mild temperature (31.3 ºC) and pressure (7.4 MPa). When CO_2 reaches the supercritical condition, it presents the low viscosity and low surface tension typical for a gas, which enhances the mass transfer ratio, but on the other side, it has a density of 0.470 g/mL, similar to liquid densities, and therefore, the solubility is improved respect to the gas state [70]. That means that $scCO_2$ presents a combination behavior between a gas and a liquid. In this way, $scCO_2$ penetrates easily into the material having excellent access to the interesting compounds and improving the mass transfer ratio [71]. After the extraction, as mentioned above, the $scCO_2$ is decompressed and relieved from the extract, giving rise to a pure extract free of solvent [68, 69].

$ScCO_2$ can be used as a unique extraction solvent for the extraction of non-polar compounds, and the polarity of the extraction can be adapted by modifying the extraction conditions (pressure, temperature, and flow rate) due to the density, viscosity, fluency and, therefore, the solubility of the solvent is determined by the operating conditions [30]. On the other hand, the polarity of the extracted compounds can be further increased by the addition of polar solvents (typically in a maximum ratio of $scCO_2$/solvent of 90:10). These organic solvents are called co-solvents or modifiers. Ethanol is the organic solvent most used as a modifier.

All these strategies contribute to performing a selective green extraction that can be adapted to different target compounds with a wide polarity range [68, 69].

The low polarity of $scCO_2$ makes that this green solvent has been typically used without the addition of co-solvents for the extraction of non-polar compounds. The extraction of low polar compounds using $scCO_2$ as extraction solvent has been reported for the extraction of non-polar lipids from coffee beans, linseed [72, 73], fatty acids from fruit pomaces [74, 75], essential oils from *Humulus Lupulus* and *Datura Stramonium* [76], lipids and pigments from algae [77, 78], terpenes from *Asteraceae* and *Lamiaceae* plants [79], wax from miscanthus [80], or hydrocarbons from microalgae [81].

Due to the high selectivity of $scCO_2$ for the non-polar fraction, its use in biorefinery processes provides important advantages for the fractionation of the valuable compounds of a particular matrix. Usually, the $scCO_2$ extraction is the first step of a downstream process in those biorefinery approaches. This first step aims to extract a pure non-polar fraction leaving a residue rich in more polar compounds like phenolic compounds, polar lipids, sugars, or proteins that are sequentially extracted with other extraction techniques [75, 77, 82 - 84]. For example, Sánchez Camargo *et al.*, used this strategy for the valorization of the alga *Nannochlorpsis gaditana* using $scCO_2$ for the recovery of non-polar lipids and carotenoids with subsequent extraction of the residue with PLE for the extraction of other bioactive compounds [77]. Three and four-step approaches using $scCO_2$ as a first fractionation level have been used for the downstream process of the algae *Scenedesmus obliquus* and *Isochrysis galbana* [82, 84] as well as for *Moringa oleifera* leaves [83].

However, $scCO_2$ can be also used for the extraction of more polar compounds by adding the appropriate modifier or co-solvent. This strategy has been used mainly for the extraction of phenolic compounds of food by-products, like the aerial parts of beetroot [85]. In this work, not only ethanol was used as a co-solvent, but also the mixture of different concentrations of ethanol/water was evaluated. The results showed that $scCO_2$ + 7% of ethanol/water (90:10, *v/v*) was the optimal solvent for the extraction of phenolic compounds from the leaves, while 10% of the co-solvent ethanol/water (90:10, *v/v*) was optimum for the stems. Other applications using co-solvent have been employed for the extraction of phenolic compounds from black mulberry leaves [86] or from onion skin by-products [87]. Also, carotenoids have been extracted using co-solvent from carrots [88], vegetable waste matrices [89], or spinach by-products [90]. Besides ethanol and water, other green solvents can be employed as co-solvents like bio-based solvents limonene, ethyl acetate, or ethyl lactate [91, 92].

Carbon Dioxide Expanded Liquids (CXLs)

Gas-expanded liquids (GXLs) are specific solvents able to dissolve a compressed gas (CO_2 or ethane) in high concentrations [93]. The most common GXLs are carbon dioxide expanded liquids (CXLs). In contrast to $scCO_2$ with the addition of co-solvent, where the organic solvent modifies the properties of the $scCO_2$, in CXLs the CO_2 is the modifier. Therefore, GXLs in general, and CXLs in particular, combine the beneficial properties of the two constituents that make up the solvent [94]. The physicochemical properties of the solvent where CO_2 is dissolved in (typically ethanol), like the dielectric constant of the solvent or the miscibility, acidity, and hydrogen-bonding ability, are altered and consequently, the polarity coverage of CXEs is increased in comparison to pure ethanol. Besides, the viscosity and the interfacial tension of the liquid are reduced at the same time that the diffusivity of the solvent is increased. These effects improve the transport rate and the mass transfer ratio during the extraction. The equilibrium phase between ethanol and CO_2 and the physicochemical properties of the CXL can be easily modified by adjusting the operating conditions of temperature, pressure, and CO_2 mole fraction [94]. In addition, the use of CXLs is associated with a reduction in energy costs, since the CXLs require even lower pressure than the $scCO_2$ extractions [93, 95].

The potential of CXLs has been proven for the extraction of a wide variety of compounds with different polarity properties. Reyes *et al.* carried out the study of the extraction of the carotenoid astaxanthin from the alga *Haematoccocus pluvialis* by $scCO_2$ with ethanol as a co-solvent and compare these results with the extraction using CXEs. They could observe that the temperature and the concentration of ethanol had a significant effect on the extraction yield of astaxanthin and its antioxidant activity. CXLs with 50% ethanol and 50% CO_2 at 45 °C provided the best recovery of this valuable carotenoid [96].

In combination with $scCO_2$ extractions and other green extraction techniques such as PLE or SWE, CXLs are used as one of the downstream steps involved in biorefinery processes. Usually, CXLs extraction is an intermediate step between $scCO_2$ extraction and more polar extraction steps like PLE with ethanol and SWE. Gilbert-López *et al.* used CXLs containing 45% ethanol for the extraction of the carotenoid fucoxanthin and polar lipids from the residue obtained after a previous $scCO_2$ extraction from the alga *Isochrysis galbana* [82]. In another downstream process for the valorization of the alga *Scenedesmus obliquus*, the carotenoids lutein and β-carotene were preferable extracted in the CXLs step consisting of

25% CO_2 dissolved in 75% ethanol [84]. Li *et al.* studied the properties and differences of the binary system ethanol+CO_2 under the subcritical and supercritical regions for the extraction of lipids from microalgae. The results showed that subcritical CXL was able to extract the complete lipid fraction with higher content of docosahexaenoic acid (DHA) and higher antioxidant activity [94].

Other applications of CXLs as green solvents have been optimized to extract different compounds with a wide variety of polarities in one unique extraction process. For example, the extraction of a complex extract, in terms of chemical composition, consisting of ascorbic acid, phenolic compounds, and carotenoids from acerola was carried out by Poletto *et al.* using 50% of ethanol in the CLX mixture [97]. On the other hand, a complex extract was also obtained by Fuentes-Gandara *et al.* for the valorization of sunflower leaves, but in this case, the optimal composition of the CXL was a mixture of CO_2 and water (50%). The characterization of the extract showed a mixture of fatty acids, terpenoids, flavonoids, lignans, and apocarotenoids [98].

In Table **4**, the most recent applications of SFE using scCO_2 as well as CXLs are presented.

Dimethyl Ether (DME)

DME is an alkyl ether with low molecular weight (C_2H_6O, MW= 46 g/mol). At room temperature and pressure, it maintains the gas state [102]. Traditionally, DME has been used to produce liquefied petroleum gas and small hydrocarbons. More recently, it has appeared as a green alternative to fuel due to its lower toxicity and lower production and transportation costs [103].

Table 4. Applications of scCO_2 and CXLs for environmentally friendly extractions of natural ingredients.

SFE			
Solvent	**Food or Ingredient Under Study**	**Extraction Method**	**Refs.**
scCO_2	Yarrow, marigold, balm and marjoram	Flow rate: 70 g/min Time: 180 min Temperature: 40 ºC Pressure: 14 MPa	[79]
scCO_2	Non-polar lipids from the alga *Nannochloropsis gaditana*	Flow rate: 10 mL/min Time: 270 min Temperature: 55 ºC Pressure: 40 MPa	[77]

(Table 4) cont.....

$scCO_2$	Fatty acids from *Moringa oleifera*	Flow rate: 2 mL/min Time: 60 min Temperature: 50 ºC Pressure: 10 MPa	[83]
$scCO_2$ + co-solvent	Waste products from beetroot (leaves and stems)	Flow rate: 8.3 g/min 10% of ethanol/water (90:10, *v/v*) (leaves) $scCO_2$ + 7% of ethanol/water (90:10, *v/v*) (stems) Time: 240 min Temperature: 40 ºC Pressure: 25 MPa	[85]
$scCO_2$ + ethanol	Peel and seed citrus by-products	Flow rate: 27 g/min $scCO_2$ + 1 mL/min ethanol Time: 120 min Temperature: 45 ºC Pressure: 20-30 MPa	[99]
$scCO_2$ + ethanol	Garden cress (*Lepidium sativum*)	Flow rate: 4 mL/min + 1 mL/min ethanol Time: 60 min static time + 10 min dynamic extraction Temperature: 50 ºC	[100]
$scCO_2$ + ethanol	Carnosic acid and carnosol enriched fraction of rosemary	Flow rate: 60 g/min Time: 60 min Temperature: 40 °C Pressure: 15 or 30 MPa	[101]
$scCO_2$ + ethanol	Carnosic acid and carnosol enriched fraction of rosemary	Flow rate: 60 g/min + 7% ethanol Time: 120 min, Temperature: 40 °C Pressure: 15 MPa	[101]
$scCO_2$ + ethanol	Microalga *Isochrysis galbana*	Flow rate: 5000 mL/min Time: 60 min Temperature: 50 °C Pressure: 30 MPa	[82]
$scCO_2$ + ethyl lactate	Caffeine from green tea (*Camellia sinensis*) leaves	Flow rate: 150 g/min + 2% ethyl lactate Time: 210 min Temperature: 70 °C Pressure: 30 MPa	[92]
CXL			
Solvent	**Food or Ingredient Under Study**	**Extraction Method**	**Refs.**
Ethanol + $scCO_2$	Carotenoids and chlorophylls from microalga *Scenedesmus obliquos*	Ethanol + 25% CO_2 Time: 150 min Temperature: 50 ºC Pressure: 7 MPa	[84]

(Table 4) cont.....

Ethanol + $scCO_2$	Caraotenoids from microalga *Isochrysis galbana*	Ethanol + 55% CO_2, 60 min, 50 °C, 7 MPa.	[82]
Ethanol + $scCO_2$	Ascorbic acid, phenolic compounds and carotenoids from acerola	Ethanol + 50% CO_2 Time: 90 min (pomace); 60 min (bagasse) Temperature: 40 °C Pressure: 7 MPa	[97]
Water + $scCO_2$	Fatty acids, terpenoids, flavonoids and lignans from sunflower leaves	Water + 50 CO_2 Time: 240 min Temperature: 55 °C Pressure: 40 MPa	[98]
d-Limonene + $scCO_2$	Fucoxanthin from microalga *Phaeodactylum tricornutum*	1.6 mL/min CO_2 + 50% d-limonene, 180 min, 40 °C, 30 MPa	[91]

CXL, carbon dioxide expanded liquids; $scCO_2$, supercritical carbon dioxide; SFE, supercritical fluid extraction.

Besides the new application of DME as a fuel alternative, it has shown good potential as a green solvent recognized by EFSA (European Food and Safety Authority) as safe for its application in the food industry [104]. Regarding its potential as a green solvent, DME can be easily liquefied at 0.51-0.59 MPa [105] and it presents a low boiling point (-24.8 °C) that allows achieving supercritical conditions at very low temperatures, even below 0 °C. Moreover, the low boiling point also contributes to easily releasing the solvent from the extract under very mild conditions, which ensures the maintenance of the native composition of the extract without damaging the compounds [103, 104]. DME offers affinity to hydrophobic matrices like oil, so it is a good candidate for the extraction of non-polar compounds, but, in addition, unlike the rest of liquefied gases, DME presents partial solubility in water. Therefore, DME can be used for the extraction of wet matrices without previous drying steps of the material, which reduces time and energy expenses. Besides, during the extraction, DME can act as a proper dehydrating agent because it can absorb water in a wet matrix, at the same time that it carries out the extraction of target compounds [67, 102, 105]. The partial solubility on water also affects the polarity of the compounds that can be extracted with DME, then DME is the only liquefied gas able to extract relatively polar compounds [67]. Another interesting property of DME is that in contrast to other alkyl ethers, DME is resistant to auto-oxidation [104]. Therefore, DME is considered an environmentally friendly, non-toxic, and low-cost solvent.

The extraction of natural compounds using DME as solvent presents big advantages over the rest of liquefied gases due to the partial solubility of water in DME. This ability reduces the pretreatment sample preparation steps like the dehydration of the sample and, therefore, reduces the time and the costs of the process. For example, in a study that compared the effect of $scCO_2$ and DME on the extraction of carotenoids from algae, the extraction with DME allowed carrying out the process with a raw material containing 93.2% of water. The final residue of the algae obtained after the extraction was almost completely dehydrated. Although both extraction techniques were successful for the extraction of carotenoids, DME reduces significantly energy costs [106].

The extraction of other wet microalgae (*Chlorococcum humicola*) with DME was compared with conventional solid-liquid extractions using acetone as an extraction solvent [107]. The results of this comparison showed that DME achieved a higher extraction yield than acetone and higher selectivity for less polar carotenoids. The effect of the water content on the lipid extraction from wet material using DME was evaluated by Wang *et al.* [105]. They tested different water contents in the raw material (65, 75, and 85%) and observed that in 20 min of extraction, the complete dewatering of the sample was possible, and in 50 min, they achieved the highest extraction yield of lipids. The optimal water content of the alga was set at 75%.

DME has been also used for the extraction of pigments (chlorophylls and carotenoids) from fresh vegetables like squash, spinach leaves, and carrots [103], aroma components from fresh orange peels and dry lavender flowers [67], monoterpenoid phenols like thymol from black cumin seeds [108], lipids and phospholipids from oat flakes [109], or phenolic compounds from mangosteen [110]. Moreover, DME has been also used for the production of biofuels from wet algae [111, 112] and spent coffee grounds, tea leaves, orange peels, gramineous weeds [113], and other food waste [114].

n-Butane

n-butane (C_4H_{10}, MW= 58 g/mol) is another green alternative for the extraction of non-polar compounds. Although it is commonly used as gas for cooking or as a compound of fuel and gasoline, *n*-butane can work as a liquefied gas under soft temperature and pressure conditions and it is authorized as a solvent in food production [66, 115]. When *n*-butane is used as liquefied gas, it presents attractive extraction characteristics like its low boiling point (-1 ºC), low viscosity as well as low-pressure procedures [116].

As mentioned above, *n*-butane presents hydrophobic properties so it is ideal for the extraction of non-polar compounds. Nevertheless, the selectivity of the target compounds can be slightly adapted by modifying the pressure of the extraction process [117]. In this sense, *n*-butane has similar polarity to $scCO_2$ due to both present similar dielectric constants (ε $scCO_2$= 1.8 vs ε *n*-butane= 1.7) [116, 117].

However, *n*-butane requires lower pressures to work as extraction liquefied gas in comparison to $scCO_2$. Besides, *n*-butane is very accessible and cheap. All these characteristics become *n*-butane an interesting alternative to $scCO_2$ [117]. Nonetheless, there are some disadvantages to the use of *n*-butane as an extraction solvent related to safety issues, since it is highly inflammable. Then, when this solvent is chosen for extraction procedures, high precautions should be taken during the whole process [66].

n-butane has been applied for the recovery of a wide variety of valuable compounds. Capeletto *et al.* compared the extraction efficiency and bioactivity of $scCO_2$ and *n*-butane extracts of phenolic compounds and terpenes from guavirova seeds. Both extracts were rich in terpenoids, but the extract obtained using *n*-butane presented a much higher concentration in phenolic compounds, showing a higher total phenolic content (four times higher than the $scCO_2$ extract), and higher antioxidant activity. On the other hand, the $scCO_2$ extract showed better antimicrobial activity [116]. In another comparison between $scCO_2$ and *n*-butane for the recovery of lipids and carotenoids from krill, *n*-butane provided higher lipid extraction yield and higher recovery of astaxanthin and tocopherol. Besides, the *n*-butane lipid extract was more resistant to lipid oxidation and presented higher nutritional quality [118].

Other different compounds extracted with *n*-butane are phenylpropanoids from grapefruit peel [119], carrot oleoresins, caraway aromas, and sunflower oils [66]. Moreover, a complex extract from cinnamon consisting of alcohols, aldehydes, esters, carboxylic acids, alkanes, and ketenes was achieved with *n*-butane extraction [120]. Fatty acids from andiroba seeds [117], pecan nuts [121], or wheat germen [122] as well as phenolic compounds from litchi and oat seeds [123] have been also extracted using *n*-butane.

An interesting application of *n*-butane was developed by Zhang *et al.* In this work, *n*-butane was used to eliminate pesticides from black tea with little loss of the compounds responsible for the quality of the tea, like catechin, theanine, or caffeine. However, this process slightly affected the aroma of the tea due to the hydrophobicity of some of the aroma compounds which allowed the solubility of these compounds in *n*-butane [124].

Hydrofluoroolefins (HFOs) and Hydrofluorocarbons (HFCs)

HFOs and HFCs are used as thermal fluids for refrigeration, they produce a very low impact on the environment since they are non-toxic, non-flammable, and do not produce greenhouse effects due to their short atmospheric lifetime [125, 126]. These environmentally friendly characteristics increased the interest in HFOs and HFCs as green solvents.

Regarding the HFOs as extraction solvents, they present high density (1163 kg m^{-3}) and very low viscosity (0.20 mPa/s), six times lower than other conventional organic solvents like ethanol. Another characteristic of these solvents in common with the rest of liquefied gases is their low boiling point [127]. Regarding the polarity of HFOs, they are slightly polar compared to *n*-hexane, but they do not present complete solubility in water [125]. There are two main HFOs used as liquefied gases for the extraction of natural products, HFO-1234yf and HFO-1234ze ($C_3H_2F_4$, MW = 114 g/mol) [125, 127]. HFO-1234yf and HFO-1234ze have very similar chemical structures, so their extraction capacities and solvation power are very similar. The main difference between them is their price, being HFO-1234yf five times more expensive than HFO-1234ze [125].

On the other hand, HFCs present a fluorine-based structure, very similar to HFOs, but the main difference between these two solvents is that, while the HFOs presented an unsaturated structure (carbon-carbon double bond), HFCs are saturated molecules, therefore, HFCs are chemically more stable. This makes HFOs have a shorter atmospheric lifetime. 1,1,1,2 tetrafluoroethane, also known as R-134a, is the most common HFC used as a green solvent. This solvent is non-flammable at room temperature and pressure, and its boiling point is achieved at -26 ºC. When it is pressurized to maintain the liquid state, it presents a density of 1300 kg m^{-3} and its viscosity and surface tension are lower than other conventional solvents, which promotes the penetration of R-134a in the solid matrices. It is not toxic and considered as a GRAS extraction solvent in the food industry and it is considered environmentally friendly due to its low cost, low toxicity, and low ozone depletion potential [126]. At this point, it is important to remark that working with liquefied gases always involves the danger of leaks in the reservoir tank. These kinds of accidents can lead to asphyxiation due to the large volume of gas that is generated in atmospheric conditions. For example, 1 L of HFO in the liquefied state can produce 200 L of gas [125].

The ability of HFOs as green extraction solvents has been employed for the recovery of oils from rapeseeds, carotenoids from carrots, and phenolic compounds and terpenoids from rosemary and olive leaves [125], as well as for the extraction of the biopharmaceutical compound artemisinin from *Artemisia annua* [127]. Rapinel *et al.* carried out a study for the comparison of three green solvents against n-hexane for the extraction of aromatic compounds from lavender flowers and orange peels. The three solvents were *n*-butane, DME, and HFO-1234ze. The extracts obtained by *n*-butane and HFO-1234ze presented similar extraction yields and chemical composition to the ones obtained with n-hexane.

However, the partial solubility of more polar compounds in DME, allowed the extract obtained by DME showed more variety of extracted analytes. Regarding olfactory perception, there was no significant difference in the extracts obtained by the four different techniques. Concerning the physicochemical properties of the three green solvents tested in this study, HFO-1234ze required lower energy to be evaporated from the final extract (0.05 kW/h) than DME and *n*-butane (0.11 and 0.10 kW/h, respectively). Besides, the density of HFO-1234ze was higher (1163 kg m^{-3}) in comparison to the density of DME and *n*-butane (661 and 573 kg m^{-3}, respectively) and the three of them showed low viscosity (0.20, 0.12, and 0.16 mPa/s for HFO-1234ze, DME and *n*-butane, respectively) [67].

HFCs have been used for the extraction of lipids from spent coffee grounds [128], bilberry seed waste [129], astaxanthin from the krill *Euphausia pacifica* [130], and active compounds from *Artemisia annua* [131]. Table **5** depicts the applications of DME, *n*-butane, and HFOs/HFCs as green solvents for the extraction of valuable ingredients from natural matrices.

Table 5. Applications of other liquefied gases for the environmentally friendly extraction of natural ingredients.

DME		
Food or Ingredient Under Study	**Extraction Method**	**Refs.**
Carotenoids from algae	Wet material Flow rate: 10 cm^3/min Time: 43 min Temperature25 °C Pressure: 0.59 MPa	[106]
Carotenoids and chlorophylls from micro alga *Chlorococcum humicola*	Wet material L/S ratio: 45:1 Time:20 min Temperature: 41 °C Pressure: 0.59 MPa	[103]
Thymol from black cumin	L/S ratio: 20 mL/g Time: 100 min Temperature: 25 °C Pressure: 0.06 MPa	[108]
Lipids and phospholipids from oat	L/S ratio: 10:1 Time: 80 min Temperature: 60 °C	[109]
Phenolic compounds from mangosteen	Wet material Flow rate: 7 mL/min Temperature: 35 °C Pressure: 0.8 MPa	[110]

(Table 5) cont.....

***n*-butane**		
Food or Ingredient Under Study	**Extraction Method**	**Refs.**
Oxyprenilated phenylpropanoids from grapefruit	Volume: 80 mL Time: 1440 min Temperature: 15 °C Pressure: 0.4-0.7 MPa	[119]
Oleoresins from carrots, aroma from caraway and oil from sunflowers	Volume: 1500 mL Time: 180 min Temperature: 20 °C Pressure: 0.2 MPa	[66]
Lipids from krill	L/S ratio: 5 mL/g Time: 120 min Temperature: 40 °C Pressure: 1 MPa	[118]
Alcohols, aldehydes, esters, carboxylic acids, alkanes and ketones from cinnamon	L/S ratio: 2:1 Time: 45 min Temperature: 40 °C	[120]
Fatty acids from pecan nuts	Flow rate: 3 mL/min Time: 180 min Temperature: 55 °C Pressure: 4 MPa	[121]
Aromatic compounds from lavender and orange peel	Volume: 900 mL Time: 240 min Temperature: 25 °C	[67]
Pesticides from tea	Volume: 1000 mL Time: 30 min Temperature: 40 °C	[124]
HFOs		
Food or Ingredient Under Study	**Extraction Method**	**Refs.**
Artemisinin from *Artemisia annua*	R1234yg Volume: 20 mL Time: 150 min Temperature: 20 °C Pressure: 0.45 mPa	[127]
HFCs		
Food or Ingredient Under Study	**Extraction Method**	**Refs.**
Lipids from bilberry seeds	R-134a Time: 270 min Temperature: 30-45 °C Pressure: 0.8-1 MPa	[129]

(Table 5) cont.....

Astaxanthin from krill *Euphausia pacific*	R-134a Flow rate: 6 g/min Time: 30 min Temperature: 60 ºC Pressure: 10 MPa	[130]
Bioactive compounds from *Artemisia annua*	R-134a L/S ratio: 1:20 Time: 60 min Temperature: 30 ºC Pressure: 1 MPa	[131]

DME: dimethyl ether; HFOs: Hydrofluoroolefins; HFCs: hydrofluorocarbons; L/S ratio: liquid to sample ratio.

ETHANOL

Ethanol obtained from biomass is a bio-based alcohol often called bioethanol. It can be used either as biofuel, as a solvent itself, or as an intermediate for the production of other solvents. Due to its potential use as biofuel, the development of biorefineries for ethanol production attracts a lot of interest. Bioethanol can be obtained from the fermentation of sugars of different biomass resources and there are three generations distinguished, based on the nature of the feedstock. The so-called first generation of bioethanol is obtained from crops rich in sucrose (*e.g.* sugarcane, sugar beet, sweet sorghum) or starch (*e.g.* corn, cassava, wheat). Sucrose may be directly fermented to produce ethanol, while starch requires the previous hydrolysis into simple sugars [132]. Currently, second-generation bioethanol is produced from lignocellulosic biomass, which relieves some problems associated with the use of food crops, such as the increase in food price. Nevertheless, the conversion of lignocellulosic biomass into bioethanol involves a more complicated process, especially the pretreatment to make cellulose and hemicellulose polymers available for the subsequent hydrolysis into simple sugars (glucose and xylose) [133]. The third generation of bioethanol is based on the development of biorefineries from algae, which do not compromise food supplies or arable land [134]. Due to its EHS properties, ethanol has been included in the list of recommended solvents for the CHEM21 ranking.

Regarding the physicochemical properties of ethanol (CH_3CH_2OH, MW= 46 g/mol), it is polar (ε= 22.4 at 20 ºC) [24], very soluble in water, and miscible in a big variety of organic solvents. To remove this solvent from the final extract, frequently, rotary evaporation is used, due to its boiling point (78 ºC). In comparison to methanol, a very common organic solvent used for natural product extraction, ethanol is slightly less polar and a little bit more expensive. Besides, methanol is easier to remove from the extract due to its lower boiling point. However, ethanol is preferable in green approaches due to methanol has a negative environmental impact, and therefore, ethanol and/or mixtures of ethanol/water are substituting its applications.

The use of ethanol as an extraction solvent in emergent green extraction techniques is an attractive and very common practice for the extraction of polar to medium polar compounds. The most noteworthy green extraction techniques where ethanol is commonly used for PLE, MAE, and UAE. Besides, as mentioned in Section "Supercritical CO_2 ($scCO_2$)", it is one of the most used modifiers in $scCO_2$ extraction [68, 69]. These extraction techniques improve the extraction yield and reduce the extraction time in comparison to conventional solid-liquid extractions. This is an advantage for the extraction of compounds that are unstable under high temperatures and long extraction times [135 - 137].

As in the case of SWE, PLE consists of the use of pressure to maintain solvents in the liquid state even at high temperatures (subcritical conditions). Pressures around 3.5-20 MPa and temperatures up to 200 ºC are the typical parameters used in PLE extraction of natural compounds. In PLE conditions, the viscosity of the solvent is reduced and, therefore, both, the solubility and the mass transfer efficiency are improved [135 - 137]. In MAE, non-ionizing electromagnetic waves in the frequency band of 300 MHz and 300 GHz are applied for carrying out the extraction. These microwaves create the rotation of the dipoles of polar solvents and matrixes, which produces increments of temperature in short times and hence, a superior mass transfer ratio [138]. On the other hand, UAE uses power ultrasounds (20 kHz – 100 MHz) which produce high acoustic pressure able to create extremely fast micro-bubbles that collapse near the matrix, generating shock waves that disrupt the wall cells and allow the easy penetration of the solvent in the matrix [139].

In the last years, the use of ethanol as a green solvent has suffered an increased interest (Table **6**). The physicochemical characteristics and polarity of ethanol make this solvent an excellent extraction agent for phenolic compounds. In particular, mixtures of ethanol and water are the most conventional solvents for the extraction of these interesting antioxidant compounds present in natural sources. In MAE, this strategy has been used for the extraction of phenolic compounds from diverse foods and spices like peppermint [140], asparagus [141], chokeberries [142], grapes [143], brown algae [144], or thymus [145]. UAE extraction of phenolic compounds from microalgae [146] with ethanol/water mixture results optimal as well.

Table 6. Green extraction methods in food analysis using ethanol.

Solvent	Food or Ingredient Under Study	Extraction Method	Refs.
	MAE		
Ethanol 50%	Phenolic compounds from thymus	L/S ratio: 20:1 Time: 9.5 min Temperature: 150 °C Power: 850 W	[145]
Ethanol 67.4%	Phenolic compounds, saponins and caffeic acid from asparragus	L/S ratio: 42:1 Time: 1.15 min Power:440 W	[141]
Ethanol	Carotenoids from microalga *Heterocapsa triquetra*	L/S ratio: 5:20 Time: 5 min Temperature: 50 °C Power: 5-400 W	[165]
Ethanol	Carotenoids from microalga *Phaeodactylum tricornutum*	(A) MAE L/S ratio: 10:0.5 Time: 2 min Temperature: 30 °C Power: 850 W max (B) PLE 1.0 g sample Time: 20 min Temperature: 50 °C Pressure: 10 MPa	[166]
	UAE		
Ethanol 50%	Phenolic compounds, carotenoids, vitamin C from orange peel	Time: 30 min Power: 400 W	[156]
Ethanol 53%	Polyphenolic compounds from red sorghum (*Sorghum bicolor* L.) bran	L/S ratio: 52:1 Time: 21 min Power: 200 W	[44]
Ethanol 55%	Phenolic compounds from potato by-products	L/S ratio: 10:1 Time: 35 min Temperature: 35 °C	[150]
Ethanol 55.4%	Phenolic compounds from microalgae	Time: 59.6 min Temperature: 74.4 °C Power: 700 W	[146]
Ethanol 57.38%	Antioxidants from *Thelephora ganbajun* mushroom	UAE L/S ratio: 70.15:1 Time: 10.58 min Temperature: 40 °C Power: 500 W	[122]

(Table 6) cont.....

Solvent	Food or Ingredient Under Study	Extraction Method	Refs.
Ethanol 65%	Phenolic acids, flavonoids, anthocyanins, and vitamin E from Thai riceberry bran	UAE L/S ratio: 20:1 Time: 40 min Temperature: 51 °C Frequency: 43 kHz	[167]
Ethanol	Rosmarinic acid and carnosic acid from rosemary	L/S ratio: 10:1 Time: 10 min Power: 140 W	[160]
PLE			
Ethanol 50%	Phenolic compounds from grape marc	Time: 54 min Temperature: 40 ºC Pressure: 10 MPa	[151]
Ethanol 80%	Rosmarinic acid, carnosic acid and carnosol from rosemery	Time: 3 min Temperature: 183 ºC Pressure: 13 MPa	[158]
Ethanol 86%	Phenolic compounds from Goji berries	Time: 20 min Temperature: 180 °C Pressure: 10 MPa	[168]
Ethanol	Phenolic compounds from feijoa leaves	Time: 50 min Temperature: 40 ºC Pressure: 10 MPa	[162]
Ethanol	Fatty acids from chia seeds	Time: 10 min Temperature: 60 ºC Pressure: 10.3 MPa	[159]

MAE: microwave-assisted extraction; L/S ratio: liquid to sample ratio (*v/w*); PLE: pressurized liquid extraction; UAE: ultrasound-assisted extraction.

Besides, the extraction of waste products from the food industry has been performed using different percentages of ethanol/water mixture, for instance, 78% ethanol was optimal for the extraction of phenolic compounds from apple skin [147]. A percentage of 50% was optimal for the extraction from pomegranate peels [148] using MAE, and from cold-pressed oil residues of almonds [149], or by-products produced in the industry of potatoes [150] with UAE extraction methods. PLE has been also extensively used for the recovery of phenolic compounds from food by-products using ethanol/water mixtures, for example, from grape marcs [151, 152], olive pomace [153], beetroot waste [154], or grape pomace from the wine industry [155]. Other kinds of compounds that have been extracted with these alternative extraction techniques using mixtures of ethanol/water are carotenoids and vitamin C [156], fatty acids from edible insects with UAE [157], carnosic acid, rosmarinic acid, and carnosol from rosemary [158], and ω-3 fatty acid-rich extract from chia seeds [159] using PLE.

On the other hand, 100% ethanol has been used for the extraction of natural compounds as well. For example, pure ethanol was the optimal extraction solvent for the MAE and UAE processes of rosmarinic acid and carnosic acid from rosemary [160], for the UAE extraction of saponins from edible seeds [161], and for PLE extraction of phenolic compounds from feijoa leaves [162], fatty acids from brow algae [163], oil from cocoa bean shells [164], and phenolic compounds from cranberry pomace [75].

OTHER BIO-BASED SOLVENTS

Ethyl Lactate and Ethyl Acetate

Ethyl lactate ($C_5H_{10}O_3$, MW= 118.13 g/mol) is the product of the esterification between ethanol and lactic acid. This solvent is considered bio-based due to its precursors (ethanol and lactic acid) are obtained from renewable sources by fermentation of sugars from biomass. Ethyl lactate has a low dielectric constant (ε= 15.7 at 20 ºC), is soluble and miscible in water and other less polar organic solvents (ketones, esters, and hydrocarbons). Its use in the food industry is allowed since it is recognized as GRAS by the USFDA and accepted by EFSA as a food additive [169]. Besides, ethyl lactate has been described as an alternative to halogenated solvents such as methylene chloride (dichloromethane) or chloroform because it is completely biodegradable, non-corrosive, non-hazardous, and non-toxic [170]. The main inconvenience to the use of ethyl lactate as a green solvent for extraction is related to the high boiling point of this solvent (154 ºC). This high temperature implies the use of long evaporation times and large energy consumption to remove the solvent from the final extract, and it can induce the degradation of compounds with low stability at high temperatures [171]. Nevertheless, due to its characteristic physical and thermodynamic properties (surface tension, viscosity, heat capacity, density), its use as an extraction solvent in natural matrices has gained attention in the last years [170, 172].

One of the most successful applications of ethyl lactate for the extraction of natural matrices is the decaffeination of green tea using this solvent in SFE and PLE extraction processes [92, 173, 174]. Using PLE, ethyl lactate showed a high selectivity for the extraction of caffeine minimizing the coextraction of catechin in the tea leaves, which is an important phenolic compound present in tea. Besides, a magnetic stirring extraction using a mixture of ethyl lactate and water (70:30, *v/v*) has been also used for the extraction of curcuminoids from turmeric [175]. The mixture of ethyl lactate and ethanol (60:40, *v/v*) provided good recoveries of carotenoids and tocols from crude palm olein [176].

On the other hand, ethyl acetate is also obtained from ethanol by different synthetic routes: i) esterification between acetic acid and ethanol, ii) partial oxidation of ethanol to acetic acid and esterification, iii) transformation of ethanol into acetaldehyde, followed by the addition of acetaldehyde to ethanol plus an additional dehydrogenation to form ethyl acetate. Besides these reactions, ethyl acetate can be also produced from biomass by the esterification of glucose [177]. In comparison to ethyl lactate, ethyl acetate ($CH_3COOC_2H_5$, MW = 88.11 g/mol; ε (20 ºC) = 6.08) is more volatile (boiling point = 77 ºC) and it is soluble but not miscible in water, but it is miscible with oxygenated and chlorinated solvents like ethanol and chloroform (PubChem database). Equally than ethyl lactate, ethyl acetate is considered GRAS by the USFDA. As an extraction solvent, ethyl acetate has been used as part of the extraction mixture for the recovery of bioactives from goldenberry. In this work, PLE using ethyl acetate/ethanol (25:75, *v/v*) provided the best extraction conditions for the extraction of ergostane-type steroids with potential bioactive properties [178]. Besides, pure ethyl acetate in PLE optimized methods showed high potential for the extraction of fatty acids from brown algae [163, 179], providing oils enriched in long-chain fatty acids like oleic acid, arachidonic acid, and eicosapentaenoic acid. Other PLE applications that used ethyl acetate for the extraction of key compounds from food sources are the recovery of carotenoids from microalgae [91], essential oils [180], or fungicide contaminants [181]. Mixtures of ethyl lactate/ethyl acetate used in UAE showed good selectivity for lycopene extracted from tomato by-products [182].

d-Limonene

(+)-Limonene or d-Limonene ($C_{10}H_{16}$, MW= 136.24 g/mol) is a natural terpene found in the essential oil of citrus peel. It is responsible for the characteristic odor of these essential oils. The production of this natural compound is closely related to biorefinery processes since it is obtained from the waste produced in the orange juice industry by a cold process consisting of centrifugal separation or steam distillation [183]. Besides, d-limonene can be also obtained by solvent-free extraction with reduced processing time [184].

As a solvent, d-Limonene has been recognized by the USFDA as GRAS for the extraction of natural compounds from food and plant material. It offers low-polar properties (ε (20 °C) = 2.3; log Kow= 4.57) so it is presented as another alternative to non-polar solvents (hexane, toluene, dichloromethane) [183, 184]. However, as ethyl lactate, d-limonene has a high boiling point (176 ºC), which increases the difficulty of its evaporation in the extraction process.

Diverse food applications using d-limonene for the extraction of compounds with nutritional interest have been described in the last years. The use of d-limonene in marine and fish oils provides good quality oils, rich in ω-3 fatty acids. Moreover, d-limonene acted as a protective agent against the oxidation of these oils due to the high antioxidant properties of this solvent [185, 186]. Also, d-limonene has been used for the extraction of fatty acids from grape seeds [187]. UAE using d-limonene has also shown good recoveries for carotenoids from citrus waste [188] and gamma-oryzanol from rice bran [189]. D-limonene has been used as solvent in other emergent extraction techniques like PLE and CXLs. Sánchez-Camargo *et al.* used d-limonene as a selective solvent in PLE and CXLs using 50% of d-limonene in the expanded solvent for the extraction of fucoxanthin from the microalgae *Phaeodactylum* [91]. Table **7** depicts the most recent applications of ethyl acetate, ethyl lactate, and d-limonene for the extraction of food ingredients.

Table 7. Green extraction methods in food analysis using bio-based solvents.

Solvent	Food or Ingredient Under Study	Extraction Method	Refs.
Ethyl acetate	Essential oil components in feedstuff additives	PLE 0.1-0.5 g additive sample, 1.4 g feedstuff, Time: 5 min Temperature: 50 °C Pressure: 10 MPa	[180]
Ethyl acetate	Fungicide residues in Matcha	PLE S/L ratio 130:1 Time: 30 min Temperature: 100 °C Pressure: 10 MPa	[181]
Ethyl acetate	Fatty acids from algae	PLE Time: 10 min Temperature: 120 °C Pressure: 10 MPa	[179]
Ethyl acetate 58%	Capsaicin in pepper	UAE Time: 10 min Temperature: 40 °C Power: 200 W	[190]
Ethyl acetate/Ethyl lactate (30:70, *v/v*)	Lycopene from tomato waste	UAE L/S ratio: 100:1 Time: 20 min Temperature: 63.4 °C	[182]
Ethyl lactate	Caffein from tee	PLE Time: 10 min Temperature: 100 °C Pressure: 10 MPa	[174]
Ethyl lactate 70%	Curcuminoids from turmeric	L/S ratio; 1:1 Time: 30 min Temperature: 30 min	[175]

(Table 7) cont.....

Solvent	Food or Ingredient Under Study	Extraction Method	Refs.
d-Limonene, or ethyl acetate	Fucoxanthin from microalga *Phaeodactylum tricornutum*	PLE 1.0 g sample, Time: 20 min Temperature: 40 °C Pressure: 10 MPa	[91]
d-Limonene/ ethanol (50:50)	Lipids from microalgae (*Spirulina, Phormidium, Anabaena, Stigeoclonium*)	PLE 2.0 g sample, Time: 15 min Temperature: 200 °C Pressure: 20.7 MPa	[191]
d-Limonene	gamma-Oryzanol compounds from Thai riceberry bran	UAE L/S ratio: 5:1 Time: 60 min Temperature: 60 °C Frequency: 43 kHz	[167]
d-Limonene	Carotenoids from citrus waste	UAE L/S ratio: 10:1 Time: 5 min Temperature: 20 °C Ultrasonic intensity: 208 W cm^{-2}	[188]

MAE: microwave-assisted extraction; L/S ratio: liquid to sample ratio (*v/w*); PLE: pressurized liquid extraction; UAE: ultrasound-assisted extraction.

AMPHIPHILIC SOLVENTS

An amphiphilic compound is a chemical with both behaviors, hydrophilic and hydrophobic. In the case of amphiphilic solvents, they are composed of a polar or ionic head group (hydrophilic character), in combination with long hydrocarbon chains (the responsible for hydrophobic properties). Amphiphilic solvents have a unique array of physical-chemical properties that render them very attractive to replace organic solvents in analytical extractions. Among the amphiphilic solvents commonly used in microextraction techniques, surfactants are the most widely used. In recent years, the design and development of amphiphilic solvents have an important role to make Green Chemistry readily available [23, 26]. Depending on their hydrophilic moiety, they may be divided into three groups:

- Non-ionic: do not ionize in solution and typically have a hydrophilic group composed of alcohol, phenol, ether, ester, or amide [192].
- Ionic (anionic or cationic): surfactants formed by an amphiphilic cation and an anion in an aqueous solution or vice versa. Often this class of surfactants contains nitrogen compounds such as fatty amine salts of quaternary ammoniums linked.

- Zwitterionic: as all zwitterionic substances, it is a molecule that contains an equal number of positively and negatively charged functional groups [193].

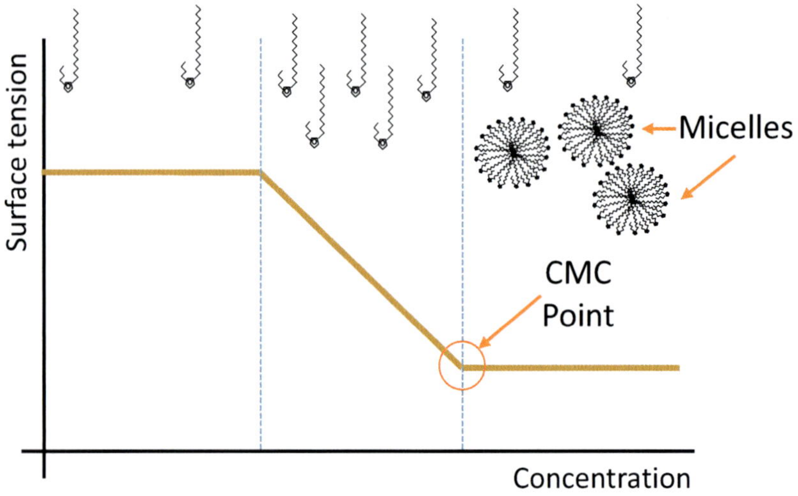

Fig. (3). CMC scheme for micelle based on surfactant formation.

One of the main characteristics of these solvents is their capacity to decrease the surface tension between two phases (liquid-liquid, solid-liquid, or gas-liquid) [194]. This dynamic is crucial for the use of amphiphilic solvent as an extraction solvent in sample preparation. Surface-active agents have the ability to alter the physical properties of interfaces. Surfactants can be adsorbed or aggregated, and it plays a significant role in several applications. Both phenomena are associated with a significant reduction of free energy (Gibbs energy) resulting from the hydrophobic effect. An important characteristic of surfactants is that they can form spontaneous aggregates in aqueous solutions at concentration levels above their critical micelle concentration (CMC). Micelle formation, or micellization, is an alternative mechanism to adsorption at the interfaces for removing hydrophobic groups from contact with water, thereby reducing the free energy of the system. The micelles have relevant importance in chemistry, because of their ability to dissolve different kinds of compounds. Compounds with low solubility in water, or even insoluble, can be dissolved and dispersed by the micelles in aqueous solutions [195, 196]. To rise the CMC value, it is important to know the surface tension of the surfactant. As the concentration of a surfactant increases in an immiscible medium, its surface tension changes as shown in Fig. (**3**). When CMC is reached, (when aggregates begin to be generated), the surface tension becomes constant again, but with a lower value. After that, the addition of surfactants will no longer affect the surface tension and is, therefore, unnecessary. A typical micelle in aqueous solution forms with the hydrophilic head regions in contact with the water and the hydrophobic aliphatic tail regions buried in the inner portion of the micelle [197]. A type of micelle based on surfactant widely used in recent years in the food analysis field is the reverse micelles (RMs). The general method of fabricating RMs is involved with making surfactant completely dissolved in a non-polar solvent. The particularity of RMs is that they have the ability to contain encapsulated water molecules as inner cores within a bulk nonpolar solvent [198]. The size, shape, and internal structure of aqueous cores, aggregation number, and micro viscosity are important parameters to characterize a common RM system.

Another important amphiphilic group of compounds is the supramolecular solvents (SUPRASs). SUPRASs are nanostructured liquids produced in colloidal solutions of amphiphilic compounds by spontaneous, sequential phenomena of self-assembly and coacervation. They are composed of long-chain alkyl alcohols and long-chain carboxylic acids. One of the advantages of SUPRASs is that they can be prepared *in situ* during the extraction process [199]. As surfactants, they present sensitivity to external environmental stimulation. This is because the forces involved in the self-assembly and coacervation process are non-covalent and therefore, reversible. SUPRAS have exceptional intrinsic properties for their use as extraction solvents. Among their properties, could be highlighted: their capacity for simultaneous solubilization of polar, non-polar, and amphiphilic compounds; their versatile structures that provide excellent interactions; they are low-cost substances; their easy synthesis, which facilitates their use both in laboratories and in the industry; their low volatility and flammability; and their tailor to conventional extraction methods [200]. These solvents are used in the same mode as surfactants in cloud-point extraction (CPE). It is important to highlight that the main difference between SUPRAS and surfactants is that surfactants have an ionic origin.

In general, amphiphilic solvents present efficient extraction properties. Both, SUPRASs and surfactants, at a concentration higher than CMC owing to the different polarity regions present in the supramolecular aggregates that offer mixed mechanisms for their solubilization. As a result, solutes in a wide polarity range can be simultaneously and efficiently extracted. Thus, high extraction efficiencies can be achieved using low volumes of these amphiphilic solvents [201, 202]. The utilization of surfactants is most frequently implemented in aqueous systems. However, there are many cases of non-aqueous systems.

The use of surfactants, SUPRASs, or other amphiphilic solvents as extractant solvent is quite simple. Surfactants are directly added to the sample and the surfactant-analyte interactions are usually increased by the use of an external force, ultrasound or microwaves (UAE or MAE). Once the analytes are concentrated in the micelle, they can proceed to the separation. The separation of the surfactant-rich phase can be forced by a change in the physical-chemical properties of the media. These external stimulations can be achieved in various modes: changes in pH, temperature, or ionic strength [203]. Regarding this, there are two main surfactant-based extraction methods: CPE, when the temperature of the environment is changed to favor the separation; and coacervative extraction (CAE), which involves an ionic surfactant. CPE is the most widely used. CPE employs neutrally charged surfactants (non-ionic or zwitterionic surfactants), while CAE involves a phase separation caused by changes in other parameters (*e.g.* addition of salt or organic solvents, pH change, *etc*) [204, 205]. Dispersive liquid-liquid microextraction (DLLME) is the most extended liquid-phase microextraction (LPME) technique used [206]. It has several variants, however, all of them concur in the dispersion of a few microliters of extractant in a liquid sample. The conventional DLLME has characteristics of CPE, however, it also resembles homogeneous LPME. It consists of three elements: an aqueous sample containing analytes, a few microliters of a non-miscible amphiphilic solvent, and a dispersive solvent miscible with both. The mixing of those elements is a successful approach because of the high contact established between the surface of fine droplets and analytes. That fact speeds up the mass-transfer processes of analytes. Nowadays several DLLME modes imply the use of external energy instead of dispersive solvent to favor the mass transfer. The most common are ultrasound-assisted, microwave-assisted, or vortex-assisted DLLME [207]. Despite the liquid state of surfactants, they have been utilized also in μSPE techniques. This is possible when the surfactants are supported onto a solid, forming composites or hybrid materials. The solid supports most widely used for surfactants and SUPRASs are magnetic nano-particles (MNPs). MNPs are a group of nano-particles that present superparamagnetic properties. These properties typically come from iron minerals and magnetic iron oxides such as magnetite (Fe_3O_4) and maghemite (γ-Fe_2O_3) [208]. Once the amphiphilic compounds have coated the MNPs, repulsive forces are generated in the nano-particle surface, and in consequence, a stable magnetic composite. The resulting composite is then used in magnetic-dispersive miniaturized solid-phase extraction (M-D-μSPE) strategies. M-D-μSPE replaces the employ of centrifugation or decantation steps to separate the phases. In this case, the microextraction procedure is developed as follows: MNPs are added to a liquid sample matrix. After a proper equilibration time, an external magnetic field is applied [209]. Several studies using amphiphilic solvents as extractant solvents are included in Table **8** [201 - 215].

Table 8. Use of surfactants as extraction solvents for agri-food analysis.

Surfactant	Analyte	Matrix	LPME Technique	Refs.
Hydroalcolic solution	Bisphenol A	Wine	DLLME	[210]
SDS/CTAB	Metals	Tomato paste	DLLME	[211]
SBA-15	Pb, Cd and Ni	Fish and water	DLLME	[212]
Tetradecanoate + MNPs	Bisphenol A	Sorft drink	M-D-μSPE	[213]
CTAB + MNPs	Xanthohumol	Beer	M-D-μSPE	[214]
AC-OFX + MNPs	Herbicides	Milk and rice	M-D-μSPE	[215]

DLLME: dispersive liquid-liquid microextraction; M-D-μSPE: magnetic dispersive liquid-liquid miniaturize solid-phase extraction.

IONIC LIQUIDS

Ionic liquids and their derivatives (ILs) are non-conventional solvents composed of salts (they are composed completely of ions). They represent a sustainable alternative to common organic solvents. ILs are non-molecular solvents with melting points below 100 °C [216, 217]. They present a wide variety of physicochemical properties, including low to negligible vapor pressure at room temperature, and high chemical and electrochemical stability. They are considered green solvents, and a great alternative to conventional organic solvents. However, these properties that ensure low emissions may cause persistence of ILs in the environment and the assessment of the toxicity of such a wide range of ILs compositions represents a big challenge [218]. ILs who have a melting point lower than room temperature are also denominated as room temperature ionic liquids (RTILs). Due to their versatile structure, a simple modification in them implies a huge modification in their intrinsic properties (such as water solubility or viscosity). The incorporation of polar or non-polar moieties as precursors during the synthesis can promote different interactions between analytes with different natures. Both aspects, among others, make ILs suitable extraction solvents. ILs are prepared by the combination of a bulky organic cation, commonly containing nitrogen or phosphorous atoms, and an anion of either, organic or inorganic nature [207]. Depending on the components and their ratios, ILs may present different properties that make them tailored for specific applications. Thus, ILs are also called, designer solvents. The most representative anion and cations for ILs designing are included in Fig. (**4**). One basic method for ILs preparation involves metathesis (or double displacement reaction) of a halide salt of the desired cation with a salt containing the desired anion, which results in halide-containing ILs [23]. Therefore, new processes are under investigation in order to obtain less toxic, renewable, and biodegradable ILs [219, 220].

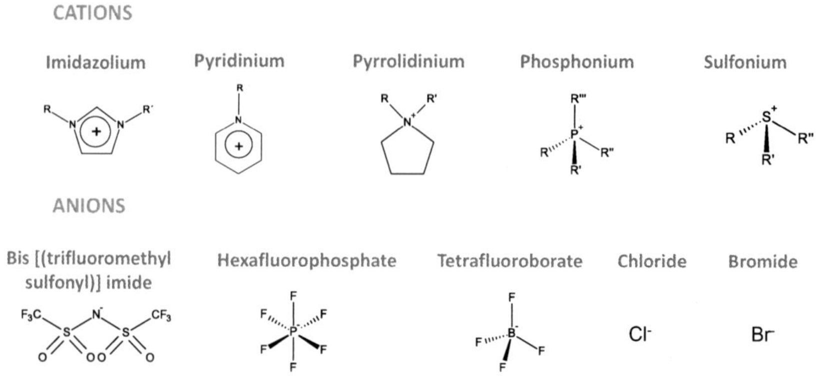

Fig. (4). Most representative cations and anions in ILs structures.

In Analytical Chemistry, not only ILs are interesting, but also their derivatives. There are different interesting ILs derivatives described, such as polymeric ionic liquids (PILs), IL-based surfactants (ILBSs), and magnetic ionic liquids (MILs) [221]. These materials combine the inherent properties of ILs with those derived from polymeric, surfactant, and paramagnetic properties, respectively. PILs have better mechanical, and thermal stability than ILs. They are obtained by the polymerization of ILs monomers [222]. ILBSs act as conventional surfactants by forming micellar aggregates when dissolved in water above their CMC. The main difference is that the CMC is much lower for ILBSs, therefore they will be required for specific applications compared with surfactants. At last, MILs are the most popular IL derivative in recent years. MILs are designed to incorporate paramagnetic metal ions or metal clusters in IL structures [223].

Aside from the aforementioned properties, ILs can be also designed to exhibit exceptional selectivity toward specific groups of compounds. It is evident that the utilization of microextraction procedures together with previously described materials based on ILs is a good choice in Analytical Chemistry. To date, ILs have been applied in a wide number of microextraction procedures. ILs and derivatives have been used in a variety of LPME approaches in food analysis, such as single-drop microextraction (SDME), hollow-fiber liquid-phase microextraction (HF-LPME), DLLME, and magnetic dispersive liquid-liquid extraction (M-DLLME). However, these solvents have also been used as supported extractant solvents in μSPE, for example in solid-phase microextraction (SPME), M-D-μSPE, and stir bar solid extraction (SBSE) [224]. Table **9** includes several examples of the use of ILs and derivatives in LPME and μSPE strategies [225 - 232].

Table 9. Use of ILs as an extraction solvent for agri-food analysis.

IL	Analyte	Matrix	LPME Technique	Refs.
[HMIM][NTF_2]	Fungicides	Juice	DLLME	[225]
[C_6MIM][PF_6]	Sulfonamides	Water, milk and honey	DLLME assisted by MW	[226]
[$P_{6,6,6,14}^+$] [Mn(hfacac)$_3^-$]	Fatty acids	Milk	Magnetic SDME (MIL)	[227]
[OMIM][PF_6]	Trihalomethanes	Water	SDME	[228]
[HMIM][BF_4]	Fatty acid	Juice	SDME	[229]
Poly([ViBzHDIM][TFSI])	Aroma	Coffee	SPME (PIL)	[230]
Poly([ViHDIM][PF_6])	Pyrethroids	Vegetables	SPME (PIL)	[231]
([DABCO–C_3OH]CL@MNPS	Antibiotics	Milk	M-D-µSPE	[232]

DLLME: dispersive liquid-liquid extraction; M-D-µSPE: magnetic-dispersive miniaturized solid-phase extraction; MW: microwaves; SDME: single drop microextraction; PIL: polymeric ionic liquids.

SDME is a LPME strategy that requires only a few microliters of a suspended extractant solvent. This technique employs the needle of a syringe to support the droplet, which is exposed to the sample solution containing analytes. It may be developed in two modes, head-space (HS-) or direct immersion (DI-) mode [233]. As a novelty, and taking advantage of the magnetic properties of MILs, SDME has been developed with a slim strong magnet instead of a syringe needle [227]. The basis of the method involves the use of a rod magnet to aid in maintaining a small microdroplet of MIL during headspace-single drop microextraction (HS-SDME). As a surfactant, ILs have been extensively used in DLLME processes [217]. It is important to highlight that due to the possible hydrophobic or hydrophilic properties of ILs, they can be used as extractants and dispersive solvents during a DLLME [195]. However, this chapter will focus only on the use of IL as an extractant solvent for food analysis. Due to the possibility of separating MILs from a mixture by the use of an external magnetic field, they have been employed in liquid phase microextraction methods such as magnetic dispersive liquid-liquid microextraction (M-DLLME) or SDME [26, 234]. HF-LPME allows the use of hollow semipermeable membranes composed of inert porous materials. The hollow membrane normally is made of polypropylene. The easiest configuration is the attachment of the hollow fiber to a syringe needle. In the HF-LPME configuration, the polypropylene fiber is filled with IL and immersed into the sample. The diffusion and the partition of the analytes from the sample through the polypropylene fiber, favor its preconcentration in only a few microliters of extractant IL [235].

The possible interaction generated by ILs permits their use in µSPE strategies when they are supported onto a fiber, a magnetic stir bar, or MNPs. ILs and PILs have been used as coating materials onto SPME fibers [236]. However, PILs have demonstrated excellent extraction performance due to their robustness compared with ILs [237]. Specifically, a metal wire acts as support and then is coated by PILs, getting an SPME fiber. The coating material (in this case PILs) can be designed to improve the fundamental features of SPME: sensitivity, loading capacity, and stability. ILs could be also be supported onto MNPs, and be used in M-D-µSPE. The strategy follows the same mechanism as explained in the above section. The last strategy which will be discussed in this chapter for ILs is the SBSE. This technique has been used with two ILs derivatives, MNP@ILs and MILs. The magnetic characteristics of these composites permit their attachment onto a stir bar [220].

DEEP EUTECTIC SOLVENTS (DESS)

Deep eutectic solvents (DESs) are considered a greener alternative to ILs, being and important advance of DESs that they are biodegradable. Despite their similar characteristic and properties, DESs are completely different solvents. As mentioned before, ILs are obtained by the combination of a cation and an anion, while a DES is the proper combination of two solid substances through hydrogen bonds. They may also be defined as a eutectic mixture of Lewis or Brønsted acids and bases that contain various cationic and/or anionic molecules [238]. DESs present a low vapor pressure, high viscosity, and melting points between -50 to 150 °C, and they commonly degrade before their boiling point [239]. Nevertheless, the basis of these solvents is the eutectic point. This principle indicates that a homogeneous mixture of substances melts at a temperature lower than the melting point of any of the constituents. All these advantages open alternative routes in their applications; however, it must be noted that DESs are potentially reactive chemicals, and their auto association by hydrogen bonds limits their reactivity. DESs are formed from mixtures of a hydrogen bond acceptor (HBA) and a hydrogen bond donor (HBD). In Fig. (**5d**), several examples of HBA and HBD are included. Another important advantage, apart from its low toxicity, is the easy synthesis of DESs [240]. Practically, a proper ratio of HBA:HBD is mixed into a flask and after heating and stirring, a colorless liquid is obtained. Also, an interesting twin screw extrusion (TSE) process has been successfully proposed for the synthesis of DESs, which overcomes problems with handling and thermal decomposition [23]. Depending on the ratio, the physicochemical properties of the final DESs varied. DESs are viscous as ILs and this is often a drawback since it causes problems with handling and mass transfer. Sometimes cosolvents such as water may be used to reduce viscosity, although they should be added in an appropriate ratio to avoid the disruption of the hydrogen bonding networks of DESs. It is important to mention that DESs synthesis is environmentally friendly because has no waste products or emissions [241].

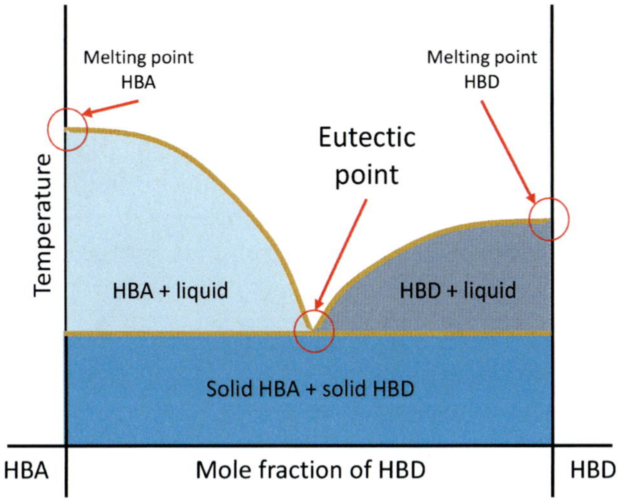

Fig. (5). Deep eutectic solvent phase diagram.

The most commonly used organic salts are quaternary ammonium or phosphonium salts, and HBDs used include alcohols, carboxylic acids, and amide. One of the HBAs most used to produce DESs is choline chloride (ChCl). ChCl is a cheap, biodegradable, and nontoxic ammonium salt [242]. DESs can be classified into four types depending on the composition:

- Type I: organic salt and metal salt // MCl_x, with M = Zn, Sn, Fe, Al or, Ga
- Type II: organic salt and metal salt hydrate // MCl_x-nH_2O, with M = Cr, Co, Cu, Ni, Fe, Ca, Mg, or La
- Type III: organic salt and organic compound as HBD // HBD with functional groups = $-CONH_2$, -COOH, -OH
- Type IV: metal salt and organic compound as HBD // in this case the metal salt acts as HAD, HBD like type III

The polarity of DESs can be designed by a proper combination between HBA and HBD. The hydrophobic DESs should contain HBD with long acidic chains or terpenes, and tetra-*n*-butyl ammonium bromide as organic salt. These polar DESs are used in the extraction of fatty acids, small alcohols, metals, or pesticides.

Regarding hydrophilic DESs, their structures are composed of biocompatible HBD, such as urea, carboxylic acid, or amino acids, and ChCl as HBA. They present higher toxicity than hydrophobic DESs, but their water solubility limits their use in extraction techniques.

DESs have evolved over the years with the aim to improve their properties. Choi *et al.* present the natural deep eutectic solvents (NADESs) based on both, the diversity of redundant metabolites in natural resources, and the occurrence of certain natural eutectic mixtures [243]. NADESs are composed of natural monomers obtained from renewable sources, which play a biological or physiological role in nature. Between them, it could be included: sugars, sugar alcohols, amino acids, or choline derivatives. However, it is important to notice that this natural origin does not automatically means that they are nontoxic solvents [23, 244]. NADESs can be classified into four groups [245, 246]:

- Group A: derivative from organic acids
- Group B: derivate from choline chloride
- Group C: based on sugar mixtures
- Group D: other combinations

Considering green and miniaturized techniques, DESs have been used in a variety of LPME techniques in agri-food analysis. Table **10** includes recent studies based on DESs as extractant sorbent [246 - 253]. DLLME with different assistances (ultrasound or vortex) is the most widely used when a hydrophobic DESs is employed, due to its simplicity and efficiency. However, it should be noted that in the case of DESs, it is very common to use salting-out to favor phase separation after the dispersion. Salting-out is also used in DLLME based on ILs, but many authors take special emphasis on the case of DESs. That occurs because the salting-out effect could break the phase separation membrane and promote the target to be fully enriched in the DES phase [254]. They have been also employed as magnetic-based materials in magnetic-assisted strategies.

Table 10. Use of DESs as extractant solvent for agri-food analysis.

DES	Analyte	Matrix	Lpme Technique	Refs.
Menthol:lauric acid:decanoic acid (3:1:1)	Testosterone and methyltestosterone	Milk	Acid based induced LLME	[246]
ChCl:phenol ChCl:butyric acid	Herbicidest	Wheat	DLLME	[247]
N-decanoic acid:methyltrioctyl ammonium bromide (2:1)	Fluoroquinolones	Milk, honey and water	DLLME salting out	[248]

(Table 10) cont.....

DES	Analyte	Matrix	Lpme Technique	Refs.
Tetra-n-octylphosphoniumbromide salt:ph	Antiepileptic drugs	Water, breast milk	DLLME	[249]
ChCl:4-chlorophenol	Pesticides	Vegetables and juices	SDME + salting out	[250]
Tetra-n-octylphosphoniumbromide salt:dodecanol (1:2)	Terpenes	Spieces	SDME	[251]
ChCl:2-chlorophenol (1:2) + MMWCNTS	Btex	Water	M-D-μSPE	[252]
Proline:propylene glycol (1:3) + MMWCNTS	Pesticides	Vegatables and friuts	M-D-μSPE	[253]

DLLME: dispersive liquid-liquid microextraction; LLME: liquid-liquid microextraction; M-D-μSPE: magnetic-dispersive miniaturized solid-phase extraction; MMWCNT: magnetic multi-walled carbon nanotubes; SDME: single-drop microextraction.

SWITCHABLE SOLVENTS

Switchable solvents are the result of the continuous efforts of Green Chemistry to find environmentally friendly solvents. Those solvents can reversibly change their physicochemical properties. As amphiphilic solvents, switchable solvents have reversible reactions in response to external stimulation. Changing temperature or adding/removing gas, polarity, hydrophilicity, or ionic strength may be affected. On the other hand, when they are exposed to another chemical trigger, they can change their chemical properties (acidic, basic, and oxidant or reduction properties). Because of the reversibility of the reaction, the changed solvent can easily be brought back to its original state [255, 256]. The switchable solvents can be considered as a system in which a non-ionic liquid is altered to an ionic liquid in the presence of CO_2, and then returns to its non-ionic structure as the gas is purged from the solution. They offer good capabilities as extraction solvents, since a solute can be extracted in one form of the solvent and then released by precipitation or by phase separation after the polarity switch triggered by CO_2. Switchable solvents can be classified depending on the addition of CO_2 into the system at atmospheric pressure [257]:

- Switchable polarity solvents (SPS): which undergo a polarity switch (change in the dielectric constant) upon CO_2 addition.
- Switchable hydrophilicity solvents (SHS): which change in polarity allows a hydrophobic solution to be miscible with water upon CO_2 addition.
- Switchable water (SW): wherein an aqueous solution undergoes a change in ionic strength.

SPSs are a mix between a nucleophilic compound (alcohol or amine) and an organic base (amine, amidine, or guanidine) [95, 257]. The addition of a CO_2 current implies the polarity switch, from low polar to high polar. Similar to traditional ILs, the properties of SPS systems turn by altering the structures such as side groups or alcohol chain lengths. One advantage is that the procedure requires a low amount of energy, and CO_2 is also cheap and safe to handle [258]. In these systems, the nucleophilic group of the solvent reacts with the introduced CO_2 molecules to produce ILs with alkyl carbonate ($[RCO_3]^-$) or alkyl carbamate ($[R_2NCO_2R']^-$) anions. The reaction is reversed by heating, sonication, or by bubbling with an inert gas such as N_2, processes that lead to the removal of CO_2 from the media [227].

SHS may be considered a sub-class of SPS. In these cases, CO_2 reacts with the molecules to produce a water-miscible form. SHSs are generally composed of hydrophobic organic bases (*e.g.*: alkylated amidines, secondary and tertiary amines) that form a biphasic mixture with water. Upon addition of CO_2, carbonic acid is generated, which protonates the organic base and liberates bicarbonate, both polar water-soluble compounds that give a single phase [259]. This is of great use because it can separate dissolved products from the solvent. In regular extraction processes, the solvent must be evaporated, which involve high energy needs. However, some of these SHSs have health and safety concerns associated, such as toxicity, volatility, or flammability, which would make them less desirable for their use. The polarity switching feature of these solvents provides a simple, reliable, and fast extraction method for scientists by decreasing the number of extraction steps and eliminating toxic solvents.

SW born to replace the resource-demanding method of separating organic solutes from water, the salting-out method. Salting-out involves the addition of salt to water media (such as NaCl) to increase the ionic strength. In this sense, the dissolved organic solutes are no longer soluble in polar aqueous media [260]. Although the method is quite effective, the addition of salt to water is not readily reversible. SW commonly refers to an aqueous solution of an uncharged organic base (amine or polyamine) that increases its ionic strength upon the addition of CO_2. As SPS and SHS, SW would involve the reversible reaction of a nitrogenous base additive dissolved in water with CO_2.

Different applications of switchable solvents in food safety have been reported, such as the extraction of bisphenols contaminants from drinks [261], sulfonamides in chicken meat [262], or contaminants from different classes in water [263]. Nevertheless, all of these applications used chemical triggers, and some work is still to be done in order to develop greener approaches. Regarding SW, it is already a quite new solvent, and its application in agri-food analysis has to be developed. In general, all switchable solvents are quite new and more studies about their toxicity, life cycle, and performance have to be made in the future in order to make comparisons to other solvents [213].

CONCLUDING REMARKS AND FUTURE TRENDS

The use of environmentally friendly solvents is a key aspect in the development of analytical methods for food analysis within the scope of Green Analytical Chemistry and their use is increasing in the last years. Nevertheless, they constitute a piece in the process of method development, which should be considered as a whole from a more holistic perspective. Therefore, current trends involve the assessment of the greenness of the full method using specific tools or chemometric analysis. Finally, it should be noted that a balance between greenness and practical aspects of method development has to be pursued. In this context has emerged the concept of White Analytical Chemistry (WAC), which is presented as a compromise between the greenness of an analytical method and its functionality, constituting a closer approach to sustainable development than GAC [264].

Solvents such as water and carbon dioxide are frequently used in pressurized green extraction techniques like SWE and SFE, respectively, especially to extract bioactive compounds from natural sources. However, these techniques may involve a high energy consumption to maintain the solvents under high pressures and temperatures. Concerning waste generation, $scCO_2$ and liquefied gases provide solvent-free extracts, since they return to their gas state and are released from the extracts during the depressurization step at the end of the process. On the other hand, if water has to be removed from the extract, it requires a process of lyophilization, which is not energetically favorable.

Bio-based solvents have also been used in combination with pressurized techniques, such as PLE, and other assisted techniques such as MAE and UAE. In general, these techniques involve a reduction in solvent consumption and/or extraction time, compared to conventional SLE or LLE procedures. Nevertheless, a solvent labeled as bio-based due to its generation from renewable sources does not directly assure a low EHS impact or economic viability.

Regarding the neoteric environmentally friendly solvents, LPME techniques take advantage of the properties of surfactants, ILs, and DES. However, the viscosity of DES and ILs is often a drawback, since it causes handling and mass transfer problems. Also, the toxicity of DESs and ILs is controversial. Finally, the reported applications so far on the use of switchable solvents used chemical triggers, and therefore there is some room for improvement in the greenness of analytical methods.

LIST OF ABBREVIATIONS

CAE	Coacervative Extraction
ChCl	Choline Chloride
CMC	Critical Micelle Concentration
COSMO-RS	Conductor-Like Screening Model for Real Solvent
CPE	Cloud-Point Extraction
CTAB	Cetrimonium Bromide
CXL	Carbon Dioxide Expanded Liquid
DES	Deep Eutectic Solvent
DFT	Density Functional Theory
DHA	Docosahexaenoic Acid
DI	Direct Immersion
DLLME	Dispersive Liquid-Liquid Microextraction
DME	Dimethyl Ether
EFSA	European Food and Safety Authority
EHS	Environmental Healh and Safety
GAC	Green Analytical Chemistry
GRAS	Generally Recognized As Safe
GXL	Gas Expanded Liquid
HBA	Hydrogen Bond Acceptor
HBD	Hydrogen Bond Donor
HFC	Hydrofluorocarbon
HF-LPME	Hollow-Fiber Liquid-Phase Microextraction
HFO	Hydrofluoroolefin
HS	Head Space
HSP	Hansen Solubility Parameters
IL	Ionic Liquid
ILBS	Ionic Liquid-Based Surfactant
LCA	Life Cycle Assessment
LLE	Liquid-Liquid Extraction
LPME	Liquid-Phase Microextraction
LTTM	Low Transition Temperature Mixtures
MAE	Microwave-Assisted Extraction
M-D-µSPE	Magnetic-Dispersive Miniaturized Solid-Phase Extraction

MIL	Magnetic Ionic Liquid
MMWCNT	Magnetic Multi-Walled Carbon Nanotube
MNP	Magnetic Nano-Particle
NADES	Natural Deep Eutectic Solvent
PCA	Principal Component Analysis
PHWE	Pressurized Hot Water Extraction
PIL	Polymeric Ionic Liquid
PLE	Pressurized Liquid Extraction
RM	Reverse Micelle
RTIL	Room Temperature Ionic Liquid
S/L	Solid to Liquid Ratio
SBSE	Stir Bar Solid Extraction
SDME	Single-Drop Microextraction
SFE	Supercritical Fluid Extraction
SHS	Switchable Hydrophilicity Solvent
SHWE	Superheated Water Extraction
SLE	Solid-Liquid Extraction
SPME	Solid-Phase Microextraction
µSPE	Miniaturized Solid-Phase Extraction
SPS	Switchable Polarity Solvent
SUPRA	Supramolecular Solvent
SW	Switchable Water
SWE	Subcritical Water Extraction
TSE	Twin Screw Extrusion
UAE	Ultrasound-Assisted Extraction
USFDA	United States Food and Drug Administration
WAT	White Analytical Chemistry

CONSENT FOR PUBLICATION

Not applicable.

CONFLICT OF INTEREST

The authors declare no conflict of interest, financial or otherwise.

ACKNOWLEDGEMENTS

B.G.L. thanks the Spanish Ministerio de Ciencia e Innovación for her Ramón y Cajal postdoctoral research contract (ref. RYC2019-026581-I). P.R.B. thanks to FEDER UJA for the support (ref. 1264521).

REFERENCES

[1] Gu Y, Jérôme F. Bio-based solvents: an emerging generation of fluids for the design of eco-efficient processes in catalysis and organic chemistry. Chem Soc Rev 2013; 42(24): 9550-70. [http://dx.doi.org/10.1039/c3cs60241a] [PMID: 24056753]

[2] Gałuszka A, Migaszewski Z, Namieśnik J. The 12 principles of green analytical chemistry and the SIGNIFICANCE mnemonic of green analytical practices. Trends Analyt Chem 2013; 50: 78-84. [http://dx.doi.org/10.1016/j.trac.2013.04.010]

[3] Płotka-Wasylka J, Rutkowska M, Owczarek K, Tobiszewski M, Namieśnik J. Extraction with environmentally friendly solvents. Trends Analyt Chem 2017; 91: 12-25. [http://dx.doi.org/10.1016/j.trac.2017.03.006]

[4] Molina-Díaz A, Beneito-Cambra M, Moreno-González D, Gilbert-López B. Ambient mass spectrometry from the point of view of Green Analytical Chemistry. Curr Opin Green Sustain Chem 2019; 19: 50-60. [http://dx.doi.org/10.1016/j.cogsc.2019.05.005]

[5] Castro-Puyana M, Mendiola JA, Ibañez E. Strategies for a cleaner new scientific discipline of green foodomics. Trends Analyt Chem 2013; 52: 23-35. [http://dx.doi.org/10.1016/j.trac.2013.06.013]

[6] Gilbert-López B, Mendiola JA, Ibáñez E. Green foodomics. Towards a cleaner scientific discipline. Trends Analyt Chem 2017; 96: 31-41. [http://dx.doi.org/10.1016/j.trac.2017.06.013]

[7] Armenta S, Garrigues S, de la Guardia M. The role of green extraction techniques in Green Analytical Chemistry. Trends Analyt Chem 2015; 71: 2-8. [http://dx.doi.org/10.1016/j.trac.2014.12.011]

[8] Armenta S, Garrigues S, Esteve-Turrillas FA, de la Guardia M. Green extraction techniques in green analytical chemistry. Trends Analyt Chem 2019; 116: 248-53. [http://dx.doi.org/10.1016/j.trac.2019.03.016]

[9] Prat D, Wells A, Hayler J, *et al.* CHEM21 selection guide of classical- and less classical-solvents. Green Chem 2016; 18(1): 288-96. [http://dx.doi.org/10.1039/C5GC01008J]

[10] Capello C, Fischer U, Hungerbühler K. What is a green solvent? A comprehensive framework for the environmental assessment of solvents. Green Chem 2007; 9(9): 927-34. [http://dx.doi.org/10.1039/b617536h]

[11] Kralisch D, Ott D, Gericke D. Rules and benefits of Life Cycle Assessment in green chemical process and synthesis design: a tutorial review. Green Chem 2015; 17(1): 123-45. [http://dx.doi.org/10.1039/C4GC01153H]

[12] Gałuszka A, Migaszewski ZM, Konieczka P, Namieśnik J. Analytical Eco-Scale for assessing the greenness of analytical procedures. Trends Analyt Chem 2012; 37: 61-72. [http://dx.doi.org/10.1016/j.trac.2012.03.013]

[13] Nowak PM, Kościelniak P. What color is your method? Adaptation of the RGB additive color model to analytical method evaluation. Anal Chem 2019; 91(16): 10343-52. [http://dx.doi.org/10.1021/acs.analchem.9b01872] [PMID: 31305064]

[14] Pena-Pereira F, Wojnowski W, Tobiszewski M. AGREE—Analytical GREEnness Metric Approach and Software. Anal Chem 2020; 92(14): 10076-82.
[http://dx.doi.org/10.1021/acs.analchem.0c01887] [PMID: 32538619]

[15] Kalinowska K, Bystrzanowska M, Tobiszewski M. Chemometrics approaches to green analytical chemistry procedure development. Curr Opin Green Sustain Chem 2021; 30: 100498.
[http://dx.doi.org/10.1016/j.cogsc.2021.100498]

[16] Hansen CM. Hansen solubility parameters: a user's handbook. CRC press 2007.
[http://dx.doi.org/10.1201/9781420006834]

[17] Sánchez-Camargo AP, Bueno M, Parada-Alfonso F, Cifuentes A, Ibáñez E. Hansen solubility parameters for selection of green extraction solvents. Trends Analyt Chem 2019; 118: 227-37.
[http://dx.doi.org/10.1016/j.trac.2019.05.046]

[18] Jessop PG, Jessop DA, Fu D, Phan L. Solvatochromic parameters for solvents of interest in green chemistry. Green Chem 2012; 14(5): 1245-59.
[http://dx.doi.org/10.1039/c2gc16670d]

[19] Jessop PG. Searching for green solvents. Green Chem 2011; 13(6): 1391-8.
[http://dx.doi.org/10.1039/c0gc00797h]

[20] Moity L, Durand M, Benazzouz A, Pierlot C, Molinier V, Aubry JM. Panorama of sustainable solvents using the COSMO-RS approach. Green Chem 2012; 14(4): 1132-45.
[http://dx.doi.org/10.1039/c2gc16515e]

[21] Kerton F M, Marriott R. Alternative solvents for green chemistry. Royal Society of chemistry: 2013.

[22] https://www.fda.gov/food/food-ingredients-packaging/generally-recognized-safe-gras

[23] Clarke CJ, Tu WC, Levers O, Bröhl A, Hallett JP. Green and sustainable solvents in chemical processes. Chem Rev 2018; 118(2): 747-800.
[http://dx.doi.org/10.1021/acs.chemrev.7b00571] [PMID: 29300087]

[24] Smallwood I. Handbook of organic solvent properties. Butterworth-Heinemann 2012.

[25] https://pubchem.ncbi.nlm.nih.gov/

[26] Pacheco-Fernández I, Pino V. Green solvents in analytical chemistry. Curr Opin Green Sustain Chem 2019; 18: 42-50.
[http://dx.doi.org/10.1016/j.cogsc.2018.12.010]

[27] Durand E, Lecomte J, Villeneuve P. From green chemistry to nature: The versatile role of low transition temperature mixtures. Biochimie 2016; 120: 119-23.
[http://dx.doi.org/10.1016/j.biochi.2015.09.019] [PMID: 26391220]

[28] Filly A, Fabiano-Tixier AS, Louis C, Fernandez X, Chemat F. Water as a green solvent combined with different techniques for extraction of essential oil from lavender flowers. C R Chim 2016; 19(6): 707-17.
[http://dx.doi.org/10.1016/j.crci.2016.01.018]

[29] Hartonen K, Riekkola M-L. Water as the first choice green solvent. In: Pena-Pereira F, Tobiszewski M (Eds.) The application of green solvents in separation processes. Elsevier 2017; pp. 19-55.
[http://dx.doi.org/10.1016/B978-0-12-805297-6.00002-4]

[30] Vian M, Breil C, Vernes L, Chaabani E, Chemat F. Green solvents for sample preparation in analytical chemistry. Curr Opin Green Sustain Chem 2017; 5: 44-8.
[http://dx.doi.org/10.1016/j.cogsc.2017.03.010]

[31] Abdelmoez W, Abdelfatah R. Therapeutic compounds from plants using subcritical water technology.Water extraction of bioactive compounds. Elsevier 2017; pp. 51-68.
[http://dx.doi.org/10.1016/B978-0-12-809380-1.00002-4]

[32] Castro-Puyana M, Marina ML, Plaza M. Water as green extraction solvent: Principles and reasons for

its use. Curr Opin Green Sustain Chem 2017; 5: 31-6.
[http://dx.doi.org/10.1016/j.cogsc.2017.03.009]

[33] Teo CC, Tan SN, Yong JWH, Hew CS, Ong ES. Pressurized hot water extraction (PHWE). J Chromatogr A 2010; 1217(16): 2484-94.
[http://dx.doi.org/10.1016/j.chroma.2009.12.050] [PMID: 20060531]

[34] Zakaria SM, Kamal SMM. Subcritical water extraction of bioactive compounds from plants and algae: applications in pharmaceutical and food ingredients. Food Eng Rev 2016; 8(1): 23-34.
[http://dx.doi.org/10.1007/s12393-015-9119-x]

[35] Plaza M, Turner C. Pressurized hot water extraction of bioactives. Trends Analyt Chem 2015; 71: 39-54.
[http://dx.doi.org/10.1016/j.trac.2015.02.022]

[36] Gilbert-López B, Plaza M, Mendiola JA, Ibáñez E, Herrero M. Subcritical water extraction and neoformation of antioxidants. In: Dominguez González H, González Muñoz MJ, Eds. Water extraction of bioactive compounds. Elsevier 2017; pp. 109-30.
[http://dx.doi.org/10.1016/B978-0-12-809380-1.00004-8]

[37] Zhang J, Wen C, Gu J, Ji C, Duan Y, Zhang H. Effects of subcritical water extraction microenvironment on the structure and biological activities of polysaccharides from Lentinus edodes. Int J Biol Macromol 2019; 123: 1002-11.
[http://dx.doi.org/10.1016/j.ijbiomac.2018.11.194] [PMID: 30468806]

[38] Ma X, Jing J, Wang J, Xu J, Hu Z. Extraction of low methoxyl pectin from fresh sunflower heads by subcritical water extraction. ACS Omega 2020; 5(25): 15095-104.
[http://dx.doi.org/10.1021/acsomega.0c00928] [PMID: 32637782]

[39] Saravana PS, Tilahun A, Gerenew C, *et al.* Subcritical water extraction of fucoidan from Saccharina japonica: optimization, characterization and biological studies. J Appl Phycol 2018; 30(1): 579-90.
[http://dx.doi.org/10.1007/s10811-017-1245-9]

[40] Alboofetileh M, Rezaei M, Tabarsa M, You S, Mariatti F, Cravotto G. Subcritical water extraction as an efficient technique to isolate biologically-active fucoidans from Nizamuddinia zanardinii. Int J Biol Macromol 2019; 128: 244-53.
[http://dx.doi.org/10.1016/j.ijbiomac.2019.01.119] [PMID: 30684576]

[41] Švarc-Gajić J, Cvetanović A, Segura-Carretero A, Linares IB, Mašković P. Characterisation of ginger extracts obtained by subcritical water. J Supercrit Fluids 2017; 123: 92-100.
[http://dx.doi.org/10.1016/j.supflu.2016.12.019]

[42] Xu H, Wang W, Liu X, Yuan F, Gao Y. Antioxidative phenolics obtained from spent coffee grounds (Coffea arabica L.) by subcritical water extraction. Ind Crops Prod 2015; 76: 946-54.
[http://dx.doi.org/10.1016/j.indcrop.2015.07.054]

[43] Cvetanović A, Švarc-Gajić J, Zeković Z, *et al.* Subcritical water extraction as a cutting edge technology for the extraction of bioactive compounds from chamomile: Influence of pressure on chemical composition and bioactivity of extracts. Food Chem 2018; 266: 389-96.
[http://dx.doi.org/10.1016/j.foodchem.2018.06.037] [PMID: 30381203]

[44] Luo X, Cui J, Zhang H, Duan Y. Subcritical water extraction of polyphenolic compounds from sorghum (Sorghum bicolor L.) bran and their biological activities. Food Chem 2018; 262: 14-20.
[http://dx.doi.org/10.1016/j.foodchem.2018.04.073] [PMID: 29751901]

[45] Koyu H, Kazan A, Ozturk TK, Yesil-Celiktas O, Haznedaroglu MZ. Optimizing subcritical water extraction of *Morus nigra* L. fruits for maximization of tyrosinase inhibitory activity. J Supercrit Fluids 2017; 127: 15-22.
[http://dx.doi.org/10.1016/j.supflu.2017.03.007]

[46] Zhang Y, Zhang Y, Taha AA, *et al.* Subcritical water extraction of bioactive components from ginseng roots (*Panax ginseng* C.A. Mey). Ind Crops Prod 2018; 117: 118-27.

[http://dx.doi.org/10.1016/j.indcrop.2018.02.079]

[47] Cheigh CI, Yoo SY, Ko MJ, Chang PS, Chung MS. Extraction characteristics of subcritical water depending on the number of hydroxyl group in flavonols. Food Chem 2015; 168: 21-6. [http://dx.doi.org/10.1016/j.foodchem.2014.07.047] [PMID: 25172678]

[48] Kheirkhah H, Baroutian S, Quek SY. Evaluation of bioactive compounds extracted from Hayward kiwifruit pomace by subcritical water extraction. Food Bioprod Process 2019; 115: 143-53. [http://dx.doi.org/10.1016/j.fbp.2019.03.007]

[49] Pinto D, Vieira EF, Peixoto AF, *et al.* Optimizing the extraction of phenolic antioxidants from chestnut shells by subcritical water extraction using response surface methodology. Food Chem 2021; 334: 127521. [http://dx.doi.org/10.1016/j.foodchem.2020.127521] [PMID: 32693333]

[50] Erşan S, Güçlü Üstündağ Ö, Carle R, Schweiggert RM. Subcritical water extraction of phenolic and antioxidant constituents from pistachio (*Pistacia vera* L.) hulls. Food Chem 2018; 253: 46-54. [http://dx.doi.org/10.1016/j.foodchem.2018.01.116] [PMID: 29502842]

[51] Kim SW, Ko MJ, Chung MS. Extraction of the flavonol quercetin from onion waste by combined treatment with intense pulsed light and subcritical water extraction. J Clean Prod 2019; 231: 1192-9. [http://dx.doi.org/10.1016/j.jclepro.2019.05.280]

[52] Ahmadian-Kouchaksaraie Z, Niazmand R, Najafi MN. Optimization of the subcritical water extraction of phenolic antioxidants from *Crocus sativus* petals of saffron industry residues: Box-Behnken design and principal component analysis. Innov Food Sci Emerg Technol 2016; 36: 234-44. [http://dx.doi.org/10.1016/j.ifset.2016.07.005]

[53] Ko MJ, Kwon HL, Chung MS. Pilot-scale subcritical water extraction of flavonoids from satsuma mandarin (*Citrus unshiu* Markovich) peel. Innov Food Sci Emerg Technol 2016; 38: 175-81. [http://dx.doi.org/10.1016/j.ifset.2016.10.008]

[54] Lachos-Perez D, Baseggio AM, Torres-Mayanga PC, *et al.* Sequential subcritical water process applied to orange peel for the recovery flavanones and sugars. J Supercrit Fluids 2020; 160: 104789. [http://dx.doi.org/10.1016/j.supflu.2020.104789]

[55] Gonçalves Rodrigues LG, Mazzutti S, Vitali L, Micke GA, Ferreira SRS. Recovery of bioactive phenolic compounds from papaya seeds agroindustrial residue using subcritical water extraction. Biocatal Agric Biotechnol 2019; 22: 101367. [http://dx.doi.org/10.1016/j.bcab.2019.101367]

[56] Jokić S, Gagić T, Knez Ž, Šubarić D, Škerget M. Separation of active compounds from food by-product (cocoa shell) using subcritical water extraction. Molecules 2018; 23(6): 1408. [http://dx.doi.org/10.3390/molecules23061408] [PMID: 29891762]

[57] Giombelli C, Iwassa IJ, da Silva C, Bolanho Barros BC. Valorization of peach palm by-product through subcritical water extraction of soluble sugars and phenolic compounds. J Supercrit Fluids 2020; 165: 104985. [http://dx.doi.org/10.1016/j.supflu.2020.104985]

[58] Vardanega R, Carvalho PIN, Santos DT, Meireles MAA. Obtaining prebiotic carbohydrates and beta-ecdysone from Brazilian ginseng by subcritical water extraction. Innov Food Sci Emerg Technol 2017; 42: 73-82. [http://dx.doi.org/10.1016/j.ifset.2017.05.007]

[59] Muñoz-Almagro N, Valadez-Carmona L, Mendiola JA, Ibáñez E, Villamiel M. Structural characterisation of pectin obtained from cacao pod husk. Comparison of conventional and subcritical water extraction. Carbohydr Polym 2019; 217: 69-78. [http://dx.doi.org/10.1016/j.carbpol.2019.04.040] [PMID: 31079687]

[60] Liew SQ, Teoh WH, Tan CK, Yusoff R, Ngoh GC. Subcritical water extraction of low methoxyl pectin from pomelo (*Citrus grandis* (L.) Osbeck) peels. Int J Biol Macromol 2018; 116: 128-35.

[http://dx.doi.org/10.1016/j.ijbiomac.2018.05.013] [PMID: 29738869]

[61] Wu H, Li C, Li Z, *et al.* Simultaneous extraction of oil and tea saponin from *Camellia oleifera* Abel. seeds under subcritical water conditions. Fuel Process Technol 2018; 174: 88-94. [http://dx.doi.org/10.1016/j.fuproc.2018.02.014]

[62] Ahmed R, Chun BS. Subcritical water hydrolysis for the production of bioactive peptides from tuna skin collagen. J Supercrit Fluids 2018; 141: 88-96. [http://dx.doi.org/10.1016/j.supflu.2018.03.006]

[63] Hu S, Fan X, Qi P, Zhang X. Identification of anti-diabetes peptides from *Spirulina platensis*. J Funct Foods 2019; 56: 333-41. [http://dx.doi.org/10.1016/j.jff.2019.03.024]

[64] Lachos-Perez D, Baseggio AM, Mayanga-Torres PC, *et al.* Subcritical water extraction of flavanones from defatted orange peel. J Supercrit Fluids 2018; 138: 7-16. [http://dx.doi.org/10.1016/j.supflu.2018.03.015]

[65] Yara-Varón E, Fabiano-Tixier AS, Balcells M, Canela-Garayoa R, Bily A, Chemat F. Is it possible to substitute hexane with green solvents for extraction of carotenoids? A theoretical *versus* experimental solubility study. RSC Advances 2016; 6(33): 27750-9. [http://dx.doi.org/10.1039/C6RA03016E]

[66] Rapinel V, Rombaut N, Rakotomanomana N, Vallageas A, Cravotto G, Chemat F. An original approach for lipophilic natural products extraction: Use of liquefied n-butane as alternative solvent to n-hexane. Lebensm Wiss Technol 2017; 85: 524-33. [http://dx.doi.org/10.1016/j.lwt.2016.10.003]

[67] Rapinel V, Santerre C, Hanaei F, *et al.* Potentialities of using liquefied gases as alternative solvents to substitute hexane for the extraction of aromas from fresh and dry natural products. C R Chim 2018; 21(6): 590-605. [http://dx.doi.org/10.1016/j.crci.2018.04.006]

[68] Herrero M, Cifuentes A, Ibañez E. Sub- and supercritical fluid extraction of functional ingredients from different natural sources: Plants, food-by-products, algae and microalgae. A review. Food Chem 2006; 98(1): 136-48. [http://dx.doi.org/10.1016/j.foodchem.2005.05.058]

[69] Sanchez-Camargo ADP, Mendiola JA, Ibáñez E, Herrero M. Supercritical fluid extraction. In: Reference Module in Chemistry, Molecular Sciences and Chemical Engineering. 2014. [http://dx.doi.org/10.1016/B978-0-12-409547-2.10753-X]

[70] Chemat F, Abert Vian M, Ravi HK, *et al.* Review of alternative solvents for green extraction of food and natural products: Panorama, principles, applications and prospects. Molecules 2019; 24(16): 3007. [http://dx.doi.org/10.3390/molecules24163007] [PMID: 31430982]

[71] Cabeza LF, de Gracia A, Fernández AI, Farid MM. Supercritical CO_2 as heat transfer fluid: A review. Appl Therm Eng 2017; 125: 799-810. [http://dx.doi.org/10.1016/j.applthermaleng.2017.07.049]

[72] Cornelio-Santiago HP, Gonçalves CB, de Oliveira NA, de Oliveira AL. Supercritical CO_2 extraction of oil from green coffee beans: Solubility, triacylglycerol composition, thermophysical properties and thermodynamic modelling. J Supercrit Fluids 2017; 128: 386-94. [http://dx.doi.org/10.1016/j.supflu.2017.05.030]

[73] Abrahamsson V, Rodriguez-Meizoso I, Turner C. Supercritical fluid extraction of lipids from linseed with on-line evaporative light scattering detection. Anal Chim Acta 2015; 853: 320-7. [http://dx.doi.org/10.1016/j.aca.2014.09.052] [PMID: 25467475]

[74] Campalani C, Amadio E, Zanini S, *et al.* Supercritical CO_2 as a green solvent for the circular economy: Extraction of fatty acids from fruit pomace. J. CO_2 Util. 2020, 41, 101259.

[75] Tamkutė L, Liepuoniūtė R, Pukalskienė M, Venskutonis PR. Recovery of valuable lipophilic and

polyphenolic fractions from cranberry pomace by consecutive supercritical CO_2 and pressurized liquid extraction. J Supercrit Fluids 2020; 159: 104755.
[http://dx.doi.org/10.1016/j.supflu.2020.104755]

[76] Campalani C, Chioggia F, Amadio E, *et al.* Supercritical CO_2 extraction of natural antibacterials from low value weeds and agro-waste. J. CO_2 Util. 2020, 40, 101198.

[77] Sánchez-Camargo AP, Pleite N, Mendiola JA, *et al.* Development of green extraction processes for *Nannochloropsis gaditana* biomass valorization. Electrophoresis 2018; 39(15): 1875-83.
[http://dx.doi.org/10.1002/elps.201800122] [PMID: 29683520]

[78] Feller R, Matos ÂP, Mazzutti S, *et al.* Polyunsaturated ω-3 and ω-6 fatty acids, total carotenoids and antioxidant activity of three marine microalgae extracts obtained by supercritical CO_2 and subcritical n-butane. J Supercrit Fluids 2018; 133: 437-43.
[http://dx.doi.org/10.1016/j.supflu.2017.11.015]

[79] Mouhid L, Gómez de Cedrón M, Vargas T, *et al.* Identification of antitumoral agents against human pancreatic cancer cells from *Asteraceae* and *Lamiaceae* plant extracts. BMC Complement Altern Med 2018; 18(1): 254.
[http://dx.doi.org/10.1186/s12906-018-2322-6] [PMID: 30223811]

[80] Attard TM, McElroy CR, Gammons RJ, *et al.* Supercritical CO_2 extraction as an effective pretreatment step for wax extraction in a Miscanthus biorefinery. ACS Sustain Chem& Eng 2016; 4(11): 5979-88.
[http://dx.doi.org/10.1021/acssuschemeng.6b01220]

[81] Yildiz-Ozturk E, Yesil-Celiktas O. Supercritical CO_2 extraction of hydrocarbons from *Botryococcus braunii* as a promising bioresource. J Supercrit Fluids 2017; 130: 261-6.
[http://dx.doi.org/10.1016/j.supflu.2017.06.019]

[82] Gilbert-López B, Mendiola JA, Fontecha J, *et al.* Downstream processing of *Isochrysis galbana:* a step towards microalgal biorefinery. Green Chem 2015; 17(9): 4599-609.
[http://dx.doi.org/10.1039/C5GC01256B]

[83] Rodríguez-Pérez C, Mendiola JA, Quirantes-Piné R, Ibáñez E, Segura-Carretero A. Green downstream processing using supercritical carbon dioxide, CO_2-expanded ethanol and pressurized hot water extractions for recovering bioactive compounds from Moringa oleifera leaves. J Supercrit Fluids 2016; 116: 90-100.
[http://dx.doi.org/10.1016/j.supflu.2016.05.009]

[84] Gilbert-López B, Mendiola JA, van den Broek LAM, *et al.* Green compressed fluid technologies for downstream processing of *Scenedesmus obliquus* in a biorefinery approach. Algal Res 2017; 24: 111-21.
[http://dx.doi.org/10.1016/j.algal.2017.03.011]

[85] Lasta HFB, Lentz L, Mezzomo N, Ferreira SRS. Supercritical CO_2 to recover extracts enriched in antioxidant compounds from beetroot aerial parts. Biocatal Agric Biotechnol 2019; 19: 101169.
[http://dx.doi.org/10.1016/j.bcab.2019.101169]

[86] Nastić N, Borrás-Linares I, Lozano-Sánchez J, Švarc-Gajić J, Segura-Carretero A. Optimization of the extraction of phytochemicals from black mulberry (*Morus nigra* L.) leaves. J Ind Eng Chem 2018; 68: 282-92.
[http://dx.doi.org/10.1016/j.jiec.2018.07.055]

[87] Campone L, Celano R, Lisa Piccinelli A, *et al.* Response surface methodology to optimize supercritical carbon dioxide/co-solvent extraction of brown onion skin by-product as source of nutraceutical compounds. Food Chem 2018; 269: 495-502.
[http://dx.doi.org/10.1016/j.foodchem.2018.07.042] [PMID: 30100465]

[88] de Andrade Lima M, Charalampopoulos D, Chatzifragkou A. Optimisation and modelling of supercritical CO_2 extraction process of carotenoids from carrot peels. J Supercrit Fluids 2018; 133: 94-102.
[http://dx.doi.org/10.1016/j.supflu.2017.09.028]

[89] de Andrade Lima M, Kestekoglou I, Charalampopoulos D, Chatzifragkou A. Supercritical fluid extraction of carotenoids from vegetable waste matrices. Molecules 2019; 24(3): 466. [http://dx.doi.org/10.3390/molecules24030466] [PMID: 30696092]

[90] Derrien M, Aghabararnejad M, Gosselin A, Desjardins Y, Angers P, Boumghar Y. Optimization of supercritical carbon dioxide extraction of lutein and chlorophyll from spinach by-products using response surface methodology. Lebensm Wiss Technol 2018; 93: 79-87. [http://dx.doi.org/10.1016/j.lwt.2018.03.016]

[91] Sánchez-Camargo AP, Pleite N, Herrero M, Cifuentes A, Ibáñez E, Gilbert-López B. New approaches for the selective extraction of bioactive compounds employing bio-based solvents and pressurized green processes. J Supercrit Fluids 2017; 128: 112-20. [http://dx.doi.org/10.1016/j.supflu.2017.05.016]

[92] Bermejo DV, Ibáñez E, Reglero G, Fornari T. Effect of cosolvents (ethyl lactate, ethyl acetate and ethanol) on the supercritical CO_2 extraction of caffeine from green tea. J Supercrit Fluids 2016; 107: 507-12. [http://dx.doi.org/10.1016/j.supflu.2015.07.008]

[93] Jessop PG, Subramaniam B. Gas-Expanded Liquids. Chem Rev 2007; 107(6): 2666-94. [http://dx.doi.org/10.1021/cr040199o] [PMID: 17564482]

[94] Li PL, Li HN, Jing KJ, David A, Lin JR, Deng G. Evaluation of lipid extraction from microalgae based on different phase regions of CO_2-expanded ethanol. Chem Eng Process 2019; 138: 1-6. [http://dx.doi.org/10.1016/j.cep.2019.02.011]

[95] Herrero M, Mendiola JA, Ibáñez E. Gas expanded liquids and switchable solvents. Curr Opin Green Sustain Chem 2017; 5: 24-30. [http://dx.doi.org/10.1016/j.cogsc.2017.03.008]

[96] Reyes FA, Mendiola JA, Ibañez E, del Valle JM. Astaxanthin extraction from *Haematococcus pluvialis* using CO_2-expanded ethanol. J Supercrit Fluids 2014; 92: 75-83. [http://dx.doi.org/10.1016/j.supflu.2014.05.013]

[97] Poletto P, Álvarez-Rivera G, López GD, *et al.* Recovery of ascorbic acid, phenolic compounds and carotenoids from acerola by-products: An opportunity for their valorization. Lebensm Wiss Technol 2021; 146: 111654. [http://dx.doi.org/10.1016/j.lwt.2021.111654]

[98] Fuentes-Gandara F, Torres A, Fernández-Ponce MT, *et al.* Selective fractionation and isolation of allelopathic compounds from *Helianthus annuus* L. leaves by means of high-pressure techniques. J Supercrit Fluids 2019; 143: 32-41. [http://dx.doi.org/10.1016/j.supflu.2018.08.004]

[99] Ndayishimiye J, Lim DJ, Chun BS. Antioxidant and antimicrobial activity of oils obtained from a mixture of citrus by-products using a modified supercritical carbon dioxide. J Ind Eng Chem 2018; 57: 339-48. [http://dx.doi.org/10.1016/j.jiec.2017.08.041]

[100] Rafińska K, Pomastowski P, Rudnicka J, *et al.* Effect of solvent and extraction technique on composition and biological activity of Lepidium sativum extracts. Food Chem 2019; 289: 16-25. [http://dx.doi.org/10.1016/j.foodchem.2019.03.025] [PMID: 30955598]

[101] Sánchez-Camargo AP, Valdés A, Sullini G, *et al.* Two-step sequential supercritical fluid extracts from rosemary with enhanced anti-proliferative activity. J Funct Foods 2014; 11: 293-303. [http://dx.doi.org/10.1016/j.jff.2014.10.014]

[102] Li P, Makino H. Liquefied dimethyl ether: An energy-saving, green extraction solvent. Alternative solvents for natural products extraction. Springer 2014; pp. 91-106. [http://dx.doi.org/10.1007/978-3-662-43628-8_5]

[103] Noriyasu A, Furukawa H, Kikuchi A, *et al.* Use of liquefied cold temperature dimethyl ether for

extraction of pigments from fresh vegetable tissues. Hortic Sci (Prague) 2015; 29(1): 48-52.

[104] Kanda H, Li P, Makino H. Production of decaffeinated green tea leaves using liquefied dimethyl ether. Food Bioprod Process 2013; 91(4): 376-80. [http://dx.doi.org/10.1016/j.fbp.2013.02.001]

[105] Wang Q, Oshita K, Takaoka M, Shiota K. Influence of water content and cell disruption on lipid extraction using subcritical dimethyl ether in wet microalgae. Bioresour Technol 2021; 329: 124892. [http://dx.doi.org/10.1016/j.biortech.2021.124892] [PMID: 33676356]

[106] Goto M, Kanda H, Wahyudiono , Machmudah S. Extraction of carotenoids and lipids from algae by supercritical CO_2 and subcritical dimethyl ether. J Supercrit Fluids 2015; 96: 245-51. [http://dx.doi.org/10.1016/j.supflu.2014.10.003]

[107] Eghbali Babadi F, Boonnoun P, Nootong K, Powtongsook S, Goto M, Shotipruk A. Identification of carotenoids and chlorophylls from green algae *Chlorococcum humicola* and extraction by liquefied dimethyl ether. Food Bioprod Process 2020; 123: 296-303. [http://dx.doi.org/10.1016/j.fbp.2020.07.008]

[108] Subratti A, Lalgee LJ, Jalsa NK. Efficient extraction of black cumin (*Nigella sativa* L.) seed oil containing thymol, using liquefied dimethyl ether (DME). J Food Process Preserv 2019; 43(4): e13913. [http://dx.doi.org/10.1111/jfpp.13913]

[109] Li Y, Obadi M, Qi Y, *et al.* Extraction of oat lipids and phospholipids using subcritical propane and dimethyl ether: experimental data and modeling. Eur J Lipid Sci Technol 2021; 123(1): 2000092. [http://dx.doi.org/10.1002/ejlt.202000092]

[110] Nerome H, Hoshino R, Ito S, *et al.* Functional ingredients extraction from *Garcinia mangostana* pericarp by liquefied dimethyl ether. Chem Eng J 2016; 20(4): 155-62.

[111] Kanda H, Hoshino R, Murakami K, Wahyudiono, Zheng Q, Goto M. Lipid extraction from microalgae covered with biomineralized cell walls using liquefied dimethyl ether. Fuel 2020; 262: 116590. [http://dx.doi.org/10.1016/j.fuel.2019.116590]

[112] Sakuragi K, Li P, Aoki N, Otaka M, Makino H. Oil recovery from wet *Euglena gracilis* by shaking with liquefied dimethyl ether. Fuel Process Technol 2016; 148: 184-7. [http://dx.doi.org/10.1016/j.fuproc.2016.03.005]

[113] Li P, Kanda H, Makino H. Simultaneous production of bio-solid fuel and bio-crude from vegetal biomass using liquefied dimethyl ether. Fuel 2014; 116: 370-6. [http://dx.doi.org/10.1016/j.fuel.2013.08.020]

[114] Sakuragi K, Li P, Otaka M, Makino H. Recovery of bio-oil from industrial food waste by liquefied dimethyl ether for biodiesel production. Energies 2016; 9(2): 106. [http://dx.doi.org/10.3390/en9020106]

[115] Union E. Directive 2009/28/EC of the European Parliament and of the Council of 23 April 2009 on the promotion of the use of energy from renewable sources and amending and subsequently repealing Directives 2001/77/EC and 2003/30/EC. Official Journal of the European Union 2009; 5: 2009.

[116] Capeletto C, Conterato G, Scapinello J, *et al.* Chemical composition, antioxidant and antimicrobial activity of guavirova (Campomanesia xanthocarpa Berg) seed extracts obtained by supercritical CO_2 and compressed n-butane. J Supercrit Fluids 2016; 110: 32-8. [http://dx.doi.org/10.1016/j.supflu.2015.12.009]

[117] Novello Z, Scapinello J, Magro JD, *et al.* Extraction, chemical characterization and antioxidant activity of andiroba seeds oil obtained from pressurized *n*-butane. Ind Crops Prod 2015; 76: 697-701. [http://dx.doi.org/10.1016/j.indcrop.2015.07.075]

[118] Sun D, Cao C, Li B, *et al.* Antarctic krill lipid extracted by subcritical n-butane and comparison with supercritical CO_2 and conventional solvent extraction. Lebensm Wiss Technol 2018; 94: 1-7. [http://dx.doi.org/10.1016/j.lwt.2018.04.024]

[119] Genovese S, Epifano F, Preziuso F, *et al.* A novel and efficient subcritical butane extraction method and UHPLC analysis of oxyprenylated phenylpropanoids from grapefruits peels. J Pharm Biomed Anal 2020; 184: 113185.
[http://dx.doi.org/10.1016/j.jpba.2020.113185] [PMID: 32113120]

[120] Liang Y, Li Y, Sun A, Liu X. Chemical compound identification and antibacterial activity evaluation of cinnamon extracts obtained by subcritical *n*-butane and ethanol extraction. Food Sci Nutr 2019; 7(6): 2186-93.
[http://dx.doi.org/10.1002/fsn3.1065] [PMID: 31289667]

[121] Scapinello J, Magro JD, Block JM, Di Luccio M, Tres MV, Oliveira JV. Fatty acid profile of pecan nut oils obtained from pressurized *n*-butane and cold pressing compared with commercial oils. J Food Sci Technol 2017; 54(10): 3366-9.
[http://dx.doi.org/10.1007/s13197-017-2771-9] [PMID: 28974822]

[122] Xu B, Han J, Zhou S, Wu Q, Ding F. Quality characteristics of wheat germ oil obtained by innovative subcritical butane experimental equipment. J Food Process Eng 2016; 39(1): 79-87.
[http://dx.doi.org/10.1111/jfpe.12201]

[123] Paliga M, Novello Z, Dallago RM, *et al.* Extraction, chemical characterization and antioxidant activity of *Litchi chinensis* Sonn. and *Avena sativa* L. seeds extracts obtained from pressurized *n*-butane. J Food Sci Technol 2017; 54(3): 846-51.
[http://dx.doi.org/10.1007/s13197-016-2485-4] [PMID: 28298700]

[124] Zhang Y, Gu L, Wang F, Kong L, Qin G. Effective subcritical butane extraction of bifenthrin residue in black tea. Molecules 2017; 22(4): 560.
[http://dx.doi.org/10.3390/molecules22040560] [PMID: 28358319]

[125] Rapinel V, Breil C, Makerri C, *et al.* Feasibility of using liquefied gas HFO-1234ze (trans-1,3,3,-tetrafluoroprop-1-ene) as an alternative to conventional solvents for solid–liquid extraction of food ingredients and natural products. Lebensm Wiss Technol 2017; 83: 225-34.
[http://dx.doi.org/10.1016/j.lwt.2017.05.027]

[126] Corr S. Chapter 3: 1, 1, 1, 2-tetrafluoroethane (R-134a): a selective solvent for the generation of flavor and fragrance ingredients. In: Frey C, Rouseff R (Eds.) Natural Flavors and Fragrances. ACS Publications 2005: Vol 908: pp- 41-59.
[http://dx.doi.org/10.1021/bk-2005-0908.ch003]

[127] Suberu J, Yamin P, Cornell R, Sam A, Lapkin A. Feasibility of using 2,3,3,3-tetrafluoropropene (r1234yf) as a solvent for solid–liquid extraction of biopharmaceuticals. ACS Sustain Chem& Eng 2016; 4(5): 2559-68.
[http://dx.doi.org/10.1021/acssuschemeng.5b01721]

[128] Colucci Cante R, Garella I, Gallo M, Nigro R. Effect of moisture content on the extraction rate of coffee oil from spent coffee grounds using Norflurane as solvent. Chem Eng Res Des 2021; 165: 172-9.
[http://dx.doi.org/10.1016/j.cherd.2020.11.002]

[129] Colucci Cante R, Prisco I, Garella I, Gallo M, Nigro R. Extracting the lipid fraction from waste bilberry seeds with a hydrofluorocarbon solvent. Chem Eng Res Des 2020; 157: 174-81.
[http://dx.doi.org/10.1016/j.cherd.2020.02.032]

[130] Han Y, Ma Q, Wang L, Xue C. Extraction of astaxanthin from *Euphausia pacific* using subcritical 1, 1, 1, 2-tetrafluoroethane. J Ocean Univ China 2012; 11(4): 562-8.
[http://dx.doi.org/10.1007/s11802-012-1948-0]

[131] Lelono A, Simanungkalit S, Umarudin I, Herdiawan H. Screening of active compounds from *Artemisia annua* using HFC-134a subcritic extraction system. J Trop Pharm Chem 2018; 4(3): 114-21.
[http://dx.doi.org/10.25026/jtpc.v4i3.149]

[132] Manochio C, Andrade BR, Rodriguez RP, Moraes BS. Ethanol from biomass: A comparative

overview. Renew Sustain Energy Rev 2017; 80: 743-55.
[http://dx.doi.org/10.1016/j.rser.2017.05.063]

[133] Zabed H, Sahu JN, Suely A, Boyce AN, Faruq G. Bioethanol production from renewable sources: Current perspectives and technological progress. Renew Sustain Energy Rev 2017; 71: 475-501.
[http://dx.doi.org/10.1016/j.rser.2016.12.076]

[134] Dalena F, Senatore A, Iulianelli A, Di Paola L, Basile M, Basile A. Ethanol from biomass: future and perspectives.Ethanol. Elsevier 2019; pp. 25-59.
[http://dx.doi.org/10.1016/B978-0-12-811458-2.00002-X]

[135] Herrero M, Castro-Puyana M, Mendiola JA, Ibañez E. Compressed fluids for the extraction of bioactive compounds. Trends Analyt Chem 2013; 43: 67-83.
[http://dx.doi.org/10.1016/j.trac.2012.12.008]

[136] del Pilar Sánchez-Camargo A, Ibáñez E, Cifuentes A, Herrero M. Bioactives obtained from plants, seaweeds, microalgae and food by-products using pressurized liquid extraction and supercritical fluid extraction. Book: *Green Extraction Techniques: Principles*. Advances and Applications 2017; 76: 27.

[137] Wianowska D, Gil M. Critical approach to PLE technique application in the analysis of secondary metabolites in plants. Trends Analyt Chem 2019; 114: 314-25.
[http://dx.doi.org/10.1016/j.trac.2019.03.018]

[138] Routray W, Orsat V. Microwave-assisted extraction of flavonoids: a review. Food Bioprocess Technol 2012; 5(2): 409-24.
[http://dx.doi.org/10.1007/s11947-011-0573-z]

[139] Picó Y. Ultrasound-assisted extraction for food and environmental samples. Trends Analyt Chem 2013; 43: 84-99.
[http://dx.doi.org/10.1016/j.trac.2012.12.005]

[140] Pavlić B, Kaplan M, Bera O, *et al.* Microwave-assisted extraction of peppermint polyphenols Artificial neural networks approach. Food Bioprocess Technol 2019; 118: 258-69.

[141] Zhang H, Birch J, Ma ZF, *et al.* Optimization of microwave-assisted extraction of bioactive compounds from New Zealand and Chinese *Asparagus officinalis* L. roots. J Food Sci Technol 2019; 56(2): 799-810.
[http://dx.doi.org/10.1007/s13197-018-3540-0] [PMID: 30906038]

[142] Simić VM, Rajković KM, Stojičević SS, *et al.* Optimization of microwave-assisted extraction of total polyphenolic compounds from chokeberries by response surface methodology and artificial neural network. Separ Purif Tech 2016; 160: 89-97.
[http://dx.doi.org/10.1016/j.seppur.2016.01.019]

[143] Garrido T, Gizdavic-Nikolaidis M, Leceta I, *et al.* Optimizing the extraction process of natural antioxidants from chardonnay grape marc using microwave-assisted extraction. Waste Manag 2019; 88: 110-7.
[http://dx.doi.org/10.1016/j.wasman.2019.03.031] [PMID: 31079623]

[144] Dang TT, Bowyer MC, Van Altena IA, Scarlett CJ. Optimum conditions of microwave-assisted extraction for phenolic compounds and antioxidant capacity of the brown alga *Sargassum vestitum*. Sep Sci Technol 2018; 53(11): 1711-23.
[http://dx.doi.org/10.1080/01496395.2017.1414845]

[145] Nabet N, Gilbert-López B, Madani K, Herrero M, Ibáñez E, Mendiola JA. Optimization of microwave-assisted extraction recovery of bioactive compounds from *Origanum glandulosum* and *Thymus fontanesii*. Ind Crops Prod 2019; 129: 395-404.
[http://dx.doi.org/10.1016/j.indcrop.2018.12.032]

[146] Gam DH, Kim SY, Kim JW. Optimization of ultrasound-assisted extraction condition for phenolic compounds, antioxidant activity, and epigallocatechin gallate in lipid-extracted microalgae. Molecules 2020; 25(3): 454.

[http://dx.doi.org/10.3390/molecules25030454] [PMID: 31973236]

[147] Casazza AA, Pettinato M, Perego P. Polyphenols from apple skins: A study on microwave-assisted extraction optimization and exhausted solid characterization. Separ Purif Tech 2020; 240: 116640. [http://dx.doi.org/10.1016/j.seppur.2020.116640]

[148] Kaderides K, Papaoikonomou L, Serafim M, Goula AM. Microwave-assisted extraction of phenolics from pomegranate peels: Optimization, kinetics, and comparison with ultrasounds extraction. Chem Eng Process 2019; 137: 1-11. [http://dx.doi.org/10.1016/j.cep.2019.01.006]

[149] Tungmunnithum D, Elamrani A, Abid M, *et al.* abra, A.; Addi, M.; Hano, C., A quick, green and simple ultrasound-assisted extraction for the valorization of antioxidant phenolic acids from moroccan almond cold-pressed oil residues. Appl Sci (Basel) 2020; 10(9): 3313. [http://dx.doi.org/10.3390/app10093313]

[150] Riciputi Y, Diaz-de-Cerio E, Akyol H, *et al.* Establishment of ultrasound-assisted extraction of phenolic compounds from industrial potato by-products using response surface methodology. Food Chem 2018; 269: 258-63. [http://dx.doi.org/10.1016/j.foodchem.2018.06.154] [PMID: 30100433]

[151] Tamires Vitor Pereira D, Vollet Marson G, Fernández Barbero G, *et al.* Concentration of bioactive compounds from grape marc using pressurized liquid extraction followed by integrated membrane processes. Separ Purif Tech 2020; 250: 117206. [http://dx.doi.org/10.1016/j.seppur.2020.117206]

[152] Pereira DTV, Tarone AG, Cazarin CBB, Barbero GF, Martínez J. Pressurized liquid extraction of bioactive compounds from grape marc. J Food Eng 2019; 240: 105-13. [http://dx.doi.org/10.1016/j.jfoodeng.2018.07.019]

[153] Cea Pavez I, Lozano-Sánchez J, Borrás-Linares I, Nuñez H, Robert P, Segura-Carretero A. Obtaining an extract rich in phenolic compounds from olive pomace by pressurized liquid extraction. Molecules 2019; 24(17): 3108. [http://dx.doi.org/10.3390/molecules24173108] [PMID: 31461900]

[154] Battistella Lasta HF, Lentz L, Gonçalves Rodrigues LG, Mezzomo N, Vitali L, Salvador Ferreira SR. Pressurized liquid extraction applied for the recovery of phenolic compounds from beetroot waste. Biocatal Agric Biotechnol 2019; 21: 101353. [http://dx.doi.org/10.1016/j.bcab.2019.101353]

[155] Huaman-Castilla NL, Martínez-Cifuentes M, Camilo C, Pedreschi F, Mariotti-Celis M, Pérez-Correa JR. The impact of temperature and ethanol concentration on the global recovery of specific polyphenols in an integrated HPLE/RP process on Carménère pomace extracts. Molecules 2019; 24(17): 3145. [http://dx.doi.org/10.3390/molecules24173145] [PMID: 31470596]

[156] Montero-Calderon A, Cortes C, Zulueta A, Frigola A, Esteve MJ. Green solvents and Ultrasound-Assisted Extraction of bioactive orange (*Citrus sinensis*) peel compounds. Sci Rep 2019; 9(1): 16120. [http://dx.doi.org/10.1038/s41598-019-52717-1] [PMID: 31695137]

[157] Otero P, Gutierrez-Docio A, Navarro del Hierro J, Reglero G, Martin D. Extracts from the edible insects *Acheta domesticus* and *Tenebrio molitor* with improved fatty acid profile due to ultrasound assisted or pressurized liquid extraction. Food Chem 2020; 314: 126200. [http://dx.doi.org/10.1016/j.foodchem.2020.126200] [PMID: 31972408]

[158] Hirondart M, Rombaut N, Fabiano-Tixier AS, Bily A, Chemat F. Comparison between pressurized liquid extraction and conventional Soxhlet extraction for rosemary antioxidants, yield, composition, and environmental footprint. Foods 2020; 9(5): 584. [http://dx.doi.org/10.3390/foods9050584] [PMID: 32380668]

[159] Villanueva-Bermejo D, Calvo MV, Castro-Gómez P, Fornari T, Fontecha J. Production of omega 3-rich oils from underutilized chia seeds. Comparison between supercritical fluid and pressurized liquid

extraction methods. Food Res Int 2019; 115: 400-7.
[http://dx.doi.org/10.1016/j.foodres.2018.10.085] [PMID: 30599958]

[160] Bellumori M, Innocenti M, Binello A, Boffa L, Mulinacci N, Cravotto G. Selective recovery of rosmarinic and carnosic acids from rosemary leaves under ultrasound- and microwave-assisted extraction procedures. C R Chim 2016; 19(6): 699-706.
[http://dx.doi.org/10.1016/j.crci.2015.12.013]

[161] Navarro del Hierro J, Herrera T, García-Risco MR, Fornari T, Reglero G, Martin D. Ultrasound-assisted extraction and bioaccessibility of saponins from edible seeds: quinoa, lentil, fenugreek, soybean and lupin. Food Res Int 2018; 109: 440-7.
[http://dx.doi.org/10.1016/j.foodres.2018.04.058] [PMID: 29803470]

[162] Santos PH, Kammers JC, Silva AP, Oliveira JV, Hense H. Antioxidant and antibacterial compounds from feijoa leaf extracts obtained by pressurized liquid extraction and supercritical fluid extraction. Food Chem 2021; 344: 128620.
[http://dx.doi.org/10.1016/j.foodchem.2020.128620] [PMID: 33223301]

[163] Otero P, López-Martínez MI, García-Risco MR. Application of pressurized liquid extraction (PLE) to obtain bioactive fatty acids and phenols from *Laminaria ochroleuca* collected in Galicia (NW Spain). J Pharm Biomed Anal 2019; 164: 86-92.
[http://dx.doi.org/10.1016/j.jpba.2018.09.057] [PMID: 30359842]

[164] Okiyama DCG, Soares ID, Toda TA, Oliveira AL, Rodrigues CEC. Effect of the temperature on the kinetics of cocoa bean shell fat extraction using pressurized ethanol and evaluation of the lipid fraction and defatted meal. Ind Crops Prod 2019; 130: 96-103.
[http://dx.doi.org/10.1016/j.indcrop.2018.12.063]

[165] Haguet Q, Bonnet A, Bérard JB, *et al.* Antimelanoma activity of *Heterocapsa triquetra* pigments. Algal Res 2017; 25: 207-15.
[http://dx.doi.org/10.1016/j.algal.2017.04.034]

[166] Gilbert-López B, Barranco A, Herrero M, Cifuentes A, Ibáñez E. Development of new green processes for the recovery of bioactives from *Phaeodactylum tricornutum*. Food Res Int 2017; 99(Pt 3): 1056-65.
[http://dx.doi.org/10.1016/j.foodres.2016.04.022] [PMID: 28865617]

[167] Peanparkdee M, Yamauchi R, Iwamoto S. Characterization of antioxidants extracted from Thai riceberry bran using ultrasonic-assisted and conventional solvent extraction methods. Food Bioprocess Technol 2018; 11(4): 713-22.
[http://dx.doi.org/10.1007/s11947-017-2047-4]

[168] Tripodo G, Ibáñez E, Cifuentes A, Gilbert-López B, Fanali C. Optimization of pressurized liquid extraction by response surface methodology of Goji berry (*Lycium barbarum L.*) phenolic bioactive compounds. Electrophoresis 2018; 39(13): 1673-82.
[http://dx.doi.org/10.1002/elps.201700448] [PMID: 29314152]

[169] https://www.efsa.europa.eu/en

[170] Pereira CS, Rodrigues AE. Ethyl lactate main properties, production processes, and applications.Alternative Solvents for Natural Products Extraction. Springer 2014; pp. 107-25.
[http://dx.doi.org/10.1007/978-3-662-43628-8_6]

[171] Kua YL, Gan S, Morris A, Ng HK. Ethyl lactate as a potential green solvent to extract hydrophilic (polar) and lipophilic (non-polar) phytonutrients simultaneously from fruit and vegetable by-products. Sustain Chem Pharm 2016; 4: 21-31.
[http://dx.doi.org/10.1016/j.scp.2016.07.003]

[172] Villanueva-Bermejo D, Reglero G, Fornari T. Recent advances in the processing of green tea biomolecules using ethyl lactate. A review. Trends Food Sci Technol 2017; 62: 1-12.
[http://dx.doi.org/10.1016/j.tifs.2016.12.009]

[173] Villanueva-Bermejo D, Mendiola JA, Ibáñez E, Reglero G, Fornari T. Pressurized liquid extraction of

caffeine and catechins from green tea leaves using ethyl lactate, water and ethyl lactate + water mixtures. Food Bioprod Process 2015; 96: 106-12.
[http://dx.doi.org/10.1016/j.fbp.2015.07.008]

[174] Villanueva Bermejo D, Ibáñez E, Reglero G, Turner C, Fornari T, Rodriguez-Meizoso I. High catechins/low caffeine powder from green tea leaves by pressurized liquid extraction and supercritical antisolvent precipitation. Separ Purif Tech 2015; 148: 49-56.
[http://dx.doi.org/10.1016/j.seppur.2015.04.037]

[175] D'Archivio AA, Maggi MA, Ruggieri F. Extraction of curcuminoids by using ethyl lactate and its optimisation by response surface methodology. J Pharm Biomed Anal 2018; 149: 89-95.
[http://dx.doi.org/10.1016/j.jpba.2017.10.042] [PMID: 29102796]

[176] Kua YL, Gan S, Morris A, Ng HK. Optimization of simultaneous carotenes and vitamin E (tocols) extraction from crude palm olein using response surface methodology. Chem Eng Commun 2018; 205(5): 596-609.
[http://dx.doi.org/10.1080/00986445.2017.1407760]

[177] Löser C, Urit T, Bley T. Perspectives for the biotechnological production of ethyl acetate by yeasts. Appl Microbiol Biotechnol 2014; 98(12): 5397-415.
[http://dx.doi.org/10.1007/s00253-014-5765-9] [PMID: 24788328]

[178] Ballesteros-Vivas D, Álvarez-Rivera G, Ibáñez E, Parada-Alfonso F, Cifuentes A. A multi-analytical platform based on pressurized-liquid extraction, *in vitro* assays and liquid chromatography/gas chromatography coupled to high resolution mass spectrometry for food by-products valorisation. Part 2: Characterization of bioactive compounds from goldenberry (*Physalis peruviana* L.) calyx extracts using hyphenated techniques. J Chromatogr A 2019; 1584: 144-54
[http://dx.doi.org/10.1016/j.chroma.2018.11.054] [PMID: 30579639]

[179] Otero P, Quintana S, Reglero G, Fornari T, García-Risco M. Pressurized liquid extraction (PLE) as an innovative green technology for the effective enrichment of Galician algae extracts with high quality fatty acids and antimicrobial and antioxidant properties. Mar Drugs 2018; 16(5): 156.
[http://dx.doi.org/10.3390/md16050156] [PMID: 29748479]

[180] Jiménez-Salcedo M, Tena MT. Determination of cinnamaldehyde, carvacrol and thymol in feedstuff additives by pressurized liquid extraction followed by gas chromatography–mass spectrometry. J Chromatogr A 2017; 1487: 14-21.
[http://dx.doi.org/10.1016/j.chroma.2017.01.042] [PMID: 28129937]

[181] Raina-Fulton R, Mohamad A. Pressurized solvent extraction with ethyl acetate and liquid chromatography—tandem mass spectrometry for the analysis of selected *Conazole fungicides* in matcha. Toxics 2018; 6(4): 64.
[http://dx.doi.org/10.3390/toxics6040064] [PMID: 30366422]

[182] Silva YPA, Ferreira TAPC, Celli GB, Brooks MS. Optimization of lycopene extraction from tomato processing waste using an eco-friendly ethyl lactate–ethyl acetate solvent: a green valorization approach. Waste Biomass Valoriz 2019; 10(10): 2851-61.
[http://dx.doi.org/10.1007/s12649-018-0317-7]

[183] Ciriminna R, Lomeli-Rodriguez M, Demma Carà P, Lopez-Sanchez JA, Pagliaro M. Limonene: a versatile chemical of the bioeconomy. Chem Commun (Camb) 2014; 50(97): 15288-96.
[http://dx.doi.org/10.1039/C4CC06147K] [PMID: 25341412]

[184] Aissou M, Chemat-Djenni Z, Yara-Varón E, Fabiano-Tixier AS, Chemat F. Limonene as an agro-chemical building block for the synthesis and extraction of bioactive compounds. C R Chim 2017; 20(4): 346-58.
[http://dx.doi.org/10.1016/j.crci.2016.05.018]

[185] Scurria A, Fabiano Tixier AS, Lino C, *et al.* High Yields of Shrimp Oil Rich in Omega-3 and Natural Astaxanthin from Shrimp Waste. ACS Omega 2020; 5(28): 17500-5.
[http://dx.doi.org/10.1021/acsomega.0c01978] [PMID: 32715235]

[186] Ciriminna R, Scurria A, Fabiano-Tixier AS, *et al.* Omega-3 extraction from anchovy fillet leftovers with limonene: chemical, economic, and technical aspects. ACS Omega 2019; 4(13): 15359-63. [http://dx.doi.org/10.1021/acsomega.9b01168] [PMID: 31572834]

[187] Rubio L, Lamas JP, Lores M, Garcia-Jares C. Matrix solid-phase dispersion using limonene as greener alternative for grape seeds extraction, followed by GC-MS analysis for varietal fatty acid profiling. Food Anal Methods 2018; 11(11): 3235-42. [http://dx.doi.org/10.1007/s12161-018-1300-4]

[188] Boukroufa M, Boutekedjiret C, Chemat F. Development of a green procedure of citrus fruits waste processing to recover carotenoids. Resource-Efficient Technologies 2017; 3(3): 252-62. [http://dx.doi.org/10.1016/j.reffit.2017.08.007]

[189] Trinovita E, Chany Saputri F, Mun'im A. Potential gastroprotective activity of rice bran (*Oryza sativa* L.) extracted by ionic liquid-microwave-assisted extraction against ethanol-induced acute gastric ulcers in rat model. Sci Pharm 2018; 86(3): 35. [http://dx.doi.org/10.3390/scipharm86030035] [PMID: 30205486]

[190] Vázquez-Espinosa M, González de Peredo AV, Ferreiro-González M, *et al.* Optimizing and comparing ultrasound-and microwave-assisted extraction methods applied to the extraction of antioxidant capsinoids in peppers. Agronomy (Basel) 2019; 9(10): 633. [http://dx.doi.org/10.3390/agronomy9100633]

[191] Golmakani MT, Mendiola JA, Rezaei K, Ibáñez E. Pressurized limonene as an alternative bio-solvent for the extraction of lipids from marine microorganisms. J Supercrit Fluids 2014; 92: 1-7. [http://dx.doi.org/10.1016/j.supflu.2014.05.001]

[192] Lombardo D, Kiselev M A, Magazù S, Calandra P. Amphiphiles self-assembly: basic concepts and future perspectives of supramolecular approaches. Adv. Condens. Matter Phys. 2015, 2015. [http://dx.doi.org/10.1155/2015/151683]

[193] Sanchez-Fernandez A, Moody GL, Murfin LC, *et al.* Self-assembly and surface behaviour of pure and mixed zwitterionic amphiphiles in a deep eutectic solvent. Soft Matter 2018; 14(26): 5525-36. [http://dx.doi.org/10.1039/C8SM00755A] [PMID: 29926037]

[194] Rebello S, Asok AK, Mundayoor S, Jisha MS. Surfactants: toxicity, remediation and green surfactants. Environ Chem Lett 2014; 12(2): 275-87. [http://dx.doi.org/10.1007/s10311-014-0466-2]

[195] Trujillo Rodríguez, González-Hernández P, Pino V. Analytical applications of ionic liquid-base surfactants in separation science ionic liquid-based surfactant science: formulation. Characterization, and Applications 2015; pp. 475-501. [http://dx.doi.org/10.1002/9781118854501.ch23]

[196] Sosa Ferrera Z, Padrón Sanz C, Mahugo Santana C, Santana Rodríguez JJ. The use of micellar systems in the extraction and pre-concentration of organic pollutants in environmental samples. Trends Analyt Chem 2004; 23(7): 469-79. [http://dx.doi.org/10.1016/S0165-9936(04)00732-0]

[197] Perinelli DR, Cespi M, Lorusso N, Palmieri GF, Bonacucina G, Blasi P. Surfactant self-assembling and critical micelle concentration: one approach fits all? Langmuir 2020; 36(21): 5745-53. [http://dx.doi.org/10.1021/acs.langmuir.0c00420] [PMID: 32370512]

[198] Sun X, Bandara N. Applications of reverse micelles technique in food science: A comprehensive review. Trends Food Sci Technol 2019; 91: 106-15. [http://dx.doi.org/10.1016/j.tifs.2019.07.001]

[199] Ballesteros-Gómez A, Sicilia MD, Rubio S. Supramolecular solvents in the extraction of organic compounds. A review. Anal Chim Acta 2010; 677(2): 108-30. [http://dx.doi.org/10.1016/j.aca.2010.07.027] [PMID: 20837178]

[200] Burakham R, Srijaranai S. Surfactant-based materials. In: Guardia M, Esteve-Turrillas FA (Eds.)

Handbook of smart materials in analytical chemistry. John Wiley & Sons Ltd. 2019, 103-158.

[201] Kislik V. Advances in development of solvents for liquid-liquid extraction. Solvent extraction. Elsevier Amsterdam 2012; pp. 451-81. [http://dx.doi.org/10.1016/B978-0-444-53778-2.10013-5]

[202] Ballesteros-Gómez A, Lunar L, Sicilia MD, Rubio S. Hyphenating supramolecular solvents and liquid chromatography: tips for efficient extraction and reliable determination of organics. Chromatographia 2019; 82(1): 111-24. [http://dx.doi.org/10.1007/s10337-018-3614-1]

[203] Yazdi AS. Surfactant-based extraction methods. Trends Analyt Chem 2011; 30(6): 918-29. [http://dx.doi.org/10.1016/j.trac.2011.02.010]

[204] Samaddar P, Sen K. Cloud point extraction: A sustainable method of elemental preconcentration and speciation. J Ind Eng Chem 2014; 20(4): 1209-19. [http://dx.doi.org/10.1016/j.jiec.2013.10.033]

[205] Melnyk A, Wolska L, Namieśnik J. Coacervative extraction as a green technique for sample preparation for the analysis of organic compounds. J Chromatogr A 2014; 1339: 1-12. [http://dx.doi.org/10.1016/j.chroma.2014.02.082] [PMID: 24666936]

[206] Rezaee M, Assadi Y, Milani Hosseini MR, Aghaee E, Ahmadi F, Berijani S. Determination of organic compounds in water using dispersive liquid–liquid microextraction. J Chromatogr A 2006; 1116(1-2): 1-9. [http://dx.doi.org/10.1016/j.chroma.2006.03.007] [PMID: 16574135]

[207] Rocío-Bautista P, Pino V. Extraction methods facilitated by the use of magnetic nanoparticles. Analytical Separation Science 2015; pp. 1681-724. [http://dx.doi.org/10.1002/9783527678129.assep061]

[208] Wang Z, Guo H, Yu Y, He N. Synthesis and characterization of a novel magnetic carrier with its composition of Fe_3O_4/carbon using hydrothermal reaction. J Magn Magn Mater 2006; 302(2): 397-404. [http://dx.doi.org/10.1016/j.jmmm.2005.09.044]

[209] Faraji M, Yamini Y, Saleh A, Rezaee M, Ghambarian M, Hassani R. A nanoparticle-based solid-phase extraction procedure followed by flow injection inductively coupled plasma-optical emission spectrometry to determine some heavy metal ions in water samples. Anal Chim Acta 2010; 659(1-2): 172-7. [http://dx.doi.org/10.1016/j.aca.2009.11.053] [PMID: 20103121]

[210] Cinelli G, Cuomo F, Ambrosone L, Venditti F, Lopez F. Determination of bisphenol A in red wine using a double vortex-ultrasound-assisted microextraction assay: Role of the interfacial properties. Biotechnol Prog 2019; 35(3): e2780. [http://dx.doi.org/10.1002/btpr.2780] [PMID: 30697978]

[211] Kahe H, Chamsaz M, Zavar MHA. A novel supramolecular aggregated liquid–solid microextraction method for the preconcentration and determination of trace amounts of lead in saline solutions and food samples using electrothermal atomic absorption spectrometry. RSC Advances 2016; 6(54): 49076-82. [http://dx.doi.org/10.1039/C6RA03874C]

[212] Sadeghi M, Shiri F, Kordestani D, Mohammadi P, Alizadeh A. SBA-15/Metformin as a novel sorbent combined with surfactant-assisted dispersive liquid–liquid microextraction (SA-DLLME) for highly sensitive determination of Pb, Cd and Ni in food and environmental samples. J Indian Chem Soc 2018; 15(4): 753-68. [http://dx.doi.org/10.1007/s13738-017-1272-0]

[213] Yazdinezhad SR, Ballesteros-Gómez A, Lunar L, Rubio S. Single-step extraction and cleanup of bisphenol A in soft drinks by hemimicellar magnetic solid phase extraction prior to liquid chromatography/tandem mass spectrometry. Anal Chim Acta 2013; 778: 31-7.

[http://dx.doi.org/10.1016/j.aca.2013.03.025] [PMID: 23639396]

[214] Ding J, Zhao Q, Sun L, Ding L, Ren N. Magnetic mixed hemimicelles solid-phase extraction of xanthohumol in beer coupled with high-performance liquid chromatography determination. J Sep Sci 2011; 34(12): 1463-8.
[http://dx.doi.org/10.1002/jssc.201000930] [PMID: 21548085]

[215] Mohd NI, Gopal K, Raoov M, *et al.* Evaluation of a magnetic activated charcoal modified with non-ionic silicone surfactant as a new magnetic solid phase extraction sorbent with triazine herbicides as model compounds in selected milk and rice samples. Talanta 2019; 196: 217-25.
[http://dx.doi.org/10.1016/j.talanta.2018.12.043] [PMID: 30683354]

[216] Welton T. Room-temperature ionic liquids. Solvents for synthesis and catalysis. Chem Rev 1999; 99(8): 2071-84.
[http://dx.doi.org/10.1021/cr980032t] [PMID: 11849019]

[217] Ho TD, Zhang C, Hantao LW, Anderson JL. Ionic liquids in analytical chemistry: fundamentals, advances, and perspectives. Anal Chem 2014; 86(1): 262-85.
[http://dx.doi.org/10.1021/ac4035554] [PMID: 24205989]

[218] Bubalo MC, Radošević K, Redovniković IR, Slivac I, Srček VG. Toxicity mechanisms of ionic liquids. Arh Hig Rada Toksikol 2017; 68(3): 171-9.
[http://dx.doi.org/10.1515/aiht-2017-68-2979] [PMID: 28976886]

[219] Marullo S, D'Anna F, Rizzo C, Billeci F. Ionic liquids: "normal" solvents or nanostructured fluids? Org Biomol Chem 2021; 19(10): 2076-95.
[http://dx.doi.org/10.1039/D0OB02214D] [PMID: 33606870]

[220] Santos JI, Gonçalves AMM, Pereira JL, *et al.* Environmental safety of cholinium-based ionic liquids: assessing structure–ecotoxicity relationships. Green Chem 2015; 17(9): 4657-68.
[http://dx.doi.org/10.1039/C5GC01129A]

[221] Singh SK, Savoy AW. Ionic liquids synthesis and applications: An overview. J Mol Liq 2020; 297: 112038.
[http://dx.doi.org/10.1016/j.molliq.2019.112038]

[222] Green O, Grubjesic S, Lee S, Firestone MA. The design of polymeric ionic liquids for the preparation of functional materials. Polym Rev (Phila Pa) 2009; 49(4): 339-60.
[http://dx.doi.org/10.1080/15583720903291116]

[223] Santos E, Albo J, Irabien A. Magnetic ionic liquids: synthesis, properties and applications. RSC Advances 2014; 4(75): 40008-18.
[http://dx.doi.org/10.1039/C4RA05156D]

[224] Trujillo-Rodríguez MJ, Rocío-Bautista P, Pino V, Afonso AM. Ionic liquids in dispersive liquid-liquid microextraction. Trends Analyt Chem 2013; 51: 87-106.
[http://dx.doi.org/10.1016/j.trac.2013.06.008]

[225] You X, Chen X, Liu F, Hou F, Li Y. Ionic liquid-based air-assisted liquid–liquid microextraction followed by high performance liquid chromatography for the determination of five fungicides in juice samples. Food Chem 2018; 239: 354-9.
[http://dx.doi.org/10.1016/j.foodchem.2017.06.074] [PMID: 28873579]

[226] Xu X, Su R, Zhao X, *et al.* Ionic liquid-based microwave-assisted dispersive liquid–liquid microextraction and derivatization of sulfonamides in river water, honey, milk, and animal plasma. Anal Chim Acta 2011; 707(1-2): 92-9.
[http://dx.doi.org/10.1016/j.aca.2011.09.018] [PMID: 22027124]

[227] Trujillo-Rodríguez MJ, Pino V, Anderson JL. Magnetic ionic liquids as extraction solvents in vacuum headspace single-drop microextraction. Talanta 2017; 172: 86-94.
[http://dx.doi.org/10.1016/j.talanta.2017.05.021] [PMID: 28602308]

[228] Aguilera-Herrador E, Lucena R, Cárdenas S, Valcárcel M. Ionic liquid-based single drop

microextraction and room-temperature gas chromatography for on-site ion mobility spectrometric analysis. J Chromatogr A 2009; 1216(29): 5580-7.
[http://dx.doi.org/10.1016/j.chroma.2009.05.071] [PMID: 19523638]

[229] Jiang C, Wei S, Li X, *et al.* Ultrasonic nebulization headspace ionic liquid-based single drop microextraction of flavour compounds in fruit juices. Talanta 2013; 106: 237-42.
[http://dx.doi.org/10.1016/j.talanta.2013.01.028] [PMID: 23598122]

[230] López-Darias J, Anderson JL, Pino V, Afonso AM. Developing qualitative extraction profiles of coffee aromas utilizing polymeric ionic liquid sorbent coatings in headspace solid-phase microextraction gas chromatography–mass spectrometry. Anal Bioanal Chem 2011; 401(9): 2965-76.
[http://dx.doi.org/10.1007/s00216-011-5394-4] [PMID: 21935597]

[231] Zhang Y, Wang X, Lin C, Fang G, Wang S. A novel SPME fiber chemically linked with 1-vinyl-3-hexadecylimidazolium hexafluorophosphate ionic liquid coupled with GC for the simultaneous determination of pyrethroids in vegetables. Chromatographia 2012; 75(13-14): 789-97.
[http://dx.doi.org/10.1007/s10337-012-2244-2]

[232] Sahebi H, Konoz E, Ezabadi A, Niazi A, Ahmadi SH. Simultaneous determination of five penicillins in milk using a new ionic liquid-modified magnetic nanoparticle based dispersive micro-solid phase extraction followed by ultra-performance liquid chromatography-tandem mass spectrometry. Microchem J 2020; 154: 104605.
[http://dx.doi.org/10.1016/j.microc.2020.104605]

[233] Jeannot MA, Cantwell FF. Solvent microextraction into a single drop. Anal Chem 1996; 68(13): 2236-40.
[http://dx.doi.org/10.1021/ac960042z] [PMID: 21619310]

[234] Trujillo-Rodríguez MJ, Anderson JL. In situ formation of hydrophobic magnetic ionic liquids for dispersive liquid-liquid microextraction. J Chromatogr A 2019; 1588: 8-16.
[http://dx.doi.org/10.1016/j.chroma.2018.12.032] [PMID: 30600165]

[235] Qiu X, Zhang Y, Zhou Y, Li G, Feng X. Progress in pretreatment and analysis of organic Acids: An update since 2010. Food Chem 2021; 360: 129977.
[http://dx.doi.org/10.1016/j.foodchem.2021.129977] [PMID: 34023712]

[236] Yu H, Ho TD, Anderson JL. Ionic liquid and polymeric ionic liquid coatings in solid-phase microextraction. Trends Analyt Chem 2013; 45: 219-32.
[http://dx.doi.org/10.1016/j.trac.2012.10.016]

[237] Patinha DJS, Silvestre AJD, Marrucho IM. Poly(ionic liquids) in solid phase microextraction: Recent advances and perspectives. Prog Polym Sci 2019; 98: 101148.
[http://dx.doi.org/10.1016/j.progpolymsci.2019.101148]

[238] Smith EL, Abbott AP, Ryder KS. Deep eutectic solvents (DESs) and their applications. Chem Rev 2014; 114(21): 11060-82.
[http://dx.doi.org/10.1021/cr300162p] [PMID: 25300631]

[239] De Oliveira Vigier K, Jérôme F. Synthesis and properties deep eutectic solvents: synthesis. Properties, and applications 2019; pp. 1-23.

[240] Makoś P, Słupek E, Gębicki J. Hydrophobic deep eutectic solvents in microextraction techniques–A review. Microchem J 2020; 152: 104384.
[http://dx.doi.org/10.1016/j.microc.2019.104384]

[241] Deetlefs M, Seddon KR. Assessing the greenness of some typical laboratory ionic liquid preparations. Green Chem 2010; 12(1): 17-30.
[http://dx.doi.org/10.1039/B915049H]

[242] Ünlü AE, Arıkaya A, Takaç S. Use of deep eutectic solvents as catalyst: A mini-review. Green Processing and Synthesis 2019; 8(1): 355-72.
[http://dx.doi.org/10.1515/gps-2019-0003]

[243] Choi YH, van Spronsen J, Dai Y, *et al.* Are natural deep eutectic solvents the missing link in understanding cellular metabolism and physiology? Plant Physiol 2011; 156(4): 1701-5. [http://dx.doi.org/10.1104/pp.111.178426] [PMID: 21677097]

[244] Vanda H, Dai Y, Wilson EG, Verpoorte R, Choi YH. Green solvents from ionic liquids and deep eutectic solvents to natural deep eutectic solvents. C R Chim 2018; 21(6): 628-38. [http://dx.doi.org/10.1016/j.crci.2018.04.002]

[245] Espino M, de los Ángeles Fernández M, Gomez FJV, Silva MF. Natural designer solvents for greening analytical chemistry. Trends Analyt Chem 2016; 76: 126-36. [http://dx.doi.org/10.1016/j.trac.2015.11.006]

[246] Torbati M, Farajzadeh MA, Afshar Mogaddam MR, Torbati M. Development of microwave-assisted liquid-liquid extraction combined with lighter than water in syringe dispersive liquid-liquid microextraction using deep eutectic solvents: Application in extraction of some herbicides from wheat. Microchem J 2019; 147: 1103-8. [http://dx.doi.org/10.1016/j.microc.2019.04.044]

[247] Li X, Yuan T, Zhao T, Wu X, Yang Y. An effective acid–base-induced liquid–liquid microextraction based on deep eutectic solvents for determination of testosterone and methyltestosterone in milk. J Chromatogr Sci 2020; 58(9): 880-6. [http://dx.doi.org/10.1093/chromsci/bmaa051] [PMID: 32798218]

[248] Yu K, Yue ME, Xu J, Jiang TF. Determination of fluoroquinolones in milk, honey and water samples by salting out-assisted dispersive liquid-liquid microextraction based on deep eutectic solvent combined with MECC. Food Chem 2020; 332: 127371. [http://dx.doi.org/10.1016/j.foodchem.2020.127371] [PMID: 32622188]

[249] Ismailzadeh A, Masrournia M, Es'haghi Z, Bozorgmehr MR. Microextraction and gas chromatography–flame ionization determination of five antiepileptic drugs in biological samples using amino acid-based deep eutectic ionic liquids. J Mol Liq 2020; 317: 113979. [http://dx.doi.org/10.1016/j.molliq.2020.113979]

[250] Abolghasemi MM, Piryaei M, Imani RM. Deep eutectic solvents as extraction phase in head-space single-drop microextraction for determination of pesticides in fruit juice and vegetable samples. Microchem J 2020; 158: 105041. [http://dx.doi.org/10.1016/j.microc.2020.105041]

[251] Triaux Z, Petitjean H, Marchioni E, Boltoeva M, Marcic C. Deep eutectic solvent–based headspace single-drop microextraction for the quantification of terpenes in spices. Anal Bioanal Chem 2020; 412(4): 933-48. [http://dx.doi.org/10.1007/s00216-019-02317-9] [PMID: 31927601]

[252] Yousefi SM, Shemirani F, Ghorbanian SA. Enhanced headspace single drop microextraction method using deep eutectic solvent based magnetic bucky gels: Application to the determination of volatile aromatic hydrocarbons in water and urine samples. J Sep Sci 2018; 41(4): 966-74. [http://dx.doi.org/10.1002/jssc.201700807] [PMID: 29193809]

[253] Zhao J, Meng Z, Zhao Z, Zhao L. Ultrasound-assisted deep eutectic solvent as green and efficient media combined with functionalized magnetic multi-walled carbon nanotubes as solid-phase extraction to determine pesticide residues in food products. Food Chem 2020; 310: 125863. [http://dx.doi.org/10.1016/j.foodchem.2019.125863] [PMID: 31835214]

[254] Ma Y, Wang Q, Zhu T. Comparison of hydrophilic and hydrophobic deep eutectic solvents for pretreatment determination of sulfonamides from aqueous environments. Anal Methods 2019; 11(46): 5901-9. [http://dx.doi.org/10.1039/C9AY02244A]

[255] Pollet P, Eckert CA, Liotta CL. Switchable solvents. Chem Sci (Camb) 2011; 2(4): 609-14. [http://dx.doi.org/10.1039/c0sc00568a]

[256] Jessop PG, Heldebrant DJ, Li X, Eckert CA, Liotta CL. Reversible nonpolar-to-polar solvent. Nature 2005; 436(7054): 1102-2.
[http://dx.doi.org/10.1038/4361102a] [PMID: 16121169]

[257] Jessop PG, Mercer SM, Heldebrant DJ. CO_2-triggered switchable solvents, surfactants, and other materials. Energy Environ Sci 2012; 5(6): 7240-53.
[http://dx.doi.org/10.1039/c2ee02912j]

[258] Ezoddin M, Abdi K, Lamei N. Development of air assisted liquid phase microextraction based on switchable-hydrophilicity solvent for the determination of palladium in environmental samples. Talanta 2016; 153: 247-52.
[http://dx.doi.org/10.1016/j.talanta.2016.03.018] [PMID: 27130115]

[259] Vanderveen JR, Durelle J, Jessop PG. Design and evaluation of switchable-hydrophilicity solvents. Green Chem 2014; 16(3): 1187-97.
[http://dx.doi.org/10.1039/C3GC42164C]

[260] Mercer SM, Jessop PG. "Switchable water": aqueous solutions of switchable ionic strength. ChemSusChem 2010; 3(4): 467-70.
[http://dx.doi.org/10.1002/cssc.201000001] [PMID: 20186910]

[261] Wang X, Gao M, Zhang Z, *et al.* u, H.; Liu, T.; Yu, N.; Wang, X.; Wang, H., Development of CO_2-mediated switchable hydrophilicity solvent-based microextraction combined with HPLC-UV for the determination of bisphenols in foods and drinks. Food Anal Methods 2018; 11(8): 2093-104.
[http://dx.doi.org/10.1007/s12161-018-1187-0]

[262] Di X, Wang X, Liu Y, Guo X, Di X. Microwave assisted extraction in combination with solid phase purification and switchable hydrophilicity solvent-based homogeneous liquid-liquid microextraction for the determination of sulfonamides in chicken meat. J Chromatogr B Analyt Technol Biomed Life Sci 2019; 1118-1119: 109-15.
[http://dx.doi.org/10.1016/j.jchromb.2019.04.036] [PMID: 31030103]

[263] Chormey DS, Bodur S, Baskın D, Fırat M, Bakırdere S. Accurate and sensitive determination of selected hormones, endocrine disruptors, and pesticides by gas chromatography-mass spectrometry after the multivariate optimization of switchable solvent liquid-phase microextraction. J Sep Sci 2018; 41(14): 2895-902.
[http://dx.doi.org/10.1002/jssc.201800223] [PMID: 29774656]

[264] Nowak PM, Wietecha-Posłuszny R, Pawliszyn J. White Analytical Chemistry: An approach to reconcile the principles of Green Analytical Chemistry and functionality. Trends Analyt Chem 2021; 138: 116223.
[http://dx.doi.org/10.1016/j.trac.2021.116223]

CHAPTER 4

Recent Applications of Deep Eutectic Solvents

Ruth Rodríguez-Ramos[1], Álvaro Santana-Mayor[1], Bárbara Socas-Rodríguez[1], Antonio V. Herrera-Herrera[2] and Miguel Ángel Rodríguez-Delgado[1,*]

[1] *Departamento de Química, Unidad Departamental de Química Analítica, Facultad de Ciencias, Universidad de La Laguna (ULL). Avenida Astrofísico Francisco Sánchez, s/nº. 38206 San Cristóbal de La Laguna (Tenerife), España*

[2] *Instituto Universitario de Bio-Orgánica Antonio González. Universidad de La Laguna (ULL). Avenida Astrofísico Francisco Sánchez, 2. 38206 San Cristóbal de La Laguna (Tenerife), España*

Abstract: Among the different strategies applied in recent years for the development of green extraction techniques in food analysis, the design and use of deep eutectic solvents (DESs) have aroused the utmost attention due to the advantages provided by these materials in terms of sustainability and versatility. Different types of DESs have been applied in this field including hydrophilic and hydrophobic mixtures, natural DESs, or polymeric-DESs. In this sense, the great availability of components and the wide range of possible combinations constitute potential tools to increase the selectivity and enhance the extraction capacity of the procedures, which is an important concern when complex food samples are tackled. This broad spectrum of possibilities has allowed the extraction of diverse compounds including not only contaminants such as pesticides, plastic migrants, heavy metals, or pharmaceuticals, among others, but also the extraction of biomolecules from food and food by-products. However, despite the advantages of these materials, there are important drawbacks like their high viscosity and low volatility that limit their application. In this context, an important effort has been carried out by the study of different combinations and the development of numerous approaches. In this chapter, the most relevant applications of DESs in the last five years in food analysis have been compiled and discussed in order to provide a global view of the advantages and limitations of the application of these green extraction solvents in the field. Additionally, the current trends and future perspectives in the use of DESs in food analysis are also pointed out.

Keywords: Biomolecule, Bioactive compound, Extraction technique, Food safety, Green solvent, Sustainability.

[*] **Corresponding Author Miguel Ángel Rodríguez Delgado:** Departamento de Química, Unidad Departamental de Química Analítica, Facultad de Ciencias, Universidad de La Laguna (ULL). Avenida Astrofísico Francisco Sánchez, s/nº. 38206 San Cristóbal de La Laguna (Tenerife), España; E-mail: mrguez@ull.edu.es.

Merichel Plaza & Maria Luisa Marina (Eds.)

INTRODUCTION

Solvents represent the largest proportion of substances used in chemical protocols. Thus, it is not surprising that, within the framework of green chemistry, several efforts have done to develop green solvents [1] due to the contaminant power and typical intrinsic toxicity of the traditional ones obtained from fossil feedstocks. In this regard, apart from water and CO_2, several bio-based molecular solvents [1] (substances available from low value and extensively accessible biomass resources), supramolecular solvents [2] (nanostructured liquids composed of three-dimensional amphiphilic aggregates), and low transition temperature mixtures [3] (LTTMs, solvents made from mixtures of compounds with a low transition temperature between liquid and solid-state) have been proposed.

Different families of substances can be classified as LTTMs [3], including ionic liquids (ILs) and DESs (from the Greek word "ευτύπος" *eutēktos* -easily melted). Generally speaking, LTTMs share similar physicochemical properties (*e.g.*, low vapor pressure, high density, low volatility, *etc.*) and components. This is the reason why some of these solvents are sometimes classified as DESs or ILs [3], depending on the particular application. Although some of them are formed by a similar process (and for this reason, there is some confusion in the nomenclature), ILs are normally recognized as molten salt liquid at room temperature. Meanwhile, DESs can be defined as mixtures of at least two constituents (a hydrogen bond acceptor (HBA) and a hydrogen bond donor (HBD)) that remain liquid over a wide range of temperatures (also room temperature) [4]. The main force driving the formation of DESs is strong hydrogen bond interaction, but other forces, such as Van der Waals, dipole, alkyl-alkyl, halogen bonds, and other electrostatic interactions, are also involved [4]. As a result of those interactions (and the corresponding charge delocalization), the eutectic temperature is significantly lower than the melting point of the specific constituents and also much lower than expected from the known enthalpies of fusion of the pure components. Strictly speaking, the eutectic point is only achieved at a particular molar ratio of the individual constituents (Fig. **1**) and, thus, the term DESs should be only employed for this mixture. However, different mixtures present melting temperatures lower than the predictable, which are frequently included in this category (although some authors prefer the term "eutectic mixtures"). It should be noted that certain particular DESs (for example those containing $SnCl_2$, $AlCl_3$, and $FeCl_3$ mixed with imidazolium chlorides) have two different eutectic points [5].

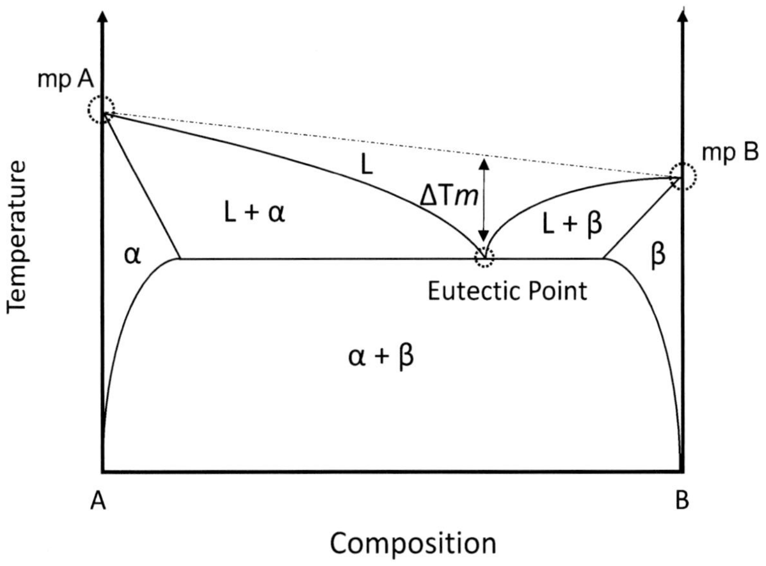

Fig. (1). Schematic representation of the phase diagram for a binary eutectic mixture. L: liquid phase, α: solid rich in component A, β: solid rich in component B.

Typical HBDs are alcohols, amides, amines, azole, carboxylic acids, imidazole, phenol, thiourea, urea, and water, whereas conventional HBAs comprise dication-based, imidazolium-based, quaternary phosphonium, and quaternary ammonium salts, inner salts, and molecular imidazole and its analogs.

Archetypally, DESs can be described by the formula C^+X^-zY, where C^+ is a cation of a salt, X^- refers to a Lewis base (such as the halide anion of the salt), and z is the number of units of a Y Lewis or Brønsted acid or HBD molecule [5]. Initially, the vast majority of DESs were classified into four different types [5], attending to the nature of the components. This way, type I DESs are those formed by a salt and metal halide; type II are composed of salts and metal halide hydrates; type III by salts and HBDs; and type IV contain metal chloride hydrates and HBDs. Moreover, the search for non-ionic DESs, that overcome the issues associated with salt-based DESs, led to the emergence of type V, which are characterized by a mixture of molecular components. It should be mentioned that in those DESs formed by non-ionic components, it is not easily distinguishable which component acts as HBD and which one as HBA.

Apart from the nature of the components, DESs can also be classified according to their hydrophobicity and to the origin of their components. As the name suggests, the hydrophobic DESs (HDESs) are those that cannot be solubilized in water, whereas the hydrophilic ones are dissolved by water [6]. Although the label HDES was used for all the DESs containing any water-insoluble component, it should be only employed for those DESs in which all the components are insoluble. In this regard, Florindo *et al.* [7] demonstrated that certain DESs typically considered HDES (the combination of quaternary ammonium salts, monoterpenes, or fatty acids with short chains- acetic, pyruvic, levulinic, or butyric acids) are unstable in contact with water since the short carboxylic acid can leach to the water phase. Another interesting type that has gained increasing popularity is the so-called natural DESs (NaDESs). Generally speaking, these DESs can be defined as those containing primary or secondary metabolites and other abundant natural compounds present in cells to solubilize, store, and transport some important water-immiscible molecules. In the pharmaceutical field, some particular DESs (the therapeutic DESs (TheDESs)) containing at least one active pharmaceutical ingredient have also been introduced [8].

DESs are characterized by particular physicochemical properties that make them very interesting for different applications. These characteristics include, among others, simple production and structural design, low cost, low volatility, nonflammability, high viscosity, high dissolution power, environmental friendliness, limited toxicity, and good biodegradability compared with conventional solvents. It should be noted that these properties are highly correlated with the individual components of DESs (and their molar ratios). As previously stated, some of these properties are shared by ILs and DESs. However, the cost of the raw materials for DESs production is lower and they are easily prepared and scaled-up. Moreover, their environmental friendliness is higher [9] and their inherent toxicity toward organisms [10], and their conductivities are lower [11]. Some previously published review articles [5, 12 - 14] compiled and analyzed available data about DESs properties. For this reason, the properties will not be extensively described in this chapter. However, some ideas, which are important for food analysis, should be considered.

The unusual solvation properties of DESs are related to their high capability of donating and accepting protons or electrons, which makes them ideal for the extraction of a wide variety of analytes [4]. High solubility of metal salts and compounds (including protic solvents) was early observed [15]. Their immiscibility with aprotic solvents allows the back-extraction of molecules of low or medium polarity. The use of this extra solvent would allow the reuse of the DESs but, in contrast, would increase the polluting potential. Apart from the dissolution power, other physicochemical properties influence the extraction capacity of DESs, such as viscosity and polarity. The viscosity of eutectic mixtures (between ≈50 and 100,000 cP [16]) is normally higher than other solvents, with the lowest possible value at the eutectic point or in its vicinity [17]. Frequently, the viscosity of DESs is reduced by the addition of an extra solvent (including water) [16, 18]. However, that addition should be carefully studied since the hydrogen bond could be broken [19]. Furthermore, viscosity decreases as temperature increases (this phenomenon is also observed for other properties such as density [11]). Thus, temperatures higher than ambient can be required for certain applications.

Other properties are affected by the particular HBA and HBD combination and thus DESs can be tuned to obtain fit-for-purpose characteristics. In this regard, the polarity, although also affected by the HBA, is frequently directly related to the HBD [20] (considering that a great majority use choline chloride (ChCl) as HBA) and is decreased with the temperature [21]. The acidity of DES depends on the nature of the individual components and acid, neutral and basic DESs can be distinguished. This acidity can be significantly modified by dissolved gasses [22] but water and temperature have a negligible effect on it [22, 23]. Typically, the conductivity of DESs (<2 mS/cm at room temperature [12]) is lower, whereas the surface tension is higher compared to those of conventional solvents.

Some of the typical constituents of DESs are halogenated products and/or metal salts and thus their toxicity could not be disregarded. Also, certain combinations could produce exacerbated toxicity. As an example, Hayyan *et al.* [24, 25] observed that DESs made with ChCl [24] or phosphonium-based salts [25] and glycerine, ethylene glycol (EtGly), tryethylene glycol, or urea showed increased toxicity on bacteria (in comparison with the individual components) due to a disruption of the cell walls produced by the delocalization of charges in these solvents. Thus, the exploitation of certain DESs in fields like pharmacology, medicine, and, particularly, the agri-food industry can be limited [26, 27].

Although eutectic mixtures were used in some research fields before (for instance as substrates for enzymatic reactions [28, 29] or transdermal administration of pharmaceuticals [30]), it was not until 2003 when Abbot *et al.* [15] introduced the term DESs. They observed that ChCl forms homogenous liquids with amides capable of forming hydrogen bonds with chloride ions. Also, the combination of quaternary ammonium salts with urea produced ambient temperature liquids (and as the symmetry of the cation decreased the melting point was reduced). Moreover, they observed that some of their interesting properties were directly influenced by hydrogen bonding. After this publication, the interest in these solvents rapidly expanded in chemistry, and particularly in analytical chemistry. This is evidenced by the multiple review articles published since then [4, 5, 12, 14, 16 - 18, 27, 31, 32]. Due to their limited toxicity and biodegradability, their use in food analysis has become one of their most popular applications [31, 32].

However, some particular aspects of DESs can have a negative effect on food analysis (*e.g.*, need to combine them with hazardous substances for extraction or with pollutant solvents before instrumental analysis, energy, and time-consuming procedures, *etc.*) and should be carefully studied before its application.

RECENT APPLICATIONS OF DESS AS EXTRACTION MATERIALS IN FOOD SAFETY

In recent years, food trade globalization and the increasing interest of the population in the effect of diet on health have pointed out the importance of ensuring the quality and safety of food products. This fact has given rise to the development of new analytical methodologies that allow the reliable evaluation of this kind of matrices and the accomplishment of current legislation on the subject. Besides, the important concern about the negative impact of different human activities on the environment has also influenced the search for sustainable procedures in this field [33]. Based on that, the scientific community not only works on the analytical quality of the developed procedures but also the principles of green chemistry. Among the variety of strategies, the use of green extraction solvents and particularly DESs, has become one of the most important approachcs in food analysis. As can be seen in Fig. (**2**), in which the main applications of DESs to the extraction of contaminants from food matrices have been presented, diverse groups of analytes have been evaluated and varied approaches based on DESs have been performed with this aim, as it is thoroughly discussed below.

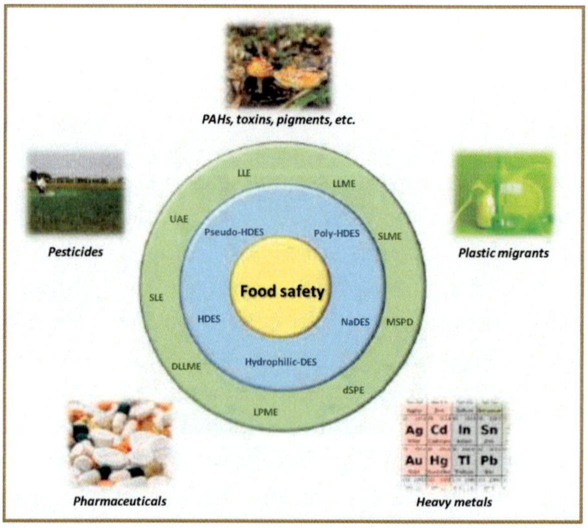

Fig. (2). Scheme of the main applications of DESs as extraction solvents in food safety including the type of solvents and techniques used in the last five years, as well as the contaminants mostly evaluated.

Pesticides

Among the different applications of DESs as extraction solvents in food safety assessment, the evaluation of pesticides has been the most widely developed. They have been applied to different types of matrices including fruit and vegetables, tea, oil, nuts, milk, honey, and, especially, for the evaluation of fruit and vegetable juices as it is shown in Table **1**, in which some relevant examples have been compiled [34 - 43]. However, despite the extended application of DESs in this area, the number and variety of addressed matrices so far are quite reduced and the evaluation of multi-residue analysis of pesticides has not been tackled up to date.

Liquid matrices have been more commonly studied than solid food due to the complexity of working with this kind of product using DESs. The particular characteristics of these solvents provide numerous advantages; however, their high viscosity and low volatility constitute important drawbacks for their use as extraction solvents and their application in chromatographic systems, especially those combined with mass spectrometry (MS) as analytical techniques [4]. In this sense, the application of DESs to the extraction of solid matrices has been carried out so far by the combination of different techniques including liquid and solid-based extraction procedures using conventional or nanomaterial sorbents [34, 39]. However, the combination of various procedures has also been a recurrent strategy for liquid samples [36 - 38, 40]. In general terms, the combination of two extractions involves two different objectives, improving clean-up efficiency and increasing the preconcentration capacity of the protocol in order to enhance the sensitivity. A good example of this procedure is the work carried out by Khanehzar *et al.* [39] for the evaluation of two neonicotinoid pesticides (acetamiprid and imidacloprid) in pistachio by the combination of a modified QuEChERS (*Quick, Easy, Cheap, Effective, Rugged, and Safe*) approach and dispersive liquid-liquid microextraction (DLLME) using a DES constituted for trioctylmethylammonium chloride (TOMAC) and isoamyl alcohol at a molar ratio 1:4. With this aim, 5 g of pistachio was subjected to a citrate-buffered QuEChERS using graphitized carbon black (GCB) as clean-up sorbent. Afterward, the supernatant obtained from this step was used as a disperser for the DLLME. The combination of both techniques clearly improved the sensitivity obtained when only QuEChERS was applied, which allowed authors to reach the limits of quantification (LOQs) required and reduce the presence of interferences. Additionally, the authors also compared the efficiency of different DESs as extraction solvents for the DLLME using combinations of TOMAC or benzyltriethylammonium chloride (BTEAC) as HBAs with ethyl hexanoic and octanoic acids, isoamyl alcohol, and thymol as HBDs. Results indicated that TOMAC and isoamyl alcohol DES provided the best extraction efficiency, which was attributed to the higher possibility of this DES to establish hydrogen bonds with the target analytes, as well as its higher hydrophobicity in comparison with the rest of the combinations, which favors its separation from the aqueous phase.

Table 1. Some examples of recent applications of DESs for the extraction of pesticides from food matrices.

Sample (Amount)	Analytes	DES (Molar Ratio; Amount)	Extraction Procedure	Analytical Technique	Recovery %	LODs	Comments	References
Apple, pear, carrot, cucumber (0.2 g)	7 pesticides	Proline:propylene glycol (1:3; 2 mL)	Vortex-UAE-m-µ-dSPE	HPLC-DAD	76-98	0.02-0.05 mg/L	- m-MWCNTs were used as sorbents for SPE. - ChCl based DES with urea, EtGly, glycerol, lactic acid, phenol, 1,4 butanediol, acetic acid, and propylene glycol, as well as citric acid:glucose, proline:EtGly and proline propylene glycol were also tested. - Experiment design was used for optimization. - ACN was used as a desorption solvent.	[34]
Honey (2 mL)	8 pesticides	PChCl:dichloroacetic acid:decanoic acid (1:1:1; 60 µL)	DLLME	GC-(Q)-MS	82-92	0.19-0.36 µg/kg	- ACN was used as a dispersive solvent.	[35]
Tea beverages (5 mL)	4 Pyrethroid pesticides	TOAB:1-dodecanol: decanoic acid (1:2:2; 60 mg)	ME-LPME	HPLC-UV/Visible	89-94	0.56-1.24 µg/L	- Dodecanoic and octanoic acid, as well as 1-dodecanol were tested as HBDs. - TBTDPB was used as emulsifier.	[36]
Edible oils (2.5 mL)	4 OPPs	ChCl:3,3–dimethyl butyric acid (1:2; 15 µL)	dSPE-SFOD-DLLME	GC-NPD	68-77	0.06-0.24 µg/L	- Saponification was done before the extraction. - PSA was used as sorbent for dSPE. - Acetone was used as disperser. - Pivalic and pyruvic acids were also evaluated as HBDs. - DLLME and SFOD-DLLME comparison.	[37]
Fruit juices (25 mL)	5 Pesticides	ChCl:p-chlorophenol (1:2; 74 µL)	PSA-based SBSE-DLLME	GC-(Q)-MS	74-92	14-29 ng/L	- Derivatization was applied. -Hydrophilic DES was applied as elution solvent in the SBSE (ChCl: EtGly, 1:2). - Glycol and phenol were tested as HBDs for hydrophilic DES. - m-chlorophenol and α-naphthol were tested as HBDs for hydrophobic DESs.	[38]

(Table 1) cont.....

Sample (Amount)	Analytes	DES (Molar Ratio; Amount)	Extraction Procedure	Analytical Technique	Recovery %	LODs	Comments	References
Pistachio (5 g)	Acetamiprid, imidacloprid	TOMAC:isoamyl alcohol (1:4; 100 µL)	QuEChERS-DLLME	HPLC-UV	96-99	1.5-3.0 µg/kg	- Benzyltriethylammonium chloride was tested as HBA. - Ethyl hexanoic, octanoic acids isoamyl alcohol, and thymol were tested as HBDs. - Different ratios were tested. - ACN was used as disperser.	[39]
Milk (5 mL)	7 pesticides	Menthol:decanoic acid (1:2; 63 µL)	LLE-VA-UA-SFOD-DLLME	GC-FID	64-89	0.9-3.9 µg/L	- An hydrophilic DES was used in LLE (*in-situ* preparation) and as disperser in the DLLME (ChCl:EtGly). -Different molar ratios were tested.	[40]
Tea (5 g)	6 Herbicides	ChCl:butyric acid (1:2, 180 µL)	In-syringe-DLLME	GC-(Q)-MS	70-89	2.6-8.4 ng/kg	- ChCl:phenol (1:3; 2 mL) was applied as extraction solvent prior to DLLME. - Dichloroacetic acid or glycerol were tested as HBDs for hydrophilic DES. - Pivalic acid and propionic acid were tested as HBDs for hydrophobic. - Different molar ratios were tested.	[41]
Fruit juices (1.6 mL) and vegetable juices (1 mL)	5 Pesticides	ChCl:p-chlorophenol (1:2; 142 µL)	Temperature controlled-DLLME	GC-FID	56-93	0.13-0.31 µg/L	- Diazinon was determined in tomato samples. - DLLME was carried out at 70 ºC and then cooled down.	[42]
Fruit and vegetable juices (5 mL)	10 Pesticides	ChCl:pivalic acid (1:2; 46 µL)	Gas assisted-LPME	GC-(Q)-MS	79-96	3-26 ng/L	- Diazinon and bromopropylate were determined in some samples. - Butyric and 3,3-dimethyl butyric acid were tested as HBDs. - Different molar ratios were tested.	[43]

ACN: acetonitrile; ChCl: choline chloride; DAD: diode array detector; DES: deep eutectic solvent; DLLME: dispersive liquid-liquid microextraction; dSPE: dispersive solid-phase extraction; EtGly: ethylene glycol; FID: flame ionization detector; GC: gas chromatography; HBA: hydrogen bond acceptor; HBD: hydrogen bond donor; HPLC: high-performance liquid chromatography; LLE: liquid-liquid extraction; LOD: limit of detection; LPME: liquid-phase microextraction; m-µ-dSPE: magnetic-micro-dispersive solid phase extraction; ME: membrane emulsification; m-MWCNT: magnetic-multi walled carbon nanotubes; MS: mass spectrometry; NPD: nitrogen phosphorous detector; OCP: organochloride pesticide; OPP: organophosphorus pesticide; PChCl: phosphocholine chloride; PSA: primary secondary amine; Q: single quadrupole; QuEChERS: *Quick, Easy, Cheap, Effective, Rugged and Safe*; SBSE: stir bar sorptive extraction; SFOD: solidification of floating organic droplet; SPE: solid-phase extraction; TBTDPB: tributyltetradecylphosphonium bromide; TOAB: tetraoctylammonium bromide; TOMAC: trioctylmethylammonium chloride; UA: ultrasound assisted; UAE ultrasound-assisted extraction; UV: ultraviolet; VA: vortex-assisted.

As can be seen in Table **1**, when the objective of the study involves liquid matrices, the application of DESs becomes easier and simpler, especially when the applied DES is immiscible in the evaluated sample, which favors phase separation [35, 43, 44]. Otherwise, the use of emulsifier agents such as tetrahydrofuran is needed to carry out the effective separation [45]. Another strategy applied to solve this problem is the modification of the extraction conditions (temperature, pH, *etc.*) to change the miscibility of the extraction solvent in the sample along the procedure as it was demonstrated in the work developed by Farajzadeh *et al.* [42], in which a temperature controlled-DLLME was applied for the extraction of 5 pesticides from fruit and vegetable juices using 142 µL of a ChCl:p-chlorophenol DES with a molar ratio 1:2. The extraction was carried out at 70 °C, at which the extraction solvent resulted to be miscible in water. However, for phase separation, the system was cooled down in order to decrease such miscibility prior to drop collection and determination by gas chromatography (GC)-flame ionization detector (FID). The procedure showed great suitability in obtaining recovery values in the range of 56-93% and limits of detection (LODs) between 0.13 and 0.31 µg/L.

Although hydrophilic DESs have been applied for the extraction of pesticides [34], the application of hydrophobic ones has been mostly performed. In these cases, the separation from aqueous samples does not result in a problem, however, the interaction the DES-analyte could suffer. The applications of microwaves, ultrasounds, vortex or air assistance are suitable approaches [40, 43], although others like the use of surfactants has demonstrated to be an interesting alternative. This approach was applied by Quian *et al.* [36]. In this case, a membrane emulsification (ME)-liquid-phase microextraction (LPME) was applied for the extraction of 4 pyrethroid pesticides from tea. With this aim, 5 mL of sample was submitted to the extraction using a ternary DESs (TDESs). Initially, the surfactant, tributyltetradecylphosphonium bromide (TBTDPB), was dissolved in the aqueous sample at 45 °C and the solution was introduced in a syringe that was connected to another one containing 60 mg of the DESs by a porous ceramic membrane. A complete emulsion was obtained by manually pushing the aqueous sample into the syringe containing the DES and the porous membrane. The process was repeated six times to ensure that the resulting emulsion was uniform. It was loaded into a plastic tube, and 160 mg of NaCl was added for demulsification. The resultant turbid solution was centrifuged for phase separation and the DES drop was diluted with methanol before injection in the high-performance liquid chromatography (HPLC)-UV/Visible system for pesticides determination. Although different HBDs such as dodecanoic and octanoic acids as well as 1-dodecanol were tested, TOAB:1-dodecanol:decanoic acid at a molar ratio 1:2:2 was the one that provided the best results with recovery values of 89-94% and LODs in the range 0.56-1.24 µg/L.

In other cases, with the same aim of searching approaches that allow simplifying the procedure maintaining the extraction efficiency, the application of different DESs with varied hydrophobic nature along the extraction procedure has also been explored [40, 41]. These approaches constitute an interesting strategy to favor the separation of phases at each step and avoid the use of organic solvents, especially when the analysis of solid samples must be afforded and the direct application of microextraction procedures such as DLLME results in making it very complex and difficult to be applied.

As previously indicated, a great number of water-immiscible DESs have been applied for the evaluation of pesticides from food samples. However, it should be remarked that the ones strictly defined as HDEs have been scarcely used in the area [40]. In most cases, pseudo-HDESs, in which a halogenated component was introduced to increase the density of the extraction agents and favor its immiscibility in water, were used [35, 38, 42]. Regarding the type of DES applied to the extraction of pesticides, it should be remarked that a considerable number of halogenated components has been used [35, 36, 38, 39, 41, 42], although other pseudo-HDESs were also tested [37, 41, 43]. With respect to the application NaDESs, ChCl has been the HBA most commonly applied in combination with short-chain fatty acids [37, 41, 43], although amino acids and terpenes such as menthol have been applied as HBAs [34, 46] combined with alcohols or fatty acids.

Finally, it should be remarked the considerable number of recent articles in which the combinations of liquid chromatography (LC) and GC with conventional detectors, ultraviolet (UV), UV/Visible, FID or nitrogen phosphorous detector (NPD), among others, have been chosen for pesticides determination, whereas the use of MS detector is reduced to the combination of GC with single quadrupole (Q) systems [35, 38, 41, 43].

Plastic Migrants

Even though most applications related to food safety have been focused on the analysis of pesticide residues, other important groups of compounds with a potential health risk have been studied using DESs as extraction agents. As can be seen in (Table **2**) [47 - 51], this is the case of plastic migrants, which comprise important families of compounds such as phthalic acid esters (PAEs), bisphenols and alkylphenols, UV filters, or fluorescent whiteners, as well as antioxidants, among others. Although most studies have been mainly destined for the analysis of bisphenols and phthalates, interesting works can be found in the determination of fluorescent brighteners or UV filters. In this regard, it is worth mentioning the work developed by Jiaqi [50], in which three UV absorbents, including UV-531,

326, and 328, were determined in food plastic bags. A cloud point extraction, also called acid-base induced extraction, was developed and combined with HPLC-UV detection. Considering that at least one of the DES components is a saturated fatty acid, the authors took advantage of the formation of micelles with the increase of the pH to improve the solubility of the synthesized HDESs into the aqueous solution, and thus, increase the extraction efficiency. Then the pH was readjusted to return the HDES to its neutral state allowing phase separation. In this regard, the authors tested mixtures of medium-chain fatty acids (octanoic, nonanoic, decanoic, undecanoic, and dodecanoic acids) or menthol at molar ratios of 2:1 for binary mixtures of fatty acids, 3:1:1 for their ternary mixtures, and 3:1 for menthol:fatty acid HDESs, finding that menthol:nonanoic acid achieved the best results. The extraction procedure, consisting of a vortex-assisted (VA)-liqui--liquid microextraction (LLME), was carried out using 500 μL of HDES and 500 μL of NaOH 1M to change DES into its anionic form and give rise to a micellar solution. After mixing, the pH was readjusted using 500 μL of HCl 1M, the solution was centrifuged, and the aqueous phase was discarded. The analytes-enriched extract was diluted with methanol and injected into the HPLC-UV system for separation and determination. Recovery values between 84% and 101%, and LODs of 0.05-0.17 mg/L were obtained.

In general terms, for the extraction of plastic migrants, the use of LPME procedures and determination by conventional detectors such as UV [50], diode array detector (DAD) [48, 51], or fluorescence detector (FD) [47] have predominated. For compounds' separation, HPLC [48 - 51] stands out compared to GC, possibly due to the well-known low volatility of DESs. Regarding the types of DESs used, it seems that there is not a clear trend for this family of compounds since both hydrophilic DESs [48] and HDESs [50, 51] have been used depending on whether the matrices or sample solutions are aqueous or oily, or if the sample is solid. In this sense, menthol [50, 51] or quaternary ammonium salts, such as ChCl [48, 49], as hydrophobic and hydrophilic HBAs, respectively, and fatty [49 - 51] and organic acids, or even alcohols [48], as HBDs, have been the most used.

Another interesting application that has recently been exploited is the so-called *in-situ* formation of DES or extraction *via* forming DES [52 - 54]. In this case, taking advantage of the formation of hydrogen bonds between a donor/acceptor pair of the analyte(s) of interest and another component that acts as a counterpart, not only the formation of a DES is achieved, but also the extraction of the target compounds. As an example, Shi *et al.* [47] developed an extraction method based on the *in-situ* formation of an HDES between a fluorescent whitening agent, fluorescent brightener 52, as the target analyte, and 2-ethylhexyl 4-hydroxybenzoate, a low toxic paraben, in different food products such as noodles, fish balls, and mushrooms, although paper cups were also analyzed. For the extraction process, 100 mg of 2-ethylhexyl 4-hydroxybenzoate was added into 10 mL of sample solution (derived from 1 g of sample), vortexed and centrifuged. After discarding the aqueous phase, 2 mL of absolute ethanol was added to the extract for dilution, and the concentration of the target compound was determined by a FD, obtaining a very good sensitivity with a LOD of 0.045 µg/L and recovery of 82-113%. During the extraction step, the interactions between the hydroxyl (OH) group of the paraben and the carbonyl (CO) group of fluorescent brightener 52 gave rise to DES formation. With the aim of studying the optimum extractant, different 4-hydroxybenzoic acid esters were evaluated, finding that the selected one provided the best results probably due to the increased alkyl chain length and the higher hydrophobicity. Besides, a selectivity study was carried out by investigating the effect of organic and inorganic interferents such as glucose, starch, sucrose, sulfadiazine sodium salt, methyl orange, hydroquinone, NaCl, $CaCl_2$, Na_2SO_4, $NaNO_3$, $FeCl_3{\cdot}6H_2O$, and $MgSO_4$. In all cases, the extraction efficiency was above 90% even at interferent concentrations 500-2000 times higher.

Table 2. Some examples of recent applications of DESs for the extraction of plastic migrants from food matrices.

Sample (amount)	Analytes	DES (molar ratio; amount)	Extraction procedure	Analytical technique	Recovery %	LODs	Comments	References
Noodles, fish balls, and mushroom (1 g)	Fluorescent brightener 52	2-ethylhexyl-4-hydroxybenzoate:analyte (-;100 mg)	VA-microextraction	FD	82-113	0.045 µg/L	- Powder samples were added into 10 mL water and the resulting solution was used for extraction. -HDES was formed *in-situ* with the analyte. - Other three hydroxybenzoates were tested as HBDs. - A selectivity study was carried out by adding different coexisting inorganic salts and common organic compounds. - The method was applied to the analysis of real samples. Paper cups were also analyzed.	[47]
Tea, apple-based beverage, and pineapple juice (10 mL)	8 PAEs	ChCl:phenol (1:2; 440 µL)	VA-DLLME	HPLC-DAD-(QqQ)-MS/MS	84-120	5.1-17.8 µg/L	- Glycerol, EtGly, urea, and acetic acid were also tested as HBDs. - THF was used as an emulsifier. - MS/MS was used for analytes confirmation. - An experimental design was used for extraction conditions optimization. - The method was applied to the analysis of real samples.	[48]

(Table 2) cont.....

Sample (amount)	Analytes	DES (molar ratio; amount)	Extraction procedure	Analytical technique	Recovery %	LODs	Comments	References
Doogh and mineral water (25-50 mL)	Irgaphos 168 and irganox 1010	ChCl:oleic acid (1:2; 3.49 g)	VA-UA-LLE	HPLC-VWD	74-83	0.03-0.09 µg/L	- NaDES was formed *in-situ* in the sample. - Myristic and palmitic acids were also tested as HBDs.	[49]
Food packaging bags (2 g)	UV-531, UV-326, and UV-328	Menthol:nonanoic acid (3:1; 500 µL)	Acid-base induced VA-LLME	HPLC-UV	82-103	0.05-0.17 mg/L	- Sample was pre-treated with ACN and diluted with water up to 50 mL. - NaOH and HCl were used in the extraction procedure for the cloud point extraction method. - Octanoic, decanoic, undecanoic, and dodecanoic acids were also tested as DESs or HBDs.	[50]
Orange juice, tea drink, and soda (40 mL)	BPA, BPAF, TBBPA, and 4-TOP	Menthol:undecanoic acid (3:1; 440 µL)	Acid-base induced VA-LLME	HPLC-DAD	84-101	6-11 ng/L	- Fe_3O_4@MgAl-layered-double hydroxide nanoparticles were used for sample clean-up pre-treatment. - NH_3 and HCl were used in the extraction procedure for the cloud point extraction method. - Octanoic to dodecanoic acids were tested as HBDs. - An experimental design was used for extraction conditions optimization. - The method was applied to the analysis of real samples.	[51]

4-TOP: 4-tert-octylphenol; ACN: acetonitrile; BPA: bisphenol A; BPAF: bisphenol AF; ChCl: choline chloride; DAD: diode array detector DES: deep eutectic solvent; DLLME: dispersive liquid-liquid microextraction; EtGly: ethylene glycol; FD: fluorescence detector; HBD: hydrogen bond donor; HDES: hydrophobic deep eutectic solvent; HPLC: high-performance liquid chromatography; LLE: liquid-liquid extraction; LLME: liquid-liquid microextraction; LOD: limit of detection; MS/MS: tandem mass spectrometry; NaDES: natural deep eutectic solvent; PAE: phthalic acid ester; QqQ: triple quadrupole; TBBPA: tetrabromobisphenol A; THF: tetrahydrofuran; UA: ultrasound-assisted; UV: ultraviolet; VA: vortex-assisted; VWD: variable wavelength detector.

Pharmaceuticals

Although studied to a lesser extent compared with pesticide residues or plastic migrants, pharmaceuticals represent another priority family of organic pollutants in food safety since their presence not only represents a potential risk for public health but also reveals an inappropriate or excessive use of them. Unlike plastic migrants, to extract these compounds both liquid-phase and solid-phase extraction or microextraction techniques are used, as can be seen in Table **3** [55 - 59]. This includes LLME [55, 56, 58] and DLLME, as well as solid-phase extraction (SPE) [59], although applications of its miniaturized [60] or magnetic micro-dispersive (m-µ-dSPE) [61] approaches can also be found.

Regarding the compounds determined, synthetic antibiotics belonging to the families of fluoroquinolones [55, 59] and sulphonamides [56 - 58], and to a lesser extent tetracyclines [59], have been the most analyzed in food, although there is a study in which carbamazepine and triclocarban [58] were extracted. As far as food matrices are concerned, there is very little variety, practically limited to the analysis of milk [55] and other animal-derived food, such as chicken meat [57], fish oil [58], or honey [55], although cereals [59] and fruit products [56] have also been analyzed.

DESs most used to carry out the extraction of drugs in foodstuff are those formed by quaternary ammonium compounds, such as HBA, ChCl [58], betaine [59], trioctylmethylammonium bromide (TOMAB) [55] or TOMAC [56], and tetrabutylammonium bromide (TBAB) [57]. As HBDs, medium-chain fatty acids [55, 57] and different alcohols [56, 58, 59] have been the most employed. Depending on whether the state of the sample is solid or liquid, and whether its nature is aqueous or oily, hydrophilic DESs [58] or HDESs [55, 56] are used. This is the case of the work developed by Liu *et al.* [58], in which a hydrophilic DES composed of ChCl and glycerol at a molar ratio of 2:1 was applied in a reversed-phase LLME to extract pharmaceuticals and one personal care product (sulfamethazine, sulfamethoxazole, carbamazepine, and triclocarban) from fish oil samples. In the extraction procedure, 2 g of fish oil diluted with 1 mL of n-hexane was extracted with 75 µL of DES. After two extraction cycles, the extracts were merged and diluted 3 times to reduce their viscosity and, consequently, their inference in the MS signal. Finally, the extract was filtered and injected into the ultra-high-performance liquid chromatography (UHPLC)-tandem mass spectrometry (MS/MS) system, providing good recovery values and low LODs.

Table 3. Some examples of recent applications of DESs for the extraction of pharmaceuticals from food matrices.

Sample (amount)	Analytes	DES (molar ratio; amount)	Extraction procedure	Analytical technique	Recovery %	LODs	Comments	References
Milk (5 mL) and honey (5 g)	6 Fluoroquinolones	TOMAB:decanoic acid (1:2; 200 µL)	VA-LLME	MEKC-UV	88-114	0.006-0.010 mg/L	- HCl/ethanol (125/25, v/v) mixture was used as back-extraction solvent. - Yoghurt, tap and river waters were also analyzed.	[55]
Fruit juices (5 mL)	Sulfapyridine, sulfamethazine, and sulfadimethoxine	TOMAC:2-octanol (1:2; 800 µL)	VA-UA-LLME	HPLC-UV	88-98	0.02-0.05 mg/L	-TOMAC:octanoic, decanoic, or oleic acids, and TBAB:octanoic acid were also tested. - Apple, grape, peach, and pear juices and black tea were analyzed.	[56]

(Table 3) cont.....

Sample (amount)	Analytes	DES (molar ratio; amount)	Extraction procedure	Analytical technique	Recovery %	LODs	Comments	References
Chicken meat (1 g)	Sulfamethazine and sulfamethoxazole	TBAB:malonic acid:hexanoic acid (1:1:1; 2.0 g)	UAE/DLLME	HPLC-UV	86-109	3-7 μg/kg	- Meat samples were lyophilized. - Heptanoic, octanoic, and decanoic acids were also tested as HBDs. - Formic, acetic, and malonic acids, and glycerol, EtGly, and butanediol were evaluated as the third component. - A selectivity study was carried out by the addition of other antibiotics (tetracyclines and fluoroquinolones). - The method was applied to the analysis of real samples.	[57]
Fish oil (2 g)	Sulfamethazine, sulfamethoxazole, carbamazepine, and triclocarban	ChCl:glycerol (1:2; 75 μL)	UA-LLME	UHPLC-(QqQ)-MS/MS	87-94	16.7-33.0 ng/L	- Sample was previously diluted with n-hexane. - Urea, EtGly, and malonic acid were also tested as HBDs. - An experimental design was used for extraction conditions optimization. - The method was applied to the analysis of real samples.	[58]
Millet (5 g)	Levofloxacin and tetracycline	Betaine:EtGly:water-based polymer (1:2:1; 1.5 mmol)	SPE	HPLC-UV	96-100	0.01-0.04 mg/L	- Sample was hydrated with 50 mL water and 1 mL was used as sample solution. - 200 mg of sorbent was used. - APTES-methacrylic acid (3.3 mmol), TEOS (1 mL), AIBN (0.3 mmol), and EGDMA (25 mmol) were used for MIPs synthesis. - NIPs, MIPs, and DES-based MIPs were compared.	[59]

AIBN: 2,2-azobisisobutyronitrile; APTES: (3-aminopropyl)triethoxysilane; DES: deep eutectic solvent; DLLME: dispersive liquid-liquid microextraction; EGDMA: ethylene glycol dimethacrylate; EtGly: ethylene glicol; HBD: hydrogen bond donor; HPLC: high-performance liquid chromatography; LLME: liquid-liquid microextraction; LOD: limit of detection; MEKC: micellar electrokinetic capillary chromatography; MIP: molecularly imprinted polymer; MS/MS: tandem mass spectrometry; NIP: non-imprinted polymers; QqQ: triple quadrupole; SLE: solid-liquid extraction; SPE: solid-phase extraction; TBAB: tetrabutylammonium bromide; TEOS: tetraethoxysilane; TOMAB: trioctylmethylammonium bromide; TOMAC: trioctylmethylammonium chloride; UA: ultrasound-assisted; UAE: ultrasound-assisted extraction; UHPLC: ultra-high-performance liquid chromatography; UV: ultraviolet; VA: vortex-assisted.

In almost all cases, analytical techniques used for separation and determination of pharmaceuticals are based on LC systems, including both HPLC [56, 57, 59] and UHPLC [58], and UV detectors [55 - 57, 59], although some applications of MS [58] detectors can be found. As an exception, Yu *et al.* [55] developed an LLME method based on TOMAB:decanoic acid (1:2) for the extraction of 6 fluoroquinolones from milk, honey, and yogurt before their determination by micellar electrokinetic capillary chromatography (MEKC)-UV detection. To couple the extraction procedure to the capillary electrophoresis system, a back-extraction procedure was required using 125 μL of HCl 0.5 M and 25 μL of ethanol, and sodium dodecyl sulfate solution for a concentration of 5 mM was added to the aqueous extract and injected into the chromatographic system.

An interesting application of DESs in the field of food safety is the work carried out by Li and Row [59] for the extraction of levofloxacin and tetracycline from millet using a molecularly imprinted polymer (MIP) based on DESs prior to their determination by HPLC-UV. In this case, DES formed by betaine, EtGly, and water at a molar ratio of 1:2:1, and acting as a functional monomer, improves the affinity capacity and selectivity of the MIP. Additionally, the polar character of the DES enhanced the aqueous compatibility of MIPs. For conventional MIP or non-imprinted polymer (NIP) synthesis, (3-aminopropyl)triethoxysilane (APTES)-methacrylic acid, as functional monomer, tetraethoxysilane (TEOS), 2,2-azobisisobutyronitrile (AIBN) as initiator, and ethylene glycol dimethacrylate (EGDMA) as cross-linker, were used, including the addition of templates in the case of MIPs. In the case of DES-based MIP, betaine:EtGly was used as an auxiliary monomer. For antibiotics purification, 200 mg of each sorbent was suitably packed and conditioned in an SPE cartridge, and 1 mL of sample solution (5 g millet/50 mL solution) was loaded into the SPE column. With the aim to select the most adequate sorbent, NIP, MIPs, and DES-based MIPs for levofloxacin and tetracycline were compared. DES-based MIPs not only achieved a lower loss rate of analytes compared to conventional NIP and MIPs but also showed better selectivity for target compounds. Besides, they presented the best adsorption capacity when compared with NIP and MIPs. DES-based MIPs showed the best selectivity and extraction efficiency for levofloxacin and tetracycline, reaching recovery higher than 95.7% and 97.2%, respectively, whereas their recovery values for the other antibiotics, that are tetracycline and levofloxacin, were lower than 40%. Apart from that, all sorbents were able to eliminate the typical chromatographic interferences of millet, but DES-based MIPs obtained the best results in terms of recovery.

Interestingly, DESs are not always used as extraction solvents per se but, taking advantage of their physical properties such as the hydrophobic/hydrophilic character of their components, they can be used as extraction assistants. This is the case of the curious methodology developed by Shishov *et al.* [57] for the extraction of the synthetic antibacterial drugs sulfamethazine and sulfamethoxazole from chicken meat in an unconventional DLLME of solid samples using a TDES in which each component plays a key role. The procedure consisted of an initial ultrasound-assisted extraction (UAE) of 1 g of sample with 2 g of DES composed of hydrophilic and hydrophobic components (TBAB:malonic acid:hexanoic acid at a molar ratio of 1:1:1) at 50 ºC for 30 min. Then, a DLLME was carried by adding 10 mL of water to the tube containing the DES/sample mixture. At this step, DES was decomposed, dispersing hexanoic acid (as extraction agent) in the aqueous solution, while TBAB and malonic acid were dissolved into the sample. In this way, the problem of compatibility of DESs with chromatographic systems is reduced and the need for dilution is avoided as a result of the lower viscosity of hexanoic acid compared with DESs. The extraction protocol developed, very similar to the conventional DLLME but that uses a single solvent (extractant and disperser are part of DES), provided good results with recovery values in the range 86-109% and LODs between 3 and 7 µg/kg. During the optimization procedure, a preliminary study was carried out using a DLLME in which DESs composed of TBAB, as HBA, and different medium-chain fatty acids (hexanoic to decanoic acids) were tested at a molar ratio of 1:2, showing that hexanoic acid provided better results due to its higher polarity (recovery of 75%). However, when this DES was evaluated in a similar procedure for meat samples, this value decreased below 30% as a result of the high viscosity of DES. Thus, common carboxylic acids such as formic, acetic, and malonic acids, and alcohols, including glycerol, EtGly, and butanediol were evaluated as DES third component at a molar ratio of 1:1:1. Introducing an additional component, not only reduces the viscosity of DES but also improves the mass transfer. In this case, using malonic acid the best results were obtained since it provided a very acidic medium for the adequate extraction of sulphonamides. This DES was characterized by Fourier-transform infrared spectroscopy showing that both acids act as HBDs and are responsible for the DES formation. TBAB as water-soluble salt acts as salting-out and disperser agent allowing not only the dispersion of hexanoic acid but also the mass transfer between aqueous and organic phases, whereas malonic acid provided lower viscosity and adequate medium for analytes extraction and, additionally, can improve the dispersion of the extractant.

Other Organic Contaminants

In addition to the main groups of organic pollutants evaluated in food matrices by using DESs as extraction solvents (*i.e.*, pesticide residues, plastic migrants, and pharmaceuticals), other substances harmful to humans and the environment have been studied. As can be seen in Table (**4**) [62 - 71], among all the possible families of compounds, synthetic dyes [64, 66, 69, 70] and toxins, including aflatoxins [65] and patulin [67], have been by far the most evaluated. However, although to a lesser extent, a few articles that determine phenols [62, 71] and polycyclic aromatic hydrocarbons (PAHs) [63, 68], among others, can be found in the literature. These organic contaminants have been evaluated in a wide variety of samples, including edible oils [71], cereals [65], smoked food [62], honey [63], spices [64], and jellies [66, 70]. Besides, different beverages, such as carbonated drinks [66, 69], fruit juices [67], coffee [68], tea [69], and bacteria beverage [69], and sweets, including chocolate dragee [66], fruity pastel [70], smarties [70], ice cream [70], and candy [70], have been studied using DESs as the extraction agents.

Regarding DESs used to carry out the extraction procedure, both hydrophilic DESs [67, 71] and HDESs [63, 64, 68, 70], or pseudo-HDESs [65, 66, 69] (those formed of hydrophilic and hydrophobic components) have been applied with this aim. In this sense, quaternary ammonium salts such as ChCl [62, 71], tetrabutylammonium chloride (TBAC) [65, 67, 69] and TOMAC [70], or even some terpenes like menthol [63] or camphor [68], have been the most used as HBAs. On the contrary, several medium-chain fatty acids [63, 69, 70] and alcohols [65, 67, 71], or terpenes such as thymol [64, 66], menthol [68], and to a lesser extent borneol [68] have been employed as HBDs. In particular, menthol has been used in both ways, although in many cases it is difficult to determine the donor and the acceptor component in those DESs formed by non-ionic compounds.

As described for plastic migrant's determination, extraction *via* forming DESs has also been applied for other contaminants. An interesting example is the work developed by Shishov *et al.* [62], in which a novel ChCl hydrophilic membrane-based microextraction was used for the extraction of phenols, including phenol, o-cresol, p-cresol, guaiacol, eugenol, and isoeugenol (the last three are bioactive compounds present in essential oils), from smoked sausage and fish through the *in-situ* formation of DES between analytes and ChCl. With this aim, 200 mg of smoked sample was mixed with 500 μL of hexane, and a syringe with a poly(vinylidene fluoride-co-tetrafluoroethylene) (PVDF-co-PTFE) membrane, impregnated with 10 μL of 200 g/L ChCl solution, was immersed. After 15 min of shaking, the membrane was taken off, dried, and placed into a *via* l containing 500 μL of water for the back-extraction due to DES decomposition, and the aqueous extract was injected into the HPLC-FD system. Afterward, the impregnated membrane was dried in an incubator at 100 °C for 5 min, since when a wet membrane was used, recovery values lower than 10% were obtained. This can be explained because hydrophilic DESs are not formed in aqueous solutions. In these conditions, good recovery values (70-80%) and LODs in the order of μg/kg were obtained.

Table 4. Some examples of recent applications of DESs for the extraction of other organic contaminants from food matrices.

Sample (amount)	Analytes	DES (molar ratio; amount)	Extraction procedure	Analytical technique	Recovery %	LODs	Comments	References
Smoked sausage and fish (0.2 g)	6 Phenols	ChCl:analyte (-; 2 mg)	Membrane-based microextraction	HPLC-FD	70-80	0.3-1.0 μg/kg	- Sample was previously mixed with hexane. - PVDF-c--PTFE membrane was used. - Cellulose acetate, nylon, and polyethersulfone membranes were also tested. - DES was formed *in-situ* between ChCl and analytes. - The method was applied to the analysis of real samples.	[62]
Honey (5 mL)	16 PAHs	Menthol:decanoic acid (1:2; 65 μL)	SFOD-DLLME	GC-(Q)-MS	76-93	14-52 ng/kg	- Sample was previously diluted with water. - Octanoic and nonanoic acids were also tested as HBDs. - A sugar cube was used as disperser. - The method was applied to the analysis of real samples.	[63]
Chili peppers, paprika, cumin, and sumac (0.1 g)	Sudan dyes (I–IV)	Coumarin:thymol (1:1; 200 μL)	UA-SLME	HPLC-UV	86-99	0.25-0.35 mg/kg	- The method was applied to the analysis of real samples.	[64]

(Table 4) cont.....

Sample (amount)	Analytes	DES (molar ratio; amount)	Extraction procedure	Analytical technique	Recovery %	LODs	Comments	References
Millet, peanut, and heepseed (1 g)	Aflatoxins B1, B2, G1, and G2	TBAC:1-hexanol (1:2; 200 μL)	MSPD	HPLC-FD	-	0.03-0.10 μg/kg	- 1-dodecanol and hexanoic acid were also tested as HBDs. - Silica gel for millet and diatomite for peanut and hempseed samples were used as disperser. Al_2O_3 was also evaluated. - ACN was used as elution solvent. - Pyridinium tribromide was used as post-column derivatizing agent.	[65]
Carbonated beverage (0.4 mL), jelly (0.2 g) and chocolate dragee (2 g)	Amaranth, ponceau 4R, allura red, azorubine, and erythrosine	BTEAC:thymol (4:1; 300 μL)	VA-LLME	HPLC-UV	94-101	0.01-0.08 μg/L	- Samples were previously diluted with water. - ChCl was also tested as HBA. - The method was applied to the analysis of real samples.	[66]
Fruit juices (5 mL)	Patulin	TBAC:2,3-butanediol (1:3; 500 μL)	UA-DLLME	UV/Visible	90-107	2.2 μg/L	- ChCl and TOAC as HBAs, and EtGly, glycerol, 1,3- and 1,4-butanediol as HBDs, were also tested. - Mg(II) solution was used as complexing agent. - Different metal ions were tested as complexing agents. - Acetone was used as emulsifier solvent. - The method was applied to the analysis of apple, orange, peach, apricot, grape, kiwi, cherry, and mango juices.	[67]

(Table 4) cont.....

Sample (amount)	Analytes	DES (molar ratio; amount)	Extraction procedure	Analytical technique	Recovery %	LODs	Comments	References
Coffee (8 mL)	14 PAHs	Menthol:borneol:camphor (5:1:4; 40 mg)	m-μ-dSPE	HPLC-FD, GC-(Q)-MS	91-121	0.31-5.9 ng/L	- Fe_3O_4@DES was used as extraction material. - ACN was used as back-extraction solvent. - Menthol, borneol, camphor, and thymol were tested as HDES components. - A computational approach was used for the selection of optimal HDES. - Selectivity was studied against PAEs. - GC-MS was used for confirmation. - The method was applied to the analysis of real samples.	[68]
Beverages (1-10 mL)	8 Synthetic pigments	TBAC:octanoic acid (1:2; 100 μL)	VA-LLME	HPLC-DAD	75-103	0.016-1.12 μg/L	- TBAC and TOMAC with octanoic or decanoic acids were tested as DES components. - Tea beverage, carbonated drinks, fruit juices, and *Lactobacillus* beverages were analyzed.	[69]
Jelly, fruity pastel, smarties, candy (2-5 g), and ice cream (10 mL)	Sunset Yellow and Brilliant Blue FCF	TOMAC:decanoic acid (1:2; 150 μL)	DLLME	UV/Visible	98-103	2.0-2.9 μg/L	- Oleic acid and ibuprofen were also tested as HBDs. - Acetic acid and sodium bicarbonate were used to produce effervescence.	[70]

(Table 4) cont.....

Sample (amount)	Analytes	DES (molar ratio; amount)	Extraction procedure	Analytical technique	Recovery %	LODs	Comments	References
							- A selectivity study was carried out using possible interfering organic compounds and inorganic ions. - The method was applied to the analysis of real samples.	
Vegetable oil (1 mL)	4 Chlorophenols and 2 phenolics	ChCl:1,6-hexanediol (1:2; 90 µL)	UA-DLLME	HPLC-UV/Visible	77-92	0.05-0.10 mg/L	- EtGly, glycerol, 1,2- and 1,4-butanediol, urea, and acetic acid were also tested as HBDs. - Acetone was used as disperser.	[71]

ACN: acetonitrile; BTEAC: benzyltriethylammonium chloride; ChCl: choline chloride; DAD: diode array detector; DES: deep eutectic solvent; DLLME: dispersive liquid-liquid microextraction; EtGly: ethylene glycol; FD: fluorescence detector; GC: gas chromatography; HBA: hydrogen bond acceptor; HBD: hydrogen bond donor; HDES: hydrophobic deep eutectic solvent; HPLC: high-performance liquid chromatography; LLME: liquid-liquid microextraction; LOD: limit of detection; m-µ-dSPE: magnetic-micro-dispersive solid-phase extraction; MS: mass spectrometry; MSPD: matrix solid-phase dispersion; PAE: phthalic acid ester; PAH: polycyclic aromatic hydrocarbon; PVDF-co-PTFE: poly(vinylidene fluoride-co-tetrafluoroethylene); Q: single quadrupole; SFOD: solidified floating organic droplet; SLME: solid-liquid microextraction; TBAC: tetrabutylammonium chloride; TOAC: trioctylammonium chloride; TOMAC: trioctylmethylammonium chloride; UA: ultrasound-assisted; UV: ultraviolet; VA: vortex-assisted.

Regarding separation and detection systems, HPLC [62, 64 - 66, 68, 69, 71] and conventional detectors, such as FD [62, 65, 68], UV/Visible [67, 70, 71], UV [64, 66], or DAD [69], have been the most widely used, although GC [63, 68] and MS [63, 68] were employed in some cases. For the extraction of the mentioned organic contaminants, both solvent and sorbent-based extraction procedures have been applied. In this regard, DLLME, LLME, solid-liquid microextraction (SLME), matrix solid-phase dispersion (MSPD), and m-µ-dSPE, among others, have been used. As an example, Fan *et al.* [68] proposed a natural terpene-based HDES as the carrier solvent for the m-µ-dSPE of 16 PAHs, including naphthalene, acenaphthylene, acenaphthene, fluorene, phenanthrene, anthracene, fluoranthene, pyrene, chrysene, benzo[b]fluoranthene, benzo[a]pyrene, dibenzo[a,h]anthracene, benzo[g,h,i]perylene, and indeno [1,2,3-c,d]pyrene from coffee samples. In this study, 1 g of the ternary HDES composed of menthol, borneol, and camphor at a molar ratio of 5:1:4 was mixed with 1 g of Fe_3O_4 magnetic nanoparticles under sonication to obtain a ferrofluid. Then, 80 mg of the nanofluid was added to 8 mL of coffee sample, and after manual agitation and magnetic isolation, 40 µL of acetonitrile (ACN) was used for the re-extraction of analytes with vortex agitation and the extract was injected into the HPLC-FD system for analysis. GC-MS was only used for samples confirmation. Initially, for HDES selection, binary mixtures of menthol or thymol with borneol or camphor, at a molar ratio of 1:1, were evaluated by computational studies through COnductor-like Screening Model for Segment Activity Coefficient (COSMO-SAC). Predicted results were confirmed by experimental assays showing that HDES composed of menthol:camphor not only provided the best extraction efficiency for most PAHs, but also the highest selectivity (against PAEs). However, given that menthol:borneol HDES showed the best recovery values for naphthalene, the mixture of both DESs, which also present a common component (menthol), improved the whole extraction performance. As previously mentioned, the addition of a third component enhances the properties of the synthesized materials such as viscosity, selectivity, and extraction capacity. These parameters provided excellent results in terms of LODs (0.31-5.9 ng/L) and recovery values (91-121%). Besides, the extraction mechanism was studied through quantum chemical calculations showing that the interactions between HDESs and PAHs become stronger as a result of a reduction in the bandgap between the highest occupied and the lowest unoccupied molecular orbitals.

Another point is that most of the developed extraction procedures are ultrasound (UA) [64, 67, 71], VA [66, 69], temperature [63], or even magnetically assisted [68], which enhances the extraction efficiency due to the dispersion of the extraction material in the sample, giving rise to simple, fast, and more efficient methods. It is interesting the microextraction method developed by Ravandi *et al.* [70], in which the dispersion of the DES into the aqueous sample solution was carried out by effervescence. In this case, an HDES of TOMAC and decanoic acid at a molar ratio of 1:2 was applied for the extraction of the synthetic colorants Sunset Yellow and Brilliant Blue FCF from different sweets (fruity pastel, smarties, ice cream, candy, and jelly) prior to their determination by UV/Visible spectrophotometry. However, the novelty of this work resides in the production of an *in-situ* effervescence as the dispersion force of HDES into the sample. With this aim, 8 mL of sample was mixed with 2 mL of $NaHCO_3$ 1.5 M, and 600 µL of HDES/acetic acid (1/3, v/v) extraction mixture was added at the bottom of the flask. At this moment, acetic acid, as a proton donor, and sodium bicarbonate, as an effervescent compound, react and an effervescent process took place. The result was the dispersion of DES in the aqueous solution, avoiding the addition of organic solvents and assistance techniques. After centrifuging, the DES extract was 2-fold diluted with methanol for the determination of the target analytes, obtaining recovery values of 98-103% and LODs of 2.9 µg/L for Sunset Yellow and 2.0 µg/L for Brilliant Blue FCF. Finally, an interference study was carried out through the evaluation of the effect of some organic compounds and inorganic species in the correct determination of the synthetic dyes. It was found that concentrations between 16 and 2000 times higher than the ones found for the target compounds cause a relative error lower than 5%. With these results, the suitability of the developed method was demonstrated for synthetic dyes determination.

Given the difficulties on many occasions of the collection of the droplet in microextraction procedures of complex food matrices, the solidification of the droplet has been postulated as a good alternative. This variant called solidified floating organic droplet (SFOD), consists of freezing the final extract, generally placing the sample after extraction in an ice bath. Then, the solid is taken out with a spatula or tweezers and either dissolved in an adequate solvent or melted before analysis. As an example, Fazaieli *et al.* [63] developed an SFOD-DLLME method using menthol:decanoic acid HDES at a molar ratio of 1:2 for the extraction of 16 PAHs from honey samples and GC-MS for their determination. In order to avoid the use of conventional organic solvents and in an approach towards more sustainable procedures, a sugar cube was used as the dispersing agent. The optimized extraction procedure consisted of the extraction of 5 mL of honey, diluted up to 15 mL, heated at 45 ºC for 2 min, with 65 μL of HDES previously added into the sugar cube (0.5 g). After 30 s of manual shaking, the sugar cube was dissolved, and the extraction solvent dispersed into the sample. Then, it was placed in an ice bath for 2 min, and the solid droplet of DES was collected and dissolved in 10 μL of ACN before GC-MS analysis. It is possible that the sugar cube not only acted as a disperser, but also as a sugaring-out agent, similar to salts in the salting-out process. The methodology performed is an interesting alternative that seeks to follow the principles of green chemistry. Besides, it provided good results with recovery values of 76-93% and low LODs between 14 and 52 ng/kg.

In general terms, for the evaluation of organic contaminants in food samples, several DESs and (pseudo-)HDESs have been used as extraction materials. Compounds belonging to the families of pesticides, plastic migrants, and pharmaceuticals constituted the organic contaminants most evaluated. However, other minority groups have also been studied, such as dyes and toxins, among others. Besides, a wide variety of food samples have been analyzed, including different beverages and animal-derived products, fruits, cereals, or edible oils, *etc.* Regarding the extraction techniques, solid and liquid-phase (micro)extraction techniques have been applied coupled, in most cases, to LC systems and conventional detector systems, since the drawbacks associated with GC and the difficulties of using MS detectors or other instruments required the volatilization of solvents.

Heavy Metals

The current trends in green chemistry not only have influenced the use of DESs for the extraction of organic contaminants, but also for inorganic metal species, allowing the reduction of volatile and toxic organic solvents used in this field. As can be seen in some examples of the last five years compiled in (Table **5**) [72 - 81], the extraction of heavy metals has been carried out from a wide variety of foodstuffs, including vegetables, fruits, beverages, animal-derived products, and dairy products, among others. With this aim, DESs consisted of ChCl as HBA combined with different HBDs such as phenol [74, 76, 80, 81], carboxylic acids [72, 79], and thioamides [78], have been the most commonly applied. Likewise, other quaternary ammonium salts such as betaine and TBAB combined with sugars and carboxylic acids have also been used for the extraction of metal species from food samples [75, 77]. Apart from that, the development of TDESs for the extraction of metal ions from food matrices has also been reported [73]. Regarding the extraction techniques, solvent-based extractions have been by far the most used with these green solvents for the analysis of heavy metals in food samples [72 - 74, 76 - 81], although, to a lesser extent, the use of DESs in sorbent-based extraction techniques has also been explored [75].

Table 5. Some examples of recent applications of DESs for the extraction of heavy metals from food matrices.

Sample (amount)	Analytes	DES (molar ratio; amount)	Extraction procedure	Analytical technique	Recovery %	LODs	Comments	References
Tomato sauce, black and green tea, and dark chocolate (0.5-3 g)	Co(II)	ChCl:n-phenyliminodiacetic acid (1:2; 20 µL)	UA-LPME	UV/Visible spectrophotometry	94-105	5.23 µg/L	- Acid digestion was previously carried out. - The DES has a dual role: chelating agent and extraction solvent. - The method was applied to the analysis of food and environmental water samples.	[72]
Dairy products (5 mL)	Pb(II), Cd(II) and Cu(II)	Menthol:sorbitol:mandelic acid (1:2:1; 100 µL)	SFOD-DLLME	FAAS	89-104	0.38-0.42 µg/L	- A protein precipitation step was previously carried out. - The NaDES has a dual role: chelating agent and extraction solvent. - Methanol (1.5 mL) was used as disperser.	[73]

(Table 5) cont.....

Sample (amount)	Analytes	DES (molar ratio; amount)	Extraction procedure	Analytical technique	Recovery %	LODs	Comments	References
							- Other binary DESs formed by ChCl and menthol as HBAs and vanillin, phthalic acid, mandelic acid, and sorbitol as HBDs, at molar ratios 1:1 and 1:2, were also investigated. - Other molar ratios were also studied for the TDES, including 4:2:1, 1:1:1, and 2:1:1. - The method was applied to the analysis of milk with different fat content, local milk and milkshake samples.	
Fruit juice (5 mL)	Co(II) and Ni(II)	ChCl:4-amylphenol (1:1; 125 mg)	DLLME	FAAS	96-97	0.22-0.30 µg/L	- The DES has a dual role: chelating agent and extraction solvent. - Methanol (1 mL) was used as disperser. - Other molar ratios were also studied, including 2:1 and 1:2. - The method was applied to the analysis of grape and peach juices and environmental water samples.	[74]
Vegetables (-)	Pb(II)	Poly(TBAB:acrylic acid) (1:2; 12 mg)	µ-dSPE	FAAS	92-106	2.0 µg/L	- Acid digestion was previously carried out. - A back-extraction step was carried out with nitric acid (1 mL). - The sorbent could be reused for 16 adsorption/ desorption cycles. - The method was applied to the analysis of onion, celery, carrot, tomato, and environmental water samples.	[75]

(Table 5) cont.....

Sample (amount)	Analytes	DES (molar ratio; amount)	Extraction procedure	Analytical technique	Recovery %	LODs	Comments	References
Mushroom and soybean (2 g)	Cr(III) and Cr(VI)	ChCl:phenol (1:2; 350 µL)	UA-DLLME	FAAS	60	0.4 µg/L	- Acid digestion was previously carried out. - PAN was used as a chelating agent. - THF (400 µL) was used as emulsifier. - A comparison with VA-DLLME technique was performed. - Other HBDs were also investigated, including urea, glycerol and EtGly. - Other molar ratios were also studied for ChCl:phenol DES, including 1:1, 1:3 and 1:4. - The method was applied to the analysis of food and environmental waters samples.	[76]
Fish (0.5 g)	MeHg, total Hg (Hg(II)) and iHg (Hg^0, Hg(I) and Hg(II))	Betaine:sorbitol (1:3; 600 µL)	UA-DLLME	UV/Visible spectrophotometry	92-99	0.25-0.92 µg/L	- Different sample pre-treatments were carried out. - AzB (300 µL) was used as a chelating agent. - ACN (375 µL) was used as emulsifier. - Other HBAs such as ChCl and HBDs such as glycerol, sorbitol, glucose, lactic acid, and sucrose were also investigated. - Other molar ratios were also studied for betaine:sorbitol DES, including 1:1, 2:1, 1:2, 2:3 and 3:2. - The NaDES contained 10% water (v/v). - The method was applied to the analysis of tuna, salmon, trout, mackerel, whiting, anchovy and shrimp, and environmental water samples.	[77]

(Table 5) cont.....

Sample (amount)	Analytes	DES (molar ratio; amount)	Extraction procedure	Analytical technique	Recovery %	LODs	Comments	References
Foodstuff (walnut, rice, tomato paste, spinach, orange juice, black tea, and water) (1-2 g)	Pb(II), As(III), Cd(II) and Cu(II)	ChCl:thioacetamide@m-MWCNTs (1:2; 40 µL)	m-AA-LLME	ETAAS	94-101	3.0-4.2 ng/L	- Acid digestion was previously carried out. - Other HBDs were also investigated, including urea and thiourea. - A back-extraction step was carried out with nitric acid (75 µL). - The method was applied to the analysis of food samples.	[78]
Bottled water, honey, and rice (0.2 g, 3-50 mL)	As(III,V) and Sb(III,V)	ChCl:oxalic acid (1:3; 700 µL)	VA-DLLME	HG-AAS	94-96	7.5-15.6 ng/L	- UA digestion was previously carried out. - Dithizone (600 µL) was used as a chelating agent. - THF (300 µL) was used as emulsifier. - Other HBAs such as TBAC and TOAC, and HBDs such as phenol, glycerol, and decanoic acid, were also investigated. - Other molar ratios were also studied for ChCl:oxalic acid DES, including 1:1, 1:2, 1:4, and 2:1. - The method was applied to the analysis of food and environmental water samples.	[79]

(Table 5) cont.....

Sample (amount)	Analytes	DES (molar ratio; amount)	Extraction procedure	Analytical technique	Recovery %	LODs	Comments	References
Foodstuff (tomato, cucumber, banana, black tea, apple, cabbage, egg, chicken meat, cultivated mushroom, spinach, honey, canned fish, cow meat, bean, coffee, cheese, red wine, white wine, cow milk and ice tea) (1 g or 1 mL)	V(V)	ChCl:phenol (1:4; 1 mL)	UA-DLLME	ETAAS	>96	0.025 μg/L	- Microwave-assisted digestion was previously carried out. - HMBATSC (1 mL) was used as a chelating agent. - THF (1 mL) was used as emulsifier. - Other HBDs were also investigated, including urea, oxalic acid, and lactic acid. - Other molar ratios were also studied for ChCl:phenol DES, including 1:1, 1:2 and 1:3. - An experimental design was carried out for the optimization step. - The method was applied to the analysis of food samples.	[80]
Foodstuff (cow milk, mixed fruit juice, orange juice, grapefruit, sheep milk, yogurt, honey, egg, canned fish and edible mushroom) (1 g or 1 mL)	Se(IV) and Se(VI)	ChCl:phenol (1:3; 500 μL)	UA-DLLME	ETAAS	97	4.61 ng/L	- Microwave-assisted digestion was previously carried out. - DAB was used as a chelating agent. - THF (500 μL) was used as emulsifier. - Decanoic acid as HBD with TOAC and TBAC as HBAs, at a molar ratio 1:2, were also investigated. - Other molar ratios were also studied for ChCl:phenol DES, including 1:1 and 1:2. - The method was applied to the analysis of food and environmental water samples.	[81]

μ-dSPE: micro-dispersive solid-phase extraction; ACN: acetonitrile; AzB: azure B (3-(dimethylamino)-7-(methylamino)phenothiazin-5-ium chloride); ChCl: choline chloride; DAB: 3,3′-diaminobenzidine; DES: deep eutectic solvent; DLLME: dispersive liquid-liquid microextraction; ETAAS: electrothermal atomic absorption spectrometry; EtGly: ethylene glycol; FAAS: flame atomic absorption spectrometry; HBA: hydrogen bond acceptor; HBD: hydrogen bond donor; HG-AAS: hydride generation atomic absorption spectrometry; HMBATSC: 2-hydroxy-3-methoxybenzaldehyde thiosemicarbazone; iHg: inorganic mercury; LLME: liquid-liquid microextraction; LOD: limit of detection; m-AA: magnetic-air-assisted; MeHg: methylmercury; m-MWCNT: magnetic-multi-walled carbon nanotubes; NaDES: natural deep eutectic solvent; PAN: 1-(2-Pyridylazo)-2-naphthol; SFOD: solidified floating organic droplet; TBAB: tetrabutylammonium bromide; TBAC: tetrabutylammonium chloride; TDES: ternary deep eutectic solvent; THF: tetrahydrofuran; TOAC: trioctylammonium chloride; UA: ultrasound-assisted; UV: ultraviolet; VA: vortex assisted.

As can be noticed from (Table **5**), the determination of metal ions usually requires the use of chelating agents to form stable complexes with the target analytes and, consequently, achieve an efficient extraction. However, Sorouraddin *et al.* [73] prepared a DES which acted simultaneously as a complexing agent and extraction solvent for the extraction and preconcentration of Pb(II), Cd(II) and Cu(II) from dairy samples prior to their determination by flame atomic absorption spectrometry (FAAS) analysis. In this way, both the need for an expensive chelating agent and a toxic organic solvent could be avoided. Besides, the authors also proposed the SFOD-DLLME as extraction procedure since solidifying the DES droplets containing the analytes significantly facilitated their separation from the aqueous phase. For this, the melting point of the DES had to be close to room temperature so that it could easily solidify. According to that, the authors tested different blends of DESs, including ChCl and menthol based-DESs combined with vanillin, phthalic acid, mandelic acid, and sorbitol as HBDs, at different molar ratios (1:1, 1:2, and 1:2:1), finding that only the TDESs formed by the natural components, menthol, sorbitol, and mandelic acid at a molar ratio 1:2:1, showed the mentioned characteristics. Additionally, the authors affirmed the TDES is less soluble in water than the individual components since the hydrogen bond interactions between their polar groups decrease the polarity. After optimizing the previous protein precipitation step and SFOD-DLLME procedure, recovery values in the range 89-104% and LODs of 0.38-0.42 μg/L were obtained using only 5 mL of sample solution at pH 5, 100 μL of TDES, and 1.5 mL of methanol as disperser solvent. Finally, the method was applied to the analysis of different samples, including milk products with different fat content and milkshakes, showing the suitability of the TDES for its purpose.

Following the trend of simplifying the extraction procedure, Shirani *et al.* [78] implemented a magnetic-air-assisted (m-AA)-LLME for the extraction and preconcentration of Pb(II), As(III), Cd(II), and Cu(II) from different foods, including walnut, rice, tomato paste, spinach, orange juice, and black tea, as well as water samples, prior to their quantification by electrothermal atomic absorption spectrometry (ETAAS) analysis. The m-AA-LLME was based on a mixture of magnetic-multi-walled carbon nanotubes (m-MWCNTs) and a ChCl-based DES as the extraction phase of the target compounds. The AA-LLME led to low consumption of extraction solvent volume and no disperser solvent was needed, since pulling in and pushing out the mixture of sample solution and extraction phase several times favored the formation of a stable cloudy solution and, consequently, the extraction of the analytes. Besides, the magnetic behavior of the nanofluid allowed it to be easily separated from the media after extraction with an external magnet and, thus, additional centrifugation step was avoided. In order to find the ChCl-based DES with the highest affinity to metal ions, the authors investigated different HBDs, including urea, thiourea, and thioacetamide, at a molar ratio of 1:2. The best extraction efficiency was provided by thioacetamide and the worst by urea, since sulfur atoms are softer than nitrogen atoms, and therefore are more polarizable, interacting more easily with analytes. Under the optimized conditions, 48 mL of sample (previously acid digested) at pH 6.5, 40 μL of magnetic nanofluid, and 75 μL of nitric acid as back-extraction solvent, the developed methodology achieved recovery values in the range 94-101% and LODs of 3.0-4.2 ng/L, demonstrating its high suitability to determine heavy metals in the selected matrices.

Another interesting work to mention is the one developed by Abdolhosseini *et al.* [75], in which a polymeric DES (poly-DES) based on TBAB and acrylic acid, at a molar ratio 1:2, has been synthesized, and then applied as an SPE sorbent for the micro-dispersive solid-phase extraction (μ-dSPE) of Pb(II) from vegetable and environmental water samples, followed by its determination by FAAS analysis. In this study, the poly-DES was obtained through a photopolymerization process without the need for solvents, which significantly reduced energy consumption and costs. The authors discussed that the prepared sorbent simultaneously possessed both the advantages of DES and those of porous materials. In this way, the presence of functional groups, including OH, carboxyl (COOH) and CO, and the large surface area and high porosity enabled chemical interactions with Pb(II) ions, as well as provided sites for the adsorption of the analyte increasing the adsorption capacity. Additionally, the negatively charged surface of DES allowed electrostatic interactions with Pb(II) ions. The authors also evaluated the reusability of the prepared sorbent, finding that it could be reused up to 16 times without changes in extraction efficiency. Finally, the method achieved good results in terms of recovery values (92-106%) and LOD (2.0 μg/L) after optimizing the process. Moreover, no significant differences were found compared to inductively coupled plasma (ICP)-MS results in the analysis of onion, celery, carrot, and tomato, and other non-food samples.

In summary, the extraction of heavy metals from food matrices by DESs has been mainly carried out by quaternary ammonium salt based-DESs applied in LLME techniques, which usually employed small volumes of emulsifier or disperser solvents such as THF, methanol, or ACN, among others, to promote the separation of the DES phase from the media after the extraction process. However, some novel applications, in which phase separation has been favored by repeated operations of pull in and push out the mixture of the sample solution and extraction solvent can also be found in the literature. Likewise, the combination of DESs with magnetic nanomaterials has also improved the extraction performance, since no centrifugation step was needed to obtain the DES phase containing the analytes. Another remarkable improvement has been the preparation of multifunctional DESs acting as complexing agents and extraction solvents simultaneously, which avoided the need for costly ligands to chelate metal ions. According to that, the use of DESs in this field has given rise to green analytical methodologies, characterized by energy-saving and cost-effective extraction procedures.

RECENT APPLICATIONS OF DESs FOR THE EXTRACTION OF BIOMOLECULES

The growing interest of the population in natural bioactive substances and food composition in recent years has led to the development of extraction methods for the evaluation of biomolecules with promising applications in the food, pharmaceutical, and cosmetic industries. As occurs in other areas of analytical chemistry, the application of green procedures using DESs, which not only provide efficient extraction, but also accomplish the principles of sustainability, have attracted a great interest. As can be seen in Fig. (**3**), DESs have been mainly applied for the extraction of different types of biomolecules including phenolic compounds, terpenes, and terpenoids, proteins, amino acids, or carbohydrates, although other organic compounds have also been evaluated using this kind of solvents as it is discussed in the following sections.

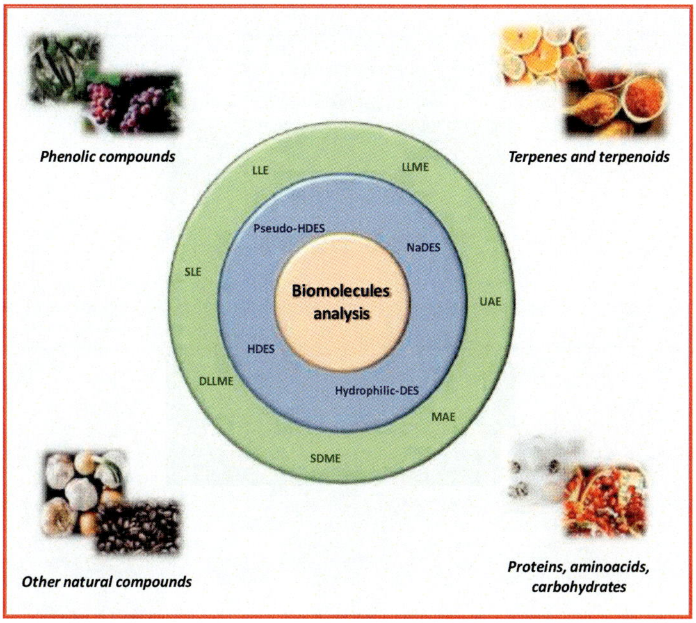

Fig. (3). Scheme of the main applications of DESs for the extraction of biomolecules in food analysis including the type of solvents and techniques used in the last five years, as well as the compounds mostly evaluated.

Phenolic Compounds

Regarding the application of DESs for the extraction of phenolic compounds, which have shown several benefits for health, due to their antioxidant, antiproliferative, anti-inflammatory, antimicrobial, and neuro-protective properties, among others [16]. Table (**6**) [82 - 91] summarizes some examples of the most relevant approaches developed in the last years (2016-present). Different classes of analytes have been evaluated, including phenolic acids [83, 85, 86, 88, 89, 91], phenolic aldehydes (vanillin) [88], phenolic alcohols (simple phenols and poly-phenols) [82, 88, 89, 91], flavonoids [82, 83, 85, 86, 88 - 91], anthocyanins [84, 88], and other phenolic metabolites [82, 87, 89, 91], from both food matrices (olive oil [82, 91], fruits [83, 84, 86, 90], vegetables [90] and spices [89, 90]) and food by-products [85, 88, 89] by UAE [83, 84, 86, 88, 89], liquid-liquid extraction (LLE) [82, 85, 91], and solid-liquid extraction (SLE) [87, 90] techniques, followed by LC coupled to conventional detectors [82 - 84, 86, 88 - 90], as well as MS [91] and MS/MS systems [84, 85, 87, 90].

Table 6. Some examples of recent applications of DESs for the extraction of phenolic compounds from food matrices.

Sample (amount)	Analytes	DES (molar ratio; amount)	Extraction Procedure	Analytical Technique	Recovery %	LODs	Comments	References
Olive oil (14 g)	2 Phenolic alcohols, 2 flavonoids, and 6 phenolic metabolites	ChCl:xylitol:water (2:1:3; 14 g)	LLE	HPLC-DAD	-	-	- The effect of removing the NaDES using a resin prior to analysis was studied. - The extraction temperature was set at 40 ºC.	[82]
Sapodilla pulp (0.2 g)	7 Phenolic acids and 4 flavonoids	ChCl based-DESs (1:1; 7 mL)	UAE	HPLC-UV	71-86	-	- Three different types of HBDs carboxylic acids, alcohols, and urea were investigated. - DESs contained 25% water (v/v). - The extraction temperature was set at 70 ºC. - Antioxidant content was characterized by TPC, TFC, and TAC. - Antioxidant and antimicrobial activities were measured using different methods. - A comparison with a conventional extraction method was performed. - The DES was reused up to three cycles.	[83]

(Table 6) cont.....

Sample (amount)	Analytes	DES (molar ratio; amount)	Extraction Procedure	Analytical Technique	Recovery %	LODs	Comments	References
Blueberries (0.1 g)	6 Anthocyanins	ChCl:glycerol:citric acid (1:4:1; 12 mL)	UAE	HPLC-DAD-(Q-ToF)-MS/MS	76	0.02 mg/L	- Different molar ratios of the ternary NaDES and individual pure components were studied. - An experimental design was carried out for the optimization of NaDES composition. - The NaDESs contained 25% water (v/v). - The extraction was carried out at room temperature. - A comparison with conventional extraction methods was performed. - A purification step by SPE was performed.	[84]
Tea seed oil (6 g)	10 Phenolic acids, 12 flavonoids and 3 carboxylic acids	ChCl:glycerol (1:2; 6 g)	LLE	UHPLC-(QqQ)-MS/MS	-	0.1-24.3 µg/L	- Other HBDs were also investigated, including xylitol (with water addition), malonic acid, EtGly, and propylene glycol. - Different molar ratios were studied, including 1:1, 1:2, 2:1 and 1:1:1. - The extraction temperature was set at 50 ºC. - A comparison with a conventional method was performed. - The removal of DES was carried out prior to analysis. - 31 phenolic compounds were profiled in the tea seed oil samples. - UHPLC-(Q-To-)-MS/MS was used for the characterization of extracts.	[85]
Bael pulp (0.5 g)	10 Phenolic acids, 6 flavonoids and cinnamic acid	ChCl:oxalic acid (1:1; 25 mL)	UAE	HPLC-DAD	>60	-	- Other molar ratios were also studied, including 1:2 and 2:1. - The NaDES contained 25% water (v/v). - The extraction temperature was set at 80 ºC. - A comparison with conventional extraction methods was performed. - The removal of NaDES was carried out prior to analysis.	[86]

(Table 6) cont.....

Sample (amount)	Analytes	DES (molar ratio; amount)	Extraction Procedure	Analytical Technique	Recovery %	LODs	Comments	References
Eucalyptus leaves (2 g)	26 Phenolic compounds (tentatively identified)	ChCl:EtGly (1:2; 20 mL)	SLE	UHPLC-(ToF)-MS/MS	-	-	- EAE, MAE, and UAE methods were also evaluated. - Other DESs were also investigated, including ChCl:xylitol (5:1), ChCl:glucose (1:1), and citric acid:glucose (1:1). - DESs contained 20% water (v/v). - The extraction temperature was set at 50 ºC. - A comparison with a conventional extraction method was performed. - TPC and TFC were determined. - The antioxidant capacity was evaluated.	[87]
Grape and olive pomace (0.5 g)	2 Phenolic acids, 2 phenolic alcohols, vanillin (phenolic aldehyde), 2 flavonoids, 9 anthocyanins, and pinoresinol (phenolic lignan)	ChCl:citric acid (2:1; 10 mL)	UAE/MAE	HPLC-DAD	-	-	- Biological activities of extracts were investigated. - The NaDES contained 30% water (v/v). - A comparison with ethanol extracts was performed.	[88]
Olive, onion, tomato, and pear food by-products (-)	4 phenolic acids, 2 phenolic alcohols, 6 flavonoids, oleuropein (phenolic secoiridoids), and cinnamic acid	Lactic acid:glucose (5:1; -)	UAE	HPLC-DAD	82-110	0.6-89.1 μg/kg	- Other NaDESs formed by citric acid and glucose or fructose, at a molar ratio 1:1, were also investigated. - An experimental design was carried out for the optimization step. - A sample to solvent ratio of 75 mg/mL was used. - The NaDES contained 15% water (v/v). - The extraction temperature was set at 40 ºC. - A stability test of phenolics in NaDES, water, and methanol extracts was carried out. - A comparison to conventional extraction solvents was performed.	[89]

(Table 6) cont.....

Sample (amount)	Analytes	DES (molar ratio; amount)	Extraction Procedure	Analytical Technique	Recovery %	LODs	Comments	References
Fruits, vegetables and spices (0.2 g)	8 Flavonoids	Acetyl ChCl:lactic acid (2:1; 600 μL)	SLE (miniaturized)	UHPLC-UV-(Q-Trap)-MS/MS	70-94	-	- Other DESs formed by ChCl and derivatives, betaine and carnitine as HBAs and citric acid, malic acid, tartaric acid and lactic acid as HBDs, at different molar ratios, were also investigated. - Acetyl ChCl:lactic acid and ChCl:lactic acid DESs were studied at different molar ratios, including 1:1, 1:2 and 2:1.	[90]
						-	- The NaDES contained 30% water (v/v). - An experimental design was carried out for the optimization step. - The extraction temperature was set at 60 °C. - A comparison with UAE and conventional extraction methods was performed. - The LOQ was found to be 0.5 μg/g. - The method was applied to the analysis of cranberry, fruits of *Lycium barbarum L.*, grape, plum and orange peel, onion and broccoli, and mustard, rosemary, and black pepper.	
Olive oil (14 g)	4 Phenolic acids and alcohols, 2 flavonoids, 6 phenolic metabolites, and elenolic acid	ChCl-based DESs (-; 14 g)	LLE	HPLC-DAD and HPLC-(Q-ToF)-MS	-	-	- Sugars, alcohols, carboxylic acids, and urea were investigated as HBDs. - Different molar ratios were studied, including 1:1, 1:2, 1:5, 2:1, 4:1 and 1:1:1. - The extraction temperature was set at 40 °C. - The removal of DES was carried out before analysis. - A comparison with a conventional extraction method was performed.	[91]

ChCl: choline chloride; DAD: diode array detector; DES: deep eutectic solvent; EAE: enzyme-assisted extraction; EtGly: ethylene glycol; HBA: hydrogen bond acceptor; HBD: hydrogen bond donor; HPLC: high-performance liquid chromatography; LLE: liquid-liquid extraction; LOD: limit of detection; LOQ: limit of quantification; MAE: microwave-assisted extraction; MS/MS: tandem mass spectrometry; MS: mass spectrometry; NaDES: natural deep eutectic solvent; Q: single quadrupole; QqQ: triple quadrupole; Q-Trap: triple quadrupole-linear ion trap; Q-ToF: quadrupole-time-of-flight; SLE: solid-liquid extraction; SPE: solid-phase extraction; TAC: total anthocyanin content; TDES: ternary deep eutectic solvent; TFC: total flavonoid content; ToF: time-of-flight; TPC: total phenolic content; UAE: ultrasound-assisted extraction; UHPLC: ultra-high-performance liquid chromatography; UV: ultraviolet.

In general terms, the extraction of phenolic compounds has been carried out using hydrophilic ChCl-based NaDESs combined with different HBDs, including sugars [82, 87, 91], carboxylic acids [83, 84, 86 - 88, 90, 91], alcohols [83 - 85, 87, 91] and urea [83, 91]. Likewise, mixtures of DESs prepared from carboxylic acids and sugars have also been explored [89]. Besides, DESs have sometimes been diluted by adding water to reduce their viscosities and, consequently, to facilitate the mass transfer of analytes from sample solution to the DES phase [82 - 90].

As an example of the extraction of phenolic compounds from food samples by DESs, García *et al.* [91] proposed different combinations of ChCl-based DESs mixed with sugars (sucrose, glucose, fructose, and xylitol), alcohols (glycerol, 1,4-butanediol, and 1,2-propanediol), organic acids (lactic acid and malonic acid), and urea as HBDs, at different molar ratios (1:1, 1:2, 1:5, 2:1 and 4:1), as well as a mixture of three sugars (sucrose:glucose:fructose, 1:1:1 molar ratio), for the extraction of phenolic compounds present in virgin olive oil, followed by their separation and quantification in an HPLC-DAD system. With the objective of evaluating the extraction efficiency of synthesized DESs, the authors compared the extraction yields obtained to those achieved by the most commonly used method for the extraction of natural compounds from food matrices, finding that, under the same conditions (extraction temperature of 40 °C, agitation time of 1 hour and a recovery step using a macro-porous resin), almost all the studied DESs showed a better performance for the extraction of phenolic compounds from olive oil than the conventional mixture of methanol/water (80:20, v/v). In this sense, organic acid-based DESs showed a good capacity to extract phenolic alcohols, hydroxytyrosol, and tyrosol, while alcohol-based DESs were the best option to extract the most abundant phenolic metabolites found in olive oil, oleacein, oleocanthal, and oleuropein aglycon, as well as to extract less polar compounds, flavonoids, luteolin, and apigenin, and the phenolic lignan, acetoxipinoresinol. In relation to sugar-based DESs, high extraction yields were achieved for almost all of the phenolic compounds, except in the case of sucrose:glucose:fructose DES for which lower yields were obtained. Finally, urea-based DESs were able to extract oleacein and oleocanthal, although phenolic alcohols were not efficiently extracted. Besides, most of the chromatographic profiles obtained showed a reasonable similarity between them for a total of 13 representative bioactive compounds of olive oil, whose identification was confirmed by MS spectra. In this way, the authors proved the successful capacity of prepared hydrophilic DESs as extraction solvents of a wide range of compounds of different polarities from olive oil.

Following this trend, the same research group has recently prepared a ChCl:xylitol:water NaDES (water was added to reduce its viscosity), at a molar ratio 2:1:3, for the LLE of 10 phenolic compounds from olive oil samples, prior to

HPLC-DAD analysis [82]. In this case, the authors carried out the optimization of the process by evaluating different extraction affecting parameters (temperature, agitation time, and sample:DES ratio), as well as recovery conditions (resin height, wash-water, and eluent volumes and pH). Besides, the effect of removing the NaDES using a resin before HPLC analysis was also investigated. The results showed that the chromatograms of both operational modes, eliminating the NaDES and directly injecting it in the chromatographic system, were comparable and no statistically significant differences were found for most of the phenolic compounds analyzed in terms of concentrations, except for hydroxytyrosol. The losses of the latter were attributed to water washing during the recovery step due to its high polarity and solubility in aqueous media. Although avoiding the recovery step saves time, it is still needed when concentrated polyphenolic extracts are required for industrial applications, and at the same time allows to reuse the extraction solvent.

Another interesting work to mention is the one developed by Saha and Chakraborty [83], in which ChCl-based DESs were proposed for the UAE of sapodilla pulp phenolic compounds, 7 phenolic acids, and 4 flavonoids followed by their determination by HPLC-UV analysis. Among the different HBDs tested, oxalic acid, malonic acid, lactic acid, and urea provided the highest extraction efficiencies, since their high polarities favored the interaction with target compounds and, in the case of carboxylic acids, the low pH values also enhanced the extraction of polar phenolic compounds present in sapodilla. DESs formed by glycerol, which contains three OH groups, and malic acid, with more than two COOH/OH groups, as HBDs, showed steric hindrance and high viscosity that reduced the availability of sites for hydrogen-bonding interaction with OH groups of phenolic compounds and, consequently, led to lower extraction yields of the target analytes. From several studies, the authors demonstrated that DESs could be successfully used for the extraction of sapodilla phenolic compounds, which can be further utilized in food applications. In this sense, the ChCl:oxalic acid extract showed the highest antioxidant activity, whereas the ChCl:malonic acid extract had the highest reducing capacity, which explains that the variation in the free radical scavenging activity of DES extracts depends on the total phenolic content (TPC). Besides, phenolic compounds proved to be more stable in DES extracts than in an aquorganic mixture, since the strong interactions established with the analytes by hydrogen bonds and the high viscosity of DESs diminished the interaction with oxygen and lowered the mass transfer of oxygen at the DES-air interface, that resulted in a decrease in phenolic compounds degradation. Finally, the authors found that the inherent solvent characteristics were the main responsible for the antimicrobial activity of DES extracts, being phenolic extracts obtained by organic acid-based DESs the best inhibitors, especially, against *Bacillus* and *Aspergillus* organisms.

Apart from that, DESs have also been applied for the extraction of phenolic compounds from food by-products with the aim of both reducing the costs of food waste treatments and revaluing the food industry [88]. In this way, Panić *et al.* [88] developed a method based on ChCl:citric acid DES (molar ratio 2:1) as extraction solvent of phenolic compounds from grape and olive pomace samples by ultrasound and microwave treatments. The authors found that phenolic content was generally higher for both pomaces in NaDES extracts than in conventional solvents, being grape pomace richer in phenolic compounds than olive pomace. The biological activity of the obtained extracts was assessed by the study of the antioxidant capacity, antiproliferative activity, cell death and cell cycle arrest, oxidative stress, and presence of DNA/RNA active compounds. In this regard, grape and olive pomaces extracted by NaDESs showed stronger cytotoxicity on two human tumor cell lines than those obtained by ethanol, which the authors attributed to both the higher TPC and the higher antioxidant capacity determined in the NaDES extracts. Besides, the ChCl:citric acid DES exhibited antioxidant activity by itself thanks to its reactive oxygen species with scavenging effects. Although an increase in the number of dead cells was not observed, the cell proliferation was delayed in NaDES extracts. Likewise, grape pomace extract showed the ability to protect cells from oxidative stress while olive pomace did not present this behavior, which the authors also attributed to its higher content in phenolic compounds. Finally, the binding study with DNA and RNA evidenced that grape pomace extracts contained DNA/RNA active compounds since it stabilized both biomacromolecules, while the constituents of the NaDES, ChCl and citric acid, did not have negative effects on them. Considering all data obtained, the authors confirmed both the good extractability of the prepared NaDES and the great potential of food by-products on food and pharmaceutical applications, especially those obtained from grapes.

Although it is still limited, the application of DESs for the extraction of this group of biologically active substances has shown higher extraction capacity and enhanced biological activity compared to conventional extraction methods based on toxic and volatile organic solvents.

Terpenes and Terpenoids

Another interesting group of phytochemicals to mention are terpenes and terpenoids, which have also attracted much attention for practical applications due to their antioxidant, anticancer, antimicrobial, and antimalarial properties, among others [16]. In search of green extraction methods for this group of volatile chemicals, conventional organic solvent extraction has been set aside in favor of the exploration of DESs. As can be seen in some examples of the last five years compiled in (Table 7) [92 - 96], DESs have been utilized for the extraction of terpenes and terpenoids from both food samples, such as spices [93] and fruit juices [94], and food by-products, such as essential oils [92, 96] and fruit pomaces [95], by different extraction techniques involving heating-stirring extraction [92], UAE [95] and LLE [96], as well as microextraction techniques such as single-drop microextraction (SDME) [93] and LLME [94], followed by GC [92, 93, 96] and LC [94] coupled to conventional detectors [94] and MS systems [93, 96], as well as by spectrophotometric methods [95].

Table 7. Some examples of recent applications of DESs for the extraction of terpenes and terpenoids from food matrices.

Sample (amount)	Analytes	DES (molar ratio; amount)	Extraction procedure	Analytical technique	Recovery %	LODs	Comments	References
Citrus essential oil (-)	Linalool	TBAC:analyte (-; -)	Heating and stirring	GC	99	-	- Other seven quaternary ammonium salts were tested as HBAs. - A deterpenation process from orange essential oil (composed of limonene and linalool) was carried out. - The TBAC:linalool mass ratio was fixed at 20:1. - The extraction temperature was set at 65 °C. - A back-extraction step was carried out. - A COSMO-RS study was performed.	[92]
Spices (cinnamon, cumin, fennel, clove, thyme and nutmeg) (50 mg)	29 terpenes	TBAB:dodecanol (1:2; 1.5 μL)	HS-SDME	GC-(IT)-MS	-	0.47-86.40 μg/g	- Other binary DESs formed by ChCl and TOMAC as HBAs and alcohols, hexanoic acid, lactic acid, and urea as HBDs,	[93]

(Table 7) cont.....

Sample (amount)	Analytes	DES (molar ratio; amount)	Extraction procedure	Analytical technique	Recovery %	LODs	Comments	References
							at different molar ratios, were also investigated. - An experimental design was carried out for the optimization step. - The extraction temperature was set at 80 °C. - More than 40 terpenes were extracted and identified in nutmeg. - The method was applied to the analysis of spice samples.	
Fruit juice (-)	β-carotene and lycopene	Nonanoic acid:decanoic acid:undecanoic acid (2:1:1; 600 μL)	Acid-base induced-LLME	HPLC-UV	90-103	2.0-50 μg/L	- NH_4OH (400 μL) and HCl were used as emulsifiers and demulsifier agents, respectively. - The method was applied to the analysis of watermelon, grapefruit, tomato, and guava juice samples.	[95]

(Table 7) cont.....

Sample (amount)	Analytes	DES (molar ratio; amount)	Extraction procedure	Analytical technique	Recovery %	LODs	Comments	References
Tomato pomace (40 mg)	Lycopene	DL-menthol:lactic acid (8:1; 5 mL)	UAE	UV/Visible spectrophotometry	-	-	- An experimental design was carried out for the optimization step. - ChCl:levulinic acid DES was also investigated. - The binary DESs were prepared at different molar ratios, including 2:1, 1:1, and 1:2. - The extraction temperature was set at 60 ºC. - A comparison with a conventional solvent extraction was carried out.	[95]
Citrus essential oil (-)	Linalool	ChCl:glycerol (1:2; -)	LLE	GC-(Q)-MS	-	-	- A deterpenation process from orange essential oil (composed of limonene and linalool) was carried out. - The addition of water was evaluated. - The extraction was carried out at room temperature. - Different sample to solvent ratios were investigated (1:2, 1:1, 2:1, 3:1, 4:1, and 5:1 wt). - A COSMO-RS study was performed.	[96]

ChCl: choline chloride; COSMO-RS: COnductor-like Screening MOdel for Real Solvents; DES: deep eutectic solvent; GC: gas chromatography; HBA: hydrogen bond acceptor; HBD: hydrogen bond donor; HPLC: high-performance liquid chromatography; HS-SDME: headspace single-drop microextraction; IT: ion trap; LLE: liquid-liquid extraction; LLME: liquid-liquid microextraction; LOD: limit of detection; MS: mass spectrometry; Q: single quadrupole; TBAB: tetrabutylammonium bromide; TBAC: tetrabutylammonium chloride; TOMAC: trioctylmethylammonium chloride; UAE: ultrasound-assisted extraction; UV: ultraviolet.

In this regard, the extraction of terpenes and terpenoids has been based mainly on the use of hydrophilic DESs composed of quaternary ammonium salts and alcohols [93, 96], carboxylic acids [93], and urea [93]. HDESs synthesized by long-chain carboxylic acids have also been applied as extraction materials, overcoming the limitation of hydrophilic DESs that disintegrate in aqueous samples [94]. Likewise, the synthesis of pseudo-HDESs [95], the *in-situ* formation of the DES with the target analyte [92], and the addition of water to reduce the viscosity of DESs [96] have also been explored in this application.

The industrial applications of essential oils have been limited to some extent due to the presence of terpenes that eventually undergo oxidation processes that give rise to unpleasant flavors [96]. This is why some research groups have investigated the separation of limonene from linalool in citrus essential oils using DESs. As an example, Li *et al.* [92] proposed to carry out the deterpenation process of the citrus essential oil by the *in-situ* formation of the DES (HBA:analyte). In this way, the authors evaluated the interactions of several quaternary ammonium salts with linalool and limonene, including ChCl and others formed by alkyl chains varying from C_2 to C_5 carbon units with chloride or bromide halide ions. The two liquid phases obtained after extraction with TBAC showed its efficiency to selectively form hydrogen bonds with the OH group of the linalool and, at the same time, have weak interactions with limonene. These experimental results were consistent with those previously obtained from quantum chemistry-based statistical thermodynamics studies, (COnductor-like Screening MOdel for Realistic Solvents, COSMOS-RS), since long alkyl chain quaternary ammonium salts showed strong interactions with both linalool and limonene (no phase separation was obtained) and the DESs based on short alkyl chain quaternary ammonium salts and linalool were still solid-liquid mixtures. After investigating the parameters affecting the associative extraction procedure, a mass ratio of 20:1 of TBAC to linalool and stirring and settling temperatures of 65 ºC and 30 ºC, respectively, were enough to obtain a good extraction performance. Likewise, it was necessary to perform a back-extraction in two steps in order to first eliminate the terpene completely from the DES phase and then, recycle the organic salt, finally obtaining the purified terpenoid extract. After evaluating sixteen organic solvents, n-hexane was used for the first step and a mixture of water and *n*-hexane was utilized for the second one. The suitability of these solvents was also confirmed with COSMO-RS studies. The whole procedure showed a satisfactory recovery value of 99% for the target analyte. Ozturk *et al.* [96], for their part, previously investigated glycerol-based DESs for the LLE of linalool from citrus essential oil before GC-MS analysis. In this case, ChCl combined with glycerol at 1:2 and 1:3 molar ratios, without dilution and diluted with 30 wt% of water, were evaluated as extraction solvents. The authors found that all DES mixtures were suitable for extracting linalool *versus* limonene from essential oil model mixtures, as the higher polarity of the first one favored its hydrogen-bond interactions with DESs. Although DESs diluted with water showed lower viscosity and higher selectivity towards linalool extraction, they also reduced the solubility of the citrus essential oil, which resulted in a lower concentration of the terpenoid in the DES phase. Therefore, pure DESs, especially ChCl:glycerol (1:2), provided the highest extraction capacities, which was consistent with the results obtained from COSMO-RS studies.

Another interesting application of DESs in terpenes and terpenoids extraction is the work developed by Li *et al.* [94], in which fatty acid-based DESs were prepared and applied for the extraction of β-carotene and lycopene from fruit juices by acid-base induced-LLME, followed by HPLC-UV determination. Among the different combinations of DESs evaluated, the ternary mixture of nonanoic acid:decanoic acid:undecanoic acid, at molar ratio 2:1:1, showed the highest extraction efficiencies due to its hydrophobic behavior that favored the interaction with the target analytes. Likewise, the changes in the sample pH using NH_4OH solution as emulsifier agent formed a milky solution that significantly improved the phase separation, and, at the same time, avoided the use of the toxic THF. Under the optimized conditions, 600 μL of DES, 400 μL of NH_4OH, 30 s of vortex agitation, HCl as demulsification agent, and centrifugation at 4000 rpm for 5 min, recovery values between 90% and 103% and LODs in the range 2.0-50 μg/L were achieved. These results proved the feasibility of the green method developed to be applied to the analysis of watermelon, grape, tomato, and guava juices.

Other Natural Compounds

In addition to the extraction of phenolic compounds, as well as terpenes and terpenoids, DESs have also been explored as extraction materials of other functional natural compounds. Table (**8**) [97 - 101] compiles some examples of these applications in the last five years, including the extraction of caffeine [97], folic acid [98], lignin [99], hydroxyapatite [100], and lignans from coffee [97], wheat flour [98], vegetable wastes [99], fish wastes [100] and sesame oils [101], by UAE/microwave-assisted extraction (MAE) [99] and SLE [100] techniques, as well as microextraction techniques involving DLLME [97] and its modes, VA-DLLME [98] and UA-LLME [101], mainly followed by LC coupled to conventional detectors [97, 98, 101]. In this way, DESs formed by ChCl and alcohols [97, 99 - 101], as well as pseudo-HDESs prepared from TOMAC and isoamyl alcohol [98] have been investigated as extraction materials of biomolecules with practical applications.

Table 8. Some examples of recent applications of DESs for the extraction of other natural compounds from food matrices.

Sample (amount)	Analytes	DES (molar ratio; amount)	Extraction Procedure	Analytical Technique	Recovery %	LODs	Comments	References
Turkish coffee (3 g)	Caffeine	ChCl:phenol (1:3; 400 μL)	DLLME	HPLC-UV	91-100	0.12 μg/mL	- Other molar ratios were also investigated, including 1:1, 1:2, and 1:4. - THF (800 μL) was used as emulsifier. - Vortex agitation and pH adjustment were not needed.	[97]
Wheat flour (5 g)	Folic acid (vitamin B9)	TOMAC:iso-amyl alcohol (1:4; 150 μL)	VA-DLLME	HPLC-UV	92-100	1.0 ng/g	- SLE was previously carried out. - BTEAC-based DESs with thymol and octanoic acid as HBDs, at a molar ratio 1:2, were also investigated. - Other molar ratios were investigated for TOMAC:isoamyl alcohol DES, including 1:1, 1:2, 1:3, 1:4, and 1:5. - Methanol (250 μL) was used as disperser solvent. - The method was applied to the analysis of four wheat flour samples.	[98]

(Table 8) cont.....

Sample (amount)	Analytes	DES (molar ratio; amount)	Extraction Procedure	Analytical Technique	Recovery %	LODs	Comments	References
Garlic skin and green onion root (-)	Lignin	ChCl:glycerol: $AlCl_3 \cdot 6H_2O$ (1:2:0.2; -)	UAE/MAE	Gravimetric analysis	90-92	-	- Other TDESs formed by ChCl as HBA, urea, and oxalic acid as HBDs, and different metal chlorides ($FeCl_3$ and $CrCl_3$) were also investigated. - The binary DESs were also investigated at different molar ratios, including 1:1 and 2:1. - Other heating methods were evaluated. - A sample to solvent ratio of 1:10 wt was used. - Enzymatic hydrolysis was carried out. - Hemicellulose and cellulose contents were also determined.	[99]
Fish scales (0.3 g)	Hydroxyapatite	ChCl:glycerol (1:2; 4.5 g)	SLE	FTIR, XRD, EDS, ICP-OES, SEM, TG and particle size distribution analysis	48	-	- Citric and acetic acids were also investigated as HBDs of ChCl-based DES at molar ratios 2:1 and 1:2, respectively. - An orthogonal test was carried out for the optimization step. - The extraction temperature was set at 70 ºC. - A purification step with NaOH was carried out after extraction. - A hemolysis test was carried out.	[100]

(Table 8) cont.....

Sample (amount)	Analytes	DES (molar ratio; amount)	Extraction Procedure	Analytical Technique	Recovery %	LODs	Comments	References
Sesame oil (0.2 g)	3 lignans	ChCl:*p*-cresol (1:2; 400 μL)	UA-LLME	HPLC-UV	97-120	0.3-0.5 mg/kg	- Other HBDs were also investigated, including phenol, *o*-cresol, *m*-cresol, catechol, and 4-methoxyphenol. - A comparison with polyol-based DESs was carried out. - The extraction temperature was set at 50 ºC. - A comparison with conventional LLE method was performed. - The method was applied to the analysis of sesame and blend oil samples.	[101]

BTEAC: benzyltriethylammonium chloride; ChCl: choline chloride; DES: deep eutectic solvent; DLLME: dispersive liquid-liquid microextraction; EDS: energy-dispersive X-ray spectrometry; FTIR: Fourier transform infrared spectroscopy; HBA: hydrogen bond acceptor; HBD: hydrogen bond donor; HPLC: high-performance liquid chromatography; ICP: inductively coupled plasma; LLE: liquid-liquid extraction; LLME: liquid-liquid microextraction; LOD: limit of detection; MAE: microwave-assisted extraction; OES: optical emission spectroscopy; SEM: scanning electron microscopy; SLE: solid-liquid extraction; TDES: ternary deep eutectic solvent; TG: thermogravimetric; THF: tetrahydrofuran; TOMAC: trioctylmethylammonium chloride; UA: ultrasound-assisted; UAE: ultrasound-assisted extraction; UV: ultraviolet; VA: vortex-assisted; XRD: X-ray diffraction.

A representative example is the extraction of alkaloids from food matrices using DESs. In this regard, Sivrikaya [97] developed a method for the extraction of caffeine from coffee by DLLME based on ChCl:phenol DES as extraction solvent, before HPLC-UV determination. In this study, the effect of the solvent molar ratio on the extraction performance was evaluated varying the concentration of the HBD (1:1; 1:2; 1:3 and 1:4), finding that the extraction efficiency increased with the molar ratio up to 1:3. In this way, the keto-enol tautomerism of caffeine led to establishing π-π interactions with the prepared DES. After that, the optimum volumes of the ChCl:phenol DES and THF (emulsifier solvent) were found at 400 μL and 800 μL, respectively. Under optimal conditions, the method achieved recovery values in the range 91-100% and a LOD of 0.12 μg/mL, which proved fit for its purpose. Therefore, it was applied to the analysis of different commercial Turkish coffee samples. It is noteworthy that the use of THF in this work was necessary to promote phase separation since ChCl:phenol DES decomposed in aqueous sample solutions, being the organic phase mainly constituted by THF, phenol, and the analyte [102].

Apart from organic species, DESs have also been applied as extraction solvents of natural inorganic species. As an example, Liu *et al.* [100] developed a method for the extraction of hydroxyapatite from fish-scale waste by DESs, composed of ChCl and different HBDs, including glycerol, citric acid, and acetic acid, at molar ratios 1:2, 2:1 and 1:2. In this regard, the high viscosity of ChCl:citric acid DES did not allow the dissolution of collagen of fish scales matrix and consequently the DES did not extract the analyte, whereas ChCl:acetic acid DES probably decomposed the analyte since after the extraction process no precipitate of hydroxyapatite was obtained. However, ChCl:glycerol DES showed properties, such as good stability, moderate viscosity, high collagen solubility, and no decomposition process, that were suitable for the extraction of hydroxyapatite. Techniques such as Fourier transform infrared spectroscopy (FTIR) analysis and thermogravimetric (TG) analysis confirmed these results. Once the most suitable DES was selected, the optimization step gave rise to the following conditions: extraction temperature of 65 ºC, extraction time of 2.5 h, and sample to solvent ratio of 1/15 (g/g) (0.3 g of fish scales and 4.5 g of ChCl:glycerol DES). That resulted in an extraction rate of 66%, which after a purification step of the obtained inorganic solutes corresponded to a 48% of hydroxyapatite. The authors also employed other characterization techniques in order to confirm the elimination of impurities from hydroxyapatite. Likewise, the hemolysis test showed that both the hydroxyapatite extract and pure ChCl:glycerol DES had good biocompatibility for biological applications. For all of this, the method proved to efficiently and respectfully with environment recover hydroxyapatite from fish wastes.

Proteins, Carbohydrates, Amino Acids, Lipids

Apart from the mentioned biomolecules, amino acids, proteins, and carbohydrates, have also been extracted by using DESs in the last years, as it can be seen in (Table **9**), in which some examples of these types of studies have been compiled [103 - 107]. The nature of the evaluated matrices is varied, including vegetables and fruit-based by-products [103, 106], plants [104], eggs [107], or milk [105], among others. However, the procedures applied generally involved conventional extractions, whereas the application of miniaturized techniques has been rarely applied. In this sense, Balaraman and Rathnasamy [107] developed a methodology based on a UA-LLME at 35 ºC, combined with preparative HPLC for the extraction and purification of immunoglobulins from quail eggs before separation and analysis by size exclusion chromatography (SEC)-UV. With this aim, five specifically designed DESs were evaluated (*i.e.* benzyltri butylammonium bromide (BTBAB):glycerol, TBAB:glycerol, benzyltrimethyl ammonium chloride (BTMAC):glycerol and benzyltributylammonium chloride (BTBAC):glycerol at a ratio 1:2 and ChCl:glycerol (1:1)). The best results were obtained when 500 µL of BTBAC:glycerol at a ratio of 1:2 was applied, obtaining a recovery of 85% and a LOD of 8 g/L. Authors attributed this result to the higher density of the selected DES, which enhanced its hydrophobicity favoring phases separation, although other parameters such as ionic strength also played a key role in the process.

Regarding the type of DESs commonly applied for amino acids, proteins, and carbohydrates extraction, the use of quaternary ammonium salts-based DESs was preferred in most cases. Among them, ChCl has been the most applied in combination with short-chain fatty acids to form pseudo-HDESs for the extraction of protein from bamboo by-products [103] or combined with EtGly or acetic acid to obtain hydrophilic NaDESs [104, 106]. However, organic acids have also been used as HBAs or HBDs to prepare other hydrophilic NaDESs. As an example, López *et al.* [105] evaluated the extraction capacity of different mixtures constituted by lactic and citric acid with glucose or fructose for the UAE of 3 seleno amino acids from powdered and lyophilized milk prior to their determination by HPLC-ICP-MS. All NaDES were diluted with water 18% (v/v). Results indicated that lactic acid:glucose at a molar ratio of 5:1 provided the highest extraction efficiency, even higher than using SDS-Tris, commonly used for the extraction of amino acids. After optimization, recovery values in the range 90-109% and LODs of 37-9.64 µg/kg were obtained.

Table 9. Some examples of recent applications of DESs for the extraction of proteins and carbohydrates from food matrices.

Sample (amount)	Analytes	DES (molar ratio; amount)	Extraction Procedure	Analytical Technique	Recovery %	LODs	Comments	References
Tip, basal and sheath bamboo shoots (-)	Proteins	ChCl:levulinic acid (1:6; 40% water; -)	SLE	Spectrophotometry	39[(a)]	-	- By-products valorization. - Higher or similar extraction yields than using conventional solvents. - SLE (50 min, 80 ºC). - Sample to solvent ratio (30-70 mg/mL)	[103]

(Table 9) cont.....

Sample (amount)	Analytes	DES (molar ratio; amount)	Extraction Procedure	Analytical Technique	Recovery %	LODs	Comments	References
Camellia oleifera Abel. seed cake (0.1 g)	Polysaccharides	ChCl:EtGly (1:2; 3 mL)	UAE	Spectrophotometry	85	-	- Re-extraction in EOPO copolymer. - Seventeen DES were tested. - 30% water was added into DESs. - Antioxidant activity was also evaluated. - UAE (50 min, 50 ºC, pH 5).	[104]
Powdered and lyophilized milk (0.94 g)	3 Seleno amino acids	Lactic acid:glucose (5:1; 3.09 mL)	UAE	HPLC-ICP-MS	90-109	7.37-9.64 µg/kg	- Citric acid: glucose (1:1) and fructose:citric acid (1:1) were also tested. - 25% (v/v) water was added into DESs. - UAE (34 min).	[105]
Pomegranate peel (150 mg)	Proteins and other bioactive compounds	ChCl:acetic acid:water (1:1.10; 5 mL)	UAE	HPLC-(Q-ToF)-MS	-	-	- Comparison with PLE using ethanol:water (70:30 v/v) was carried out.	[106]
Quail egg (-)	Immunoglobulins	BTBAC:glycerol (1:2; 500µl)	UA-LLME-preparative HPLC	SEC-UV	85	8 g/L	- BTBAB, TBAB, BTMAC and ChCl were tested as HBAs. - Sample to solvent ratio (85% v/v). - UA-LLME (35ºC, 12 min).	[107]

(a) yield g of analytes /g of matrix; BTBAB: benzyltributylammonium bromide; BTBAC: benzyltributylammonium chloride; BTMAC: benzyltrimethylammonium chloride; ChCl: choline chloride; DES: deep eutectic solvent; EOPO: ethylene oxide-propylene oxide; EtGly: ethylene glycol; HBA: hydrogen bond acceptor; HBD: hydrogen bond donor; HPLC: high-performance liquid chromatography; ICP: inductively coupled plasma; LLME: liquid-liquid microextraction; LOD: limit of detection; MS: mass spectrometry; PLE: pressurized liquid extraction; Q-ToF: quadrupole-time-of-flight; SEC: size exclusion chromatography; SLE: solid-liquid extraction; TBAB: tetrabutylammonium bromide; UA: ultrasound-assisted; UAE: ultrasound-assisted extraction; UV: ultraviolet.

In the same vein, other authors have also compared the suitability of DESs with the one presented by other solvents usually applied for the extraction of these kinds of biomolecules obtaining similar results, which demonstrates the great potential of these new green materials in the field. An interesting example is the work carried out by Lin *et al.* [103]. In this case, the authors applied a low cost, environmentally sustainable, non-toxic NaDES constituted by ChCl:levulinic acid at a molar ratio 1:6 and diluted with water, 40% (v/v), for the extraction of proteins from bamboo shoots and its processing wastes by SLE at 80ºC during 50 min. Results were compared with a conventional SLE using 0.2% (w/w) NaOH solution at 40 ºC for 10 min. The extraction yield achieved with DESs was higher (39.16 g/g dw) than the one obtained with NaOH solution (23.88 g/g dw). Although the energy and time consumption requirements are higher when NaDES was used, the authors remarked that less toxicity and safer handling in relatively high concentrations of the prepared solvent make the procedure more sustainable in this case. Additionally, they also highlighted the possibility of reusing the remaining solid residues after NaDES extraction of protein in a biorefinery approach for valorization as a source of dietary fiber and other compounds, such as xylan, xylooligosaccharide, xylitol, and sodium carboxymethyl cellulose that could be further extracted as potential functional food ingredients, an aspect that enhances the green character of this approach.

Finally, it should be pointed out that, although there is still a wide field of study in the application of DESs for the extraction of biomolecules that need to be explored, these solvents have been demonstrated to be a promising alternative to the use of conventional toxic solvents. In this sense, DESs not only allow the development of more sustainable methodologies by using greener components, but also the application of NaDESs with non-toxic effects allows the simplification of the procedures and the possibility of directly using the extracts obtained in products intended for human or animal consumption. This aspect constituted an important advance for the food industry and the development of natural products.

CONCLUDING REMARKS

The remarkable extraction ability and the versatility of DESs have propitiated their intensive use in food analysis. This capacity depends on inherent factors such as the individual constituents of these solvents, their concentration and water content, and also on extrinsic aspects such as the extraction method employed. As expected, DESs and extraction methods were selected considering the analytes and the physical state of the food samples (liquid, solid or viscous). The polarity of the DESs plays a predominant role. This way, HDESs were predominantly employed for the extraction by LLE or LPME of both organic and inorganic compounds from aqueous matrices. Meanwhile, hydrophilic DESs were mainly used for the extraction of analytes from non-aqueous samples or water-based samples but using an additional step for the separation of the phases. Regarding the extraction protocols, SLE (assisted by microwave, ultrasound, heating, *etc.*) was employed for solid samples whereas liquid samples were extracted through LLE and LPME (also assisted by microwave, ultrasound, vortex, heating, and gases or surfactants), although in some cases the combination of both types of approaches was carried out. The need for this assistance can be easily explained by the high viscosity, which hinders the intimate contact between the sample and DESs and makes difficult mass transference. Due to the intrinsic low volatility of DESs, the mentioned extraction techniques were frequently coupled to LC or spectrometry protocols, although techniques such as GC have also been employed. Although DESs are mainly used in the mL or g scale, their inherent scalability is fundamental for future industrial agri-food applications.

A wide diversity of organic and inorganic contaminants (pesticides, plastic migrants, pharmaceuticals, metals, *etc.*) and naturally occurring and bioactive compounds (phenolic compounds, terpenes and terpenoids, proteins, carbohydrates, amino acids, lipids, and other biomolecules) have been extracted from food by these solvents. However, its industrial application is still limited and requires further research. Elucidation of the extraction mechanism is fundamental to facilitate the widespread use of DESs in food analysis, particularly for the recovery of bioactive compounds. Also, other aspects such as toxicity, stability, and operational cost should be evaluated with this objective.

Although DESs have a limited contaminant power, the possible solvent regeneration or recycling (which also decreases the cost) can increase the greenness of this kind of mixture. Due to the low vapor pressure, their separation from analytes by vaporization is problematic, if not impossible. Thus, alternative procedures (such as SLE, LLE, ultrafiltration, or back-extraction) already proposed in biomass fractionation [108] will be indispensable in industrial agri-food applications. An interesting option for these recycling methods would be the use of the extract (without recovering the target compounds) when NaDESs are employed, given their low toxicity. This kind of DESs is particularly interesting for the heavily regulated agri-food industry.

It should be mentioned that, despite the advantages and the good results provided by DESs, they are not massively commercialized and maybe this is the reason why these solvents have not been routinely implemented.

LIST OF ABBREVIATIONS

µ-dSPE	micro-dispersive solid-phase extraction
4-TOP	4-tert-octylphenol
ACN	acetonitrile

AIBN	2,2-azobisisobutyronitrile
APTES	(3-aminopropyl)triethoxysilane
AzB	azure B (3-(dimethylamino)-7-(methylamino)phenothiazin-5-ium chloride)
BPA	bisphenol A
BPAF	bisphenol AF
BTBAB	benzyltributylammonium bromide
BTBAC	benzyltributylammonium chloride
BTEAC	benzyltriethylammonium chloride
BTMAC	benzyltrimethylammonium chloride
ChCl	choline chloride
COSMO-RS	COnductor-like Screening MOdel for Realistic Solvents
COSMO-SAC	COnductor-like Screening MOdel for Segment Activity Coefficient
DAB	3,3′-diaminobenzidine
DAD	diode array detector
DES	deep eutectic solvent
DLLME	dispersive liquid-liquid microextraction
dSPE	dispersive solid-phase extraction
EAE	enzyme-assisted extraction
EDS	energy-dispersive X-ray spectrometry
EGDMA	ethylene glycol dimethacrylate
EOPO	ethylene oxide-propylene oxide
ETAAS	electrothermal atomic absorption spectrometry
EtGly	ethylene glycol
FAAS	flame atomic absorption spectrometry
FD	fluorescence detector
FID	flame ionization detector
FTIR	Fourier transform infrared spectroscopy
GC	gas chromatography
GCB	graphitized carbon black
HBA	hydrogen bond acceptor
HBD	hydrogen bond donor
HDES	hydrophobic deep eutectic solvent
HG-AAS	hydride generation atomic absorption spectrometry
HMBATSC	2-hydroxy-3-methoxybenzaldehyde thiosemicarbazone
HPLC	high performance liquid chromatography

HS-SDME	headspace single-drop microextraction
ICP	inductively coupled plasma
iHg	inorganic mercury
IL	ionic liquid
IT	ion trap
LC	liquid chromatography
LLE	liquid-liquid extraction
LLME	liquid-liquid microextraction
LOD	limit of detection
LOQ	limit of quantification
LTTM	low transition temperature mixture
LPME	liquid-phase microextraction
m-µ-dSPE	magnetic-micro-dispersive solid phase extraction
m-AA	magnetic-air-assisted
MAE	microwave-assisted extraction
ME	membrane emulsification
MeHg	methylmercury
MEKC	micellar electrokinetic capillary chromatography
MIP	molecularly imprinted polymer
m-MWCNT	magnetic-multi-walled carbon nanotube
MS/MS	tandem mass spectrometry
MS	mass spectrometry
MSPD	matrix solid-phase dispersion
NaDES	natural deep eutectic solvent
NIP	non-imprinted polymer
NPD	nitrogen phosphorous detector
OCP	organochloride pesticide
OES	optical emission spectroscopy
OPP	organophosphorous pesticide
PAE	phthalic acid ester
PAH	polycyclic aromatic hydrocarbon
PAN	1-(2-Pyridylazo)-2-naphthol
PChCl	phosphocholine chloride
PLE	pressurized liquid extraction
Poly-DES	polymeric-deep eutectic solvent

PSA	primary secondary amine
PVDF-co-PTFE	poly(vinylidene fluoride-co-tetrafluoroethylene)
Q	single quadrupole
QqQ	triple quadrupole
Q-ToF	quadrupole-time-of-flight
Q-Trap	triple quadrupole-linear ion trap
QuEChERS	Quick, Easy, Cheap, Effective, Rugged and Safe
SBSE	stir bar sorptive extraction
SDME	single-drop microextraction
SEC	size exclusion chromatography
SEM	scanning electron microscopy
SFOD	solidified floating organic droplet
SLE	solid-liquid extraction
SLME	solid-liquid microextraction
SPE	solid-phase extraction
TAC	total anthocyanin content
TBAB	tetrabutylammonium bromide
TBAC	tetrabutylammonium chloride
TBBPA	tetrabromobisphenol A
TBTDPB	tributyltetradecylphosphonium bromide
TDES	ternary deep eutectic solvent
TEOS	tetraethoxysilane
TFC	total flavonoid content
TG	thermogravimetric
TheDES	therapeutic deep eutectic solvent
THF	tetrahydrofuran
TOAB	tetraoctylammonium bromide
TOAC	trioctylammonium chloride
ToF	time-of-flight
TOMAB	trioctylmethylammonium bromide
TOMAC	trioctylmethylammonium chloride
TPC	total phenolic content
UA	ultrasound-assisted
UAE	ultrasound-assisted extraction
UHPLC	ultra-high performance liquid chromatography

UV	ultraviolet
VA	vortex-assisted
VWD	variable wavelength detector
XRD	X-ray diffraction

CONSENT FOR PUBLICATION

Not applicable.

CONFLICT OF INTEREST

The authors declare no conflict of interest, financial or otherwise.

ACKNOWLEDGEMENTS

This work has been supported by the Spanish Ministry of Economy and Competitiveness (project AGL2017-89257-P). R.R.-R. and A.S.-M. would like to thank the Agency of Economy, Industry, Trade and Knowledge of the Government of the Canary Islands for the FPI fellowships (co-financed with an 85% from the European Social Funds).

REFERENCES

[1] Pena-Pereira F, Kloskowski A, Namieśnik J. Perspectives on the replacement of harmful organic solvents in analytical methodologies: A framework toward the implementation of a generation of eco-friendly alternatives. Green Chem 2015; 17(7): 3687-705. [http://dx.doi.org/10.1039/C5GC00611B]

[2] Rubio S. Twenty years of supramolecular solvents in sample preparation for chromatography: Achievements and challenges ahead. Anal Bioanal Chem 2020; 412(24): 6037-58. [http://dx.doi.org/10.1007/s00216-020-02559-y] [PMID: 32206847]

[3] Durand E, Lecomte J, Villeneuve P. From green chemistry to nature: The versatile role of low transition temperature mixtures. Biochimie 2016; 120: 119-23. [http://dx.doi.org/10.1016/j.biochi.2015.09.019] [PMID: 26391220]

[4] Santana-Mayor Á, Rodríguez-Ramos R, Herrera-Herrera AV, Socas-Rodríguez B, Rodríguez-Delgado MÁ. Deep eutectic solvents. the new generation of green solvents in analytical chemistry. TrAC Trends Anal Chem 2020; 34: 116108.

[5] Smith EL, Abbott AP, Ryder KS. Deep eutectic solvents (DESs) and their applications. Chem Rev 2014; 114(21): 11060-82. [http://dx.doi.org/10.1021/cr300162p] [PMID: 25300631]

[6] Makoś P, Słupek E, Gębicki J. Hydrophobic deep eutectic solvents in microextraction techniques–A review. Microchem J 2020; 152: 104384. [http://dx.doi.org/10.1016/j.microc.2019.104384]

[7] Florindo C, Branco LC, Marrucho IM. Quest for green-solvent design: from hydrophilic to hydrophobic (Deep) eutectic solvents. ChemSusChem 2019; 12(8): 1549-59. [http://dx.doi.org/10.1002/cssc.201900147] [PMID: 30811105]

[8] Rahman MS, Roy R, Jadhav B, Hossain MN, Halim MA, Raynie DE. Formulation, structure, and

applications of therapeutic and amino acid-based deep eutectic solvents: An overview. J Mol Liq 2021; 321: 114745.
[http://dx.doi.org/10.1016/j.molliq.2020.114745]

[9] Paiva A, Craveiro R, Aroso I, Martins M, Reis RL, Duarte ARC. Natural deep eutectic solvents – solvents for the 21st century. ACS Sustain Chem& Eng 2014; 2(5): 1063-71.
[http://dx.doi.org/10.1021/sc500096j]

[10] Hou XD, Liu QP, Smith TJ, Li N, Zong MH. Evaluation of toxicity and biodegradability of cholinium amino acids ionic liquids. PLoS One 2013; 8(3): e59145.
[http://dx.doi.org/10.1371/journal.pone.0059145] [PMID: 23554985]

[11] Hou Y, Yao C, Wu W. Deep eutectic solvents: Green solvents for separation applications. Wuli Huaxue Xuebao 2018; 34(8): 873-85.
[http://dx.doi.org/10.3866/PKU.WHXB201802062]

[12] Zhang Q, De Oliveira VK, Royer S, Jérôme F. Deep eutectic solvents: Syntheses, properties and applications. Chem Soc Rev 2012; 41(21): 7108-46.
[http://dx.doi.org/10.1039/c2cs35178a] [PMID: 22806597]

[13] Kovács A, Neyts EC, Cornet I, Wijnants M, Billen P. Modeling the physicochemical properties of natural deep eutectic solvents. ChemSusChem 2020; 13(15): 3789-804.
[http://dx.doi.org/10.1002/cssc.202000286] [PMID: 32378359]

[14] Liu Y, Friesen JB, McAlpine JB, Lankin DC, Chen SN, Pauli GF. Natural deep eutectic solvents: Properties, applications, and perspectives. J Nat Prod 2018; 81(3): 679-90.
[http://dx.doi.org/10.1021/acs.jnatprod.7b00945] [PMID: 29513526]

[15] Abbott AP, Capper G, Davies DL, Rasheed RK, Tambyrajah V. Novel solvent properties of choline chloride/urea mixtures. Chem Commun 2003; (1): 70-1.
[http://dx.doi.org/10.1039/b210714g] [PMID: 12610970]

[16] Zainal-Abidin MH, Hayyan M, Hayyan A, Jayakumar NS. New horizons in the extraction of bioactive compounds using deep eutectic solvents: A review. Anal Chim Acta 2017; 979: 1-23.
[http://dx.doi.org/10.1016/j.aca.2017.05.012] [PMID: 28599704]

[17] Durand E, Lecomte J, Villeneuve P. Deep eutectic solvents: Synthesis, application, and focus on lipase-catalyzed reactions. Eur J Lipid Sci Technol 2013; 115(4): 379-85.
[http://dx.doi.org/10.1002/ejlt.201200416]

[18] Shishov A, Pochivalov A, Nugbienyo L, Andruch V, Bulatov A. Deep eutectic solvents are not only effective extractants. Trends Analyt Chem 2020; 129: 115956.
[http://dx.doi.org/10.1016/j.trac.2020.115956]

[19] Dai Y, Witkamp GJ, Verpoorte R, Choi YH. Tailoring properties of natural deep eutectic solvents with water to facilitate their applications. Food Chem 2015; 187: 14-9.
[http://dx.doi.org/10.1016/j.foodchem.2015.03.123] [PMID: 25976992]

[20] Pandey A, Rai R, Pal M, Pandey S. How polar are choline chloride-based deep eutectic solvents? Phys Chem Chem Phys 2014; 16(4): 1559-68.
[http://dx.doi.org/10.1039/C3CP53456A] [PMID: 24305780]

[21] Pandey A, Pandey S. Solvatochromic probe behavior within choline chloride-based deep eutectic solvents: Effect of temperature and water. J Phys Chem B 2014; 118(50): 14652-61.
[http://dx.doi.org/10.1021/jp510420h] [PMID: 25418894]

[22] Li W, Zhang Z, Han B, *et al.* Switching the basicity of ionic liquids by CO2. Green Chem 2008; 10(11): 1142-5.
[http://dx.doi.org/10.1039/b811624e]

[23] Kareem MA, Mjalli FS, Hashim MA, AlNashef IM. Phosphonium-based ionic liquids analogues and their physical properties. J Chem Eng Data 2010; 55(11): 4632-7.
[http://dx.doi.org/10.1021/je100104v]

[24] Hayyan M, Hashim MA, Hayyan A, *et al.* Are deep eutectic solvents benign or toxic? Chemosphere 2013; 90(7): 2193-5.
[http://dx.doi.org/10.1016/j.chemosphere.2012.11.004] [PMID: 23200570]

[25] Hayyan M, Hashim MA, Al-Saadi MA, Hayyan A, AlNashef IM, Mirghani MES. Assessment of cytotoxicity and toxicity for phosphonium-based deep eutectic solvents. Chemosphere 2013; 93(2): 455-9.
[http://dx.doi.org/10.1016/j.chemosphere.2013.05.013] [PMID: 23820537]

[26] Dai Y, Witkamp GJ, Verpoorte R, Choi YH. Natural deep eutectic solvents as a new extraction media for phenolic metabolites in Carthamus tinctorius L. Anal Chem 2013; 85(13): 6272-8.
[http://dx.doi.org/10.1021/ac400432p] [PMID: 23710664]

[27] Fernández MÁ, Boiteux J, Espino M, Gomez FJV, Silva MF. Natural deep eutectic solvents-mediated extractions: The way forward for sustainable analytical developments. Anal Chim Acta 2018; 1038: 1-10.
[http://dx.doi.org/10.1016/j.aca.2018.07.059] [PMID: 30278889]

[28] Gill I, Vulfson E. Enzymic catalysis in heterogeneous eutectic mixtures of substrates. Trends Biotechnol 1994; 12(4): 118-22.
[http://dx.doi.org/10.1016/0167-7799(94)90088-4] [PMID: 7764807]

[29] López-Fandiño R, Gill I, Vulfson EN. Protease-catalyzed synthesis of oligopeptides in heterogenous substrate mixtures. Biotechnol Bioeng 1994; 43(11): 1024-30.
[http://dx.doi.org/10.1002/bit.260431105] [PMID: 18615513]

[30] Stott P, Williams AC, Barry BW. Transdermal delivery from eutectic systems: Enhanced permeation of a model drug, ibuprofen. J Control Release 1998; 50(1-3): 297-308.
[http://dx.doi.org/10.1016/S0168-3659(97)00153-3] [PMID: 9685897]

[31] Chen J, Li Y, Wang X, Liu W. Application of deep eutectic solvents in food analysis: A review. Molecules 2019; 24(24): 4594.
[http://dx.doi.org/10.3390/molecules24244594] [PMID: 31888138]

[32] Lu W, Liu S, Wu Z. Recent application of deep eutectic solvents as green solvent in dispersive liquid–liquid microextraction of trace level chemical contaminants in food and water. Crit Rev Anal Chem 2020; 0(0): 1-15.
[PMID: 32845172]

[33] Socas-Rodríguez B, Herrera-Herrera AV, Asensio-Ramos M, Rodríguez-Delgado MÁ. Recent applications of magnetic nanoparticles in food analysis. Processes 2020; 8(9): 1140.
[http://dx.doi.org/10.3390/pr8091140]

[34] Zhao J, Meng Z, Zhao Z, Zhao L. Ultrasound-assisted deep eutectic solvent as green and efficient media combined with functionalized magnetic multi-walled carbon nanotubes as solid-phase extraction to determine pesticide residues in food products. Food Chem 2020; 310(310): 125863.
[http://dx.doi.org/10.1016/j.foodchem.2019.125863] [PMID: 31835214]

[35] Mokhtari N, Torbati M, Farajzadeh MA, Afshar Mogaddam MR. Synthesis and characterization of phosphocholine chloride-based three-component deep eutectic solvent: Application in dispersive liquid–liquid microextraction for determination of organothiophosphate pesticides. J Sci Food Agric 2020; 100(6): 2364-71.
[http://dx.doi.org/10.1002/jsfa.10203] [PMID: 31853973]

[36] Qian H, Yang Q, Qu Y, Ju Z, Zhou W, Gao H. Hydrophobic deep eutectic solvents based membrane emulsification-assisted liquid-phase microextraction method for determination of pyrethroids in tea beverages. J Chromatogr A 2020; 1623: 461204.
[http://dx.doi.org/10.1016/j.chroma.2020.461204] [PMID: 32505292]

[37] Zahiri E, Khandaghi J, Farajzadeh MA, Afshar Mogaddam MR. Combination of dispersive solid phase extraction with solidification organic drop–dispersive liquid–liquid microextraction based on deep

eutectic solvent for extraction of organophosphorous pesticides from edible oil samples. J Chromatogr A 2020; 1627: 461390.
[http://dx.doi.org/10.1016/j.chroma.2020.461390] [PMID: 32823096]

[38] Mogaddam MRA, Farajzadeh MA, Mohebbi A, Nemati M. Stir bar sorptive extraction combined with deep eutectic solvent-based dispersive liquid–liquid microextraction: Application in simultaneous derivatisation and extraction of acidic pesticides. Int J Environ Anal Chem 2020; 00(00): 1-13.

[39] Khanehzar H, Faraji M, Nezhadali A, Yamini Y. Combining of modified QuEChERS and dispersive liquid–liquid microextraction as an efficient sample preparation method for extraction of acetamiprid and imidacloprid from pistachio samples. J Indian Chem Soc 2021; 18(3): 641-9.
[http://dx.doi.org/10.1007/s13738-020-02050-6]

[40] Jouyban A, Farajzadeh MA, Afshar Mogaddam MR. In matrix formation of deep eutectic solvent used in liquid phase extraction coupled with solidification of organic droplets dispersive liquid-liquid microextraction; application in determination of some pesticides in milk samples. Talanta 2020; 206(206): 120169.
[http://dx.doi.org/10.1016/j.talanta.2019.120169] [PMID: 31514834]

[41] Torbati M, Farajzadeh MA, Mogaddam MRA, Torbati M. Deep eutectic solvent based homogeneous liquid–liquid extraction coupled with in-syringe dispersive liquid–liquid microextraction performed in narrow tube; application in extraction and preconcentration of some herbicides from tea. J Sep Sci 2019; 42(9): 1768-76.
[http://dx.doi.org/10.1002/jssc.201801016] [PMID: 30809926]

[42] Farajzadeh MA, Shahedi HA, Afshar MMR. Development of a new temperature-controlled liquid phase microextraction using deep eutectic solvent for extraction and preconcentration of diazinon, metalaxyl, bromopropylate, oxadiazon, and fenazaquin pesticides from fruit juice and vegetable samples fo. J Food Compos Anal 2017; 2018(66): 90-7.

[43] Jouyban A, Farajzadeh MA, Afshar MMR. A lighter-than-water deep eutectic-solvent-based dispersive liquid-phase microextraction method in a u-shaped homemade device. New J Chem 2018; 42(12): 10100-10.
[http://dx.doi.org/10.1039/C8NJ00597D]

[44] Farajzadeh MA, Abbaspour M, Kazemian R, Afshar MMR. Preparation of a new three-component deep eutectic solvent and its use as an extraction solvent in dispersive liquid–liquid microextraction of pesticides in green tea and herbal distillates. J Sci Food Agric 2020; 100(5): 1904-12.
[http://dx.doi.org/10.1002/jsfa.10200] [PMID: 31825526]

[45] Li X, Wang M, Zhao J, *et al.* Ultrasound-assisted emulsification liquid phase microextraction method based on deep eutectic solvent as extraction solvent for determination of five pesticides in traditional Chinese medicine. J Pharm Biomed Anal 2019; 166: 213-21.
[http://dx.doi.org/10.1016/j.jpba.2019.01.018] [PMID: 30660036]

[46] Shishov A, Terno P, Moskvin L, Bulatov A. In-syringe dispersive liquid-liquid microextraction using deep eutectic solvent as disperser: Determination of chromium (VI) in beverages. Talanta 2020; 206(206): 120209.
[http://dx.doi.org/10.1016/j.talanta.2019.120209] [PMID: 31514865]

[47] Shi Y, Li X, Shang Y, Li T, Zhang K, Fan J. Effective extraction of fluorescent brightener 52 from foods by *in situ* formation of hydrophobic deep eutectic solvent. Food Chem 2020; 311: 125870.
[http://dx.doi.org/10.1016/j.foodchem.2019.125870] [PMID: 31740270]

[48] Santana-Mayor Á, Socas-Rodríguez B, Rodríguez-Ramos R, Rodríguez-Delgado MÁ. A green and simple procedure based on deep eutectic solvents for the extraction of phthalates from beverages. Food Chem 2020; 312: 125798.
[http://dx.doi.org/10.1016/j.foodchem.2019.125798] [PMID: 31887621]

[49] Naebi M, Jamshidi MA, Farajzadeh MA, Abolhassni J, Mogaddam MRA. In-process prepared deep eutectic solvent based homogeneous liquid-liquid microextraction for the determination of irgaphos

168 and irganox 1010 in polypropylene packed drinks. J Sep Sci 2020; 43(14): 2850-7. [http://dx.doi.org/10.1002/jssc.202000148] [PMID: 32363747]

[50] Wen J, Lu Y, Shi L, Yang Y. A novel cloud point extraction based on fatty acid deep eutectic solvent combined with high-performance liquid chromatography for determination of ultraviolet absorbent in food packaging bags. Microchem J 2020; 153: 104466. [http://dx.doi.org/10.1016/j.microc.2019.104466]

[51] Yang D, Wang Y, Li H, Yang Y. Acid-base-governed deep eutectic solvent-based microextraction combined with magnetic solid-phase extraction for determination of phenolic compounds. Mikrochim Acta 2020; 187(2): 124. [http://dx.doi.org/10.1007/s00604-020-4109-y] [PMID: 31932975]

[52] Pang K, Hou Y, Wu W, Guo W, Peng W, Marsh KN. Efficient separation of phenols from oils *via* forming deep eutectic solvents. Green Chem 2012; 14(9): 2398-401. [http://dx.doi.org/10.1039/c2gc35400d]

[53] Qin L, Li J, Cheng H, Chen L, Qi Z, Yuan W. Association extraction for vitamin E recovery from deodorizer distillate by *in situ* formation of deep eutectic solvent. AIChE J 2017; 63(6): 2212-20. [http://dx.doi.org/10.1002/aic.15606]

[54] Shishov AY, Chislov MV, Nechaeva DV, Moskvin LN, Bulatov AV. A new approach for microextraction of non-steroidal anti-inflammatory drugs from human urine samples based on in-situ deep eutectic mixture formation. J Mol Liq 2018; 272: 738-45. [http://dx.doi.org/10.1016/j.molliq.2018.10.006]

[55] Yu K, Yue ME, Xu J, Jiang TF. Determination of fluoroquinolones in milk, honey and water samples by salting out-assisted dispersive liquid-liquid microextraction based on deep eutectic solvent combined with MECC. Food Chem 2020; 332: 127371. [http://dx.doi.org/10.1016/j.foodchem.2020.127371] [PMID: 32622188]

[56] Ji Y, Meng Z, Zhao J, Zhao H, Zhao L. Eco-friendly ultrasonic assisted liquid–liquid microextraction method based on hydrophobic deep eutectic solvent for the determination of sulfonamides in fruit juices. J Chromatogr A 2020; 1609: 460520. [http://dx.doi.org/10.1016/j.chroma.2019.460520] [PMID: 31519407]

[57] Shishov A, Gorbunov A, Baranovskii E, Bulatov A. Microextraction of sulfonamides from chicken meat samples in three-component deep eutectic solvent. Microchem J 2020; 158: 105274. [http://dx.doi.org/10.1016/j.microc.2020.105274]

[58] Liu C, Liu D, Liu X, *et al.* Deep eutectic solvent-based liquid phase microextraction for the determination of pharmaceuticals and personal care products in fish oil. New J Chem 2017; 41(24): 15105-9. [http://dx.doi.org/10.1039/C7NJ03350H]

[59] Li X, Row KH. Purification of antibiotics from the millet extract using hybrid molecularly imprinted polymers based on deep eutectic solvents. RSC Advances 2017; 7(28): 16997-7004. [http://dx.doi.org/10.1039/C7RA01059A]

[60] Tang W, Gao F, Duan Y, Zhu T, Ho Row K. Exploration of deep eutectic solvent-based molecularly imprinted polymers as solid-phase extraction sorbents for screening chloramphenicol in milk. J Chromatogr Sci 2017; 55(6): 654-61. [http://dx.doi.org/10.1093/chromsci/bmx011] [PMID: 28334778]

[61] Alipanahpour Dil E, Ghaedi M, Asfaram A, Tayebi L, Mehrabi F. A ferrofluidic hydrophobic deep eutectic solvent for the extraction of doxycycline from urine, blood plasma and milk samples prior to its determination by high-performance liquid chromatography-ultraviolet. J Chromatogr A 2020; 1613: 460695. [http://dx.doi.org/10.1016/j.chroma.2019.460695] [PMID: 31740033]

[62] Shishov A, Gagarionova S, Bulatov A. Deep eutectic mixture membrane-based microextraction: HPLC-FLD determination of phenols in smoked food samples. Food Chem 2020; 314: 126097.

[http://dx.doi.org/10.1016/j.foodchem.2019.126097] [PMID: 31945551]

[63] Fazaieli F, Afshar Mogaddam MR, Farajzadeh MA, Feriduni B, Mohebbi A. Development of organic solvents-free mode of solidification of floating organic droplet–based dispersive liquid–liquid microextraction for the extraction of polycyclic aromatic hydrocarbons from honey samples before their determination by gas chromatography–mass spectrometry. J Sep Sci 2020; 43(12): 2393-400. [http://dx.doi.org/10.1002/jssc.202000136] [PMID: 32145048]

[64] Sivrikaya OS, Yılmaz Y. Ultrasound-assisted hydrophobic deep eutectic solvent based solid-liquid microextraction of Sudan dyes in spice samples. Spectrochim Acta A Mol Biomol Spectrosc 2020; 236: 118353. [http://dx.doi.org/10.1016/j.saa.2020.118353] [PMID: 32320916]

[65] Wu X, Zhang X, Yang Y, Liu Y, Chen X. Development of a deep eutectic solvent-based matrix solid phase dispersion methodology for the determination of aflatoxins in crops. Food Chem 2019; 291: 239-44. [http://dx.doi.org/10.1016/j.foodchem.2019.04.030] [PMID: 31006465]

[66] Faraji M. Determination of some red dyes in food samples using a hydrophobic deep eutectic solvent-based vortex assisted dispersive liquid-liquid microextraction coupled with high performance liquid chromatography. J Chromatogr A 2019; 1591: 15-23. [http://dx.doi.org/10.1016/j.chroma.2019.01.022] [PMID: 30651206]

[67] Altunay N, Elik A, Gürkan R, Novel A. A novel, green and safe ultrasound-assisted emulsification liquid phase microextraction based on alcohol-based deep eutectic solvent for determination of patulin in fruit juices by spectrophotometry. J Food Compos Anal 2019; 82: 103256. [http://dx.doi.org/10.1016/j.jfca.2019.103256]

[68] Fan C, Cao X, Han T, *et al.* Selective microextraction of polycyclic aromatic hydrocarbons using a hydrophobic deep eutectic solvent composed with an iron oxide-based nanoferrofluid. Mikrochim Acta 2019; 186(8): 560. [http://dx.doi.org/10.1007/s00604-019-3651-y] [PMID: 31338683]

[69] Zhu S, Zhou J, Jia H, Zhang H. Liquid–liquid microextraction of synthetic pigments in beverages using a hydrophobic deep eutectic solvent. Food Chem 2018; 243: 351-6. [http://dx.doi.org/10.1016/j.foodchem.2017.09.141] [PMID: 29146348]

[70] Ghorbani Ravandi M, Fat'hi MR. Green effervescence assisted dispersive liquid–liquid microextraction based on a hydrophobic deep eutectic solvent for determination of Sunset Yellow and Brilliant Blue FCF in food samples. New J Chem 2018; 42(18): 14901-8. [http://dx.doi.org/10.1039/C8NJ00782A]

[71] Ma W, Tang W, Row KH. Characterization of deep eutectic solvents for dispersive liquid–liquid microextraction for phenolics. Anal Lett 2017; 50(14): 2177-88. [http://dx.doi.org/10.1080/00032719.2016.1277533]

[72] Torabi A, Shirani M, Semnani A, Akbari A. Deep eutectic solvent-based ligandless ultrasound-assisted liquid-phase microextraction for extraction of cobalt ions from food samples prior to spectrophotometric determination. J Iran Chem Soc 2020; 4: 893-902.

[73] Sorouraddin SM, Farajzadeh MA, Dastoori H. Development of a dispersive liquid-liquid microextraction method based on a ternary deep eutectic solvent as chelating agent and extraction solvent for preconcentration of heavy metals from milk samples. Talanta 2020; 208: 120485. [http://dx.doi.org/10.1016/j.talanta.2019.120485] [PMID: 31816735]

[74] Sorouraddin SM, Farajzadeh MA, Okhravi T. Development of dispersive liquid-liquid microextraction based on deep eutectic solvent using as complexing agent and extraction solvent: Application for extraction of heavy metals. Sep Sci Technol 2020; 55(16): 2955-66. [http://dx.doi.org/10.1080/01496395.2019.1666874]

[75] Abdolhosseini M, Shemirani F, Yousefi SM. Poly (deep eutectic solvents) as a new class of sustainable sorbents for solid phase extraction: Application for preconcentration of Pb (II) from food

and water samples. Mikrochim Acta 2020; 187(11): 602.
[http://dx.doi.org/10.1007/s00604-020-04564-5] [PMID: 33034749]

[76] Fasihi M, Rajabi M, Barfi B, Sajjadi SM. Deep eutectic-based vortex-assisted/ultrasound-assisted liquid-phase microextractions of chromium species. J Indian Chem Soc 2020; 17(7): 1705-13.
[http://dx.doi.org/10.1007/s13738-020-01890-6]

[77] Altunay N, Elik A, Gürkan R. Natural deep eutectic solvent-based ultrasound-assisted-microextraction for extraction, pre-concentration and analysis of methylmercury and total mercury in fish and environmental waters by spectrophotometry. Food Addit Contam Part A Chem Anal Control Expo Risk Assess 2019; 36(7): 1079-97.
[http://dx.doi.org/10.1080/19440049.2019.1619939] [PMID: 31140933]

[78] Shirani M, Habibollahi S, Akbari A. Centrifuge-less deep eutectic solvent based magnetic nanofluid-linked air-agitated liquid–liquid microextraction coupled with electrothermal atomic absorption spectrometry for simultaneous determination of cadmium, lead, copper, and arsenic in food samples and non-alcoholic beverages. Food Chem 2019; 281: 304-11.
[http://dx.doi.org/10.1016/j.foodchem.2018.12.110] [PMID: 30658761]

[79] Altunay N, Elik A, Gürkan R. Innovative and practical deep eutectic solvent based vortex assisted microextraction procedure for separation and preconcentration of low levels of arsenic and antimony from sample matrix prior to analysis by hydride generation-atomic absorption spectrometry. Food Chem 2019; 293: 378-86.
[http://dx.doi.org/10.1016/j.foodchem.2019.05.019] [PMID: 31151625]

[80] Zounr RA, Tuzen M, Khuhawar MY. Novel ultrasonic-assisted deep eutectic solvent-based dispersive liquid–liquid microextraction for determination of vanadium in food samples by electrothermal atomic absorption spectrometry: A multivariate study. Appl Organomet Chem 2018; 32(3): e4144.
[http://dx.doi.org/10.1002/aoc.4144]

[81] Panhwar AH, Tuzen M, Kazi TG. Ultrasonic assisted dispersive liquid-liquid microextraction method based on deep eutectic solvent for speciation, preconcentration and determination of selenium species (IV) and (VI) in water and food samples. Talanta 2017; 175: 352-8.
[http://dx.doi.org/10.1016/j.talanta.2017.07.063] [PMID: 28842002]

[82] Rodríguez-Juan E, Rodríguez-Romero C, Fernández-Bolaños J, Florido MC, Garcia-Borrego A. Phenolic compounds from virgin olive oil obtained by natural deep eutectic solvent (NADES): Effect of the extraction and recovery conditions. J Food Sci Technol 2021; 58(2): 552–61.
[PMID: 33568848]

[83] Saha SK, Chakraborty R. Effect of deep eutectic solvent's characteristics on extraction and bioactivity of polyphenols from sapodilla pulp. Chem Pap 2020; 75: 691-702.

[84] Silva DT, Pauletto R, Cavalheiro SS, *et al.* Natural deep eutectic solvents as a biocompatible tool for the extraction of blueberry anthocyanins. J Food Compos Anal 2020; 89: 103470.
[http://dx.doi.org/10.1016/j.jfca.2020.103470]

[85] Wang X, Jia W, Lai G, Wang L, Mar Contreras M, Yang D. Extraction for profiling free and bound phenolic compounds in tea seed oil by deep eutectic solvents. J Food Sci 2020; 85(5): 1450-61.
[http://dx.doi.org/10.1111/1750-3841.15019] [PMID: 32249418]

[86] Saha SK, Dey S, Chakraborty R. Effect of choline chloride-oxalic acid based deep eutectic solvent on the ultrasonic assisted extraction of polyphenols from Aegle marmelos. J Mol Liq 2019; 287: 110956.
[http://dx.doi.org/10.1016/j.molliq.2019.110956]

[87] Gullón B, Muñiz-Mouro A, Lú-Chau TA, Moreira MT, Lema JM, Eibes G. Green approaches for the extraction of antioxidants from eucalyptus leaves. Ind Crops Prod 2019; 138: 111473.
[http://dx.doi.org/10.1016/j.indcrop.2019.111473]

[88] Panić M, Radić SM, Kraljić K, *et al.* Ready-to-use green polyphenolic extracts from food by-products. Food Chem 2019; 283: 628-36.
[http://dx.doi.org/10.1016/j.foodchem.2019.01.061] [PMID: 30722921]

[89] Fernández MÁ, Espino M, Gomez FJV, Silva MF. Novel approaches mediated by tailor-made green solvents for the extraction of phenolic compounds from agro-food industrial by-products. Food Chem 2018; 239: 671-8.
[http://dx.doi.org/10.1016/j.foodchem.2017.06.150] [PMID: 28873620]

[90] Bajkacz S, Adamek J. Development of a method based on natural deep eutectic solvents for extraction of flavonoids from food samples. Food Anal Methods 2018; 11(5): 1330-44.
[http://dx.doi.org/10.1007/s12161-017-1118-5]

[91] García A, Rodríguez-Juan E, Rodríguez-Gutiérrez G, Rios JJ, Fernández-Bolaños J. Extraction of phenolic compounds from virgin olive oil by deep eutectic solvents (DESs). Food Chem 2016; 197(Pt A): 554-61.
[http://dx.doi.org/10.1016/j.foodchem.2015.10.131] [PMID: 26616988]

[92] Li J, Wang J, Wu M, Cheng H, Chen L, Qi Z. Deep deterpenation of citrus essential oils intensified by *in situ* formation of a deep eutectic solvent in associative extraction. Ind Eng Chem Res 2020; 59(19): 9223-32.
[http://dx.doi.org/10.1021/acs.iecr.0c00442]

[93] Triaux Z, Petitjean H, Marchioni E, Boltoeva M, Marcic C. Deep eutectic solvent–based headspace single-drop microextraction for the quantification of terpenes in spices. Anal Bioanal Chem 2020; 412(4): 933-48.
[http://dx.doi.org/10.1007/s00216-019-02317-9] [PMID: 31927601]

[94] Li H, Zhao C, Tian H, Yang Y, Li W. Liquid–liquid microextraction based on acid–base-induced deep eutectic solvents for determination of β-carotene and lycopene in fruit juices. Food Anal Methods 2019; 12(12): 2777-84.
[http://dx.doi.org/10.1007/s12161-019-01639-9]

[95] Silva YPA, Ferreira TAPC, Jiao G, Brooks MS. Sustainable approach for lycopene extraction from tomato processing by-product using hydrophobic eutectic solvents. J Food Sci Technol 2019; 56(3): 1649-54.
[http://dx.doi.org/10.1007/s13197-019-03618-8] [PMID: 30956346]

[96] Ozturk B, Esteban J, Gonzalez-Miquel M. Deterpenation of citrus essential oils using glycerol-based deep eutectic solvents. J Chem Eng Data 2018; 63(7): 2384-93.
[http://dx.doi.org/10.1021/acs.jced.7b00944]

[97] Sivrikaya S. A deep eutectic solvent based liquid phase microextraction for the determination of caffeine in Turkish coffee samples by HPLC-UV. Food Addit Contam Part A Chem Anal Control Expo Risk Assess 2020; 37(3): 488-95.
[http://dx.doi.org/10.1080/19440049.2020.1711972] [PMID: 31917645]

[98] Faraji M, Mahmoodi-Maymand M, Dastmalchi F. Green, fast and simple dispersive liquid-liquid microextraction method by using hydrophobic deep eutectic solvent for analysis of folic acid in fortified flour samples before liquid chromatography determination. Food Chem 2020; 320: 126486.
[http://dx.doi.org/10.1016/j.foodchem.2020.126486] [PMID: 32208186]

[99] Ji Q, Yu X, Yagoub AEGA, Chen L, Zhou C. Efficient removal of lignin from vegetable wastes by ultrasonic and microwave-assisted treatment with ternary deep eutectic solvent. Ind Crops Prod 2020; 149: 112357. [Ji, S].
[http://dx.doi.org/10.1016/j.indcrop.2020.112357]

[100] Liu Y, Li J, Wang D, *et al.* Enhanced extraction of hydroxyapatite from bighead carp (*Aristichthys nobilis*) scales using deep eutectic solvent. J Food Sci 2020; 85(1): 150-6.
[http://dx.doi.org/10.1111/1750-3841.14971] [PMID: 31877234]

[101] Liu W, Zhang K, Yang G, Yu J. A highly efficient microextraction technique based on deep eutectic solvent formed by choline chloride and p-cresol for simultaneous determination of lignans in sesame oils. Food Chem 2019; 281: 140-6.
[http://dx.doi.org/10.1016/j.foodchem.2018.12.088] [PMID: 30658740]

[102] Shishov A, Gorbunov A, Moskvin L, Bulatov A. Decomposition of deep eutectic solvents based on choline chloride and phenol in aqueous phase. J Mol Liq 2020; 301: 112380. [http://dx.doi.org/10.1016/j.molliq.2019.112380]

[103] Lin Z, Jiao G, Zhang J, Celli GB, Brooks MSL. Optimization of protein extraction from bamboo shoots and processing wastes using deep eutectic solvents in a biorefinery approach. Biomass Convers. Biorefinery 2020; 11: 2763-74.

[104] Gao C, Cai C, Liu J, *et al.* Extraction and preliminary purification of polysaccharides from camellia oleifera abel. seed cake using a thermoseparating aqueous two-phase system based on eopo copolymer and deep eutectic solvents. Food Chem 2020; 313(313): 126164. [http://dx.doi.org/10.1016/j.foodchem.2020.126164] [PMID: 31935662]

[105] López R, D'Amato R, Trabalza-Marinucci M, *et al.* Green and simple extraction of free seleno-amino acids from powdered and lyophilized milk samples with natural deep eutectic solvents. Food Chem 2020; 326(326): 126965. [http://dx.doi.org/10.1016/j.foodchem.2020.126965] [PMID: 32413755]

[106] Hernández-Corroto E, Plaza M, Marina ML, García MC. Sustainable extraction of proteins and bioactive substances from pomegranate peel (*Punica granatum L.*) using pressurized liquids and deep eutectic solvents. Innov Food Sci Emerg Technol 2020; 60(February): 102314. [http://dx.doi.org/10.1016/j.ifset.2020.102314]

[107] Balaraman HB, Rathnasamy SK. High selective purification of IgY from quail egg: Process design and quantification of deep eutectic solvent based ultrasound assisted liquid phase microextraction coupled with preparative chromatography. Int J Biol Macromol 2020; 146: 253-62. [http://dx.doi.org/10.1016/j.ijbiomac.2019.12.242] [PMID: 31917210]

[108] Ruesgas-Ramón M, Figueroa-Espinoza MC, Durand E. Application of deep eutectic solvents (DES) for phenolic compounds extraction: Overview, challenges, and opportunities. J Agric Food Chem 2017; 65(18): 3591-601. [http://dx.doi.org/10.1021/acs.jafc.7b01054] [PMID: 28414232]

CHAPTER 5

Ionic Liquids

Alfonso Jiménez[1], **Carlos Javier Pelegrín**[1] and **María Carmen Garrigós**[1,*]

[1] *Department of Analytical Chemistry, Nutrition & Food Sciences, University of Alicante, Campus San Vicente, ES-03690, San Vicente del Raspeig (Alicante), Spain*

Abstract: The significant potential of ionic liquids (ILs) in the extraction and separation of valuable products from food samples is deeply discussed in this chapter, where the main studies on the application of ionic liquids to food analysis are presented. The novel extraction strategies reviewed in this chapter have the potential to significantly enhance the extraction yield, in particular when the combination of ionic liquids with accelerated and green extraction techniques, such as microwave-assisted extraction (MAE), ultrasound-assisted extraction (UAE) or subcritical water extraction (SBWE) are used. ILs are considered environmentally-friendly solvents and they offer some advantageous properties which are particularly relevant in extraction systems in food matrices, such as their low toxicity and volatility and different polarity, hydrophobicity and selectivity. A particular section is devoted to microextraction techniques with ionic liquids, which have shown great performance in the extraction of valuable compounds for a variety of food samples. This chapter summarizes and gives an overview of the latest developments and applications of ILs in the extraction of bioactive compounds from food.

Keywords: Bioactive compounds, Food analysis, Green extraction techniques, Ionic liquids, Microextraction, Sustainable solvents.

INTRODUCTION

One of the main reasons for the gradual implementation of green extraction techniques in analytical procedures is the increasing awareness of the use of high amounts of potentially toxic and non-sustainable organic solvents in traditional extraction methods. The replacement of solvents with emerging green alternatives, such as deep eutectic solvents, ionic liquids, and non-ionic surfactants, is in line with sustainable development strategies [1]. Ionic liquids (ILs) have been proposed as environmentally-friendly alternatives in their use as extraction solvents in food analysis. Their use in combination with advanced and sustainable extraction techniques, such as microwave-assisted extraction (MAE),

* **Corresponding Author María Carmen Garrigós:** Department of Analytical Chemistry, Nutrition & Food Sciences, University of Alicante, Campus San Vicente, ES-03690, San Vicente del Raspeig (Alicante), Spain; Tel: +34-965903529; Fax: +34-965903697; E-mail: mc.garrigos@ua.es

Merichel Plaza & Maria Luisa Marina (Eds.)

ultrasound-assisted extraction (UAE) and supercritical/subcritical extraction techniques is considered an adequate alternative to traditional solid-liquid methods. In particular, their use in food matrices can be considered a very promising possibility with multiple applications. ILs have been the subject of a number of interesting and recent reviews and they have raised great attention since their introduction during the 90s [2]. Most of these reviews deal with general issues in the use of ILs under different extraction conditions [3 - 5] and specific applications, including food analysis [6 - 9].

It is noteworthy that ILs can also be considered as one of the most fascinating "green" solvents for food analysis owing to their high thermal stability (higher than 300 °C in most of the cases) along with very low vapour pressure and the huge number of possibilities in their application, including their use in miniaturized systems [10]. ILs are organic salts with low melting temperature (below 100 °C), consisting of large organic cations and organic or inorganic anions. They exhibit unique characteristics, such as high ionic conductivity, low flammability and non-volatility among others, making them environmentally-friendly alternatives to traditional organic solvents commonly used in the extraction of food components. Additionally, their high viscosity and surface tension lead to the formation of stable phase boundaries, enhancing the easy extraction of analytes.

These outstanding properties make ionic liquids the ideal candidates as green solvents and hardeners in many extraction processes. In particular, halogen-free ILs are more advantageous for environmental reasons due to their low toxicity, being excellent solvents for food analysis. As an example, Husson *et al.* [11] indicated that acidic ionic liquids are good solvents for fast esterification of many food components, particularly lipids, without the need of any catalyst. Hence, the use of ILs as green solvents for the selective extraction of polyacids from food samples can modify the food structure without any additional catalyst in the esterification process [12]. The main aim of this chapter is to perform a general overview of the use of ionic liquids in food analysis, particularly in the extraction of valuable components coupled with advanced and sustainable extraction methods.

CHEMICAL STRUCTURE AND PROPERTIES OF IONIC LIQUIDS

Ionic liquids are liquid organic salts with low melting point, which are usually made up of a large organic cation and a smaller inorganic or organic anion [4, 6]. Those ILs with melting points at or below room temperature are defined as room temperature ionic liquids (RTILs) [13] and they are highly valuable since their use leads to significant energy savings during extraction. The origin of ILs dates back to 1914, when Walden [14] succeeded in the synthesis of ethylammonium nitrate ($[C_2NH_3][NO_3]$); with a melting point of 12 °C [6]. However, the interest in the use of ILs as alternative to conventional solvents has raised only in the last three decades and their research and development have become a milestone in different scientific applications, including their use as catalysts in organic synthesis and separation processes, as part of lithium-ion batteries and sensors, thermal fluids, biomass solvents, lubricant additives or as a CO_2 absorption medium [15].

Some of the most common organic cations that make up the structure of ILs are those based on imidazolium, pyridinium, pyrrolidinium, triazolium, thiazolium, piperidinium, sulfonium, phosphonium or ammonium. In the same way, anions can be both inorganic and organic, including magnetic anions. Inorganic anions include halides (such as Cl^-, Br^- and I^-), tetrafluoroborate (BF_4^-) and hexafluorophosphate (PF_6^-). Organic anions include acetate, bis(trifluoromethylsulfonyl)imide ($[NTf_2]^-$), trifluoro-methanesulfonate ($[OTf]^-$), trifluoroacetate, bis(pentafluoroethylsulfonyl) imide ($[BETI]^-$) and many others. In addition, the incorporation of paramagnetic components, including metal ions (Mn, Fe, Co and Ni) and rare earths (Gd and Dy), into the IL structure has raised some interest in recent years [16, 17]. Some examples include tetrachloroferrate, tetrachloromanganate^{2-} and tetrachlorocobaltate^{2-}. The structure and nomenclature of the most common cations and anions in ILs are shown in Fig. (**1**).

The large number of potential combinations of cations and anions in the formation of ILs results in their unique and characteristic properties, including their broad range of working temperatures, high polarities, variable viscosities, low charge densities, reasonable ionic conductivities and low electrical conductivities [18]. Their miscibility with water and organic solvents can be controlled by the different cation/anion combination, their ratio or by incorporating certain functional groups to the IL structure, resulting in tailor-made compounds for specific applications. In addition, their very low vapour pressure (associated with low combustibility and high thermal stability), as well as their non-explosive and minimally corrosive character, give ILs a clear environmentally-friendly nature [4]. However, not all ILs can be classified as fully sustainable, since some of them have been reported as toxic when incorporated into the environment [19 - 22].

A summary of those unique properties of ILs, which make them especially valuable as extraction media is shown in Fig. (**2**). As mentioned above, one of the most important characteristics of ILs is their negligible vapour pressure, making unnecessary the inclusion of distillation processes in the overall extraction methods. However, it has been observed that vapour pressure should not be considered negligible at very low pressures and temperatures above 200 or 300

°C, which is the case in ultrasound- and microwave-assisted extraction techniques [6, 16].

Cations	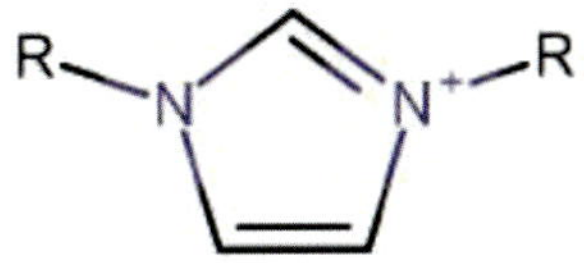 Imidazolium based	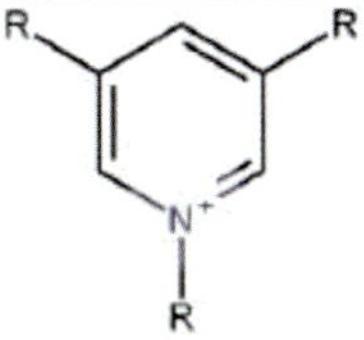 Pyridinium based	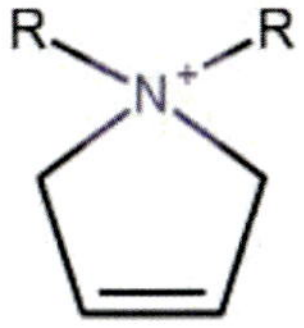 Pyrrolidinium based
	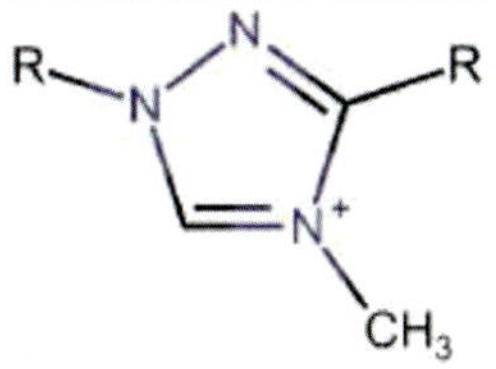 1,2,4-Triazolium based	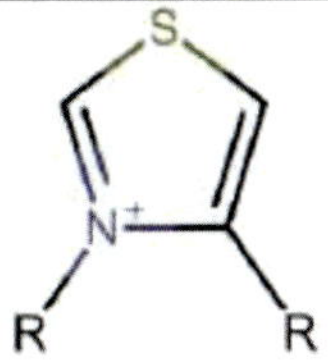 Thiazolium based	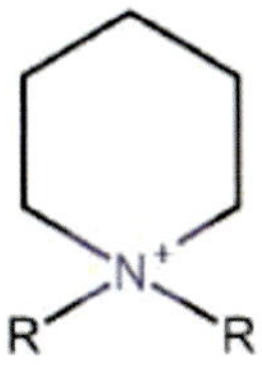 Piperidinium based
	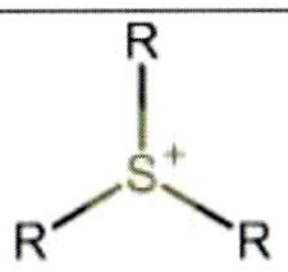 Sulfonium based	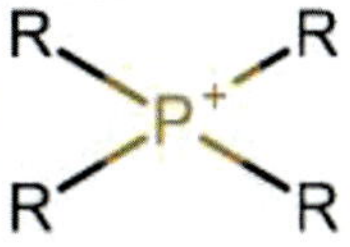 Phosphonium based	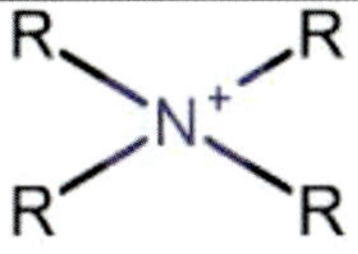 Ammonium based
Inorganic anions	X^- Cl^- Br^- I^- Halide	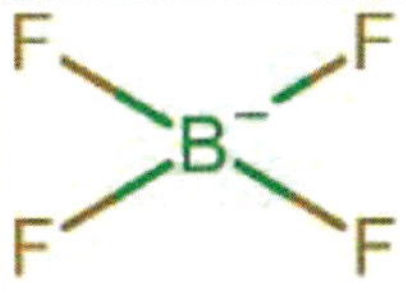 Tetrafluoroborate $[BF_4]^-$	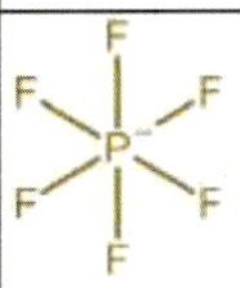 Hexafluorophosphate $[PF_6]^-$
Organic anions	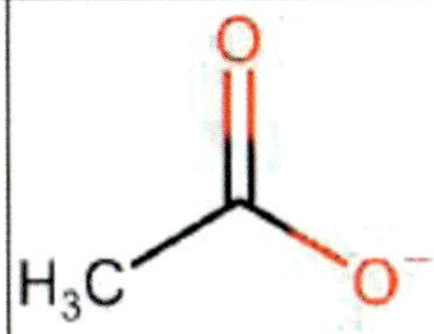 Acetate $[OAc]^-$	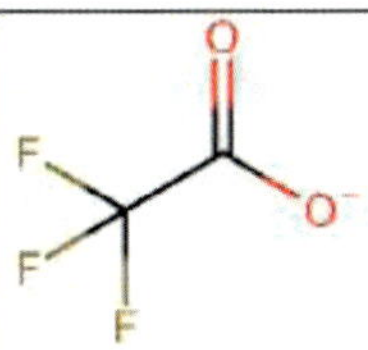 Trifluoroacetate	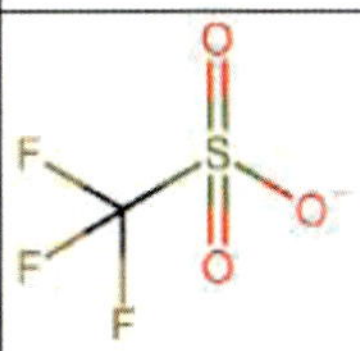 Trifluoro-methanesulfonate $[OTf]^-$

(Fig. 1) contd.....

	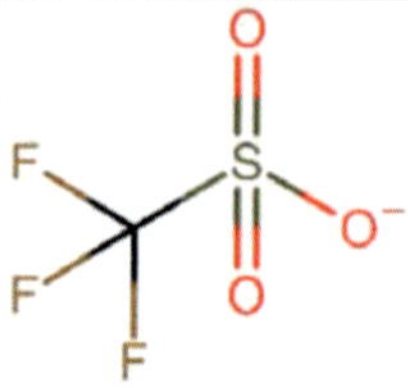 bis(trifluoromethylsulfonyl)imide $[NTf_2]^-$	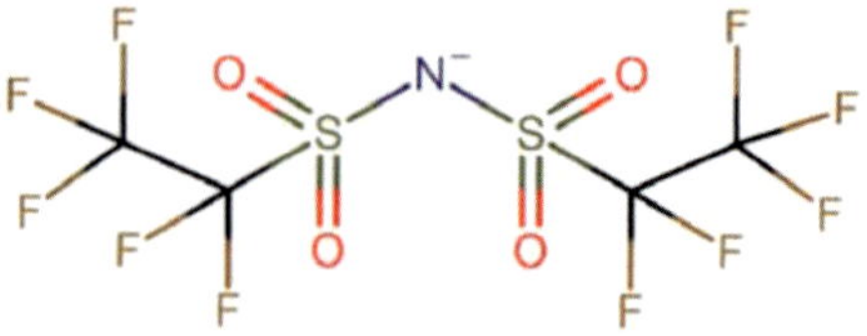 bis(pentafluoroethylsulfonyl)imide $[BETI]^-$	
Magnetic anions	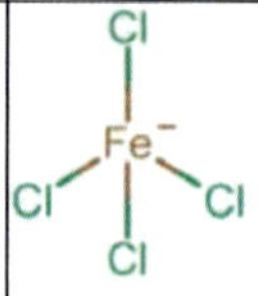 Tetrachloroferrate	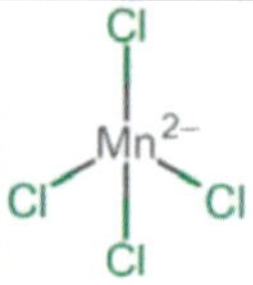 Tetrachloromanganate (2-)	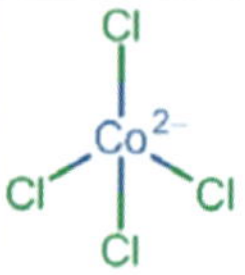 Tetrachlorocobaltate (2-)

Fig. (1). Structure and nomenclature of some of the most common cations and anions in ILs.

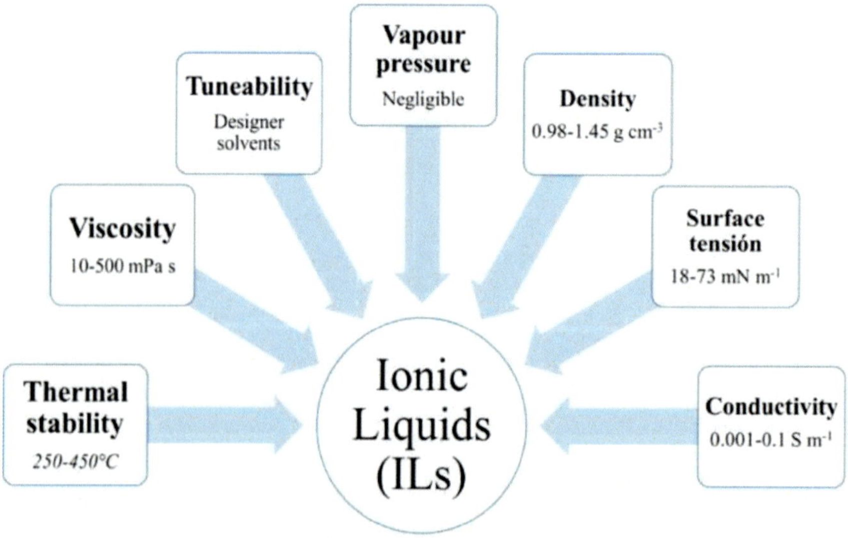

Fig. (2). Main physicochemical properties of ILs as extracting media (adapted from Reference [4]).

ILs are also able to allow solvation interactions with both polar and non-polar molecules, a property that can be finely tuned according to the IL structure, making them optimal solvents in different extraction techniques [4]. In general terms, when compared with conventional organic solvents, ILs are shifted towards polar compounds. It has been observed that the polarity of ILs decreases with the anion size and the length of the alkyl chain in the cation [6, 16, 23].

The thermal stability of ILs is also an important issue since it could limit the maximum operating temperature in some applications. ILs have been reported to decompose at relatively high temperatures (well above 200 °C) and in some cases at temperatures higher than 450 °C [6, 16]. This parameter is closely related to the structure of the anionic part of ILs. In general terms, ILs with organic anions are more stable at high temperatures than those with inorganic anions in their structure, resulting in the following sequence $[Cl]^- < [Br]^- < [I]^- < [BF_4]^- < [OTf]^- < [NTf_2]^- < [PF_6]^-$. The relatively low thermal stability of halide-based ILs is probably associated with their nucleophilic character [4, 18]. On the other hand, it has been reported that the length of the alkyl chain in the cation does not have a significant influence on the thermal stability of ILs [6, 16]. However, it has been reported that the decomposition temperatures of ILs can be ranked according to the cation as ammonium-based < imidazolium-based < phosphonium based [23].

RTILs are ILs with low melting temperatures and they are gaining some space in their use as suitable solvents by their easy manipulation in the liquid form. The melting point is a complex property that reflects the intermolecular forces and crystal lattice arrangement of ILs; depending largely on the cation asymmetric chemical structure that hinders crystallization and consequently reduces melting temperature [18]. This is particularly relevant in the case of imidazolium-based ILs, since the long and cross-linked substituent results in low melting points, regardless of the anion. Nevertheless, it is difficult to define a clear relationship between the anion structure and the IL-melting point. In general terms, ILs containing halides show higher melting points than those containing $[NTf_2]^-$ or $[PF_6]^-$ due to the hydrogen bonds formed in the IL structure. This effect is particularly significant in fluorine-based ILs [6].

One of the major drawbacks in the use of ILs is their viscosity, considerably higher than water and traditional organic solvents, which could affect the extraction process under some specific conditions. As an example, the viscosities of hexane, benzene and water at room temperature are 0.3, 0.6 and 0.9 mPa s, respectively, while common viscosities of ILs range from 10 to 500 mPa s [6, 16]. The viscosity of ILs is conditioned by several parameters, the most important being the increase with their concentration and the decrease at high temperatures and with the anion asymmetry. It is worth noting that the cations present in the IL structure are mainly responsible for the variation in viscosity due to their different nature and size, conditioning changes in the intermolecular interactions, including hydrogen bonds and Van der Waals forces [4]. The introduction of branched and/or long alkyl groups will result in some increase in the IL viscosity. In the case of the anion, ILs with large, non-symmetric anions, such as the bis(trifluoromethanesulfonyl) imide anion ($[TFSI]^-$), have lower viscosity than ILs with non-polar, symmetric anions [6, 16].

Surface tension is another important parameter to evaluate the performance of ILs as extraction solvents, giving an accurate estimation of the surface cohesion energy, which is closely related to the strength of the interactions between anions and cations in the IL structure. The surface tension values of ILs are slightly higher than those of traditional organic solvents, but not as high as in water [23]. The anion type and size show the greatest influence on the surface tension of ILs, decreasing with the anion size ($[BF_4]^- > [PF_6]^- > [OTf]^- > [TFSI]^-$). In turn, the length of the alkyl chain leads to the reduction in the IL surface tension as the result of the dilution of the ionic charge and the lower strength of the hydrogen bonds between anion and cation. The introduction of alkyl substituents on the cation leads to the increase in the ILs surface tension [6, 16, 18].

The miscibility of ILs with water is one of the main parameters conditioning their use in different extraction techniques [6, 16, 18]. ILs can be either hydrophobic or hydrophilic and their solubility depends on the anion size, with the hydrophobic nature increasing with the length of the alkyl chain attached to the cation: $[Br]^- \sim [Cl]^- < [BF_4]^- < [PF_6]^- < [NTf_2]^- < [BETI]^-$ [4]. The water solubility of ILs is also dependent on the temperature and polarity. It is well known that long alkyl chains increase the hydrophobicity of ILs [4, 6, 23].

Finally, the density of ILs ranges from 0.98-1.45 g cm^{-3}, slightly higher than that of water and conventional organic solvents. The correlation of the IL type with their density value is not a simple task, but it seems that it is mostly dependent on the anion $[OAc]^- < [Cl]^- < [BF_4]^- < [Br]^- < [PF_6]^- < [TFSI]^-$.

Polymeric Ionic Liquids

Polymeric ionic liquids (PILs) are a group of polyelectrolytes that can be synthesized by polymerization of ionic liquid monomers (ILMs) or by post-polymerization processes [24, 25]. These compounds have been known for more than fifty years, when Hoover [26] used them for the very first time. However, they are gaining interest in the scientific field, in particular by their broad range of applications in microextraction techniques, as it will be discussed later in this chapter. PILs can be produced from ILMs resulting in the combination of some of the valuable properties of ILs as solvents with the mechanical and thermal resistance of polymers [24], achieving unique properties, such as high thermal and chemical stability, durability and mechanical robustness, as well as a longer service life than ordinary ILs. In addition, PILs also exhibit processability and plasticity similar to those of traditional polymers [25]. PILs can be synthesized to achieve linear, helical, hyper-branched or star-shaped structures, such as common polymers, and they can also be synthesized by including other non-ionic monomers, creating random or block copolymers [24]. Considering these properties, PILs have emerged as promising materials for the synthesis of different sorbent surfaces used in microextraction techniques.

PILs can be divided into three different families according to their repeating units (cationic, anionic and zwitterionic PILs), as shown in Fig. (**3**). In cationic PILs, the cationic elements are covalently bound to the polymer matrix. The most common cationic PILs are formed on the basis of imidazolium, pyridinium, pyrrolidinium and ammonium. On the other hand, anionic groups, such as sulphonate, carboxylate and phosphate, are covalently bonded to the polymer backbones to obtain anionic PILs. Finally, in the case of zwitterionic PILs, both elements are covalently bonded to polymer chains [25].

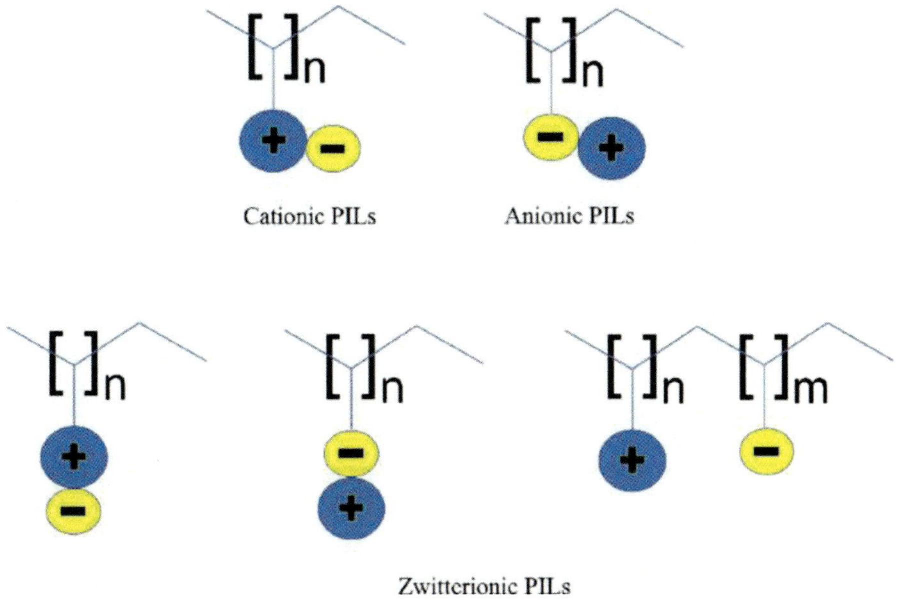

Fig. (3). Structure and types of PILs (adapted from Reference [25]).

PILs can be obtained by polymerization of ILMs. Examples of such techniques could be free radical polymerization (homo- or copolymerization), atom transfer polymerization, ring-opening metathesis polymerization, electropolymerization and chemically induced oxidative polymerization [24]. Free radical polymerization uses monomers prone to be chemically modified, resulting in PILs with similar properties and structure to the starting compounds, but higher thermal and mechanical resistance. Although they show the advantage of maintaining the original molecular weight, making them less viscous and easier to manipulate, they also show certain disadvantages such as reduced chemical tunability and difficulties in achieving high conversions due to possible steric hindrances. These free radical-synthesized PILs have been mainly used in electrochemistry or in membranes for gas separation. On the other hand, the atom transfer polymerization is based on the synthesis and polymerization of specific ILMs, as proposed in 2000 by Hirao *et al.* [27], who synthesized different N-vinyl imidazolium salts and performed their subsequent thermal-initiated free polymerization to be further used in applications requiring high thermal conductivity. It should be noted that, although the polymerizable group can be located either in the cation or in the anion, the former has been the most studied and used [24].

The use of PILs as solid-phase microextraction (SPME) coating phases has recently attracted considerable attention [28 - 36], since the pioneering work by Zhao *et al.* in 2008 [31]. The use of PILs in SPME will be discussed in more detail later in this chapter, but in brief, it is an excellent tool to easily manipulate significant properties of SPME materials by improving their sensitivity, charge capacity, selectivity, and stability. The use of long alkyl chains or hydroxyl groups in the ILM cation, as well as the use of halides as anions to improve the affinity of polar target compounds, are some examples of simple modifications. It is well known that PILs show high extraction efficiency for a wide range of compounds since they are rich in functional groups that can interact in multiple ways with the target analytes [24, 25].

Bio-Based Ionic Liquids

It is well known that one of the main disadvantages in the use of ILs is their potential toxicity and poor biodegradability in some cases, particularly in ILs with alkyl chains of imidazolium and benzimidazolium [37]. Therefore, some research has been recently carried out to get ILs obtained from renewable raw materials, referred to as bio-based ILs (B-ILs). Several biomolecules can be used as precursors for B-ILs: amino acids and amino alcohols; sugars from cellulose, chitin, starch and other polysaccharides; aromatic aldehydes from lignin and a diverse group of other compounds, such as fatty acids from algal or vegetable derived oils [38]. It has been reported that (2-hydroxyethyl)-ammonium lactate has the highest biodegradability levels (95%) and choline-based B-ILs show the lowest toxicity with high biodegradability [39]. Fig. (**4**) shows a graphical overview of the main sources of B-ILs.

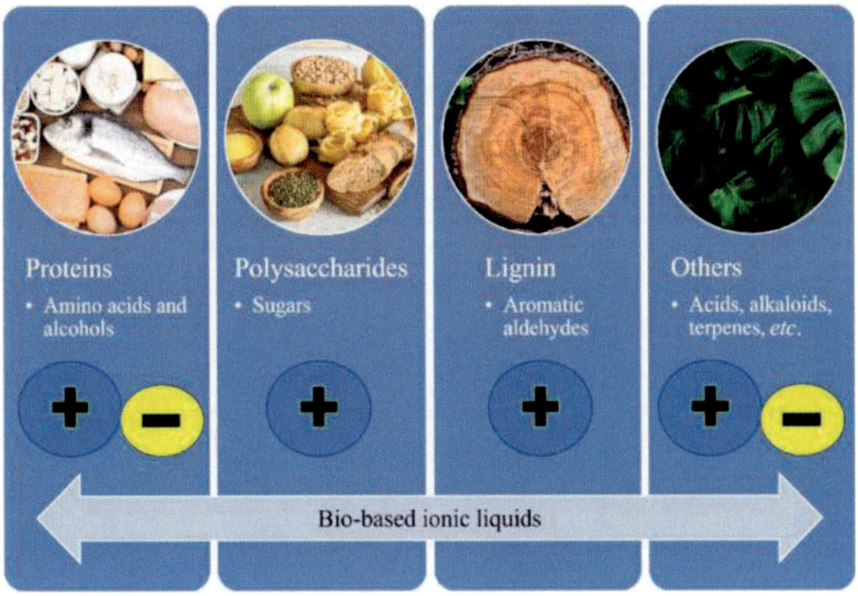

Fig. (4). Main synthesis pathways to B-ILs. For each class of renewable resource, typical derived monomers are depicted, divided into anions (-) or cations (+) (adapted from Reference [38]).

Most of the ongoing research on the synthesis of B-IL cations and anions is based on the use of protein-derived amino acids, obtained by hydrolysis followed by the amino acids cleavage. In general terms, the synthesis of cations involves a series of steps, whereas anions are usually obtained by simple acid-base neutralization reactions. For example, DeStefano *et al.* [40] developed four different B-ILs, consisting of tetramethyl guanidinium (TMG) as the cation with different amino acid anions (serine, threonine, aspartate and glutamate). Shaikh *et al.* [41], developed some B-ILs using two amino acid ILs (tetramethyl phosphonium glycinate, $[P_{1111}]$[Gly], and tetrabutyl phosphonium glycinate, $[P_{4444}]$[Gly]) with potential in atmospheric CO_2 capture. Choline-amino acid ionic liquids (ChoAA-ILs) have also been of great interest since they were introduced by Liu *et al* [42]. Their structure and properties have been discussed by Gontrani [43]. Tao *et al.* [44] also used amino acids and their derivatives as the most abundant natural source of quaternary nitrogen precursors to synthesize IL cations. Aslam *et al.* [45] selected three amino acid esters based ionic liquids (L-Phenyl Alanine methyl ester, L-Leucine methyl ester and L-Alanine methyl ester) such as cation and different anions to get materials with high resistance to corrosion. Another work by Ohno *et al.* [46] used 1-ethyl-3-methyl imidazole ($[Emim]^+$) as cation and twenty natural amino acids as anions to synthesize this type of B-ILs, while Kagimoto *et al.* [47] reported the combined use of tetrabutyl phosphonium cation ($[C_{16}H_{36}P]^+$) in such synthesis processes.

B-ILs based on sugars are currently under research. They are usually synthesized by depolymerization of polysaccharides, either glucose and derivatives obtained from cellulose or fructose extracted from plants. It should be noted that most of the outstanding research on sugar-based B-ILs has focused on the synthesis of cations. Chiappe *et al.* [48] described the synthesis and applications of B-ILs derived from natural sugars, like arabinose, $_D$-fructose, methyl α-$_D$-glucopyranoside, isosorbide, isomannide, 2,3,5-tri-O-benzyl $_D$-arabinose and $_D$-xylose. Glucose-linked 1,2,3-triazolium ionic liquids have been effectively synthesized by Jha *et al.* [49].

It is well known that lignin is abundantly available in nature, as it is one of the main components of woody plants. Some work has been reported on the development of lignin-based ILs. For example, Socha *et al.* [50] replaced imidazolium-based ILs with a series of tertiary amine-based B-ILs from aromatic aldehydes obtained from lignin and hemicelluloses.

Finally, it should be noted that other molecules of biological origin, in particular fatty acids, alkaloids and terpenes can be used to synthesize ILs. For example, six tartaric acid-based ILs containing different cation species were designed and reported by Zhang *et al.* [51]. Furthermore, Kwan *et al.* [52] reported the synthesis of different lipid-based B-ILs with the use of methyl oleate and methyl stearate in combination with imidazolium substituents. Different authors succeeded in the synthesis of new B-ILs from fatty acids used either as anions or cations [53 - 59]. On the other hand, Heckel *et al.* [60] described the synthesis of ILs based on (-)-nicotine *via* quaternization of the pyridine ring. Ammonium and imidazolium-based ILs have been also designed and synthesized using isosorbide as a bio-based and renewable substrate by Buu *et al.* [61] and Ferraz *et al.* [62] used ampicillin, an active pharmaceutical anionic ingredient, to prepare novel ILs with different organic cation hydroxides.

IONIC LIQUIDS COMBINED WITH GREEN EXTRACTION TECHNIQUES

The adequate combination of advanced and sustainable extraction methods with environmentally-friendly solvents, particularly with ionic liquids, has recently proved as successful in many different applications, including food analysis [6, 17, 63]. Whatever the extraction method used, the chemical nature of the extraction solvent is of primary importance to favour the solubility of target compounds, resulting in improvements in the extraction yield and selectivity, which is of outstanding relevance in the overall process for the target analytes obtained from natural sources, including food components.

One of the main features of ILs, compared to traditional organic solvents, is their ability to incorporate functional groups within their structure, which can selectively interact with specific analytes present in the food structure. This is particularly relevant in the case of vegetables, since many authors have reported the high ability of ILs to react with and finally dissolve cellulose and other cell-wall components, which is clearly improved when using intensive extraction techniques, such as MAE or UAE. ILs can produce the specific rupture of the lignin and hemicellulose network phases resulting in further loss of crystallinity of the cellulose portion of the cell walls in vegetables [64]. This property can be exploited in many applications in food analysis, for instance, for the pre-treatment of vegetables prior to hydrodistillation, at laboratory scale, to increase production of valuable essential oils [65]. Some examples of the combination of ILs with advanced and sustainable extraction techniques are discussed below.

IL-MAE

As it will be fully described in another chapter in this volume, MAE is increasing their uses and applications as a fast and reliable extraction technique. MAE provides a rapid heat transfer by the use of microwaves while the possibility to modify pressures inside the extraction containers results in high yields for many targeted compounds extracted from biological matrices, including food. Since MAE operating conditions are quite hard in some cases, ILs can be considered adequate solvents based on their high thermal and chemical resistance, which are important parameters that govern the suitability and long-term stability of the extraction process. In particular, highly polar ILs show a very fast and homogeneous heating profile under microwave irradiation, so their extracting ability is clearly improved and they are considered as excellent microwave absorbers [65]. The high thermal stability of ILs make them very useful for MAE in plants and vegetables, resulting in a rapid heating rate of the solvent-sample mixture, cell walls and/or membranes disruption and fast procedures, which combined with the negligible vapour pressure of ILs may increase the extraction rates.

The huge possibilities offered by ILs due to the tunability of their structure and, consequently, of their properties result in successful combinations with MAE that can be applied to a large number of extractions, including food analysis [66]. In particular this combination could be applied to both, metabolomics studies and the selective extraction of specific endogenous and exogenous compounds in plants and vegetables. In this context, ILs can be exploited for the simultaneous determination of volatile and non-volatile components [67, 68] with different approaches, which is particularly relevant in the case of extraction of target analytes from food samples. In the first case, an efficient system based on the use of ILs in microwave-assisted hydrodistillation was used in the pre-treatment step with further liquid-liquid extraction of purified fractions of essential oils and the simultaneous extraction of a set of non-volatile flavonoids from several vegetables [69]. The obtained extracts were collected in two separation columns, the first recovering the volatile components, while the second collecting the non-volatile compounds after the addition of an organic solvent.

The use of ionic liquid-based microwave-assisted extraction (IL-MAE) to obtain different analytes from vegetable food matrices has been widely reported. One example is the extraction of essential oils from dried fruits of *Illicium verum Hook. f.* and *Cuminum cyminum L* [70]. A more recent work [71] also reported the successful IL-MAE combination to extract verbascoside from *Rehmannia* root, which is widely used as functional food in East Asia. Authors used 10 different imidazolium-based ILs where different anions and cations were combined to evaluate their performance in the extraction of this analyte, taking into consideration some of the most important properties of ILs as extractive compounds, like viscosity, hydrophobic character and chemical structure.

The IL-MAE combination has been widely applied to extract active compounds from natural sources, providing high yields and extraction rates while the amounts consumed of solvent and energy are minimal. Different experimental conditions have been proposed for MAE with ILs, including microwave power ranging from 280 to 700 W, temperatures from 50 to 100 ºC irradiation times from 2 to 40 min, solid/liquid ratio from 1:7 to 1:40, while other operating conditions are considerably reduced when compared to conventional solid/liquid extraction. The increase in solvent volume could result in lower extraction efficiencies and potential sample agglomeration with difficulties to get high extraction yields [72].

Among all the experimental conditions to be optimized in MAE, power must be considered a key parameter. Some studies on the influence of the microwave power in the extraction yield when using ILs have been recently reported. For example, Chen *et al.* [73] successfully extracted trans-resveratrol from tree peony seed oil-extracted residues. Authors reported that, by changing the microwave irradiation power, the yield of the resveratrol extraction could be modified. They increased the power from 120 to 385 W with a clear improvement in yield (from 1.2 μmol/g to 2.9 μmol/g) with high purity level in both cases, not lower than 80% of trans-reservatrol in the extracts. This can be due to the increase in the analyte diffusion from the food matrix to the solvent, induced by the microwave irradiation energy [74]. However, values higher than 385 W resulted in a clear decrease in the yield due to the internal over-heating and further degradation of the product structure.

On the other hand, Panic *et al.* [75] developed a very innovative and selective method for the extraction of anthocyanins from grapes with ILs and natural deep eutectic solvents (NADES). These authors proposed the use of MAE, UAE and a combination of both extraction techniques in order to maximize anthocyanin extraction yields. The best results were obtained by the simultaneous ultrasound/microwave-assisted extraction (UMAE) (with microwave power at 300 W, and ultrasound power at 50 W) and good yields (1.77 mg/g in dry weight of anthocyanins), which were also efficiently recovered from all the solvents used in this work resulting in purified extracts which could be further up-scaled.

Another important application of the IL-MAE combination is the extraction of polysaccharides from different biomass resources, including lignocellulosic materials and food residues. Particularly, the ability of ILs to extract the most abundant polysaccharides from natural sources, namely cellulose, chitin, starch, hemicelluloses and pectins, has been largely studied in the last few years and some excellent reviews have been recently published [66, 76]. The interaction mechanisms between ILs and polysaccharides are complex, but a detailed study for each individual case would result in the design of selective extraction processes. A deep discussion of the work developed for each polysaccharide as well as the innovation degree and the development stage of dissolution and extraction technologies is currently on-going and some studies have reported nice results. For example, Badgujar & Bhanage discussed the different possibilities offered by ILs in the extraction of different polysaccharides from lignocellulosic biomass, including food [77]. Authors studied various MAE process parameters and their influence for biomass dissolution by ILs and concluded with the great potential shown by this combination to selectively extract polysaccharides from complex biomass structures. MAE-ILs could also provide a good condition for the homogeneous reaction of biopolymers by the dissolution, regeneration and modification of cellulose, starch, chitosan, lignin and proteins [78].

IL-UAE

UAE has been proposed as a valid alternative to other extraction techniques, including MAE, that are energy-consuming. UAE has been largely studied for the extraction of thermolabile compounds in plants and food residues at room temperature, since ultrasounds are able to break down cell walls releasing volatile analytes with no risk of thermal degradation. Ultrasonic waves promote solvent penetration and swelling of the natural matrix by the repeated formation and subsequent collapse of gas bubbles, causing mechanical tension in cells, with increase in the number and size of pores and finally the cell tissue rupture, facilitating in this way the release of the target compounds [72]. The ability to extract simultaneously hydrophilic and hydrophobic non-volatile compounds is also very interesting and could enhance the use of ILs in combination with UAE [79]. For instance, an ionic-liquid-based UAE method was proposed to simultaneously extract gingerols and polysaccharides from ginger (*Zingiber officinale Roscoe*) with high yields and reduced extraction times in both cases [80].

This ionic liquid ultrasound-assisted extraction (IL-UAE) combination was firstly proposed for extraction of bioactive compounds from food matrices in 2007, and it has been the object of raising interest in the last 10 years, as it has been recently reviewed [81, 82]. As an example, Murador *et al.* [83] reported the use of IL-UAE for the extraction of carotenoids from orange peels. These authors succeeded in developing an effective, rapid and environmentally-friendly method after optimization of some key parameters, such as ultrasound power, time, IL concentration, soaking time, solid-liquid ratio and the number of extraction cycles. IL-UAE provided higher extraction efficiency and reduced significantly time and energy consumption in comparison with similar extractions with organic solvents (acetone) and ILs without UAE. Authors concluded that this pioneering work could be extended for the successful extraction of a wide range of active compounds from a variety of vegetable food matrices.

Another interesting study where the IL-UAE combination was applied to extraction and further determination of heavy metals (Cr and Cd) in vegetables was performed by Elik *et al.* [84]. Authors reported low detection and quantification limits, high accuracy in their method, with low relative standard deviations (RSD) in all cases, high enhancement factor and wide linear working area when compared with conventional methods for the analysis of Cr and Cd by flame atomic absorption spectroscopy [85]. Otu *et al.* reported a very interesting study where ILs were used in combination with ultrasound irradiation to extract polysaccharides from leaves with short times and high extraction yields [86]. Authors showed the main structure-function relationship for extracted and purified polysaccharides, that were reported to be very useful for the food and pharmaceutical industries.

The IL-UAE combination is very useful for the specific extraction of potential contaminants in food samples. Bani *et al.* developed a method based on an *in situ* derivatization combined with ionic liquid-based ultrasound-assisted dispersive liquid-liquid microextraction (IL-USA-DLLME) for the extraction of seven biogenic amines (BAs) namely putrescine, cadaverine, histamine, tryptamine, tyramine, spermidine, and spermine in foods [87]. This method showed a wide linear interval (1-100 mg/L), low detection and quantification limits and high recoveries, with application to a broad range of foods (from fruits to canned fish) with a sensitive, rapid and cost-effective performance.

Protic ionic liquids (PIL) coupled to UAE could be used as alternative to conventional organic solvents for the extraction of alkaloids from foods, such as theobromine and caffeine from coffee and cocoa grains [88]. The optimum conditions for UAE were reported as 100-300 W power, 1:6 solid/liquid ratio, PIL concentration 4.5 M. It was reported that the solid/liquid ratio influenced greatly the extraction process of alkaloids, while the ultrasonic power and other UAE parameters did not change significantly the extraction yields. Similar ILs have been also used in the UAE and further determination of lipids from grape seeds [89].

Another work by Medina-Torres *et al.* [90] reported the isolation of bioactive phenols-loaded chitosan nanoparticles by ionic gelation from Persian lemon (*Citrus latifolia*) waste (PLW) and chitosan nanoparticles with UAE. Authors used Response Surface Methodology (RSM) to optimize the UAE conditions in terms of Total Phenolic Content (TPC) and extraction yields. A synergic effect between phenolic compounds from PLW and chitosan nanoparticles was observed in antioxidant and antibacterial activity, showing the high potential of these compounds as substrates for the production of value-added compounds with potential application in the food industry.

IL Coupled to Supercritical/Subcritical Extraction Techniques

Subcritical water refers to high-temperature and high-pressure water. A unique and useful characteristic of subcritical water is that polarity can be dramatically decreased at high temperatures. Therefore, subcritical water can behave similarly to methanol or ethanol. This makes subcritical water a green extraction fluid used for a variety of organic species with the development of specific methods for subcritical water extraction (SBWE) which have been recently reviewed [91].

A very interesting study was developed by Vo Dinh *et al.* [92], who explored the combination of ILs with SBWE methods to obtain different phenolic compounds from the brown seaweed *Saccharina japonica*. These authors used the imidazolium-based IL 1-butyl-3-methylimidazolium tetrafluoroborate as the catalyst in reactions to selectively extract the target analytes. This method provided a clear improvement in phenolics extraction yields, with identification and quantification of gallic, chlorogenic, gentisic, protocatechuic, caffeic, and syringic acids in extracted products with yields 4-5 times higher than those obtained without the use of ILs. However, p-hydroxybenzoic acid showed slightly lower values in the presence of the IL and vanillic acid easily decomposed under the experimental conditions.

Another interesting application of the IL-SBWE combination is the conversion of hemicellulose to the corresponding monosaccharides at ultra-high hydrolysis rate

[93]. Authors reported that the hemicellulose removal rate exceeded the values reported for most acid, alkali, or deep eutectic solvent treatments, as well as results obtained for ILs without the combination with subcritical water. In another study, Yu *et al.* reported that the extraction of lipids from microalgae was improved by using IL-SBWE [94]. Authors reported that the reaction between ILs and cellulose and hemicellulose in cell walls was the most important step to determine the lipids extraction yield.

Supercritical CO_2 fluid extraction (SFE) is a technique that uses the properties of CO_2 over its critical point to selectively extract soluble components from raw materials [95]. In addition, supercritical CO_2 is recognised as an ideal solvent for extracting bioactive compounds because it is non-toxic, non-explosive and readily available, not causing major disruptions in biocompounds [96]. In particular, the combination of SFE with ILs has been reported for specific extractions, such as the fractionation of polysaccharides where ILs are used for purification of the extracts after SFE [97].

IONIC LIQUIDS IN MICROEXTRACTION TECHNIQUES

Sample preparation is a key step in analytical processes, where parameters such as accuracy, sensitivity, repeatability or selectivity should be improved [17]. Natural samples are usually very complex in composition and they require several steps prior to the direct application of the selected analytical method for target analytes. However, in recent years, some research has been carried out to ensure that all pre-treatment processes are as simple as possible, while maintaining possibilities for miniaturization and automation, with safety and sustainability [98]. In this sense, the use of new extractant phases, such as ILs, has shown raising interest in the last few years.

It is well known that ILs can be designed by the combination of many different organic and inorganic anions, cations or alkyl substituents of cations, which give them different properties and a broad range of applications [16, 99]. ILs can be used in combination with different sample preparation techniques: liquid-liquid microextraction (LLME), dispersive liquid-liquid microextraction (DLLME), dispersive liquid phase microextraction (DLPME), single drop microextraction (SDME), solid phase microextraction (SPME), stir bar sorptive extraction (SBSE) or ultrasonic assisted microextraction (UAME).

The exponential growth in research work dealing with the use of ILs as sorption materials in the preparation of food samples for further chemical analysis has been observed in the last two decades (Fig. **5**). The application of ILs in microextraction-related techniques is due to their unique properties [16]. This section will discuss some of the ILs applications in microextraction techniques used for the identification and quantification of different analytes in food samples.

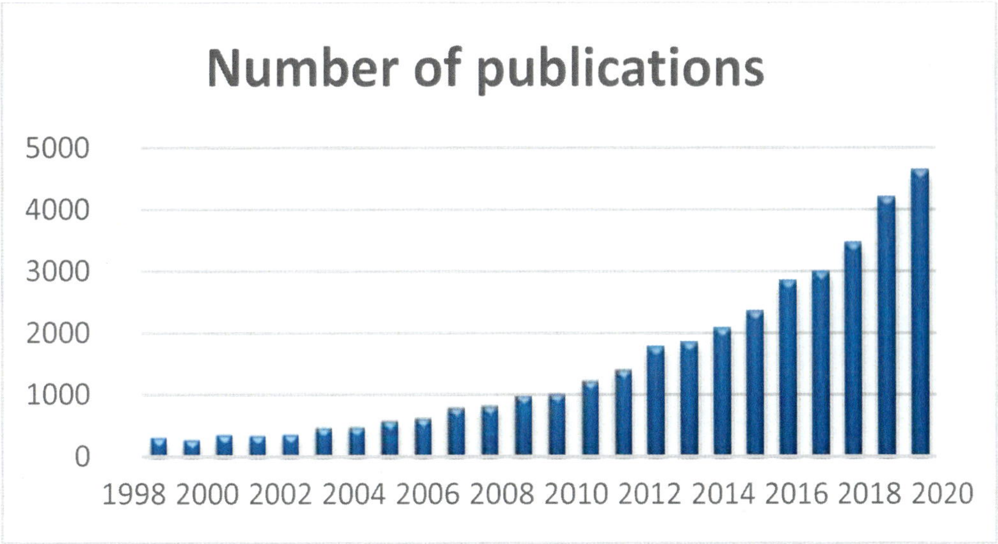

Fig. (5). Evolution of articles related with ionic liquids as sorption media in sample preparation techniques for food analysis (ScienceDirect database, 2021; keywords: ionic liquids; food analysis; sample preparation).

Liquid-Liquid Microextraction (LLME)

The use of ILs in liquid-liquid extraction (LLE) has emerged as a potential alternative to traditional solvents. Although their high viscosity and cost in some cases could be considered a disadvantage, the possibilities of miniaturization and automation are clear advantages in microextraction processes [16, 17]. Some examples of the use of ILs in LLE procedures for the determination of different analytes in food samples are shown in Table **1** and they are discussed below.

Vortex-assisted liquid-liquid microextraction methods (VA-LLME) are adequate for the determination of metals in food samples. Ali *et al.* [100] developed a VA-LLME method for the pre-concentration and identification of Se in water and fifteen different food samples with further detection by electrothermal atomic absorption spectroscopy (ETAAS). Authors suggested the use of ammonium pyrrolidine dithiocarbamate (APDC) to form a hydrophobic complex with the metal cation, while Triton X-114 and 1-butyl-3-methylimidazolium hexafluorophosphate ionic liquid [C_4mim][PF_6], were used in the dispersant medium. Similarly, Castro Grijalba *et al.* [101] developed a VA-LLME-based method for the speciation of Se (IV) and Se (VI) in Allium and Brassica vegetables. In this case, trihexyl(tetradecyl)-phosphonium decanoate $[P_{6,6,6,14}][C_{10}H_{19}O_2]$ was used, resulting in the extraction of the Se(IV)-APDC complex followed by determination by ETAAS. Chamsaz *et al.* [102] developed a rapid and simple method based on VA-LLME for the determination of trace Cd in spinach leaves, using 1-hexyl-3-methylimidazolium hexafluorophosphate $[Hmim][PF_6]$ as the most adequate solvent and flame atomic absorption spectrometry (FAAS) as the detection technique.

On the other hand, Yang *et al.* [103] reported a method based on ionic liquid-assisted liquid-liquid microextraction based on the solidification of floating organic droplets (ILSFOD-LLME). They used trihexyl-tetradecy--hexafluorophosphate ($[P_{6,6,6,14}][PF_6]$) as the most adequate solvent for seven benzoylurea insecticides in fruit juice. You *et al.* [104] also proposed a method for fruit juices based on ionic liquid-based air-assisted liquid-liquid microextraction (IL-AALLME). In this method, three different ionic liquids ($[Hmim][NTf_2]$, $[Omim][PF_6]$ and $[Hmim][PF_6]$) were tested as pesticides extractants. Finally, a pre-treatment based on in-syringe temperature-controlled liquid-liquid microextraction based on solidified floating ionic liquid (in-syringe TC-LLM--SFIL) combined with HPLC/DAD detection was used for the quantification of five triazine and phenylurea herbicides in vegetable protein drinks [105]. Authors selected $[P_{4,4,4,12}][BF_4]$ as the extractant medium and the final advantages of the whole procedure were the high efficiency, simplicity and possible classification within green chemistry. Finally, a LLME method was developed by Sun *et al.* [106] for the extraction of four Sudan dyes from red wine and fruit juice with the use of 1-hexyl-3-methylimidazolium hexafluorophosphate ($[C_6mim][PF_6]$) and HPLC/DAD detection.

Dispersive Liquid-Liquid Microextraction (DLLME)

Dispersive liquid-liquid microextraction (DLLME) was firstly developed in 2006, when Rezaee *et al.* [107] succeeded in using effectively this method to determine Polycyclic Aromatic Hydrocarbons (PAHs) in water samples. This technique is based on the use of three components consisting of an aqueous phase, a non-polar extraction solvent and a dispersant. This is usually an organic solvent that is miscible in both, the extraction solvent and the aqueous phase, thereby promoting the dispersion of the acceptor phase and increasing the extraction efficiency of the analytes in complex matrices [17, 99]. The most important feature of this technique lies in achieving the quantitative extraction of the analytes from just some millilitres of the aqueous sample into microliters of the extraction solvent [18], providing a great advantage in sustainability by consuming a small volume of solvent as well as by being a very fast extraction technique. Traditionally, chlorinated organic compounds such as dichloromethane, tetrachloroethylene, chlorobenzene, carbon tetrachloride or chloroform have been used as extraction solvents in DLLME while the dispersant was based on organic solvents such as acetone, acetonitrile, methanol or ethanol. However it is well known that many of them are highly toxic and environmentally unfriendly [6], so despite of the fact that their use in DLLME is reduced to small quantities, ILs have emerged as strong and ideal alternatives due to their advantages, as well as their tunability to be miscible or immiscible with the dispersant depending on whether they are designed to be hydrophobic or hydrophilic [99]. Consequently, ionic liquids dispersive liquid-liquid microextraction (IL-DLLME) has emerged in a multitude of applications. Table **1** shows a summary of the analytical characteristics of the selected IL-DLLME applications for the determination of various analytes in food samples.

IL-DLLME has been applied to the extraction of inorganic compounds, most of them requiring complexation in different matrices. Abdolmohammad-Zadeh *et al.* [108] developed a methodology for the pre-concentration of zinc in water and milk samples based on the application of 1-hexylpyridinium hexafluorophosphate [HPy][PF_6] as extracting solvent and 8-hydroxyquinoline as complexing agent, prior to FAAS determination. Similarly, they also developed a method with the same IL and complexing agent to evaluate the Al concentration in water, fruit juices and food samples [109]. In this case, stopped flow spectrofluorometry (SFS) was used for detection. Tuzen *et al.* [110] developed an IL-DLLME method using pyrocatechol violet as the complexing agent and 1-hexyl-3-methylimidazolium bis(trifluoromethylsulfonyl)-imide [C_6mim][NTf_2] as the extracting solvent for the detection of Sn in twenty-four food samples by ETAAS.

Other authors developed simple and effective methods based on vortex-assisted ionic liquid dispersion liquid-liquid microextraction (VA-IL-DLLME) [111 - 113] and ultrasound-assisted ionic liquid dispersive liquid-liquid phase microextraction (UA-IL-DLLME) [114]. Altunay *et al.* [111, 112] developed methods based on VA-IL-DLLME to identify and quantify Al and Cr in vegetables by FAAS [111]; and inorganic Se species (total Se, Se(IV) and Se(VI)) in water, beverages and food samples by HGAAS [112]. The ILs used in these studies were 1-butyl-3-methylimidazolium bis(trifluorosulfonyl)imide [C_4mim][NTf_2] and 1-n-octy--3-methylimidazolium bis(trifluoromethane)-sulfonamide [C_8mim][NTf_2], respectively. Elik *et al.* [113] used 1-butyl-3-methylimidazolium-chloride [C_4mim][Cl] in VA-DLLME followed by FAAS for the pre-concentration and determination of trace Mn and Cd in vegetables, cereals and nuts. On the other hand, Biata *et al.* [114] developed a fast and simple method based on UA-I--DLLME for the pre-concentration of Sb and Sn in beverages by optimizing the experimental factors using RSM based on a small central composite design

(SCCD), in which the interactions between the IL (Aliquat 336) and butyl acetate were evaluated. Finally, the concentrations of both metals were quantified using Inductively Coupled Plasma Optical Emission Spectrometry (ICP-OES).

Recently, magnetic ionic liquids (MILs) have gained great importance and, due to their paramagnetic properties and tunability, have been widely applied to microextraction methods, since they simplify tedious and time-consuming steps, such as centrifugation or filtration by applying magnetic separation [6, 17]. Fiorentini *et al.* [115 - 117] developed a dispersive liquid-liquid microextraction method based on the application of a magnetic ionic liquid (MIL-DLLME) for the determination of As [115], Cr [116] and Pb [117] in honey and bee products by ETAAS. Authors used trihexyl-(tetradecyl) phosphonium tetrachloroferrate (III) $[P_{6,6,6,14}][FeCl_4]$ for As and Cr; and 1-trihexyl-(tetradecyl) phosphonium tetrachloromanganate (II) $([P_{6,6,6,14}]_2MnCl_4)$ for Pb. A new method was developed by Beiraghi *et al.* [118] and it was based on the application of a new task-specific magnetic polymeric ionic liquid (TSMPIL) for the pre-concentration of K in olive and sesame oils and subsequent determination by FAES.

IL-DLLME has been also used for the extraction of organic compounds in different food matrices. Rivera-Vera *et al.* [119] developed an IL-DLLME method with $[C_6mim][PF_6]$), and ethyl-dimethyl-(2-methoxyethyl) ammonium tris (pentafluoroethyl) trifluorophosphate ([MOEDEA][FAP]) for the determination of six carcinogenic PAHs in tea by high-performance liquid chromatography coupled to fluorescence detection (HPLC/FLD). Zhao *et al.* [120] reported the extraction, purification and determination by HPLC/FLD of four aflatoxins (AFs) in wheat by IL-DLLME. Six ILs including 1-alkyl-3-methyl imidazolium bis[(trifluoromethyl) sulfonyl] imide ($[C_nmim][NTf_2]$, n = 6 and 8), 1-alkyl-3-methyl imidazolium hexafluorophosphate ($[C_nmim][PF_6]$, n = 6, 8 and 10) and 1-hexyl-3-methyl imidazolium tetrafluoroborate ($[C_6mim][BF_4]$) were used. Wang *et al.* [121] and Ravelo-Pérez *et al.* [122] also used $[C_6mim][PF_6]$ for the extraction and quantification by HPLC/DAD of five triazines and eight pesticides in honey and bananas, respectively. Fan *et al.* [123] proposed the extraction of four phthalate esters in forty-one alcoholic beverages by using IL-DLLME. Samples were extracted with microvolumes of $[C_4mim][PF_6]$, $[C_6mim][PF_6]$ and $[C_8mim][PF_6]$ in the presence of a certain amount of ethanol and 10% (w/v) sodium chloride. These extracts were further analyzed by HPLC/DAD. Zhang *et al.* [124] developed an *in-situ* IL-DLLME method for the determination of polychlorinated biphenyls (PCBs) and acrylamide at trace levels in milk and coffee with 1-butyl-3-methyl imidazolium bromide ([Bmim][Br]) and 1-(--hydroxyethyl)-3-methyl imidazolium chloride ([HeOHmim][Cl]), respectively.

Quantification was performed using headspace gas chromatography with electron capture and mass spectrometry detection (HS-GC/ECDMS).

Some modifications of this technique have been recently proposed. For example, Piao *et al.* [44] used effervescence-assisted DLLME for the determination of five triazine herbicides in tea beverages. In this study [C_4mim][HSO_4] was used not only as extractant, but also as Brønsted acid. Su *et al.* [125] developed a novel method called microwave-assisted ionic liquid/ionic liquid dispersive liquid-liquid microextraction (MA-IL/IL-DLLME) for the analysis of seven triazine herbicides in fruit juices using [C_4mim][PF_6], [C_6mim][PF_6] and [C_8mim][PF_6]. Altunay [126] used VA-DLLME based in 1-ethyl-3-methyl imidazolium ethylsulfate [Emim][$EtSO_4$] coupled to ultraviolet-visible spectrophotometry (UV-vis) for the determination of vanillin in six food products. These authors also developed another UA-IL-DLLME method with [C_6mim][BF_4] for the extraction of melamine in dairy products followed by UV-Vis detection [127]. Qu *et al.* [128] developed and optimized the integration of a conventional IL extractant [C_6mim][PF_6] with [C_4mim-SH][Br] as a disperser, for the UA-IL-DLLME of three bisphenols (BPs) in juice and milk samples, as well as their subsequent quantification by HPLC/FLD.

Furthermore, it is important to highlight the use of the QuEChERS (Quick, Easy, Cheap, Effective, Rugged & Safe) technique combined with ionic liquid dispersive liquid-liquid microextraction (QuEChERS-IL-DLLME). Zhang *et al.* [129] used this method for the identification of six triazole fungicides in pears, apples, and grapefruit by high-performance liquid chromatography coupled with photodiode array detection (HPLC-PDA). The ILs used in this study were 1-alky--3-methyl imidazolium hexafluorophosphate ([$C_n$mim][PF_6], n = 4, 6 and 8). Lawal *et al.* [130] used QuEChERS-IL-DLLME based on [C_6mim][PF_6] for the HPLC/MS detection of eight pesticides in fruits and vegetables.

In addition, MILs have been also applied with DLLME for the determination of organic compounds in different complex food matrices. For instance, Wang *et al.* [131] developed a MIL-DLLME method to extract six triazine herbicides from vegetable oils using 1-hexyl-3-methyl imidazolium tetrachloroferrate ([C_6mim][$FeCl_4$]), as the microextraction solvent. The same authors also used MILs for the extraction of six triazine herbicides from oilseeds by combining solid-phase dispersion with magnetic dispersive ionic liquid-liquid microextraction (MSPD-MIL-DLLME) [132]. Yao *et al.* [133] synthesized and characterized four different types of MILs ([$C_n$mim-TEMPO][Cl], n = 2, 3, 4, 5). They used them to develop a new MIL-DLLME method coupled with high-performance liquid chromatography-time-of-flight mass spectrometry (HPLC/ToF/MS) to separate, pre-concentrate and quantify five different sulphonamides in milk. Yih Hui *et al.*

[134] developed poly(β-cyclodextrin-ionic liquid) grafted magnetic nanoparticles (poly(βCD-IL)@Fe_3O_4) that combined with 1-octanol as supramolecular solvent was used in the microextraction of seven representative PAHs in water, rice and tea beverages. Cao *et al.* developed a new method for the identification of BAs based on *in situ* derivatization combined with the MIL-DLLME method [135, 136]. The MILs used in this study were trihexyl-tetradecyl-phosphonium tetrachlorocobalt (II) ($[P_{6,6,6,14}^{+}][CoCl_4^{2-}]$) and benzyl-trioctyl-ammonium thiocyanatecobalt (II) ($[N_{8,8,8,B}^{+}]_2[Co(SCN)_4^{2-}]$), for wine and fish [135], and beer and milk [136], respectively.

Another application of ILs in combination with DLLME is the so-called temperature-controlled ionic liquid dispersive liquid-liquid microextraction (TC-IL-DLLME). The main feature of this technique is that no dispersant is required because the dispersion process is not chemically achieved, but it is thermally assisted. Kamarei *et al.* [137] used this strategy for the pre-concentration and determination of chlorobenzenes in water by using $[C_4mim][PF_6]$ and HPLC/DAD for detection. This combination was also used by Elik [138] who developed an analytical procedure for the determination of histamine in canned fish products using TC-IL-DLLME. The experimental variables were optimized by using a RSM based on a central composite design (CCD) while the IL was 1-hexyl-3-methyl imidazolium tetrafluoroborate $[Hmim][BF_4]$.

Single Drop Microextraction (SDME)

In SDME the analyte uptake occurs in a liquid droplet suspended in the tip of a microsyringe, either directly immersed in a stirred solution (DI-SDME) or in the headspace (HS-SDME) for volatile compounds, thus avoiding possible matrix interferences [99]. SDME is a cheap and sensitive technique, requiring very small volumes of extractant. However, despite the advantages shown by SDME, some experimental problems, such as the instability of the droplet formed or losses due to evaporation or solubilization in the sample matrix should be taken into account [17, 18]. In this sense, properties of ILs, such as their high thermal stability, surface tension and viscosity or their low vapour pressure, have made them strong alternatives to conventional organic solvents for SDME. Moreover, the tunability of ILs directly influences properties such as their miscibility, a very important parameter in SDME [17, 99]. Both hydrophobic and hydrophilic ILs have been widely used in recent years with DI-SDME and HS-SDME [16] in the ionic liquid based SDME (IL-SDME). Table **1** shows a summary of the analytical characteristics of some IL-SDME applications for the determination of various analytes in food samples.

Wen *et al.* developed several methods for the determination of Cu [139] and Cd [140] in food samples using [C_4mim][PF_6] as extractant in IL-SDME and UV-Vis spectrophotometry and ETAAS as methods of quantification, respectively. Liu *et al.* [141] determined formaldehyde in shiitake mushrooms by using [C_4mim][PF_6] in IL-SDME coupled to HPLC/PDA determination. Moreover, a HS-SDME method was developed by Zhao *et al.* [142] for the GC/FID determination of four volatile phenols in waters using [C_4mim][PF_6] and [C_8mim][PF_6].

Some of the most recent applications of the use of ILs coupled to SDME have focused on the use of MILs, where a rod magnet achieves suspension of the MIL directly in the aqueous sample or in the headspace. As an example, Jahromi *et al.* [143] synthesized a hydrophobic MIL based on the tetrachloromanganate (II) anion and used it for the extraction of ascorbic acid from orange juice with subsequent quantification by voltammetry.

Solid Phase Microextraction (SPME)

SPME is a technique developed several decades ago by Pawliszyn *et al.* [144] and stands out, like most microextraction techniques, as a simple, fast and cheap procedure, requiring small volumes of solvents or no solvent at all. SPME covers two different and consecutive stages. The first one is based on the partition of analytes between the sample and an immobilized stationary phase consisting of a solid or a liquid phase dispersed in a solid. The second step consists of the desorption of those analytes that have been impregnated on the fibre surface and their release for subsequent quantification. More specifically, samples are firstly placed in a vial in which a reagent can be added if necessary, and the vial is further sealed hermetically with a silicone membrane cap. The fibre is then directly immersed in the sample or exposed to the headspace for a certain time and an equilibrium adsorption of the analytes onto the fibre takes place. After extraction, the analytes are thermally or chemically desorbed for identification and quantification [18, 99].

The continuous research into the SPME technique and its application to different real matrices has encouraged the development of new sorbents, which can replace the limited catalogue of commercially available fibre coatings [16]. In this sense, IL-based sorbent coatings have been gaining relevance in recent years, since they can be designed with selected affinity towards specific analytes [6]. Their negligible vapour pressure, good thermal stability, as well as their adjustable viscosity and miscibility with water and organic solvents are the properties of ILs that make them promising coatings for application in SPME fibres [99]. In this context, ILs could help to solve or minimize some of the well-described disadvantages of the SPME technique, such as low chemical and thermal resistance of the sorbents, low selectivity for certain analytes, fragility or limited shelf-life. Table **1** shows a summary of the analytical characteristics of selected IL-SPME combinations for the determination of different analytes in food samples.

Direct immersion-solid phase microextraction (DI-SPME) was firstly developed since no complex procedures for sample preparation are required. Erdem *et al.* [145] synthesized a SPME fibre coated with montmorillonite (MMT) clay intercalated with carboxymethyl cellulose (CMC) and a dicationic imidazolium-based ionic liquid (Im-DIL) with detection by gas chromatography-mass spectrometry (GC/MS) for the determination of six PAEs in water and beverages. Haghi *et al.* [146] developed an IL-based sol-gel coated SPME fibre on a porous copper substrate using silica nanoparticles (PSG-ILs-silica NPs) for the extraction of five pesticides from tomato with subsequent GC/FID quantification. Liu *et al.* [147] used DI-SPME for the extraction of four sulfonylurea herbicides in tea samples in combination with HPLC/UV-Vis. In this case, ILs were immobilized on magnetic nanoparticles (Fe_3O_4@SiO_2@ILs MNPs) by a microwave-assisted method. Tang *et al.* [148] synthesized 1-vinyl-3-(4-vinyl-benzyl) imidazolium chloride (Poly([ViViPIm][Cl])) with further polymerization in the presence of azobisobutyronitrile to be physically coated onto stainless steel wires. Subsequently, these fibres were applied to wine matrices for the determination of six organic acids by GC/MS. Wu *et al.* [149] developed a hybrid material using 1-(triethoxy silyl)-propyl-3-aminopropyl imidazole hexafluorophosphate to be used as SPME fibre for the extraction of eight organochlorine pesticides (OCPs) from fruit juices and milk. Finally, Wu *et al.* [150] developed an ionic liquid polymer composite functionalized with multi-walled carbon nanotubes by electrodeposition. This composite was coated onto a commercial fluorinated polymer (Nafion®) and subsequently used for the analysis of four carbamate pesticides in apples and lettuces by DI-SPME-GC/FID.

HS-SPME has been widely used as a successful extraction technique in food analysis. For example, Martín-Calero *et al.* [151] studied the use of several ILs (1-Butyl-3-methyl imidazolium tetrafluoroborate ([Bmim][BF_4]), [C_6mim][BF_4] and 1-Methyl-3-octylimidazolium-tetrafluoroborate ([Omim][BF_4])) for the separation, identification and quantification by HPLC/PLD of six heterocyclic aromatic amines in meat-based infant foods. Pang *et al.* [152] developed an ionogel-based IL (1-hexadecyl-3-methyl imidazolium bis(trifluoromethyl sulfonyl) imide [C_{16}mim][NTf_2]) for HS-SPME and it was successfully applied to four organophosphorus pesticides from wine and juice samples. Identification and quantification of these analytes was performed by gas chromatography with flame photometric detector (GC/FPD). Li *et al.* [153] used 1-hydroxyethyl-3-methyl imidazolium tetrafluoroborate ([HOEmim] BF_4), multi-walled carbon nanotubes (MWCNTs) and reduced graphene oxide (rGO) to synthesize a new material (MWCNTs-rGO-IL) which was subsequently co-electrodeposited with polyaniline on stainless steel wires by cyclic voltammetry. This material was then used as SPME fibre for the identification and quantification of six alcohols in soft drinks followed by GC-FID determination.

PILs have raised large interest in SPME in recent years, as they show the extracting capabilities of ILs but also additional mechanical stability, making them very robust and adequate for a large number of applications [25]. As an example, González-Álvarez *et al.* [28] synthesized two different types of PILs [poly(VHIM-NTf2)] and [poly(VHIM-taurate)] by a free radical polymerization reaction to be further used as coatings for SPME. These materials were applied to the extraction and quantification of eight fragrances in lemon beer by GC/MS. Sadeghi *et al.* [29] developed a highly innovative method for the determination of sulfathiazole in milk and honey samples with a membrane synthesized *via* copolymerization of 1-hexyl-3-vinyl imidazolium bromide ([MMA-HVIm][Br]) and methyl-methacrylate (MMA) with further determination by fluorescence spectroscopy. Trujillo-Rodríguez *et al.* [30] developed four PIL-SPME coatings and they compared their performance with nine commercial SPME coatings for the extraction and quantification by GC/FID of sixteen volatile compounds (two free fatty acids, two aldehydes, two ketones and ten phenols) in cheeses. Zhao *et al.* [31] synthesized three polymeric imidazolium-based IL coatings (poly([ViHim][NTf_2]), poly([ViDDim][NTf_2]) and poly([ViHDim][NTf_2])) by free radical polymerization and they used them to extract esters and Fatty Acid Methyl Esters (FAMEs) in wine with subsequent quantification by GC/FID. Pelit *et al.* [32] mixed ILs with MMT to be co-deposited with polythiophene (PTh) and electrochemically coated onto a SPME fibre for subsequent use in the analysis of pesticides in grape juice. In this case, the ILs used for the synthesis of the PTh-MMT-IL fibre were 1-methyl-3-octyl imidazolium bromide ([C_8mim][Br]), 1-methyl-3-undecyl imidazolium bromide ([C_{12}mim][Br]) and 1-methyl-3-octadecyl imidazolium bromide ([C_{18}mim][Br]). Moreover, two KSF-IL coating materials for SPME (KSF-[H_2N-C_3bim][Br] and KSF-[H_2N-C_3bim][BF_4]) were successfully prepared and characterized by Tashakkori *et al.* [154]. They used these materials to analyze sixteen phenolic compounds in fruit juice by SPME-GC/MS and SPME-HPLC/DAD. Gionfriddo *et al.* [33] evaluated the performance of two PILs used in SPME coatings for the separation, identification and quantification of OPPs and fruit metabolites in grapes. The two different PIL coatings used in this work were poly(1-4-vinylbenzyl-3-hexadecyl imidazolium) bis[(trifluoromethyl)sulfonyl] imide (poly ([VBHDIM][NTf_2]) and N,N-didecy--N-methyl-d-glucaminium poly(2-methyl-acrylic acid 2-[1-(3-{2-[2-(--trifluoromethanesulfonylamino-propoxy)-ethoxy]-ethoxy}-propylaino)-vinylamino]-ethyl ester) (poly([DDMGlu][MTFSI]). Feng *et al.* [34, 35] also used HS-SPME for the identification and quantification of five pesticides and 2-naphthol in tea beverages and fruit, respectively. In the first case, ILs used were [C_8mim][Br], [C_{12}mim][Br] and [C_{18}mim][Br] with quantification by GC/ECD. However, for the quantification of 2-naphthol, a PIL/MWCNT was synthesized *in situ* on a functionalized stainless-steel wire, with subsequent quantification by GC/FID. The SPME-coated PIL/MWCNT fibres were synthesized from 1-Viny--3-ethyl imidazolium bromide (VEImBr) and 1-vinyl-3-octyl imidazolium bromide (VOImBr). Chen *et al.* [36] developed a new molecular imprinted polymer (MIP) using 1-allyl-3-vinyl imidazolium chloride as the functional monomer. The final synthesized template was used as an extraction medium for five phenolic acids in multiple monolithic fibre solid phase microextraction (MMF-SPME). Finally, Pacheco-Fernández *et al.* [155] developed different PIL-based sorbent coatings based on zwitterionic IL monomers and dicationic IL crosslinkers for the determination of short chain free fatty acids in wine.

Table 1. Summary of the use of ILs in microextraction techniques for the determination of different analytes in foods.

Analytes	Sample	Ionic Liquid	Sample Treatment	Analytical Method	LOD (µg/L)	RSD (%)	Linear Range (µg/L)	Enhan. Factor	Recovery	Refs.
LLME										
Se	water and 15 food samples	[C_4mim][PF_6]	VA-LLME	ETAAS	0.07	3.51	0.24-1.75	98.7	-	[100]
7 BUs	fruit juice	[$P_{6,6,6,14}$][PF_6]	SFOD-LLME	HPLC/UV-Vis	0.03-0.28	0.1-7.3	0.5-500	160-246	75.6-113.9	[103]
5 fungicides	fruit juice	[Hmim][NTf_2] [Omim][PF_6] [Hmim][PF_6]	AALLME	HPLC/UV-Vis	0.4-1.8	4.2-6.2	5-400	-	74.9-115.4	[104]
Cd	spinach leaves	[Hmim][PF_6]	VA-LLME	FAAS	1.1	4.3	5-150	31	97.3-98	[102]
5 triazine and phenylurea pesticides	vegetable protein drinks	[$P_{4,4,4,12}$][BF_4]	in-syringe TC-LLME	HPLC/DAD	0.25-2.59	< 8.17	7.8-1000	-	81.26-118.42	[105]
Se species	vegetables	[$P_{6,6,6,14}$][$C_{10}H_{19}O_2$]	VA-LLME	ETAAS	0.005	4.9	-	100	94.0-102	[101]
4 azo dyes	wine and fruit juices	[C_6mim][PF_6]	LLME	HPLC/DAD	0.428-4.846	1-6	2-1000	-	68.54-108.28	[106]
DLLME										
Vanillin	6 food samples	[Emim][$EtSO_4$]	VA-DLLME	UV-vis spectroscopy	0.15	3.7-4.1	0.5-300	-	92.1-103	[126]
Al, Cr	8 food samples	[C_4mim][NTf_2]	VA-DLLME	FAAS	0.02 0.05	1.1-2.3 1.1-1.8	0.07-100 0.2-80	95 135	96.7-103.2 97.1-103.8	[111]
Zn	water and milk	[HPy][PF_6]	DLLME	FAAS	0.22	1.92	0.8-33	71	-	[108]
Se species	water, beverages and 6 food samples	[C_8mim][NTf_2]	VA-DLLME	HGAAS	0.0015	2.4-3.5	0.005-0.5	120	92.7-103.4	[112]
Mn Cd	12 food samples	[C_4mim][Cl]	VA-DLLME	FAAS	0.15 0.25	1.0-2.5 1.0-2.1	0.5-300 1.0-225	150 110	94.2-103.7 95.1-104.2	[113]

(Table 1) cont.....

Analytes	Sample	Ionic Liquid	Sample Treatment	Analytical Method	LOD (μg/L)	RSD (%)	Linear Range (μg/L)	Enhan. Factor	Recovery	Refs.
As	honey	$[P_{6,6,6,14}][FeCl_4]$	DLLME	ETAAS	0.012	3.9	0.02-5	110	96-102	[115]
Cr	honey	$[P_{6,6,6,14}][FeCl_4]$	DLLME	ETAAS	0.005	3.5	0.015-0.18	105	94-101	[116]
Pb	bee products	$[P_{6,6,6,14}]_2[MnCl_4]$	DLLME	ETAAS	0.003	3.6	0.01-3	107	94.8-101	[117]
5 triazine herbicides	tea beverages	$[C_4mim][HSO_4]$	effervescence-assisted DLLME	HPLC/UV-Vis	$8{\cdot}10^{-5}$	1.1-9.3	0.002-0.2	-	76.3-135.9	[44]
6 PAHs	tea beverages	$[C_6mim][PF_6]$ [MOEDEA][FAP]	DLLME	HPLC/FLD	0.002-0.004	2.02-28.6	1-500	57-94	57-94	[119]
Sn	24 food samples	$[C_6mim][NTf_2]$	UA-DLLME	ETAAS	0.0034	4.5	0.02-3	200	96-98	[110]
6 triazine herbicides	vegetable oils	$[C_6mim][FeCl_4]$	DLLME	HPLC/UV-Vis	1.31-1.49	1.1-7.7	5-1000	-	81.8-110.7	[131]
6 triazine herbicides	oilseeds	$[C_4mim][FeCl_4]$	MSPD-DLLME	HPLC/UV-Vis	1.20-2.72	< 7.7	8-1000	-	82.9-113.7	[132]
5 sulfonamides	milk	$[C_nmim$-TEMPO][Cl]	DLLME	HPLC/ToF/MS	0.534-0.891	0.73-4.17	3-500	42-47	95-105	[133]
4 Afs	wheat	$[C_nmim][NTf_2]$ $[C_nmim][PF_6]$ $[C_6mim][BF_4]$	DLLME	HPLC/FLD	0.015-0.20	0.8-4.7	0.15-100	-	80.8-112.5	[120]
6 triazole fungicides	fruit	$[C_4mim][PF_6]$ $[C_6mim][PF_6]$ $[C_8mim][PF_6]$	QuEChERS-DLLME	HPLC/PDA	3.4-26.8	1.1-12.6	10-10000	-	63.8-119.1	[129]
7 triazine herbicides	fruit juice	$[C_4mim][PF_6]$ $[C_6mim][PF_6]$ $[C_8mim][PF_6]$	MA-DLLME	HPLC	1.01-1.57	1.8-5.6	5-250	-	82.4-104.1	[125]
8 pesticides	fruit and vegetables	$[C_6mim][PF_6]$	QuEChERS-SPE-DLLME	HPLC/MS	0.01-0.54	0-25	5-300	-	89-138	[130]
PCBs and acrylamide	milk and coffee	[Bmim][Br] [HeOHmim][Cl]	in-situ DLLME	HS-GC/ECDMS	0.005-0.025	3.9-13.8	0.025-5	-	-	[124]
5 triazines	honey	$[C_6mim][PF_6]$	DLLME	HPLC/DAD	5.31-8.59	1.7-16.5	250-11400	125-225	60.1-94.6	[121]
4 PAEs	41 alcoholic beverages	$[C_4mim][PF_6]$ $[C_6mim][PF_6]$ $[C_8mim][PF_6]$	DLLME	HPLC/DAD	1.5-4.2	0.7-8	10-1000	51-107	88.5-104.6	[123]
Al	waters, fruit juice and food samples	$[Hpy][PF_6]$	DLLME	SFS	0.05	1.7	0.06-15	100	92.7-101.5	[109]
6 BAs	wine and fish	$[P_{6,6,6,14}^{+}][CoCl_4^{2-}]$	DLLME	HPLC/UV-Vis	1.2-3.9	1.7-4.9	10-1000	-	93.2-103.1	[135]

Other Techniques

Stir bar sorptive microextraction (SBSE) emerged as a novel technique where analytes are coated onto a polymer layer with a moving sorptive element, similar to a magnetic stir bar with further elution with a suitable solvent or thermal desorption [156]. Since not many polymers are adequate for such processes, there has been growing interest in the possibilities of ILs in this application as it was reported by Fan *et al.* [157], who developed a novel SBSE coating layer for the sorption of three non-steroidal anti-inflammatory drugs (NSAIDs) in water and milk. This layer was formed by 1-allylimidazole tetrafluoroborate $[AIM][BF_4]$ as the active substance, which was bound to the surface of the stir bar by g-(methacryl-oyloxy-propyl) trimethoxysilane. The main result of this new development was the higher extraction yield of NSAIDs when compared to other conventional methods.

PILs have been also used as monolithic cakes in a modified version of solid phase extraction, called stir cake sorptive extraction (SCSE). PIL-based SCSE (using 1-allyl-3-vinyl imidazolium chloride for the PIL synthesis), has been described by Chen *et al.* [158], who efficiently determined three organic acids used as preservatives in beverages. Lin *et al.* [159] developed a method for simultaneous determination of sorbic acid, benzoic acid and cinnamic acid in juices and soft drinks by SCSE coupled to HPLC/DAD, using a polymeric ionic liquid-based monolith (PILM) as the extractive medium. At the same time, Wang *et al.* [160] also used a PILM in SCSE-HPLC/DAD to monitor benzimidazole anthelmintics (BAs) in water, milk and honey samples.

FUTURE CHALLENGES AND LIMITATIONS

As it has been stated in the previous sections of this chapter, ILs are very attractive for industrial food extraction processes, as they are able to integrate chemical processes at temperatures lower than those used with conventional solvents. These temperatures lead to reduced energy costs, less thermal injuries and minimum loss of solvents by evaporation, which is important for the activity of nutraceutical compounds and sensorial profiles in food [161]. ILs have been intensely explored in the extraction of selected compounds from food matrices, especially from vegetables because of their high ability to dissolve cellulose by deconstructing its intramolecular structure and interacting with their hydroxyl groups [162]. ILs solvation ability also allows working with a broad range of food matrices which would be the most straightforward access to the target analytes [163]. Nevertheless, there are still some challenging issues related to the use of ILs in food analysis that need to be solved. ILs have been used as surfactants, extractants for the purification of biomolecules, stabilizers for enzymes and in biosensor applications in food analysis. Regardless of the lack of data on the toxicity of food products with ILs residues, low doses in *in vitro* tests did not lead to cell damage. Moreover, several ILs have shown stabilizing effects on biomolecules such as enzymes and proteins.

Although the large number of outstanding properties that make ILs one of the most important developments in extraction technologies in the last decades, there are still some important points to be solved for their complete implementation in food analysis. One of the main issues still pending is their potential toxicity in some cases that could result in environmental problems after extraction. Although they were originally considered to be green chemicals, some concerns about their potential environmental impact have been raised in recent years [19, 164]. In fact, most of the ILs are obtained from fossil sources and show poor degradability. In this respect, the current trend is to design ILs with higher degradability under natural conditions while showing no toxicity. In particular, ILs with fluorine-free anions and with more biodegradable cations, such as pyridinium or guanidinium have been developed [165]. It has also been stated that shorter side chains result in ILs with lower toxicity [166]. As an example, guanidinium-based ILs have been applied in the extraction of flavonoids from *Mangifera spp.* and *Passiflora spp.* leaves with high potential in their use as food additives [167].

As already described in this chapter, the use of anionic or cationic components fully or partly derived from natural compounds would represent a more sustainable approach for the development of a new generation of bio-renewable ILs (B-ILs). Phenolic acids, flavonoids, volatile compounds as well as primary metabolites (sugars, proteins, *etc.*) were successfully extracted from different food matrices using B-ILs [168]. These solvents were able to provide extracts with higher yields of target compounds in shorter times, getting high advantages in the possibilities of use in these applications. Unfortunately, these B-ILs are still not widely applied to food analysis, but hopefully some on-going research would be successful and consequently the range of applications of B-ILs in food analysis should increase in the next few years. Future advancements in the development of B-ILs are therefore expected and their application in the extraction of food matrices should be explored.

Another drawback in the use of ILs in the extraction of specific analytes from food samples is the difficulty in recovering the extracted compounds because of the negligible vapour pressure of most of the ILs. This limit mostly affects their use in the isolation of natural compounds, resulting in serious problems in their further analytical applications [169]. In addition, the poor volatility of ILs also means that their compatibility with LC or GC-MS systems can be also a challenging issue. As a result, a back-extraction of the analytes from the IL phase is normally required before their application to extraction of target analytes from food samples.

The new generations of ILs also include 3rd generation ILs that are derived from pharmaceutical ingredients. These compounds are gaining some space in current research, particularly in the extraction of target analytes from microalgae [170]. It is expected that these advantageous properties of 3rd generation ILs will be applied to food analysis in the next future. Due to that merit and the low toxicity reported, they can be certainly suitable candidates for extraction, particularly for those compounds with possibilities to be further isolated to get chemicals with active properties [171].

Another possibility for the future application of ILs by the food industry is their use in the fabrication of active biomaterials. For example, the development of gelatin-based ionogel films was reported by Metha and Kumar [172]. They used choline salicylate, a bio-based IL, to introduce antioxidant and antimicrobial properties to the film, making it able to be used in active packaging formulations. These films displayed high antimicrobial and antioxidant properties, outstanding mechanical strength and UV shielding properties, showing excellent applicability to avoid oxidation in packaged apples. In another study by Wang *et al.* [173], long-chain cellulose esters were used in biodegradable films for food packaging. Catalyzed synthesis of long-chain cellulose esters using lipase in a mixture of ILs, [C_4mim][Cl] and [Bmim][BF_4] was reported. Not only a green approach for biodegradable plastic synthesis was provided, but these films also enhanced the activity of *Candida rugosa* lipase in the IL and reduced the time required for the cellulose ester formation.

Food adulteration is another important issue in food analysis, since either adulterant substances are deliberately added to food for economic profit or essential substances in food are removed to decrease the production costs. ILs have been proposed as adequate solvents to detect these illegal practices in high value added food. For example, Calvano *et al.* [174] developed a technique to detect adulteration in extra virgin olive oil using tributylammonium α-cyano-4-hydroxycinnamate for liquid-liquid pre-extraction followed by mass spectrometry detection. This method was based on the determination of phospholipids, which is highly efficient to detect samples with 1% of contaminants.

CONCLUDING REMARKS

The use of ionic liquids as green solvents for the selective and efficient extraction and separation of specific high-added value compounds from food samples has been the object of a large number of studies in recent years. ILs have shown good performance as powerful alternatives to conventional organic solvents due to their unique physico-chemical properties, excellent tunability and possibilities to cover a broad range of analytes, from hydrophobic to hydrophilic under different extraction conditions for food by-products, particularly but not only polyphenols, essential oils, lipids, glycosides and alkaloids. The possibilities of ILs in combination with advanced extraction techniques such as microwave-assisted extraction (IL-MAE), ultrasound-assisted extraction (IL-UAE), and subcritical water extraction (IL-SBWE) and their application in food analysis have shown great potential for quantitative and reliable extraction of target analytes from complex matrices, such as food. Nevertheless, there is still much work to do, on one side the extension of the use of ILs as adequate solvents for food matrices and on the other hand to overcome the drawbacks still inherent to their use, such as potential toxicity in some particular cases. The application of ILs combined with green extraction techniques has not been completely exploited, and new applications on the extraction of specific analytes from food samples are expected in the coming years. It is expected that these techniques and the use of ILs for extraction could be up-scaled from lab-scale to industrial environment to give new tools for the food industry.

LIST OF ABBREVIATIONS

AFs	Aflatoxins
BAs	Biogenic amines
B-ILs	Bio-based ILs
BPs	bisphenols
CCD	Central composite design
ChoAA-ILs	Choline-amino acid ionic liquids
CMC	Carboxymethyl cellulose
DI-SDME	Direct immersion-solid phase microextraction
DLLME	Dispersive liquid-liquid microextraction
DLPME	Dispersive liquid phase microextraction
ETAAS	Electrothermal atomic absorption spectroscopy
FAAS	Flame atomic absorption spectrometry
FAES	Flame atomic emission spectrometry
FAMEs	Fatty acid methyl esters
GC/FID	Gas chromatography-flame ionization detection
GC/FPD	Gas chromatography with flame photometric detection
GC/MS	Gas chromatography-mass spectrometry
HGAAS	Hydride generation atomic absorption spectrometry
HPLC/DAD	High-performance liquid chromatography coupled to diode array detection
HPLC/FLD	High-performance liquid chromatography coupled with fluorescence detection
HPLC-PDA	High-performance liquid chromatography coupled with photodiode array detection
HPLC/ToF/MS	High-performance liquid chromatography-time-of-flight mass spectrometry
HS-GC/ECDMS	Headspace gas chromatography with electron capture and mass spectrometry detection
HS-SDME	Headspace-solid phase microextraction

ICP-OES	Inductively coupled plasma optical emission spectrometry
IL-AALLME	Ionic liquid-based air-assisted liquid-liquid microextraction
IL-DLLME - I	onic liquid dispersive liquid-liquid microextraction
IL-MAE	Ionic liquid-based microwave-assisted extraction
ILMs	Ionic liquid monomers
ILs	Ionic liquids
IL-SDME	Ionic liquid-based SDME
ILSFOD-LLME	Ionic liquid-assisted liquid-liquid microextraction based on the solidification of floating organic droplets
IL-UAE - Ionic liquid ultrasound-assisted extraction	
IL-USA-DLLME	Ionic liquid-based ultrasound-assisted dispersive liquid-liquid microextraction
Im-DIL	dicationic imidazolium-based ionic liquid
LLE - Liquid-liquid extraction	
LLME	Liquid-liquid microextraction
MAE	Microwave-assisted extraction
MIL-DLLME	Dispersive liquid-liquid microextraction method based on the application of a magnetic ionic liquid
MILs	Magnetic ionic liquids
MIP	Molecular imprinted polymer
MMA	Methyl-methacrylate
MMF- SPME - Multiple monolithic SPME	
MMT	Montmorillonite
MSPD-MIL-DLLME	Solid-phase dispersion with magnetic dispersive ionic liquid-liquid microextraction
MWCNTs	Multi-walled carbon nanotubes
NADES	Natural deep eutectic solvents
NSAIDs	Non-steroidal anti-inflammatory drugs
OCPs	Organochlorine pesticides
PAHs	Polycyclic Aromatic Hydrocarbons
PCBs	Polychlorinated biphenyls
PIL	Protic ionic liquids
PILM	Polymeric ionic liquid-based monolith
PILs	Polymeric ionic liquids
PLW	Persian lemon (*Citrus latifolia*) waste

poly(βCD-IL)@Fe_3O_4)	poly(β-cyclodextrin-ionic liquid) grafted magnetic nanoparticles
PTh	Polythiophene
QuEChERS	Quick Easy, Cheap, Effective, Rugged & Safe technique
QuEChERS-IL-DLLME	QuEChERS technique combined with ionic liquid dispersive liquid-liquid microextraction
rGO	Reduced graphene oxide
RSD	Relative standard deviation
RSM	Response surface methodology
RTILs	Room temperature ionic liquids
SBSE	Stir bar sorptive microextraction
SBWE	Subcritical water extraction
SCCD	Small central composite design
SCSE	Stir cake sorptive extraction
SFE	Supercritical CO_2 fluid extraction
SFS	Stopped flow spectrofluorometry
SDME - Single drop microextraction	
SPME	Solid-phase microextraction
TC-IL-DLLME	Temperature-controlled ionic liquid dispersive liquid-liquid microextraction
TC-LLME-SFIL	Temperature-controlled liquid-liquid microextraction based on solidified floating ionic liquid
TPC	Total Phenolic Content
TSMPIL	Task-specific magnetic polymeric ionic liquid
UAE	Ultrasound-assisted extraction
UAME	Ultrasonic-assisted microextraction
UMAE	Simultaneous ultrasound/microwave-assisted extraction
VA-LLME	Vortex-assisted liquid-liquid microextraction

CONSENT FOR PUBLICATION

Not applicable.

CONFLICT OF INTEREST

The authors declare no conflict of interest, financial or otherwise.

ACKNOWLEDGEMENT

Declared none.

REFERENCES

[1] Picot-Allain C, Mahomoodally MF, Ak G, Zengin G. Conventional *versus* green extraction techniques — a comparative perspective. Curr Opin Food Sci 2021; 40: 144-56. [http://dx.doi.org/10.1016/j.cofs.2021.02.009]

[2] Welton T. Ionic liquids: A brief history.Biophys Rev. 2018; 10: pp. 691-706. [http://dx.doi.org/10.1007/s12551-018-0419-2]

[3] Agrawal A, Keçili R, Ghorbani-Bidkorbeh F, Hussain CM. Green miniaturized technologies in analytical and bioanalytical chemistry. Trends Analyt Chem 2021; 143: 116383. [http://dx.doi.org/10.1016/j.trac.2021.116383]

[4] Ramos M, Jiménez A, Garrigós MC. Il-based advanced techniques for the extraction of value-added compounds from natural sources and food by-products. Trends Analyt Chem 2019; 119: 115616. [http://dx.doi.org/10.1016/j.trac.2019.07.027]

[5] Fomo G, Madzimbamuto TN, Ojumu TV. Applications of nonconventional green extraction technologies in process industries: Challenges, limitations and perspectives. Sustainability 2020; 12(13): 5244. [http://dx.doi.org/10.3390/su12135244]

[6] Delińska K, Yavir K, Kloskowski A. Ionic liquids in extraction techniques: Determination of pesticides in food and environmental samples. Trends Analyt Chem 2021; 143: 116396. [http://dx.doi.org/10.1016/j.trac.2021.116396]

[7] Huelsmann RD, Will C, Carasek E. Determination of bisphenol A: Old problem, recent creative solutions based on novel materials. J Sep Sci 2021; 44(6): 1148-73. [http://dx.doi.org/10.1002/jssc.202000923] [PMID: 33006433]

[8] Souza Mesquita LM, Martins M, Pisani LP, Ventura SPM, Rosso VV. Insights on the use of alternative solvents and technologies to recover bio-based food pigments. Compr Rev Food Sci Food Saf 2021; 20(1): 787-818. [http://dx.doi.org/10.1111/1541-4337.12685] [PMID: 33325108]

[9] Musarurwa H, Tavengwa NT. Emerging green solvents and their applications during pesticide analysis in food and environmental samples. Talanta 2021; 223(Pt 1): 121507. [http://dx.doi.org/10.1016/j.talanta.2020.121507] [PMID: 33303178]

[10] Freire MG, Cláudio AFM, Araújo JMM, *et al.* Aqueous biphasic systems: A boost brought about by using ionic liquids. Chem Soc Rev 2012; 41(14): 4966-95. [http://dx.doi.org/10.1039/c2cs35151j] [PMID: 22692625]

[11] Husson E, Hulin L, Hadad C, Boughanmi C, Stevanovic T, Sarazin C. Acidic ionic liquid as both solvent and catalyst for fast chemical esterification of industrial lignins: performances and regioselectivity. Front Chem 2019; 7: 578. [http://dx.doi.org/10.3389/fchem.2019.00578] [PMID: 31475140]

[12] Bernardo J, Gírio F, Łukasik R. The effect of the chemical character of ionic liquids on biomass pre-treatment and posterior enzymatic hydrolysis. Molecules 2019; 24(4): 808. [http://dx.doi.org/10.3390/molecules24040808] [PMID: 30813398]

[13] Martín-Calero A, Pino V, Afonso AM. Ionic liquids as a tool for determination of metals and organic compounds in food analysis. Trends Analyt Chem 2011; 30(10): 1598-619. [http://dx.doi.org/10.1016/j.trac.2011.04.023]

[14] Walden P. Molecular weights and electrical conductivity of several fused salts. Bull Acad Imper Sci 1914; 1800: 405-22.

[15] Nanda B, Sailaja M, Mohapatra P, Pradhan RK, Nanda BB. Green solvents: A suitable alternative for

sustainable chemistry. Mater Today Proc 2021; 47: 1234-40.
[http://dx.doi.org/10.1016/j.matpr.2021.06.458]

[16] Marcinkowska R, Konieczna K, Marcinkowski Ł, Namieśnik J, Kloskowski A. Application of ionic liquids in microextraction techniques: Current trends and future perspectives. Trends Analyt Chem 2019; 119: 115614.
[http://dx.doi.org/10.1016/j.trac.2019.07.025]

[17] Llaver M, Oviedo MN, Fiorentini EF, Quintas PY, Wuilloud RG. Analytical developments and applications of ionic liquids for environmental studies. Anal Chem 2021; 31: e00131.

[18] Nawała J, Dawidziuk B, Dziedzic D, Gordon D, Popiel S. Applications of ionic liquids in analytical chemistry with a particular emphasis on their use in solid-phase microextraction. Trends Analyt Chem 2018; 105: 18-36.
[http://dx.doi.org/10.1016/j.trac.2018.04.010]

[19] Cho CW, Pham TPT, Zhao Y, Stolte S, Yun YS. Review of the toxic effects of ionic liquids. Sci Total Environ 2021; 786: 147309.
[http://dx.doi.org/10.1016/j.scitotenv.2021.147309] [PMID: 33975102]

[20] Abdelghany TM, Leitch AC, Nevjestić I, *et al.* Emerging risk from "environmentally-friendly" solvents: Interaction of methylimidazolium ionic liquids with the mitochondrial electron transport chain is a key initiation event in their mammalian toxicity. Food Chem Toxicol 2020; 145: 111593.
[http://dx.doi.org/10.1016/j.fct.2020.111593] [PMID: 32777338]

[21] Cho CW, Yun YS. Application of general toxic effects of ionic liquids to predict toxicities of ionic liquids to Spodoptera frugiperda 9, Eisenia fetida, Caenorhabditis elegans, and Danio rerio. Environ Pollut 2019; 255(Pt 1): 113185.
[http://dx.doi.org/10.1016/j.envpol.2019.113185] [PMID: 31522005]

[22] Leitch AC, Abdelghany TM, Probert PM, *et al.* The toxicity of the methylimidazolium ionic liquids, with a focus on M8OI and hepatic effects. Food Chem Toxicol 2020; 136: 111069.
[http://dx.doi.org/10.1016/j.fct.2019.111069] [PMID: 31883992]

[23] Yavir K, Konieczna K, Marcinkowski Ł, Kloskowski A. Ionic liquids in the microextraction techniques: The influence of ILs structure and properties. Trends Analyt Chem 2020; 130: 115994.
[http://dx.doi.org/10.1016/j.trac.2020.115994]

[24] Patinha DJS, Silvestre AJD, Marrucho IM. Poly(ionic liquids) in solid phase microextraction: Recent advances and perspectives. Prog Polym Sci 2019; 98: 101148.
[http://dx.doi.org/10.1016/j.progpolymsci.2019.101148]

[25] Mei M, Huang X, Chen L. Recent development and applications of poly (ionic liquid)s in microextraction techniques. Trends Analyt Chem 2019; 112: 123-34.
[http://dx.doi.org/10.1016/j.trac.2019.01.003]

[26] Hoover MF. Cationic quaternary polyelectrolytes-a literature review. J Macromol Sci Part A - Chem 1970; 4: 1327-418.

[27] Hirao M, Ito K, Ohno H. Preparation and polymerization of new organic molten salts; N-alkylimidazolium salt derivatives. Electrochim Acta 2000; 45(8-9): 1291-4.
[http://dx.doi.org/10.1016/S0013-4686(99)00334-5]

[28] González-Álvarez J, Blanco-Gomis D, Arias-Abrodo P, *et al.* Analysis of beer volatiles by polymeric imidazolium-solid phase microextraction coatings: Synthesis and characterization of polymeric imidazolium ionic liquids. J Chromatogr A 2013; 1305: 35-40.
[http://dx.doi.org/10.1016/j.chroma.2013.07.012] [PMID: 23891206]

[29] Sadeghi S, Oliaei S. Microextraction of sulfathiazole from milk and honey samples using a polymeric ionic liquid membrane followed by fluorometric determination. J Food Compos Anal 2021; 97: 103774.
[http://dx.doi.org/10.1016/j.jfca.2020.103774]

[30] Trujillo-Rodríguez MJ, Yu H, Cole WTS, *et al.* Polymeric ionic liquid coatings *versus* commercial solid-phase microextraction coatings for the determination of volatile compounds in cheeses. Talanta 2014; 121: 153-62.
[http://dx.doi.org/10.1016/j.talanta.2013.12.046] [PMID: 24607122]

[31] Zhao F, Meng Y, Anderson JL. Polymeric ionic liquids as selective coatings for the extraction of esters using solid-phase microextraction. J Chromatogr A 2008; 1208(1-2): 1-9.
[http://dx.doi.org/10.1016/j.chroma.2008.08.071] [PMID: 18805539]

[32] Pelit FO, Pelit L, Dizdaş TN, *et al.* A novel polythiophene – ionic liquid modified clay composite solid phase microextraction fiber: Preparation, characterization and application to pesticide analysis. Anal Chim Acta 2015; 859: 37-45.
[http://dx.doi.org/10.1016/j.aca.2014.12.043] [PMID: 25622604]

[33] Gionfriddo E, Souza-Silva ÉA, Ho TD, Anderson JL, Pawliszyn J. Exploiting the tunable selectivity features of polymeric ionic liquid-based SPME sorbents in food analysis. Talanta 2018; 188: 522-30.
[http://dx.doi.org/10.1016/j.talanta.2018.06.011] [PMID: 30029408]

[34] Feng J, Sun M, Xu L, Wang S, Liu X, Jiang S. Novel double-confined polymeric ionic liquids as sorbents for solid-phase microextraction with enhanced stability and durability in high-ionic-strength solution. J Chromatogr A 2012; 1268: 16-21.
[http://dx.doi.org/10.1016/j.chroma.2012.10.037] [PMID: 23127811]

[35] Feng J, Sun M, Li L, Wang X, Duan H, Luo C. Multiwalled carbon nanotubes-doped polymeric ionic liquids coating for multiple headspace solid-phase microextraction. Talanta 2014; 123: 18-24.
[http://dx.doi.org/10.1016/j.talanta.2014.01.030] [PMID: 24725859]

[36] Chen L, Huang X. Preparation and application of a poly (ionic liquid)-based molecularly imprinted polymer for multiple monolithic fiber solid-phase microextraction of phenolic acids in fruit juice and beer samples. Analyst 2017; 142(21): 4039-47.
[http://dx.doi.org/10.1039/C7AN01186E] [PMID: 28960220]

[37] Jadhav NR, Bhosale SP, Bhosale SS, Mali SD, Toraskar PB, Kadam TS. Ionic liquids: Formulation avenues, drug delivery and therapeutic updates. J Drug Deliv Sci Technol 2021; 65: 102694.
[http://dx.doi.org/10.1016/j.jddst.2021.102694]

[38] Hulsbosch J, De Vos DE, Binnemans K, Ameloot R. Biobased ionic liquids: Solvents for a green processing industry? ACS Sustain Chem& Eng 2016; 4(6): 2917-31.
[http://dx.doi.org/10.1021/acssuschemeng.6b00553]

[39] Pavlovica S, Zicmanis A, Gzibovska E, Klavins M, Mekss P. (2-Hydroxyethyl) ammonium lactates-highly biodegradable and essentially non-toxic ionic liquids. Green and Sustainable Chemistry 2011; 1(3): 103-10.
[http://dx.doi.org/10.4236/gsc.2011.13017]

[40] DeStefano I, DeStefano G, Paradis NJ, *et al.* Thermodynamic destabilization of azurin by four different tetramethylguanidinium amino acid ionic liquids. Int J Biol Macromol 2021; 180: 355-64.
[http://dx.doi.org/10.1016/j.ijbiomac.2021.03.090] [PMID: 33744247]

[41] Shaikh AR, Ashraf M, AlMayef T, Chawla M, Poater A, Cavallo L. Amino acid ionic liquids as potential candidates for CO_2 capture: Combined density functional theory and molecular dynamics simulations. Chem Phys Lett 2020; 745: 137239.
[http://dx.doi.org/10.1016/j.cplett.2020.137239]

[42] Liu QP, Hou XD, Li N, Zong MH. Ionic liquids from renewable biomaterials: Synthesis, characterization and application in the pretreatment of biomass. Green Chem 2012; 14(2): 304-7.
[http://dx.doi.org/10.1039/C2GC16128A]

[43] Gontrani L. Choline-amino acid ionic liquids: Past and recent achievements about the structure and properties of these really “green” chemicals. Biophys Rev 2018; 10(3): 873-80.
[http://dx.doi.org/10.1007/s12551-018-0420-9] [PMID: 29687272]

[44] Piao H, Jiang Y, Qin Z, *et al.* Development of a novel acidic task-specific ionic liquid-based effervescence-assisted microextraction method for determination of triazine herbicides in tea beverage. Talanta 2020; 208: 120414.
[http://dx.doi.org/10.1016/j.talanta.2019.120414] [PMID: 31816681]

[45] Aslam R, Mobin M, Huda , Obot IB, Alamri AH. Ionic liquids derived from α-amino acid ester salts as potent green corrosion inhibitors for mild steel in 1M HCl. J Mol Liq 2020; 318: 113982.
[http://dx.doi.org/10.1016/j.molliq.2020.113982]

[46] Ohno H, Fukumoto K. Amino acid ionic liquids. Acc Chem Res 2007; 40(11): 1122-9.
[http://dx.doi.org/10.1021/ar700053z] [PMID: 17970600]

[47] Kagimoto J, Fukumoto K, Ohno H. Effect of tetrabutylphosphonium cation on the physico-chemical properties of amino-acid ionic liquids. Chem Commun 2006; 21(21): 2254-6.
[http://dx.doi.org/10.1039/b600771f] [PMID: 16718320]

[48] Chiappe C, Marra A, Mele A. Synthesis and applications of ionic liquids derived from natural sugars. Top Curr Chem 2010; 295: 177-95.
[http://dx.doi.org/10.1007/128_2010_47] [PMID: 21626744]

[49] Jha AK, Jain N. Synthesis of glucose-tagged triazolium ionic liquids and their application as solvent and ligand for copper(I) catalyzed amination. Tetrahedron Lett 2013; 54(35): 4738-41.
[http://dx.doi.org/10.1016/j.tetlet.2013.06.114]

[50] Socha AM, Parthasarathi R, Shi J, *et al.* Efficient biomass pretreatment using ionic liquids derived from lignin and hemicellulose. Proc Natl Acad Sci 2014; 111(35): E3587-95.
[http://dx.doi.org/10.1073/pnas.1405685111] [PMID: 25136131]

[51] Zhang Q, Ren S, Li A, Zhang J, Xue S, Sun X. Tartaric acid-based ionic liquid-type chiral selectors: Effect of cation species on their enantioseparation performance in capillary electrophoresis. Separ Purif Tech 2021; 275: 119228.
[http://dx.doi.org/10.1016/j.seppur.2021.119228]

[52] Kwan ML, Mirjafari A, McCabe JR, *et al.* Synthesis and thermophysical properties of ionic liquids: Cyclopropyl moieties *versus* olefins as Tm-reducing elements in lipid-inspired ionic liquids. Tetrahedron Lett 2013; 54(1): 12-4.
[http://dx.doi.org/10.1016/j.tetlet.2012.09.101]

[53] Sernaglia M, Blanco D, Battez AH, Viesca JL, González R, Bartolomé M. Two fatty acid anion-based ionic liquids - part I: Physicochemical properties and tribological behavior as neat lubricants. J Mol Liq 2020; 305: 112827.
[http://dx.doi.org/10.1016/j.molliq.2020.112827]

[54] Sernaglia M, Blanco D, Hernández Battez A, González R, Fernández-González A, Bartolomé M. Two fatty acid anion-based ionic liquids - part II: Effectiveness as an additive to a polyol ester. J Mol Liq 2020; 310: 113158.
[http://dx.doi.org/10.1016/j.molliq.2020.113158]

[55] Ramajo B, Blanco D, Rivera N, Viesca JL, González R, Battez AH. Long-term thermal stability of fatty acid anion-based ionic liquids. J Mol Liq 2021; 328: 115492.
[http://dx.doi.org/10.1016/j.molliq.2021.115492]

[56] Khan A, Gusain R, Sahai M, Khatri OP. Fatty acids-derived protic ionic liquids as lubricant additive to synthetic lube base oil for enhancement of tribological properties. J Mol Liq 2019; 293: 111444.
[http://dx.doi.org/10.1016/j.molliq.2019.111444]

[57] Rivera N, Blanco D, Viesca JL, Fernández-González A, González R, Battez AH. Tribological performance of three fatty acid anion-based ionic liquids (FAILs) used as lubricant additive. J Mol Liq 2019; 296: 111881.
[http://dx.doi.org/10.1016/j.molliq.2019.111881]

[58] Parmentier D, Metz SJ, Kroon MC. Tetraalkylammonium oleate and linoleate based ionic liquids:

Promising extractants for metal salts. Green Chem 2013; 15(1): 205-9.
[http://dx.doi.org/10.1039/C2GC36458A]

[59] Blanco D, Rivera N, Oulego P, Díaz M, González R, Battez AH. Novel fatty acid anion-based ionic liquids: Contact angle, surface tension, polarity fraction and spreading parameter. J Mol Liq 2019; 288: 110995.
[http://dx.doi.org/10.1016/j.molliq.2019.110995]

[60] Heckel T, Winkel A, Wilhelm R. Chiral ionic liquids based on nicotine for the chiral recognition of carboxylic acids. Tetrahedron Asymmetry 2013; 24(18): 1127-33.
[http://dx.doi.org/10.1016/j.tetasy.2013.07.021]

[61] Van Buu ON, Aupoix A, Hong NDT, Vo-Thanh G. Chiral ionic liquids derived from isosorbide: Synthesis, properties and applications in asymmetric synthesis. New J Chem 2009; 33(10): 2060-72.
[http://dx.doi.org/10.1039/b902956g]

[62] Ferraz R, Branco LC, Marrucho IM, *et al.* Development of novel ionic liquids based on ampicillin. MedChemComm 2012; 3(4): 494-7.
[http://dx.doi.org/10.1039/c2md00269h]

[63] Mastellone G, Marengo A, Sgorbini B, Rubiolo P, Cagliero C. New phases for analytical scale extraction from plants: Current and future trends. Trends Analyt Chem 2021; 141: 116288.
[http://dx.doi.org/10.1016/j.trac.2021.116288]

[64] Brandt A, Gräsvik J, Hallett JP, Welton T. Deconstruction of lignocellulosic biomass with ionic liquids. Green Chem 2013; 15(3): 550-83.
[http://dx.doi.org/10.1039/c2gc36364j]

[65] Lanari D, Marcotullio M, Neri A. A design of experiment approach for ionic liquid-based extraction of toxic components-minimized essential oil from myristica fragrans houtt. Fruits †. Molecules 2018; 23(11): 2817.
[http://dx.doi.org/10.3390/molecules23112817] [PMID: 30380729]

[66] Morais ES, Lopes AMC, Freire MG, Freire CSR, Coutinho JAP, Silvestre AJD. Use of ionic liquids and deep eutectic solvents in polysaccharides dissolution and extraction processes towards sustainable biomass valorization. Molecules 2020; 25(16): 3652.
[http://dx.doi.org/10.3390/molecules25163652] [PMID: 32796649]

[67] Li C, Zhang J, Zhao C, *et al.* Separation of the main flavonoids and essential oil from seabuckthorn leaves by ultrasonic/microwave-assisted simultaneous distillation extraction. R Soc Open Sci 2018; 5(7): 180133.
[http://dx.doi.org/10.1098/rsos.180133] [PMID: 30109070]

[68] Gil KA, Tuberoso CIG. Crucial challenges in the development of green extraction technologies to obtain antioxidant bioactive compounds from agro-industrial by–products. Chem Biochem Eng Q 2021; 35(2): 105-38.
[http://dx.doi.org/10.15255/CABEQ.2021.1923]

[69] Liu X, Jing X, Li G. A process to acquire essential oil by distillation concatenated liquid-liquid extraction and flavonoids by solid-liquid extraction simultaneously from Helichrysum arenarium (L.) Moench inflorescences under ionic liquid-microwave mediated. Separ Purif Tech 2019; 209: 164-74.
[http://dx.doi.org/10.1016/j.seppur.2018.07.028]

[70] Lu C, Wang H, Lv W, *et al.* Ionic liquid-based ultrasonic/microwave-assisted extraction combined with uplc for the determination of anthraquinones in rhubarb. Chromatographia 2011; 74(1-2): 139-44.
[http://dx.doi.org/10.1007/s10337-011-2023-5]

[71] Fan Y, Xu C, Li J, *et al.* Ionic liquid-based microwave-assisted extraction of verbascoside from Rehmannia root. Ind Crops Prod 2018; 124: 59-65.
[http://dx.doi.org/10.1016/j.indcrop.2018.07.063]

[72] Bogdanov MG. Ionic liquids as alternative solvents for extraction of natural products.Alternative

Solvents for Natural Products Extraction, Green Chemistry and Sustainable Technology. Berlin, Heidelberg: Springer-Verlag 2014.
[http://dx.doi.org/10.1007/978-3-662-43628-8_7]

[73] Chen F, Zhang X, Zhang Q, *et al.* Simultaneous synergistic microwave–ultrasonic extraction and hydrolysis for preparation of trans-resveratrol in tree peony seed oil-extracted residues using imidazolium-based ionic liquid. Ind Crops Prod 2016; 94: 266-80.
[http://dx.doi.org/10.1016/j.indcrop.2016.08.048]

[74] Liu T, Sui X, Zhang R, *et al.* Application of ionic liquids based microwave-assisted simultaneous extraction of carnosic acid, rosmarinic acid and essential oil from Rosmarinus officinalis. J Chromatogr A 2011; 1218(47): 8480-9.
[http://dx.doi.org/10.1016/j.chroma.2011.09.073] [PMID: 21999917]

[75] Panić M, Gunjević V, Cravotto G, Radojčić I. Enabling technologies for the extraction of grape-pomace anthocyanins using natural deep eutectic solvents in up-to-half-litre batches extraction of grape-pomace anthocyanins using NADES. Food Chem 2019; 300: 125185.
[http://dx.doi.org/10.1016/j.foodchem.2019.125185] [PMID: 31326673]

[76] Alokika , Anu , Kumar A, Kumar V, Singh B. Cellulosic and hemicellulosic fractions of sugarcane bagasse: Potential, challenges and future perspective. Int J Biol Macromol 2021; 169: 564-82.
[http://dx.doi.org/10.1016/j.ijbiomac.2020.12.175] [PMID: 33385447]

[77] Badgujar KC, Bhanage BM. Factors governing dissolution process of lignocellulosic biomass in ionic liquid: Current status, overview and challenges. Bioresour Technol 2015; 178: 2-18.
[http://dx.doi.org/10.1016/j.biortech.2014.09.138] [PMID: 25451772]

[78] Yang X, Qiao C, Li Y, Li T. Dissolution and resourcfulization of biopolymers in ionic liquids. React Funct Polym 2016; 100: 181-90.
[http://dx.doi.org/10.1016/j.reactfunctpolym.2016.01.017]

[79] Mena-García A, Ruiz-Matute AI, Soria AC, Sanz ML. Green techniques for extraction of bioactive carbohydrates. Trends Analyt Chem 2019; 119: 115612.
[http://dx.doi.org/10.1016/j.trac.2019.07.023]

[80] Kou X, Ke Y, Wang X, *et al.* Simultaneous extraction of hydrophobic and hydrophilic bioactive compounds from ginger (Zingiber officinale Roscoe). Food Chem 2018; 257: 223-9.
[http://dx.doi.org/10.1016/j.foodchem.2018.02.125] [PMID: 29622202]

[81] Benvenutti L, Zielinski AAF, Ferreira SRS. Which is the best food emerging solvent: IL, DES or NADES? Trends Food Sci Technol 2019; 90: 133-46.
[http://dx.doi.org/10.1016/j.tifs.2019.06.003]

[82] Khadhraoui B, Ummat V, Tiwari BK, Fabiano-Tixier AS, Chemat F. Review of ultrasound combinations with hybrid and innovative techniques for extraction and processing of food and natural products. Ultrason Sonochem 2021; 76: 105625.
[http://dx.doi.org/10.1016/j.ultsonch.2021.105625] [PMID: 34147916]

[83] Murador DC, Braga ARC, Martins PLG, Mercadante AZ, de Rosso VV. Ionic liquid associated with ultrasonic-assisted extraction: A new approach to obtain carotenoids from orange peel. Food Res Int 2019; 126: 108653.
[http://dx.doi.org/10.1016/j.foodres.2019.108653] [PMID: 31732025]

[84] Elik A, Karakoy T, Çilesiz Y, Altunay N. Flame atomic absorption spectrometric determination of total chromium and cadmium in bean samples after ultrasonic-assisted microextraction using ionic liquid Aliquat-336. J Indian Chem Soc 2021; 18(1): 117-27.
[http://dx.doi.org/10.1007/s13738-020-02009-7]

[85] Tuzen M, Elik A, Altunay N. Ultrasound-assisted supramolecular solvent dispersive liquid-liquid microextraction for preconcentration and determination of Cr(VI) in waters and total chromium in beverages and vegetables. J Mol Liq 2021; 329: 115556.
[http://dx.doi.org/10.1016/j.molliq.2021.115556]

[86] Otu PNY, Osae R, Abdullateef MT, Cunshan Z, Xiaojie Y, Azumah BK. Characterization of*MORINGA OLEIFERA* leaf polysaccharides extracted by coupling ionic liquid separation system with ultrasound irradiation. J Food Process Eng 2020; 43(7): e13417. [http://dx.doi.org/10.1111/jfpe.13417]

[87] Bani SM, Saaid M, Saad B. An *in situ* dansylation ultrasound-assisted dispersive liquid-liquid microextraction based on ionic liquid for determination of biogenic amines in foods. Food Anal Methods 2020; 13(2): 568-77. [http://dx.doi.org/10.1007/s12161-019-01656-8]

[88] De Oliveira NS, Carlos AS, Mattedi S, *et al.* Ionic liquid-based ultrasonic-assisted extraction of alkaloids from cacao (theobroma Cacao). Chem Eng Trans 2018; 64: 49-54.

[89] Zhang X, Li X, He Q, *et al.* Ionic liquid based ultrasonic-assisted extraction of oleanolic acid from grape seeds. Open Access Libr J 2017; 4: 1-16. [http://dx.doi.org/10.4236/oalib.1104190]

[90] Medina-Torres N, Espinosa-Andrews H, Trombotto S, *et al.* Ultrasound-assisted extraction optimization of phenolic compounds from *citrus latifolia* waste for chitosan bioactive nanoparticles development. Molecules 2019; 24(19): 3541. [http://dx.doi.org/10.3390/molecules24193541] [PMID: 31574952]

[91] Cheng Y, Xue F, Yu S, Du S, Yang Y. Subcritical water extraction of natural products. Molecules 2021; 26(13): 4004. [http://dx.doi.org/10.3390/molecules26134004] [PMID: 34209151]

[92] Vo Dinh T, Saravana PS, Woo HC, Chun BS. Ionic liquid-assisted subcritical water enhances the extraction of phenolics from brown seaweed and its antioxidant activity. Separ Purif Tech 2018; 196: 287-99. [http://dx.doi.org/10.1016/j.seppur.2017.06.009]

[93] Xu Y, Wang P, Xue S, Kong F, Ren H, Zhai H. Green biorefinery — the ultra-high hydrolysis rate and behavior of *Populus tomentosa* hemicellulose autohydrolysis under moderate subcritical water conditions. RSC Advances 2020; 10(32): 18908-17. [http://dx.doi.org/10.1039/D0RA02350G] [PMID: 35518329]

[94] Yu Z, Chen X, Xia S. The mechanism of lipids extraction from wet microalgae Scenedesmus sp. by ionic liquid assisted subcritical water. J Ocean Univ China 2016; 15(3): 549-52. [http://dx.doi.org/10.1007/s11802-016-2831-1]

[95] Barbosa JR, S Freitas MM, Oliveira LC, *et al.* Obtaining extracts rich in antioxidant polysaccharides from the edible mushroom Pleurotus ostreatus using binary system with hot water and supercritical CO_2. Food Chem 2020; 330: 127173. [http://dx.doi.org/10.1016/j.foodchem.2020.127173] [PMID: 32569930]

[96] Baldino L, Zuppolini S, Cardea S, *et al.* Production of biodegradable superabsorbent aerogels using a supercritical CO_2 assisted drying. J Supercrit Fluids 2020; 156: 104681. [http://dx.doi.org/10.1016/j.supflu.2019.104681]

[97] Zhao HK, Wei XY, Xie YM. Supercritical CO_2 extraction, structural analysis and bioactivity of polysaccharide from Grifola frondosa. J Food Compos Anal 2021; 102: 104067. [http://dx.doi.org/10.1016/j.jfca.2021.104067]

[98] Lucena R, Cárdenas S. Ionic liquids in sample preparation.Comprehensive Analytical Chemistry. Elsevier 2017; 76: pp. 203-24.

[99] Aguilera-Herrador E, Lucena R, Cárdenas S, Valcárcel M. The roles of ionic liquids in sorptive microextraction techniques. Trends Analyt Chem 2010; 29(7): 602-16. [http://dx.doi.org/10.1016/j.trac.2009.11.009]

[100] Ali J, Tuzen M, Feng X, Kazi TG. Determination of trace levels of selenium in natural water, agriculture soil and food samples by vortex assisted liquid-liquid microextraction method: Multivariate

techniques. Food Chem 2021; 344: 128706.
[http://dx.doi.org/10.1016/j.foodchem.2020.128706] [PMID: 33267987]

[101] Castro Grijalba A, Martinis EM, Wuilloud RG. Inorganic selenium speciation analysis in allium and brassica vegetables by ionic liquid assisted liquid-liquid microextraction with multivariate optimization. Food Chem 2017; 219: 102-8.
[http://dx.doi.org/10.1016/j.foodchem.2016.09.117] [PMID: 27765204]

[102] Chamsaz M, Eftekhari M, Atarodi A, Asadpour S, Ariani M. Preconcentration procedure using vortex agitator system for determination of trace levels of cadmium by flame atomic absorption spectrometry. J Braz Chem Soc 2012; 23(9): 1630-5.
[http://dx.doi.org/10.1590/S0103-50532012005000026]

[103] Yang M, Zhang P, Hu L, *et al.* Ionic liquid-assisted liquid-phase microextraction based on the solidification of floating organic droplets combined with high performance liquid chromatography for the determination of benzoylurea insecticide in fruit juice. J Chromatogr A 2014; 1360: 47-56.
[http://dx.doi.org/10.1016/j.chroma.2014.07.076] [PMID: 25124227]

[104] You X, Chen X, Liu F, Hou F, Li Y. Ionic liquid-based air-assisted liquid–liquid microextraction followed by high performance liquid chromatography for the determination of five fungicides in juice samples. Food Chem 2018; 239: 354-9.
[http://dx.doi.org/10.1016/j.foodchem.2017.06.074] [PMID: 28873579]

[105] Xu W, Li J, Feng J, Wang Z, Zhang H. In-syringe temperature-controlled liquid–liquid microextraction based on solidified floating ionic liquid for the simultaneous determination of triazine and phenylurea pesticide in vegetable protein drinks. J Chromatogr B Analyt Technol Biomed Life Sci 2021; 1174: 122721.
[http://dx.doi.org/10.1016/j.jchromb.2021.122721] [PMID: 33957354]

[106] Sun S, Wang Y, Yu W, *et al.* Determination of sudan dyes in red wine and fruit juice using ionic liquid-based liquid-liquid microextraction and high-performance liquid chromatography. J Sep Sci 2011; 34(14): 1730-7.
[http://dx.doi.org/10.1002/jssc.201100037] [PMID: 21656676]

[107] Rezaee M, Assadi Y, Milani Hosseini MR, Aghaee E, Ahmadi F, Berijani S. Determination of organic compounds in water using dispersive liquid–liquid microextraction. J Chromatogr A 2006; 1116(1-2): 1-9.
[http://dx.doi.org/10.1016/j.chroma.2006.03.007] [PMID: 16574135]

[108] Abdolmohammad-Zadeh H, Sadeghi GH. A novel microextraction technique based on 1-hexylpyridinium hexafluorophosphate ionic liquid for the preconcentration of zinc in water and milk samples. Anal Chim Acta 2009; 649(2): 211-7.
[http://dx.doi.org/10.1016/j.aca.2009.07.040] [PMID: 19699396]

[109] Abdolmohammad-Zadeh H, Sadeghi GH. Combination of ionic liquid-based dispersive liquid–liquid micro-extraction with stopped-flow spectrofluorometry for the pre-concentration and determination of aluminum in natural waters, fruit juice and food samples. Talanta 2010; 81(3): 778-85.
[http://dx.doi.org/10.1016/j.talanta.2010.01.012] [PMID: 20298853]

[110] Tuzen M, Uluozlu OD, Mendil D, *et al.* A simple, rapid and green ultrasound assisted and ionic liquid dispersive microextraction procedure for the determination of tin in foods employing ETAAS. Food Chem 2018; 245: 380-4.
[http://dx.doi.org/10.1016/j.foodchem.2017.10.115] [PMID: 29287384]

[111] Altunay N, Yıldırım E, Gürkan R. Extraction and preconcentration of trace Al and Cr from vegetable samples by vortex-assisted ionic liquid-based dispersive liquid–liquid microextraction prior to atomic absorption spectrometric determination. Food Chem 2018; 245: 586-94.
[http://dx.doi.org/10.1016/j.foodchem.2017.10.134] [PMID: 29287413]

[112] Altunay N, Elik A, Katin K. Optimization of vortex-assisted ionic liquid dispersive liquid–liquid microextraction by experimental design prior to hydride generation atomic absorption spectrometry for

determination of selenium species in food, beverage and water samples. J Food Compos Anal 2021; 99: 103871.
[http://dx.doi.org/10.1016/j.jfca.2021.103871]

[113] Elik A, Altunay N, Gürkan R. Microextraction and preconcentration of Mn and Cd from vegetables, grains and nuts prior to their determination by flame atomic absorption spectrometry using room temperature ionic liquid. J Mol Liq 2017; 247: 262-8.
[http://dx.doi.org/10.1016/j.molliq.2017.09.121]

[114] Biata NR, Nyaba L, Ramontja J, Mketo N, Nomngongo PN. Determination of antimony and tin in beverages using inductively coupled plasma-optical emission spectrometry after ultrasound-assisted ionic liquid dispersive liquid-liquid phase microextraction. Food Chem 2017; 237: 904-11.
[http://dx.doi.org/10.1016/j.foodchem.2017.06.058] [PMID: 28764084]

[115] Fiorentini EF, Canizo BV, Wuilloud RG. Determination of as in honey samples by magnetic ionic liquid-based dispersive liquid-liquid microextraction and electrothermal atomic absorption spectrometry. Talanta 2019; 198: 146-53.
[http://dx.doi.org/10.1016/j.talanta.2019.01.091] [PMID: 30876542]

[116] Fiorentini EF, Oviedo MN, Wuilloud RG. Ultra-trace cr preconcentration in honey samples by magnetic ionic liquid dispersive liquid-liquid microextraction and electrothermal atomic absorption spectrometry. spectrochim. Acta - Part B - At Spectrosc 2020; 169: 105879.

[117] Fiorentini EF, Botella MB, Wuilloud RG. A simple preconcentration method for highly sensitive determination of Pb in bee products by magnetic ionic liquid dispersive liquid-liquid microextraction and electrothermal atomic absorption spectrometry. J Food Compos Anal 2021; 95: 103661.
[http://dx.doi.org/10.1016/j.jfca.2020.103661]

[118] Beiraghi A, Shokri M. A novel task specific magnetic polymeric ionic liquid for selective preconcentration of potassium in oil samples using centrifuge-less dispersive liquid-liquid microextraction technique and its determination by flame atomic emission spectroscopy. Talanta 2018; 178: 616-21.
[http://dx.doi.org/10.1016/j.talanta.2017.08.080] [PMID: 29136871]

[119] Rivera-Vera C, Lasarte-Aragonés G, Bravo MA, Muñoz-Lira D, Salazar R, Toledo-Neira C. Ionic liquids-based dispersive liquid-liquid microextraction for determination of carcinogenic polycyclic aromatic hydrocarbons in tea beverages: Evaluation of infusion preparation on pollutants release. Food Control 2019; 106: 106685.
[http://dx.doi.org/10.1016/j.foodcont.2019.06.011]

[120] Zhao R, An J, Sun Y, He L, Jiang X, Zhang S. A simple and low-cost sample preparation for the effective extraction, purification and enrichment of aflatoxins in wheat by combining with ionic liquid-based dispersive liquid–liquid microextraction. Microchem J 2021; 164: 106036.
[http://dx.doi.org/10.1016/j.microc.2021.106036]

[121] Wang Y, You J, Ren R, *et al.* Determination of triazines in honey by dispersive liquid–liquid microextraction high-performance liquid chromatography. J Chromatogr A 2010; 1217(26): 4241-6.
[http://dx.doi.org/10.1016/j.chroma.2010.03.031] [PMID: 20382395]

[122] Ravelo-Pérez LM, Hernández-Borges J, Asensio-Ramos M, Rodríguez-Delgado MÁ. Ionic liquid based dispersive liquid–liquid microextraction for the extraction of pesticides from bananas. J Chromatogr A 2009; 1216(43): 7336-45.
[http://dx.doi.org/10.1016/j.chroma.2009.08.012] [PMID: 19700165]

[123] Fan Y, Liu S, Xie Q. Rapid determination of phthalate esters in alcoholic beverages by conventional ionic liquid dispersive liquid–liquid microextraction coupled with high performance liquid chromatography. Talanta 2014; 119: 291-8.
[http://dx.doi.org/10.1016/j.talanta.2013.11.023] [PMID: 24401417]

[124] Zhang C, Cagliero C, Pierson SA, Anderson JL. Rapid and sensitive analysis of polychlorinated biphenyls and acrylamide in food samples using ionic liquid-based *in situ* dispersive liquid-liquid

microextraction coupled to headspace gas chromatography. J Chromatogr A 2017; 1481: 1-11. [http://dx.doi.org/10.1016/j.chroma.2016.12.013] [PMID: 28017564]

[125] Su R, Li D, Wu L, *et al.* Determination of triazine herbicides in juice samples by microwave-assisted ionic liquid/ionic liquid dispersive liquid-liquid microextraction coupled with high-performance liquid chromatography. J Sep Sci 2017; 40(14): 2950-8. [http://dx.doi.org/10.1002/jssc.201700270] [PMID: 28493534]

[126] Altunay N. Development of vortex-assisted ionic liquid-dispersive microextraction methodology for vanillin monitoring in food products using ultraviolet-visible spectrophotometry. Lebensm Wiss Technol 2018; 93: 9-15. [http://dx.doi.org/10.1016/j.lwt.2018.03.021]

[127] Altunay N, Elik A, Kaya S. A simple and quick ionic liquid-based ultrasonic-assisted microextraction for determination of melamine residues in dairy products: Theoretical and experimental approaches. Food Chem 2020; 326: 126988. [http://dx.doi.org/10.1016/j.foodchem.2020.126988] [PMID: 32447162]

[128] Qu J, Li Y, Gao M, *et al.* Development and optimization of a thiol imidazolium-based ionic liquid for ultrasonic assisted liquid-liquid microextraction combined with HPLC-FLD for determination of bisphenols in milk and juice samples. Lebensm Wiss Technol 2019; 111: 653-62. [http://dx.doi.org/10.1016/j.lwt.2019.05.096]

[129] Zhang Y, Zhang Y, Nie J, Jiao B, Zhao Q. Determination of triazole fungicide residues in fruits by quechers combined with ionic liquid-based dispersive liquid-liquid microextraction: Optimization using response surface methodology. Food Anal Methods 2016; 9(12): 3509-19. [http://dx.doi.org/10.1007/s12161-016-0548-9]

[130] Lawal A, Wong RCS, Tan GH, Abdulra'uf LB. Determination of pesticide residues in fruit and vegetables by high-performance liquid chromatography–tandem mass spectrometry with multivariate response surface methodology. Anal Lett 2019; 52(2): 231-48. [http://dx.doi.org/10.1080/00032719.2018.1459655]

[131] Wang Y, Sun Y, Xu B, *et al.* Magnetic ionic liquid-based dispersive liquid–liquid microextraction for the determination of triazine herbicides in vegetable oils by liquid chromatography. J Chromatogr A 2014; 1373: 9-16. [http://dx.doi.org/10.1016/j.chroma.2014.11.009] [PMID: 25464995]

[132] Wang Y, Sun Y, Xu B, *et al.* Matrix solid-phase dispersion coupled with magnetic ionic liquid dispersive liquid–liquid microextraction for the determination of triazine herbicides in oilseeds. Anal Chim Acta 2015; 888: 67-74. [http://dx.doi.org/10.1016/j.aca.2015.07.028] [PMID: 26320960]

[133] Yao T, Du K. Simultaneous determination of sulfonamides in milk: In-situ magnetic ionic liquid dispersive liquid-liquid microextraction coupled with HPLC. Food Chem 2020; 331: 127342. [http://dx.doi.org/10.1016/j.foodchem.2020.127342] [PMID: 32590266]

[134] Yih Hui B, Mohamad Zain NN, Mohamad S, Varanusupakul P, Osman H, Raoov M. Poly(cyclodextrin-ionic liquid) based ferrofluid: A new class of magnetic colloid for dispersive liquid phase microextraction of polycyclic aromatic hydrocarbons from food samples prior to GC-FID analysis. Food Chem 2020; 314: 126214. [http://dx.doi.org/10.1016/j.foodchem.2020.126214] [PMID: 31972404]

[135] Cao D, Xu X, Xue S, Feng X, Zhang L. An *in situ* derivatization combined with magnetic ionic liquid-based fast dispersive liquid-liquid microextraction for determination of biogenic amines in food samples. Talanta 2019; 199: 212-9. [http://dx.doi.org/10.1016/j.talanta.2019.02.065] [PMID: 30952249]

[136] Cao D, Xu X, Feng X, Zhang L. Designed multifunctional visual observation of magnetic ionic liquid coupling with microwave-assisted derivatization for determination of biogenic amines. Food Chem 2020; 333: 127518.

[http://dx.doi.org/10.1016/j.foodchem.2020.127518] [PMID: 32712545]

[137] Kamarei F, Ebrahimzadeh H, Yamini Y. Optimization of temperature-controlled ionic liquid dispersive liquid phase microextraction combined with high performance liquid chromatography for analysis of chlorobenzenes in water samples. Talanta 2010; 83(1): 36-41. [http://dx.doi.org/10.1016/j.talanta.2010.08.035] [PMID: 21035640]

[138] Elik A. Response surface methodology based on central composite design for optimizing temperature-controlled ionic liquid-based microextraction for the determination of histamine residual in canned fish products. J Food Compos Anal 2021; 98: 103807. [http://dx.doi.org/10.1016/j.jfca.2021.103807]

[139] Wen X, Deng Q, Guo J. Ionic liquid-based single drop microextraction of ultra-trace copper in food and water samples before spectrophotometric determination. Spectrochim Acta A Mol Biomol Spectrosc 2011; 79(5): 1941-5. [http://dx.doi.org/10.1016/j.saa.2011.05.095] [PMID: 21697004]

[140] Wen X, Deng Q, Wang J, Yang S, Zhao X. A new coupling of ionic liquid based-single drop microextraction with tungsten coil electrothermal atomic absorption spectrometry. Spectrochim Acta A Mol Biomol Spectrosc 2013; 105: 320-5. [http://dx.doi.org/10.1016/j.saa.2012.12.040] [PMID: 23318776]

[141] Liu J, Peng J, Chi Y, Jiang G. Determination of formaldehyde in shiitake mushroom by ionic liquid-based liquid-phase microextraction coupled with liquid chromatography. Talanta 2005; 65(3): 705-9. [http://dx.doi.org/10.1016/j.talanta.2004.07.037] [PMID: 18969856]

[142] Zhao FQ, Li J, Zeng BZ. Coupling of ionic liquid-based headspace single-drop microextraction with GC for sensitive detection of phenols. J Sep Sci 2008; 31(16-17): 3045-9. [http://dx.doi.org/10.1002/jssc.200800308] [PMID: 18704999]

[143] Jahromi Z, Mostafavi A, Shamspur T, Mohamadim M. Magnetic ionic liquid assisted single-drop microextraction of ascorbic acid before its voltammetric determination. J Sep Sci 2017; 40(20): 4041-9. [http://dx.doi.org/10.1002/jssc.201700664] [PMID: 28841257]

[144] Arthur CL, Pawliszyn J. Solid phase microextraction with thermal desorption using fused silica optical fibers. Anal Chem 1990; 62(19): 2145-8. [http://dx.doi.org/10.1021/ac00218a019]

[145] Erdem P, Altınışık Tağaç A, Tashakkori P, Seyhan Bozkurt S, Merdivan M. Biopolymer-imidazolium based dicationic ionic liquid modified clay bionanocomposite coating for solid-phase microextraction of phthalate esters. Microchem J 2021; 170: 106648. [http://dx.doi.org/10.1016/j.microc.2021.106648]

[146] Haghi JN, Husain SW, Azar PA, Tehran MS. Fabrication of silica nanoparticle-peg-ionic liquid spme fiber for determination of pesticide residues in tomato. IIOAB J 2017; 8: 11-7.

[147] Liu G, Su P, Yang L, Yang Y. Preparation of novel ionic-liquid-modified magnetic nanoparticles by a microwave-assisted method for sulfonylurea herbicides extraction. J Sep Sci 2015; 38(22): 3936-44. [http://dx.doi.org/10.1002/jssc.201500477] [PMID: 26354900]

[148] Tang Z, Duan Y. Fabrication of porous ionic liquid polymer as solid-phase microextraction coating for analysis of organic acids by gas chromatography – mass spectrometry. Talanta 2017; 172: 45-52. [http://dx.doi.org/10.1016/j.talanta.2017.05.032] [PMID: 28602302]

[149] Wu M, Chen G, Liu P, Zhou W, Jia Q. Preparation of porous aromatic framework/ionic liquid hybrid composite coated solid-phase microextraction fibers and their application in the determination of organochlorine pesticides combined with GC-ECD detection. Analyst 2016; 141(1): 243-50. [http://dx.doi.org/10.1039/C5AN01372K] [PMID: 26579991]

[150] Wu M, Wang L, Zeng B, Zhao F. Ionic liquid polymer functionalized carbon nanotubes-doped poly(3,4-ethylenedioxythiophene) for highly-efficient solid-phase microextraction of carbamate

pesticides. J Chromatogr A 2016; 1444: 42-9.
[http://dx.doi.org/10.1016/j.chroma.2016.03.074] [PMID: 27036210]

[151] Martín-Calero A, Pino V, Ayala JH, González V, Afonso AM. Ionic liquids as mobile phase additives in high-performance liquid chromatography with electrochemical detection: Application to the determination of heterocyclic aromatic amines in meat-based infant foods. Talanta 2009; 79(3): 590-7.
[http://dx.doi.org/10.1016/j.talanta.2009.04.032] [PMID: 19576417]

[152] Pang L, Yang P, Pang R, *et al.* Ionogel-based ionic liquid coating for solid-phase microextraction of organophosphorus pesticides from wine and juice samples. Food Anal Methods 2018; 11(1): 270-81.
[http://dx.doi.org/10.1007/s12161-017-0997-9]

[153] Li L, Wu M, Feng Y, Zhao F, Zeng B. Doping of three-dimensional porous carbon nanotube-graphene-ionic liquid composite into polyaniline for the headspace solid-phase microextraction and gas chromatography determination of alcohols. Anal Chim Acta 2016; 948: 48-54.
[http://dx.doi.org/10.1016/j.aca.2016.11.020] [PMID: 27871609]

[154] Tashakkori P, Tağaç AA, Merdivan M. Fabrication of montmorillonite/ionic liquid composite coated solid-phase microextraction fibers for determination of phenolic compounds in fruit juices by gas chromatography and liquid chromatography. J Chromatogr A 2021; 1635: 461741.
[http://dx.doi.org/10.1016/j.chroma.2020.461741] [PMID: 33253998]

[155] Pacheco-Fernández I, Trujillo-Rodríguez MJ, Kuroda K, Holen AL, Jensen MB, Anderson JL. Zwitterionic polymeric ionic liquid-based sorbent coatings in solid phase microextraction for the determination of short chain free fatty acids. Talanta 2019; 200: 415-23.
[http://dx.doi.org/10.1016/j.talanta.2019.03.073] [PMID: 31036204]

[156] Baltussen E, Sandra P, David F, Cramers C. Stir bar sorptive extraction (SBSE), a novel extraction technique for aqueous samples: Theory and principles. J Microcolumn Sep 1999; 11(10): 737-47.
[http://dx.doi.org/10.1002/(SICI)1520-667X(1999)11:10<737::AID-MCS7>3.0.CO;2-4]

[157] Fan W, Mao X, He M, Chen B, Hu B. Development of novel sol–gel coatings by chemically bonded ionic liquids for stir bar sorptive extraction—application for the determination of NSAIDS in real samples. Anal Bioanal Chem 2014; 406(28): 7261-73.
[http://dx.doi.org/10.1007/s00216-014-8141-9] [PMID: 25234304]

[158] Chen L, Huang X. Preparation of a polymeric ionic liquid-based adsorbent for stir cake sorptive extraction of preservatives in orange juices and tea drinks. Anal Chim Acta 2016; 916: 33-41.
[http://dx.doi.org/10.1016/j.aca.2016.02.030] [PMID: 27016436]

[159] Lin F, Nong S, Huang X, Yuan D. Sensitive determination of organic acid preservatives in juices and soft drinks treated by monolith-based stir cake sorptive extraction and liquid chromatography analysis. Anal Bioanal Chem 2013; 405(6): 2077-81.
[http://dx.doi.org/10.1007/s00216-012-6646-7] [PMID: 23307124]

[160] Wang Y, Zhang J, Huang X, Yuan D. Preparation of stir cake sorptive extraction based on polymeric ionic liquid for the enrichment of benzimidazole anthelmintics in water, honey and milk samples. Anal Chim Acta 2014; 840: 33-41.
[http://dx.doi.org/10.1016/j.aca.2014.06.039] [PMID: 25086891]

[161] Toledo Hijo AAC, Maximo GJ, Costa MC, Batista EAC, Meirelles AJA. Applications of ionic liquids in the food and bioproducts industries. ACS Sustain Chem& Eng 2016; 4(10): 5347-69.
[http://dx.doi.org/10.1021/acssuschemeng.6b00560]

[162] Pinkert A, Marsh KN, Pang S, Staiger MP. Ionic liquids and their interaction with cellulose. Chem Rev 2009; 109(12): 6712-28.
[http://dx.doi.org/10.1021/cr9001947] [PMID: 19757807]

[163] Ventura SPM, e Silva FA, Quental MV, Mondal D, Freire MG, Coutinho JAP. Ionic-liquid-mediated extraction and separation processes for bioactive compounds: Past, present, and future trends. Chem Rev 2017; 117(10): 6984-7052.
[http://dx.doi.org/10.1021/acs.chemrev.6b00550] [PMID: 28151648]

[164] Khan MI, Mubashir M, Zaini D, *et al.* Cumulative impact assessment of hazardous ionic liquids towards aquatic species using risk assessment methods. J Hazard Mater 2021; 415: 125364. [http://dx.doi.org/10.1016/j.jhazmat.2021.125364] [PMID: 33740721]

[165] Pacheco-Fernández I, Pino V, Ayala JH, Afonso AM. Guanidinium ionic liquid-based surfactants as low cytotoxic extractants: Analytical performance in an in-situ dispersive liquid–liquid microextraction method for determining personal care products. J Chromatogr A 2018; 1559: 102-11. [http://dx.doi.org/10.1016/j.chroma.2017.04.061] [PMID: 28487122]

[166] Trush MM, Semenyuta IV, Hodyna D, *et al.* Functionalized imidazolium-based ionic liquids: Biological activity evaluation, toxicity screening, spectroscopic, and molecular docking studies. Med Chem Res 2020; 29(12): 2181-91. [http://dx.doi.org/10.1007/s00044-020-02631-3]

[167] Moučková K, Pacheco-Fernández I, Ayala JH, Bajerová P, Pino V. Evaluation of structurally different ionic liquid-based surfactants in a green microwave-assisted extraction for the flavonoids profile determination of *mangifera* sp. and *passiflora* sp. leaves from canary islands. Molecules 2020; 25(20): 4734. [http://dx.doi.org/10.3390/molecules25204734] [PMID: 33076463]

[168] Chemat F, Abert Vian M, Ravi HK, *et al.* Review of alternative solvents for green extraction of food and natural products: Panorama, principles, applications and prospects. Molecules 2019; 24(16): 3007. [http://dx.doi.org/10.3390/molecules24163007] [PMID: 31430982]

[169] Musarurwa H, Tavengwa NT. Deep eutectic solvent-based dispersive liquid-liquid micro-extraction of pesticides in food samples. Food Chem 2021; 342: 127943. [http://dx.doi.org/10.1016/j.foodchem.2020.127943] [PMID: 33041169]

[170] Imbimbo P, D'Elia L, Liberti D, Olivieri G, Monti DM. Towards green extraction methods from microalgae learning from the classics. Appl Microbiol Biotechnol 2020; 104(21): 9067-77. [http://dx.doi.org/10.1007/s00253-020-10839-x] [PMID: 32960292]

[171] Martins PLG, Braga AR, de Rosso VV. Can ionic liquid solvents be applied in the food industry? Trends Food Sci Technol 2017; 66: 117-24. [http://dx.doi.org/10.1016/j.tifs.2017.06.002]

[172] Mehta MJ, Kumar A. Ionic liquid assisted gelatin films: Green, uv shielding, antioxidant, and antibacterial food packaging materials. ACS Sustain Chem& Eng 2019; 7(9): 8631-6. [http://dx.doi.org/10.1021/acssuschemeng.9b00423]

[173] Wang F, Zhao G, Lang X, Li J, Li X. Lipase-catalyzed synthesis of long-chain cellulose esters using ionic liquid mixtures as reaction media. J Chem Technol Biotechnol 2017; 92(6): 1203-10. [http://dx.doi.org/10.1002/jctb.5109]

[174] Calvano CD, Ceglie CD, D'Accolti L, Zambonin CG. MALDI-TOF mass spectrometry detection of extra-virgin olive oil adulteration with hazelnut oil by analysis of phospholipids using an ionic liquid as matrix and extraction solvent. Food Chem 2012; 134(2): 1192-8. [http://dx.doi.org/10.1016/j.foodchem.2012.02.154] [PMID: 23107747]

CHAPTER 6

Supramolecular Solvents

Noelia Caballero-Casero[1,*] and **Soledad Rubio**[1]

[1] *Department of Analytical Chemistry. Institute of Chemistry for Energy and Environment. Edificio Anexo Marie Curie. Campus de Rabanales, University of Córdoba, Córdoba 14071, Spain*

Abstract: Supramolecular solvents (SUPRASs) are becoming more and more demanded for sample preparation in food analysis. Their inherent properties (*e.g.* different polarity microenvironments, multiple binding sites, discontinuous nature, easy tailoring of their properties, *etc.*) make them highly efficient for the extraction of single- and multi-class contaminants in food matrices. Likewise, they offer numerous opportunities for the development of innovative sample treatment platforms not attainable by conventional solvents. In this chapter, the fundamentals underlying the production of SUPRASs and their more relevant properties regarding their application to the extraction of food contaminants are discussed. An overview of representative developments in this field is given based on the different types of SUPRASs applied so far in food analysis. Major achievements attained, mainly related to the extraction of single- and multi-components prior to their quantification by liquid chromatography coupled to different detection systems, are critically presented. The main challenges to be faced in order to get SUPRAS-based methodologies that meet European requirements for screening/quantification of contaminants in food and promote their use in food control labs are discussed.

Keywords: Coacervates, Cloud point, Extraction, Food analysis, Liquid chromatography-mass spectrometry, Supramolecular solvents.

INTRODUCTION

Supramolecular solvents (SUPRASs), tailored and green nanostructured liquids consisting of self-assembled amphiphilic aggregates, have gained momentum in recent years for sample treatment in analytical chemistry [1]. Major driving forces for their increasing demand in this field are the potential of SUPRASs for facing the challenges that the detection/quantification of contaminants have in areas such as food quality assurance [2], environmental analysis [3], anti-doping control [4], and so on.

* **Corresponding Author Noelia Caballero Casero:** Department of Analytical Chemistry. Institute of Chemistry for Energy and Environment. Edificio Anexo Marie Curie. Campus de Rabanales, University of Córdoba, Córdoba 14071, Spain; Tel: +34 957218644; E-mail: a42caasn@uco.es

Merichel Plaza & Maria Luisa Marina (Eds.)

Developing effective sample treatments in food analysis has to cope with highly demanding requirements. Among them, the high chemical structural variability of the analytes, the required low detection limits, the need for high specificity to avoid false-positive and negative findings; and the high number of samples to be analysed while delivering results in a short time, are worth noting. Thus, regulated and unregulated chemicals in food are very diverse and originate from different sources along the food chain (crop, transport, storage, processing, packaging, *etc.*). Regulated chemicals in food include a variety of contaminants such as mycotoxins, dioxins, polychlorinated biphenyls (PCBs), polycyclic aromatic hydrocarbons (PAHs), residues of veterinary medicines, hormones, pesticides and so on. Unregulated food contaminants are not routinely controlled, despite they could exert a risk to public health. One of the most important sources of unknown potentially toxic food contaminants is food contact materials (FCMs) [5], which constitute a hot topic and a priority for the European Food Safety Authority (EFSA) [6]. Global lists for FCMs do not usually take into account the so-called non-intentionally added substances (NIAS), impurities, by-products or degradation products that originated during food processing or come from recycling and that could constitute the majority of chemicals present in a product [7]. So, effective food quality control requires *multi-methods* for dealing with the huge number of chemicals under regulation, as well as proactive approaches to cope with the potential harm of unknown chemicals and fraudulent practices.

Nowadays, analytical method requirements for food control of regulated substances are well established [8, 9]. Mass spectrometry (MS), combined with liquid chromatography (LC) and gas chromatography (GC), has become the primary technique for confirmatory methods in food control since they provide suitable identification points for legislated chemicals. For this purpose, LC-MS/MS using electrospray (ESI) or atmospheric pressure chemical ionisation (APCI) and GC-MS using electron impact (EI) are routinely used, whose selection mainly depends on the compound polarity and technique availability [10]. GC-head space (HS) injectors have gained popularity in routine analysis, given their potential to perform clean-up in an easy automated way through the selective desorption of analytes from the matrix.

Suspect and non-target screening analysis is nowadays of great interest for identifying emerging food safety issues at their early stages. To address this interest, LC-MS with time-of-flight (TOF, QTOF) and Orbitrap analysers have become the most suitable techniques due to their high capacity for compound identification on the basis of mass accuracy and isotopic pattern [11, 12]. On the other hand, ambient mass spectrometry (AMS) is an emerging approach based on new ion sources capable of desorbing the analytes directly from the sample with minimal or no sample treatment and of ionising chemicals under ambient conditions. AMS is gaining interest in food fraud detection given the possibility of generating results quicker than conventional MS techniques [13, 14].

On the whole, sophisticated processes are mostly required for food quality control. So, there is a need for quick but accurate screening and quantification methods in order to make them affordable and not hamper trade [15]. In this respect, comprehensive MS-based screening and confirmation analytical platforms able to identify and quantify as many as possible *single- or multi-class analytes* in a run should be developed. To fulfil this demand, suitable sample treatments for efficient isolation of a wide polarity range of chemicals while providing also effective sample clean-up should be developed [16]. In short, there is a need for greener, simple and quick matrix- and analyte-independent sample treatments, which are compatible with LC- and GC- combined with MS, and emerging techniques such as AMS. In this respect, green SUPRASs, tailored to meet specific functionalities, represent a great promise for developing innovative sample treatments.

This book chapter deals with some fundamentals and characteristics of SUPRASs relevant for their application to the extraction of contaminants in food, the achievements attained so far in this area, and the ongoing challenges that need to be confronted to make SUPRAS competitive extractants in food analysis.

SYNTHESIS OF SUPRAMOLECULAR SOLVENTS

The synthesis of SUPRASs occurs through a bottom-up approach in which amphiphilic molecules in colloidal solutions undergo self-assembly and coacervation under environmental conditions where amphiphile-amphiphile interactions predominate over amphiphile-solvent interactions [17]. The general synthesis procedure comprises two steps (Fig. **1**). First, amphiphilic molecules are in an aqueous or organic solution above the critical aggregation concentration (*cac*), forming a colloidal suspension. The aggregates in suspension are usually aqueous or reversed micelles or vesicles. Second, the size of the supramolecular aggregates can be controlled by changing the environmental conditions of the colloidal suspension through the action of a coacervation-inducing agent (*e.g.* pH or temperature modification, the addition of inorganic and organic salts or poor solvents for the amphiphile). As a result, the aggregates grow into oily droplets, which spontaneously associate in clusters of individual droplets. The density of these clusters differs from that of the synthesis solution, so they are separated into a new liquid phase (SUPRAS) by flocculation or precipitation. The SUPRAS is not a continuous phase but is composed of coacervate droplets clearly visible by optical microscopy. This phase is in equilibrium with the solution from which it formed (bulk solution), which contains the amphiphile at the *cac* [18].

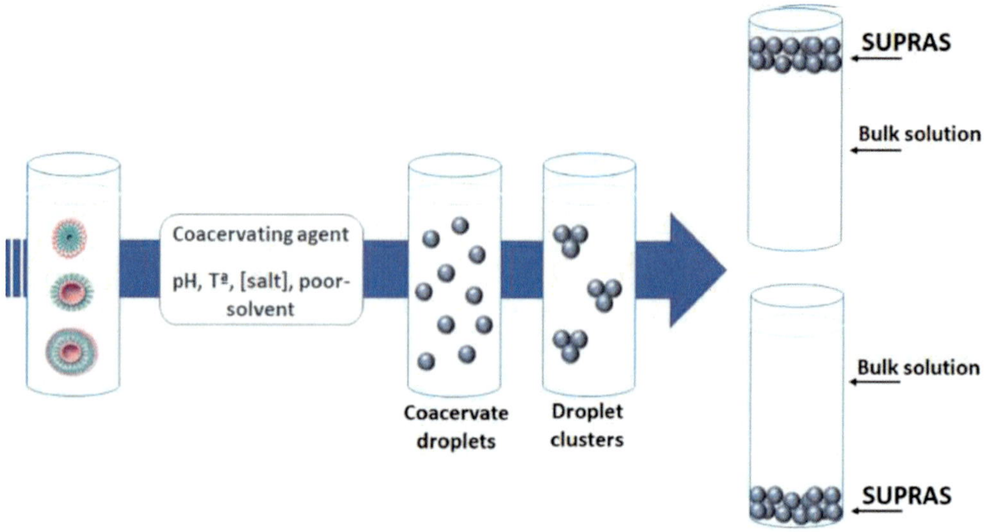

Fig. (1). General scheme of SUPRAS synthesis process.

Coacervation only occurs when the size of colloidal aggregates increases. Mostly, solvophobicity guides the self-assembly of amphiphiles to give colloidal suspensions while the aggregate size is commonly stopped by electrostatic repulsions or steric hindrance among amphiphilic head groups [19]. This implies that to obtain coacervates, environmental conditions allowing higher packing of head groups have to be fostered.

Based on the predominant character of the head group of the amphiphile (ionic/neutral), aggregate growth occurs according to two basic mechanisms [20]. In ionic systems, an effective strategy is the addition of inorganic or organic salts or amphiphilic counter ions in order to reduce the repulsion forces among head groups through charge neutralisation. Similarly, acid concentrations on requirements for ionisable amphiphiles neutralisation [21] promote their coacervation. Regarding nonionic amphiphiles, the reduction of the number of solvent molecules that can solvate head groups of amphiphiles is the best approach to promote their coacervation. This strategy can be accomplished in different ways such as an increment of the temperature [22] or the addition of a poor solvent for the supramolecular aggregates and miscible with the solvation solvent [23]. The increment of temperature removes also some solvent molecules, which reduces the area per head group and the neighbour micelles overlap, triggering SUPRAS formation [24].

For non-ionic amphiphiles bearing zwitterionic head groups, SUPRAS formation is promoted by the reduction of the temperature below a critical value [25]. While it is expectable that electrostatic interactions are involved in both intra- and inter-aggregate interactions, these electrostatic interactions are short-range and, consequently, their effects are qualitatively different from those of ionic surfactants.

Phase diagrams are constructed for delimiting the experimental conditions under which SUPRASs are synthesised [17]. The extent of the SUPRAS region is mainly dependent on the amphiphile and the agent driving the coacervation. In the case of SUPRAS prepared directly in samples, the SUPRAS region may be broadened or shortened by the action of matrix components [26, 27]. An example of this influence is shown in Fig. (**2A**) for SUPRAS prepared from reverse micelles of decanoic acid in THF under the addition of water. Figs. (**2A** and **2B**) depict the phase diagrams for the SUPRAS prepared in distilled water and different beverages, respectively. Except for red tea infusion, sample matrix components moved the upper boundary in the phase diagrams (lines 1,3,4,5 in Fig. **2B**) compared to that obtained in distilled water (Fig. **2A**). Ethanol in beer and wine and sugar in lemon soft drinks accounted for these changes in the extent of SUPRAS region [26].

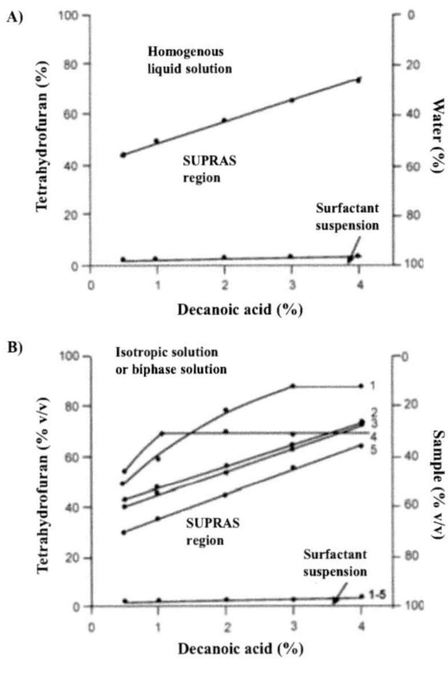

Fig. (2). Phase diagrams for decanoic acid in binary mixtures of (A) THF and water and (B) THF and (1) lemon soft drink, (2) red tea infusion, (3) beer, (4) must and (5) wine (Reproduced with permission from Ballesteros-Gómez *et al.* [26]).

Besides *supramolecular solvents*, a term coined by our research group in 2009 [26], surfactants undergoing liquid phase separation have been also named in the analytical literature as *cloud point* [28, 29] and *coacervates* [30, 31]. These terms refer to the temperature at which non-ionic surfactants become clouding in aqueous solutions, and the spontaneous phase separation (*i.e.* coacervation), respectively. Although the term SUPRAS will be used throughout this book chapter, applications of cloud point extraction and coacervative extraction in food analysis will be also included in the discussion.

CHARACTERISTICS OF SUPRASS RELEVANT FOR EXTRACTION OF CONTAMINANTS IN FOOD

The volume of SUPRAS used in most of the proposed analytical methodologies for the extraction of contaminants in food analysis is usually below 1 mL, so most of the applications in this area are in fact microextractions. Compared to molecular solvents, SUPRAS have unique features that confer outstanding properties for extraction processes [32, 33]. Below the most significant features are discussed.

Supramolecular Aggregates Offer Microenvironments of Different Polarity

Main ingredients in SUPRASs are amphiphiles and a solvent (water or hydro-organic mixtures [23]). Additionally, it may contain salts (organic or inorganic) that are used as coacervation-inducing agents [34].

Ordered structures of amphiphiles in the SUPRAS provide regions of different polarity; polar in the head groups and nonpolar in the hydrocarbon chains. As a result, solutes covering a wide polarity range can be simultaneously solubilised in the distinct regions of the amphiphilic aggregates, which fosters their application in multicomponent analysis. This property, along with the fact that solutes can be solubilised by mixed-mode mechanisms, both in the polar groups (*e.g.* ionic, hydrogen and halogen bonding, dipole-dipole, *etc.*) and hydrocarbon chains (*e.g.* dispersion, π-π, polar hydrophobicity, *etc.*) makes SUPRAS particularly suitable for the development of analyte-independent sample treatments.

Nonpolar analytes are solubilised in the nonpolar microenvironment of the amphiphilic aggregates according to their octanol-water distribution constants. Despite the high extraction efficiencies reported for these types of compounds (*e.g.* extraction of polycyclic aromatic hydrocarbons from smoked meat and fish, bivalve molluscs and processed cereal-based food for infants [35]). However, the most outstanding applications of SUPRASs are related to the extraction of polar compounds (*e.g.* the extraction of contaminants in liquid foods [26, 29]), an area in which conventional organic solvents are inefficient. A great variety of amphiphiles, made up of a diversity of polar groups including neutral (carboxylic acids, polyethene oxides), anionic (carboxylates, sulfates, sulfonates) and cationic (ammonium, pyridinium) have been used for extraction of polar compounds. Among interactions, those providing high energy bonds (*e.g.* ionic and hydrogen bonds) have shown the highest extraction efficiency.

Despite the scarcity of theoretical studies on the mechanisms governing the extraction of solutes in SUPRASs [36], it is widely accepted that mixed-mode interactions are largely responsible for the high extraction efficiencies commonly observed. The establishment of mixed-mode interactions has been clearly proved in the extraction of amphiphilic compounds such as pharmaceutical drugs in freshwater fish [37]. These analytes interact in the hydrophobic and hydrophilic SUPRASs regions and may give mixed analyte-SUPRAS aggregates.

The suitability of SUPRASs to extract analytes in a wide range of polarities from a variety of samples makes them especially suitable for application to the determination of multi-components and the development of generalised sample treatment in food analysis [38]. This ability has been applied to the development of interesting analytical strategies for the determination of multiresidues of sulfonamides in meat [39] and bisphenols and their derivatives in canned food [40].

Supramolecular Aggregates Provide Multiple Binding Sites

Amphiphiles in the SUPRASs are highly packed, sometimes reaching concentrations as high as 1 mg/μL. As a result, there are a huge number of binding sites for solutes, which enable high extraction efficiencies using low SUPRAS volumes. This characteristic renders them highly suitable for microextraction.

The packing degree of amphiphiles in the SUPRAS greatly depends on their structure. Thus, the bigger the size of polar groups or hydrocarbon chains, the lower the concentration of amphiphile in the SUPRAS. For instance, the concentration of alcohol ethoxylate in temperature-induced SUPRASs goes from 0.34 to 1.35 M as the length of the hydrocarbon chain is reduced from 16 to 12 carbon atoms [41]. The packing of amphiphiles is also dependent on the value set for the coacervation-inducing agent (*e.g.* electrolyte concentration, temperature, water content, *etc.*) in the colloidal suspension. Thus, the higher the value for this agent, the greater the concentration of amphiphile in the SUPRAS. As an example, the concentration of Triton X-114 in temperature-induced SUPRASs varies from 0.22 to 0.72 M as the temperature goes from 30 to 60 ºC [42]. Likewise, the concentration of decanoic acid in water-induced SUPRASs goes from 1.22 to 3.19 M as water in the colloidal suspension varies from 70 to 90% [23].

SUPRASs are Discontinuous Liquid Phases

SUPRASs are not continuous liquid phases, as it is clearly observable by microscopy (Fig. **3**). The size of the coacervate droplets making up them is commonly in the interval from nm to μm. So, as SUPRASs are used as extractant phases, extraction equilibria are quickly reached on the basis of the large surface area provided by the coacervate droplets.

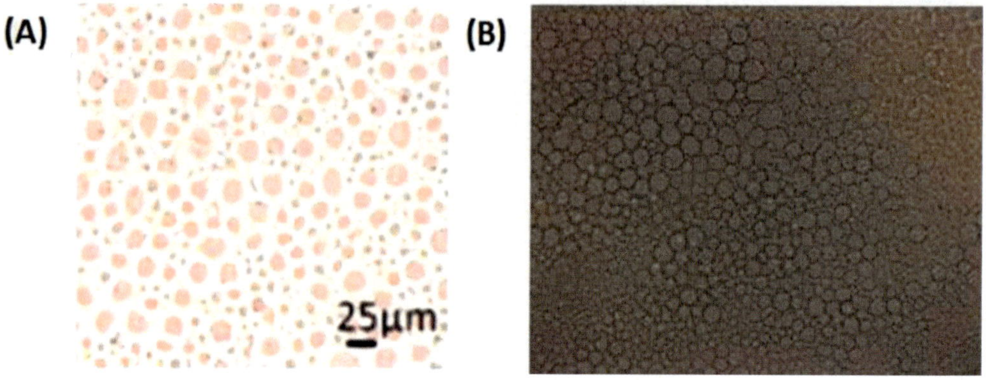

Fig. (3). Micrograph of coacervate droplets of (A) alkanol-based SUPRAS (Reproduced with permission from Ballesteros-Gómez *et al.* [36]) and (B) rhamnolipids in hydrochloric acid.

The Properties and Functionalities of SUPRASs can be Tailored

The ability to design SUPRASs with specific composition, properties and nanostructures is one of their most interesting features [20, 38]. SUPRAS tailoring can be easily achieved by the selection of the amphiphile and/or coacervation conditions.

Selection of the amphiphile enables the maximization of solute-SUPRAS chemical interactions, and consequently, the enhancement of extraction efficiency. Given the large availability of both natural and synthetic amphiphiles, their selection should be only guided by a good knowledge of the mechanisms governing the solubilisation of targeted solutes.

Selection of the coacervation conditions enables the control of amphiphile self-assembly, which offers great opportunities for the tailoring of SUPRAS composition and nanostructures [43]. So, in this respect, SUPRAS are considered environment-responsive solvents.

Environment-responsive SUPRAS allow the development of matrix-independent methods by achieving simultaneously analyte extraction and sample cleanup. SUPRAS made up of carboxylic acids [23] or alkanols [36], organic solvent and water (usually water–tetrahydrofuran or water-ethanol) are interesting examples of environment-responsive SUPRASs. As it is depicted in Fig. (**4A**), these SUPRAS are formed by inverted hexagonal-shaped supramolecular aggregates in which the head groups of the amphiphile are arranged to form an aqueous cavity inside the aggregate, while the hydrocarbon chains are arranged outwards and dispersed in the organic solvent (*i.e.* tetrahydrofuran). The environmental conditions allow tailoring the size of the aqueous cavities of the aggregates (the increment of organic solvent to water ratio in the synthesis led to an increment of the aqueous cavities size (Fig. **4A**), whereby SUPRASs with restricted access properties (SUPRAS-RAM) [36] have been synthesized. Fig. (**4B**) shows the micrograph of the whole SUPRAS, a single coacervate droplet and its internal structure, respectively, obtained by cryo-scanning electron microscopy (cryo-SEM). The three micrographs show the hexagonal-shaped aggregates at both the nano- and micro-scale.

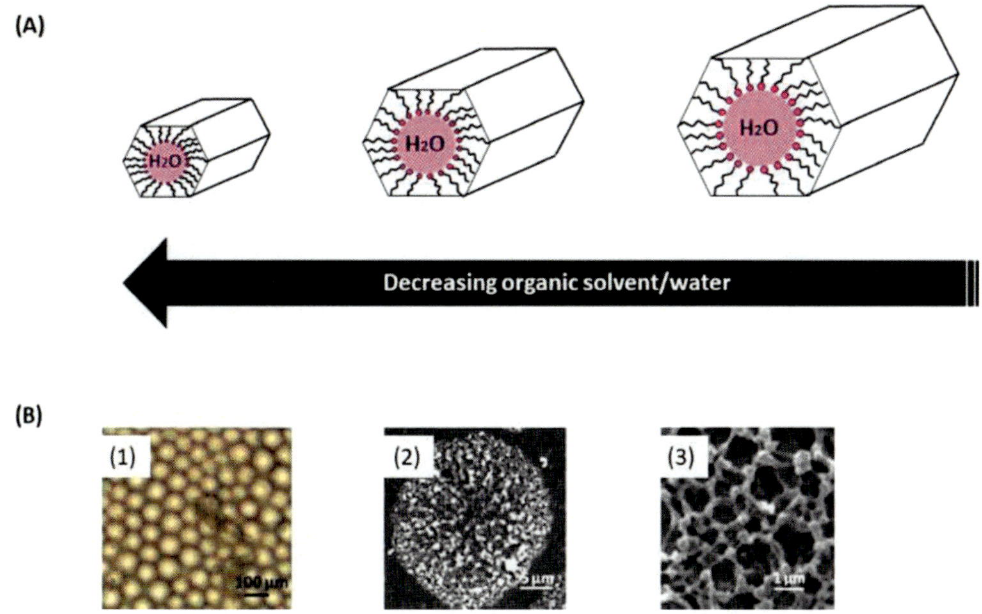

Fig. (4). (A) Picture of the inverted hexagonal arrangement of aggregates and the effect of environmental conditions on the SUPRAS structure. (B) Micrographs of (1) alkanol-based SUPRAS, (2) coacervate droplets, and (3) coacervate droplet surface obtained by microscopy (Reproduced with permission from Ballesteros-Gómez *et al.* [36]).

SUPRAS-RAM do not extract polar macromolecules such as polysaccharides or humic acids because SUPRAS aqueous cavities are not large enough for their incorporation (Fig. **5A**). On the other hand, SUPRAS-RAM do not also extract proteins because these macromolecules denaturalise and flocculate by the action of SUPRAS components (Fig. **5B**).

The ability of SUPRAS-RAM to simultaneously solubilise analytes in a wide range of polarity and exclude polar macromolecules allows the development of general sample treatment platforms that enable analytes' extraction and sample cleanup in a single stage. Interesting examples of that are the application of SUPRAS-RAM for the extraction of both ochratoxin A in spices subject to EU regulation (*i.e. Capsicum* spp., *piper* spp., nutmeg, ginger and turmeric) [44] and *Fusarium* toxins in cereals [45]. Conventional SUPRAS potentially produce strong matrix effects as combined with LC-MS [46]. However, the use of SUPRAS-RAM allowed external calibration-based quantitation even for the direct analysis of SUPRAS extract by LC-MS/MS [45] or LC-FL [44].

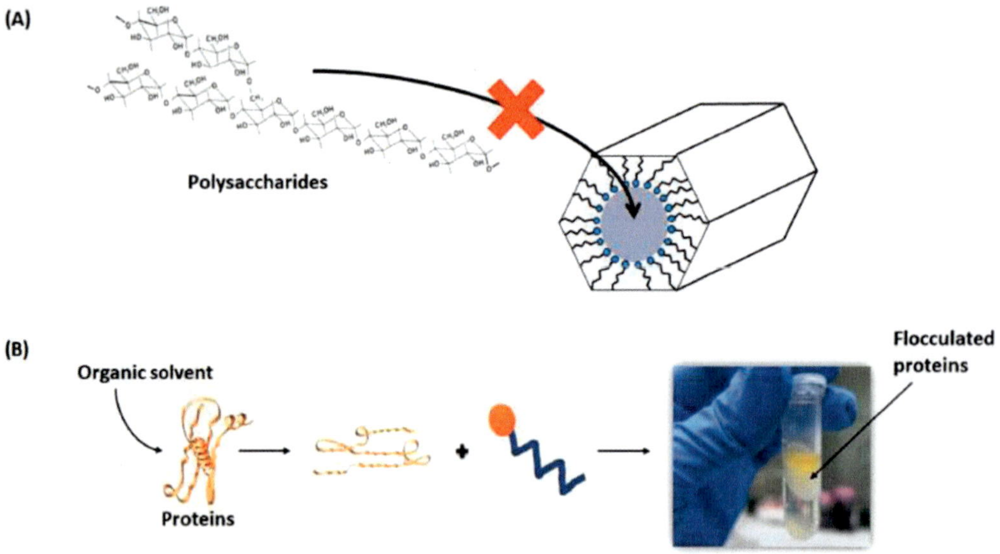

Fig. (5). Restricted access properties of SUPRASs for interferences removal by (A) physical and (B) chemical mechanisms.

The SUPRAS-RAM's ability to reduce matrix effects can be insufficient in the case of samples with high lipid content. Lipids from the matrix are extracted by SUPRAS through the formation of mixed aggregates with the amphiphiles making up them and may produce matrix effects. A strategy to deal with this interference is the use of SUPRAS-RAM synthesised from volatile amphiphiles [27]. After extraction, the SUPRAS extract is evaporated to dryness, producing the precipitation of matrix lipids. Then, the analytes are solubilised in a low volume of hydro-organic solution where lipidic components are not soluble. Based on these types of amphiphiles, it has been recently reported the determination of 21 bisphenols and derivatives (log K_{ow} from 1.254 to 6.564) in a variety of canned food and beverages [47]. Treatment of both liquid and solid samples was based on the use of low volumes of SUPRAS (190-200 µL). High absolute recoveries (70-114%) along with low matrix effects (signal suppression or enhancement in the range 78-116%) were obtained for food samples encompassing a wide range of macromolecules content (*e.g.* protein, fat, carbohydrates, *etc.*). Fig. (**6**) shows a schematic of the whole procedure. Quantification was carried out by liquid chromatography and electrospray tandem mass spectrometry using external calibration.

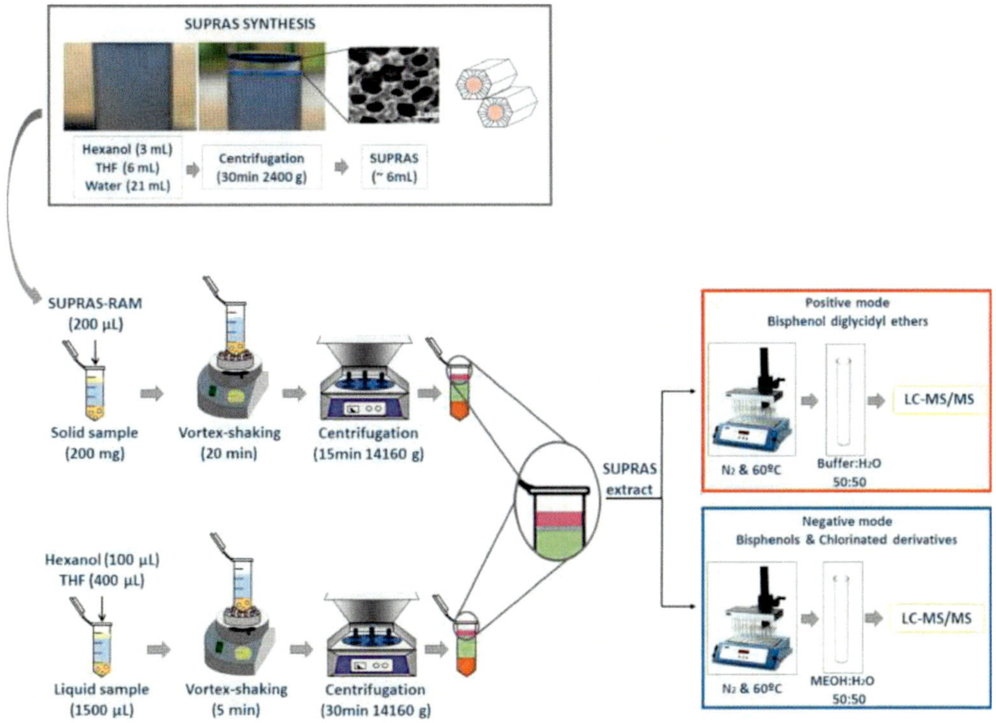

Fig. (6). Overview of the SUPRAS synthesis and its structure (top) for the extraction/cleanup of solid samples (middle), and the SUPRAS synthesis into the liquid sample and simultaneous extraction/cleanup (bottom); and LC-ESI-MS/MS sequential analysis of bisphenol diglycidyl ethers and bisphenols/chlorinated bisphenols (right) (Reproduced with permission from Caballero-Casero and Rubio.[47]).

SUPRAS Match Most of the Criteria Set for Green Solvents

A valuable asset of SUPRASs is that their compliance with the twelve criteria set for green solvents can be easily reached [48]. Thus their *synthesis* occurs through an energy-saving approach based on self-assembly processes that are highly selective and efficient. On the other hand, their *performance* regarding extraction processes compares favourably with those of organic solvents in terms of applicability, extraction yields and tailoring possibilities. Regarding *storage*, SUPRAS nanostructures are highly stable once separated from the solution where coacervation occurred and storage conditions will mainly depend on SUPRAS composition. Nevertheless, storage is not necessary for applications involving liquid-liquid extraction since SUPRASs are mostly prepared *in situ* in the sample.

The nine remaining criteria are all related to the properties (*toxicity, biodegradability, stability, flammability,* and *recyclability)* and market status (*grade, price, availability,* and *renewability)* of the selected amphiphiles since they constitute the major component in SUPRASs. In this respect, it is worth noting that the market of natural amphiphiles obtained from renewable sources (*i.e.*bio-amphiphiles) at competitive prices is becoming increasingly important. Thus, bioSUPRASs made up of rhamnolipids which are produced by microorganisms through fermentation processes, have been recently reported [49, 50].

An aspect that remains largely unexplored in this context is SUPRAS recyclability, mainly due to the minute volume of solvent used in analytical microextractions and the fact that SUPRASs are directly prepared in the sample. However, recyclability is desirable in applications where higher SUPRAS volumes are consumed (*e.g.* in the extraction of bioactives from vegetal biomass). In this case, both the amphiphile in the SUPRAS and the coacervation-inducing agents remaining in the equilibrium solution after SUPRAS separation can be recycled. The strategy for the recovery of amphiphiles will depend on their chemical and physical properties. On the other hand, the equilibrium solution containing the coacervation-inducing agents can be reused for new syntheses of SUPRAS after the addition of the amphiphile and some replenishment of these agents whenever necessary.

SUPRAS Extractions Adapt to Different Formats

SUPRASs have been applied to the extraction of contaminants in both liquid and solid food samples [2, 26, 29, 51]. In the case of liquid food, the SUPRAS is formed *in situ* in the sample. The synthesis process consists of the addition of the amphiphile to the liquid sample and establishing the proper experimental conditions to promote coacervation. SUPRAS formation is a spontaneous and instantaneous process, giving a non-continuous phase since coacervate droplets maintain their integrity. As a consequence, the extraction equilibrium is rapidly reached, being usually required only a few minutes of sample stirring. Moreover, most of the procedures include a subsequent step of centrifugation in order to accelerate the separation of SUPRAS from the bulk solution. Once the extraction is performed, the SUPRAS extract or an aliquot is submitted for analysis.

Calculation of the volume of SUPRAS produced in the sample as a function of the concentration of amphiphile and/or coacervation-inducing agents used for

SUPRAS synthesis can be easily performed through previously derived equations [33]. When no equation has been derived, there are different strategies to measure the SUPRAS volume obtained. In the case of SUPRAS with lower density than the sample, such as those made up of ionic amphiphiles, alkyl carboxylic acids, alkanols, *etc*.; the volume can be easily calculated through the cylindrical volume equation $\pi r^2 h$, where *h* is the height of the SUPRAS column in the centrifuge tube and *r* is the radius of the tube. Both parameters can be measured with a digital calliper. To facilitate the measurement, especially when low volumes of solvent (microlitres) are produced [52], is advisable to use centrifuges tubes of narrow neck shape (*e.g*. Fig. **7A**) in the SUPRAS synthesis. A scheme of the overall methodology proposed for the extraction of mycotoxin in red wine [52] is shown in Fig. (**7A**). For SUPRAS denser than the sample (*e.g*. many non-ionic and zwitterionic surfactants), the measurement and withdrawal of SUPRAS can be facilitated by increasing the viscosity of the SUPRAS and decanting the supernatant. For this purpose, after the centrifugation step, the SUPRAS is cooled down. Once the SUPRAS has been separated from the bulk solution, the volume can be measured (*e.g*. with a microsyringe) or calculated by weight.

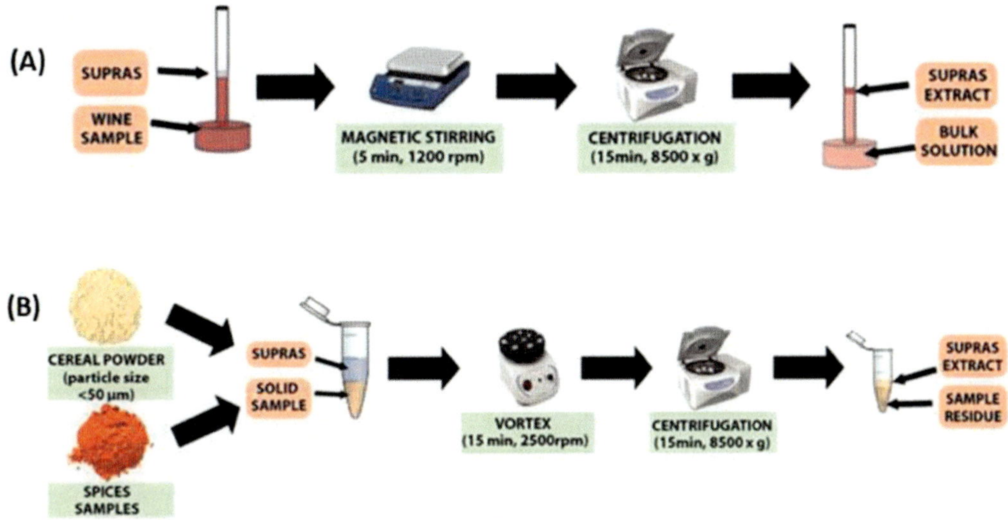

Fig. (7). Illustrative scheme of the mycotoxins SUPRAS-based extraction from (A) red wine, a liquid sample, and (B) solid foodstuffs. (Reproduced with permission from García-Fonseca *et al.* [52]).

Regarding the extraction of analytes from solid food samples, SUPRAS is synthesised in a previous stage. Typically, SUPRAS is produced and stored until its use. The SUPRAS volume obtained is enough to treat multiple food solid samples, usually 30-50 samples. Two strategies have been followed for the application of SUPRAS to the extraction of solid food. The first approach is based on the direct addition of SUPRAS to the solid food sample. Typically, low sample amount (*e.g.* 01-05 g) is extracted with low SUPRAS volume (*e.g.* 0.2-0.8 mL). Then, the analyte extraction is facilitated by vortexing the mixture and subsequently is submitted to centrifugation for accelerating phase separation. A schematic of this strategy [52] is shown in Fig. (**7B**). This approach is suitable for multianalyte extraction within a wide range of polarity, considering the highly polar fraction. An example is the application of inverted hexagonal aggregates of decanoic acid-based [53] SUPRAS to the extraction of veterinary residues belonging to the group of quinolone antibiotics (log K_{ow} = 0.94-1.6) from farmed fish and shellfish, achieving high extraction efficiencies for all the target analytes (recoveries of 99-102%). In the second strategy, the equilibrium solution produced during the coacervation process is added together with the SUPRAS to the solid sample, resulting in three phases: the solid residue at the bottom, the SUPRAS extract in the upper phase and the equilibrium solution in between. The equilibrium solution has two functions, it reduces the required volume of SUPRAS by wetting the sample and improves the selectivity by entrapping polar matrix components. It is also worth noting that very polar analytes can distribute between the SUPRAS and the equilibrium solution, decreasing the extraction efficiency. Thus, the use of the equilibrium solution is not recommended for the extraction of very polar analytes. This second strategy has been applied to the decanoic acid vesicles SUPRAS-based extraction of benzimidazole fungicides in fruits and vegetables [54].

Recently, SUPRASs combining magnetic nanoparticles (MNPs) have attracted attention since they allow separating the SUPRAS from the liquid sample without a centrifugation stage [55 - 57]. With this aim, the SUPRAS is mixed with the MNPs employing two different ways. The first option proposes the formation of the SUPRAS *in situ* in the liquid sample continued by the addition of the MNPs. While in the second, SUPRAS and MNPs are mixed to form a ferrofluid, which is added to the liquid sample. The remainder of the process is the same in both cases, the mixture is stirred to facilitate the extraction after which SUPRAS extract is separated from the liquid sample by means of a magnet. The sample can be discarded by decantation and the SUPRAS extract is solubilised in an aliquot of organic solvent. SUPRASs combining MNPs have been applied to the extraction of pesticides from water and fruit juices [55 - 57].

SUPRASs are Compatible with Chromatographic Techniques

From the very beginning, SUPRASs have shown high compatibility with liquid chromatography (LC), particularly with reversed-phase LC[38]. The main reasons for this fruitful coupling are that both are able to deal with compounds covering a wide polarity range and that the hydro-organic mobile phases used in this LC mode break down the supramolecular aggregates into amphiphile monomers. Compatibility has been also proved for other chromatographic modes including micellar liquid chromatography (MLC) [58] or chiral LC [59].

In most applications, SUPRAS extracts are directly injected into the chromatographic system and they do not commonly influence the chromatographic profile of analytes. A particular case is the use of SUPRAS made up of highly water-insoluble amphiphiles in chromatographic determinations where the mobile phase contains high water proportion (*e.g.* >80%). Under these conditions, SUPRAS do not break down into monomers but in colloidal aggregates, thus constituting a pseudophase where solute may partition. As a result, chromatographic profiles may be modified and the use of analyte standards in SUPRAS is imperative for their accurate identification and/or quantification. As the viscosity of the SUPRAS is high, the extract is diluted with a polar solvent such as methanol or acetonitrile (usually twice) before injection [41].

The compatibilisation of SUPRAS with the most used LC detectors (UV-visible, fluorescence, and mass spectrometry) mainly depends on the characteristics of the amphiphile used for its synthesis. Initially, SUPRASs were mostly made up of amphiphiles belonging to the Triton X (polyoxyethylene (n) tert-octylphenyl ethers) and PONPE (polyoxyethylene (n) nonyl phenyl ethers) series, which undergo coacervation above a specific temperature. These amphiphiles are commercially available as a mixture of homologs and oligomers and many of them show high absorbance and fluorescence signals. So, although they were combined with LC coupled to UV-visible [60] and fluorescence [61] detectors, applications were restricted to analytes with absorption or fluorescence characteristics different to those of amphiphiles [62, 63]. A step forward in this respect was the synthesis of SUPRAS from pure monomers of amphiphiles (to avoid the chromatographic coelution produced by homologs and oligomers) that absorb below 200 nm. An example was the SUPRASs consisting of alkylammonium sulfate [64] and alkyl sulfates and sulfonates [21]. These and many other SUPRAS have enabled the development of many interesting applications based on LC coupled to UV-visible and fluorescence detection [2, 51].

A major advance in the combination of LC with SUPRASs has been their use in MS [33, 38]. Although amphiphiles are usually selected to give low ionisation efficiency in the ionisation source (*e.g.* alkyl alcohols in ionisation by electrospray, ESI), the amphiphilic monomers leaving the chromatographic column are automatically sent to waste before entering the ionisation source and, in this respect, they are considered as matrix components.

SUPRASs compatibility with gas chromatography (GC) has been less straightforward owing to the low volatility of amphiphiles. Strategies to increment compatibility have mainly focused on removing the amphiphile before chromatographic analysis by back-extraction assisted by ultrasonic or microwaves with a polar solvent [65] or by retention on an ion exchange sorbent [66].

SUPRAS APPLICATIONS IN FOOD ANALYSIS: ACHIEVEMENTS AND CHALLENGES AHEAD

Table **1** shows representative applications of SUPRAS to sample treatment in food analysis. While this list is not exhaustive, it sets out a reference framework for understanding the achievements made so far in this field as well as the challenges ahead. Applications have been organised according to the type of inducing agents used for amphiphile coacervation (*e.g.* temperature, water, salt or acid) since they are closely linked to the scope of SUPRAS application. Within each type of SUPRAS, applications have been grouped according to the type of food contaminant (*e.g.* pesticides, PAHs, veterinary residues, mycotoxins, packaging material residues, emerging contaminants, *etc.*) or additive (*e.g.* dyes, antioxidants, *etc.*).

The ***temperature-induced SUPRASs***, made up of nonionic amphiphiles, were the only ones available for analytical extractions until around the decade of the 2000s. Currently, although many nonionic amphiphiles are commercially available, the alkylphenol ethoxylate Triton X-114 continues as the preferred one, followed by Triton X-100 (Table **1**). The applications developed so far [60, 67 - 72] have extensively proved that the different microenvironments existing in these SUPRASs made them effective extractants of a wide variety of nonpolar and polar contaminants (*e.g.* pesticides, antibiotics, hormones, mycotoxins), and additives (*e.g.* dyes) in food. Thus, excellent extraction recoveries have been obtained in most of the applications (Table **1**). On the other hand, the high number of binding sites provided by these SUPRASs has permitted their application at low solvent/sample ratios. As a result, detection limits at the $\mu g\ L^{-1}$ or $\mu g\ kg^{-1}$ level have been generally reached.

Nevertheless, despite these valuable assets, temperature-induced SUPRASs feature two important operational drawbacks that have hindered their use in current food control labs. Firstly, extractions have to be carried out at temperatures above the cloud point in order to maintain SUPRAS integrity (see Table **1**). This requirement adds complexity to the sample treatment process and excludes the extraction of volatile and thermolabile compounds. A strategy frequently used for reducing the working extraction temperature is to add salts to the colloidal suspension (*e.g.* Na_2SO_4, NaCl, Na_2CO_3), since the salts decrease the cloud point of nonionic amphiphiles (Table **1**).

On the other hand, as commented above, the nonionic amphiphiles undergoing temperature-induced coacervation are commercially available as a mixture of homologs and oligomers. This fact constitutes a serious drawback for chromatographic analysis owing to their co-elution with the targeted food contaminants. As a result, food methods based on extractions with these types of SUPRAS are mostly limited to the quantification of a few contaminants not co-eluting with the homologs/oligomers of the amphiphile (Table **1**). Likewise, because it is very difficult to create many chromatographic segments to send all the homologues/oligomers to waste before their detection, the use of mass spectrometry is hindered and most of the applications in this field have been based on LC-UV (Table **1**). This fact is a great limiting factor of the cloud point technique in food analysis, given the requirements of current food legislation for the identification of contaminants, which make MS necessary. So, the methods developed so far using temperature-induced SUPRAS could only serve, at best, as screening methods. Regarding dyes, given that selective detection can be achieved in many cases using diode array detectors, these methods could serve as quantification methods.

The development of ***water-induced SUPRASs*** using alkyl carboxylic acids [23] and alkanols [36] solved most of the operational problems related to temperature-induced SUPRAS while maintaining and improving their valuable extraction characteristics. So, many applications closer to the needs of current food control labs started being developed [33, 73 - 82].

Water-induced SUPRASs are synthesised at room temperature, and although their chemical properties and nanostructures do not change in a wide range of temperature values, most of the proposed applications have been developed without fixing this variable (Table **1**), thus adding simplicity to the extraction process. On the other hand, both alkyl carboxylic acids and alkanols are commercially available as pure compounds. So, they give a single peak in chromatographic analysis, thus facilitating their no co-elution with food contaminants. This, along with the fact that these amphiphiles do not absorb above 230 nm and do not give a fluorescence signal, has allowed the quantification of a higher number of contaminants in a single run (*e.g.* 16 PAHs in apple peel [83]) compared to temperature-induced SUPRASs. Particular attention has been given to the determination of contaminants by LC-FD owing to the higher sensitivity and selectivity of this technique compared to LC-UV. In this respect, interesting applications have been developed for PAHs [26, 83], veterinary residues [39, 53], mycotoxins [26, 44, 84, 85] or bisphenols [40, 86, 87] in a variety of liquid and solid food samples (Table **1**).

Table 1. Applications of supramolecular solvents to the extraction of contaminants and additives in food.

Compounds	Type of food	SUPRAS components	Extraction recoveries (%)	Instrumental analysis	LOD (µg/L or µg/Kg)	Refs.
Temperature-Induced SUPRAS						
Pesticides						
Diquat, paraquat	milk	Triton X-114/60°C	95-98	CZE-UV	4-18	[69]
Fenitrothion, chlorpyrifos, parathion, methidathion	honey	Triton X-114/85°C	90-107	GC-MS	0.03-0.47	[68]
Methomyl, propoxur, carbofuran, carbaryl, isoprocarb, promecarb	Apple, pear, guava	Triton X-114/NaCl/45°C	80-107	LC-UV	0.1-1	[60]
Aprocarb, carbofuran, isoprocarb, fenobucarb	Corn	Triton X-100/Na_2SO_4/25°C	84-93	LC-UV	0.2-0.5	[72]
Carbofuran	Rice	Triton X-100/NaCl/ $(NH_4)_2SO_4$/50°C	88-92	LC-UV	0.5	[67]
Dichlorvos, methamidophos, acephate, diazinon, dimethoate, chlorpyrifos, parathion-methyl, malathion and parathion-ethyl	Fruit juice	PEG 6000/ Na_2SO_4/20°C Back extraction with ethyl acetate	72-95	GC-FPD	0.5-3.0	[71]
Metsulfuron-methyl, chlorsulfuron, bensulfuron-methyl	Rice	Triton X-114/ Na_2SO_4/50°C	82-91	LC-UV	4-6	[70]
Veterinary Residues						
Ampicillin	Bovine milk	Triton X-114/Cetyltrimethyl ammonium/Na_2SO_4/40 °C	79-99	LC-UV	2-3	[102]
Sulfafurazole, sulfa-methoxazole, sulfamethoxypyridazine, sulfaquinoxaline sulfachloropyrid-azine, sulfadimethoxine, Sulfamonom-ethoxine	Milk	Triton X-114/ Na_2SO_4/50 °C	67-106	LC-UV	2.2-9.8	[103]
Hormones						
Beclometasone dipropionate, hydrocortisone butyrate, nandrolone phenylpropionate	Milk	Tergitol TMN-6/NaCl/35 °C	95-98	LC-DAD	0.01-0.017	[104]
Mycotoxins						
Ergotamine, ergonovine	Cereals (grains and flours)	Water/buffer tetraborate/PONPE 7,5 +ethanol	>92	CZE-UV	2.2-2.6	[100]
Dyes						
Sudan I, II, III, IV	Chilli powder	Triton X-100/Na_2CO_3/70 °C	81-85	LC-UV	2-4	[105]
Allura red	Candy, drinks, pastel, jelly	Triton X-100/Cetyltrimethyl ammonium/KCl/	96-106	UV-Vis spectrophotometry	7.8	[106]
Carmoisine, Brilliant blue FCF	Raspberry jelly, fruity candy, smarties	Triton X-100/NaCl/76 °C	96-104	UV-Vis spectrophotometry	16-17	[107]
Quercetin	Onion, tomato, apple and orange juice	Triton X-114/Na_2SO_4/40 °C	87-94	UV-Vis spectrophotometry	2.2	[108]
Antioxidants						
Xanthohumol	Beer	Triton X-114/NaCl/70 °C	91-102	LC-UV	3	[109]

(Table 1) cont.....

Compounds	Type of food	SUPRAS components	Extraction recoveries (%)	Instrumental analysis	LOD (µg/L or µg/Kg)	Refs.
Water-Induced SUPRAS						
Pesticides						
Monuron, linuron, isoproturon	Water, rice	Decanoic acid/THF	80-99	LC-UV	10-30	[33]
Ethion, phosalone, diazinon, chlorpyrifos	Rice, cucumber, tomatoes	Decanoic acid/THF/NaCl	102-178	LC-UV	0.05-0.2	[73]
Permethrin enantiomers (1S-trans, 1R-trans, 1S-cis, 1R-cis)	Tomato, green bean, cabbage, apple, orange, melon, green pepper	Hexanol/THF	93-105	LC-MS/MS	0.2-0.3	[78]
Carbaryl	Water, fruits, vegetables	Heptanol/THF	90-122	LC-MS/MS	30	[76]
Malathion	Water	Heptanol/THF	Non reported	LC-MS	1.4	[75]
Malathion	Water, eggplant, tomatoes	Heptanol/THF	89-104	LC-MS	1.4	[75]
Triadimefon, triadimenol	Beer	Decanol/THF	84-100	LC-MS/MS	0.24-0.98	[74]
Diuron, hexazinone, ametryn, tebuthiuron	Natural water	Decanol/THF/NaCl	95-111	LC-DAD	0.13-1.45	[81]
Carbendazim, fipronil, picoxystrobin	Natural water	Decanol/THF/NaCl	93-110	LC-DAD	0.23-0.45	[82]
Chlorpyrifos, diazinon, phosalone	Orange juice, tap water	Undecanol/THF	>94	LC-UV	0.5-1.3	[80]
Pendimethalin, trifluralin	Water, onion	Undecanol/THF	91–98	LC-UV	1.3-2.1	[77]
Perfluorooctanoic acid, perfluoroundecanoic acid, perfluorododecanoic acid, perfluorotridecanoic acid, flusilazole, flubendiamide, fipronil, triflumizole, flufenoxuron	Drinking and environmental water	Undecanol/THF/NaCl	81-107	LC-MS	0.42-0.84	[79]
PAHs						
Benzo[a]pyrene	Tea and coffee infusions	Decanoic acid/THF	90-96	LC-FD	0.11–0.13 ng L^{-1}	[26]
Naphthalene, acenaphthylene, acenaphthene, fluorene, phenanthrene anthracene, fluoranthene, pyrene, benzo(a)anthracene, chrysene, benzo(b)fluoroanthene, benzo(k)fluoroanthene, benzo(a)pyrene, dibenz(a,h)anthracene, benzo(g,h,i)perylene, indeno(1,2,3-cd)pyrene	Apple peel	Decanoic acid/THF after supercritical fluid extraction	60-78	LC-FD	0.34-1.27	[83]
Veterinary Residues						
Flumequine, oxolinic acid	Fish and shellfish	Decanoic acid/THF	99–102	LC-FD	1.5-2.7	[53]
Sulfadiazine, sulfamerazine, sulfamethoxypyridazine, sulfachloropyridazine, sulfadoxine, sulfamethoxazole, sulfadimethoxine and sulfaquinoxaline	Muscle tissues of pork, chicken, turkey, lamb and beef	Decanoic acid/THF	98-109	LC-FD	3.4-13.1	[39]

(Table 1) cont.....

Compounds	Type of food	SUPRAS components	Extraction recoveries (%)	Instrumental analysis	LOD (µg/L or µg/Kg)	Refs.
Ibuprofen, naproxen, ketoprofen	Freshwater fish	Decanoic acid/THF	97-104	LC-MS/MS	1.7-3.3	[37]
Albendazole, Albendazole sulfone, Albendazole sulfoxide, Albendazole-2-aminosulfone, Febendazole, Fenbendazole sulfone, Oxfendazole, Febantel	Cow milk	Hexanol/THF	80-110	LC-MS/MS	0.03-0.14	[110]
Narasin A, Salinomycin, Lasalocid A, Maduramicin, Monensin A, Semduramicin	Cow milk	Hexanol/THF	71-112	LC-MS/MS	0.004-0.07	[111]
Mycotoxins						
Ochratoxin A	Wheat	Decanoic acid/THF	84-95	LC-FD	0.5	[84]
Ochratoxin A	White, red and rose wine	Decanoic acid/THF	85-100	LC-FD	4.5-15 ng L^{-1}	[85]
Ochratoxin A	Water, wine, vinegar, must, beer	Decanoic acid/THF	79-93	LC-FD	4-9 ng L^{-1}	[26]
Ochratoxin A	Hot and sweet paprika, black pepper, nutmeg, ginger, turmeric	Decanoic acid/THF	81-101	LC-FD	1.2-3	[44]
Ochratoxin A, Aflatoxin B1	Wine, spices, cereals	Tetradecanoic acid/THF	75-96%	Enzyme-linked Immunosorbent assay (ELISA)	100-3100	[52]
Deoxynivalenol, Fumonisin B1, Fumonisin B2, Zearalenone	Wheat, maize	Oleic acid/THF	87-105	LC-ESI-Ion trap/MS	8-15	[45]
Dyes						
Bixin, Norbixin	Snack, Ice-cream, crème caramel, cheese, margarine	Octanoic acid/THF	78-113	LC-DAD	63-76	[112]
Sudan I, II, III, IV	Chilli-containing sauces	Decanoic acid/THF	86-108	LC-UV	2.7-7.4	[113]
Curcumin, demethoxycurcumin, bis-demethoxycurcumin	Snacks, gelatine, yoghurt, mayonnaise, butter, candy, fish	Decanoic acid/THF	82-106	LC-UV	0.95-3.5	[114]
Rhodamine B	Chilli powder	Pentanol/THF	95-117	UV-Vis Spectrophotometry	8	[115]
Curcumin	Turmeric, tea	Undecanol/THF/NaCl	94-104	Smartphone digital image colourimetry	200-900	[116]
Packaging material residues						
Bisphenol A	Canned fruits and vegetables	Decanoic acid/THF	81-96	LC-FD	1.3	[86]
Bisphenol A	Canned fatty food (fish and meat)	Decanoic acid/THF	90–99	LC-FD	9	[87]
Bisphenol A	Water, white soda	Decanoic acid/THF	90-96	LC-ESI-MS/MS	0.20-0.21	[26]

(Table 1) cont.....

Compounds	Type of food	SUPRAS components	Extraction recoveries (%)	Instrumental analysis	LOD (μg/L or μg/Kg)	Refs.
Bisphenols A (BPA), bisphenol B (BPB), bisphenol E (BPE), bisphenol F (BPF), bisphenol P (BPP), bisphenol S (BPS), bisphenol Z (BPZ), bisphenol AF (BPAF), bisphenol AP (BPAP), monochloro-BPA (MCBPA), dichloro-BPA (DCBPA), trichloro-BPA (TCBPA), tetrachloro BPA (TeCBPA), bisphenol A diglycidyl ether (BADGE) and their derivatives (BADGE.H_2O, BADGE.2H_2O, BADGE.HCl, BADGE.2HCl, BADGE.HCl.H_2O), bisphenol F diglycidyl ether (BFDGE), and the derivative BFDGE.2H_2O	Sweet corn, lentils, chickpeas, Sausages, meatballs, tonic, bottled water, tea, cola, beer	Hexanol/THF	70-114	LC-ESI-MS/MS	0.019-0.19 (liquid food) 0.06-0.81 (solid food)	[47]
Bisphenol A, bisphenol AF, tetrabromobisphenol A	Lactic acid milk drinks, carbonated drinks, milk	Octanol/THF	91-105	LC-DAD	0.14–0.32	[117]
Bisphenol A, bisphenol B, bisphenol F, bisphenol E, bisphenol F diglycidyl ether (BFDGE), bisphenol A diglycidyl ether (BADGE)] and their derivatives [BADGE·2H2O, BADGE·H2O, BADGE·HCl·H2O, BADGE·HCl, BADGE·2HCl and BFDGE·2HCl]	Canned foodstuffs (vegetables, legumes, fruits, fish and other seafood, meat products and grains)	Tetradecanol/THF	80-110	LC-FD	0.3-1.2	[40]
		Emerging contaminants				
Stereoisomers of hexabromocyclododecane (HBCD): (-)-α-HBCD, (+)-- -HBCD, (-)-β-HBCD, (+)-- -HBCD, (-)-γ-HBCD, (+)-- -HBCD,	Fish muscle from hake, cod, sole,panga, whiting and sea bass	Decanoic acid/THF	87-114	LC-MS/MS	0.15-1.9	[90]
		Salt-Induced SUPRAS				
		Pesticides				
2,4-dichlorophenoxyacetic acid (2,4-D), 2-methyl-4-chlorophenoxyacetic acid (MCPA), 2-(2,- -dichlorophenoxy)propanoic acid (Dichloprop), 2- (4-Chloro-2-methylphenoxy) propanoic acid (Mecoprop), 4-(2,- -dichlorophenoxy)butanoic acid (2,4-DB) and 4- (4-Chloro-2-methylphenoxy) butanoic acid (MCPB)	Water, rice	Dodecyl sulfate/tetrabutyl ammonium bromide	81-110	LC-UV	1-2	[94]
Fenitrothion, guthion, parathion, methidathion, chlorpyrifos	Water	Dodecyl sulfate/:$Al_2(SO_4)_3$	>81	LC-UV	0.7-2.5	[93]
Fenitrothion, phosalone, chlorpyrifos	Water, apple, pineapple, peach	Decanoic acid/tetrabutyl ammonium hydroxide (molar ratio 2:1)	92-110	LC-UV	0.1-0.35	[55]

(Table 1) cont.....

Compounds	Type of food	SUPRAS components	Extraction recoveries (%)	Instrumental analysis	LOD (μg/L or μg/Kg)	Refs.
Carbendazim, thiabendazole, fuberidazole	Fruits (oranges, tangerines, lemons, limes, grapefruits, apples, pears and bananas) and vegetables (potatoes and lettuces)	Decanoic acid/tetrabutyl ammonium hydroxide (molar ratio 2:1)	93–102	LC-FD	0.03–14.0	[54]
PAHs						
Benzo[a]anthracene, benzo[a]pyrene, chrysene, bezo[b]fluoranthene	Smoked meat and fish, bivalve molluscs and processed cereal food	Octanoic acid/tetrabutyl ammonium hydroxide (molar ratio 2:1)	92-103	LC-FD	0.3-0.7	[35]
Benzo[a]pyrene	Instant noodles, biscuits, rice crust and fried dough sticks.	Octanol/tetrabutyl ammonium bromide	90-100	LC-FD	0.11	[95]
Veterinary residues						
Tetracycline, oxytetracycline, chlortetracycline, methacycline, doxycycline	Milk, egg, honey	Dodecyltrimethylammonium/ didodecyldimethylammonium/NaCl	70-122	LC-UV	0.7-3.4	[92]
Mycotoxins						
Ochratoxin A	Dried vine fruits	Decanoic acid/tetrabutyl ammonium hydroxide (molar ratio 2:1)	95-101	LC-FD	1.8	[91]
Acid-Induced SUPRAS						
Pesticides						
Etofenprox	Tap water, fountain, lake water, beer	Dodecane sulphonate/5.4M HCl	94-100	LC-UV	4	[64]
Carbaryl	Tap and drinking water, vegetables	Dodecyl sulfate/6M HCl	85-103	LC-UV	50	[98]
PAHs						
Acenaphtene, anthracene, benzo[a]anthracene benzo[a]pyrene, chrysene, dibenzo[a,h]anthracene, phenanthrene pyrene, fluorine, fluoranthene	Tap water	Dodecyl sulfate/6M HCl	67-95	LC-FD	0.01-20 nM	[97]
Magnetic SUPRAS						
Pesticides						
Diazinon, metalaxyl	Orange, peach, grape and apple juices	Ferrofluid-mediated calcined layered double hydroxide@tannic acid-based SUPRAS	85-97	GC-FID	0.2-0.8	[99]
Cyanazine, simazine, prometon, propazine	Tap, river and spring water	Decanoic acid-MNP /tetrabutyl ammonium	90-106	LC-UV	300-500	[56]
Fenitrothion, Phosalone, Chlorpyrifos	Water and fruit juice	Oleic acid-MNP/decanoic acid/tetrabutyl ammonium	92-110	LC-UV	0.1-0.35	[55]

In terms of detection, the turning point for hyphenating SUPRASs with LC-MS has been the development of water-induced SUPRASs based on alkanols. These amphiphiles have negligible signal in LC-ESI-MS/MS since they do not ionise in ESI [88], the widest ionisation source used in food analysis. As a result, many applications based on the combination of alkanol-based SUPRASs with LC-ES--MS/MS have been developed in the last few years (Table **1**). This hyphenation has already proved successful for the detection and quantification of contaminants, including pesticides, veterinary residues, packaging material residues and emerging contaminants (Table **1**). All these methods have been validated following the EU requirements set for food analysis.

Regarding chemical composition and nanostructure, both the alkanol- and alkyl carboxylic acid-based SUPRASs consist of densely packed inverted hexagonal aggregates with a high number of binding sites (up to 1 mg/μL) that show restricted access properties. These features have made them suitable for the development of analyte- and matrix-independent methods [89], although their advantages in this respect have been scarcely exploited in food analysis so far (*e.g.* determination of 21 bisphenols and derivatives in a variety of foodstuffs [47]).

An interesting field explored with the hyphenation of water-induced SUPRAS and LC-MS/MS has been the quantification of enantiomers from pesticides [78] and emerging contaminants [90], where SUPRASs were proved to be compatible with chiral liquid chromatography. Although no EU requirements have been still set for differentiated analysis of isomers in food, their different toxicity claims for this evaluation in order to have an accurate assessment of human health risks.

The application of ***salt-induced SUPRASs*** to food analysis suffers from many of the problems already mentioned for temperature-induced SUPRASs, although the causes for these problems are a bit different. As a result, the applications developed so far involve only a few contaminants and are mostly based on LC-UV or LC-FD detection [35, 54, 55, 91 - 95].

The main salt-induced SUPRASs used in food analysis are those synthesised from carboxylic acids [35, 54, 55, 91], alkanols [95] or dodecyl sulfate [94], under the action of tetraalkylammonium salts. Also, SUPRASs synthesised from alkyl trimethyl and alkyl dimethyl ammonium mixtures [92] or dodecyl sulfate [93], in the presence of an inorganic salt, have been applied in this field. Many of these SUPRASs are made of vesicles with high kinetic stability, which difficult their solubilisation in typical mobile phases used in reversed-phase chromatography. As a result, retention times may shift by the partition of solutes to the vesicular phase, which may hinder contaminant separation. Nevertheless, the main problem derives from the presence of high amounts of alkylammonium salts in the chromatographic system, which are strongly retained and cause background contamination in MS as working in positive ESI mode.

Unfortunately, these operational problems have limited their use in food analysis, despite they offer a great variety of mixed-mode mechanisms for solute solubilisation. Thus, it is common that the polar region includes ionic, hydrogen bonds, polar and halogen bonds, while the nonpolar region offers dispersion interactions, so they have the potential for efficient extraction in multiresidue analysis in food.

The ***acid-induced SUPRASs,*** synthesised from alkyl sulfates and sulfonates under strongly acidic conditions [21] were the first ones obtained from pure amphiphiles and, given their ionic character, they were well separated in chromatographic analysis from very different polar and nonpolar analytes. This feature allowed outstanding applications in environmental analysis, particularly in the extraction of analytes strongly retained on solid matrices [96]. However, the application of acid-induced SUPRASs in food has been very limited and mainly focused on water samples [64, 97, 98]. The major limitation of these SUPRASs in the analysis of liquid samples is the high concentration of hydrochloric acid required for coacervation (5-6 M).

Recently, SUPRASs have been combined with magnetic nanoparticles (MNPs) in order to facilitate SUPRAS retrieval after extraction [55, 56, 99]. Once separated the ***magnetic SUPRAS*** from the sample, the analytes are recovered in a proper organic solvent. A major advantage of this approach is that the layer covering the MNPs may add an additional mechanism for analyte extraction. Nevertheless, it also adds a new step (*i.e.* solvent back extraction) which may decrease recoveries, particularly in multiresidue analysis, and precision. The applications of magnetic SUPRAS in food analysis so far have been scarce and limited to a few analytes and simple detectors (UV and FD in LC and FID in GC), so more investigation is required to prove their potential in current food analysis.

As Table **1** clearly shows, SUPRASs application to food analysis is predominantly carried out by LC, since the combination of SUPRASs with GC involves the back-extraction of analytes with a proper solvent in order to remove the amphiphile [71]. On the other hand, satisfactory results can be obtained by combining SUPRASs with capillary zone electrophoresis (CZE) provided that non-aqueous media are used to avoid surfactant adsorption onto the wall of the capillary. This strategy has proved successful in the determination of herbicides [69] and mycotoxins [100]. As aqueous buffers are used, high surfactant adsorption in the bare fused silica capillary occurs thus leading to a marked loss of efficiency and reproducibility both in the migration times and in the areas of the electrophoretic peaks [101].

According to the discussion above, the advances achieved in the last two decades in the application of SUPRASs to food analysis have been outstanding. However, the most distinctive features of SUPRASs that make them highly competitive in this field remain largely unexplored. This is particularly true regarding their potential to extract a large number of structurally unrelated compounds as required for the screening/confirmation of both legislated and unlegislated contaminants in food. All types of SUPRASs included in Table **1** have this potential, because of the different polarity microenvironments existing in their nanostructures. So, major general challenges facing SUPRASs application in food analysis refer to their compatibility with MS, which is the reference detector in this field.

Regarding temperature-induced SUPRASs, they do not require organic solvents for their synthesis, which is a valuable asset. Nevertheless, their successful application in food analysis will undoubtedly require the search for pure amphiphiles able to undergo coacervation at room temperature and not ionise in ESI sources.

Concerning salt-induced SUPRASs, compatibility with MS will require the search for amphiphiles undergoing coacervation under the action of inorganic salts and having low ionisation efficiency in ESI sources. On the other hand, for acid-induced SUPRAS, the coacervation of amphiphiles bearing weak acids as functional groups (*e.g.* carboxylic acids) will require less acidic conditions to be neutralised, which will render in mild extraction conditions. As an example, it has been recently reported that the coacervation of rhamnolipids occurs at concentrations of hydrochloric acid as low as 0.01M [50].

Water-induced SUPRASs, particularly alkanol-based SUPRASs, have already proved to be compatible with LC-MS/MS and, a priori, are the ones with the greatest potential in food analysis. Nevertheless, applications related to single- and multi-class analytes involving a variety of matrices and using not only low-resolution MS, but also high-resolution (LC-HRMS), and other emerging techniques such as ambient MS are lacking.

CONCLUDING REMARKS

While it is true that many applications of SUPRASs in food analysis have been reported and that SUPRASs exhibit the potential of being used in the routine control of food contaminants, new developments should now focus on addressing the current needs of food labs. SUPRASs surpass organic solvents in this field in terms of their ability for extracting efficiently a great number of compounds covering a wide polarity range while simultaneously removing major matrix components. Likewise, they surpass solid-phase extraction in terms of simplicity and cost. These great advantages have not been conveniently exploited and efforts in developing methods that can be directly transferable from research labs to food control labs by closely working with the different agents involved should be a priority.

LIST OF ABBREVIATIONS

AMS	Ambient mass spectrometry
APCI	Atmospheric pressure chemical ionisation
BioSUPRASs	SUPRASs made up of bioamphiphiles
CAC	Critical aggregation concentration
Cryo-SEM	Cryo-scanning electron microscopy
CZE	Capillary zone electrophoresis
EFSA	European Food Safety Authority
EI	Electron impact
ESI	Electrospray ionisation
EU	European Union
FCMs	Food contact materials
FD	Fluorescence detection
FID	Flame ionisation detector
GC	Gas chromatography
HRMS	High-resolution mass spectrometry
HS	GC-head space injector
LC	Liquid chromatography
MLC	Micellar liquid chromatography
MNPs	Magnetic nanoparticles
MS	Mass spectrometry
NIAS	Non-intentionally added substances
PAHs	Polycyclic aromatic hydrocarbons
PCBs	Polychlorinated biphenyls
QTOF	Quadrupole time-of-flight mass spectrometry
SUPRASs	Supramolecular solvents
SUPRAS-RAM	SUPRASs with restricted access properties
THF	Tetrahydrofuran
TOF	Time-of-flight mass spectrometry

UV Ultraviolet detection

CONSENT FOR PUBLICATION

Not applicable.

CONFLICT OF INTEREST

The authors declare no conflict of interest, financial or otherwise.

ACKNOWLEDGEMENTS

This work was supported by the Andalusian Department of Knowledge, Innovation and University (P18-RT-2654). Dr Caballero-Casero acknowledges their post-doctoral contract from the Andalusian Government (Ref. Doc_00289).

REFERENCES

[1] Hansen FA, Pedersen-Bjergaard S. Emerging Extraction Strategies in Analytical Chemistry. Anal Chem 2020; 92(1): 2-15.
[http://dx.doi.org/10.1021/acs.analchem.9b04677] [PMID: 31625733]

[2] Musarurwa H, Tavengwa NT. Supramolecular solvent-based micro-extraction of pesticides in food and environmental samples. Talanta 2021; 223(Pt 1): 121515.
[http://dx.doi.org/10.1016/j.talanta.2020.121515] [PMID: 33303131]

[3] Moradi M, Yamini Y, Feizi N. Development and challenges of supramolecular solvents in liquid-based microextraction methods. Trends Analyt Chem 2021; 138: 116231.
[http://dx.doi.org/10.1016/j.trac.2021.116231]

[4] Accioni F, Nieddu M, Corona P, Boatto G. Hexanol-Based Supramolecular Solvents Tool for the Determination of 11 Illicit Phenethylamines in Oral Fluid by LC-MS/MS. J Anal Toxicol 2020; 44(1): 15-21.
[PMID: 31095695]

[5] Grob K. The European system for the control of the safety of food-contact materials needs restructuring: a review and outlook for discussion. Food Addit Contam Part A Chem Anal Control Expo Risk Assess 2017; 34(9): 1643-59.
[http://dx.doi.org/10.1080/19440049.2017.1332431] [PMID: 28627317]

[6] https://www.efsa.europa.eu/en/topics/topic/food-contact-materials

[7] Grob K. Work plans to get out of the deadlock for the safety assurance of migration from food contact materials? A proposal. Food Control 2014; 46: 312-8.
[http://dx.doi.org/10.1016/j.foodcont.2014.05.044]

[8] European Commission. SANCO/3029/99 Rev. 4 (11/07/2000) Residues: Guidance for Generating and Reporting Methods of Analysis in Support of Pre-Registration Data Requirements for Annex II (Part A, Section 4) and Annex III (Part A, Section 5) of Directive 91/414. SANCO/3029/99 rev.4 2000, 26.

[9] European Commission. 2002/657/EC COMMISSION DECISION of 12 August 2002 Implementing Council Directive 96/23/EC Concerning the Performance of Analytical Methods and the Interpretation of Results. 2002/657/EC Comm. Decis. 2002..

[10] Locatelli M, Celia C, Eds. Analytical Chemistry: Developments, Applications and Challenges in Food Analysis. Nova Science Publishers 2017.

[11] Aceña J, Stampachiacchiere S, Pérez S, Barceló D. Advances in liquid chromatography–high-resolution mass spectrometry for quantitative and qualitative environmental analysis. Anal Bioanal Chem 2015; 407(21): 6289-99.
[http://dx.doi.org/10.1007/s00216-015-8852-6] [PMID: 26138893]

[12] Knolhoff AM, Zweigenbaum JA, Croley TR. Nontargeted Screening of Food Matrices: Development of a Chemometric Software Strategy To Identify Unknowns in Liquid Chromatography–Mass Spectrometry Data. Anal Chem 2016; 88(7): 3617-23.
[http://dx.doi.org/10.1021/acs.analchem.5b04208] [PMID: 26938581]

[13] Porcari AM, Fernandes GD, Barrera-Arellano D, Eberlin MN, Alberici RM. Food quality and authenticity screening *via* easy ambient sonic-spray ionization mass spectrometry. Analyst (Lond) 2016; 141(4): 1172-84.
[http://dx.doi.org/10.1039/C5AN01415H] [PMID: 26820190]

[14] Black C, Chevallier OP, Elliott CT. The current and potential applications of Ambient Mass Spectrometry in detecting food fraud. Trends Analyt Chem 2016; 82: 268-78.
[http://dx.doi.org/10.1016/j.trac.2016.06.005]

[15] Núñez O, Lucci P, Eds. New Trends in Sample Preparation Techniques for Food Analysis. Nova Science Publishers 2016.

[16] Lehotay SJ, Chen Y. Hits and misses in research trends to monitor contaminants in foods. Anal Bioanal Chem 2018; 410(22): 5331-51.
[http://dx.doi.org/10.1007/s00216-018-1195-3] [PMID: 29951771]

[17] Ballesteros-Gómez A, Sicilia MD, Rubio S. Supramolecular solvents in the extraction of organic compounds. A review. Anal Chim Acta 2010; 677(2): 108-30.
[http://dx.doi.org/10.1016/j.aca.2010.07.027] [PMID: 20837178]

[18] IUPAC Compendium of Chemical Terminology, (the "Gold Book")-Online Version. 2nd ed., Blackwell Scientific Publications 2019.

[19] Steed W, Turner R. j. W. Core Concepts in Supramolecular Chemistry and Nanochemistry; John Wiley, 2007; Vol. 129.

[20] Evans DF, Wennerström H. The Colloidal Domain: Where Physics. 2nd ed., Chemistry, Biology, and Technology Meet 1999.

[21] Casero I, Sicilia D, Rubio S, Pérez-Bendito D. An Acid-Induced Phase Cloud Point Separation Approach Using Anionic Surfactants for the Extraction and Preconcentration of Organic Compounds. Anal Chem 1999; 71(20): 4519-26.
[http://dx.doi.org/10.1021/ac990106g]

[22] Hinze WL, Pramauro E, Poole CF. A Critical Review of Surfactant-Mediated Phase Separations (Cloud-Point Extractions): Theory and Applications. Crit Rev Anal Chem 1993; 24(2): 133-77.
[http://dx.doi.org/10.1080/10408349308048821]

[23] Ruiz FJ, Rubio S, Pérez-Bendito D. Water-induced coacervation of alkyl carboxylic acid reverse micelles: phenomenon description and potential for the extraction of organic compounds. Anal Chem 2007; 79(19): 7473-84.
[http://dx.doi.org/10.1021/ac0708644] [PMID: 17764154]

[24] Glatter O, Fritz G, Lindner H, *et al.* Nonionic Micelles near the Critical Point: Micellar Growth and Attractive Interaction. Langmuir 2000; 16(23): 8692-701.
[http://dx.doi.org/10.1021/la000315s]

[25] Saitoh T, Hinze WL. Concentration of hydrophobic organic compounds and extraction of protein using alkylammoniosulfate zwitterionic surfactant mediated phase separations (cloud point extractions). Anal Chem 1991; 63(21): 2520-5.
[http://dx.doi.org/10.1021/ac00021a023] [PMID: 1763809]

[26] Ballesteros-Gómez A, Rubio S, Pérez-Bendito D. Potential of supramolecular solvents for the extraction of contaminants in liquid foods. J Chromatogr A 2009; 1216(3): 530-9. [http://dx.doi.org/10.1016/j.chroma.2008.06.029] [PMID: 18603255]

[27] Salatti-Dorado JÁ, Caballero-Casero N, Sicilia MD, Lunar ML, Rubio S. The use of a restricted access volatile supramolecular solvent for the LC/MS-MS assay of bisphenol A in urine with a significant reduction of phospholipid-based matrix effects. Anal Chim Acta 2017; 950: 71-9. [http://dx.doi.org/10.1016/j.aca.2016.11.026] [PMID: 27916133]

[28] Stalikas CD. Micelle-mediated extraction as a tool for separation and preconcentration in metal analysis. Trends Analyt Chem 2002; 21(5): 343-55. [http://dx.doi.org/10.1016/S0165-9936(02)00502-2]

[29] Arya SS, Kaimal AM, Chib M, Sonawane SK, Show PL. Novel, energy efficient and green cloud point extraction: technology and applications in food processing. J Food Sci Technol 2019; 56(2): 524-34. [http://dx.doi.org/10.1007/s13197-018-3546-7] [PMID: 30906010]

[30] Melnyk A, Namieśnik J, Wolska L. Theory and recent applications of coacervate-based extraction techniques. Trends Analyt Chem 2015; 71(71): 282-92. [http://dx.doi.org/10.1016/j.trac.2015.03.013]

[31] Pacheco-Fernández I, González-Martín R, e Silva FA, Freire MG, Pino V. Insights into coacervative and dispersive liquid-phase microextraction strategies with hydrophilic media – A review. Anal Chim Acta 2021; 1143: 225-49. [http://dx.doi.org/10.1016/j.aca.2020.08.022] [PMID: 33384121]

[32] Caballo C, Sicilia MD, Rubio S. Supramolecular Solvents for Green Chemistry.The Application of Green Solvents in Separation Processes. Elsevier 2017; pp. 379-414. [http://dx.doi.org/10.1016/B978-0-12-805297-6.00005-X]

[33] Rubio S. Twenty years of supramolecular solvents in sample preparation for chromatography: achievements and challenges ahead. Anal Bioanal Chem 2020; 412(24): 6037-58. [http://dx.doi.org/10.1007/s00216-020-02559-y] [PMID: 32206847]

[34] Ruiz F J, Rubio S, Pérez-Bendito D. Tetrabutylammonium-Induced Coacervation in Vesicular Solutions of Alkyl Carboxylic Acids for the Extraction of Organic Compounds 2006; 78(20): 7229-39. [http://dx.doi.org/10.1021/ac060427+]

[35] López-Jiménez FJ, Ballesteros-Gómez A, Rubio S. Determination of polycyclic aromatic hydrocarbons (PAH4) in food by vesicular supramolecular solvent-based microextraction and LC–fluorescence detection. Food Chem 2014; 143: 341-7. [http://dx.doi.org/10.1016/j.foodchem.2013.07.136] [PMID: 24054249]

[36] Ballesteros-Gómez A, Rubio S. Environment-responsive alkanol-based supramolecular solvents: characterization and potential as restricted access property and mixed-mode extractants. Anal Chem 2012; 84(1): 342-9. [http://dx.doi.org/10.1021/ac2026207] [PMID: 22103906]

[37] Caballo C, Sicilia MD, Rubio S. Enantioselective analysis of non-steroidal anti-inflammatory drugs in freshwater fish based on microextraction with a supramolecular liquid and chiral liquid chromatography–tandem mass spectrometry. Anal Bioanal Chem 2015; 407(16): 4721-31. [http://dx.doi.org/10.1007/s00216-015-8675-5] [PMID: 25869485]

[38] Ballesteros-Gómez A, Lunar L, Sicilia MD, Rubio S. Hyphenating Supramolecular Solvents and Liquid Chromatography: Tips for Efficient Extraction and Reliable Determination of Organics. Chromatographia 2019; 82(1): 111-24. [http://dx.doi.org/10.1007/s10337-018-3614-1]

[39] Costi EM, Sicilia MD, Rubio S. Multiresidue analysis of sulfonamides in meat by supramolecular solvent microextraction, liquid chromatography and fluorescence detection and method validation according to the 2002/657/EC decision. J Chromatogr A 2010; 1217(40): 6250-7.

[http://dx.doi.org/10.1016/j.chroma.2010.08.017] [PMID: 20810118]

[40] Alabi A, Caballero-Casero N, Rubio S. Quick and simple sample treatment for multiresidue analysis of bisphenols, bisphenol diglycidyl ethers and their derivatives in canned food prior to liquid chromatography and fluorescence detection. J Chromatogr A 2014; 1336: 23-33. [http://dx.doi.org/10.1016/j.chroma.2014.02.008] [PMID: 24594089]

[41] Carabias-Martínez R, Rodríguez-Gonzalo E, Moreno-Cordero B, Pérez-Pavón JL, García-Pinto C, Fernández Laespada E. Surfactant cloud point extraction and preconcentration of organic compounds prior to chromatography and capillary electrophoresis. J Chromatogr A 2000; 902(1): 251-65. [http://dx.doi.org/10.1016/S0021-9673(00)00837-2] [PMID: 11192158]

[42] Taechangam P, Scamehorn JF, Osuwan S, Rirksomboon T. Effect of nonionic surfactant molecular structure on cloud point extraction of phenol from wastewater. Colloids Surf A Physicochem Eng Asp 2009; 347(1-3): 200-9. [http://dx.doi.org/10.1016/j.colsurfa.2009.04.005]

[43] Lehn JM. Toward complex matter: Supramolecular chemistry and self-organization. Proc Natl Acad Sci USA 2002; 99(8): 4763-8. [http://dx.doi.org/10.1073/pnas.072065599] [PMID: 11929970]

[44] Caballero-Casero N, García-Fonseca S, Rubio S. Restricted access supramolecular solvents for the simultaneous extraction and cleanup of ochratoxin A in spices subjected to EU regulation. Food Control 2018; 88: 33-9. [http://dx.doi.org/10.1016/j.foodcont.2018.01.003]

[45] García-Fonseca S, Rubio S. Restricted access supramolecular solvents for removal of matrix-induced ionization effects in mass spectrometry: Application to the determination of Fusarium toxins in cereals. Talanta 2016; 148: 370-9. [http://dx.doi.org/10.1016/j.talanta.2015.11.014] [PMID: 26653462]

[46] Kojro G, Rudzki PJ, Pisklak DM, Giebułtowicz J. Matrix effect screening for cloud-point extraction combined with liquid chromatography coupled to mass spectrometry: Bioanalysis of pharmaceuticals. J Chromatogr A 2019; 1591: 44-54. [http://dx.doi.org/10.1016/j.chroma.2019.01.031] [PMID: 30679045]

[47] Caballero-Casero N, Rubio S. Comprehensive supramolecular solvent-based sample treatment platform for evaluation of combined exposure to mixtures of bisphenols and derivatives by liquid chromatography-tandem mass spectrometry. Anal Chim Acta 2021; 1144: 14-25. [http://dx.doi.org/10.1016/j.aca.2020.11.057] [PMID: 33453791]

[48] Gu Y, Jérôme F. Bio-based solvents: an emerging generation of fluids for the design of eco-efficient processes in catalysis and organic chemistry. Chem Soc Rev 2013; 42(24): 9550-70. [http://dx.doi.org/10.1039/c3cs60241a] [PMID: 24056753]

[49] Romera-García E, Ballesteros-Gómez A, Rubio S. Supramolecular biosolvents made up of self-assembled rhamnolipids: synthesis and characterization. Green Chem 2020; 22(18): 6115-26. [http://dx.doi.org/10.1039/D0GC02078H]

[50] Romera-García E, Ballesteros-Gómez A, Rubio S. An environmentally stable supramolecular biosolvent: Characterization and study of its potential for the elimination of polar toxic substances in water. J Clean Prod 2021; 321: 128975. [http://dx.doi.org/10.1016/j.jclepro.2021.128975]

[51] Musarurwa H, Tavengwa NT. Emerging green solvents and their applications during pesticide analysis in food and environmental samples. Talanta 2021; 223(Pt 1): 121507. [http://dx.doi.org/10.1016/j.talanta.2020.121507] [PMID: 33303178]

[52] García-Fonseca S, Ballesteros-Gómez A, Rubio S. Restricted access supramolecular solvents for sample treatment in enzyme-linked immuno-sorbent assay of mycotoxins in food. Anal Chim Acta 2016; 935: 129-35. [http://dx.doi.org/10.1016/j.aca.2016.06.042] [PMID: 27543022]

[53] Costi EM, Sicilia MD, Rubio S. Supramolecular solvents in solid sample microextractions: Application to the determination of residues of oxolinic acid and flumequine in fish and shellfish. J Chromatogr A 2010; 1217(9): 1447-54.
[http://dx.doi.org/10.1016/j.chroma.2009.12.073] [PMID: 20060977]

[54] Moral A, Sicilia MD, Rubio S. Determination of benzimidazolic fungicides in fruits and vegetables by supramolecular solvent-based microextraction/liquid chromatography/fluorescence detection. Anal Chim Acta 2009; 650(2): 207-13.
[http://dx.doi.org/10.1016/j.aca.2009.07.056] [PMID: 19720194]

[55] Zohrabi P, Shamsipur M, Hashemi M, Hashemi B. Liquid-phase microextraction of organophosphorus pesticides using supramolecular solvent as a carrier for ferrofluid. Talanta 2016; 160: 340-6.
[http://dx.doi.org/10.1016/j.talanta.2016.07.036] [PMID: 27591622]

[56] Safari M, Yamini Y, Tahmasebi E, Ebrahimpour B. Magnetic nanoparticle assisted supramolecular solvent extraction of triazine herbicides prior to their determination by HPLC with UV detection. Mikrochim Acta 2016; 183(1): 203-10.
[http://dx.doi.org/10.1007/s00604-015-1607-4]

[57] Yang D, Li X, Meng D, Wang M, Yang Y. Supramolecular solvents combined with layered double hydroxide-coated magnetic nanoparticles for extraction of bisphenols and 4-tert-octylphenol from fruit juices. Food Chem 2017; 237: 870-6.
[http://dx.doi.org/10.1016/j.foodchem.2017.06.063] [PMID: 28764080]

[58] Kukusamude C, Quirino JP, Srijaranai S. A coacervative extraction based on single-chain and double-chain cationic surfactants. J Chromatogr A 2016; 1472: 10-5.
[http://dx.doi.org/10.1016/j.chroma.2016.10.030] [PMID: 27776773]

[59] Lara AB, Caballo C, Sicilia MD, Rubio S. Speeding up the extraction of hexabromocyclododecane enantiomers in soils and sediments based on halogen bonding. Anal Chim Acta 2018; 1027: 47-56.
[http://dx.doi.org/10.1016/j.aca.2018.05.002] [PMID: 29866269]

[60] Santalad A, Srijaranai S, Burakham R, Glennon JD, Deming RL. Cloud-point extraction and reversed-phase high-performance liquid chromatography for the determination of carbamate insecticide residues in fruits. Anal Bioanal Chem 2009; 394(5): 1307-17.
[http://dx.doi.org/10.1007/s00216-009-2663-6] [PMID: 19242683]

[61] Ferrer R, Beltrán JL, Guiteras J. Use of cloud point extraction methodology for the determination of PAHs priority pollutants in water samples by high-performance liquid chromatography with fluorescence detection and wavelength programming. Anal Chim Acta 1996; 330(2-3): 199-206. [http://dx.doi.org/10.1016/0003-2670(96)00176-6]

[62] Xie S, Paau M C, Li C F, Xiao D, Choi M M F. Separation and Preconcentration of Persistent Organic Pollutants by Cloud Point Extraction 2010.
[http://dx.doi.org/10.1016/j.chroma.2009.11.075]

[63] Moradi M, Yamini Y. Surfactant roles in modern sample preparation techniques: A review. J Sep Sci 2012; 35(18): 2319-40.
[http://dx.doi.org/10.1002/jssc.201200368] [PMID: 22887709]

[64] Jia G, Bi C, Wang Q, Qiu J, Zhou W, Zhou Z. Determination of Etofenprox in environmental samples by HPLC after anionic surfactant micelle-mediated extraction (coacervation extraction). Anal Bioanal Chem 2006; 384(6): 1423-7.
[http://dx.doi.org/10.1007/s00216-005-0273-5] [PMID: 16491345]

[65] Sikalos TI, Paleologos EK. Cloud point extraction coupled with microwave or ultrasonic assisted back extraction as a preconcentration step prior to gas chromatography. Anal Chem 2005; 77(8): 2544-9.
[http://dx.doi.org/10.1021/ac048267u] [PMID: 15828791]

[66] Faria AM, Dardengo RP, Lima CF, Neves AA, Queiroz MELR. Determination of disulfoton in surface water samples by cloud-point extraction and gas chromatography. Int J Environ Anal Chem 2007;

87(4): 249-58.
[http://dx.doi.org/10.1080/03067310601068841]

[67] Chen J, Zhao W, Liu W, Zhou Z, Yang M. Cloud point extraction coupled with derivative of carbofuran as a preconcentration step prior to HPLC. Food Chem 2009; 115(3): 1038-41.
[http://dx.doi.org/10.1016/j.foodchem.2008.12.035]

[68] Fontana AR, Camargo AB, Altamirano JC. Coacervative microextraction ultrasound-assisted back-extraction technique for determination of organophosphates pesticides in honey samples by gas chromatography–mass spectrometry. J Chromatogr A 2010; 1217(41): 6334-41.
[http://dx.doi.org/10.1016/j.chroma.2010.08.021] [PMID: 20817191]

[69] Kukusamude C, Srijaranai S, Kato M, Quirino JP. Cloud point sample clean-up and capillary zone electrophoresis with field enhanced sample injection and micelle to solvent stacking for the analysis of herbicides in milk. J Chromatogr A 2014; 1351: 110-4.
[http://dx.doi.org/10.1016/j.chroma.2014.05.050] [PMID: 24908154]

[70] Wu YJ, Fu XW, Yang H. Cloud point extraction with Triton X-114 for separation of metsulfuron-methyl, chlorsulfuron, and bensulfuron-methyl from water, soil, and rice and analysis by high-performance liquid chromatography. Arch Environ Contam Toxicol 2011; 61(3): 359-67.
[http://dx.doi.org/10.1007/s00244-010-9626-y] [PMID: 21127849]

[71] Zhao W, Sun X, Deng X, Huang L, Yang M, Zhou Z. Cloud point extraction coupled with ultrasonic-assisted back-extraction for the determination of organophosphorus pesticides in concentrated fruit juice by gas chromatography with flame photometric detection. Food Chem 2011; 127(2): 683-8.
[http://dx.doi.org/10.1016/j.foodchem.2010.12.122] [PMID: 23140719]

[72] Zhou Z, Chen J, Zhao D, Yang M. Determination of four carbamate pesticides in corn by cloud point extraction and high-performance liquid chromatography in the visible region based on their derivatization reaction. J Agric Food Chem 2009; 57(19): 8722-7.
[http://dx.doi.org/10.1021/jf901644c] [PMID: 19807151]

[73] Gorji S, Biparva P, Bahram M, Nematzadeh G. Rapid and Direct Microextraction of Pesticide Residues from Rice and Vegetable Samples by Supramolecular Solvent in Combination with Chemometrical Data Processing. Food Anal Methods 2019; 12(2): 394-408.
[http://dx.doi.org/10.1007/s12161-018-1371-2]

[74] Zhao W, Zhao J, Zhao H, Cao Y, Liu W. Supramolecular solvent-based vortex-mixed microextraction: Determination of chiral triazole fungicide in beer samples. Chirality 2018; 30(3): 302-9.
[http://dx.doi.org/10.1002/chir.22798] [PMID: 29314267]

[75] ALOthman ZA, Yilmaz E, Habila MA, *et al.* Supramolecular solvent microextraction and ultra-performance liquid chromatography-tandem mass spectrometry combination for the preconcentration and determination of malathion in environmental samples. Desalination Water Treat 2019; 144: 166-71.
[http://dx.doi.org/10.5004/dwt.2019.23574]

[76] ALOthman ZA, Yilmaz E, Habila MA, Alhenaki B, Soylak M, Ahmed AYBH, Alabdullkarem EA. Development of Combined-Supramolecular Microextraction with Ultra-Performance Liquid Chromatography-Tandem Mass Spectrometry Procedures for Ultra-Trace Analysis of Carbaryl in Water, Fruits and Vegetables. Int J Environ Anal Chem 2022; 102(7): 1491-501.

[77] Amir S, Jan MR, Shah J. Efficient environment friendly method for microextraction of dinitroaniline herbicides using supramolecular solvent. Int J Environ Anal Chem 2021; 101(10): 1393-407.
[http://dx.doi.org/10.1080/03067319.2019.1683550]

[78] Lara AB, Caballo C, Sicilia MD, Rubio S. Quick and Sensitive Enantioselective Determination of Permethrin in Fruits and Vegetables by Combining Supramolecular Solvents and Chiral Liquid Chromatography-Tandem Mass Spectrometry. J Agric Food Chem 2020; 68(33): 9014-23.
[http://dx.doi.org/10.1021/acs.jafc.0c02533] [PMID: 32687353]

[79] Deng H, Wang H, Liang M, Su X. A novel approach based on supramolecular solvent microextraction

and UPLC-Q-Orbitrap HRMS for simultaneous analysis of perfluorinated compounds and fluorine-containing pesticides in drinking and environmental water. Microchem J 2019; 151: 104250. [http://dx.doi.org/10.1016/j.microc.2019.104250]

[80] Peyrovi M, Hadjmohammadi M. Alkanol-based supramolecular solvent microextraction of organophosphorus pesticides and their determination using high-performance liquid chromatography. J Indian Chem Soc 2017; 14(5): 995-1004. [http://dx.doi.org/10.1007/s13738-017-1049-5]

[81] Scheel GL, Tarley CRT. Feasibility of supramolecular solvent-based microextraction for simultaneous preconcentration of herbicides from natural waters with posterior determination by HPLC-DAD. Microchem J 2017; 133: 650-7. [http://dx.doi.org/10.1016/j.microc.2017.03.007]

[82] Scheel GL, Teixeira Tarley CR. Simultaneous microextraction of carbendazim, fipronil and picoxystrobin in naturally and artificial occurring water bodies by water-induced supramolecular solvent and determination by HPLC-DAD. J Mol Liq 2020; 297: 111897. [http://dx.doi.org/10.1016/j.molliq.2019.111897]

[83] Falsafi Z, Raofie F, Ariya PA. Supercritical fluid extraction followed by supramolecular solvent microextraction as a fast and efficient preconcentration method for determination of polycyclic aromatic hydrocarbons in apple peels. J Sep Sci 2020; 43(6): 1154-63. [http://dx.doi.org/10.1002/jssc.201900886] [PMID: 31883198]

[84] García-Fonseca S, Ballesteros-Gómez A, Rubio S, Pérez-Bendito D. Supramolecular solvent-based microextraction of ochratoxin a in raw wheat prior to liquid chromatography-fluorescence determination. J Chromatogr A 2010; 1217(16): 2376-82. [http://dx.doi.org/10.1016/j.chroma.2009.10.085]

[85] García-Fonseca S, Ballesteros-Gómez A, Rubio S, Pérez-Bendito D. Coacervative extraction of Ochratoxin A in wines prior to liquid chromatography/fluorescence determination. Anal Chim Acta 2008; 617(1-2): 3-10. [http://dx.doi.org/10.1016/j.aca.2007.11.002] [PMID: 18486634]

[86] García-Prieto A, Lunar L, Rubio S, Pérez-Bendito D. Decanoic acid reverse micelle-based coacervates for the microextraction of bisphenol A from canned vegetables and fruits. Anal Chim Acta 2008; 617(1-2): 51-8. [http://dx.doi.org/10.1016/j.aca.2008.01.061] [PMID: 18486640]

[87] Pérez Bendito MD, Rubio Bravo S, Lunar Reyes ML, García Prieto A. Determination of bisphenol A in canned fatty foods by coacervative microextraction, liquid chromatography and fluorimetry. Food Addit Contam Part A Chem Anal Control Expo Risk Assess 2009; 26(2): 265-74. [http://dx.doi.org/10.1080/02652030802368740] [PMID: 19680898]

[88] Qi W, Wang Y, Cao Y, *et al.* Simultaneous Analysis of Fatty Alcohols, Fatty Aldehydes, and Sterols in Thyroid Tissues by Electrospray Ionization-Ion Mobility-Mass Spectrometry Based on Charge Derivatization. Anal Chem 2020; 92(13): 8644-8. [http://dx.doi.org/10.1021/acs.analchem.0c01292] [PMID: 32574041]

[89] Accioni F, García-Gómez D, Girela E, Rubio S. SUPRAS extraction approach for matrix-independent determination of amphetamine-type stimulants by LC-MS/MS. Talanta 2018; 182: 574-82. [http://dx.doi.org/10.1016/j.talanta.2018.02.039] [PMID: 29501195]

[90] Lara AB, Caballo C, Sicilia MD, Rubio S. Enantiomer-specific determination of hexabromocyclododecane in fish by supramolecular solvent-based single-step sample treatment and liquid chromatography–tandem mass spectrometry. Anal Chim Acta 2012; 752: 62-8. [http://dx.doi.org/10.1016/j.aca.2012.09.039] [PMID: 23101653]

[91] Caballero-Casero N, García-Fonseca S, Rubio S. Vesicular aggregate-based solventless microextraction of Ochratoxin A in dried vine fruits prior to liquid chromatography and fluorescence detection. Talanta 2012; 89: 377-82.

[http://dx.doi.org/10.1016/j.talanta.2011.12.046] [PMID: 22284505]

[92] Gissawong N, Boonchiangma S, Mukdasai S, Srijaranai S. Vesicular supramolecular solvent-based microextraction followed by high performance liquid chromatographic analysis of tetracyclines. Talanta 2019; 200: 203-11.
[http://dx.doi.org/10.1016/j.talanta.2019.03.049] [PMID: 31036174]

[93] Mammana SB, Berton P, Camargo AB, Lascalea GE, Altamirano JC. Coprecipitation-assisted coacervative extraction coupled to high-performance liquid chromatography: An approach for determining organophosphorus pesticides in water samples. Electrophoresis 2017; 38(9-10): 1334-43.
[http://dx.doi.org/10.1002/elps.201600335] [PMID: 28176350]

[94] Seebunrueng K, Phosiri P, Apitanagotinon R, Srijaranai S. A new environment-friendly supramolecular solvent-based liquid phase microextraction coupled to high performance liquid chromatography for simultaneous determination of six phenoxy acid herbicides in water and rice samples. Microchem J 2020; 152: 104418.
[http://dx.doi.org/10.1016/j.microc.2019.104418]

[95] Yang S, Tang T, Tan Y, *et al.* Determination of benzo(*a*)pyrene in fried and baked foods by HPLC combined with vesicular coacervative supramolecular solvent extraction. J Food Sci Technol 2019; 56(1): 428-35.
[http://dx.doi.org/10.1007/s13197-018-3504-4] [PMID: 30728586]

[96] Merino F, Rubio S, Pérez-Bendito D. Mixed aggregate-based acid-induced cloud-point extraction and ion-trap liquid chromatography–mass spectrometry for the determination of cationic surfactants in sewage sludge. J Chromatogr A 2003; 998(1-2): 143-54.
[http://dx.doi.org/10.1016/S0021-9673(03)00565-X] [PMID: 12862380]

[97] Goryacheva IY, Shtykov SN, Loginov AS, Panteleeva IV. Preconcentration and fluorimetric determination of polycyclic aromatic hydrocarbons based on the acid-induced cloud-point extraction with sodium dodecylsulfate. Anal Bioanal Chem 2005; 382(6): 1413-8.
[http://dx.doi.org/10.1007/s00216-005-3287-0] [PMID: 15995862]

[98] Santalad A, Srijaranai S, Burakham R, Sakai T, Deming RL. Corrigendum to "Acid-induced cloud-point extraction coupled to spectrophotometry for the determination of carbaryl residues in waters and vegetables" [Microchem. J. 90 (2008) 50–55]. Microchem J 2010; 96(2): 423. [Microchem. J. 90 (2008) 50–55].
[http://dx.doi.org/10.1016/j.microc.2010.06.009]

[99] Adlnasab L, Ezoddin M, Shabanian M, Mahjoob B. Development of ferrofluid mediated CLDH@Fe3O4@Tanic acid- based supramolecular solvent: Application in air-assisted dispersive micro solid phase extraction for preconcentration of diazinon and metalaxyl from various fruit juice samples. Microchem J 2019; 146: 1-11.
[http://dx.doi.org/10.1016/j.microc.2018.12.020]

[100] Felici E, Wang CC, Fernández LP, Gomez MR. Simultaneous separation of ergot alkaloids by capillary electrophoresis after cloud point extraction from cereal samples. Electrophoresis 2015; 36(2): 341-7.
[http://dx.doi.org/10.1002/elps.201400215] [PMID: 25257749]

[101] Wu YW, Jiang YY, Xiao TX, Zhang HL. Determination of triptonide by cloud point extraction combined with MEKC. J Sep Sci 2008; 31(5): 865-71.
[http://dx.doi.org/10.1002/jssc.200700463] [PMID: 18348329]

[102] Kukusamude C, Santalad A, Boonchiangma S, Burakham R, Srijaranai S, Chailapakul O. Mixed micelle-cloud point extraction for the analysis of penicillin residues in bovine milk by high performance liquid chromatography. Talanta 2010; 81(1-2): 486-92.
[http://dx.doi.org/10.1016/j.talanta.2009.12.029] [PMID: 20188951]

[103] Zhang W, Duan C, Wang M. Analysis of seven sulphonamides in milk by cloud point extraction and high performance liquid chromatography. Food Chem 2011; 126(2): 779-85.

[http://dx.doi.org/10.1016/j.foodchem.2010.11.072]

[104] Qin H, Yu G, Chen M, Zou Y, Yang Y. Ultrasonic-thermostatic-assisted cloud point extraction coupled to high-performance liquid chromatography for the analysis of adrenalines residues in milk. Eur Food Res Technol 2012; 234(3): 543-50.
[http://dx.doi.org/10.1007/s00217-012-1660-8]

[105] Liu W, Zhao W, Chen J, Yang M. A cloud point extraction approach using Triton X-100 for the separation and preconcentration of Sudan dyes in chilli powder. Anal Chim Acta 2007; 605(1): 41-5.
[http://dx.doi.org/10.1016/j.aca.2007.10.034] [PMID: 18022409]

[106] Pourreza N, Rastegarzadeh S, Larki A. Determination of Allura red in food samples after cloud point extraction using mixed micelles. Food Chem 2011; 126(3): 1465-9.
[http://dx.doi.org/10.1016/j.foodchem.2010.11.158] [PMID: 29243637]

[107] Pourreza N, Ghomi M. Simultaneous cloud point extraction and spectrophotometric determination of carmoisine and brilliant blue FCF in food samples. Talanta 2011; 84(1): 240-3.
[http://dx.doi.org/10.1016/j.talanta.2010.12.043] [PMID: 21315926]

[108] Khani R, Sheykhi R, Bagherzade G. An environmentally friendly method based on micro-cloud point extraction for determination of trace amount of quercetin in food and fruit juice samples. Food Chem 2019; 293: 220-5.
[http://dx.doi.org/10.1016/j.foodchem.2019.04.099] [PMID: 31151604]

[109] Chen L, Zhao Q, Jin H, *et al.* Determination of xanthohumol in beer based on cloud point extraction coupled with high performance liquid chromatography. Talanta 2010; 81(1-2): 692-7.
[http://dx.doi.org/10.1016/j.talanta.2010.01.004] [PMID: 20188983]

[110] Accioni F, Caballero-Casero N, García-Gómez D, Rubio S. Restricted Access Volatile Supramolecular Solvents for Single-Step Extraction/Cleanup of Benzimidazole Anthelmintic Drugs in Milk Prior to LC-MS/MS. J Agric Food Chem 2019; 67(1): 520-30.
[http://dx.doi.org/10.1021/acs.jafc.8b06003] [PMID: 30516988]

[111] González-Rubio S, García-Gómez D, Ballesteros-Gómez A, Rubio S. A new sample treatment strategy based on simultaneous supramolecular solvent and dispersive solid-phase extraction for the determination of ionophore coccidiostats in all legislated foodstuffs. Food Chem 2020; 326: 126987.
[http://dx.doi.org/10.1016/j.foodchem.2020.126987] [PMID: 32416421]

[112] Cardeñosa V, Lunar ML, Rubio S. Generalized and rapid supramolecular solvent-based sample treatment for the determination of annatto in food. J Chromatogr A 2011; 1218(50): 8996-9002.
[http://dx.doi.org/10.1016/j.chroma.2011.10.041] [PMID: 22071422]

[113] López-Jiménez FJ, Rubio S, Pérez-Bendito D. Supramolecular solvent-based microextraction of Sudan dyes in chilli-containing foodstuffs prior to their liquid chromatography-photodiode array determination. Food Chem 2010; 121(3): 763-9.
[http://dx.doi.org/10.1016/j.foodchem.2009.12.081]

[114] Caballero-Casero N, Ocak M, Ocak Ü, Rubio S. Quick supramolecular solvent-based microextraction for quantification of low curcuminoid content in food. Anal Bioanal Chem 2014; 406(8): 2179-87.
[http://dx.doi.org/10.1007/s00216-013-7409-9] [PMID: 24170269]

[115] Harikrishnan R, Sundraraj K, Hui BY, *et al.* Vortex-assisted supramolecular-based dispersive liquid phase microextraction for penentuan spectrophotometric determination of rhodamine b in chili powder. Malays J Anal Sci 2020; 24: 197-208.

[116] Caleb J, Alshana U. Supramolecular solvent-liquid-liquid microextraction followed by smartphone digital image colorimetry for the determination of curcumin in food samples. Sustain Chem Pharm 2021; 21: 100424.
[http://dx.doi.org/10.1016/j.scp.2021.100424]

[117] Li Y, Jiao Y, Guo Y, Yang Y. Determination of bisphenol-A, 2,4-dichlorophenol, bisphenol-AF and tetrabromobisphenol-A in liquid foods and their packaging materials by vortex-assisted

supramolecular solvent microextraction/high-performance liquid chromatography. Anal Methods 2013; 5(19): 5037-43.
[http://dx.doi.org/10.1039/c3ay40586a]

CHAPTER 7

Supercritical Fluid Extraction

Ádina L. Santana[1,*], Larry O. Chañi-Paucar[2,3], Priscilla C. Veggi[4], Juliane Viganó[5] and M. Angela A. Meireles[2]

[1] *Grain Science and Industry Department, Kansas State University, 1301 N Mid Campus Drive, Manhattan, KS 66506, USA*

[2] *School of Food Engineering, University of Campinas, Rua Monteiro Lobato, 80, Campinas, Brazil*

[3] *Escuela Profesional de Ingeniería Agroindustrial, Universidad Nacional Autónoma Altoandina de Tarma (UNAAT), Jr. Huaraz 431, 12651, Junín, Perú*

[4] *Department of Chemical Engineering, Institute of Environmental, Chemical and Pharmaceutical Sciences, Universidade Federal de São Paulo, R. São Nicolau 210, Diadema, Brazil*

[5] *Centro de Ciências da Natureza, Universidade Federal de São Carlos, Rod. Lauri Simões de Barros, km 12 – SP 189, Buri, SP 18290-000, Brazil*

Abstract: In this chapter, we highlight the basic concepts behind the use of SFE to select molecules present in food matrices, *e.g.*, carotenoids, essential oils, waxes, and phenolic compounds. Also, we highlight the SFE equipment setup, the methods for process intensification, and mass transfer mechanisms involved in the process, besides the advantages and drawbacks. Supercritical fluids have been suggested as a powerful tool to improve the performance of analytical methods in terms of reduced steps for sample preparation and waste generation, besides enhanced precision and recovery of analytes detected. The offline association of SFE with analytical detection has been elucidated for decades. Currently, many efforts have been made to reach the miniaturization of equipment as well as the online hyphenation between extraction and analytical detection with supercritical fluids as a novel method for sample preparation to detect food analytes in real time with accuracy and robustness.

Keywords: Analytical chemistry, Chromatography, Food bioactive, Overall extraction curves, Sample preparation, Supercritical CO_2.

INTRODUCTION

Supercritical fluid extraction (SFE) is a green technology widely used to extract active principles present in varied matrices for multiple applications, including foods, pharmaceuticals, and cosmetics. A significant number of works have sho-

* **Corresponding Author Ádina L. Santana:** Grain Science and Industry Department, Kansas State University, 1301 N Mid Campus Drive, Manhattan, KS 66506, USA; E-mail: adina.santana@gmail.com and adina@ksu.edu

Merichel Plaza & Maria Luisa Marina (Eds.)

wn that supercritical fluids are more effective in extracting target compounds than conventional methods that use organic solvents (methanol, acetone, and hexane) [1, 2].

Supercritical carbon dioxide (SC-CO_2) is the most used supercritical fluid due to its mild critical point (31.1 °C and 7.38 MPa), low cost, and non-flammability. Compared with conventional solvent extraction and steam distillation, CO_2 does not interact with the analytes in the sample, providing extracts of high quality in terms of target compounds, including carotenoids, phenolic compounds, and tocotrienols, which are extensively required by food industries.

In food analysis, SFE has been used as a sample preparation protocol before detecting target compounds in an analytical instrument. In chromatographic methods, the use of SC-CO_2 as mobile phase reduced retention times of compounds, and enhanced sensitivity over other methods, including HPLC and GC. Such advantages supported works on the optimization of supercritical fluid chromatography (SFC) and the implementation of direct hyphenation of SFE of matrix with the SFC or other analytical instruments for the real-time acquisition of data [3].

Efforts have been made to develop robust and accurate analytical methods to identify and quantify target compounds in food samples, including the miniaturization of SFE hyphenated with chromatographic techniques [4 - 6]. The direct hyphenation of SFE and analytical detection has been proposed to reduce the analysis time and generation of waste solvents. However, direct hyphenation represents a difficult task because of damage to the detector or lack of stability in the baseline induced by high pressure [7, 8].

In this chapter, we introduce recent findings on the use of SFE to obtain target compounds present in food samples, as well as the theoretical aspects behind SFE and the equipment used. Also, we discuss the factors that justify the selection of SFE to analyze foods, and the current developments to include supercritical technology as an analytical method to improve the reliability of methods used in food analysis.

FUNDAMENTALS ABOUT SFE

Properties and Procedure

A substance becomes a supercritical fluid when subjected to temperature and pressure above its critical values [1]. At a supercritical state, there are no distinctions between the liquid and vapor phases [9]. Some properties can be significantly tuned by increasing the temperature and/or pressure above their critical values.

Supercritical fluids have a density close to a liquid, whereas their viscosity is close to that of gas. In contrast, its diffusivity between liquid and gas values enhances transport properties, favoring solute extraction from the solid matrix [3, 4]. Likewise, the solvent density is easily tuned by temperature and pressure changes, which effectively modifies the solubility of the solute in the solvent, inducing high selectivity [10-12]. Thus, the flexibility of SFE that allows modification of the solvating power and its selectivity are the attractions of the process [13-15].

Specifically, the thermodynamic properties of carbon dioxide (CO_2) attract attention and make it the most commonly used solvent in food applications. Carbon dioxide presents a mild critical point (Tc = 31.1 °C, Pc = 7.38 MPa); it is nontoxic, nonexplosive, has low cost, commercially available, and has an environmentally friendly nature [3, 4, 7 - 9]. In addition, CO_2 was designated as a GRAS (generally recognized as safe) solvent for the food industry. Additionally, CO_2 is a gas at room temperature, and 1 atm (0.1 MPa), making it possible to obtain a solvent-free extract by depressurization, which is applicable while making bioactive compounds [16]. For these reasons, CO_2 has been used as a supercritical solvent in over 90% of SFE applications [13].

On the other hand, the physical properties of CO_2 limit its use in the extraction of compounds with low polarity or nonpolar compounds. A polar cosolvent (also known as a modifier) can be added to CO_2 in small amounts (usually 1 to 15%) [17] to improve the solubility of polar compounds in CO_2. Due to its low toxicity, anhydrous ethanol or aqueous ethanol is the most common cosolvent reported in food applications [18 - 21]. There are also reports in the literature regarding the use of water, methanol, ethyl ether, formic acid, acetic acid, acetone, and isopropanol as cosolvents [12, 14]. Likewise, cosolvents such as ethyl lactate, vegetable oils, and ethyl acetate combined with CO_2 (less than 10% of the total) have been applied [22].

Concerning the process, the SFE of solid matrices comprises two essential steps: extraction and separation. Firstly, the solvent must be tuned to a supercritical state, *i.e.*, the fluid is heated and pressurized, reaching the desired operating temperature and pressure. Once the fluid has reached the supercritical state, it flows through the extractor filled with the pretreated sample, extracting the solid matrix's solute by convection and diffusion. Afterward, the dissolved compounds pass through the separator, where they are separated from the solvent. The solute is precipitated by pressure reduction and/or temperature increase, and the supercritical fluid turns into a gas [7, 16, 17, 23].

In addition to selecting the supercritical solvent during the SFE process, other important parameters, such as extraction pressure and temperature, raw material characteristics, and others, can considerably influence both thermodynamics and transport issues [24] of the process. For this reason, a successful SFE process requires a good understanding of the supercritical fluid properties.

The thermodynamic and primary transport parameters of the process are the extracting temperature and pressure [25]. Temperature and pressure directly affect the solvent density and the solvation power. Generally, an increase in pressure for a given temperature results in a higher supercritical fluid density, improving the target compounds' solubility. In addition, an increase in temperature at constant pressure decreases the solvent density, reducing its solvating power [4, 25]. However, increasing the temperature increases the vapor pressure of solutes, promoting their transfer from the sample to the supercritical fluid [25, 26].

For these reasons, SFE can be a rapid, effective, and clean technology for extracting a range of natural compounds, such as carotenoids, alkaloids, fatty acids, essential oils, anthocyanins, flavors, vitamins, and others [15, 20].

Applications of SFE in Food Samples

Supercritical fluids have been widely used for decades due to their efficiency in obtaining a range of compounds. Their use in the extraction process is recognized for both industrial and laboratory applications. Many studies illustrate the different opportunities of SFE applied to food processes at academic and industrial levels. Some of the first large-scale uses of SFE, known since the 1970s, include the decaffeination of coffee and tea and the isolation of hop extracts for flavoring beers [11]. SFE applications in food science remain significant today. Supercritical processes are mostly used in food samples to obtain bioactive compounds from natural matrices by extraction, concentrate them by fractionation, and identify such compounds.

Pretreatment

Integrated techniques are frequently employed because of new environmentally friendly protocols, allowing higher efficiency and sustainability [11]. A sequential extraction is an approach that enables the extraction of different chemical compounds [27]. This technique consists of applying several successive extraction steps, consecutively increasing the fluid solvent power [28] and thus exploring its selectivity under different operational conditions [29].

Vardanega *et al.* [29] found high selectivity of sequential CO_2 extraction to recover geranylgeraniol and tocols from annatto seeds. Viganó *et al.* [30] explored the supercritical CO_2 selectivity to fractionate extracts from passion fruit bagasse. These authors fractionated carotenoids and fatty acids using two steps at different temperature and pressure process conditions. Many examples in the literature refer to sequential extraction processes [31–34].

Moreover, SFE has been a technology with the potential to be integrated into a biorefinery since it can be successfully used to extract high added-value products from biomass waste [35]. Therefore, SFE is integrated with other emerging technologies to accelerate and enhance bioactive compound recovery.

Several studies have applied the biorefinery concept with SFE for many purposes. SFE is an effective method for wax and lipid extraction when utilized as a part of a biorefinery, improving downstream biomass treatment [12]. Aguiar *et al.* [27] proposed a sequential extraction to recover bioactive compounds from biquinho peppers. SFE with CO_2 extracted nonpolar capsinoids while pressurized liquid extraction with aqueous ethanol extracted phenolics of high polarity. Some examples of sequential extraction methods include the extraction of tocols [36], tocols and carotenoids from passion fruit [37], flavonoids from pepper-rosmarin [38], anthocyanins and other phenolic compounds from purple corn [39], tocotrienols from annatto seeds [40], fatty oil and essential oil from fennel [41].

Extraction

Carotenoids

Carotenoids consist of lipophilic natural pigments responsible for yellow-red colors [42]. They are extensively employed as natural food colorants in the food industry and can be used as precursors of flavor or aroma compounds [25]. They are among the most important compounds with bioactive properties to human health since they have antioxidant and anticancer activities and are vitamin A precursor [43]. They are also associated with a low risk of numerous chronic diseases, such as cancer and cardiovascular disease, age-related macular degeneration, and cataracts [44]. Among all carotenoids, including α-carotene, β-carotene, β-cryptoxanthin, lycopene, lutein, and zeaxanthin, β-carotene is the most commonly studied [44].

According to Lima *et al.* [43], due to the high demand for carotenoids, the global carotenoid market tends to increase 4% from 2018 to 2023 and exceeds US$ 2 billion by 2023.

In this context, SFE methodologies have been investigated and successfully employed to obtain carotenoids from various vegetable matrices and fruit wastes [36, 37].

Sánchez-Camargo *et al.* [45] found SFE an interesting alternative to valorize mango peels. High concentrations of all-trans-β-carotene were found. β-Carotene is the primary precursor of vitamin A and usually occurs in edible plants such as acerola, carrot, nuts, and oil palm [19]. Carotenoids have also been successfully extracted from citrus fruit byproducts [46], passion fruit bagasse [30], annatto seeds [29], ripe bitter melon pericarp [47], pink guava, peach, sweet potato, apricot, green, yellow, and red bell peppers [43], pumpkin [48], camelina seeds [49], and microalgae [50].

Lycopene is a red pigment with high antioxidant activity and is among the most studied dietary antioxidants from food byproducts [52, 53]. Lycopene is a major carotenoid present in fresh tomatoes and their processing wastes and peels [51], [52]. Peels and seeds provide 48% lycopene, 53% of the total amount of phenols, 52% of flavonoids, and 52% of the tomatoes' antioxidant activity [53]. SFE has been employed to extract lycopene from tomato processing byproducts, peels, and seeds in many studies [53–55].

Phenolic Compounds

Phenolic compounds are highly attractive due to their wide range of biological properties, including antioxidant, antiviral, anti-inflammatory, antimicrobial, anticancer, and other properties, playing an essential role in human health [56]. In food applications, they are responsible for increasing the shelf-life of products due to their antioxidant activity [57].

Like carotenoids, phenolic compounds are among the main classes of bioactive compounds obtained from agricultural and food byproducts by SFE. Different samples provide various sources of phenolics groups, *i.e.,* flavonoids (flavones, anthocyanins, flavanones and others) and non-flavonoids (phenolic acids and tannins) [58]. Many studies report peels and seeds as important sources of phenolic compounds. High polyphenol contents were obtained from mango peel [45] and tomato skin byproducts [53]. Phenolic compounds were also obtained from caferana [31], radish leaves [59], fermented orange pomace [60], coffee grounds [61], orange juice bagasse [60], green tea waste [62], cacao pod husk [63], and apple pomace [64]. Hop is another source that has been continuously studied for centuries because of its high content of bioactive compounds. The phenolic fraction of hop cones has been extensively studied, and more than 1% of dried hop cones correspond to xanthohumol. Polyphenols such as catechin, prenylflavonoids, flavonols, resveratrol, and the simple phenolic ferulic acid were also identified in hop [57].

Anthocyanins are a valuable group of phenolic compounds that can be used as food colorants because of their desirable blue-red colors and nontoxicity [59, 60]. Phenolic compounds and anthocyanins were evaluated in many samples, such as genipap [65], black bean [66], juçara [67], and chokeberry peel [68]. In addition, grape seeds are an important source of catechin, epicatechin trans-resveratrol, and procyanidin B1 [69].

Essential Oils and Lipids

Essential oils from herbs comprise the most important group of compounds that SFE has extracted. The SFE with CO_2 has been used because it does not promote thermal degradation of compounds or oil contamination as extraction with organic solvents does [70].

Essential oils have lipophilic substances, such as terpenes and terpenoids, as their major components responsible for their aroma [25]. Traditionally, essential oils are applied in food, cosmetics, medicine, herbicides, fragrances, and insecticides [25]. Their use in the food industry is attractive since they promote flavor and act as a natural antimicrobial, anti-inflammatory, and antioxidant agents [71].

SFE has been successfully applied to recover essential oils from many food samples, in addition to agricultural and food byproducts. Examples include turmeric root [70], hop [72], *Piper auritum* and *Porophyllum ruderale* [73], rosemary [74], grape seeds [69], apple seed [75], winter savory, coriander, and sweet basil [71], and peppermint [76].

Regarding lipids, several studies report SFE as an outstanding alternative for their recovery. SFE has been used to obtain pure lipid extracts from various samples, analyze the fat content from foods, and obtain products with a low quantity of specific lipids or cholesterol [77].

The oil of babassu extracted by SFE was rich in essential fatty acids at high concentrations, such as lauric, myristic, oleic, and palmitic acids [78]. The seeds and peels from Avocado presented high levels of monounsaturated fatty acids (*e.g.*, palmitoleic, oleic) [79]. Researchers have shown high fatty acid contents in extracts obtained by SFE from seeds of papaya [80], apple [75], mandarin, lemon, and grapefruit [81]. Caferana [31], spent coffee grounds [61], mango peel [45], fish food [82], and the byproducts from *"horchata"* [83] were also good sources of fatty acids obtained by SFE.

Among fatty acids, omega-3 has gained attention due to its nutritional and health indication, mainly because of its beneficial effects on chronic diseases like atherosclerosis or rheumatoid arthritis [84]. Many samples are potential sources of omega-3 to be extracted by SFE, including the seeds of *Portulaca oleracea* L. [84] and *Camelina sativa* [85].

Removal of Undesirable Compounds

Another role of SFE is the extraction of undesirable compounds from food samples [13]. For example, SFE has successfully removed pesticides from food products [86].

An interesting commercial application is the removal of pesticides and heavy metals from rice. In addition, SFE helps prevent rice degradation by removing waxy layers and fatty acids [13]. Another study used SFE to determine neonicotinoid pesticides in green onions [87].

In some cases, natural compounds from foods are nontoxic but undesirable and reduce food quality. SFE provided milk and dairy products with the highest shelf life and acceptable sensorial assets, ensuring minimal quality loss [88]. The technique effectively removed cholesterol from milk [82, 83]. In addition, SFE is being used to fractionate lipids and isolate vitamins for quantification [89–91].

Analytical Tool

Another use of supercritical fluids is in analytical methods, which use showed higher efficiency in comparison with existing chemical analytical processes [13].

Several studies have been published concerning analytical methods for quantifying a wide range of bioactive compounds. In this regard, supercritical fluid chromatography (SFC) is a post extraction quantification technique, a relatively recent, highly efficient, and "green" separation technique that uses environmentally neutral CO_2 as a mobile phase [91]. A suitable coupling of SFE to the SFC system allows precise detection and separation of analytes [13].

Some analytical methods have been published for major compound quantification from hops (flowers of *Humulus lupulus*) due to their potential application in food and herbal remedies [92]. SFC was successfully employed, offering a rapid and ecologically alternative for hops quality control. Moreover, SFC is an effective alternative to determine contaminants in food samples [93]. This technique was used to estimate mycotoxins in ready-to-eat food based on cereals, legumes, vegetables, meat, and fish [93]. Additional information, including advantages and applications, can be found in the section: Applications in Food Analysis.

Advantages and Drawbacks

Supercritical fluid extraction (SFE) has been extensively applied compared to other green technologies, such as ultrasound-assisted extraction, microwave-assisted extraction, pressurized liquid extraction, mechanical pressing, *etc*.

A large number of reviews highlight SFE as a potential technique to extract many food samples. In addition, as previously discussed, SFE has also been used as a preparatory step to remove undesirable compounds.

The main advantage of SFE includes manipulating the fluid density *via* tuning the temperature and pressure, which enhances the extraction of compounds and their separation into different fractions [86].

SFE leads to a free-of-solvent extract when no cosolvent is used. Additionally, the SFE depressurization step requires no further separation and purification, decreasing the number of unit operations required [13]. In addition, SFE is suitable for extracting volatile and thermolabile biomolecules because of the possibility of operating at a low temperature [94 - 95].

Moreover, after depressurization, CO_2 can be recirculated and continuously reused [88, 89]; this step eliminates solvent removal, one of the most expensive steps of conventional processes. It can also be mentioned that it has a rapid extraction time and high productivity [25]. In addition, SFE can be directly combined with analytical chromatographic methods or SFC [86].

Despite all the advantages of using supercritical CO_2 for a wide range of applications, extraction is still the most widespread application on a large scale [24].

Many studies highlight the high investment costs of implementing the SFE process as its main disadvantage [96]. However, current works have shown the importance of research on large-scale and economic evaluation of the SFE process to allow a reliable comparison with conventional methods [25, 97].

Another disadvantage of SFE is the low CO_2 polarity, which is effective only for nonpolar molecules. This drawback can be solved by adding a modifier [95]. Moreover, SFE requires a technical understanding of process properties such as phase behavior and crossover region, requiring more trained staff to deal with it [95].

EXTRACTION PROFILE (OVERALL EXTRACTION CURVES)

Overall extraction curves (OECs) are among the most important tools in evaluating the dynamic SFE, as they allow us to observe how the extraction evolves with time. They are graphical representations that relate the mass of the extract with the extraction time [23, 133] and can be determined according to Equation 1 [9].

$$m_{total} = Q\int_0^t Y(H,t)dt \quad \textbf{(1)}$$

where m_{total} is the mass of extract, Q is the mass solvent flow rate, and $Y(H,t)$ is the ratio of solute mass to solvent mass at a given time (t) at the extractor or bed outlet (H). Alternatively, the mass of extract can be represented by the extract's yield, *i.e.*, the mass extracted per unit of raw material. However, the used mass of solvent or the mass ratio between the solvent (S) and feed (F) can be applied to replace the time.

Fig. (**1**) shows a representation of an OEC composed of the extraction yield vs. S/F. The three main periods of the whole extraction process are also presented, namely, constant extraction rate (CER), falling extraction rate (FER), and diffusion controlled (DC).

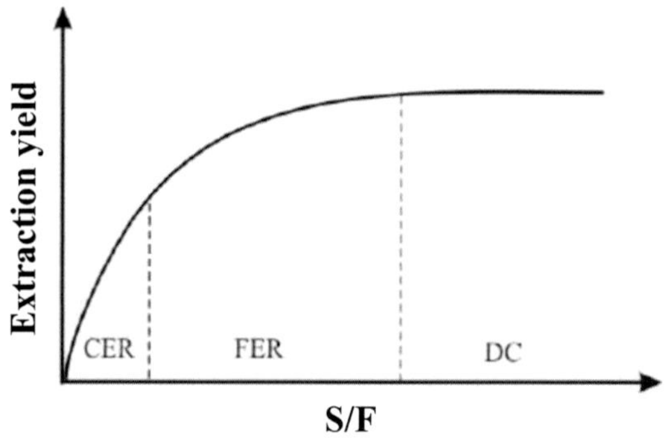

Fig. (1). Representation of an overall extraction curve and the extraction periods.

The CER period is characterized by the extraction of solutes covering the raw material's outer particle surface. This fraction is considered an easily accessible solute. The main extraction mechanism driving the process is mass convection. FER is the second extraction period and starts when there is insufficient solute to keep the extraction rate constant. The last period is DC, in which the process is governed only by solvent diffusion inside the particles and solvent/solute diffusion to their surface [98].

Several authors have reported that 50-90% of the total extract is obtained during the CER period. Therefore, the extraction ends shortly after the CER period for many applications, except when the target compound is not very soluble in the supercritical solvent or is located in a difficult access cell structure [9]. In general, for technical and economic issues, the best operational conditions will be those that produce the highest amount of extract in the shortest time. On the other hand, it is advised to lead the extraction until the DC period for analytical purposes since quantitative extraction must be performed.

The OEC evolves over time depending on the extraction kinetic properties, the raw material characteristics, and the solvation capacity of the supercritical fluid affected by the operating conditions [99]. Fig. **2** shows three possible profiles that an OEC can assume. Curves controlled by diffusion assume the shape illustrated in Fig. (**2a**), in which the extraction rate is high at the beginning of the process until the solute concentration at the particle surface drops to zero, and the rate then becomes much slower. The extraction yield eventually approaches 100%. Fig. (**2b**) presents a curve for extractions limited by solubility. The extraction rate is almost constant and only flattens off towards the end of the extraction. Fig. (**2c**) shows a typical curve affected by matrix effects, where there is some reversible interaction between extract and matrix, such as desorption. In such cases, if the maximum recovery of the extract is unknown, it is difficult to identify whether the extraction is complete [100].

Fig. (2). OEC profile for different types of extraction.

The main information given by OEC is the mass transfer rate, which, along with solute solubility in the supercritical fluid and the supercritical fluid selectivity, is essential for scale-up procedures from laboratory to pilot and industrial scales. In this sense, the mathematical modeling of OECs allows a rational approach to extraction, making it possible to generalize experimental results and extrapolate them to systems not yet studied. There are three mathematical approaches to describe OECs for supercritical processes: empirical, based on an analogy between heat and mass transfer, and those with differential mass balance [2].

OEC mathematical modeling has been widely applied to represent the extraction of several raw materials by applying different models. However, the most commonly used model is Sovová's model, which allows obtaining parameters with a physical meaning. These parameters are the mass transfer coefficient of the fluid phase, the mass transfer coefficient in the solid phase, and the solute concentration in unbroken cells [101]. Sovová [102] presented a model based on the broken and intact cell approach. The author considered that part of the solute was directly exposed to the solvent due to raw material pretreatment. The pretreatment reduces the particle size, increases the contact surface between the solvent and solid, and decreases the diffusive pathways. The easily accessed solute would cover the particle surface, while the other solute fraction would be retained within unbroken cells. Years later, Sovová [103] proposed a more complete model for SFE. It considers more phenomena than those published in 1994, especially concerning equilibrium relationships, and includes two new parameters, making the model more accurate and the applicability more complex [101]. For full access to Sovová's model, we invited the readers to access both original publications.

INSTRUMENTATION

SFE has emerged as a unit operation for the extraction of botanical matrices, micronization of pharmaceuticals, and equipment design since the first reports dated from 1879 [104].

The main application of SFE-based equipments was the extraction of bioactive compounds. With this objective, the equipment has been configured in different ways, to improve extraction efficiency of specific compounds, and the fractionation of bioactives present in the extracts. Supercritical extraction has also found other applications in the processing of food, drugs, analytical determinations, among others [105]. In this section, we describe the components of SFE equipment.

In 1879, Hannay and Hogarth observed the properties of supercritical fluids for the first time with the use of an apparatus of simple construction, consisting of a glass sight cell, a pressure generator, and a pressure measuring device. The operation of this apparatus was described by McHugh and Krukonis [104]. These first assays were developed using inorganic salts and supercritical ethanol, which resulted in different solubilities. At the end of the 1970s, supercritical fluids were oriented to obtain natural products but only applied to a small number of products [105]. Currently, there are a great variety of studies describing the supercritical extraction of natural products to obtain polar and nonpolar bioactive compounds [106].

Conventional SFE Equipment

A simple SFE process on a laboratory scale consists of the following stages: 1) preparation of the raw material, 2) packing of the extraction bed, 3) pressurization, 4) conditioning of the extraction temperature, 5) time static extraction, and 6) dynamic extraction. Laboratory-scale supercritical extraction equipment (Fig. **3**) consists of a CO_2 container, a cooler apparatus, a CO_2 compressor, a heater, an extraction vessel, an extract collection container, blocking valves located at different points in the equipment piping system, a micrometer valve, and a CO_2 flow meter [107, 108].

The operational procedure starts with the CO_2 transportation from the cylinder to the CO_2-driven pump. In the cylinder, CO_2 is at 6-7 MPa at room temperature. Afterward, CO_2 is transferred to the CO_2 pump. The head of the pump is cooled down *via* the use of a heat exchanger (Fig. **3**) to maintain CO_2 liquid during pumping. If CO_2 is pumped in the gaseous state, cavitation of the pump occurs. Consequently, the pressure does not increase, leading to a considerable increase in the number of pump compression cycles without producing an increase in the system pressure. The solvent temperature is adjusted by the heating system (oven) coupled with the extraction vessel. Static time or static extraction starts with CO_2 and the plant matrix interaction, leading to vessel pressurization up to the setup pressure [99, 103]. Afterward, dynamic extraction starts with the valve opening (V5, in Fig. **3**) just after the extraction vessel, and CO_2 flow regulation is achieved using a micro metering valve (MV, in Fig. **3**). The micro metering valve must be heated to avoid freezing due to the Joule-Thomson effect, produced by the sudden drop in pressure that allows the separation of the solvent and the recovered analytes.

Research on supercritical technology is growing, promoting the development of commercial equipment and homemade prototypes. The manufacturers of commercial SFE equipment at bench and pilot scales include TharProcess™ [109], SepareCo$_2$ Srl [110], HENAN Tradematt Industry [111], Natex [112], and ExtrateX [113]. Although commercial SFE equipment is available, homemade construction may be economically advantageous for research and projects. For instance, Johner and Meireles [114] estimated the costs of assembling a homemade SFE unit with a 0.1 L vessel extraction, which was lower than the costs of commercial units with the same characteristics. Additionally, the authors evaluated the incorporation of a fractionation system with two separators. Likewise, the cost of assembling this homemade unit was lower than the price of commercial units with the same characteristics. Depending on the implementation of the supercritical extraction apparatus, commercial equipment or homemade prototypes can be chosen. The homemade prototypes are relatively easy to assemble, cheaper, and present greater flexibility in terms of their configuration.

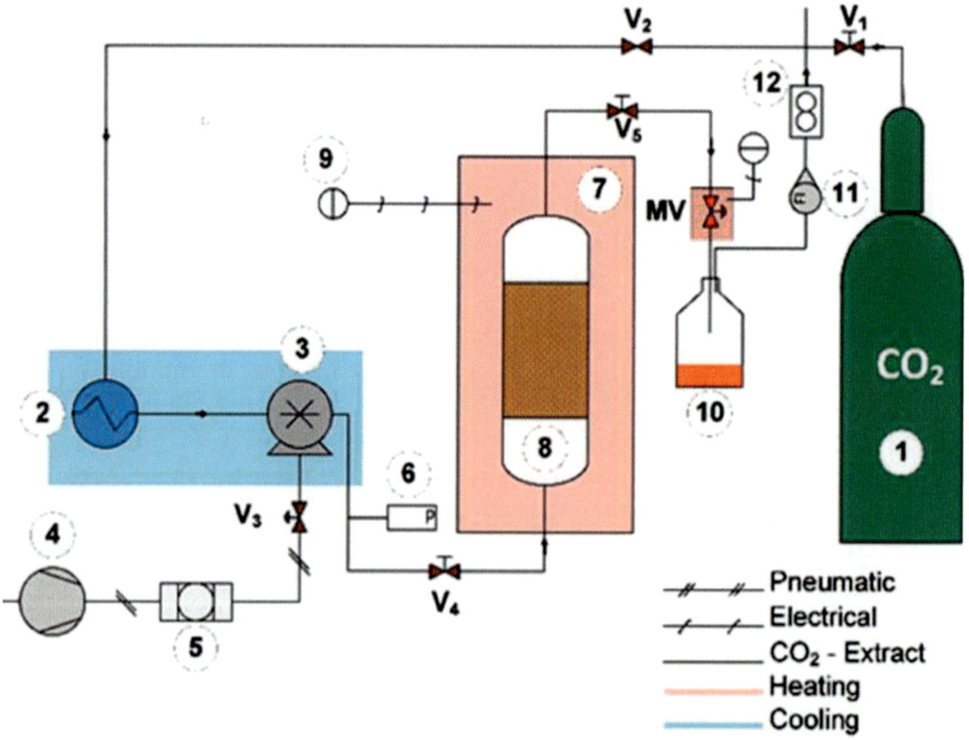

Fig. (3). Diagram of the Spe-ed SFE unit (Applied Separations, model 7071, Allentown, EUA). (1) – CO_2 reservoir, (2) – Heat exchanger, (3) – CO_2 pump, (4) – Air compressor, (5) – Air filter, (6) – Manometer, (7) – Column oven, (8) – Extraction vessel, (9) – Thermocouple, (10) – Collection flask, (11) – flowmeter, (12) – CO_2 totalizer, MV – Micrometric valve, (V3) – Control (Air pressure), and (V1, V2, V4, and V5) – Blocking valve.

SFE Equipment with Separators

The SFE of bioactive compounds using CO_2 has several advantages over conventional extraction methods [2]. The main advantage of the SFE process with

CO_2 is the possibility of continuously modulating the solvation power and selectivity of the solvent. This can be achieved by varying the pressure and temperature of the process [2, 110]. The selectivity of CO_2 under supercritical conditions was used to develop separation/fractionation processes for bioactive extracts, implemented immediately after extraction [111, 112].

Fig. (**4**) shows SFE equipment with an extraction column and two separation or fractionation vessels [114] built and validated by the Supercritical Technology Laboratory Research Group (LASEFI, Unicamp, Brazil). In this unit, the fractionation vessels were designed in series. The fractionation vessels operate based on the cascade decrease in pressure from the first to the last separator. The first separator is connected to the extraction vessel, which pressure is higher than the pressure in the last separator. The last separator's outlet pipe can release CO_2 to the environment or be connected to a recirculation system to optimize the solvent's use.

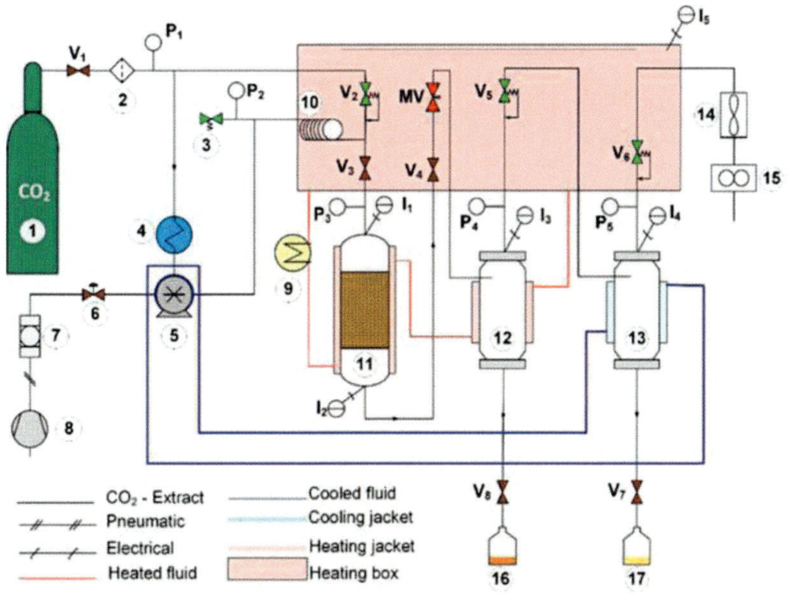

Fig. (4). Diagram of SFE-0.1L apparatus (1) – CO_2 reservoir, (2) – CO_2 filter, (3) – Safety valve, (4) – Cooling bath, (5) – Air-driven CO_2 pump, (6) – Control (air flow), (7) – Air filter, (8) – Air compressor, (9) – Heating bath, (10) – Serpentine tube, (11) – Extraction vessel, (12 and 13) – Separators, (14) – Flowmeter, (15) – Flow totalizer, (16 and 17) – Extract collecting vessel, (I1, I2, I3, I4, and I5) – Temperature indicator, (P1, P2, P3, P4, and P5) – Pressure gauge, (V1, V3, V4, V7, and V8) – Blocking valve, (V2) – Back pressure (60 MPa), (V5) – Back pressure (40 MPa), (V6) – Back pressure (10 MPa), and (MV) – Micrometering valve.

The fractionation process is convenient, both for depressurization and for the fractionation of bioactive compounds. The number of extract fractions to be obtained will depend on the number of separators, process conditions, and the extract's composition [115–118].

In the fractionation of fish oil obtained from the tuna's head, using two, three, and six fractionation glasses, it was observed that the fatty acid profile in each fraction was different [118]. Compared to the first oil fractions, the last oil fractions presented higher concentrations of PUFAs and MUFAs, with lowered SFAs. When six fractionation vessels were used, six oil fractions were obtained, observing that from the first fraction to the last oil fraction, an increase in PUFAs and MUFAs was observed, with a concomitant decrease in SFAs (Table **1**).

Table 1. Fractionation of extracts of some raw materials.

Raw-Material	Extraction Conditions	Fractions	References
Foeniculum vulgare seeds	Pressure: 20 MPa Temperature: 40 ºC Flow rate CO_2: 1.95 – 25.8 g/min. Separator 1: 8 MPa and 35 ºC Separator 2: 2 MPa and 8 ºC	Fraction 1 Yield: 3.6 g/100 g RM Fraction 2 Yield: 0.6 g/100 g RM	[41]
Ginger rhizomes	Pressure: 27.6 MPa Temperature: 40 ºC Flow rate CO_2: 1200 g/min. Time: 153 min Extractor volume: 25 L Separator 1: 17.5 MPa and 40 ºC Separator 2: 4 MPa and 40 ºC	Fraction 1 Oleoresin yield: 5.95 % w/w Fraction 2 Volatile oil yield: 2.77 % w/w	[117]
Tuna By-Product	Pressure: 40 MPa. Temperature: 65 °C. Flow rate CO_2 + Ethanol: 3 mL/min. Co-solvent: ethanol 20 %, v/v. Time: 120 min. Number of separators: 6	Fraction 1 PUFA: 10.5 %, MUFA: 23.2 %, SFA: 61.3 %, and others. Fraction 2 PUFA: 17.0 %, MUFA: 24.7 %, SFA: 53.6 %, and others. Fraction 3 PUFA: 23.9 %, MUFA: 25.5 %, SFA: 45.8 %, and others. Fraction 4 PUFA: 34.5 %, MUFA: 30.3 %, SFA: 30.5 %, and others. Fraction 5 PUFA: 45.9 %, MUFA: 32.7 %, SFA: 17.2 %, and others. Fraction 6 PUFA: 48.9 %, MUFA: 33.6 %, SFA: 13.6 %, and others.	[118]

(Table 1) cont.....

Raw-Material	Extraction Conditions	Fractions	References
Rosmarinus officinalis L. leaves	Pressure: 30 MPa Temperature: 40 °C Flow rate CO_2: 60 g/min Time: 300 min Number of separators: 2 Separator 1: 10 MPa and 40 °C Separator 2: 5 MPa and 40 °C	**Fraction 1*** 1,8-cineole: 72 % Camphor: 28 % **Fraction 2*** 1,8-cineole: 50.2 % Camphor: 21.4 % Borneol: 6.5 % α-terpineol: 6.3 % Verbenone: 4.3 % β-caryophyllene: 4 % Others (< 2 %).	[119]
Baccharis dracunculifolia leaves	Pressure: 30 MPa Temperature: 50 °C Flow rate CO_2: 16.7 g/min Time: 120 min. Separator 1: 20 MPa and 50 °C Separator 2: 10 MPa and 50 °C Separator 3: 8 MPa and 25 °C. Separator 4: 0.093 MPa and 25 °C	**Fraction 1** Yield: 0.75 % w/w Artepillin C: 238 mg/g extract Kaempferide: 40 mg/g extract **Fraction 2** Yield: 0.86 % w/w Artepillin C: 154 mg/g extract Kaempferide: 22 mg/g extract **Fraction 3** Yield: 1.58 % w/w Artepillin C: 5 mg/g extract Kaempferide: 1 mg/g extract **Fraction 4** Yield: 0.04 % w/w Artepillin C: 238 mg/g extract Kaempferide: 21 mg/g extract	[120]
Cannabis sativa L. inflorescences	Pressure: 10 MPa Temperature: 40 °C Flow rate CO_2: 50 g/min Extractor volume: 1 L Separator 1: 7 MPa and 25 °C. Separator 2: 5 MPa and 15 °C.	**Fraction 1** Waxes yield: 1.03 % w/w **Fraction 2** Volatile compounds yield: 0.67 % w/w Composition of volatile compounds*: Monoterpene hydrocarbons: 51.7 % Oxygenated monoterpenes: 2.7 % Sesquiterpene hydrocarbons: 39.1 % Oxygenated sesquiterpenes: 6.5 %	[121]

*GC peak area percentage.

The extract of *B. dracunculifolia* leaves obtained by SFE was fractionated using four vessels hyphenated in series [120]. Two fractionation conditions were studied, and the results show that the second condition (see Table **1**) is the most appropriate for obtaining rich fractions of artepilin C and kaempferide. The first two fractions presented a higher content of these two compounds. Although the last fraction was observed to have a high concentration of these two compounds, it was insignificant because the extract yield was insignificant. Therefore, the content of these two compounds was also insignificant. Similar results were obtained in the first fractionation condition (we the reader to review the reference article), but higher temperatures and pressures were used. For this reason, the first condition would be the most convenient.

The waxes are generally extracted together with the active principles of the vegetable matrices. These compounds can be separated from the bioactive compounds by applying fractionation vessels in the supercritical extraction process using CO_2. According to Baldino *et al.* [33], the waxes of the bioactive extracts of the fruits and leaves of *R. graveolens* were separated in the first fractionation vessel (20 MPa and -10 ºC) in a fractionation system with two fractionation vessels. The bioactive compounds were recovered in the second fractionation vessel (15 MPa and 24 °C). Similar results were observed in the fractionation of extracts of inflorescences of *C. sativa*; the waxes were extracted in the first separator, and the volatile compounds were concentrated in the second separator, according to the conditions shown in Table **1** [121]. Similarly, volatile compounds from *F. vulgare* seeds were concentrated in the second fractionation vessel using a unit with two separators. The extract presented a solid appearance in the first fraction, rich in fatty acids, and the second fraction had a liquid appearance, rich in essential oils [41]. The extraction and fractionation of bioactive compounds with supercritical CO_2 can be scaled up by keeping the solvent/raw material (S/F) ratio constant [117]. In this way, the extraction and fractionation of ginger rhizome extracts were increased 50 times, maintaining a constant S/F ratio of 45.9 [117]. Fractionation was carried out using two separator vessels, concentrated oleoresin was obtained in the first separator, and concentrated volatile oil was obtained in the second separator (Table **1**).

SFE Equipment with Cosolvents

The implementation of supercritical cosolvent extraction in SFE equipment is relatively straightforward. The cosolvent can be injected into the supercritical extraction system by implementing a cosolvent container connected to an HPLC pump and a mixing point before the extraction container inlet [122].

The cosolvents usually used for bioactive extraction are water and ethanol, both considered GRAS. The objective of using these solvents as cosolvents is to modify the polarity of supercritical CO_2. With this, it seeks to broaden the spectrum of bioactive compounds that can be extracted, increasing the extraction yield due to the increase in bioactive compounds' solubility [66]. Several comparative studies of supercritical extraction with CO_2, with and without cosolvents, can be found in the current literature. For example, in the extraction of bioactive compounds from *V. vinifera* leaves [122], polyunsaturated fatty acids from fish residues [123], oleoresins from dried curcuma rhizomes (*Curcuma longa* and *Curcuma amada*) [124], bioactive residues from olive processing [125], bioactive extracts of leaves and stems of *Synadenium grantii* [126], phenolic compounds of *Beta vulgaris* leaves [115], antioxidant compounds of apple residues [64], polyphenols from *Callisia fragrans* leaves [127], and bioactive compounds from dried calyces of *Hibiscus sabdariffa* [128].

Although using organic cosolvents in the supercritical extraction process increases the extraction yield of bioactive compounds, some aspects must be considered before its implementation. For example, it must be considered that the use of cosolvents in the supercritical extraction process implies the need to implement a cosolvent recovery process, which is generally carried out by evaporation [120, 125]. Cosolvent recovery can increase operational costs [123]. There is also the possibility that bioactive compounds degrade in the cosolvent recovery process and that traces of the cosolvent may remain in the final extract [123].

SFE Equipment Assisted with Ultrasound

Implementing the ultrasound extraction method in an online hyphenation of ultrasound and SFE is potentially interesting due to the increase in bioactive compounds' extraction performance compared to conventional supercritical extraction [129].

Ultrasound-assisted extraction may be used before SFE as a pretreatment of the raw material to increase the extraction yield, as was observed in the extraction of oil from *Salvia hispanica* L. seeds [130]. It can also be applied to solid waste derived from supercritical extraction. This allows increasing the extraction performance compared to other methods, as observed in the recovery of polyphenols from *Spondias tuberosa* seeds [131]. These two forms of ultrasound application with the supercritical extraction process could be more expensive and time-consuming to execute, as they are performed sequentially in the supercritical extraction process. However, this can be overcome when the two methods are integrated. That is, the two methods are executed simultaneously [108, 129, 132, 133]. Barrales *et al.* [108] describe a prototype of ultrasound-assisted supercritical extraction equipment (SFE+US). In this equipment, the ultrasound system consisted of a transducer and a power controller. The ultrasound probe was installed inside the extraction vessel. This prototype was used to extract oil from passion fruit (*Passiflora edulis*) seeds. In general terms, an increase in extraction performance was observed due to the application of ultrasound in the supercritical process, obtaining an oil rich in γ-tocopherol, γ-tocotrienol, and δ-tocotrienol, with a maximum antioxidant capacity (2.6 mg Trolox equivalent/g oil) when the oil was extracted at 160-640 Watt power, 16 MPa and 40 °C [108]. In contrast, when this same equipment was applied to extract yeast oil, yields higher than conventional extraction methods were not observed [132]. In another study with the same equipment, ultrasound-assisted supercritical extraction (SFE+US) was applied to obtain oil from *Dipteryx alata* Vogel seeds. In this case, the use of ultrasound did not affect the fatty acid composition and overall yield compared with SFE without ultrasound. However, an increase in fatty acid extraction rates was observed in the first moments of ultrasound-assisted extraction [133].

Another way the SFE+US process was implemented was by coupling an ultrasound bath to the supercritical extraction process [129]. In this case, the extraction column containing the sample is immersed in an ultrasonic bath. With this configuration of SFE+US, a yield of 8.6 mg cucurbitacin E (CuE)/g of *Iberis amara* seeds (IAS) was obtained. After 60 min of extraction, this yield was observed at 25 MPa, 50 °C, with a CO_2 flow of 2 mL/min, 12 % cosolvent concentration, and an ultrasound power of 200 W. Under these conditions, the CuE yield was higher than that obtained in the supercritical extraction process without ultrasound (7.1 mg CuE/g IAS), with higher extraction conditions (100 min dynamic extraction, 55 °C, 2.5 mL CO_2/min) for the SFE process without ultrasound [129]. The SFE+US process turns out to be advantageous for increasing the extraction yield of bioactive compounds. However, as could be observed, this effect may be different depending on the raw material. Therefore, the implementation of this process to recover bioactive compounds from raw materials not studied is conditioned on the previous study in various supercritical conditions to determine optimal extraction conditions. Likewise, pilot and industrial scale studies are necessary to determine the technical viability and economic feasibility of the SFE process.

SFE Equipment Assisted with Cold Pressing

This method was recently developed at the Laboratory of Supercritical Technology: Extraction, Fractionation, and Identification of Vegetal Extracts (LASEFI, Unicamp, Brazil). Its first application was carried out to extract oil from the pulp of *Caryocar brasiliense* [134]. In this study, the operating mode of this equipment is detailed. In summary, it consists of the following stages: 1) feeding the raw material to the extraction vessel, 2) coupling the cold pressing system to the vessel extraction, 3) cold pressing, 4) uncoupling the cold pressing system from the extraction vessel, 5) closing the extraction vessel and connecting the CO_2 supply pipes, 6) pressurizing the system, 7) conditioning the temperature of the system, 8) static extraction or static time, 9) dynamic extraction, 10) recovery and weighing of the extract obtained, and 11) depressurization of the system and cleaning. A piston constituted the cold pressing system joined to a screw that allowed pressing through torque application using a digital torque meter. Supercritical fluid extraction assisted by the cold pressing (SFEAP) method was more efficient than the SFE method in extracting oil from *C. brasiliense* pulp [134]. In this study, the oil yield increased from 53 % w/w (SFE) to 55 % w/w (SFEAP) after 2 h of extraction at 40 MPa and 60 °C. However, in the first minute of extraction with SFEAP, an increase in the yield was observed eight times compared to the SFE method. This fact is due to the increase in the CER period's mass transfer rate (called the increasing extraction rate) due to the cold pressing effect. In this period, the mass transfer is controlled mostly by convection, according to the spline model an empiric alternative to Sovová's model [134]. A similar effect was observed in extracting essential oil from clove inflorescences using the same extraction conditions (15 MPa and 40 °C) for the two methods (SFE and SFEAP). The yield increased from 21 to 22 % w/w [135]. A more significant difference in extraction yields was observed when the SFE (9.8 % w/w) and SFEAP (12.2 % w/w) methods were applied to obtain the oily extract of *Foeniculum vulgare* seeds [41].

In this context, SFE assisted with cold pressing showed to be a promising technique due to improvements in extraction yields *via* enhancement of mass transfer rates. Also, cold pressing method is relatively easy and inexpensive method.

APPLICATIONS IN FOOD ANALYSIS

Food analysis includes sampling, sample preparation, detection of the target compounds, and data handling [5, 6].

Sample preparation represents a bottleneck of current analytical chemistry procedures because it is the most time-consuming and error-prone part of the analytical scheme [136-138].

Direct contact of raw material particles with sensors/detectors may damage equipment [139], or inappropriate sample preparation may allow the flow of particles that provoke clogging. Consequently, separation processes applied to the samples (extraction, isolation, and preconcentration) are required. However, such steps are time-consuming and use large amounts of solvents.

Sample preparation includes the sample pretreatment (grinding, homogenization) and the treatments designed to decompose the matrix structure, allowing the obtainment of a multicomponent product, called extract. The extract is commonly fractionated to select the compounds of interest and transformed *via* phase exchange and derivatization reactions to increase the stability and compatibility of target compounds with the detector to improve their detectability [138].

Supercritical fluid extraction (SFE) has been proposed as a sample preparation procedure to improve the sensitivity of the analytical method [4, 6].

Compounds of low polarity, including carotenoids, cannabinoids, oleoresins, and waxes, have a high affinity with SC-CO_2. SC-CO_2 is a green alternative for replacing halogenated solvents (dichloromethane, ethyl acetate, and chloroform) due to its low toxicity, nonflammability, and reusability allowing the acquisition of extracts with no contamination.

However, when compared with solvents with an enhanced polarity such as fluorinated solvents (Freon-22), SC-CO_2 does not allow the extraction of high polar analytes, including catechins, phospholipids and alkaloids.

Otherwise, the use of polar cosolvents (modifiers or makeup solvents) may improve the extraction of polar compounds *via* modification of CO_2 solubility .The use of alcohols (methanol) and water has emerged as a green cosolvent to extract polar compounds.

Even with modifiers (cosolvents) to assist SC-CO_2 in extracting polar molecules, offline analysis requires an additional step to remove the solvent, which may allow contamination of the extract due to handling.

In this context, the analytical scale of SFE of polar compounds is underexplored because of low extraction rates attributed to SC-CO_2.

Integration of SFE with Analytical Instruments

Supercritical Technology in Chromatography

Carbon dioxide (CO_2) is the preferred solvent used in SFE because of its low toxicity, low cost, and low critical point, which is easy to set up. Similar to SFE, supercritical fluid chromatography (SFC) uses SC- CO_2 to separate nonpolar compounds, whereas the separation of polar is done with polar modifiers (cosolvents), such as methanol and water, whose concentration should be below 10% of the total CO_2 injected [22].

As an analytical method, SFC is included in the normal phase liquid chromatography (NPLC) category, because it uses nonpolar supercritical carbon dioxide as a mobile phase to separate target compounds. The NPLC uses stationary phases consisting of columns packed with polar adsorbents, and mobile phases consist of nonpolar solvents [140].

SFC provides advantages over conventional chromatographic methods, including shorter retention times and enhanced detection sensitivity. In comparison with gas chromatography, no sample derivatization is needed before injection [3].

Instrument

The SFC generally uses open-tubular and packed columns (0.05-1 μm i.d, 10-20 m height) inserted into an oven coupled with a heating source, a temperature controller and a detector. In contrast, the SFE uses the separation vessel coupled with a heating source and a temperature controller, as already mentioned in the INSTRUMENTATION section. SFC is compatible with detectors used in HPLC (ultraviolet - UV, diode array detector - DAD, charged aerosol detector - CAD), GC (flame ionization detector - FID) or mass spectrometry (triple-quadrupole mass spectrometer - QqQMS) [141].

However, when the detector is selected, the mobile phase composition, column type, flow rate, and a backpressure regulator must be considered. Also, the detector components must be able to withstand the high pressure of the mobile phase otherwise baseline should not get stable.

A temperature controller and a pump are used to keep the CO_2 in a supercritical state.

Polar cosolvents, including methanol, can be used to increase the polarity of SC-CO_2 to allow the selection of high polar molecules.

A generic SFC equipment is similar to the SFE one illustrated in Fig. **3**, regardless of the presence of an autosampler, a degasser for the modifier, a communication module, a back pressure valve to regulate the pressure and flow of mobile phases, a column installed into an oven coupled with a temperature controller, an injection valve, and a detector (Fig. **5**).

The CO_2 pump is coupled with an internal or external heat exchanger to maintain the CO_2 liquid, and a cosolvent pump is used to transport the liquid modifiers (Fig. **5**). The mobile phases are pumped to a preheating coil to increase temperature, leading the solvent mixture to reach the supercritical state. The mobile phases at the supercritical state flow through a valve, in which the sample is injected into the supercritical stream and then transferred to the column.

Offline Techniques

Most of the protocols to prepare offline SFE extracts for analysis consist of the solubilization of extracts in a solvent, followed by filtration using a membrane filter (0.22-0.44 μm).

Several offline combinations with SFE (Table **2**) used solid-phase extraction (SPE), a sorbent-based microextraction technique that fractionates the SFE extract to select the target compounds for analysis. SPE was used to fractionate and isolate the compounds present in the SFE extracts of hemp with the use of a column packed with octadecyl-modified silica gel (5g, 40–60μm) and injected into reverse-phase high-performance liquid chromatography (RP-HPLC) [5].

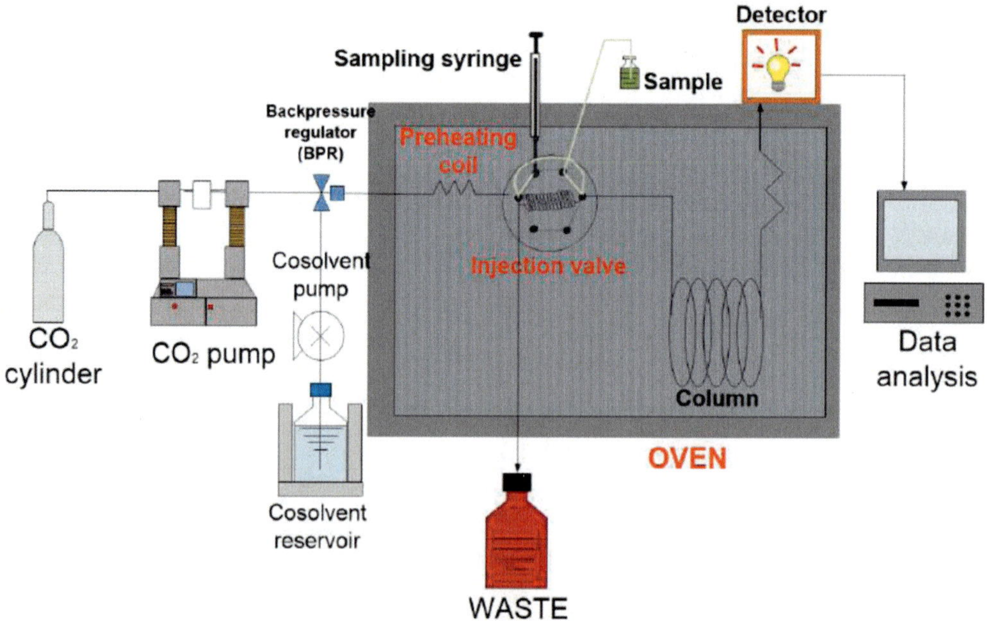

Fig. (5). Schematic diagram of the supercritical fluid chromatography experimental apparatus.

Table 2. Recently reviewed SFE conditions for offline hyphenation.

Method	Analyte Matrix	Conditions	References
SFE-MSPE-HPLC	Pyrethroids from fruit and vegetable samples	Temperature: 65 °C Pressure: 28 MPa CO_2 flow rate: 0.5 mL/min Time: 35 min	[4]

(Table 2) cont.....

Method	Analyte Matrix	Conditions	References
SFE-SPE- RP-HPLC	Tetrahydrocannabinol from *Cannabis sativa* L.	Temperature: 40-80 °C Pressure: 15-33 MPa Solvent: CO_2 Cosolvent: Ethanol Time: 240 min CO_2 flow rate:0.5 kg/h Ethanol: 0-5%	[5]
SFE-SPE-UPLC	Quinones from compost (food waste and wood chips)	Temperature: 35-55 °C Pressure: 25MPa Time: 10-40 min Solvent: CO_2 Cosolvent: methanol CO_2 flow rate:2.7 mL/min Methanol flow rate: 0.3mL/min	[142]
SFE-SPE-HPLC-UPLC	Quinones from activated sludge	Temperature: 55 °C Pressure: 25 MPa Solvent: CO_2 Cosolvent: methanol CO_2 flow rate: 2.7 L/min Methanol flow rate: 0.3 mL/min Time: 15 min	[143]

SFE-SPE-RP-HPLC – Supercritical fluid extraction coupled with solid phase extraction and reversed-phase high-performance liquid chromatography RP-HPLC.

Juliasih and co-workers [142] measured quinones from food waste compost using offline SFE and UHPLC (Ultra-High-Performance Liquid Chromatography). Before UHPLC analysis, the SFE extract was preconcentrated by SPE with hexane into a silica cartridge and filtered with a polyester filter with an integrated glass fiber prefilter membrane.

Bagheri and co-workers [4] used magnetic solid-phase extraction (MSPE) to preconcentrate SFE extracts of fruits and vegetables to further detect pyrethroid pesticides by HPLC (Table **2**).

An online fractionation of activated sludge extracts was performed by direct transportation from the SFE vessel to the C18 silica extraction cartridges to trap and collect the target fractions of quinones for further detection with UHPLC [143].

For the SFE extracts obtained with modifiers, the sample preparation starts with removing the modifier by rotary evaporation or drying under nitrogen, followed by dissolution, filtration (0.22-0.44 μm), and further resuspension in a solvent with a polarity similar to the mobile phase.

Depending on the aspects of the extract and the type of compound to be analyzed, the extract needs to be derivatized. Derivatization reactions for gas chromatography (GC) and mass spectrometry (MS), including silylation and methylation, transform the target compound into a derivative product with enhanced stability, volatility, sensitivity, greater selectivity, and improvements in peak formats and separation aspects [144].

In liquid chromatography, derivatization converts non-UV absorbing analytes into forms easily detectable at high sensitivity [145].

Yang and co-workers [146] proposed a derivatization method *via* silylation of SFE extract with *N*,*O*-bis(trimethylsilyl)trifluoroacetamide with 0.1% trimethylchlorosilane and injected for detection with gas chromatography-mass spectrometry (GC–MS).

Online Techniques

Online hyphenation involving SFE and analytical equipment has emerged from analyzing plant metabolites, human serum, and pesticides (Table **3**).

Table 3. The reviewed operational conditions for online hyphenation involving SFE.

Method	Analyte	Conditions	References
SFE-SFC	Volatile compounds from vanilla beans	Vessel packaging: 0.1g of sample loaded into the vessel Temperature: 35-55 °C Pressure: 10 -20 MPa (SFE) and 40 MPa (analysis period) Solvent: CO_2 Cosolvent: methanol (2-12%) CO_2 flow rate: 2 mL/min Valve switching: 2 backpressures	[6]
SFE-LC-FTMS	Organic matter from soil samples	Vessel packaging: 20 mg of sample was inserted into a 50 µL extraction vessel Solvent: CO_2 CO_2 flow rate: 4–6 µL/min Temperature: 31 °C Pressure: 20 MPa Valve switching: one 6-port- and one 4-port valve	[147]

(Table 3) cont.....

Method	Analyte	Conditions	References
SFE-SFC	Free fatty acids and PACs, PAHs in water	Vessel packaging:3 mL of sample were mixed with 3.4g of adsorbent Temperature: 40 °C Pressure: 22 MPa Solvent: CO_2 (99.7%) Cosolvents: methanol (1%) or water (0.3%) CO_2 flow rate: 0.85 g/min Valve switching: 2 backpressures	[148]
SFE-SFC-QqQ/MS	Apocarotenoids and apocarotenoids fatty acid esters from tamarillo	Vessel packaging: 100 mg of the sample:adsorbent mixture (1:100, w/w) was loaded into a 0.2 mL extraction vessel Solvent: CO_2 Cosolvent: ethanol (5-10%) CO_2 flow rate: 2 mL/min Temperature: 80 °C Pressure: 15 MPa Valve switching: 2 backpressures	[151]
SFE-SFC	Polyphenols, carotenoids, and chlorophylls from ivy and rosemary leaves	Vessel packaging: 1g of sample mixed with 1 g of diatomaceous earth loaded into a 5 mL vessel Solvent: CO_2 Cosolvent: ethanol (100%) or ethanol:water (50:50, v/v) at 3-23% CO_2 flow rate: 3 mL/min Cosolvent flow rate: 0.3 mL/min Temperature: 25 °C Pressure: 10-20 MPa Valve switching: two 6-port valves	[152]
SFE-SFC-MS	Benzenes from pesticide mixtures	Vessel packaging: 100 ng of standard were mixed with 1g of water-absorbent into a 5 mL vessel Solvent: CO_2 Cosolvent: 0.1% (w/v) ammonium formate in methanol at 5% CO_2 flow rate: 4 mL/min Temperature: 40 °C Pressure: 15 MPa (SFE) and 40 MPa (analysis period) Online coupling strategy: switching (one valve)	[153]

(Table 3) cont.....

Method	Analyte	Conditions	References
SFE–SFC–MS	Ketamine from urine	Vessel packaging: 20 mg of adsorbent with 20 μL of sample and 10 μL of internal standard Solvent: CO_2 Cosolvent: methanol containing 0.1% of formic acid (v/v) at 3-20% CO_2 flow rate: 0.2 mL/min Temperature: 25 °C Pressure: 15-16 MPa (SFE) and 40 MPa (analysis period) Valve switching: 2 backpressures	[154]
SFE–SFC–MS	PAHs from soil	Vessel packaging: 1 g of soil was mixed with 1g of adsorbent and 10 μL of internal standard in 5 mL stainless steel vessels. Solvent: CO_2 Cosolvent: acetonitrile (10%) CO_2 flow rate: 3 mL/min Temperature: 40 °C Pressure: 15 MPa (SFE) and 40 MPa (analysis period) Valve switching: 1 backpressure	[155]
SFE-SFC-MS/MS	Cancer biomarkers from human serum	Vessel packaging: 1μL of sample mixed with internal standards (2 μL) in a DMS card, which was dried at room temperature for 2 h Solvent: CO_2 (99.9%) Cosolvent: formic acid (0.1%) in methanol CO_2 flow rate: 0.8 g/min Temperature: 40 °C Pressure: 10 MPa Valve switching: 2 backpressures	[156]
SFE-SFC	Lignans from *Schisandra chinensis*	Vessel packaging: Solid samples (1 g) mixed with cellulose derivative powder (0.5-1 g) into a 5 mL vessel Solvent: CO_2 (99%) Cosolvent: methanol (1%) CO_2 flow rate: 1.5-3 mL/min Temperature: 50 °C Pressure: 5 MPa (SFE) and 40 MPa (analysis period) Valve switching: 2 backpressures	[157]

SFE-SFC – Supercritical fluid extraction coupled with supercritical fluid chromatography. SFE-LC-FTMS – Supercritical fluid extraction (SFE) coupled with liquid chromatography (LC) and Fourier transform mass spectrometry (FTMS) SFE-SFC-QqQ/MS - Supercritical fluid extraction (SFE) coupled with supercritical fluid chromatography with triple quadrupole mass spectrometry detection.

Such a strategy enhances the accuracy of miniaturized units for the analysis of smaller sample volumes with fully automated operations compared to the offline mode or isolated instruments [147].

The online hyphenation's other advantages include the direct separation of heat-sensitive target compounds, avoiding oxidation caused by derivatization when it is required for analysis [148].

Furthermore, online hyphenation with SFE is an answer to the demands for efficient and environmentally friendly alternatives to classical analytical methods that use high amounts of organic solvents. The time spent with online hyphenation is reduced because of the high diffusivity of SC-CO_2. Consequently, it reduces the steps involved in sample preparation, allowing the reduction of sample contamination and operator errors, improving the method's accuracy [149].

The miniaturization of instruments is useful when sample availability is limited. Additionally, miniaturization provides higher sample throughput and unattended operation, which is of interest in analytical chemistry for analyzing plant secondary metabolites and the analysis of biological and environmental materials [150].

The need for new analytical methods with enhanced sensitiveness and robustness was emphasized by Abrahamsson *et al.* [148]. The authors validated an online SFE-SFC system with diode array detector (DAD) and mass spectrometry (MS) detection to analyze biomarkers present in aqueous samples by detecting 16 compounds with lower detection limits of 0.02-40 μg/L and precision up to 13 %.

The online hyphenation of SFE with SFC with triple quadrupole mass spectrometry detection (SFE-SFC-QqQ) identified and quantified 31 carotenoids from yellow tamarillo with the use of a sub-2 micron particle C30 column. The results obtained were comparable with SFC-QqQ/MS and LC-DAD [151].

The general experimental equipment proposed for SFE-hyphenated analytical equipment consists of a CO_2 cylinder, a pump for CO_2, a pump for the liquid modifiers, an extraction vessel coupled with a temperature controller and a heating source, an analytical column inserted into an oven, an autosampler, backpressure regulators to control the flow of solvents (one located in-line before the column, and the other located in-line post-column), and detectors (Fig. **5**).

The extraction vessel is packaged with the raw material. Liquid CO_2 and cosolvents are pumped to a mixing unit and are inserted into the vessel to allow the supercritical state of the solvent mixture. Afterward, SC-CO_2 with or without cosolvent flows axially throughout the vessel.

According to Lefebvre *et al.* [152], in the case of the absence of cosolvent or when the cosolvent is present in a low amount, there is a need to insert a flow of another solvent (make-up solvent) after the extraction cell and before the backpressure regulator to avoid precipitation of the extract at atmospheric pressure.

At the outlet direction of vessel, the extract flows through an injection loop connected to a six-port rotary valve that serves as an injection system. The valve switches its position from the extraction (loading) position to the analysis position (injection). When this valve changes position, the extract present in the injection loop is then diffused in the mobile phase flow, which passes through the switching valve [152].

When the position changes, the extract present in the injection loop is carried by solvent into the downstream valve.

Upon the injection of SFE extract with a switchable system, part of the sample is transferred to the analytical column. With thc backpressure regulator (BPR1, Fig **6**) closed, an inflow of the mobile phase into the column and the extract was fractionated at the trap-column (solid-phase extraction, SPE) to allow the separation of compounds for detection [153].

The extraction vessel packaging is crucial to support the selection of target compounds. There is no standardized procedure to allow the packaging of the vessel [7].

In most of the manuscripts reviewed, the cosolvent did not act directly in the extraction but was used to saturate the raw material to allow the compounds' solubilization to be carried out by CO_2.

The saturated sample can be mixed with an adsorbent or inserted in a column packed with an adsorbent to attach nonpolar compounds to be carried by SC-CO_2. It is important to apply filtering agents (glass wool, stainless steel filters or cotton) to the outer parts of the extraction vessel to avoid potential line clogging by the carrier of particulate material by SC-CO_2.

The vessel packaging is completed by filling of empty spaces to avoid the unnecessary solvent spent and time-wasting with pressurization. If the raw material is not enough to fill the empty spaces, glass beads, silicone fillers, or even cotton [152] have been used to solve this issue.

Shen *et al.* [147] reported the direct loading of soil samples into the vessel and further treatment of extracts through a micro solid-phase extraction fused silica capillary column to enhance the separation of the compounds before injection.

Liang *et al.* [6] did not report the use of an adsorbent of any further procedure to enhance the fractionation of compounds but only the loading of 0.1 g of sample into the vessel.

For aqueous samples, the adsorbent contributes to the retention of water and consequently allows the retention of nonpolar analytes, which are easily carried by SC-CO_2.

Hofstetter *et al.* [154] validated an online SFE method to analyze urine metabolites packaged in SFE vessels by mixing 20 μL of spiked sample with 20 mg of ISOLUTE® HM-N, a modified type of diatomaceous earth.

Once the vessel is ready and the setup temperature is established, the SFE starts. The SFE consists of two periods: static and dynamic. The static period consists of the insertion of CO_2/modifier into the extraction vessel to allow pressurization to the setup pressure. For miniaturized equipment, the extraction period was reported to vary between 3 and 7 min [155, 156].

The dynamic period consists of the continuous flow of CO_2/modifier through the extraction vessel, followed by the transportation of extract to the analytical column. The dynamic period for online systems may vary between 1 and 10 min [152, 155]. Once the selective fractions of the extracts entered the analytical column, the analysis period (or separation period) started. Depending on the nature of the equipment hyphenated, the CO_2/modifier used in SFE may persist in the analysis period as a mobile phase or not.

For instance, Giuffrida and coworkers [151] extracted and analyzed carotenoids by flowing SC-CO_2 and modifier during SFE coupled with SFC. The SC-CO_2 and modifier flow was interrupted during QqQ-MS detection, from which only its mobile phase was allowed to flow the throughout the analytical column.

On the other hand, Abrahamsson and coworkers [148], who directly hyphenated SFE and SFC to analyze aqueous samples, used SC-CO_2 and modifiers for both extraction and separation.

The pressure gradient involved in the process is not standardized as well. Suzuki and co-workers [156] maintained the online SFE-SFC/MS at 10 MPa.

Some reports maintained pressure at approximately 15-16 MPa for both static and dynamic periods. The analysis starts with handling the first backpressure regulator to set up column pressure at 22 or 40 MPa [151, 157]. Once the elution and separation are finished, the second backpressure regulator is handled/programmed to equalize the pressure to the first 15-16 MPa to equilibrate the system and deliver the separated fractions for detection [154].

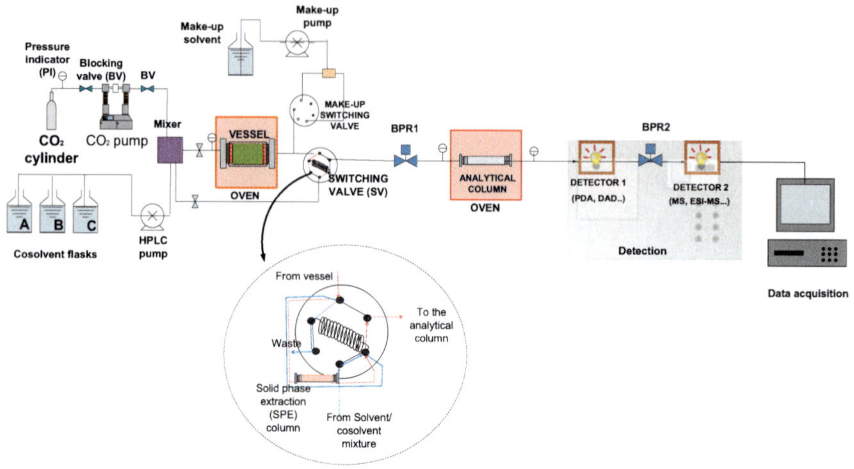

Fig. (6). A general scheme for the online hyphenation of SFE with analytical detection.

One limitation of online hyphenation reported was overpressure, which was attributed to solvent volume administrated. The valve switching strategy opens the possibilities for online coupling by connecting the extraction system with the analytical column [137].

Based on the documents mentioned, Fig. (**6**) shows a general scheme for online hyphenation of SFE and SFC.

CONCLUSION AND FUTURE REMARKS

Supercritical fluid extraction (SFE) has emerged as green process to select molecules of great utility for the food industry, including carotenoids, phenolic compounds, and fatty acids. Also, SFE showed to be feasible to select metabolites of mentioned molecules in food waste, human tissues, and soil. The optimization of SFE-based equipment to improve the extraction of substances, as well as the integration of SFE with analytical equipment has been studied on the possibility to obtain real-time identification of analytes while the food matrix is being extracted. Future directions on the use of SFE as an analytical tool include miniaturized equipment hyphenated with chromatographic techniques *via* the inclusion of backpressure valves to avoid huge pressure variations, and consequently promote baseline stability.

LIST OF ABBREVIATIONS

HPLC	High-performance liquid chromatography
SFC	Supercritical fluid chromatography
SFE	Supercritical fluid extraction
SFE-SFC	Supercritical fluid extraction coupled with supercritical fluid chromatography
SFE-LC-FTMS	Supercritical fluid extraction (SFE) coupled with liquid chromatography (LC) and Fourier transform mass spectrometry (FTMS)
SFE-SFC-QqQ/MS	Supercritical fluid extraction (SFE) coupled with supercritical fluid chromatography with triple quadrupole mass spectrometry detection
SFE-SPE-RP-HPLC	Supercritical fluid extraction coupled with solid-phase extraction and reverse-phase high-performance liquid chromatography RP-HPLC
PUFA	Polyunsaturated fatty acids
MUFA	Monounsaturated fatty acids
SFA	Saturated fatty acids

CONSENT FOR PUBLICATION

Not applicable.

CONFLICT OF INTEREST

The author(s) declare no conflict of interest, financial or otherwise.

ACKNOWLEDGEMENTS

Larry Oscar Chañi-Paucar thanks Coordenação de Aperfeiçoamento de Pessoal de Nível Superior - Brasil (CAPES) - Finance Code 001 for the doctoral fellowship. M. A. A. Meireles thanks CNPq for the productivity grant (BPQ-309825/2020-2). P. C. Veggi and J. Viganó thank financial support from Fundação de Amapro a Pesquisa do Estado de São Paulo – FAPESP (2020/7575-2; 2020/15774-5).

REFERENCES

[1] Brunner G. Supercritical fluids: technology and application to food processing. J Food Eng 2005; 67(1–2): 21-33.
[http://dx.doi.org/10.1016/j.jfoodeng.2004.05.060]

[2] Reverchon E, De Marco I. Supercritical fluid extraction and fractionation of natural matter. J Supercrit Fluids 2006; 38(2): 146-66.
[http://dx.doi.org/10.1016/j.supflu.2006.03.020]

[3] Lafont R, Dauphin-Villemant C, Warren JT, Rees H. 4 - Ecdysteroid chemistry and biochemistry.Insect Endocrinology. San Diego: Academic Press 2012; pp. 106-76.
[http://dx.doi.org/10.1016/B978-0-12-384749-2.10004-4]

[4] Bagheri H, Yamini Y, Safari M, Asiabi H, Karimi M, Heydari A. Simultaneous determination of pyrethroids residues in fruit and vegetable samples *via* supercritical fluid extraction coupled with magnetic solid phase extraction followed by HPLC-UV. J Supercrit Fluids 2016; 107: 571-80.

[http://dx.doi.org/10.1016/j.supflu.2015.07.017]

[5] Gallo-Molina AC, Castro-Vargas HI, Garzón-Méndez WF, *et al.* Extraction, isolation and purification of tetrahydrocannabinol from the Cannabis sativa L. plant using supercritical fluid extraction and solid phase extraction. J Supercrit Fluids 2019; 146: 208-16.
[http://dx.doi.org/10.1016/j.supflu.2019.01.020]

[6] Liang Y, Liu J, Zhong Q, *et al.* Determination of major aromatic constituents in vanilla using an on-line supercritical fluid extraction coupled with supercritical fluid chromatography. J Sep Sci 2018; 41(7): 1600-9.
[http://dx.doi.org/10.1002/jssc.201701097] [PMID: 29282892]

[7] Santana ÁL, Meireles MAA. Green analytical chemistry for food industries.Green Sustainable Process for Chemical and Environmental Engineering and Science: Analytical Techniques. Elsevier 2020; pp. 143-60.

[8] Jurado-Campos N, Carpio A, Zougagh M, Arce L, Arroyo-Manzanares N. Innovative coupling of supercritical fluid extraction with ion mobility spectrometry. Talanta 2018; 188: 637-43.
[http://dx.doi.org/10.1016/j.talanta.2018.06.045] [PMID: 30029424]

[9] Pereira CG, Meireles MAA. Supercritical fluid extraction of bioactive compounds: Fundamentals, applications and economic perspectives. Food Bioprocess Technol 2010; 3(3): 340-72.
[http://dx.doi.org/10.1007/s11947-009-0263-2]

[10] Silva A, Silva SA, Carpena M, *et al.* Macroalgae as a source of valuable antimicrobial compounds: Extraction and applications. Antibiotics 2020; 9(10): 642.
[http://dx.doi.org/10.3390/antibiotics9100642] [PMID: 32992802]

[11] Chemat F, Abert vian M, Fabiano-Tixier A-S, *et al.* A review of sustainable and intensified techniques for extraction of food and natural products. Green Chem 2020; 22(8): 2325-53.
[http://dx.doi.org/10.1039/C9GC03878G]

[12] Attard TM, Bukhanko N, Eriksson D, *et al.* Supercritical extraction of waxes and lipids from biomass: A valuable first step towards an integrated biorefinery. J Clean Prod 2018; 177: 684-98.
[http://dx.doi.org/10.1016/j.jclepro.2017.12.155]

[13] Anukiruthika T, Dutta S, Moses JA, Anandharamakrishnan C. Modern Applications of Supercritical Fluids Extraction in Food Toxicology. Elsevier 2021.
[http://dx.doi.org/10.1016/B978-0-08-100596-5.22939-9]

[14] Chemat F, *et al.* Review of green food processing techniques. preservation, transformation, and extraction. Innov Food Sci Emerg Technol 2017; 41(November): 357-77.
[http://dx.doi.org/10.1016/j.ifset.2017.04.016]

[15] Sánchez-Camargo AP, Parada-Alonso F, Ibáñez E, Cifuentes A. Recent applications of on-line supercritical fluid extraction coupled to advanced analytical techniques for compounds extraction and identification. J Sep Sci 2019; 42(1): 243-57.
[http://dx.doi.org/10.1002/jssc.201800729] [PMID: 30156751]

[16] Herrero M, Ibañez E. Green extraction processes, biorefineries and sustainability: Recovery of high added-value products from natural sources. J Supercrit Fluids 2018; 134: 252-9.
[http://dx.doi.org/10.1016/j.supflu.2017.12.002]

[17] Saravana PS, Getachew AT, Cho Y-J, *et al.* Influence of co-solvents on fucoxanthin and phlorotannin recovery from brown seaweed using supercritical CO_2. J Supercrit Fluids 2017; 120: 295-303.
[http://dx.doi.org/10.1016/j.supflu.2016.05.037]

[18] Brondz I, Sedunov B, Sivaraman N. Influence of modifiers on supercritical fluid chromatography (SFC) and supercritical fluid extraction (SFE), part I. Int J Anal Mass Spectrom Chromatogr 2017; 5(2): 17-39.
[http://dx.doi.org/10.4236/ijamsc.2017.52002]

[19] Xu DP, Li Y, Meng X, *et al.* Natural antioxidants in foods and medicinal plants: Extraction,

assessment and resources. Int J Mol Sci 2017; 18(1): 96.
[http://dx.doi.org/10.3390/ijms18010096] [PMID: 28067795]

[20] Al Khawli F, Pateiro M, Domínguez R, *et al.* Innovative green technologies of intensification for valorization of seafood and their by-products. Mar Drugs 2019; 17(12): 689.
[http://dx.doi.org/10.3390/md17120689] [PMID: 31817754]

[21] Manjare SD, Dhingra K. Supercritical fluids in separation and purification: A review. Mater Sci Energy Technol 2019; 2(3): 463-84.
[http://dx.doi.org/10.1016/j.mset.2019.04.005]

[22] Sánchez-Camargo AP, Parada-Alfonso F, Ibáñez E, Cifuentes A. On-line coupling of supercritical fluid extraction and chromatographic techniques. J Sep Sci 2017; 40(1): 213-27.
[http://dx.doi.org/10.1002/jssc.201601040] [PMID: 27813259]

[23] da Silva RPFF, Rocha-Santos TAP, Duarte AC. Supercritical fluid extraction of bioactive compounds. Trends Analyt Chem 2016; 76: 40-51.
[http://dx.doi.org/10.1016/j.trac.2015.11.013]

[24] Vardanega R, Náthia-Neves G, Veggi PC, de Meireless MAA. Supercritical fluid processing and extraction of food.Green Food Processing Techniques Preservation, Transformation and Extraction. 1st ed. Academic Press 2019; p. 562.

[25] Uwineza PA, Waśkiewicz A. Recent advances in supercritical fluid extraction of natural bioactive compounds from natural plant materials. Molecules 2020; 25(17): 3847.
[http://dx.doi.org/10.3390/molecules25173847] [PMID: 32847101]

[26] Belwal T, Ezzat SM, Rastrelli L, *et al.* A critical analysis of extraction techniques used for botanicals: Trends, priorities, industrial uses and optimization strategies. Trends Analyt Chem 2018; 100: 82-102.
[http://dx.doi.org/10.1016/j.trac.2017.12.018]

[27] de Aguiar AC, da Fonseca MAP, Figueiredo ACF, *et al.* Sequential high-pressure extraction to obtain capsinoids and phenolic compounds from biquinho pepper (Capsicum chinense). J Supercrit Fluids 2019; 150: 112-21.
[http://dx.doi.org/10.1016/j.supflu.2019.04.016]

[28] Perrut M, Perrut V. Towards ingredients by combining supercritical fluids with other processes. J Supercrit Fluids 2018; 134(November): 214-9.
[http://dx.doi.org/10.1016/j.supflu.2017.11.019]

[29] Vardanega R, Nogueira GC, Nascimento CDO, Faria-Machado AF, Meireles MAA. Selective extraction of bioactive compounds from annatto seeds by sequential supercritical CO_2 process. J Supercrit Fluids 2019; 150: 122-7.
[http://dx.doi.org/10.1016/j.supflu.2019.01.013]

[30] Viganó J, Coutinho JP, Souza DS, *et al.* Exploring the selectivity of supercritical CO_2 to obtain nonpolar fractions of passion fruit bagasse extracts. J Supercrit Fluids 2016; 110: 1-10.
[http://dx.doi.org/10.1016/j.supflu.2015.12.001]

[31] Fraga S, Gonçalves D, Nasário F, *et al.* Sequential high-pressure extraction of caffeine and bioactive compounds from caferana seeds (Bunchosia glandulifera). J Supercrit Fluids 2020; 165: 104958.
[http://dx.doi.org/10.1016/j.supflu.2020.104958]

[32] Romero de la Vega G, Salgado Cervantes MA, Garcia Alvarado MA, Romero-Martínez A, Hegel PE. Fractionation of vanilla oleoresin by supercritical CO_2 technology. J Supercrit Fluids 2016; 108: 79-88.
[http://dx.doi.org/10.1016/j.supflu.2015.10.022]

[33] Baldino L, Adami R, Reverchon E. Concentration of ruta graveolens active compounds using sc-co_2 extraction coupled with fractional separation. J Supercrit Fluids 2018; 131(September): 82-6.
[http://dx.doi.org/10.1016/j.supflu.2017.09.008]

[34] Baldino L, Reverchon E, Della PG. An optimized process for SC-CO_2 extraction of antimalarial

compounds from *Artemisia annua* L. J Supercrit Fluids 2017; 128(March): 89-93. [http://dx.doi.org/10.1016/j.supflu.2017.05.018]

[35] Ferro DM, Mazzutti S, Vitali L, Oliveira Müller CM, Ferreira SRS. Integrated extraction approach to increase the recovery of antioxidant compounds from sida rhombifolia leaves. J Supercrit Fluids 2019; 149: 10-9. [http://dx.doi.org/10.1016/j.supflu.2019.03.013]

[36] Hatami T, dos Santos LC, Zabot GL, *et al.* Integrated supercritical extraction and supercritical adsorption processes from passion fruit by-product: experimental and economic analyses. J Supercrit Fluids 2020; 162: 104856. [http://dx.doi.org/10.1016/j.supflu.2020.104856]

[37] dos Santos LC, Johner JCF, Scopel E, *et al.* Integrated supercritical CO_2 extraction and fractionation of passion fruit (*Passiflora edulis* Sims) by-products. J Supercrit Fluids 2021; 168(October): 105093. [http://dx.doi.org/10.1016/j.supflu.2020.105093]

[38] Garmus TT, Paviani LC, Queiroga CL, Cabral FA. Extraction of phenolic compounds from pepper-rosmarin (*Lippia sidoides* Cham.) leaves by sequential extraction in fixed bed extractor using supercritical CO_2, ethanol and water as solvents. J Supercrit Fluids 2015; 99: 68-75. [http://dx.doi.org/10.1016/j.supflu.2015.01.016]

[39] Cristianini M, Guillén Sánchez JS. Extraction of bioactive compounds from purple corn using emerging technologies: A review. J Food Sci 2020; 85(4): 862-9. [http://dx.doi.org/10.1111/1750-3841.15074] [PMID: 32237090]

[40] Moraes MN, Zabot GL, Meireles MAA. Extraction of tocotrienols from annatto seeds by a pseudo continuously operated SFE process integrated with low-pressure solvent extraction for bixin production. J Supercrit Fluids 2015; 96: 262-71. [http://dx.doi.org/10.1016/j.supflu.2014.09.007]

[41] Hatami T, Johner JCF, Meireles MAA. Extraction and fractionation of fennel using supercritical fluid extraction assisted by cold pressing. Ind Crops Prod 2018; 123(June): 661-6. [http://dx.doi.org/10.1016/j.indcrop.2018.07.041]

[42] Saini RK, Keum YS. Carotenoid extraction methods: A review of recent developments. Food Chem 2018; 240(June): 90-103. [http://dx.doi.org/10.1016/j.foodchem.2017.07.099]

[43] de Andrade Lima M, Kestekoglou I, Charalampopoulos D, Chatzifragkou A. Supercritical fluid extraction of carotenoids from vegetable waste matrices. Molecules 2019; 24(3): 466. [http://dx.doi.org/10.3390/molecules24030466] [PMID: 30696092]

[44] Prado J, Veggi P, Meireles M. Extraction methods for obtaining carotenoids from vegetables - review. Curr Anal Chem 2013; 10(1): 29-66. [http://dx.doi.org/10.2174/15734110114100010005]

[45] Sánchez-Camargo AP, Gutiérrez LF, Vargas SM, Martinez-Correa HA, Parada-Alfonso F, Narváez-Cuenca CE. Valorisation of mango peel: Proximate composition, supercritical fluid extraction of carotenoids, and application as an antioxidant additive for an edible oil. J Supercrit Fluids 2019; 152: 104574. [http://dx.doi.org/10.1016/j.supflu.2019.104574]

[46] Ndayishimiye J, Chun BS. Optimization of carotenoids and antioxidant activity of oils obtained from a co-extraction of citrus (Yuzu ichandrin) by-products using supercritical carbon dioxide. Biomass Bioenergy 2017; 106: 1-7. [http://dx.doi.org/10.1016/j.biombioe.2017.08.014]

[47] Patel AS, Kar A, Dash S, Dash SK. Supercritical fluid extraction of β-carotene from ripe bitter melon pericarp. Sci Rep 2019; 9(1): 19266. [http://dx.doi.org/10.1038/s41598-019-55481-4] [PMID: 31848376]

[48] Wang X, Wang C, Zha X, Mei Y, Xia J, Jiao Z. Supercritical carbon dioxide extraction of β-carotene and α-tocopherol from pumpkin: A box–behnken design for extraction variables. Anal Methods 2017; 9(2): 294-303.
[http://dx.doi.org/10.1039/C6AY02862D]

[49] Xie L, Cahoon E, Zhang Y, Ciftci O N. Extraction of astaxanthin from engineered camelina sativa seed using ethanol-modified supercritical carbon dioxide. J Supercrit Fluids 2019; 143(June 2018): 171-8.
[http://dx.doi.org/10.1016/j.supflu.2018.08.013]

[50] Crampon C, Nikitine C, Zaier M, *et al.* Oil extraction from enriched spirulina platensis microalgae using supercritical carbon dioxide. J Supercrit Fluids 2017; 119: 289-96.
[http://dx.doi.org/10.1016/j.supflu.2016.10.006]

[51] Vallecilla-Yepez L. Increasing *cis*lycopene content of the oleoresin from tomato processing byproducts using supercritical carbon dioxide and assessment of its bioaccessibility. 2017.Dissertations, Theses, & Student Research in Food Science and Technology. 89.

[52] Santana ÁL, Meireles MAA. Lycopene.A Centrum of Valuable Plant Bioactives. 1st ed. Cambridge: Elsevier 2021; pp. 33-58.
[http://dx.doi.org/10.1016/B978-0-12-822923-1.00032-7]

[53] Pellicanò TM, Sicari V, Loizzo MR, Leporini M, Falco T, Poiana M. Optimizing the supercritical fluid extraction process of bioactive compounds from processed tomato skin by-products. Food Sci Technol 2020; 40(3): 692-7.
[http://dx.doi.org/10.1590/fst.16619]

[54] Hatami T, Meireles MAA, Ciftci ON. Supercritical carbon dioxide extraction of lycopene from tomato processing by-products: Mathematical modeling and optimization. J Food Eng 2019; 241: 18-25.
[http://dx.doi.org/10.1016/j.jfoodeng.2018.07.036]

[55] Kehili M, Kammlott M, Choura S, *et al.* Supercritical CO_2 extraction and antioxidant activity of lycopene and β-carotene-enriched oleoresin from tomato (*Lycopersicum esculentum* L.) peels by-product of a Tunisian industry. Food Bioprod Process 2017; 102: 340-9.
[http://dx.doi.org/10.1016/j.fbp.2017.02.002]

[56] Veggi PC, Prado JM, Bataglion GA, Eberlin MN, Meireles MAA. Obtaining phenolic compounds from jatoba (*Hymenaea courbaril* L.) bark by supercritical fluid extraction. J Supercrit Fluids 2014; 89: 68-77.
[http://dx.doi.org/10.1016/j.supflu.2014.02.016]

[57] Knez Hrnčič M, Španinger E, Košir I, Knez Ž, Bren U. Hop compounds: Extraction techniques, chemical analyses, antioxidative, antimicrobial, and anticarcinogenic effects. Nutrients 2019; 11(2): 257.
[http://dx.doi.org/10.3390/nu11020257] [PMID: 30678345]

[58] Tyśkiewicz K, Konkol M, Rój E. The application of supercritical fluid extraction in phenolic compounds isolation from natural plant materials. Molecules 2018; 23(10): 2625.
[http://dx.doi.org/10.3390/molecules23102625] [PMID: 30322098]

[59] Goyeneche R, Fanovich A, Rodriguez Rodrigues C, Nicolao MC, Di Scala K. Supercritical CO_2 extraction of bioactive compounds from radish leaves: Yield, antioxidant capacity and cytotoxicity. J Supercrit Fluids 2018; 135: 78-83.
[http://dx.doi.org/10.1016/j.supflu.2018.01.004]

[60] Espinosa-Pardo FA, Nakajima VM, Macedo GA, Macedo JA, Martínez J. Extraction of phenolic compounds from dry and fermented orange pomace using supercritical CO_2 and cosolvents. Food Bioprod Process 2017; 101: 1-10.
[http://dx.doi.org/10.1016/j.fbp.2016.10.002]

[61] Muangrat R, Pongsirikul I. Recovery of spent coffee grounds oil using supercritical CO_2 : Extraction

optimisation and physicochemical properties of oil. CYTA J Food 2019; 17(1): 334-46. [http://dx.doi.org/10.1080/19476337.2019.1580771]

[62] Sökmen M, Demir E, Alomar SY. Optimization of sequential supercritical fluid extraction (SFE) of caffeine and catechins from green tea. J Supercrit Fluids 2018; 133: 171-6. [http://dx.doi.org/10.1016/j.supflu.2017.09.027]

[63] Valadez-Carmona L, Ortiz-Moreno A, Ceballos-Reyes G, Mendiola J A, Ibáñez E. Valorization of cacao pod husk through supercritical fluid extraction of phenolic compounds. J Supercrit Fluids 2018; 131(June 2017): 99-105. [http://dx.doi.org/10.1016/j.supflu.2017.09.011]

[64] Ferrentino G, Morozova K, Mosibo OK, Ramezani M, Scampicchio M. Biorecovery of antioxidants from apple pomace by supercritical fluid extraction. J Clean Prod 2018; 186: 253-61. [http://dx.doi.org/10.1016/j.jclepro.2018.03.165]

[65] Náthia-Neves G, Vardanega R, Meireles MAA. Extraction of natural blue colorant from Genipa americana L. using green technologies: Techno-economic evaluation. Food Bioprod Process 2019; 114: 132-43. [http://dx.doi.org/10.1016/j.fbp.2018.12.004]

[66] Cai ZZ, Hsieh CM. Prediction of solid solute solubility in supercritical carbon dioxide with and without organic cosolvents from PSRK EOS. J Supercrit Fluids 2020; 158: 104735. [http://dx.doi.org/10.1016/j.supflu.2019.104735]

[67] Garcia-Mendoza MP, Espinosa-Pardo FA, Baseggio AM, *et al.* Extraction of phenolic compounds and anthocyanins from juçara (Euterpe edulis Mart.) residues using pressurized liquids and supercritical fluids. J Supercrit Fluids 2017; 119: 9-16. [http://dx.doi.org/10.1016/j.supflu.2016.08.014]

[68] Woźniak Ł, Marszałek K, Skąpska S, Jędrzejczak R. The application of supercritical carbon dioxide and ethanol for the extraction of phenolic compounds from chokeberry pomace. Appl Sci 2017; 7(4): 322. [http://dx.doi.org/10.3390/app7040322]

[69] Da Porto C, Natolino A. Supercritical fluid extraction of polyphenols from grape seed (Vitis vinifera): Study on process variables and kinetics. J Supercrit Fluids 2017; 130: 239-45. [http://dx.doi.org/10.1016/j.supflu.2017.02.013]

[70] Priyanka , Khanam S. Influence of operating parameters on supercritical fluid extraction of essential oil from turmeric root. J Clean Prod 2018; 188: 816-24. [http://dx.doi.org/10.1016/j.jclepro.2018.04.052]

[71] Elgndi MA, Filip S, Pavlić B, *et al.* Antioxidative and cytotoxic activity of essential oils and extracts of Satureja montana L., Coriandrum sativum L. and Ocimum basilicum L. obtained by supercritical fluid extraction. J Supercrit Fluids 2017; 128(March): 128-37. [http://dx.doi.org/10.1016/j.supflu.2017.05.025]

[72] Kupski SC, Klein EJ, da Silva EA, Palú F, Guirardello R, Vieira MGA. Mathematical modeling of supercritical CO_2 extraction of hops (*Humulus lupulus* L.). J Supercrit Fluids 2017; 130(June): 347-56. [http://dx.doi.org/10.1016/j.supflu.2017.06.011]

[73] Conde-Hernández L A, Espinosa-Victoria J R, Guerrero-Beltrán J. Supercritical extraction of essential oils of piper auritum and porophyllum ruderale. J Supercrit Fluids 2017; 127(November): 97-102. [http://dx.doi.org/10.1016/j.supflu.2017.03.026]

[74] Conde-Hernández LA, Espinosa-Victoria JR, Trejo A, Guerrero-Beltrán JÁ. CO_2 -supercritical extraction, hydrodistillation and steam distillation of essential oil of rosemary (Rosmarinus officinalis). J Food Eng 2017; 200: 81-6. [http://dx.doi.org/10.1016/j.jfoodeng.2016.12.022]

[75] Montañés F, Catchpole O J, Tallon S, Mitchell K A, Scott D, Webby R F. Extraction of apple seed oil

by supercritical carbon dioxide at pressures up to 1300 bar. J Supercrit Fluids 2018; 141(October 2017): 128-36.
[http://dx.doi.org/10.1016/j.supflu.2018.02.002]

[76] Pavlić B, Teslić N, Zengin G, *et al.* Antioxidant and enzyme-inhibitory activity of peppermint extracts and essential oils obtained by conventional and emerging extraction techniques. Food Chem 2021; 338(February): 127724.
[http://dx.doi.org/10.1016/j.foodchem.2020.127724] [PMID: 32795878]

[77] Goto M, Askin R, Sasaki M. Supercritical Fluid Extraction in Food Analysis. In: Ötleş S (Ed.) Handbook of Food Analysis Instruments. Boca Raton: CRC Press 2008, Routledge Handbooks Online.
[http://dx.doi.org/10.1201/9781420045673.ch3]

[78] de Oliveira NA, Mazzali MR, Fukumasu H, Gonçalves CB, Oliveira AL. Composition and physical properties of babassu seed (Orbignya phalerata) oil obtained by supercritical CO_2 extraction. J Supercrit Fluids 2019; 150: 21-9.
[http://dx.doi.org/10.1016/j.supflu.2019.04.009]

[79] Páramos PRS, Granjo JFO, Corazza ML, Matos HA. Extraction of high value products from avocado waste biomass. J Supercrit Fluids 2020; 165: 104988.
[http://dx.doi.org/10.1016/j.supflu.2020.104988]

[80] Devi V, Khanam S. Development of generalized and simplified models for supercritical fluid extraction: Case study of papaya (Carica papaya) seed oil. Chem Eng Res Des 2019; 150: 341-58.
[http://dx.doi.org/10.1016/j.cherd.2019.08.006]

[81] Rosa A, Era B, Masala C, *et al.* Supercritical CO_2 extraction of waste citrus seeds: Chemical composition, nutritional and biological properties of edible fixed oils. Eur J Lipid Sci Technol 2019; 121(7): 1800502.
[http://dx.doi.org/10.1002/ejlt.201800502]

[82] Kuvendziev S, Lisichkov K, Zeković Z, Marinkovski M, Musliu Z H. Supercritical fluid extraction of fish oil from common carp (*Cyprinus carpio* L.) tissues. J Supercrit Fluids 2018; 133(December 2017): 528-34.
[http://dx.doi.org/10.1016/j.supflu.2017.11.027]

[83] Roselló-Soto E, *et al.* Evaluating the impact of supercritical-co 2 pressure on the recovery and quality of oil from 'horchata' by-products: Fatty acid profile, α-tocopherol, phenolic compounds, and lipid oxidation parameters. Food Res Int 2019; 120(July): 888-94.
[http://dx.doi.org/10.1016/j.foodres.2018.11.054]

[84] Sodeifian G, Ardestani N S, Sajadian S A, Moghadamian K. Properties of portulaca oleracea seed oil *via* supercritical fluid extraction: Experimental and optimization. J Supercrit Fluids 2018; 135(December): 34-44.
[http://dx.doi.org/10.1016/j.supflu.2017.12.026]

[85] Belayneh HD, Wehling RL, Cahoon E, Ciftci ON. Extraction of omega-3-rich oil from camelina sativa seed using supercritical carbon dioxide. J Supercrit Fluids 2015; 104: 153-9.
[http://dx.doi.org/10.1016/j.supflu.2015.06.002]

[86] Cvjetko BM, Vidović S, Radojčić RI, Jokić S. New perspective in extraction of plant biologically active compounds by green solvents. Food Bioprod Process 2018; 109: 52-73.
[http://dx.doi.org/10.1016/j.fbp.2018.03.001]

[87] Nakamura K, Otake T, Hanari N. Evaluation of supercritical fluid extraction for the determination of neonicotinoid pesticides in green onion. J Environ Sci Health B 2020; 55(7): 604-12.
[http://dx.doi.org/10.1080/03601234.2020.1747905] [PMID: 32274963]

[88] Singh SK, Pavan MS, N SP, Kant R. Applications of super critical fluid extraction in milk and dairy industry: A review. J Food Process Technol 2018; 9(12)
[http://dx.doi.org/10.4172/2157-7110.1000769]

[89] Ghosh M, Srivastava SCJ, Mishra HN. Advent of clean and green technology for preparation of low-cholesterol dairy cream powder: supercritical fluid extraction process. Food Quality and Safety 2018; 2(4): 205-11.
[http://dx.doi.org/10.1093/fqsafe/fyy012]

[90] Pourmortazavi SM, Saghafi Z, Ehsani A, Yousefi M. Application of supercritical fluids in cholesterol extraction from foodstuffs: A review. J Food Sci Technol 2018; 55(8): 2813-23.
[http://dx.doi.org/10.1007/s13197-018-3205-z] [PMID: 30065391]

[91] Schretter C, Langeder J, Freisinger V, Rollinger JM, Grienke U. Quantitative analysis of prenylated constituents in commercial hops samples using ultrahigh-performance supercritical fluid chromatography. Planta Med 2020; 86(15): 1140-7.
[http://dx.doi.org/10.1055/a-1130-0590] [PMID: 32182624]

[92] Bartolomé OA, Škerget M, Knez Ž. Separation of xanthohumol from hop extracts by supercritical fluid chromatography. Chem Eng Res Des 2016; 109: 335-45.
[http://dx.doi.org/10.1016/j.cherd.2016.02.013]

[93] Carballo D, Font G, Ferrer E, Berrada H. Evaluation of mycotoxin residues on ready-to-eat food by chromatographic methods coupled to mass spectrometry in tandem. Toxins 2018; 10(6): 243.
[http://dx.doi.org/10.3390/toxins10060243] [PMID: 29914055]

[94] Hewavitharana GG, Perera DN, Navaratne SB, Wickramasinghe I. Extraction methods of fat from food samples and preparation of fatty acid methyl esters for gas chromatography: A review. Arab J Chem 2020; 13(8): 6865-75.
[http://dx.doi.org/10.1016/j.arabjc.2020.06.039]

[95] Khaw KY, Parat MO, Shaw PN, Falconer JR. Solvent supercritical fluid technologies to extract bioactive compounds from natural sources: A review. Molecules 2017; 22(7): 1186.
[http://dx.doi.org/10.3390/molecules22071186] [PMID: 28708073]

[96] Prado JM, Veggi PC, Angela M, Meireles A. Scale-up issues and cost of manufacturing bioactive compounds by supercritical fluid extraction and ultrasound assisted extraction. In: Barbosa-Cánovas GV, Pastore GM, Candoğan K, et al. Global Food Security and Wellness. Springer: New York, NY 2017: pp. 377-433.
[http://dx.doi.org/10.1007/978-1-4939-6496-3_20]

[97] Zabot GL, Moraes MN, Angela A. Meireles M. Supercritical fluid extraction of bioactive compounds from botanic matrices: Experimental data, process parameters and economic evaluation. Recent Pat Eng 2012; 6(3): 182-206.
[http://dx.doi.org/10.2174/187221212804583204]

[98] Viganó J, Martinez J. Trends for the application of passion fruit industrial by-products: A review on the chemical composition and extraction techniques of phytochemicals. Food Public Health 2015; 5(5): 164-73.
[http://dx.doi.org/10.5923/j.fph.20150505.03]

[99] Brunner G. Gas extraction: An introduction to fundamentals of supercritical fluids and the application to separation processes. New York: Steinkopff-Verlag Heidelberg 1994.
[http://dx.doi.org/10.1007/978-3-662-07380-3]

[100] Sapkale GN, Patil SM, Surwase US, Bhatbhage PK. Supercritical fluid extraction. Int J Chem Sci 2010; 8(2): 729-43.

[101] Silva LPS, Martínez J. Mathematical modeling of mass transfer in supercritical fluid extraction of oleoresin from red pepper. J Food Eng 2014; 133: 30-9.
[http://dx.doi.org/10.1016/j.jfoodeng.2014.02.013]

[102] Sovová H. Rate of the vegetable oil extraction with supercritical CO_2—I. Modelling of extraction curves. Chem Eng Sci 1994; 49(3): 409-14.
[http://dx.doi.org/10.1016/0009-2509(94)87012-8]

[103] Sovová H. Mathematical model for supercritical fluid extraction of natural products and extraction curve evaluation. J Supercrit Fluids 2005; 33(1): 35-52. [http://dx.doi.org/10.1016/j.supflu.2004.03.005]

[104] McHugh MA, Krukonis VJ. Supercritical FIuid Extraction: Principles and Practice. 2nd ed., Boston: BUTTERWORTH-HEINEMANN 1994.

[105] Herrero M, Mendiola JA, Cifuentes A, Ibáñez E. Supercritical fluid extraction: Recent advances and applications. J Chromatogr A 2010; 1217(16): 2495-511. [http://dx.doi.org/10.1016/j.chroma.2009.12.019] [PMID: 20022016]

[106] de Melo MMR, Silvestre AJD, Silva CM. Supercritical fluid extraction of vegetable matrices: Applications, trends and future perspectives of a convincing green technology. J Supercrit Fluids 2014; 92: 115-76. [http://dx.doi.org/10.1016/j.supflu.2014.04.007]

[107] Zhang Z, Liu Y, Che L. Optimization of supercritical carbon dioxide extraction of *Eucommia ulmoides* seed oil and quality evaluation of the oil. J Oleo Sci 2018; 67(3): 255-63. [http://dx.doi.org/10.5650/jos.ess17153] [PMID: 29459511]

[108] Barrales FM, Rezende CA, Martínez J. Supercritical CO_2 extraction of passion fruit (*Passiflora edulis* sp.) seed oil assisted by ultrasound. J Supercrit Fluids 2015; 104: 183-92. [http://dx.doi.org/10.1016/j.supflu.2015.06.006]

[109] Thar Process. Scale up to thar. CO_2 extraction and purification 2020. Available from: http://www.tharprocess.com [accessed Jan. 11, 2020].

[110] SepareCo. SepareCo$_2$ 2018. Available from: http://www.separeco.com [Accessed Jan. 11, 2020].

[111] Tradematt. Tradematt (Henan) Industry Tradematt 2020. Available from: http://www.cntradematt.com [Accessed Jan. 11, 2020].

[112] Natex. natex PROZESSTECHNOLOGIE 2020. Available from: https://www.natex.at (accessed Jan. 11, 2020).

[113] ExtrateX. ExtrateX: SUPERCRITICAL FLUID INNOVATION 2020.

[114] Johner JCF, Meireles MAA. Construction of a supercritical fluid extraction (SFE) equipment: Validation using annatto and fennel and extract analysis by thin layer chromatography coupled to image. Food Sci Technol 2016; 36(2): 210-47. [http://dx.doi.org/10.1590/1678-457X.0027]

[115] Goyeneche R, Di Scala K, Ramirez CL, Fanovich MA. Recovery of bioactive compounds from beetroot leaves by supercritical CO_2 extraction as a promising bioresource. J Supercrit Fluids 2020; 155: 104658. [http://dx.doi.org/10.1016/j.supflu.2019.104658]

[116] Reverchon E. Supercritical fluid extraction and fractionation of essential oils and related products. J Supercrit Fluids 1997; 10(1): 1-37. [http://dx.doi.org/10.1016/S0896-8446(97)00014-4]

[117] Shukla A, Naik SN, Goud VV, Das C. Supercritical CO_2 extraction and online fractionation of dry ginger for production of high-quality volatile oil and gingerols enriched oleoresin. Ind Crops Prod. 2019; 130: pp. (September)352-62. [http://dx.doi.org/10.1016/j.indcrop.2019.01.005]

[118] Ferdosh S, Sarker MZI, Norulaini Nik Ab Rahman N, Haque Akanda MJ, Ghafoor K, Kadir MOA. Simultaneous extraction and fractionation of fish oil from tuna by-product using supercritical carbon dioxide (SC-CO_2). J Aquat Food Prod Technol 2016; 25(2): 230-9. [http://dx.doi.org/10.1080/10498850.2013.843629]

[119] Vicente G, García-Risco MR, Fornari T, Reglero G. Supercritical fractionation of rosemary extracts to improve the antioxidant activity. Chem Eng Technol 2012; 35(1): 176-82.

[http://dx.doi.org/10.1002/ceat.201100367]

[120] Paula JT, Sousa IMO, Foglio MA, Cabral FA. Selective fractionation of supercritical extracts from leaves of *Baccharis dracunculifolia.* J Supercrit Fluids 2017; 127(March): 62-70. [http://dx.doi.org/10.1016/j.supflu.2017.03.032]

[121] Da Porto C, Decorti D, Natolino A. Separation of aroma compounds from industrial hemp inflorescences (Cannabis sativa L.) by supercritical CO_2 extraction and on-line fractionation. Ind Crops Prod 2014; 58(August): 99-103. [http://dx.doi.org/10.1016/j.indcrop.2014.03.042]

[122] de Melo MMR, Carius B, Simões MMQ, Portugal I, Saraiva J, Silva CM. Supercritical CO_2 extraction of *V. vinifera* leaves: Influence of cosolvents and particle size on removal kinetics and selectivity to target compounds. J Supercrit Fluids 2020; 165: 104959. [http://dx.doi.org/10.1016/j.supflu.2020.104959]

[123] Melgosa R, Sanz MT, Beltrán S. Supercritical CO_2 processing of omega-3 polyunsaturated fatty acids – Towards a biorefinery for fish waste valorization. J Supercrit Fluids 2021; 169: 105121. [http://dx.doi.org/10.1016/j.supflu.2020.105121]

[124] Nagavekar N, Singhal RS. Supercritical fluid extraction of *curcuma longa* and *curcuma amada* oleoresin: Optimization of extraction conditions, extract profiling, and comparison of bioactivities. Ind Crops Prod 2019; 134(March): 134-45. [http://dx.doi.org/10.1016/j.indcrop.2019.03.061]

[125] Caballero AS, Romero-García JM, Castro E, Cardona CA. Supercritical fluid extraction for enhancing polyphenolic compounds production from olive waste extracts. J Chem Technol Biotechnol 2020; 95(2): 356-62. [http://dx.doi.org/10.1002/jctb.5907]

[126] Guedes AR, de Souza ARC, Turola Barbi RC, Nottar Escobar EL, Zanoello ÉF, Corazza ML. Extraction of *Synadenium grantii* Hook f. using conventional solvents and supercritical CO_2 + ethanol. J Supercrit Fluids 2020; 160: 104796. [http://dx.doi.org/10.1016/j.supflu.2020.104796]

[127] Phan LTM, *et al.* Supercritical fluid extraction of polyphenols from vietnamese callisia fragrans leaves and antioxidant activity of the extract. Hindawi J Chem 2020; 1-7.

[128] Pimentel-Moral S, Borrás-Linares I, Lozano-Sánchez J, Arráez-Román D, Martínez-Férez A, Segura-Carretero A. Supercritical CO_2 extraction of bioactive compounds from *Hibiscus sabdariffa.* J Supercrit Fluids 2019; 147: 213-21. [http://dx.doi.org/10.1016/j.supflu.2018.11.005]

[129] Liu X, Ou H, Gregersen H. Ultrasound-assisted supercritical CO_2 extraction of cucurbitacin E from Iberis amara seeds. Ind Crops Prod 2020; 145(83): 112093. [http://dx.doi.org/10.1016/j.indcrop.2020.112093]

[130] Fernandes SS, Tonato D, Mazutti MA, *et al.* Yield and quality of chia oil extracted *via* different methods. J Food Eng 2019; 262: 200-8. [http://dx.doi.org/10.1016/j.jfoodeng.2019.06.019]

[131] Dias JL, Mazzutti S, de Souza JAL, *et al.* Extraction of umbu (*Spondias tuberosa*) seed oil using CO_2, ultrasound and conventional methods: Evaluations of composition profiles and antioxidant activities. J Supercrit Fluids 2019; 145(Mar): 10-8. [http://dx.doi.org/10.1016/j.supflu.2018.11.011]

[132] Duarte SH, Santos P, Michelon M, Oliveira SMP, Martínez J, Maugeri F. Recovery of yeast lipids using different cell disruption techniques and supercritical CO_2 extraction. Biochem Eng J 2017; 125: 230-7. [http://dx.doi.org/10.1016/j.bej.2017.06.014]

[133] dos Santos P, de Aguiar AC, Viganó J, Boeing JS, Visentainer JV, Martínez J. Supercritical CO_2

extraction of cumbaru oil (*Dipteryx alata* Vogel) assisted by ultrasound: Global yield, kinetics and fatty acid composition. J Supercrit Fluids 2016; 107: 75-83.
[http://dx.doi.org/10.1016/j.supflu.2015.08.018]

[134] Johner JCF, Hatami T, Meireles MAA. Developing a supercritical fluid extraction method assisted by cold pressing for extraction of pequi (Caryocar brasiliense). J Supercrit Fluids 2018; 137: 34-9.
[http://dx.doi.org/10.1016/j.supflu.2018.03.005]

[135] Hatami T, Johner J C F, Zabot G L, Meireles M A A. Supercritical fluid extraction assisted by cold pressing from clove buds: Extraction performance, volatile oil composition, and economic evaluation. J Supercrit Fluids 2019; 144(September): 39-47.
[http://dx.doi.org/10.1016/j.supflu.2018.10.003]

[136] Ramos L, Ramos JJ, Brinkman UAT. Miniaturization in sample treatment for environmental analysis. Anal Bioanal Chem 2005; 381(1): 119-40.
[http://dx.doi.org/10.1007/s00216-004-2906-5] [PMID: 15602622]

[137] Liang Y, Zhou T. Recent advances of online coupling of sample preparation techniques with ultra high performance liquid chromatography and supercritical fluid chromatography. J Sep Sci 2019; 42(1): 226-42.
[http://dx.doi.org/10.1002/jssc.201800721] [PMID: 30136406]

[138] Ramos L. Critical overview of selected contemporary sample preparation techniques. J Chromatogr A 2012; 1221: 84-98.
[http://dx.doi.org/10.1016/j.chroma.2011.11.011] [PMID: 22137129]

[139] Ali I, Gupta VK, Aboul-Enein HY, Hussain A. Hyphenation in sample preparation: Advancement from the micro to the nano world. J Sep Sci 2008; 31(11): 2040-53.
[http://dx.doi.org/10.1002/jssc.200800123] [PMID: 18615807]

[140] Cooper WT. Normal-phase liquid chromatography. Encyclopedia of Analytical Chemistry. American Cancer Society 2006.
[http://dx.doi.org/10.1002/9780470027318.a5913]

[141] Takeda H, Takahashi M, Hara T, Izumi Y, Bamba T. Improved quantitation of lipid classes using supercritical fluid chromatography with charged aerosol detector. J Lipid Res 2019; 60: jlr.D094516.
[http://dx.doi.org/10.1194/jlr.D094516]

[142] Juliasih NLGR, Yuan LC, Sago Y, Atsuta Y, Daimon H. Supercritical Fluid Extraction of Quinones from Compost for Microbial Community Analysis. J Chem 2015; 2015: 1-7.
[http://dx.doi.org/10.1155/2015/717616]

[143] Hanif M, Atsuta Y, Fujie K, Daimon H. Supercritical fluid extraction and ultra performance liquid chromatography of respiratory quinones for microbial community analysis in environmental and biological samples. Molecules 2012; 17(3): 2628-42.
[http://dx.doi.org/10.3390/molecules17032628] [PMID: 22391598]

[144] van der Jagt H, Reijnders HF. Water Analysis: Overview. In: Worsfold P, Townshends A, Poole C, Eds. Encyclopedia of Analytical Science. Second Edi.. Oxford: Elsevier 2005; pp. 233-52.
[http://dx.doi.org/10.1016/B0-12-369397-7/00653-1]

[145] Choi CK, Dong MW. 5 - Sample preparation for hplc analysis of drug products.Handbook of Pharmaceutical Analysis by HPLC. Academic Press 2005; Vol. 6: pp. 123-44.
[http://dx.doi.org/10.1016/S0149-6395(05)80049-2]

[146] Yang TJ, Tsai FJ, Chen CY, Yang TCC, Lee MR. Determination of additives in cosmetics by supercritical fluid extraction on-line headspace solid-phase microextraction combined with gas chromatography–mass spectrometry. Anal Chim Acta 2010; 668(2): 188-94.
[http://dx.doi.org/10.1016/j.aca.2010.04.001] [PMID: 20493297]

[147] Shen Y, Zhao R, Tolić N, *et al.* Online supercritical fluid extraction mass spectrometry (SFE-L--FTMS) for sensitive characterization of soil organic matter. Faraday Discuss 2019; 218(0): 157-71.

[http://dx.doi.org/10.1039/C9FD00011A] [PMID: 31155623]

[148] Abrahamsson V, Henderson BL, Zhong F, Lin Y, Kanik I. Online supercritical fluid extraction and chromatography of biomarkers analysis in aqueous samples for *in situ* planetary applications. Anal Bioanal Chem 2019; 411(30): 8091-101. [http://dx.doi.org/10.1007/s00216-019-02189-z] [PMID: 31773227]

[149] Giuffrida D, Zoccali M, Mondello L. Carotenoid and apocarotenoid analysis by SFE-SFC-QqQ/MS. Methods Mol Biol 2020; 2083: 209-19. [http://dx.doi.org/10.1007/978-1-4939-9952-1_16] [PMID: 31745924]

[150] Shen Y, Zhao R, Berger SJ, Anderson GA, Rodriguez N, Smith RD. High-efficiency nanoscale liquid chromatography coupled on-line with mass spectrometry using nanoelectrospray ionization for proteomics. Anal Chem 2002; 74(16): 4235-49. [http://dx.doi.org/10.1021/ac0202280] [PMID: 12199598]

[151] Giuffrida D, Zoccali M, Arigò A, *et al.* Comparison of different analytical techniques for the analysis of carotenoids in tamarillo (Solanum betaceum Cav.). Arch Biochem Biophys 2018; 646: 161-7. [http://dx.doi.org/10.1016/j.abb.2018.03.011] [PMID: 29559322]

[152] Lefebvre T, Destandau E, Lesellier E. Sequential extraction of carnosic acid, rosmarinic acid and pigments (carotenoids and chlorophylls) from rosemary by online supercritical fluid extraction-supercritical fluid chromatography. J Chromatogr A 2020; 461709. [http://dx.doi.org/10.1016/j.chroma.2020.461709]

[153] Sakai M, Hayakawa Y, Funada Y, Ando T, Fukusaki E, Bamba T. Development of a practical online supercritical fluid extraction–supercritical fluid chromatography/mass spectrometry system with an integrated split-flow method. J Chromatogr A 2019; 1592: 161-72. [http://dx.doi.org/10.1016/j.chroma.2019.01.044] [PMID: 30712818]

[154] Hofstetter R, Fassauer GM, Link A. Supercritical fluid extraction (SFE) of ketamine metabolites from dried urine and on-line quantification by supercritical fluid chromatography and single mass detection (on-line SFE–SFC–MS). J Chromatogr B Analyt Technol Biomed Life Sci 2018; 1076: 77-83. [http://dx.doi.org/10.1016/j.jchromb.2018.01.024] [PMID: 29406031]

[155] Wicker AP, Carlton DD Jr, Tanaka K, *et al.* On-line supercritical fluid extraction—supercritical fluid chromatography-mass spectrometry of polycyclic aromatic hydrocarbons in soil. J Chromatogr B Analyt Technol Biomed Life Sci 2018; 1086: 82-8. [http://dx.doi.org/10.1016/j.jchromb.2018.04.014] [PMID: 29660666]

[156] Suzuki M, Nishiumi S, Kobayashi T, *et al.* Use of on-line supercritical fluid extraction-supercritical fluid chromatography/tandem mass spectrometry to analyze disease biomarkers in dried serum spots compared with serum analysis using liquid chromatography/tandem mass spectrometry. Rapid Commun Mass Spectrom 2017; 31(10): 886-94. [http://dx.doi.org/10.1002/rcm.7857] [PMID: 28332299]

[157] Dai Z, Xin H, Fu Q, *et al.* Exploration and optimization of conditions for quantitative analysis of lignans in *Schisandra chinensis* by an online supercritical fluid extraction with supercritical fluid chromatography system. J Sep Sci 2019; 42(14): 2444-54. [http://dx.doi.org/10.1002/jssc.201900222] [PMID: 31087615]

CHAPTER 8

Gas-Expanded Liquids Extraction

Zully J. Suárez Montenegro[1,2], Norelhouda Abderrezag[1,3], Elena Ibáñez[1,*] and **Jose A. Mendiola[1]**

[1] *Foodomics Laboratory, Institute of Food Science Research (CIAL, CSIC), Nicolas Cabrera, 9, 28049 Madrid, Spain*

[2] *Facultad de Ingeniería Agroindustrial Universidad de Nariño, C/18 Cra. 50 Ciudadela Universitaria Torobajo, 520001 Pasto, Colombia*

[3] *Laboratory of Environmental process engineering, Faculty of Process Engineering, University Salah Boubnider Constantine 3, Nouvelle Ville Constantine, 25000 Constantine, Algeria*

Abstract: Gas Expanded Liquids (GXL) are mixtures of liquid solvents (organic, water) and gases or supercritical fluids with diverse physicochemical properties halfway between pure liquids and supercritical fluids. The possibility of changing their properties by introducing small changes in pressure, temperature, and/or solvent/gas ratio, makes these solvents a very interesting and appropriate option for developing green extraction protocols for food analysis. In general, GXLs have similar densities as the solvent used in their composition, while having improved mass transfer through reduced viscosity, increased solute diffusivity, and decreased interfacial tension. Some other advantages are related to the wide range of polarities that can be obtained, depending on the liquid selected. Moreover, the substitution of a liquid fraction for a gas reduces the final use of organic solvent, thus improving the green character of GXLs. In the present chapter, the physicochemical properties of GXL are addressed along together with the description of applications in the food science and technology area.

Keywords: Biorefinery, By-products, Food analysis, Gas-expanded liquids, Green solvent, Natural compounds.

INTRODUCTION

Sustainable Development Goals (SDGs) are at present guiding the action of countries, governments, industries, and people since we have great challenges in front that are necessary to address for the future of humankind. The high levels of pollution, food waste, increasing poverty, and the accelerated degradation of natural resources are only some of the challenges we are facing; the shift towards

* **Corresponding Author Elena Ibáñez:** Foodomics Laboratory, Institute of Food Science Research (CIAL, CSIC), Nicolas Cabrera, 9, 28049 Madrid, Spain; E-mail: elena.ibanez@csic.es

Merichel Plaza & Maria Luisa Marina (Eds.)

SDGs in all areas is necessary, and food analysis can contribute by greening the different steps involved in any analytical protocol and linking them to Green Chemistry and Green Analytical Chemistry (GAC) principles, as suggested in a very recent opinion paper [1].

According to United Nations (https://www.un.org/sustainabledevelopment/), energy is the main contributor to climate change, accounting for about 60% of all global greenhouse gas emissions. In fact, in 2015, only 17.5% of final energy consumption worldwide came from renewable sources. In this sense, research has the challenge of developing increasingly cleaner and safer processes that incorporate high energy efficiency, while companies need to move forward in decreasing energy consumption or using renewable energy sources as a competitive strategy towards sustainable industrialization.

Therefore, green technologies (considered this way not only the use of green solvents but also energy-efficient practices) are at the forefront when applied to the extraction of bioactive metabolites and to food analysis.

This book chapter is structured in three parts: 1) some general features about GXL solvents such as fundamental principles, parameters, and basic equipment, 2) some applications of GXL extraction in natural sources as highly efficient alternative for obtaining bioactive metabolites from natural sources and by-products, food wastes and agricultural residues, and 3) examples of multi-fluid platforms using GXL involved in biorefinery approach that can be run at analytical or processing scale.

GREEN EXTRACTION TECHNIQUE: MEAN FEATURES

In the previous chapter, the reader has entered deeply into the Supercritical Fluid Extraction world, which is highly connected with the Gas Expanded Liquid Extraction (GXLE). Before knowing more about extraction using Gas Expanded Liquids (GXL), this type of solvents will be introduced to let the reader know more about the composition, types, physical properties, and the effects of physical parameters such as pressure, temperature, and gas molar ratio on solvents' properties.

Fundamentals of GXL

An organic solvent that dissolves, partially or totally, a compressible gas or a supercritical fluid, composes GXLs. The gas phase is commonly carbon dioxide, ethane, or ethylene, but due to the environmental and economic advantages, and safety of CO_2, it is the main fluid used in GXLs. When carbon dioxide is employed under GXLs conditions, it is named as carbon dioxide expanded-liquids

(CXL). The transformation from a pure solvent to a GXL by increasing pressure and addition of gas is shown in Fig. (**1**). GXL can be defined as *switchable solvents* since they can be easily turned back to a pure solvent just by decreasing pressure or removing the gas phase. In fact, this switchability of GXL and the common use of CO_2 have been used to define GXL as "*a type of switchable solvents that is halfway from pressurized liquids to supercritical fluids by increasing the amount of compressed* CO_2 [2-4]". They are often referred to as "enhanced fluidity solvents" when liquid phase is under subcritical conditions [5]. The advantages of using GXLs are similar to those described for supercritical fluids, such as higher gas diffusivity and enhanced transport properties in solvents, but the extraction conditions using GXL can be milder than those required for SFE. The possibility offered by GXL of introducing a gas or a fluid such as supercritical CO_2 for producing multiple immiscible phases with enhanced fluidity is also interesting.

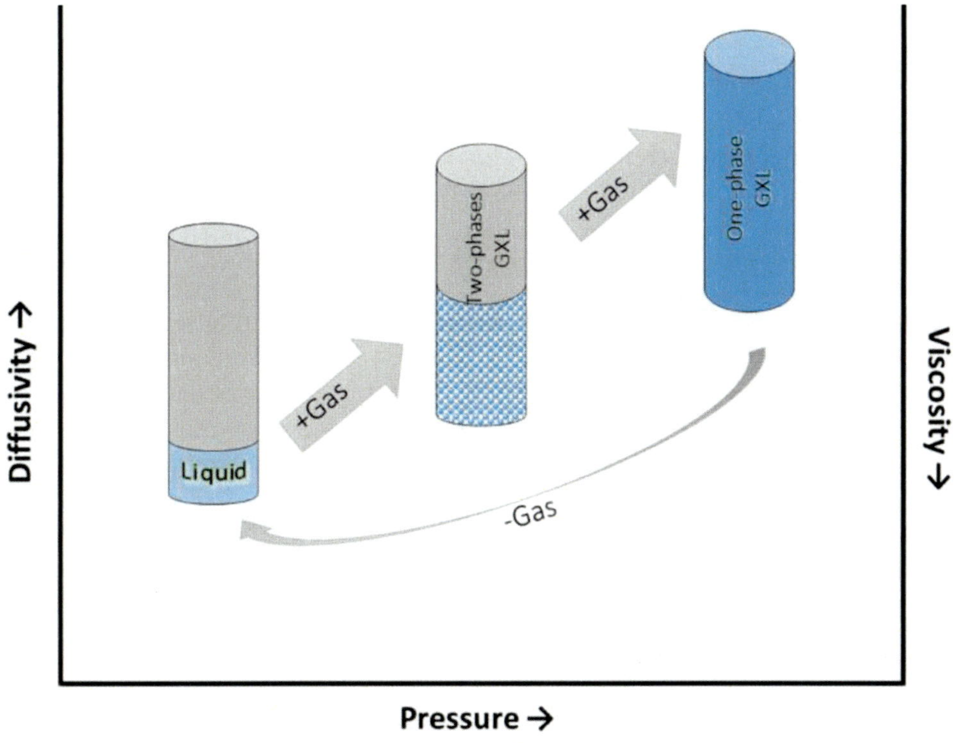

Fig. (1). Scheme of GXL formation from neat liquid and main physicochemical parameters affected.

It is important to note that in this chapter we will talk about gas phase or supercritical phase indistinctly in contrast with the liquid phase. As can be seen in

Fig. (**1**), different types of GXL can be obtained according to the conditions used which are mainly affected by pressure and gas molar fraction. In this sense, it is important to remark the partition of GXL described by Jessop & Subramaniam (2007) [6] depending on the solubility of the gas phase in the liquid phase.

a. The liquid phase can only dissolve small quantities of gas: consequently, only small changes in some properties can be observed compared to liquid solvents (polarity or volume expansion). Nevertheless, viscosity can change significantly in some liquid phases such as some liquid polymers, oils, and certain ionic liquids.
b. One-phase GXL: the gas is highly soluble in the liquid phase so its physicochemical properties exhibit enormous variations. In this regard, the most common mixtures are those with CO_2 or $scCO_2$ and ethanol, methanol, and hexane as liquid phases (compounds with high affinity to CO_2).
c. Two-phases GXL: the gas phase is not totally soluble in the liquid; in fact, there are two different phases in the mixture (some kind of tiny bubbles in the liquid). The changes in physicochemical parameters are between the other types of GXL. When these GXLs are used, they are mainly composed of water and CO_2, but in this case, the pH is highly affected so buffers can be used.

The same mixture of liquid and gas phases can be in any of the two types mentioned above, mainly depending on the temperature and pressure of the system. In fact, the study of phase equilibrium in GXLs is relatively complicated being the Monte Carlo and molecular-dynamics simulation the most accurate systems to predict changes in volume expansion, viscosity, pressure-composition, and pressure-density, as described by [7]; in some cases even better than standard PR-EOS (Peng Robinson Equation of Estate) thermodynamic modeling.

Effects of Pressure, Temperature and Percentage of CO_2 in the Mixture on the GXLs' Properties

Very important changes in GXLs' physicochemical properties can be experienced by changes in the extraction temperature and pressure and on the solvent composition (gas:liquid molar ratio). Therefore, these conditions should be carefully selected. Following, a discussion on the effects of extraction conditions on GXLs properties is presented.

Volume Expansion

It is the parameter that best defines GXLs, that is, the volume of a neat liquid solvent is expanded by adding gas phase and pressure. There is a volume increase with pressure, not like in pure gases, pure liquids, or supercritical fluids. Volume expansion at certain temperature can be defined as the "fraction V/V_0, where V_0 corresponds to the volume of 1 mole of the pure solvent at 1 atm at the chosen temperature, and V corresponds to the molar volume of the mixture" in the GXL. The change in volume depends on the solubility of the gas phase in the organic solvent. Abbot *et al.* (2009) demonstrated that the same addition of CO_2 does not cause the same expansion in all the solvents, nevertheless solvents with similar solubilities in CO_2 at 5 MPa (including acetone, ethanol, and propan-1-ol, among others) will experience the same expansion [8]. Aida *et al.* (2010) [9] associated this expansion to the hydrogen bond network. For example, comparing linear against branched alkyl alcohols, more CO_2 moles can be dissolved in branched alkyl alcohols at lower pressure because CO_2 can create larger hydrogen bond networks (even tetramer or pentamer) in the branched alkyl alcohols, which impact directly in volume expansion. On contrary, at higher pressures, the mixture is composed of a single liquid phase and the volume increases with pressure and decreases with temperature. It is caused by the dissolution of CO_2 into the liquid phase, which expands its volume. On the other hand, the effect of increasing temperature must be taken into account, since it "helps the gas to form a new phase" so higher temperatures drive to two-phase GXL formation.

Density

It is a crucial property when dealing with extraction processes since it is highly related to transport phenomena. Density is influenced by the composition of the mixture (molar ratios), pressure, and temperature. In one-phase GXL density is generally increased by increasing pressure or decreasing temperature. Nevertheless, general statements are not easy in two-phase GXL. In the same chamber volume, an expansion of the gas phase (adding gas) leads to an increase in the density of the liquid phase. However, the increasing temperature has a double effect; it expands the gas phase and decreases density in the liquid phase. On the contrary, small increments of pressure may result in decreasing density of the mixture due to increments in solubility of the gas in the liquid. The molar fraction of gas affects differently depending on the liquid, for example, in systems formed by CO_2 and alcohols (methanol, ethanol, isopropyl alcohol, tert-butyl alcohol, propanol, butanol), acetonitrile or, even, cyclohexane, the density is higher at low CO_2 molar fraction in the liquid.

Polarity

Undoubtedly, polarity is a key parameter for solubility; in fact, following the statement *similia similibus solvuntur* is normally a good point to select an extraction solvent. In GXL, polarity can be modified mainly by changing the molar ratio of gas in the liquid phase. For example, in case of CO_2, when its amount is increased in the liquid phase, the final GXL has lower polarity but higher acidity (which can be important in water and some alcohols). When dealing with one-phase GXL, the molar ratio of gas in liquid can be increased even without adding gas to the mixture; for example by decreasing temperature or increasing pressure. These variations in polarity can be calculated by means of Electronic Transition Energy (E_T, also called Reichardt's polarity) and Kamlet-Taft solvatochromic parameters. UV/vis/near-IR absorption spectra of chemical compounds are influenced by the surrounding medium (other molecules and solvents), they can change the position, intensity, and shape of absorption bands, this phenomenon is called solvatochromism [10]). These parameters, firstly introduced in 1976 [11] quantifies hydrogen bond accepting ability (β), hydrogen bond donating ability (α), and dipolarity (π*). Initially, they were obtained from the normalized UV spectra of certain solvatochromic dyes. They are very useful to correlate linearly with the logarithmic function of reaction rates and equilibria and can be calculated using COSMO-RS software [12]. For GXL Wyatt *et al.* (2005) [13] demonstrated that they can be modified by changing CO_2 molar ratios. Besides, E_T is also a solvatochromic measurement that can be used as numerical indicator of polarity [10]. It is used to calculate both polarizability/dipolarity interactions and hydrogen bonding donor interactions, which are solvation processes. As for classical liquids, E_T can be used to calculate interactions in GXL [14]. The approach of using solvatochromic measurements to calculate polarity of GXL for extraction has only been followed with single or less complex natural raw materials. However, they provide good estimations to choose the appropriate GLX for the target compounds to be extracted by GXL in natural products. Some GXL show negligible volumetric expansion, that is normally associated with gas absorption; for example in the mixtures water:CO_2 there is a CO_2 absorption associated to the carbonic acid equilibria in solution and, therefore, a change in pH. This change can be calculated using the Hunter & Savage (2008) equation [15]. This acidification of the extraction solvent, also called carbonated subcritical water, can be used to replace inorganic acids like sulphuric or hydrochloric, therefore, creating an environmentally friendly extraction or medium.

Diffusion Coefficient

It can be defined using Fick's laws of diffusion (movement of molecules from higher concentration to lower concentration region) as "*the molar flow through a surface in a unit time at a concentration gradient of unity*". Its value is determined by the interactions between solutes and solvents. Diffusion coefficients of GXL are midway liquids and supercritical fluids. For example, diffusion coefficient of pure alcohols (*i.e.* ethanol or methanol) is lower than the corresponding GXL formed by CO_2 and alcohol, which is lower than a supercritical fluid composed of them. But, as previously seen, diffusion is also altered by changing pressure, temperature, and gas:liquid molar ratio.

A good example of the effects of changing pressure, temperature, and molar ratios can be found in the interesting work by Palafox-Hernandez *et al.* (2019) [16]. They studied GXL composed of CO_2 and 1-octene or nonanal binary liquid mixtures at different pressures and temperatures, using Gibbs Ensemble Monte Carlo (GEMC) and molecular dynamics (MD) simulations to analyze associations between the phase, structural and dynamic characteristics. They found similar phase behaviors for 1-octene and nonanal despite their very different vapor phase transitions. The volume expansion for both solvents was similar with the function of CO_2 molar ratio. Interestingly, using Stokes-Einstein, they found that the volume expansion of the organic liquid, for both 1-octene and nonanal, is linearly dependent on the inverse of the shear viscosity, ηs^{-1}. These kinds of approaches have also been applied to biobased solvents by Granero-Fernandez *et al.* (2019) [17]; authors determined density and viscosity changes with CO_2 pressure. In this case, dipolarity/polarizability parameter and phase equilibrium of the expanded phase were calculated experimentally while viscosity and density were calculated through molecular dynamics. They observed that dipolarity/polarizability parameter (π^*) can change from values close to pure solvent up to negative values (pure CO_2) by changing P and T. Moreover, density and viscosity can be reduced up to 90% compared to the pure solvent. Authors also noticed that the solubility of CO_2 and therefore, variation in physicochemical properties were higher in linear organic carbonates than in the cyclic ones.

Equipment Set Up For GXL Extractions

Due to the mentioned similarities with SFE, the same equipment described in the previous chapter can be used for GXLs extraction. Nevertheless, in this particular case, it is not always necessary to reach pressures above 30 MPa; in fact, 7 MPa can be enough for many applications. The basic apparatus involves: two pumps (one for liquid and one for gas/supercritical phase, normally CO_2), heated extraction vessel, one valve to control flow and/or pressure, and, finally, a separator vessel. Besides, there are some extra materials recommended, but not mandatory for lab scales: gauges, thermocouples, and passive security system (rupture disk or safety valve).

Normally GXL are prepared by adding the liquid and the gas phase simultaneously to the extraction vessel or adding first the liquid and second the gas. For this reason, two pumps are necessary. Nevertheless, the pump for liquid can be avoided if liquid is added together with the raw material in the extraction vessel, before closing it. Mixing together liquid and gas is the main option used and it is very useful for one phase GXL. On the other hand, when working with two-phase GXL, we should consider the differences in compressibility of both, liquids and gases, and therefore, the second option is preferred.

Regarding the extraction vessel, it is commonly constructed in stainless steel; the wall size and seals must be designed considering that working pressures are commonly in the range of 7-30 MPa. Alloys with low carbon content, normally 316 grade are used for vessels and tubbing. Extraction vessels can be batch type, tubular, counter-current, sonicated, stirred, *etc.* For heating, normally ovens, jackets, or heating tapes based on electrical resistance are used, but even heat exchangers can be used for dynamic extractions.

GXL as Sustainable Solvents

One of the main advantages of using GXL for extraction is that they can be considered as green solvents according to the principles of Green Chemistry [18], Green Engineering [19], and, also, Green Analytical Chemistry [20]. Commonly, GXL are composed of CO_2 and alcohols or biobased solvents and they can fit perfectly in most of the features shown in Table **1**.

Table 1. Gas Expanded Liquids and green chemistry principles.

Principle	How GXL fit
Exists in nature without needing to be produced *ex profeso* and/or produced from byproducts sharing with other products the environmental impact	Components in GXL mixtures used for extraction are generally ethanol or biosolvents and CO_2. Normally, they are not produced to be used as solvents, they can be considered byproducts of other processes.
Easy biodegradation with no negative effects in the environment. Low toxicity for any organisms including humans	Most of the individual components of these mixtures have very high rates of biodegradability.
Low or negligible vapor pressure, so it is easy to control where the solvent is.	GXL extraction is carried out in closed vessels to maintain pressure, so this way vapor pressure is increased.
No need to add energy to be evaporated.	Some energy is necessary but, due to the expanded volume of the liquid phase thanks to the gas, the volume used is much less than in conventional extractions, so it is the energy.
High solubility towards target compounds, so, less amounts of solvents are needed in the process.	As seen in the previous subsection their polarity can be tuned so in each extraction, the solubility can be adjusted to the target compound in order to use less GXL.

To date, there are not enough studies that quantify the greenness of GXL used in extraction; in this sense, calculation methods such as Life Cycle Assessment (LCA) or Environment, Health and Safety (EHS) can be used for example to compare GXL extractions with other solvents.

APPLICATIONS: EXTRACTION OF NATURAL COMPOUNDS FROM DIFFERENT SOURCES

The wide variety of phytochemicals offered by nature go hand in hand with the growing challenge of finding new sources of functional food ingredients with pharmacological properties. However, it is important to highlight that the search for a greater bioavailability of these natural compounds is accompanied by friendly, efficient, and sustainable extraction techniques. In this sense, the role played by GXL as green and safe extraction solvents makes this extraction technology one of the greenest options to obtain bioactive molecules from natural sources and by-products, attending to criteria of effectiveness (ability to obtain compounds of interest selectively from solid or liquid matrices) and efficiency (high rate of mass transfer and low thermal degradation), among other aspects.

Current pressurized extraction and separation techniques have demonstrated to be useful for valorization of natural matrices because, together with the choice of a suitable solvent (or mixture solvents), tuneable physicochemical properties can be easily achieved by small changes in the operating conditions (density, diffusivity, viscosity, and polarity). This way, mass transfer, selectivity, and solubility of the target compounds could be increased leading to faster extractions with lower consumption of solvent and energy, lower degradation of bioactive compounds, and, in summary, with improved sustainable processes.

As mentioned earlier, this can be achieved by using GXLs extraction since the appropriate thermodynamic and physicochemical properties can be achieved at mild pressure, temperature and with a lower consumption of organic solvent (compared to liquid extraction, even with pressurized liquid extraction).

In this section, some interesting applications of GXLs to analyze and obtain bioactive compounds from natural sources such as algae, seeds, fruit, and roots will be presented targeting higher solubilities and lower thermal degradation of the bioactive analytes. Usually, carbon dioxide is the most common gas phase employed, its use implies higher diffusivity and low viscosity, promoting a higher extraction rate. Along with CO_2, various organic solvents (ethanol, methanol, *etc.*) have been used to improve the solubility of target metabolites. It is worth mentioning that when GXL is composed of carbon dioxide and ethanol, it is commonly named as carbon-dioxide expanded ethanol (CXE); this is one of the most common combinations employed in GXLs extraction.

On the other hand, some studies related to the measurement of solubility of target compounds in GXL for extraction purposes will be discussed.

Experimental Measurement of Solubility and Extraction Kinetic

Turner´s researchers group in Sweden carried out interesting studies of solubility measurement and extraction kinetics. For instance, two studies of Cunico *et al.* (2017, 2020) using carbon dioxide expanded liquids, evaluate the solubility, mass transfer resistance, partitioning coefficient, and thermal degradation of curcumin (polyphenol) [21] and quercetin (flavonoid) [22]. The CXL selected showed higher mass transfer (due to the presence of CO_2), solubility enhancement (organic solvent allows fine-tuning of physicochemical properties of the solvent), and moderate to low working temperature, which prevents thermal degradation; consequently, the extraction efficiency was enhanced.

In the solubility study of curcumin [21], CO_2-expanded ethanol was used in the following ranges of temperature (35 and 50 °C), pressure (10–30 MPa), and CO_2:EtOH mixture (10 and 50 mol%). Authors suggested that the solubility of curcumin was improved at 50 °C and the lowest value of CO_2 was added (10 mol%). Both parameters, temperature and the amount of CO_2 added in the mixture, exhibited a significant influence on the solubility, even higher than pressure. A 10 mol% of CO_2 added in the mixture is a reasonable value due to the highly polar character of curcumin and, consequently, the presence of a suitable amount of polar organic solvent favors the solubility of polyphenols.

In terms of quercetin solubility [22], two green organic solvents with different polarities (ethanol and ethyl lactate) were studied in terms of solubility and thermal degradation. The working parameters were as follows: temperature (35, 50, and 70 °C), amount of CO_2 (10, 30, and 50 mol%), and fixed pressure (10 MPa). The highest solubility was achieved for CXE at maximum values for ethanol (70 °C and 30 mol% CO_2), but minimum values for ethyl lactate (35 °C and 10% mol CO_2). The authors reported a higher solubility and thermal degradation of quercetin in CO_2-expanded ethyl lactate compared to CO_2-expanded ethanol. They suggested that the higher thermal degradation observed in CO_2-expanded ethyl lactate was due to a lower amount of CO_2 added (10% mol CO_2), so this increased thermal degradation in GXL can be applied to the extraction of other thermolabile compounds.

Al-Hamimi *et al.* (2016) studied the extraction of two volatile monoterpenes (α-pinene and cis-verbenol) from *Boswellia sacra* tree resin by using liquefied CO_2-expanded ethanol [23]. Besides, supercritical fluid extraction using CO_2 ($scCO_2$) and conventional solid-liquid extraction (SLE) were applied for comparison. In this work, $scCO_2$ extraction was performed at 55°C, 30 MPa, 5% (v/v) ethanol as modifier, and a total extraction time of 80 min. Typical method of SLE and continuous-flow SLE with ethanol was studied for resin powder extraction at 40

°C, atmospheric pressure, and 80 min extraction time in a home-built system. For a better understanding of the relationship between solubility/partitioning and desorption/diffusion rate-limiting steps, authors performed a extraction kinetics under optimal extraction conditions for CXE, $scCO_2$, and continuous-flow SLE in order to measure speed mass transfer and efficiency of the extraction processes. Hence, fractions were collected along the extraction time at different flow rates (1, 2, and 3 mL/min). Several models were applied to carry out the kinetics study of the processes' extraction rate:

1. Hansen Solubility Parameter (HSP), which is a theoretical prediction for the selection of solvents based on the group-contribution method. This tool can be used to evaluate the chemical affinity between molecular structure of analytes and solvents [24], reducing the number of experiments and the solvent waste.
2. Response surface methodology (RSM) by Box Behnken experimental design, to determine functional relationships between the extracted amount of each monoterpene as responses, and the independent variables (pressure, temperature, and molar fraction of CO_2)
3. Peleg's model, which is related to extraction rate curves in which the absorption (concentration) of analytes into a solvent at different time is a function of the initial concentration of extracted analytes at time t = 0, and the equilibrium in which the concentration of the extracted compound has its maximum.

Authors [23] demonstrated that CXE at 40 °C, 9.3 MPa, and 0.3 molar fraction of CO_2 in ethanol exhibited an extraction rate up to 10 fold faster than SFE at optimal extraction conditions. An increased solubility of analytes, a higher mass transfer rate and therefore higher efficiency were achieved in the following descending order CXE > $scCO_2$ > SLE. In fact, they concluded that initial extraction kinetics with CXE is controlled by solubility so a high flow rate is more convenient at this stage. After a certain extraction time, the flow rate is limited by internal desorption and diffusion of the analytes from the matrix to the bulk solvent and therefore a reduction in flow rate is recommended [23]. This improvement in efficiency is accompanied by relatively low pressure and moderate temperature, reducing the risk of decomposition of thermolabile compounds.

Other research works of Cunico *et al.* (2020) evaluated the efficiency of mass transfer of aqueous contaminants such as acetylsalicylic acid (aspirin), acetaminophen (paracetamol) and caffeine, by means of CO_2-expanded liquid-liquid extraction (CXLLE) In this work, a novel ternary system of CO_2/n-octanol/water was studied in terms of distribution kinetics and the data was compared with a conventional LLE utilizing n-octanol/water. Phase equilibrium of the mixture was measured by gas chromatography, for this purpose, the authors conducted an equipment consisting of a high-pressure view cell coupled on-line to GC; besides, this equipment was connected on-line to HPLC to study the potential of GXL extraction of pharmaceutical contaminants in water. They tested different pressures (7-10 MPa) and 50 °C at fixed temperature conditions to determine log D (octanol-water distribution ratio). Results (Fig. **2**) show that the efficiency of liquid-phase extraction can be improved by the addition of CO_2, contributing to reducing the viscosity and increasing the solubility of the compounds. They found faster distribution rates by adding CO_2 as viscosity-lowering entrainer, enabling 2 to 3 times shorter extraction times; moreover, since CO_2 also changed the log D (octanol-water distribution ratio) values, it can be used as a procedure for tuning the selectivity of the extraction process [25].

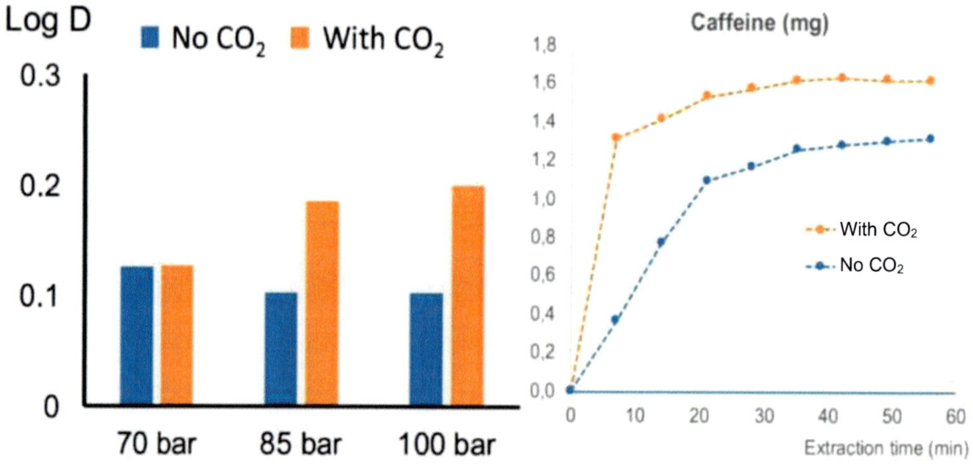

Fig. (2). Average compound log D values and extraction kinetic found for caffeine in the GXL extraction from water. Reprinted from Cunico *et al.* (2020) [25] with permission of Elsevier under the terms of the Creative Commons CC-BY license.

GXLs Extraction of Bioactives from Plants and Fruits

The vegetable kingdom is a plethora of bioactive compounds and since ancient times there has been an interest in using natural plants as a source of secondary metabolites with therapeutic properties. In this section, we highlighted some examples of the use of GXL extraction for obtaining bioactive compounds from different plants and fruits.

Two studies of natural plants from Brazil used GXLs to extract target compounds with biological properties; for instance, Santos *et al.* (2013) evaluated the antioxidant activity of Brazilian ginseng (*Pfaffia glomerata*) roots extracts obtained with $scCO_2$ with high amounts (10-100%) of modifier (ethanol), reaching CXL conditions. They investigated the effects of the independent factors (temperature, pressure, and fraction of ethanol) on response variables (extraction yield, antioxidant activity, and energy consumption per unit of manufactured extract). Using 10% (w/w) of ethanol (minimum value studied) they produced extracts with the highest antioxidant activity, while working at high temperature (90 °C) and ethanol percentage (100%) resulted in higher extraction yield and lower energy consumption per unit of manufactured product; under these conditions, pressure did not affect any of the studied response variables. The study suggested that a better extraction with reasonable antioxidant activity and reasonable energy consumption per unit of manufactured product can be obtained at the GXL range (pressure: 10 MPa, temperature: 90 °C, and mass fraction of ethanol: 50%) [26].

A second study related with *Copaifera langsdorffii* leaves was done by Costa and co-workers (2015) to obtain two representative compounds with biological properties, kaurenoic acid (diterpene) and quercitrin (glycosylated flavonol) [27]. They also investigated the effect of pressure, temperature, and percentage of co-solvent (ethanol) in the concentration of the target compounds in *C. langsdorffii*, total flavonoids content and antioxidant activity as response variables. This work determined that temperature and ethanol percentage were the main factors involved in the selectivity of the target compounds. As a result, a high amount of total flavonoids and high concentration of quercitrin were observed in the extract obtained at 10 MPa, 70 °C, and 50% (v/v) of ethanol as CO_2 modifier. This amount of ethanol in the mixture is considered within a GXL system. Opposite to the results obtained in [26], in which high antioxidant activity of *Pfaffia glomerata* was obtained at the lowest percentage (10% (v/v)) of ethanol, quercitrin exhibited a higher level of antioxidant activity at the maximum percentage of ethanol studied (50% (v/v)). Authors suggested that antioxidant activity can be due to the polar affinity of flavonoids to solvent and the co-extraction of several compounds other than flavonoids, belonging or not to the polyphenolic group with recognized antioxidant properties. As for kaurenoic acid, it exhibited antioxidant capacity at the lowest percentage of ethanol due to its non-polar nature.

David *et al.* (2019) [28] performed a study using CO_2 expanded ethanol to obtain essential oil from *Vetiveria zizanioides* (L.) Nash roots and they made an evaluation of the yield, chemical composition, antioxidant and antimicrobial activities. Supercritical CO_2 ($scCO_2$), hydrodistillation (HD), and indirect vapour distillation (IVD) were used for comparison purposes. The highest oil yield (7.42%) was obtained by CXE extraction at optimal conditions (8.4 MPa, 50 °C, 5 mL/min ethanol, and 0.22 mole fraction of CO_2), while oil yields of HD, IVD, and SFE were 0.6%, 0.3% to 0.5%, and 0.5%, respectively. The gas chromatography-mass spectrometry (GC-MS) analysis showed the presence of hydrocarbons and carbonyl compounds as the main compounds classes, besides, CXE and HD contained also alcohols and carboxylic acids while SFE showed only the presence of hydrocarbons. In terms of antimicrobial effects [28], CXE and SFE extracts exhibited moderate activity against Gram-positive bacteria such as *Staphylococcus aureus* (MIC=78 µg/mL) while a potent activity was reported for HD (MIC=39 µg/mL). With these findings, authors proved that extraction yield and target analytes, as well as biological activities, can be affected by three factors: 1) addition/removal of mole fraction of CO_2 into the liquid mixture, 2) modification of ethanol flow rate into a liquid phase promoting a wide spectrum of polarity solvents and 3) pressure, because it determines subcritical to supercritical state of CO_2 in the CXE and therefore impacts on the solubility and mass transfer properties. Fig. (**3**) shows the effect of mole fraction of CO_2 on the response factor (yield) along the extraction time.

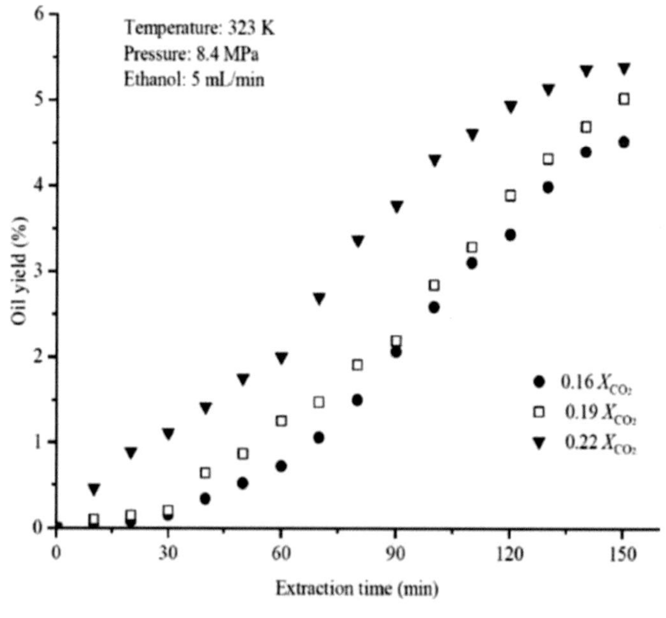

Fig. (3). Effect of CO_2 mole fraction on *Vetiveria zizanioides* (L.) Nash roots extraction yield at ethanol flow rate of 5 mL/min and constant pressure of 8.4 MPa. Reprinted from David *et al.* (2019) [28] with MDPI permission under the terms and conditions of the Creative Commons Attribution (CC BY) license.

Besides leaves and roots, fruits have also been extracted employing GXLs. For example, Sakai and coworkers (2021) applied carbon dioxide-expanded liquid (CXL) to obtain crocin-1 and crocin-2, both polar natural pigments from *Gardenia jasminoides* Ellis fruit pulp with antioxidant, anti-inflammatory, anti-cancer properties, and protective effects against Alzheimer′s disease and diabetes

[29]. In this work, the authors evaluated CXL extraction assisted (or not) by direct sonication into the liquid-CO_2 system. CXL extraction parameters were: type of liquid phase (ethanol, water, and aqueous ethanol 50 or 80% *v/v*), temperature (5–25°C), pressure (8–14 MPa), constant flow rate of 10 mL/min, and sonication time (25-200 s); response variable evaluated were the concentration of the target pigments. In order to determine the effects of different solvents (water, ethanol, and methanol as a control) on pigments extraction concentration, a conventional solid-liquid extraction (SLE) was performed giving methanol the highest extraction concentration of crocin-1 and crocin-2 (10.43 ± 0.3 and 0.37 ± 0.02 μg/mL, respectively) and pure ethanol the lowest values. However, methanol was discarded for not being a green solvent. Then, the authors analyzed the composition of the mixture of ethanol/water in the extraction phase (30, 50, or 80%) in the conventional SLE system. Due to the polar nature of crocins, pure ethanol (CXE), water (CXW), and 50 or 80% of aqueous ethanol solution (CXE50% and CXE80%, respectively) were tested. They determined that the highest concentration of pigments was reached at ethanol:water:CO_2 (64:16:20 *v/v*) because the addition of water to ethanol increases the dielectric constant and the solubility of the analytes characterized by an amphiphilic structure. Results showed that a higher amount of crocin-1 and crocin-2 (13.63 ± 0.5 and 0.51 ± 0.05 μg/mL, respectively) was found at 25 °C, 10 MPa, CXE 80% (20% CO_2 and 80% liquid phase, ethanol:water:CO_2 64:16:20 *v/v*) extraction system with direct sonication, compared to conventional solid-liquid methanol extraction (10.43 ± 0.3 and 0.37 ± 0.02 μg/mL, respectively). The time and direct sonication in CXE-80% extraction were other two factors with significant influence in pigments concentration in the extract since cavitation helps break cell membranes which improves the mass transfer rate. In that sense, 125s of direct sonication promoted the highest concentration of crocins. Authors concluded also that a direct sonication of the liquid/CO_2 two-phase system caused rapid physical mixing between the water and CO_2 phases, including micro-phase separation and cavitation, which may speed up the mass transfer at lower frequencies. Likewise, at the maximum extraction temperature tested (25 °C), there is a positive effect on pigment concentration caused by decreasing of surface tension between the solvent, solute, and matrix, which reduces viscosity and stimulates a better solvent penetration into the matrix. In terms of extraction pressure, no significant influence was noticed in the range tested (8 to 14 MPa) despite an increasing pressure increases the density of the CXL extraction system. They observed that the overall density of compressed CO_2 under these conditions was very similar and therefore little effect was noticed on liquid CO_2 solubility.

Algae and Microalgae Applications of GXL

The potential of algae and microalgae in foods, nutraceutical, cosmetic and pharmaceutical fields is widely recognized for being a rich source of essential nutrients and valuable (and unique) bioactive metabolites. Besides, another interesting field is emerging such a biofuel feedstock due to a significant content of lipid components in algae and microalgae. In this sense, interesting facts based on the technological application of GXL in *Arthrospira platensis* (Spirulina), *Haematococcus pluvialis* and others, have been reported.

Our research group (Foodomics) has published many contributions related to GXL extraction using microalgae as biomass. In 2012, Golmakani *et al.* [30] compared the extraction yield of γ-linolenic acid (GLnA)-enriched fractions from *Arthrospira platensis* (Spirulina) using CXE and pressurized ethyl lactate (PLE). Authors applied a chemometric optimization by means of statistical models based on multifactorial experimental designs to describe, at the same time, the behavior of different extraction factors as well as to predict the response variables for each extraction method. Experimental parameters of temperature, pressure, extraction time, and solvent composition (from 100% ethanol to 100% ethyl lactate) were analyzed. In terms of CXE, a total yield of lipids of 6.7% (w/w) and 24.7% of GLnA recovery were achieved at optimum conditions of 40 °C, 30 MPa, 90 min, and solvent mixture ethanol:ethyl lactate (50:50, *v/v*). Results were in accordance with their predicted responses by RSM, demonstrating the robustness of the mathematical model used. According to PLE, total lipids yield (%, w/w) and GLnA recovery were up to 3 times higher than in GXL, with values of 20.7% (w/w) and 68.3% respectively and closer to predicted responses data of 21.5% (w/w) of total yield and 63.9% of GLnA recovery at optimized conditions of 180 °C, 20.7 MPa, 15 min, and solvent mixture ethanol:ethyl lactate (50:50, *v/v*). Researchers suggested that the main factors influencing total yield and selectivity in PLE were temperature, extraction time, and solvent composition, with temperature, being the most significant factor due to an increase in the solubility of the components and a decrease in the liquid (solvent) viscosity, favoring the mass transport diffusion from the matrix to the liquid solvent. Whereas in CXE, solvent composition (% ethanol, w/w) and extraction time played an important role in response variables, having a less significant influence of temperature (see Fig. **4**). They concluded that this could be due to the use of one extracting phase, which seems to favor the recovery of a higher amount of extract, in contrast to the use of two phases as it happens with the liquid-gas binary mixture in GXL extraction, which was more selective. In this work, pressure showed a small influence on extraction and was not considered a significant factor.

Haematococcus pluvialis microalga was employed in a study published in 2014 [31] as a source for obtaining astaxanthin (xanthophyll) with powerful antioxidant activity. This molecule is characterized by having a heavy and relatively polar structure; hence a neat supercritical CO_2 extraction is limited. In this sense, the authors investigated the effect of extraction factors (pressure, temperature, and ethanol content in scCO_2) on yield, astaxanthin recovery (estimated as a percentage of the total astaxanthin extracted in each extraction method to the total astaxanthin extracted in a conventional acetone extraction), and antioxidant activity as response variables, and compared scCO_2 and CXE techniques. For this purpose, a Box–Behnken experimental design was used in order to find the relationship between extraction responses and independent extraction factors. In terms of scCO_2 extraction, the authors set conditions of pressure (from 20 to 35 MPa), temperature (from 40 to 70 °C), ethanol as modifier (from 0 to 13% (w/w)), flow of the co-solvent pump (0.04 g/min of ethanol at 20 °C and 0.08 g/min at 20 °C) to obtain mixtures containing 6.5 and 13.0% (w/w) ethanol, respectively. Values of extraction yield (73.4-282.5 mg/g), astaxanthin content (5.98-53.48 mg/g), astaxanthin recovery of (2.5-82.3%) and antioxidant activity (0.084 to 0.243 mM TE/g) were reached at optimum conditions of 20 MPa, 55 °C, 13% (w/w) of ethanol as a CO_2 modifier. In CXE experiments, the following factors and ranges were selected: temperature (30, 45, or 60 °C), pressure 7 MPa, and co-solvent pump (0.6 g/min of ethanol at 20 °C or 0.83 g/min at 20 °C) to obtain liquid mixtures containing 50 and 70% (w/w) ethanol. Results showed that the highest values of astaxanthin yield (62.57 mg/g), astaxanthin recovery (124.2%), and antioxidant activity (0.233 mM TE/g) were obtained at optimum conditions of 7 MPa, 45 °C, and 50% (w/w) ethanol. This study concluded that the overall responses of CXE extraction were better than those using SFE. In fact, ethanol content, besides the temperature, was the most representative factor contributing to the amount of astaxanthin and antioxidant activity of *H. pluvialis* extracts. Authors highlighted that antioxidant activity increased with the astaxanthin content, independently of the extraction method employed due to the presence of other bioactive compounds (such as phenols) that were co-extracted with ethanol in the mixture of the liquid solvent.

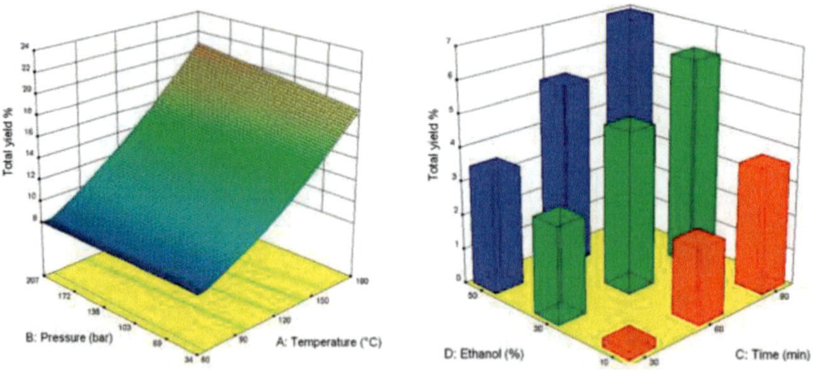

Fig. (4). Effect of significant factors in response variables for both PLE and CXE process to obtain γ-linolenic acid (GLnA)-enriched fractions from *Arthrospira platensis* (Spirulina). Reprinted from Golmakani *et al.* [30] with permission of Elsevier.

Another work, using a heterotrophic marine microalga, *Schizochytrium* sp., was published by Wang (2015) for obtaining DHA-enriched lipids by CXE [32]. The effects of pressure, temperature, ethanol, and CO_2 flow rates on the extracted lipid were analyzed for values of pressure (2.8 to 9.0 MPa) and temperature (40 to 60 °C). A total lipid yield of 87 wt% was achieved working at continuous CXE extraction for 30 min at optimal conditions (6.9 MPa, 40 °C with ethanol and a CO_2 flow rate of 1 mL/min and 6.0 mL/min, respectively). They compared the ratio of extracted lipid yields per gram of solvent employed in a conventional extraction (Soxhlet), pressurized ethanol, and CXE extraction. They found that CXE required the lowest amount of ethanol and less time to achieve the same extraction yield.

This microalga has also been examined to understand the variables (pressure, temperature, and CO_2 mole fraction) involved in different phase regions of CXE (as shown in Fig. **5**), and how they influence the extraction yield and lipids recovery [33]. For this purpose, they measured the mentioned parameters in order to find the phase region that achieved the optimum CXE lipids extraction. Besides, they compared it with conventional Soxhlet and neat $scCO_2$ extraction.

As can be observed, A, B and C are regions characterized by CXE with CO_2 mole fraction below 1.0 or critical composition (Xc). Region A is formed by biphasic fluids in which coexists a liquid-gas phase equilibrium between CO_2 as gas phase and ethanol as liquid solvent. In region B, CO_2 under subcritical conditions (below critical pressure) is mixed with ethanol. Above the critical pressure, CO_2 is compressed to liquid estate and completely dissolved in the liquid ethanol phase to form a monophasic system named enhanced fluidity liquid (EFL), as can be observed in region C. On the other side, region D is located above critical points of pressure and CO_2 mole fraction, being considered as a supercritical CO_2 with a small amount of ethanol, so we are talking about $scCO_2$ extraction with cosolvent rather than CXE extraction. Region E is characterized by CO_2 vapor. Finally, the bubble point curve is the limiting condition between subcritical and biphasic fluid regions below the critical pressure. Their results showed that CXE (subcritical) performed comprehensive extraction of lipids from microalgae, due to the enhanced mass transfer and solubility. Additionally, higher DHA content and antioxidant activity were found in CXE extracts. So, due to their mild conditions, CXE can be considered a promising alternative for extracting easily oxidized or thermolabile components such as omega-3 from natural resources.

Another important application of microalgae biomass is its possible use as biofuels, considering its high lipid content that can be useful for renewable energy based on biodiesel. Among different microalgae reported, two cases have been selected: *Chlorella vulgaris* [34] and *Botryococcus braunii* [35]. For comparative purposes, the aforementioned works studied the effect of independent extraction factors (temperature, pressure, and solvent flow rate) in total yield as the main response variable. It should be highlighted that, when research is addressed towards biodiesel industry, it requires a certain selectivity degree in terms of class of lipids that can be transesterified for biodiesel purposes, so, in both works authors studied the selectivity of the processes based on GXL processes.

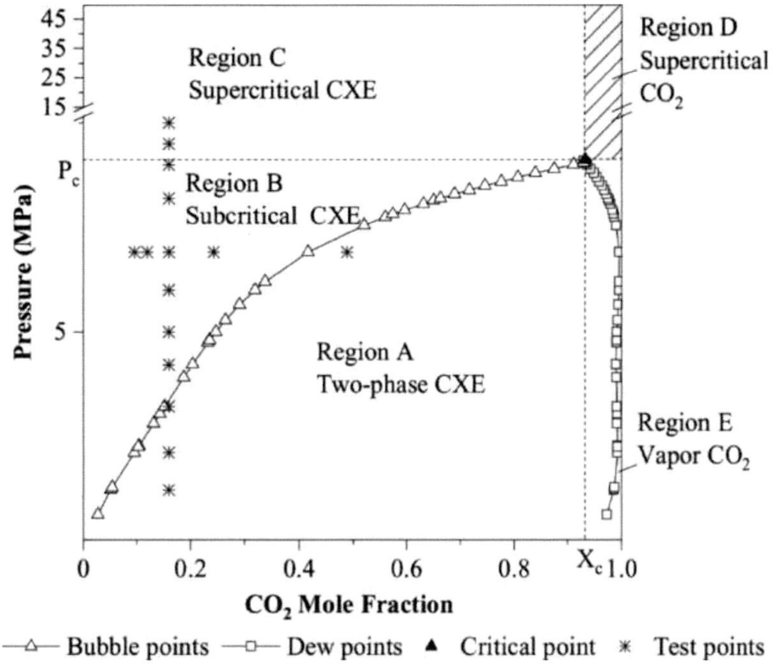

Fig. (5). Experimental design followed by Li *et al.* [33] circumscribed in different phase regions of CO_2 and ethanol binary system. Reproduced from Li *et al.* [33] with permission of Elsevier.

Yang *et al.* (2015) studied the total lipid yield from *Chlorella vulgaris* microalgae for biodiesel production on continuous mode, and compared conventional Soxhlet extraction and CXL [34]. In addition, methanol, ethanol, pressurized methanol, pressurized ethanol, CO_2 expanded methanol (CXM), and CXE were compared in the continuous extraction process in order to select suitable solvents promoting a high lipid extraction yield and adequate transesterification. As a result, CXM featuring high diffusivity and low surface tension promotes the mass transfer rate of lipids through the cell wall, thus increasing the efficiency of the lipid extraction with lower amount of solvent and shorter extraction time. Methanol also is the most often solvent used as a substrate for transesterification to produce fatty acid methyl esters (FAME). Authors determined an experimental highest extracted lipid yield of 84.8%, even better to the predicted value (82%), at optimum conditions of continuous CXM at pressure of 5.5 MPa, temperature of 85 °C, methanol flow rate of 1 mL/min, CO_2 flow rate of 3.0 mL/min (90% of the CO_2 equilibrium solubility). The total lipid content obtained by CXM was also higher than that obtained by Soxhlet extraction (20.7% of the dried biomass). Additionally, authors also evaluated the effect of changing from CO_2 to propane in the gas phase of GXL. Lipid yield of propane expanded methanol was 4% higher than CXM. However, despite the operating temperature of propane-based GXL, 60 vs 85 °C in CXE, the energy consumption of propane expanded methanol is 315% higher than CXM, besides, the high flammability of propane compared to CO_2 should be considered [34].

Paudel *et al.* [35] applied the properties of liquid CO_2 and CXM for the extraction of neutral lipids and free fatty acids from *Botryococcus braunii* sp microalgae. In CXM extraction at mild conditions, 35 °C and 7.2 MPa, lipid yield was 24 wt%, which means 5% higher than using liquid CO_2, but 8% lower than conventional extraction using $CHCl_3$/MeOH. GC-FID analysis was used to evaluate the selectivity towards neutral lipids (NL) and free fatty acids (FFA). CXM reached 78–82% selectivity for target compounds, followed by neat methanol (68%) and $CHCl_3$/MeOH mixture (46%). An interesting relationship between the yield of FFA and neutral lipids per volume of organic solvent used in the extraction process was observed, which is very important for economic and environmental cost of biodiesel production. In CXM, 21 mg of NL and FFA per mL of organic solvent was achieved, while neat methanol and $CHCl_3$/MeOH produce 3 mg of target compounds per milliliter.

GXL Extraction and Agricultural By-Products

Valorization of food by-products or waste is important not only due to the amount and bioactivity of many added-value products contained in these biomasses but also considering the collateral impacts that this valorization has in terms of scaling-up, energy consumption, economic and environmental costs, and final disposal of solvents. In order to replace the use of CO_2 within a GXL extraction, recently, Santos (2019) used nitrogen (gas phase) and water (liquid phase) to evaluate the phenolic yield from pomegranate peel in a GXL coupled to ultrasound-assisted extraction (UAE) [36]. For this purpose, they analyzed the interaction of different independent parameters such as ultrasound power, initial pressure of expansion gas, system pressure, and material particle size (0.5–1.0 mm< PS<1.0 mm) in total phenolic content. This procedure was performed in an "EXTRACT-US" system (Project FAPESP 2013/04304-4 – Patent pending), using extraction cycles of 20 min. The highest yield was reached by a combination of ultrasound (400-600W) and nitrogen (5-10 bar), while a higher phenolic yield was obtained by interaction of all factors analyzed in this study (including also the particle size).

In 2017, Chhouk and researchers [37] compared the amount of total phenolics and antioxidant activity of garlic bulk by application of CXE and conventional Soxhlet. Garlic bulk is a by-product from food processing industry composed of lignocellulosic material as a source for insoluble phenolic compounds in which authors report the presence of caffeic, p-coumaric, ferulic, and di-ferulic acids. In terms of extraction parameters, they evaluated temperature (50-200 °C) and CO_2 flow rate (0.5-1.5 mL/min) as independent variables, being the amount of total phenolics and the antioxidant activity as the response variables. Constant pressure (10 MPa) and ethanol flow rate (3mL/min) were fixed. In this study, the authors found that high temperature and low CO_2 flow rate provided the largest content of total phenolics (56.26 mgGAE/g dried sample) as well as antioxidant activity (IC_{50}=0.41 mg/mL). Probably, high temperature stimulated a disruption of the cell tissues of lignocellulosic material increasing the mass transfer rate during extraction and therefore, a higher amount of phenolic compounds could be extracted. In terms of antioxidant activity, not only the amount of phenolic compounds is involved, but also new antioxidant compounds formed as a result of Maillard reaction.

Besides common alcohols (ethanol and methanol), other liquid phases can be used in GXL extraction. In this sense, bio-based solvents offer a wide range of liquid phases produced from renewable biomass sources such as energy crops (*e.g.*, corn), forest products (*e.g.*, wood), aquatic biomass, *etc.* Hoang *et al* (2017) proposed the use of 2-methyltetrahydrofuran (MeTHF), as a renewable resource coming from furfural or levulinic acid, in a binary GXL system for biocatalysis reaction of *Candida antarctica* lipase B [38]. First, they measured the volumetric expansion of the novel liquid phase by varying CO_2 pressure at 20°C in order to find the pressure of optimum liquid expansion. In this sense, by using only 4 MPa, a higher conversion rate was achieved, showing a decrease in substrate concentration and a high concentration of CO_2. Authors suggested that these results can be explained by improved transport properties, increased hydrophobicity of the solvent systems, and a reduction of solvent inhibition due to polar MeTHF molecules. They also reported that lipase maintained a high activity and enantioselectivity in this new liquid phase [38]. As a summary, Table **2** shows the principal selected applications of GXL involved in this chapter.

MULTI-FLUID PLATFORMS USING GXL (BIOREFINERY PLATFORM)

Biorefinery is one of the most promising approaches for the sustainable exploitation of biomass sources into a wide range of high-value products. Mostly, these processes employ a single raw material source to extract a maximum product output, despite there are many considerations about environmental impact and scarcity of feedstocks [39]. Moreover, the use of compressed fluid processes with green solvents for such purpose has various advantages including less processing time, enhancement of the extraction recovery, semi-continuous operation, and reduction in energy consumption.

Table 2. Summary of main applications of GXL in natural sources and by-products.

Natural biomass	Target compound	Family	GXL method	Optimal conditions	Main feature measured	References
N.A.	Curcumin	Polyphenol	CXE	50 °C, 10 MPa 10 mol % CO_2:EtOH mixture	Solubility	[21]
N.A.	Quercetin	Flavonoid	CXE	70 °C, 10 MPa 30% mol CO_2:EtOH mixture	Solubility and thermal degradation	[22]
			CX-ethyl lactate	35 °C, 10 MPa 10 mol % CO_2:ethyl lactate		
Boswellia sacra tree resin	α-pinene cis-verbenol	Monoterpenes	CXE	40 °C, 9.3 MPa 0.31 molar fraction CO_2: EtOH	Faster mass transfer Efficiency	[23]
N.A.	Acetylsalicylic acid, Acetaminophen, Caffeine		CXLLE	50°C, 8.5 MPa	Ternary CO_2/n-octanol/water system.	[25]

(Table 2) cont.....

Natural biomass	Target compound	Family	GXL method	Optimal conditions	Main feature measured	References
Pfaffia glomerata roots	n.r.	n.r.	CXE	90 °C, 10 MPa 50% mol CO_2:EtOH mixture	Antioxidant activity, Extraction yield, energy consumption per manufactured unit	[26]
Copaifera langsdorffii leaves	kaurenoic acid	diterpene	sc CO_2	70 °C, 39 MPa Neat CO_2	Antioxidant activity, Extraction yield	[27]
	quercitrin	Glycosylated flavonol	CXE	70 °C, 10 MPa 50% mol CO_2:EtOH mixture		
Essential oil of *Vetiveriazizanioides* (L.) Nash, roots	valerenol, valerenal β-cadinene	Terpenes	CXE	50 ° C, 8.4 MPa 0.22 mol CO_2 5 mL/min EtOH	Antioxidant activity, Antimicrobial activity	[28]
Gardenia jasminoides Ellis fruit pulp	crocin-1, crocin-2	Natural pigments	CXE+sonication	25 °C, 10 MPa CXE 80% (64:16:20, EtOH:H_2O, CO_2*v/v*) 125 s, 20% amplitude	Extraction concentration	[29]
Arthrospira platensis (Spirulina) algae	γ- linolenic acid (GLnA	Fatty acid	CXE-ethyl lactate	40 °C, 30 MPa 50:50 *v/v* EtOH:ethyl lactate in liquid phase	Total yield of lipids,Maximum GLnA recovery (%, w/w)	[30]
Haematococ-cus pluvialis algae	Astaxanthin	Xanthophyll	CXE	45 °C, 7 MPa 50% w/w EtOH content	Total yield,Total astaxanthin content, Antioxidant activity	[31]
Schizochytrium sp. microalgae	Docosahexaenoic acid (DHA)	Fatty acid	CXE	40 °C, 6.9 MPa EtOH flow rate: 1mL/min, CO_2 flow rate: 6 mL/min	Total yield	[32]
Schizochytrium sp. microalgae	Docosahexaenoic acid (DHA)	Fatty acid	CXE in different phase regions	40 °C, 6.9 MPa 0.16 mol fraction CO_2	Total yield, Total amount of DHA, Antioxidant activity	[33]

(Table 2) cont.....

Natural biomass	Target compound	Family	GXL method	Optimal conditions	Main feature measured	References
Chlorella vulgaris microalgae		Lipids	CXM	85 °C, 5.5 MPa MeOH flow rate: 1 mL/min CO_2 flow rate: 3.0 mL/min	Total lipids yield, Energy efficiency	[34]
Botryococcus braunii		Neutral lipids and free fatty acids	CXM	35 °C, 7.2 MPa MeOH flow rate: 0.15 mL/min	Total lipids yield	[35]
Pomegranate (*Punica granatum L.*) peel	α-punicalagin, β-punicalagin and ellagic acid, among others	Polyphenols	GXL coupled UAE	40°C, 20 MPa N_2-Pi: 0.5-1 MPa, Ultrasound power: 400-600W	Gas phase: N_2 Liquid phase: H_2O	[36]
garlic husk	garlic acid, 4-hydrobenzoic acid, caffeic acid, p-coumaric acid among others	Phenolic	CXE	200 °C, 10 MPa CO_2 flow rate: 0.5 mL/min.		[37]
Candida antarctica lipase B		Lipase enzyme	CX- bio-based liquids	20 °C, 6 MPa MeTHF: 10% *v/v*,	Liquid phase: 2-methyltetrahydrofuran (MeTHF)	[38]

N.A. No apply n.r. no reported

As shown in Fig. (**6**), depending on the polarity of the target compound, different green compressed fluid technologies can be used. These techniques are supercritical fluid extraction (SFE), gas-expanded liquids (GXL), pressurized liquid extraction (PLE), and subcritical water extraction (SWE); they all demonstrated high efficiency and selectivity for the extraction of bioactive compounds. The order of application of these techniques can be varied depending on the expected composition of the matrix, the intensity of the conditions, and their effect on the sample during the sequential process [5].

In an interesting chapter, Srinivas and King (2018) [5] discussed the integrated platform of compressed fluid technologies such as supercritical fluid extraction or pressurized liquid extraction using ethanol or water. The application of multi-fluid processes, under controlled experimental conditions, permits not only extracting several value-added products from the same matrix but also fractionating them.

The use of $scCO_2$, depending on its density, allows extracting the non-polar solutes and when adding a co-solvent like ethanol, and also extraction of medium polar fractions such as flavonoids and carotenoids, followed by subcritical water extraction to isolate the highly polar and high molecular weight compounds such as sugars and proteins from the matrix. As previously mentioned by Herrero *et al.* (2017) in an opinion article on the properties and applications of GXL and switchable solvents (a switchable solvent can reversibly change the physical or chemical properties such as the dielectric constant, ionic strength by effect of an external trigger), GXLs can be included in a compressed fluid technologies platform [3]. In this sense, GXLs using carbon dioxide can be considered a switchable solvent, implying that the solvents can change their properties such as polarity and hydrophilicity (or hydrophobicity), by just varying the amount of CO_2. Besides, the above mentioned advantages such as low pressure (compared to SFE) and low solvent consumption (compared to PLE or SLE) make this technique really useful in a biorefinery process.

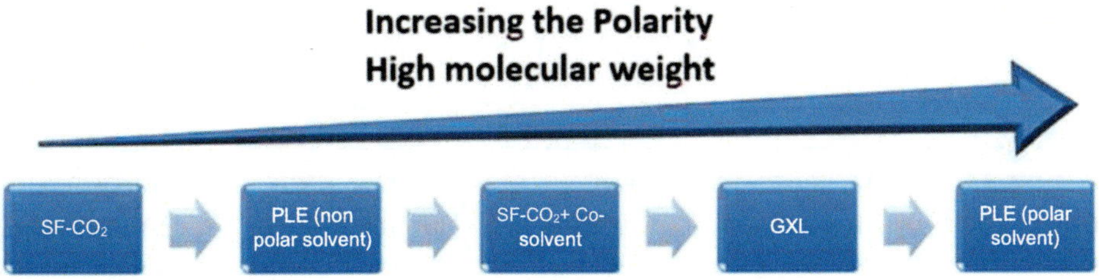

Fig. (6). Example Green Platform for Downstream processing using compressed fluids.

In terms of using GXL in a sequential and integrated extraction process, the number of publications found is still small; Fig. (7) *in vitro* represents the manuscripts published in this field. One example is the extraction of different groups of compounds from *Moringa oleifera* leaves [40]. A cascade approach was proposed, starting with the extraction of non-polar fraction using $scCO_2$. Then, CXE was performed to increase the polarity of the extract, and finally, the extraction of the most polar compounds by pressurized hot water extraction was implemented. An optimization of total recovery was carried out for each process. In addition, the antioxidant activity by TEAC assay was evaluated. In the $scCO_2$ and CXE processes, GC-MS analysis showed similar profiles, comparable in terms of qualitative and quantitative compositions, with a total of 17 and 20 compounds identified, respectively. It was observed that the total flavonoids was 2-folder higher in GXL extract than in SWE extract. Quercetin and kaempferol were the main flavonoids identified in the extracts by HPLC-ESI-QTOF-MS, which were predominantly extracted in the GXL step. The best antioxidant activity and the higher phenolic content were obtained in SWE step.

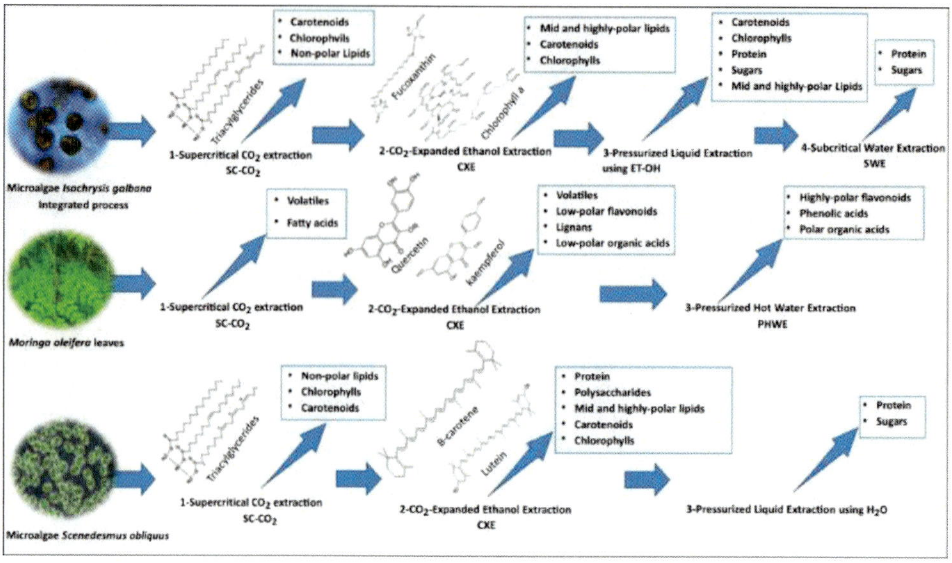

Fig. (7). Different multi-step sequential and integrated processes tested for the extraction of bioactives.

As we have seen in the previous section, microalgae are a promising source for high-value products, such as proteins, lipids, carotenoids, and carbohydrates. Therefore, the biorefinery concept is required, in accordance with the principles of Green Chemistry [18], to get a wide range of bioactive compounds. For instance, Gilbert *et al.* [39] reported a study on compressed fluids as a model process for the integrated biorefinery from *Isochrysis galbana*, a microalga well known in aquaculture. The extraction was divided into four consecutive steps to recover the maximum amount and variety of valuable compounds. This approach included the following steps: (1) SFE, (2) CXE using 75% ethanol, (3) pressurized ethanol extraction, and (4) PHWE. Results showed that SFE was more suitable for extracting triacylglycerides compared to CXE and PLE, while the medium polar compounds and polar lipids were obtained in the second and third steps. Moreover, in the CXE step, an efficient extraction of carotenoids and chlorophylls was obtained compared to acetone and methanol conventional extractions (146.58 vs. 57.19 mg per g extract and 96.56 vs. 44.48 mg per g extract, respectively, for carotenoids and chlorophylls). The most abundant carotenoid identified was fucoxanthin, which was highly extracted by CXE. The third and the fourth steps showed an important protein and carbohydrate recovery.

Another study, based on three sequential steps using scCO_2, CXE, and subcritical water to extract a wide range of bioactive molecules from *Scenedesmus oliquus* microalgae was published [41]. Results showed that triacylglycerides were mainly extracted in the first step (less polar), while the medium polar and polar lipids were detected in CXE extracts. It was concluded that for carotenoids, the best results were obtained using CXE; in this case, the main pigments extracted were lutein and β-carotene. It is worth mentioning that the content of lutein in the CXE extracts was 10-folder higher than its amount in the $scCO_2$ extracts. The second and the third step provide an enriched extract of sugars and proteins.

All the works mentioned above demonstrated the effectiveness of GXL for the extraction of carotenoids, chlorophylls, and flavonoids compared to neat $scCO_2$. In addition, the potential of GXL to extract medium polar substances has been used as an advantage in biorefinery platforms. Without any doubt, this type of platform using multi-compressed fluids, including GXLs, can be a step into the future of industrial applications in several fields such as pharmaceutical, food and agriculture, toward the greenness, sustainability, and the efficiently to fulfill the different challenges the society is facing.

CONCLUSION

In the present chapter, we have presented GXLs as a green and sustainable alternative for extracting valuable compounds from different natural raw materials such as plants, food wastes, and by-products, algae, *etc*. Fig. (**8**) summarizes the main features of GXL extraction, together with the selected applications. Advantages of the use of such technology have been discussed in terms of versatility of the liquid solvent and gases employed that can allow extraction of different polarity compounds and with different characteristics; moreover, environmental and economic costs are reduced since less amount of solvent is employed by a substantial replacement of the organic solvent with CO_2. On the other hand, GXLs can be created at relatively mild pressures compared to supercritical conditions and the different parameters can be easily modified to optimize the process in terms of yields and/or purity of the target compounds. Therefore, GXLs combine the beneficial properties of compressed gases (such as the improved mass transport) and of traditional solvents (large solvating power), leading to a new class of tunable solvents that can be tuned to become the ideal type of solvent for a given application. In this sense, although not many applications can be found in the literature, the technology is evolving targeting interesting compounds with biological activity to be employed in the food and cosmetics industries (antioxidant, antimicrobial). Moreover, the possibility of GXLs to be included in biorefinery platforms based on compressed fluids has been discussed since it is expected to become one of the main fields of application in the near future.

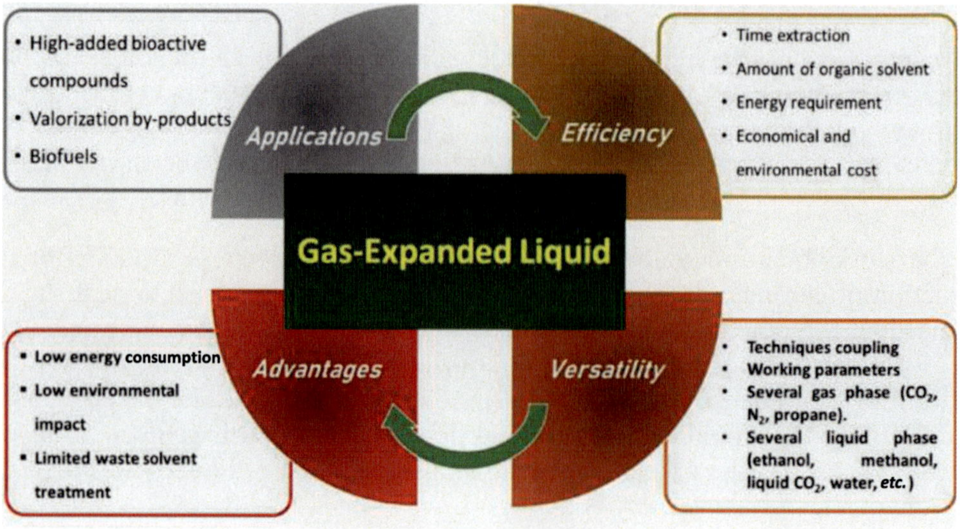

Fig. (8). Overview of the advantages and applications of GXLs.

FUTURE OUTLOOKS

Although we have tried to provide a global vision on the possibilities offered by GXLs in the extraction of high added value compounds from different sources, it is important to highlight that there is still a long way to go before implementation of these technologies becomes a reality in food analytical laboratories. It is crucial to understand that applications provided have been carried out (most of them) at laboratory scale but can set the basis for further operation at larger scale, therefore, conditions, equipment, and operation can serve as a starting point for new advances. In this sense, more studies are needed for a better understanding on the nature of GXLs and how the different factors can modify the properties of these solvents, and to take advantage of the versatility to implement new and greener methodologies in the analytical laboratories. Aspects such as the measurement of sustainability and eco-friendliness of a particular process are a goal that has to be approached through the application of different tools such as life-cycle analysis (LCA). Moreover, new applications and strategies are expected to be developed in the near future including the combination of GXLs with other extraction methodologies; for example, on-line or in sequential steps (as in compressed fluids platforms); intensification processes including ultrasounds, microwave or enzymes, testing of new bio-based solvents such as Natural Deep Eutectic Solvents (NADES) or ionic liquids (ILs), benefiting the integration of reaction and extraction processes into the same system, *etc*. Much more research is needed in this field that can help achieve some of the goals related to sustainability, focusing on the implementation of GAC principles in the laboratories and therefore addressing some important challenges the society is demanding.

LIST OF ABBREVIATIONS

CXE	Carbon Dioxide Expanded Ethanol
CXL	Carbon dioxide expanded-liquids
CXLLE	CO_2-expanded liquid-liquid extraction
CXM	Carbon Dioxide Expanded Methanol
CXW	Carbon Dioxide Expanded Water
DHA	Docosahexaenoic acid
EFL	Enhanced fluidity liquid
EHS	Environment, Health and Safety
E_T	Electronic Transition Energy
FFA	Nree fatty acids
GAC	Green Analytical Chemistry
GEMC	Gibbs Ensemble Monte Carlo
GLnA	γ-linolenic acid
GXL	Gas Expanded Liquids
GXLE	Gas Expanded Liquid Extraction
HD	Hydrodistillation
IVD	Indirect vapour distillation
LCA	Life Cycle Assessment
MD	Molecular dynamics
MeTHF	2-methyltetrahydrofuran
MIC	Minimal inhibition concentration
NL	Neutral lipids
PLE	Pressurized Liquid Extraction
PR-EOS	Peng Robinson Equation of Estate
RSM	Response surface methodology
scCO_2	Supercritical CO_2
SDGs	Sustainable Development Goals
SFE	Supercritical Fluid Extraction
SLE	Solid-liquid extraction
SWE	Subcritical Water Extraction
TEAC	Trolox equivalent antioxidant capacity

w/w	Weight per Weight
π*	Polarizability parameter

CONSENT FOR PUBLICATION

Not applicable.

CONFLICT OF INTEREST

The authors declare no conflict of interest, financial or otherwise.

ACKNOWLEDGEMENTS

Z.J.S.M. and N.A. would like to acknowledge Foodomics Laboratory (Spain). The research fellowships of Z.J.S.M. and N.A. have been financed by the University of Nariño (Colombia) and University Salah Boubnider (Algeria) respectively. This work was financed by INCGLO0019 project of Spanish Council for Higher Research (CSIC) and PID2020-113050RB-100 Ministerio de Ciencia e Innovación – Agencia Estatal de Investigación, España.

REFERENCES

[1] Ballesteros-Vivas D, Socas-Rodríguez B, Mendiola JA, Ibáñez E, Cifuentes A. Green food analysis: Current trends and perspectives. Curr Opin Green Sustain Chem 2021; 31: 100522. [http://dx.doi.org/10.1016/j.cogsc.2021.100522]

[2] Sánchez-Camargo ADP, Mendiola JA, Ibáñez E. Chapter 17: Gas Expanded-liquids. In: RSC Green Chemistry, vol. 2018, no. 57, Royal Society of Chemistry 2018: pp. 512-531

[3] Herrero M, Mendiola JA, Ibáñez E. Gas expanded liquids and switchable solvents. Curr Opin Green Sustain Chem 2017; 5: 24-30. [http://dx.doi.org/10.1016/j.cogsc.2017.03.008]

[4] Mendiola JA. Green Processes in Foodomics. Gas-Expanded Liquids Extraction of Bioactives. Compr Foodomics 2021; 2: 744-53. [http://dx.doi.org/10.1016/B978-0-08-100596-5.22927-2]

[5] Srinivas K, King J W. Chapter 6: Developments in the Processing of Foods and Natural Products Using Pressurized Fluids, RSC Green Chem., vol. 2018-January, no. 53, pp. 196-250, 2018.

[6] Jessop PG, Subramaniam B. Gas-Expanded Liquids. Chem Rev 2007; 107(6): 2666-94. [http://dx.doi.org/10.1021/cr040199o] [PMID: 17564482]

[7] Laird BB, Houndonougbo YA, Kuczera K. Phase equilibrium, structure, and transport properties of carbon-dioxide expanded liquids: A molecular simulation study ACS Symposium Series. vol. 1006: 41-65. [http://dx.doi.org/10.1021/bk-2009-1006.ch002]

[8] Abbott AP, Hope EG, Mistry R, Stuart AM. Probing the structure of gas expanded liquids using relative permittivity, density and polarity measurements. Green Chem 2009; 11(10): 1530-5. [http://dx.doi.org/10.1039/b915570h]

[9] Aida T, Aizawa T, Kanakubo M, Nanjo H. Relation between volume expansion and hydrogen bond networks for CO_2-alcohol mixtures at 40 °C. J Phys Chem B 2010; 114(43): 13628-36.

[http://dx.doi.org/10.1021/jp1017339] [PMID: 20936838]

[10] Reichardt C. Solvatochromic dyes as solvent polarity indicators. Chem Rev 1994; 94(8): 2319-58. [http://dx.doi.org/10.1021/cr00032a005]

[11] Kamlet MJ, Taft RW. The solvatochromic comparison method. I. The. β.-scale of solvent hydrogen-bond acceptor (HBA) basicities. J Am Chem Soc 1976; 98(2): 377-83. [http://dx.doi.org/10.1021/ja00418a009]

[12] Sherwood J, Granelli J, McElroy CR, Clark JH. A method of calculating the kamlet-abboud-taft solvatochromic parameters using COSMO-RS. Molecules 2019; 24(12): 2209. [http://dx.doi.org/10.3390/molecules24122209] [PMID: 31200457]

[13] Wyatt VT, Bush D, Lu J, Hallett JP, Liotta CL, Eckert CA. Determination of solvatochromic solvent parameters for the characterization of gas-expanded liquids. J Supercrit Fluids 2005; 36(1): 16-22. [http://dx.doi.org/10.1016/j.supflu.2005.03.009]

[14] Duereh A, Inomata H. Prediction of solvatochromic parameters of electronic transition energy for characterizing dipolarity/polarizability and hydrogen bonding donor interactions in binary solvent systems of liquid nonpolar-polar mixtures, CO_2-expanded liquids and supercritical carbon dioxide with cosolvent. J Mol Liq 2020; 320: 114394. [http://dx.doi.org/10.1016/j.molliq.2020.114394]

[15] Hunter SE, Savage PE. Quantifying rate enhancements for acid catalysis in CO_2-enriched high-temperature water. AIChE J 2008; 54(2): 516-28. [http://dx.doi.org/10.1002/aic.11392]

[16] Palafox-Hernandez JP, Mendis CH, Thompson WH, Laird BB. Pressure and Temperature Tuning of Gas-Expanded Liquid Structure and Dynamics. J Phys Chem B 2019; 123(13): 2915-24. [http://dx.doi.org/10.1021/acs.jpcb.8b09826] [PMID: 30848599]

[17] Granero-Fernandez E, Lacaze-Dufaure C, Condoret JS, Gerbaud V, Medina-Gonzalez Y. Controlling Solvation and Mass Transport Properties of Biobased Solvents through CO_2 Expansion: A Physicochemical and Molecular Modeling Study. Ind Eng Chem Res 2019; 58(41): 18942-64. [http://dx.doi.org/10.1021/acs.iecr.9b02218]

[18] Anastas PT, Warner JC. Green Chemistry: Theory and Practice Green Chem Theory Pract. New York: Oxford Univ. Press 1998; p. 30.

[19] Anastas PT, Zimmerman JB. Design through the 12 principles of green engineering. Environ Sci Technol 2003; 37(5): 94A-101A. [http://dx.doi.org/10.1021/es032373g] [PMID: 12666905]

[20] Tobiszewski M, Mechlińska A, Namieśnik J. Green analytical chemistry—theory and practice. Chem Soc Rev 2010; 39(8): 2869-78. [http://dx.doi.org/10.1039/b926439f] [PMID: 20502819]

[21] Cunico LP, Acosta MC, Turner C. Experimental measurements and modeling of curcumin solubility in CO_2 -expanded ethanol. J Supercrit Fluids 2017; 130: 381-8. [http://dx.doi.org/10.1016/j.supflu.2017.06.018]

[22] Cunico LP, Cobo AM, Al-Hamimi S, Turner C. Solubility and Thermal Degradation of Quercetin in CO_2-Expanded Liquids. Molecules 2020; 25(23): 5582. [http://dx.doi.org/10.3390/molecules25235582] [PMID: 33261120]

[23] Al-Hamimi S, Abellan Mayoral A, Cunico LP, Turner C. Carbon Dioxide Expanded Ethanol Extraction: Solubility and Extraction Kinetics of α-Pinene and cis-Verbenol. Anal Chem 2016; 88(8): 4336-45. [http://dx.doi.org/10.1021/acs.analchem.5b04534] [PMID: 27002237]

[24] Stefanis E, Panayiotou C. Prediction of hansen solubility parameters with a new group-contribution method. Int J Thermophys 2008; 29(2): 568-85. [http://dx.doi.org/10.1007/s10765-008-0415-z]

[25] Cunico LP, Sun M, Rui Y, *et al.* Enhanced distribution kinetics in liquid-liquid extraction by CO_2-expanded solvents. J Supercrit Fluids 2020; 163: 104874.
[http://dx.doi.org/10.1016/j.supflu.2020.104874]

[26] Santos DT, Vardanega R, Albarelli JQ, Ensinas AV, Maréchal F, Meireles MAA. Energy Consumption *Versus* Antioxidant Activity of Pressurized Fluid Extracts from *Pfaffia glomerata* roots. Chem Eng Trans 2013; 35: 1099-104.

[27] Costa ARM, Freitas LAP, Mendiola J, Ibáñez E. Copaifera langsdorffii supercritical fluid extraction: Chemical and functional characterization by LC/MS and *in vitro* assays. J Supercrit Fluids 2015; 100: 86-96.
[http://dx.doi.org/10.1016/j.supflu.2015.02.028]

[28] David A, Wang F, Sun X, *et al.* Chemical Composition, Antioxidant, and Antimicrobial Activities of *Vetiveria zizanioides* (L.) Nash Essential Oil Extracted by Carbon Dioxide Expanded Ethanol. Molecules 2019; 24(10): 1897.
[http://dx.doi.org/10.3390/molecules24101897] [PMID: 31108854]

[29] Sakai H, Ono K, Tokunaga S, Sharmin T, Aida TM, Mishima K. Extraction of natural pigments from gardenia Jasminoides J.Ellis fruit pulp using CO_2-expanded liquids and direct sonication. Separations 2020; 8(1): 1-14.
[http://dx.doi.org/10.3390/separations8010001]

[30] Golmakani MT, Mendiola JA, Rezaei K, Ibáñez E. Expanded ethanol with CO*2* and pressurized ethyl lactate to obtain fractions enriched in γ-Linolenic Acid from *Arthrospira platensis* (Spirulina). J Supercrit Fluids 2012; 62: 109-15.
[http://dx.doi.org/10.1016/j.supflu.2011.11.026]

[31] Reyes FA, Mendiola JA, Ibañez E, del Valle JM. Astaxanthin extraction from Haematococcus pluvialis using CO_2-expanded ethanol. J Supercrit Fluids 2014; 92: 75-83.
[http://dx.doi.org/10.1016/j.supflu.2014.05.013]

[32] Wang HC, Klinthong W, Yang YH, Tan CS. Continuous extraction of lipids from *Schizochytrium* sp. by CO_2-expanded ethanol. Bioresour Technol 2015; 189: 162-8.
[http://dx.doi.org/10.1016/j.biortech.2015.04.011] [PMID: 25879184]

[33] Li PL, Li HN, Jing KJ, David A, Lin JR, Deng G. Evaluation of lipid extraction from microalgae based on different phase regions of CO_2-expanded ethanol. Chem Eng Process 2019; 138: 1-6.
[http://dx.doi.org/10.1016/j.cep.2019.02.011]

[34] Yang YH, Klinthong W, Tan CS. Optimization of continuous lipid extraction from *Chlorella vulgaris* by CO_2-expanded methanol for biodiesel production. Bioresour Technol 2015; 198: 550-6.
[http://dx.doi.org/10.1016/j.biortech.2015.09.076] [PMID: 26433151]

[35] Paudel A, Jessop MJ, Stubbins SH, Champagne P, Jessop PG. Extraction of lipids from microalgae using CO_2 -expanded methanol and liquid CO_2. Bioresour Technol 2015; 184: 286-90.
[http://dx.doi.org/10.1016/j.biortech.2014.11.111] [PMID: 25537138]

[36] Santos MP, Souza MC, Sumere BR, *et al.* Extraction of bioactive compounds from pomegranate peel (*Punica granatum* L.) with pressurized liquids assisted by ultrasound combined with an expansion gas. Ultrason Sonochem 2019; 54: 11-7.
[http://dx.doi.org/10.1016/j.ultsonch.2019.02.021] [PMID: 30827900]

[37] Chhouk K, Uemori C, Wahyudiono, Kanda H, Goto M. Extraction of phenolic compounds and antioxidant activity from garlic husk using carbon dioxide expanded ethanol. Chem Eng Process 2017; 117: 113-9.
[http://dx.doi.org/10.1016/j.cep.2017.03.023]

[38] Hoang HN, Nagashima Y, Mori S, Kagechika H, Matsuda T. CO_2-expanded bio-based liquids as novel solvents for enantioselective biocatalysis. Tetrahedron 2017; 73(20): 2984-9.
[http://dx.doi.org/10.1016/j.tet.2017.04.024]

[39] Gilbert-López B, Mendiola JA, Fontecha J, *et al.* Downstream processing of *Isochrysis galbana*: a step towards microalgal biorefinery. Green Chem 2015; 17(9): 4599-609. [http://dx.doi.org/10.1039/C5GC01256B]

[40] Rodríguez-Pérez C, Mendiola JA, Quirantes-Piné R, Ibáñez E, Segura-Carretero A. Green downstream processing using supercritical carbon dioxide, CO_2-expanded ethanol and pressurized hot water extractions for recovering bioactive compounds from *Moringa oleifera* leaves. J Supercrit Fluids 2016; 116: 90-100. [http://dx.doi.org/10.1016/j.supflu.2016.05.009]

[41] Gilbert-López B, Mendiola JA, van den Broek LAM, *et al.* Green compressed fluid technologies for downstream processing of *Scenedesmus obliquus* in a biorefinery approach. Algal Res 2017; 24: 111-21. [http://dx.doi.org/10.1016/j.algal.2017.03.011]

CHAPTER 9

Pressurized Liquid Extraction (PLE)

Natalie Álvarez Alarcon[1], **Juan S. Ortiz-Ramírez**[2], **Jenny P. Ortega-Barbosa**[1], **Diego Ballesteros-Vivas**[3], **Luis I. Rodríguez-Varela**[2] and **Fabián Parada Alfonso**[1,*]

[1] *Food Chemistry Research Group, Departamento de Química, Facultad de Ciencias, Universidad Nacional de Colombia, Bogotá, Colombia*

[2] *Food Chemistry Research Group, Departamento de Ingeniería Química y Ambiental, Facultad de Ingeniería, Universidad Nacional de Colombia, Bogotá, Colombia*

[3] *Departamento de Nutrición y Bioquímica, Facultad de Ciencias, Pontificia Universidad Javeriana, Bogotá, Colombia*

Abstract: Pressurized liquid extraction (PLE) is regarded as an emergent extraction technique; it is an appropriate tool to obtain green extracts from foods or related samples. Studies on the content of contaminants in foods or food raw materials can be carried out by PLE. In the same way, studies on the obtention of bioactive extracts from classic and emerging foods and their by-products can be carried out by PLE too. Besides sequential process combinations of PLE with other innovative extraction techniques could generate benefits for the food industry. The objective of this chapter is to clearly define the role that this technique plays in food analysis, as well as the updated spectrum of some of its applications during the last lustrum.

Keywords: Bioactive extracts, Food analysis, Green extraction, Instrumental analysis techniques, Pressurized liquid extraction.

INTRODUCTION

The pressurized liquid extraction (PLE) technique was developed in 1995 by the Dionex Corporation under the name Accelerated solvent extraction (ASE®). It is also known as pressurized hot solvent extraction (PHSE) and if it uses water as the solvent, it can also be called subcritical water extraction (SWE) or pressurized hot water extraction (PHWE) [1]. PLE is used in chemical analysis to undertake exhaustive extractions from solid samples in short periods of time with small amounts of solvent [2]. With the growth of green chemistry, the use of environmentally friendly solvents under PLE allows the recovery of harmless

* **Corresponding Author Fabián Parada Alfonso:** Food Chemistry Research Group, Departamento de Química, Facultad de Ciencias, Universidad Nacional de Colombia, Bogotá, Colombia; Ext. 14448, Tel: +57-601-3165000; Fax: +57-601-3165220; E-mail: fparadaa@unal.edu.co

Merichel Plaza & Maria Luisa Marina (Eds.)

extracts and extraction by-products [3]. Fig. (**1**) shows some basic applications of PLE in food analysis, classic and emerging foods, or their by-products. These matrices can be subjected to a PLE to obtain an enriched extract with the contaminants, or the bioactive compounds from the initial sample.

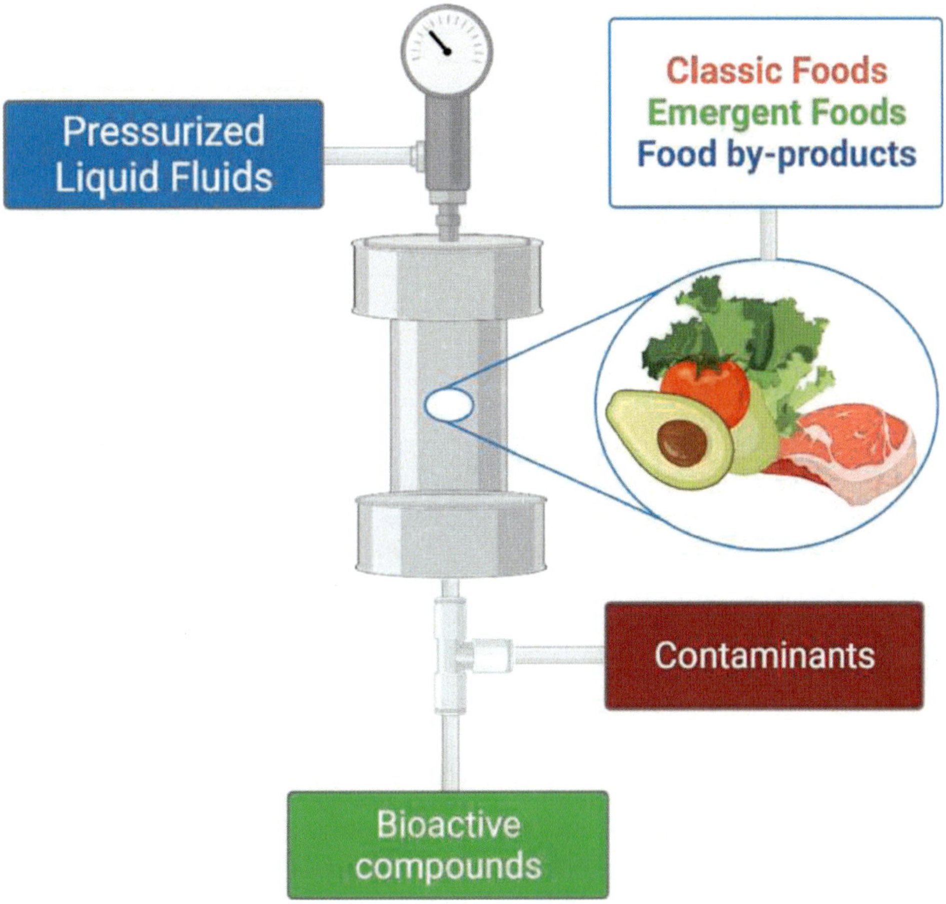

Fig. (1). PLE basic applications in food analysis. *Created with BioRender.com.*

PLE consists of exposing the sample for extraction to pressurized solvents (pressures lower than the critical pressure) and temperatures higher than those of their boiling point, thus ensuring that the solvent remains in its liquid phase [4]. This technique was initially employed in the analysis of soils contaminated with organic compounds (PCBs, PHAs) [5]. The Environmental Protection Agency (EPA) of the U.S. government has now established the use of PLE in the analysis of some contaminants in soil samples [6]. More recently the use of PLE has been growing relative to other emerging extraction techniques such as supercritical fluid extraction (SFE). As shown in Fig. (**2**), two decades ago the relationship of works that "mentioned SFE" topic to those that "mentioned PLE" topic was 2:1 approximately. Nowadays that relationship is close to 1:1. At this time, the works that mention the "PLE" topic has increased from 293 to 2414.

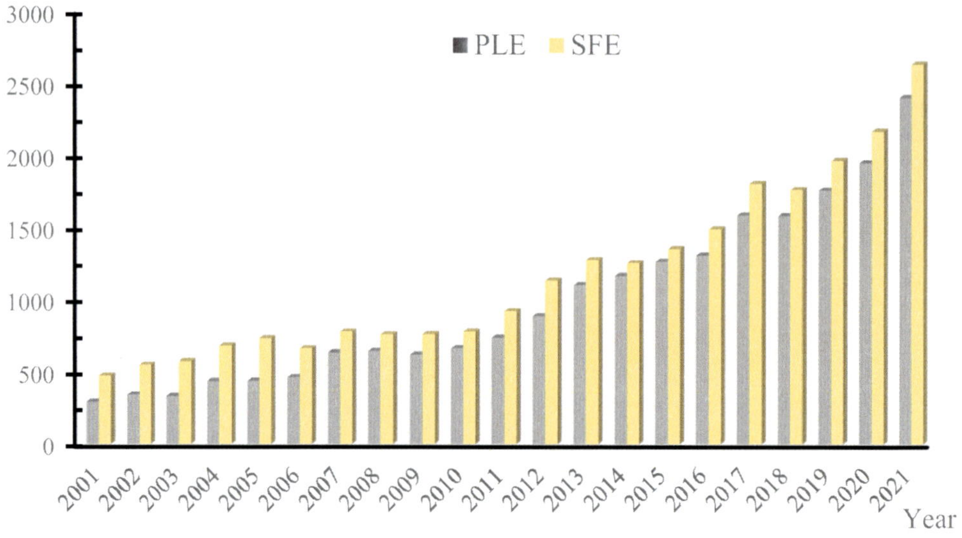

Fig. (2). Ratio of works that mention the PLE topic vs the SFE topic.

The main uses of PLE in food analysis consist of obtaining extracts from biomasses commonly used for food preparation, emerging biomasses, or their corresponding by-products. Different applications can be given to PLE using green solvents [3, 7]: Food valorization using green techniques [8], recovery of bioactive compounds or extracts from biomasses [9], obtention of antioxidants from agro-industrial by-products [10], isolation of organic contaminants from food samples or food used materials [11, 12], as is the case of ethylene polymers content [13, 14].

In this chapter, we will deal with the basic principles of PLE and its experimental conditions as well as the equipment and the extraction methods used in PLE. We will consider the combination of PLE with other extraction and analysis techniques, and we will finish by reviewing the use of PLE to obtain extracts from traditional or promising foods and from their by-products as well as from inputs used in the food industry, for example, the raw packaging materials.

PROCEDURES TO FOLLOW IN PLE - REQUIRED EQUIPMENT AND EXTRACTION CONDITIONS

The sample to analyze under PLE must be subject to the normal preparation processes, not different from those under other extraction techniques. These processes may consist of drying up the sample and grinding it to reduce particle sizes. Also, the extractant phase to be used is usually degasified to displace the remaining air in the solvent.

The following steps need to be followed for the analysis of samples under PLE:

a. Place the sample in the extraction cell. To reduce the dead volume in the cell some inert material such as sand, fiberglass, or glass beads could be added to the sample.
b. Fill the cell up with the solvent, heat it, and pressurize it. The cell can be heated up before or after filling it up completely with the solvent. Given that the extraction temperature (T_{ext}) is an important variable in PLE, the extraction system should be fitted with a thermostat to control the temperature for a constant extraction pressure (P_{ext}) and for a given extraction time (t_{ext}). The P_{ext} must permit the solvent is in the liquid phase.
c. Undertake the extraction. If the extraction is done in the static mode, the analytes migrate from the sample to the extractant phase given their solubility through diffusion. This happens at T_{ext} and P_{ext} for a given t_{ext}. If the extraction is done in the dynamic mode of the system, the extraction cell having been filled up with the solvent is pressurized and heated up until T_{ext} is reached. The system is fed continuously with fresh extractant phase and at a constant flow once the T_{ext} and P_{ext} are reached.
d. Collect the extract until the extraction is exhausted. In the static mode, once t_{ext} has elapsed, the cell is depressurized and is washed with a fresh solvent and purged with an inert gas (N_2) to achieve an exhaustive extraction. In some instances, it will be necessary to undertake several processes of static extraction to achieve exhaustive extraction. The quantity of solvent to use in the washing process corresponds to 60-70% of the empty cell capacity. The purging with N_2 will let the extraction product be free of solvent. In the dynamic mode, once T_{ext} and P_{ext} are reached, the entry and exit valves of the cell are partially opened and fresh solvent starts to be fed continuously and at a constant flow during the extraction period (the dynamic extraction time, $t_{d\text{-}ext}$). In this case, larger volumes than in the static mode of the extractant phase are necessary to ensure an exhaustive extraction. The dynamic extraction mode facilitates the work of PLE with other extraction techniques which may offer greater selectivity or with instrumental analysis techniques (*e.g.*, gas

chromatography coupled with mass-spectrometry, GC-MS; high performance liquid chromatography with diode-array detector, HPLC-DAD; HPLC coupled with mass spectrometry, HPLC-MS).

Some extraction procedures, such as static extraction and dynamic extraction, could be improved if they were performed in combination. In these cases, it is recommended to optimize the t_{ext} and $t_{d\text{-}ext}$. The response variables to use in the optimization could be the yield of the extract or the quality of the extract. The latter could be measured in terms of a higher content of a certain family of components (terpenoid, phenolic compounds, *etc.*) or in terms of a certain biological activity (antioxidant activity, larvicidal activity, cytotoxic activity, *etc.*). It is important to bear in mind that since the PLE is a technique under which exhaustive extractions can be achieved, the extracts delivered in this way may contain components of a different nature. Under these circumstances, the quality of the extract may not be commensurate with its performance.

Required Equipment into PLE

PLE can be undertaken in static or dynamic mode, the basic equipment can see in Fig. (**3**). In the static mode, a given volume of solvent is placed into contact with the sample for a given time, thus allowing solvent saturation. The basic equipment for a PLE system in static mode consists of the vessel for the solvent, a pump for liquids (usually the one found in a liquid chromatography system), a preheating spiral, an extraction cell, a pressure control valve, and a collection vessel for the extract. It should be necessary to exhaust the sample and several static extractions could be undertaken on the same sample. In the dynamic mode, the solvent flows continuously through the extraction cell which contains the sample. In this instance, the relationship between sample mass and solvent flow influences the results of the analysis. In the dynamic mode, the constant flow of fresh solvent allows for several material transfer equilibria and prevents saturation. The choice of one mode over the other depends on what is being analyzed and the properties of the equipment available.

It must be remembered that exhaustive extractions are achieved under PLE, hence extraction selectivity is limited. Not only high yield extracts are obtained, but also these extracts can have a wide polarity range [10]. Therefore, this technique is recommended for obtaining raw extracts with components having a wide polarity range. Finally, PLE has been a green technique given that it meets some of the twelve principles of Green Chemistry [15], to what has been called "the principles of Green Extraction" [16], Green extraction embraces the following: (i) the use of renewable resources in a responsible manner; (ii) the use of green solvents (water, ethanol-EtOH, and CO_2 among others) as an alternative to substitute harmful organic solvents (*e.g.* benzene, toluene, xylene, chloroform, methanol-MeOH); (iii) saving on energy consumption by intensifying processes; (iv) the use of waste products from current industrial processes and the generation of by-products from them; and (v) obtaining undenatured extracts, harmless and solvent-free, with the potential of industrial use.

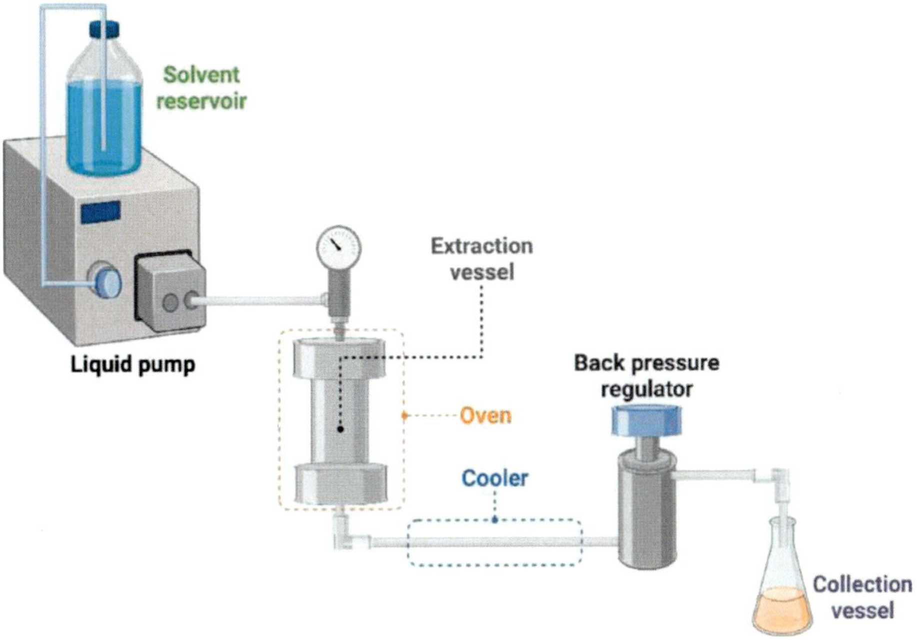

Fig. (3). Basic Equipment for a PLE system [1]. *Created with BioRender.com.*

The Effect of the Extraction Conditions on the Composition of Extracts

According to Mustafa and Turner [17], the parameters that influence the extraction process are the nature of the solvent, the extraction conditions in particular, the extraction temperature, the matrix characteristics (the interaction between the matrix and the analytes), and the extraction mode used.

As mentioned earlier, PLE consists of undertaking extractions with pressurized solvents in the liquid phase and at temperatures above boiling point and below the critical temperature. It is because of this those exhaustive extractions can be undertaken in shorter periods and smaller volumes of solvent. Normally it is possible to obtain high yielding extraction and low selectivity in terms of the type of the extract components [10].

The Nature of the Solvent

A critical factor in the PLE technique is the nature of the solvent as well as the pre-treatment that is given to it. The nature of the solvent will largely determine the selectivity of the process [18] –like its dissolving nature, the partition constants of each analyte will determine its equilibrium, see Fig. (**4**). Also, the extractant phase to be used must be degassed. This frees up the solvent of any air, thus reducing or preventing the oxidation of the analytes in the extraction process.

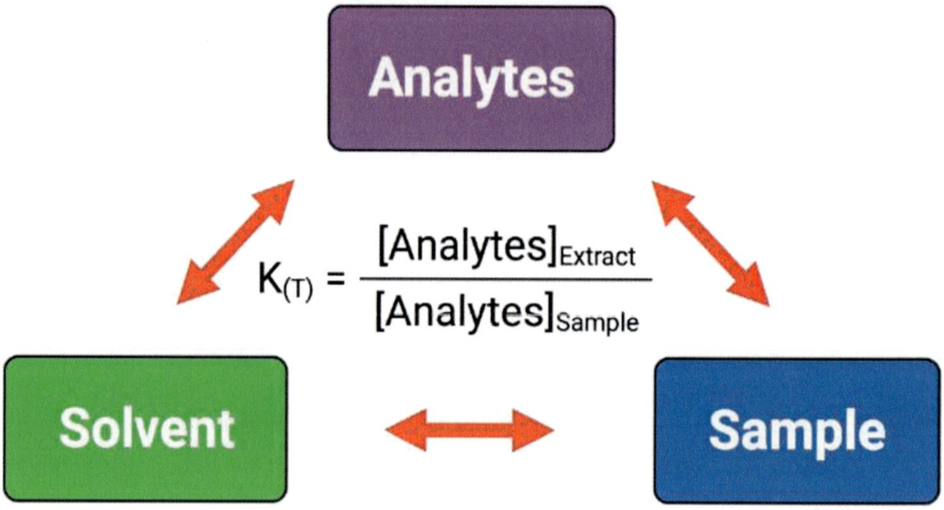

Fig. (4). Interaction between the analytes, the extractant phase, and the matrix. *Created with BioRender.com.*

The solvent commonly used in PLE is water, followed by hydroalcoholic mixtures, EtOH, esters (ethyl acetate and ethyl lactate), and limonene among others. These solvents have been classified as generally recognized as safe (GRAS) because they do not represent any risk to the environment or human health. Indeed, some of them can be obtained through chemical or biochemical transformation [16]. It is in this sense that PLE is acknowledged as a "green" extraction technique within the framework of sustainable chemistry [19].

By way of compliment to what has been said above, some esters (*e.g.*, ethyl acetate or ethyl lactate) and terpenes (*e.g.*, d-limonene) are used to extract substances of low polarity and even non-polar compounds [20]. Tables **2** and **3** summarise some PLE applications, where the most used solvents in works on classic and emerging foods, and their by-products can be seen.

Extraction Temperature (T_{ext})

In PLE, temperature is perhaps one of the most important factors since it determines the energy needed to overcome the analyte-matrix interactions (van der Waals forces, hydrogen bonding, and dipole attractions) and analyte-analyte interactions. These interactions generate the activation energy needed to start the desorption process. A high temperature reduces surface tension, increases viscosity, and improves solvent diffusivity, thus making it easier for the solvent to penetrate the sample particles and making the transportation of mass faster and more efficient [21]. All this translates into shorter extraction periods and reduced solvent use. However, the simultaneous extraction of components with different polarities makes the whole process less selective. Also, the high temperature can decompose thermolabile compounds, indeed a critical outcome in the extraction process. Pressure, on the other hand, keeps the solvent in its liquid phase despite high temperatures. The use of moderate pressure levels in PLE facilitates the penetration of the extractant phase into the structure of the matrix, thus favoring the transportation of the mass given the complete moistening of the sample. However, it has been observed that pressure does not significantly affect the yield or the selectivity of the extraction [17].

Water, a universal solvent, has been used to extract from natural sources compounds of different polarity, these sources being food products and their by-products. Indeed, the underlying reason for the use of water is because of the variations that the dielectric constant is exposed to as a function of temperature. In so far as temperature increases, water polarity diminishes, thus water being comparable to organic solvents at room temperature [22]. Fig. (**5**) shows variations in the dielectric constant of the water and of EtOH as a function of temperature and compares its polarity with that of some organic solvents. Therefore, the use of water in its liquid phase at high temperatures will allow the dissolution of hydrophobic or non-polar compounds. This is in total contrast to the property that water has in normal conditions to dissolve hydrophilic or polar compounds. This condition arises because as the temperature rises the intermolecular interactions diminish, which arise from hydrogen bonding with water molecules as well as from water molecules with those of the surrounding compounds, thus favoring the London dispersion forces (dipole-dipole forces).

On the other hand, EtOH is a versatile substance given the possibility to adjust some of its physical properties concerning pressure and temperature. For example, the Kamlet-Taft solvatochromic parameters (dipolarity/polarizability, acidity, and basicity) of EtOH decrease simultaneously as a function of temperature, its polarity diminishes together with its hydrogen bond donation and acceptor abilities [23]. Density also affects these parameters, dropping directly with temperature and increasing with pressure. This behavior implies that in a selective extraction process ethanol in the liquid phase can dissolve compounds of different polarities. Provided such solvent is subject to changes in the extraction temperature, it will become like non-polar compounds at high temperatures and like polar compounds at room temperature.

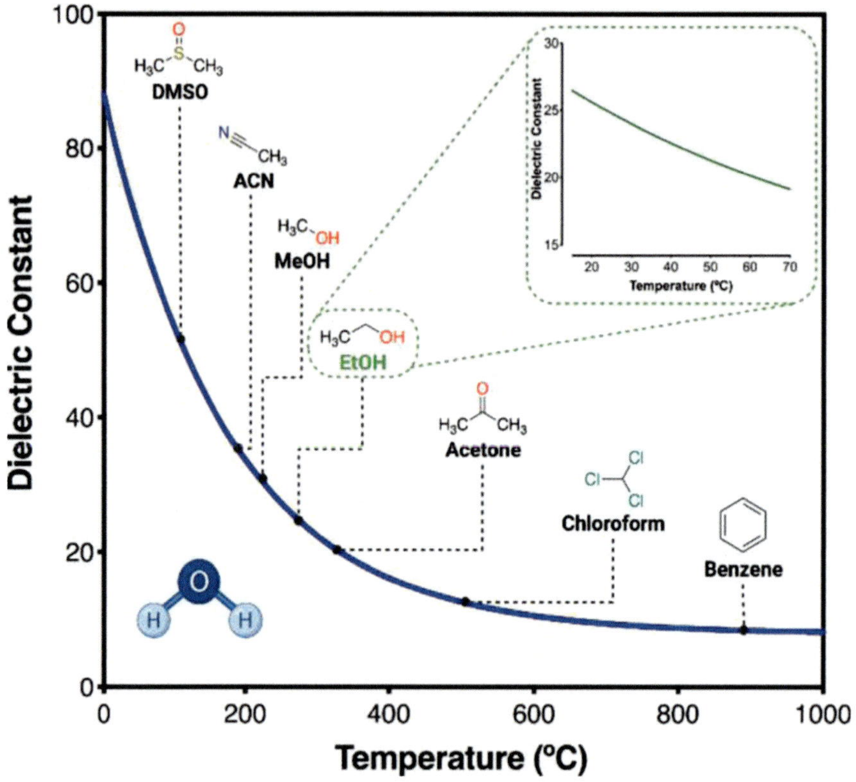

Fig. (5). Variation of the dielectric constants of water and EtOH as a function of temperature [24 - 26]. *Created with BioRender.com.*

Extraction Mode

The relationship between the required amount of solvent and the quantity of sample to be analyzed depends on the type of extraction to be undertaken. If the extraction in static mode is used (see Fig. **6**), the extract will consist of the components that will migrate from the sample to the extractant phase during the static extraction time (t_{ext}) at temperature (T_{ext}) and extraction pressure (P_{ext}), and the components that have been dragged during the washing of the extraction cell with solvent and the purge with inert gas. When standard cells of 11 mL capacity, which can accommodate samples of between 0.5 g to 5.0 g, are used, the amount of solvent per extraction batch required will normally be 20 mL.

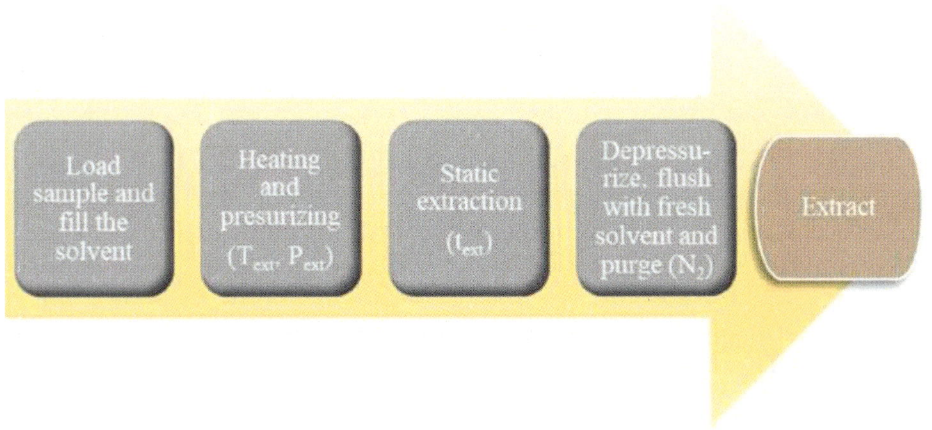

Fig. (6). Associated steps to static mode of PLE.

In a batch for extraction in the dynamic mode, using standard extractions cells, once the solvent is in contact with the sample at the suitable extraction conditions (T_{ext} and P_{ext}), the flow of fresh solvent is allowed towards the collection vial which could be at the rate of 1 to 2 mL/min. The total volume of the extract will depend on the solvent flow and the dynamic extraction time ($t_{d\text{-}ext}$) – the time during which fresh solvent flows towards the collection vial. Normally, for the same sample quantity, the volume of solvent used in the dynamic mode will be twice as much as that used in the static mode. Fig. (7) shows the steps to be taken during a dynamic-mode extraction.

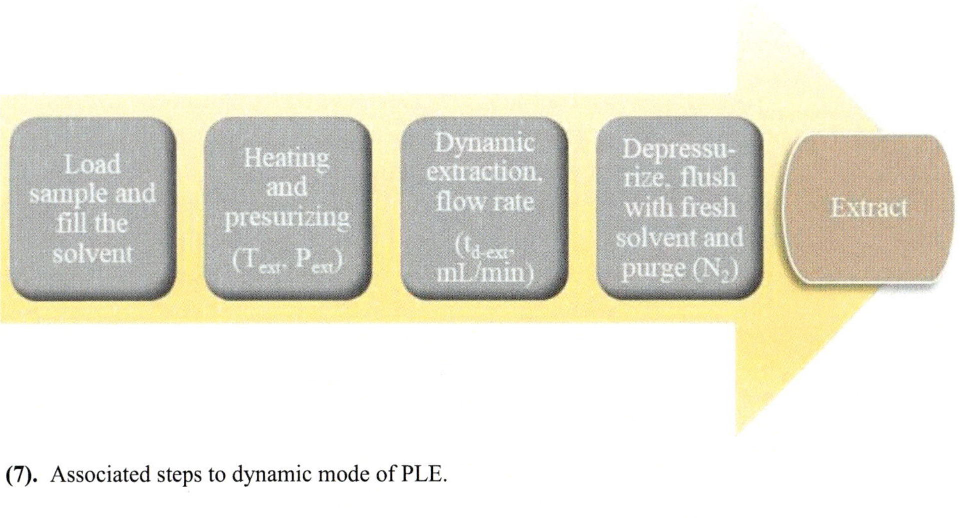

Fig. (7). Associated steps to dynamic mode of PLE.

It is possible to choose an extraction process that would alternate the extraction modes between static and dynamic to get a better extraction yield and deliver a better-quality extract.

Extraction Time (t_{ext})

The extraction time (t_{ext}) in PLE is shorter (5 to 20 min approx.) than those under classic techniques like Soxhlet extraction or maceration (3 to 10 h). In the static mode, with one single extraction batch, the concentration of each analyte will depend on the corresponding participation coefficient between the solvent and the matrix. For unknown samples, it is advisable to undertake a kinetic study to evaluate the required t_{ext} to achieve the equilibrium to deliver the extract of interest. The t_{ext} in the dynamic mode, in which the fresh solvent flows continuously to the extraction cell containing the sample, the extraction is more efficient, and the required extraction time would tend to be slightly longer than in the static mode. In the dynamic mode, the t_{ext} will affect the concentration of the analytes in the extract because of the dilution factor. This will increase as a larger volume of solvent is used in an extraction batch.

Matrix Characteristics

Extraction efficiency heavily depends on the chemical nature of the analytes (structure, size, and polarity) and their interaction with the matrix. In so far as the nature of the analytes is concerned, one of the basic principles of extraction techniques prevails – “like dissolves like”. Regarding the analyte-matrix interaction, assuming that the matrix is made up of a group of semispherical-porous particles, the compounds could be found: (i) over the surface of the matrix which would make them easily available to the extract; (ii) absorbed and/or dissolved into the micro or nanopores of the matrix which would require increased energy; and/or (iii) chemically bound in the matrix. In this latter instance, the extraction process should be more aggressive to free up the compounds from the matrix. Alternatively, the sample should previously be subject to chemical processes to break up the chemical analyte-matrix bonding. To achieve a good extraction, the analyte-matrix interaction must be overcome by the energy provided in the extraction process so that the analyte-solvent may prevail [27], the strength of this interaction depends on the affinity between the analytes and the solvent [10]. Those compounds that are found on the surface of the matrix or weakly bound to it will be transferred with greater ease by a process of controlled solubility. Instead, those compounds that interact with greater intensity with the matrix and/or are found in the porous structure of the latter are slowly dissolved and transferred to the extractant phase by controlled diffusion. These phenomena are of great interest since the extraction of the target compounds can be undertaken as a function of time. Therefore, it is important to have exhaustive extraction processes such as PLE [17].

The size of the particle in the sample is an important factor that may affect the t_{ext} and the extraction efficiency. If the extraction efficiency depends on the diffusion velocity of the analytes a smaller size particle would benefit the extraction process. However, the particle size must be appropriate to increase the contact surface evading the particle agglomeration and others channeling effects [10].

PLE APPLICATIONS INTO FOOD ANALYSIS

Investigations under PLE have been developed to study contaminants agents present in foods or food raw materials, and other studies have been developed to study traditional and emergent foods and their by-products. These topics will be treated in the next paragraphs. Finally, emphasis will be placed on some works where intensive or integrated extraction processes were carried out, in which PLE was used, to achieve better results, either obtaining higher yields or extracts with different characteristics.

Studies on Contaminants in Food or Food Raw Materials

The first PLE application was contaminant recovery from soils, it was the classic work of Richter *et al.* published at the end of the last century [5]. However, contaminant studies in foods using PLE have been made not very. Some works about contaminants in food have been published, they are shown in Table **1**. Studies have been conducted on the content of compounds harmful to health using PLE in food and food packaging. On the other hand, pesticide residues in different foods have also been determined using PLE. Finally, this extraction technique has been used to detect the exposure to contaminants from some living beings using the human placenta and bivalves.

Table 1. Studies on Contaminants in Food and Food Raw Materials.

Analyte	Matrix	Solvent (v:v)	Sample Preparation	T_{ext} (°C)	P_{ext} (MPa)	t_{ext} (min)	Refs.
PAHs	Smoked bacon	Water/ACN (8:2), 0.1% formic acid	Homogenized samples (5.0 g), diatomaceous earth (DE) (2.0 g)	75	1.9	St.: 0.17-0.25	[28]

(Table 1) cont.....

Analyte	Matrix	Solvent (v:v)	Sample Preparation	T_{ext} (°C)	P_{ext} (MPa)	t_{ext} (min)	Refs.
Polyethylene, polypropylene, polyethylene, terephthalate	Rice	DCM	0.1 g of sample	180	10	St.: 5	[29]
BFRs	Feed samples (fish oil, meal) and food samples (fish, milk, eggs, muscle, sheep liver, crustaceans)	Toluene/Acetone (7:3)	1.0 g of fatty extract	120	10	St.: 5 Cycles: 3	[30]
BFRs, PBDEs, emerging FRs	Ten species of fish and shellfish	Hexane/DCM (1:1)	1.5 g of lyophilized sample	100	10.34	St.: 10 Cycles: 2	[31]
MOSH and MOAH	Unprinted and printed recycled cardboards, printed recycled, virgin cardboards, paper bags used for bread, baking paper, baking cups	Hexane, EtOH	1.0 g of cardboard, 0.2-0.5 g paper samples	30-70	10	St.: 5-10 Cycles: 1-3	[32]
PRs	Residue tuber: sweet potato (*I. batatas*), taro (*C. esculenta*) and yam (*D. alata*)	EtOAC, ACN	3 g sample, 0.5% acetic acid, mixed with DE (1:1)	100-130	9.7	St.: 3-6	[33]
PRs	Green tea residues	DCM, water	2 g of homogenized tea sample and 5 g DE	100	14	St.: 5	[34]
PRs	Sediment (*S*), macrophytes (*Mphy*) (*U. spp.*, *Z. noltii*), aquatic animals (*Aq.an.*) (*N. diversicolor*, *S. plana*)	MeOH	1.0 g of *S*, 0.1 g of *Mphy*, 0.5 g of *Aq.an.*	40 (*Mphy*, *Aq.an.*) 80 (*S*)	10	St.: 3 Cycles: 3	[35]
OCPs and POPs	Fish tissue	Hexane/DCM (7:3)	1.5 g DE, 20-30 g of adsorbent, a 1:2 w/w mixture of sample-DE	100	10	St.: 5 Cycles: 2-3	[36]

(Table 1) cont.....

Analyte	Matrix	Solvent (v:v)	Sample Preparation	T_{ext} (°C)	P_{ext} (MPa)	t_{ext} (min)	Refs.
OPPs, OCPs, PAHs, PCBs, PBDEs, PYRs	Human placenta	ACN	1.0 g of homogenized placenta	100	10	St.: 5 Cycles: 3	[37]
Alkylphenols, Nonylphenol, Bisphenol A	Wild mussels (*M. galloprovincialis*)	MeOH	0.5 g of sample, 3 g of neutral alumina	40	–	–	[38]

PAHs: Polycyclic aromatic hydrocarbons; BFRs: Brominated flame retardants; PBDEs: Polybrominated diphenyl ether; MOSH: Mineral oils saturated hydrocarbons; MOAH: Mineral oils aromatic hydrocarbons; PRs: Pesticide residues; OCPs: Organochlorine pesticides; POPs: Persistent organic pollutants; OPPs: Organophosphorus pesticides; PCBs: Polychlorinated biphenyls; PYRs: Pyrethroid pesticides; *S*: Sediments; *Mphy*: Macrophytes; *Aq.an.*: Aquatic animals; DCM: Dichloromethane; EtOAc: Ethyl acetate; ACN: Acetonitrile; DE: Diatomaceous earth; St.: Static time.

A novel PLE procedure based on a hard cap espresso machine followed by liquid chromatography (LC) coupled with tandem mass spectrometry (LC-MS/MS) was developed to determine PAHs in smoked bacon by Merlo *et al.* (2021). The method was applied to real samples providing to be a fast and cheap approach for the efficient and reliable determination of PAHs in smoked bacon [28]. Dessì *et al.* investigated the content of some plastics in store-bought rice. Polyethylene, polyethylene terephthalate, poly-(methyl methacrylate), polypropylene, polystyrene, and polyvinyl chloride were quantified using PLE coupled to double-shot pyrolysis gas chromatography/mass spectrometry. Instant (pre-cooked) rice contained fourfold higher levels of plastics, suggesting that industrial processing potentially increases contamination [29]. Brominated Flame Retardants (BFRs) are contaminants that may enter the food chain where they mainly occur in food of animal origin, these compounds are widely used for industrial purposes, they reduce the flammability of plastics used for food packing. A PLE method able to quantify the widest range of BFRs in feed and food items was developed by Bichon *et al.*, they obtained PLE extracts using organic solvents, then the obtained extracts were analyzed by gas and liquid chromatography coupled to high resolution and tandem mass spectrometry [30]. The BFR concentrations were determined in some species of fish and shellfish widely consumed, using PLE followed by gas chromatography coupled to tandem mass spectrometry. The BFR concentrations found in fish and shellfish did not suggest significant health risks for the consumers [31]. A PLE method, followed by online LC-GC analysis, was optimized for the rapid determination of mineral oil in cardboard and paper samples. This method is useful to detected mineral oil in different cardboard and paper materials intended for contact with foods [32].

There are some dedicated manuscripts to pesticide studies using PLE. Khan *et al.* developed a multi-residue method for simultaneous analysis of a diverse range of pesticides in tuber crops, based on PLE using ethyl acetate as the solvent, followed by selective identification and quantification of the residues using gas chromatography coupled with mass spectrometry (GC-MS) selected reaction monitoring. The method was evaluated for 150 pesticides. Results showed that their limits of quantification were 0.0-0.1 ng/g, with recoveries of 70-120%. When compared to the conventional analytical techniques, such as QuEChERS and buffered ethyl acetate extraction, this method provided superior performance in terms of precision, and recovery of the spiked and incurred residues with similar productivity. The method could be applied to commercial products and regulatory residue analysis [33]. Chen *et al.* developed a methodology to study multiple pesticides from green tea into hot water *via* PLE. This study helped in the risk assessment of pesticide residues and the formulation of maximum residue levels (MRLs) in tea and its by-products [34]. Rodrigues *et al.* reported a fast, simple, sensitive, and selective analytical methodology for multi-residue determination of pesticides (atrazine, azoxystrobin, bentazon, -cyhalothrin, penoxsulam, and terbuthylazine) in sediment, macrophytes (algae and aquatic plants) and aquatic animals. They developed and validated it. The established methods were based on selective PLE followed by online solid phase extraction and ultra-performance liquid chromatography with electrospray ionization tandem mass spectrometry detection (on-line SPE-UPLC-ESI-MS/MS). The established methodology was applied on real marine samples and no pesticide concentrations above their respective method quantification limits were measured in sediments or aquatic plants [35]. A rapid automated extraction and cleanup method using selective PLE was developed and validated for 14 organochlorine pesticides in fish. The proposed method met the most important requirements of an extraction and cleanup procedure, including having a short preparation time (cleanup and concentration) and minimal sample contamination and being able to be automated [36]. The early exposure to organic pollutants-OPs (them like polychlorinated biphenyl-PCBs, polybromodiphenyl ethers-PBDEs, pyrethroids, and polycyclic aromatic hydrocarbons-PAHs) related to dietary habit or environmental exposure is a concern of growing interest in environmental health. Fernández *et al.* published relevant analytical methods for determining OPs in the human placenta. Clean-up of extracts was performed by solid-phase extraction (SPE) using EZ-POP cartridges. Detection of OPs was carried out by gas chromatography (GC) coupled to tandem mass spectrometry (MS/MS). Most of the target OPs were detected in twenty real placenta samples studied, with pyrethroids, PAHs, and OPPs representing most of the 90% of OPs with means of 13-18 ng/g *versus* PCBs, OCPs, and PBDEs with means <4.0 ng/g. The results of this work indicated a prenatal exposure to Ops [37]. Salgueiro *et al.* studied wild mussels (*Mytilus galloprovincialis*) as bioindicators of chemical contamination to evaluate the occurrence and spatial distribution of endocrine-disrupting compounds by selective PLE followed by liquid chromatography determination by tandem mass spectrometry, the results allowed to propose that the discharges of the wastewater treatment plants, and nautical, fishing and maritime transport activities were considered the main sources of pollution by alkylphenols [38].

Studies on Classic Foods and their By-products

Some of the most recent and interesting applications of PLE into the analysis of classic foods and their by-products are listed in Table **2**. In this table, the works corresponding to by-products are highlighted. The largest number of works focuses on studying phenolic-type compounds (PC) [39 - 41]. Some works studied fatty acid profile (FAP) [42], diterpenes (DT) [43], alkaloids [44], and sterols. Many authors have used PLE to obtain extracts with different kinds of components, for example, some extracts can contain PC and non-polar fatty compounds.

Table 2. Studies on Classic Foods and their By-products.

	Analyte	Matrix	Solvent (v:v)[a]	Sample Preparation	T_{ext} (°C)	P_{ext} (MPa)	t_{ext} (min)	IAT	Refs.
DT	DT: (cafestol and kahweol)	Green (*C. arabica*) coffee beans	MeOH	20 g sample, 50 mL solvent	60-100	2	165 FR: 2-3 mL/min	HPLC	[72]
PC	Isoflavones	Soybean (*G. max*) flour and protein isolate	Water	Solid liquid ratio (1-7%). 10 g sea sand	80-200	10	St.: 2-14	HPLC-UV	[39]
	PC	Rice (*O. sativa*) grains	EtOAc in MeOH (60-80%)	2.5-5 g sample	100-200	10-20	St.:5-10	HPLC-DAD	[40, 73]
	PC	White and black grape, spinach kiwi, carrot, kale, celery, beet, broccoli orange, strawberry	EtOH, (40-80%) HCl 0.1%	1 g waste sample mixed with 2 g DE	80-120	1	St.: 5-15 Cycles:1-2	HPLC-UV, HPLC-HRMS	[41]
	PC	Cocoa (*T. cacao*) bean shell	EtOH	15 g sample 15 g EtOH	70	10	St.: 20	GC-MS	[74]

(Table 2) cont.....

	Analyte	Matrix	Solvent (v:v)[a]	Sample Preparation	T_{ext} (°C)	P_{ext} (MPa)	t_{ext} (min)	IAT	Refs.
Wide polarity compounds	Volatile compounds, CGA and derivatives	Green coffee (*C. arabica*) beans	Water	2.0 g of sample, 4.0 g sea sand, 15.0 mL water	40-200	3.45-17.24	St.: 2-18	LC-q-TOF-MS, HS-SPME-GC-MS	[75]
	FAP, PC, mangiferin	Mango (*M. indica*) seed kernel	First step: Heptane 100°C Second step: EtOH 150°C	Mass/sea sand ratio 1:2	1st step:100 2nd step:150	10	St.: 20	GC-MS, UHPLC--TOF-MS/MS	[47, 48]
	FAP, PC	Watermelon (*C. lanatus*) seed oil	EtOH	Mass/solvent ratio 0.3	40-80	10	St.: 3-36 Cycles: 8 max.	–	[76]
Wide polarity compounds	DT (cafestol, kahweol), FAP, phytosterol, PC, tocopherol	Green coffee (*C. arabica*) beans	EtOH	20.0 g sample, 40.0 mL EtOH	100	10	St.: 30	HPLC	[45]
	Phytosterols, PC	Olive (*O. europea*) seeds	EtOAc (60%)	0.5 g sample, 11 g sand	40-200	10	St.: 10	GC-MS	[77]
	DT (cafestol and kahweol), PC	Green coffee (*C. arabica*) beans	EtOH	22 g, sample, 30 mL EtOH	50-70	10	St.: 4-8 Cycles: 3	GC-FID	[78]
	PC, caffeine	Spent coffee (*C. arabica*) grounds	EtOH (25-75%)	0.5-2.5 g sample, 30 mL EtOH, sea sand	40-200	3.45-17.24	St.: 5-20	HPLC	[43]
	Flavanols, alkaloids (caffeine, theophylline)	Black tea (*C. sinensis*)	Water	60 g sample	90-180	0.8	Dt.: 15-60 FR: 6-18 mL/min	HPLC	[79]
	PC, TOC, catechin, epicatechin, procyanidin B2, theobromine	Cocoa (*T. cacao*) bean shell	EtOH	Mass ratio 1:3, solid:solvent	60-90	10.3	St.: 5-50	GC, UPLC-MS/MS	[42, 44]
	PC, anthocyanins	Grape (*V. vinifera*) pomace	EtOH (50%) pH 2.0	52 g sample/11,3 kg solvent	40	10	St.: 54	MAC, FTIR, AFM	[68]
	PC, pro-anthocyanidins, vanillin	Sweet (*P. avium*) cherry pomace	Buffer phosphate pH: 6-8	5.5 g sample. Commercial enzyme (439 µL/g).	60-80	-	St.: 5-40	HPLC-SEC	[69]
	PC, anthocyanins, delphinin-3-glucoside	Black bean (*P. vulgaris*) hulls	EtOH (30-70%). pH 3-4 (citric acid 0.1 M)	5 g sample, 36 g glass spheres	30-70	10	Dt.: FR: 3-5 mL/min	HPLC	[80]
	PC, mangiferin	Mango (*M. indica*) leaves	EtOH	Lab. scale: 15 g sample Pilot scale: vessel of 5 L	100	20	Lab.:180 Pilot: 360 FR: 10-30 g/min	HPLC	[81]

(Table 2) cont.....

	Analyte	Matrix	Solvent (v:v)[a]	Sample Preparation	T_{ext} (°C)	P_{ext} (MPa)	t_{ext} (min)	IAT	Refs.
Wide polarity compounds	PC, secoiridoids, flavonoids, elenolic acids	Olive (*O. europea*) leaves	EtOH (0-100%)	4 g sample with DE 1:2	50-200	10.34	St.: 5-20	HPLC-ESI-TOF-MS	[82]
	PC (carnosic acid, carnosol, rosmarinic acid)	Rosemary (*R. officinalis*) leaves	EtOH (80%)	2 g sample	60	20	St.: 20	HPLC-MS	[83]
	PC, flavonoids, quercetin	Onions varieties	Low polarity pressurized water. 0.1% formic acid	5 g sample. 80 mL water	60	15	St.: 60	HPLC	[84]
	PC, flavonoids, quercetin	Onion (*A. cepa*) peels	EtOH	40 g sample. Teflon rod	40-60	12-20	St.: 20	HPLC	[85]
	PC, allicin	Garlic (*A. sativum*)	Water	-	40	6	-	-	[86]
	PC, mannitol, carbohydrates	Olive (*O. europea*) leaves	EtOH (60%)	3 g sample with 2 g Ottawa sand	190	11.7	St.: 5	HPLC	[87]

IAT: Instrumental analytical techniques; DT: Diterpenes; PC: Phenolic compounds; CGA: Chlorogenic acid; FAP: Fatty acid profile; TOC: Tocopherols; EtOAc: Ethyl acetate; DE: Diatomaceus earth; St.: Static time; Dt.: Dynamic time; FR: Flow rate; MAC: Monomeric anthocyanins content; FTIR: Fourier-transform infrared spectroscopy; AFM: Atomic force microscopy.

[a] If the solvent is not specified, it is water (*e.g.*, EtOH 70%: EtOH 70%, 30% water).

The by-products are highlighted with blue color.

Recently, Dong *et al.* obtained green coffee oil from Arabica coffee beans using ultrasonic/microwave-assisted extraction (UMAE), microwave-assisted extraction (UAE), ultrasound-assisted extraction (UAE), and PLE (2021) [45]. The obtained yields were 10.6, 9.1, 9.3, and 6.3%, respectively. Obtained extracts were submitted to FAMEs analysis by GC-MS, phytosterols analysis by GC, tocopherols analysis by high performance liquid chromatography (HPLC), and diterpenes analysis by HPLC. The PLE technique was ideal for obtaining oils rich in δ-tocopherol, β-sitosterol, and total phenolic compounds, these compounds are very important to the cosmetic industry. Additionally, PLE extract showed the most antioxidant activity by ABTS and FRAP methods. On the other way, a cranberry pomace study was published. Consecutively it was fractionated using supercritical CO_2 and PLE, for the recovery of lipophilic and polyphenolic fractions, respectively [46]. EtOH and water were used in PLE. PLE with EtOH effectively recovered anthocyanins and procyanidins. The strongest antioxidant extract was obtained by PLE with water. Ballesteros *et al.* proposed another fractionation strategy, which consisted of subjecting the mango seed kernel to a two-step PLE procedure. Heptane was used as a solvent in the first PLE step, obtaining the lipidic fraction. EtOH was used as a solvent in the second PLE step, obtaining the polar fraction. A comprehensive GC- and LC-q-TOF-MS/MS profiling analysis allowed the complete characterization of the lipidic and phenolic obtained fractions, demonstrating the abundance of oleic and stearic acids, as well as bioactive xanthones, phenolic acids, flavonoids, gallate derivatives, and gallotannins. Additionally, high antioxidant capacity and antiproliferative activity against human colon cancer cells in the polar fraction were detected [47, 48].

The use of green solvents was predominant in studies on classic foods and their by-products. Water, EtOH, and ethyl acetate were the most used solvents. The static extraction mode was the most used. In general, the T_{ext} values were less than 100°C, some authors included T_{ext} between 180 to 200°C in their studies. Analytical separation techniques (GC and LC) coupled with mass spectrometer techniques were fundamental tools for the characterization of the obtained extracts.

Studies on Emerging Foods and their By-products

Some studies of emerging foods and their by-products under PLE are listed in Table **3**. In this table, the works corresponding to by-products are highlighted. The largest number of works focuses on studying phenolic-type compounds (PC) [46, 49, 50]. Some works studied fatty acid profile (FAP) [51 - 53]. diterpenes (DT) [54], alkaloids [55], and sterols [56]. Some works reported obtained extracts with different types of components, for example, some extracts may contain lipophilic and hydrophilic compounds.

Table 3. Studies on emerging foods and their by-products.

	Analyte	Matrix	Solvent (v:v)[a]	Sample Preparation	T_{ext} (°C)	P_{ext} (MPa)	t_{ext} (min)	IAT	Refs.
FAP	ω-3 PUFAs	Chia (*C*), echium (*E*), camelina (*Cm*) oilseeds; *N. gaditana* (*N.g.*)	EtOAc (*C, E, Cm*) Hexane: Isopropanol (2:1) (*N.g.*)	3 g sample (*C, E Cm*); 3 g of dry equivalent *N.g.*	120	10	_	GC-MS, HPLC-ELSD	[57]
	FAP	Chia (*S. hispanica*) seeds	EtOH, DCM: MeOH (2:1)	2 g sample, 2 g sand	40-80	10	St.: 10	GC-MS	[88]
	FAP	*Isochrysis* sp. (*I. sp.*)	EtOH (80-100%)	0.5 g sample, 0.1 g DE	80	10.3	St.: 5 Cycles: 3	GC-FID, TLC-FID	[51]
	FAP	Safflower (*Carthamus tinctorius*) seeds	EtOH	5 g sample	30-40	10-20	St.: 60	GC-MS	[52]
	ω-3 PUFAs	Echium (*E. plantagineum*) seeds	EtOAc, hexane, EtOH, (10-100%)	3 g sample	EtOAc, hexane, 60-150 EtOH: 120-200	10	St.: 10	GC-MS	[89]
	MUFAs, PUFAs and SFAs	House cricket (*A. domesticus*) and mealworm (*T. molitor*)	EtOH (50-100%)	2 g of edible insect	120	10	St.: 15	GC-FID	[53]

(Table 3) cont.....

	Analyte	Matrix	Solvent (v:v)[a]	Sample Preparation	T_{ext} (°C)	P_{ext} (MPa)	t_{ext} (min)	IAT	Refs.
Carotenoids	Violaxanthin, zeaxanthin, canthaxanthin, mono, diesters	*Neochloris oleoabundans*	EtOH	0.6 g sample, 2 g sea sand	100	10.3	St.: 20	HPLC-DAD	[90]
	Lutein, β-carotene, proteins, and sugars	*Scenedesmus obliquus*	EtOH (0-100%)	0.5 g sample, 1.5 g sea sand	50-170	10	St.: 20	HPLC-ELSD, HPLC-DAD-MS/MS	[91]
	Fucoxanthin Carotenoids	*Phaeodactylum tricornutum*	EtOH (0-100%)	1 g sample, 2.5 g sea sand	50-170	13	St.: 5 (40-100°C) St.: 8 (170°C)	HPLC-DAD-MS/MS, HPLC-ELSD	[92]
Sterols	Sterols	Champignon (*A. bisporus*) bodies	EtOH	1 g sample, 4 g sea sand	50-100	10.68	St.: 5 Cycles: 5	GC-MS-FID	[56]
Withano-lides	4βHWE and WE	Goldenberry (*P. peruviana*) chalyx	EtOH/EtOAc (0-100%)	1 g samples, 2 g sand	50- 150	10	St.: 20	UPLC-q-TOF-MS	[93]
Cannabi-noids	Cannabinoids	*Cannabis sativa*	Acetone, EtOH (20-80%)	5 g sample, 5 g DE	Acetone: 30-130 EtOH 100	10.3	3 cycles Acetone: St.: 5-25 EtOH: St.: 15	HPLC-DAD, UPLC-MS, UPLC/ESI-q-TOF-MS	[94]
PC	PC, Benzoic acid, cinnamic acid derivatives	Brown algae (*C. abies-marina, U. pinnatifida, S. muticum*), red algae (*C. crispus*)	MeOH (80%)	50 mg sample, 20 mg ascorbic acid. 300 μL MeOH (90%)	130	13	St.: 10 Cycles: 2	RP-HPLC-MS/MS	[49]
	TOC	Blackcurrant (*R. nigrum*) pomace	EtOH, water	Residue after SFE-CO_2, DE	70 (EtOH) 130 (Water)	10.3	St.: 5 Cycles: 3	RP-HPLC	[95]
	TOC	*Viburnum opulus* fruits and pomace	Acetone, EtOH, water	15 g residue obtained after SFE-CO_2	70 (EtOH Acetone) 120 (Water)	10.3	St.: 5 Cycles: 3	_	[96]
	Phlorotannins	*Sargassum muticum*	EtOH (75%)	2 g sample	120	3.45	St.: 20	HPLC-DAD-MS	[97]
	PC, TOC	Sal*via* species	EtOH (75%)	5 g sample, DE (1:1)	140	10	St.: 10	UPLC-MS/ MS, UPLC--TOF, HPLC-DPPH	[98]
	PC	Goldenberry (*P. peruviana*)	EtOH	3 g sample	25	10-20	St.: 0-60	HPLC-DAD	[99]
	PC	Pink pepper (*S. terebinthifolius*) fruits	EtOH, water	5 g sample, 64 g glass spheres	EtOH: 60 Water: 130	10	St.: 30	-	[100]
	PC	Pomegranate (*P. granatum*) peels	Water, EtOH	_	40	10-13	St.: 20 Cycles: 3-5	UHPLC-MS/MS	[101]
	PC	Pomegranate (*P. granatum*) peels	EtOH (0-70%)	1 g sample, 5 g glass spheres	50-100	10	St.: 10	UHPLC-MS/MS	[102]

(Table 3) cont.....

	Analyte	Matrix	Solvent (v:v)[a]	Sample Preparation	T_{ext} (°C)	P_{ext} (MPa)	t_{ext} (min)	IAT	Refs.
PC	PC	Chontaduro (*B. gasipaes*) epicarp	EtOH	5 g sample, 40 g glass	50-100	6.6-20	St.: 20-70	HPLC-DAD	[58]
	PC	Feijoa (*A. sellowiana*) leaves	EtOH (0-100%)	3 g sample	40-80	10	Dt.: 50 St.: 25 Cycles: 2	HPLC-ESI-MS/MS	[59]
	PC	*Feijoa* (*A. sellowiana*) peels	EtOH (0-100%)	3 g sample	40-80	10	St.: 120	HPLC-ESI-MS/MS	[50]
	PC, flavonoids, anthocyanins	Blackberry (*R. fruticosus*) pomace	EtO, water	3 g residue after optimized SFE-CO_2, 1 g DE	EtOH: 50-90 H_2O: 130	10.3	3 cycles St.: 5 EtOH St.: 10 Water	UPLC-q-TOF-MS	[103]
	Anthocyanins	Blackberry, blueberry and grumixama residues	EtOH (50-70%), citric acid (pH 2.0)	5 g sample	80	10	St.: 30	UHPLC	[104]
	PC, anthocyanins	Black chokeberry (*A. melanocarpa*) pomace	Hexane, MeOH, water, acetone (80%)	7 g sample, DE (1:1)	40-130	10.3	St.: 45	UPLC-q-TOF/MS, HPLC-UV/DPPH	[105]
	PC, anthocyanins, proantho-cyanidins, flavonoids	Black chokeberries (*A. melanocarpa*) pomace	Ultrapure water, EtOH, formic acid	1 g sample remaining after SFE	50-170	10.3	St.: 20	LC×LC-DAD-MS	[106]
	PC	Passion fruit (*P. edulis*) bagasse	EtOH (75%)	40 g defatted sample	70	10	St.: 300	HPLC-DAD	[70]
	PC flavanols, flavonols, flavones, chalcones or anthocyanidins	*P. edulis-P.e.*; *P. mollissima-P.m.*; *P. ligularis-P.l.*; *P. edulis flavic.-P.e.f.* (peels)	EtOH (94%), 1% formic acid	2 g of sea sand *P.l.*, *P.e.f.*, *P.m.*: 1 g sample. *P.e.*: 0.5 g sample	99	10.3	St.: 1	HPLC-DAD-q-TOF-MS, HPLC-DAD-MS	[107]
	Flavonoids, flavanols and proantho-cyanidins	Banana passion fruit (*P. mollissima*) seeds	EtOH/EtOAc (0-100%)	Sample, sea sand (1:2 w/w)	50-150	10	St.: 10-60	UHPLC-q-TOF-MS/ MS, GC-q-TOF-MS	[108]

(Table 3) cont.....

	Analyte	Matrix	Solvent (v:v)[a]	Sample Preparation	T_{ext} (°C)	P_{ext} (MPa)	t_{ext} (min)	IAT	Refs.
PC	Isoorientin, vicenin, vitexin, orientin, and isovitexin	Passion fruit (*P. edulis*) bagasse	EtOH (70-100%). Water	3.0 g defatted sample	30-60	10	St.: 30	UPLC-MS/MS	[109]
	Piceatannol, scirpusin B	Passion fruit (*P. edulis*) bagasse	EtOH (0-100%)	3.0 g defatted sample	50-70	10	St.: 120	ESI(-)-FTICR-MS, UHPLC-MS/MS	[110]
	PC, anthocyanins	Juçara (*Euterpe edulis*)	EtOH (0-100%), citric acid	2.5 g sample	40-80	10	-	UPLC-q-TOF-MS, UPLC-UV-vis	[111]
	PC	Genipap (*Genipa americana*) peel, seeds, mesocarp, endocarp, and fruit	EtOH	3.5-6 g sample	50-80	0.2-2	St.: 5	HPLC-DAD	[112]
	PC (HCA, flavonoids), glucosinolates	Moringa (*M. oleifera*)	Water	3 g sample, 1 g DE	100	6.89	_	UHPLC-q-TOF-MS	[113]
	Alcohols, PC, flavonoids	Moringa (*M. oleifera*)	Water	1 g sample, 1 g sea sand	125	7	St.: 60	GC-MS, HPLC-ESI-q-TOF-MS	[114]
	Flavonoids, PC	*D. kotschyi*	MeOH	30 g sample	74	3.4	St.: 11 Dt.: 17	HPLC	[115]
	PC	Arrowleaf (*S. rhombifolia*) leaves	EtOH (70%)	3 g sample	80	10	St.: 60	MSEM	[116]
	Scopoletin	Noni (*M. citrifolia*)	EtOH	1 g sample	60	_	St.: 12	HPLC-DAD	[60]
	Flavonoid glycosides	Tee oil camellia (*C. oleifera*) seeds	Water	4 g sample	50-140	4.14-8.27	St.: 5-30	HPLC-ESI(-)-MS-MS	[117]
	PC	Mate (*I. paraguariensis*)	EtOH, water	Sample:solvent relation 0,02 g/mL	60	10	St.: 190	Integrated SPE-PLE x HPLC-DAD	[71]
PC	PC	Buckwheat (*F. esculentum*) hulls	Acetone (70%), EtOH (70%); water	5 g sample, DE (1:1)	EtOH, Acetone: 70 Water: 140	10.3	St.: 15	-	[118]
Sugars	Poly-saccharides	Oyster mushroom (*P. ostreatus*) and *G. lucidum*	MilliQ water	0.5 g sample, sea sand (1:8)	50-180	10.2-11.7	St.: 5-30	NMR-H HSQC, GC-MS	[119]
	Iminosugars	*Aglaonema sp.*	Water	10-160 mg sample	50-150	10	St.: 2-10 Cycle: 1-3	GC-MS	[120]
	Starchy materials, soluble fiber	Turmeric (*C. longa*) and annatto (*B. orellana*)	EtOH, water	-	60	10	-	-	[67]

(Table 3) cont.....

	Analyte	Matrix	Solvent (v:v)[a]	Sample Preparation	T_{ext} (°C)	P_{ext} (MPa)	t_{ext} (min)	IAT	Refs.
Wide range of polarity compounds	FAP, carbohydrates, hydrocarbons, and sterols	House cricket (*A. domesticus*) and mealworm (*T. molitor*)	EtOH (0-100%)	2 g edible insect	120	10	St.: 15	GC-MS	[61]
	Caffeine, PC TOC, FAP	Caferana (*B. glandulifera*)	EtOH	-	60	40	St.: 300	UPLC, GC-FID	[121]
	TOC, FAP	Raspberry (*R. idaeus*) pomace	Hexane, MeOH	Sample, DE	30-110	10.3	St.: 5-25	-	[122]
	PC, flavonoids, carotenoids	Tamarillo (*C. betacea*) peels	EtOH, water, formic acid (2%)	0.2 g sample (EtOH free), 1 g sample (EtOH)	60	10.34	St.: 20	UHPLC-q-TOF-MS/MS	[62]
	PC, carotenoids	Buriti (*M. flexuosa*) shell	EtOH, water	3 g sample, 33 g glass beads, cotton	28-71	10	St.: 90	LC-ESI-MS/MS	[123]
	Peptides, PC	Pomegranate (*P. granatum*) peels	EtOH (70%)	2 g sample, 8 g sand	120	10.34	St.: 3-12	RP-HPLC-ESI-q-TOF, HILIC-HPLC-ESI-q-TOF	[63]
	Sterols, alcohols, PC	Elephant grass (*P. purpureum*)	EtOH (50%) EtOAc	5 g sample	100	10.3	St.: 15 Cycles: 3	GC-MS, HPLC RID	[124]
	Phytosterols, TOC	Crambe (*C.* abyssinaca) seed	Methyl acetate	5.5 g sample	140	10	St.: 30	GC-FID	[125]
	Terpenoids, PC	Ora-pro nobis (*P. aculeata*)	EtOH, water	8 g sample, 60 g glass beads	50-110	10	St.: 15	GC-q-TOF-MS	[54]
Wide range of polarity compounds	TOC, sugar acids, PC, triterpenes	Sea buckthorn (*H. rhamnoides*) pomace and seeds	EtOH	25 g sample	70	10.3	St.: 15 Cycles: 3-5	UPLC-q-TOF-MS	[126]
	Withanolides (C_{28} - isoprenoids), sucrose esters, terpenoids	Goldenberry (*P. peruviana*) fruit calyx extracts	EtOH/EtOAc (0-100%)	1 g sample, 2 g sea sand	50-150	10	St.: 21	UPLC-ESI-q-TOF-MS/MS, GC-q-TOF-MS	[127]
	PC, alkaloids	Guarana (*Paullinia cupana*) seeds	EtOH (50%)	3 g sample	40-60	10	St.: 60	HPLC-DAD	[55]
	PC and SS	Chontaduro (*B. gasipaes*) waste	Water	4 g sample	100-130	6-10	St.: 30	HPLC-UV	[128]

IAT: Instrumental analytical techniques; FAP: Fatty acid profile; PC: Phenolic compounds; PUFAs: Polyunsaturated fatty acids; MUFAs: Monounsaturated fatty acids; SFAs: Saturated fatty acids; 4βHWE: 4β-Hydroxywithanolide; WE: Withanolide E; TOC: Tocopherols; HCA: Hydroxycinnamic acid; SS: Soluble sugars; *C*: Chia seeds; *E*: Echium; *Cm*: Camelina; *N.g.*: *N. gaditana*; EtOAc: Ethyl acetate; DE: Diatomaceous earth; St.: Static time; Dt.: Dynamic time; FR: Flow rate; MSEM: MatLab simulation models.
[a] If the solvent is not specified, it is water (*e.g.*, EtOH 70%: EtOH 70%, 30% water).
The by-products are highlighted with blue color.

Lipid extracts of novel omega-3 sources were obtained by PLE. Oilseeds (*Salvia hispanica*, chia; *Camelina sativa*, camelina; *Echium plantagineum*, echium) and wet biomass of microalgae (*Nannochloropsis gaditana*) were the studied samples. The obtained oils were subjected to microencapsulation. Microencapsulated oils stability was studied by GC-MS analysis. GC-MS showed that the microencapsulation process did not negatively affect their omega-3 fatty acid profile [57]. On the other way, a sequential process was used to obtain antioxidant extracts from the epicarp of chontaduro (*Bactris gasipaes*) fruit, it is a tropical palm. SFE was carried out in the first extraction stage obtaining carotenoid extract. SFE residue was submitted to microwave treatment, then the second extraction stage (PLE) was applied to obtain phenolic extract [58]. Santos *et al.* compared the efficiency of PLE and SFE techniques in obtaining antioxidant and antibacterial extracts from feijoa (*Feijoa sellowiana*) leaves. Major phenolic compounds were identified by HPLC. PLE extract obtained the highest yield, total phenolic content, and antioxidant activities, but SFE extract exhibited effectiveness against *S. aureus*, *E. coli*, and *S. typhimurium*. These extracts have important bioactivities, presenting great potential as a natural food additive [59]. Other authors used a procedure of extraction/analysis, PLE followed HPLC-DAD (PLE-HPLC-DAD), to study scopoletin content in noni (*Morinda citrifolia*). Scopoletin contributes to some noni bioactivities. The analytical method was validated, it provides a rapid and reliable routine procedure for the identification and quantification of scopoletin in noni samples [60]. A study from two edible insects (*Acheta domesticus* and *Tenebrio molitor*) under PLE showed that these extracts had antioxidant activity and inhibitory activity of lipase. The extracts characterization was made by GC-MS [61]. Suárez *et al.* studied the in-vitro neuroprotective potential of tamarillo (*Cyphomandra betacea*) epicarp, which is a tropical fruit from the Andean region of South America. This study was made under the PLE technique. The chemical characterization of obtained extracts was carried out by liquid chromatography-tandem mass spectrometry (UHPLC--TOF-MS/MS) [62]. PLE obtained extracts from pomegranate (*Punica granatum*) peels were characterized by HPLC-ESI-q-TOF/MS and their antioxidant and hypocholesterolemic activities were measured. These extracts proved to be highly antioxidant and hypocholesterolemic. Considering the above, the authors of these works state that: *"Therefore, they could be used for the development of nutraceuticals and functional foods or even in medicinal, cosmetic, and pharmaceutical preparations"* [63]. Finally, a promising and exotic food plant (Ora-pro-nobis, *Pereskia aculeata*, leaves) was subjected to different compressed fluid technologies to obtain extracts rich in antioxidants. The extracts were characterized by GC-q-TOF-MS [54].

Finally, EtOH, water, and ethyl acetate were the most used green solvents in studies on emerging foods and their by-products. Likewise, the static extraction mode was the most used. In general, the T_{ext} values ranged between 50 and 150 °C. Analytical separation techniques coupled with MS techniques were a fundamental tool for the characterization of the obtained extracts.

COMBINATION OF PLE WITH OTHER INNOVATIVE EXTRACTION TECHNIQUES - INTEGRATION AND INTENSIFICATION OF SEQUENTIAL PROCESSES

Process Intensification (PIs) and Process Integration (PI) are among the most effective strategies leading to better process designs and operations. Recently, IPs is a field receiving considerable interest. According to Stankiewicz and Moulijn, IPs correspond to: *"Any chemical engineering development that leads to substantially smaller, cleaner, safer, and more energy-efficient technology"* [64].

PIs are able to obtain the most benefit of a determined procedure, those in which multiple processing steps are integrated into a single operating unit or alternative energy sources are used (*e.g.*, sample preparation and PLE or SFE and PLE on the same vessel). PI considers the interactions between different unit operations from the outset, without optimizing them separately (*e.g.*, oil and flour obtention from soybeans) [65, 66]. Table **4** shows different examples to obtain enriched fractions from food materials and by-products under sequential extraction processes, using PLE with other innovative techniques [67].

Table 4. Studies about integration (PI) and intensification (PIs) of the extraction processes.

Technique	Matrix	Analyte	Combination Result	Refs.
SFE + Static PLE	Chia seeds	FAP	Increase phospholipids extraction yield	[88]
Static PLE + SAF	Mango seed kernel	PC with AproA	PC rich extracts with improved inhibitory effect on HT-29 colon cancer cells proliferation	[47]
Static PLE + SAF	Rosemary leaves	PC with AproA	Provide two different polarity fractions. Raffinate (Polar fraction): RA rich fraction. Extract (less polar fraction): CA+CS rich fraction. Rosmarinic acid, RA; carnosic acid + carnosol, CA+CS	[83]
SFE + Static PLE	Sal*via* species	PC	Oil extract rich in TOC. Ethanolic and water extracts rich in rosmarinic acid and caffeic acid	[98]
SFE + CXE + Static PLE	Microalga	PC, FAP, carotenoids	SFE extracts rich in triacylglygerids, CXE extracts rich in carotenoids and chlorophyll. PLE extracts rich in protein, sugars, and PC	[91]
SFE + Static PLE	Champignon bodies and by-products	Sterols	Greater sterols extraction yield. SFE extracted less material but more selective (approx. 50%) than PLE (approx. 5%)	[56]
SFE + Dynamic PLE	Pink pepper fruits	PC with AA	The defatting step by SFE increases the yield of PC and increases the AA values	[100]

(Table 4) cont.....

Technique	Matrix	Analyte	Combination Result	Refs.
SFE + Static PLE	Buckwheat hulls	PC with AA	The efficiency of EAE after PLE with acetone and EtOH considerably increased the extraction yield	[118]
SFE + Static PLE	Blackcurrant pomace	PC with AA	Approx. 67% of extractable substances was extracted. A lipophilic fraction from SFE and two polar fractions after PLE and EAE.	[95]
SFE + Static PLE	European cranberry bush pomace	PC	Consecutive re-extraction of SFE residues with pressurized solvents enabled to recover polyphenol-rich fractions with high AA.	[96]
SFE + Dynamic PLE	Moringa leaves	PC	Extraction of wide range of compounds. SFE fraction was rich in fatty acids and alkanes. CXE fraction was rich in glycosylated flavonoids and PLE fraction was rich in PC with high AA.	[114]
SFE + PLE	Cranberry pomace	FAP, PC with AA	A non-polar fraction rich in polyunsaturated fatty acids and a polar fraction rich in PC with AA was obtained.	[46]
SFE + Static PLE	Elephant grass	FAP, PC	The combination of PLE with SFE allowed non-polar compounds extraction such fatty acids and sterols with high extraction yield.	[124]
SFE + Static PLE	Cocoa been hulls	PC with AA	Defatted residue from SFE showed higher TPC and AA in PLE after SFE compared to extract from separated methods.	[74]
SFE + Static PLE + EAE	Cannabis threshing residues	PC, cannabinoids	The combined SFE, PLE and EAE reduced initial plant AA after sequential extractions.	[94]
EAE + PLE	Seaweed algae	PC, phlorotannins	The use of carbohydrases before PLE did not improve the TPC and the phlorotannins recovery. However, the study of more selective enzymes could improve the extraction yield and the phlorotannins recovery.	[19]
MF/NF/UF + Dynamic PLE	Grape pomace	PC	The extraction of grape pomace using PLE, and microfiltration process is more effective for the concentration of bioactive compounds.	[68]
UAE + Static PLE + μ-SPE	Algae samples	PC	The hyphenated combination of UAE+PLE+μ-SPE was more efficient for all analytes of interest.	[49]
UAE + PLE	Berries by-products	PC with AA	UAE+PLE using hydroethanolic solvents increases TPC and AA compared to isolated techniques.	[104]
PLE + SPE x HPLC-DAD	Mate	PC	Wide range of compound classes extracted in a single step without sample manipulation	[71]

(Table 4) cont.....

Technique	Matrix	Analyte	Combination Result	Refs.
SFE + Static PLE	Biquinho pepper	PC, capsiate	COM was 71,6% lower for the integrated process than the sum of the single-stages processes	[129]
SFE + Dynamic PLE	Caferana seeds	Fatty acids, PC	SFE extracts showed high caffeine, TOC, and fatty acids concentrations. PLE extracts showed high levels of PC	[121]
SFE + Dynamic PLE	Passion fruit bagasse	PC	A multi-stage SFE+PLE plant could extract four fractions: TOC and tocotrienols-rich extract; Fatty acids-rich extract; Carotenoids-rich extract and PC-rich extract. Scaling-up increased the process productivity and decreased the COM	[130]
Static PLE combined EAE	Sweet cherry pomace	PC, bioactive peptides	Higher PC and bioactive peptides extraction and AA than single PLE or DES	[69]
UAPLE	Passion fruit bagasse	PC, piceatannol	UAPLE increased the yields, resulting 60% more TPC and piceatannol	[70]
UAPLE	Pomegranate peel	PC	Improve the PC in terms of TPC, speed and solvent consumption. Increase the yield more than 20%	[101]
UAPLE	Pomegranate peel	PC	UAPLE demonstrated to be an efficient alternative for PC extraction specially with small particle size	[102]
Two steps process: 1. PI: SFE + LPSE. 2. PIs: SFE + PLE	Annatto, turmeric	Starchy materials, dietary fibers	PI: Two fractions in different equipment. Oil fraction, tocotrienols and TOC. Polar fraction, PC PIs: Two fractions in single multipurpose equipment. Oil fraction, terpenoids. Polar fraction, curcuminoids.	[67]

SAF: Supercritical antisolvent fractionation; CXE: Carbon-dioxide expanded ethanol; EAE: Enzyme assisted extraction; MF/NF/UF: Micro/nano/ultra-filtration; SPE: Solid phase extraction; UAE: Ultrasound assisted extraction; UAPLE: Ultrasound assisted pressurized liquid extraction; LPSE: Low polarity solvent extraction; FAP: Fatty acid profile; PC: Phenolic compounds; AproA: Antiproliferative activity; AA: Antioxidant activity; RA: Rosmarinic acid; CA: Carnosic acid; CS: Carnosol; TOC: Tocopherols; COM: Cost of manufacturing; TPC: Total phenolic content; DES: Deep euthectic solvents

Some authors had realized PI or PIs using PLE. A PI study was realized from the mango seed kernel. PLE obtained extract from the mango seed kernel was submitted to supercritical antisolvent fractionation (SAF) to attain enriched fractions with improved inhibitory effect on colon cancer cells (HT-29). HPLC---TOF-MS/MS analysis followed by a multivariate statistical analysis of bioactivity evidenced the presence of characteristic compounds which might explain the enhanced antiproliferative activity of the optimal SAF fraction [47, 48], Tamires *et al.* reported the integration of PLE with membrane separation technology to recover and concentrate monomeric anthocyanins from grape marc pomace, being emerging for obtaining functional products with high added value [68]. Monomeric anthocyanin extracts were analyzed by UHPLC-q-TOF-MS. Dominguez *et al.* reported the integration of PLE with enzyme-assisted extraction (EAE) to recover bioactive compounds from sweet cherry pomace. Size-exclusion chromatography profiles showed that PLE-EAE allowed obtaining non-extractable polyphenols (NEPs) with higher molecular weight (>8000 Da) than PLE alone [69]. On the other hand, Viganó *et al.* used PLE assisted by ultrasound (US) to intensify the extraction of phenolic compounds from defatted passion fruit bagasse. Piceatannol was quantified by HPLC-DAD [70]. Finally, the development of PI of intensified processes is possible and desirable. For example, an online system for extraction/analysis of phenolic compounds was designed, assembled, and validated, it can be applied to foods and agricultural products. The system is based on a bidimensional separation through the combination of PLE with in-line solid-phase extraction (SPE) coupled online with HPLC-DAD. The 1st dimension separation is achieved by PLE-SPE, the 2nd dimension presents the HPLC-DAD separation. The extraction/analysis process (2D PLE-SPE HPLC-DAD) can be applied to food and agricultural topics [71].

To summarize, sequential systems that include PLE let obtain better results (shorter analysis time, cheap processes, most bioactive extracts, global yield higher) than using exclusively PLE.

CONCLUDING REMARKS

This chapter presents an overview of the principles, the main factors involved, and some applications in food analysis. A general description is provided of the procedures to be followed in the PLE (*e.g.*, on the equipment required and the extraction conditions). Some applications corresponding to a wide range of analytes of interest in food samples were selected. The information provided shows some strategies on the most common parameters and conditions for the selected applications about the extraction conditions and the instrumental analysis techniques used, although the information provided is not exhaustive. Thus, recent applications for the analysis of food contaminants are presented, along with some examples for obtaining bioactive natural extracts from foods and their by-products. Analytical strategies are shown to evaluate their composition. For that reason, lipids, terpenoids, alkaloids, and polyphenols were determined as main components from different biomasses. Finally, new extraction strategies for valuable natural compounds can benefit from sequential chemical engineering processes, developing more efficient and selective systems, in the sense of sustainable development goals. Following these topics, it is worth mentioning that PLE has been successfully used in biorefinery processes.

LIST OF ABBREVIATIONS

ELSD	Evaporative light scattering detector
ESI-MS/MS	Electrospray ionization tandem mass spectrometry
GC-MS	Gas chromatography coupled with mass spectrometry
HPLC	High performance liquid chromatography
HPLC-DAD	High performance liquid chromatography with diode-array detector
HPLC-MS	High performance liquid chromatography coupled with mass spectrometry
HPLC-MS/MS	High performance liquid chromatography coupled with tandem mass spectrometry
MS/MS	Tandem mass spectrometry
P_{ext}	Extraction pressure
PHSE	Pressurized hot solvent extraction
PHWE	Pressurized hot water extraction
PI	Process Integration
PIs	Process Intensification
PLE	Pressurized liquid extraction
q-TOF	Quadrupole time-of-flight
SAF	Supercritical antisolvent fractionation
SFE	Supercritical fluid extraction
SPE	Solid phase extraction
SWE	Subcritical water extraction
t_{d-ext}	Dynamic extraction time
T_{ext}	Extraction temperature
t_{ext}	Extraction time
UHPLC	Ultrahigh-performance liquid chromatography
UPLC	Ultra-performance liquid chromatography

CONSENT FOR PUBLICATION

Not applicable.

CONFLICT OF INTEREST

The authors declare no conflict of interest, financial or otherwise.

ACKNOWLEDGEMENTS

N. Á-A., J.S. O.-R., J.P. O.-B., L.I. R.-V., and F. P.-A. thank DIEB-Universidad National de Colombia, Project number 51274 for the financial support.

REFERENCES

[1] Plaza M, Marina ML. Pressurized hot water extraction of bioactives. Trends Analyt Chem 2019; 116: 236-47.
[http://dx.doi.org/10.1016/j.trac.2019.03.024]

[2] Richter BE, Jones BA, Ezzell JL, Porter NL, Avdalovic N, Pohl C. Accelerated solvent extraction: A technique for sample preparation. Anal Chem 1996; 68(6): 1033-9.
[http://dx.doi.org/10.1021/ac9508199]

[3] Pagano I, Campone L, Celano R, Piccinelli AL, Rastrelli L. Green non-conventional techniques for the extraction of polyphenols from agricultural food by-products: A review. J Chromatogr A 2021; 1651: 462295.
[http://dx.doi.org/10.1016/j.chroma.2021.462295] [PMID: 34118529]

[4] Kettle A. Use of Accelerated Solvent Extraction to Improve Laboratory Workflow. Sunnyvale, CA, USA 2008.

[5] Richter BE, Ezzell JL, Knowles DE, *et al.* Extraction of polychlorinated dibenzo-p-dioxins and polychlorinated dibenzofurans from environmental samples using accelerated solvent extraction (ASE). Chemosphere 1997; 34(5-7): 975-87.
[http://dx.doi.org/10.1016/S0045-6535(97)00400-1]

[6] U.S. Environmental Protection Agency (EPA). Method 3545A (SW-846): Pressurized Fluid Extraction (PFE). Washington D.C.: Revisión 1 2007.

[7] El-Shamy S, Farag MA. Novel trends in extraction and optimization methods of bioactives recovery from pomegranate fruit biowastes: Valorization purposes for industrial applications. Food Chem 2021; 365: 130465.
[http://dx.doi.org/10.1016/j.foodchem.2021.130465] [PMID: 34243129]

[8] Poletto P, Alvarez-Rivera G, Torres TMS, Mendiola JA, Ibañez E, Cifuentes A. Compressed fluids and phytochemical profiling tools to obtain and characterize antiviral and anti-inflammatory compounds from natural sources. Trends Analyt Chem 2020; 129: 115942.
[http://dx.doi.org/10.1016/j.trac.2020.115942] [PMID: 32834241]

[9] Plaza M, Turner C. Pressurized hot water extraction of bioactives. Green Extraction Techniques. 2017; pp. 53-82.
[http://dx.doi.org/10.1016/bs.coac.2016.12.005]

[10] Alvarez-Rivera G, Bueno M, Ballesteros-Vivas D, Mendiola JA, Ibañez E. Pressurized liquid extraction. Liquid-Phase Extraction. Elsevier 2020; pp. 375-98.
[http://dx.doi.org/10.1016/B978-0-12-816911-7.00013-X]

[11] Andreu V, Picó Y. Pressurized liquid extraction of organic contaminants in environmental and food samples. Trends Analyt Chem 2019; 118: 709-21.
[http://dx.doi.org/10.1016/j.trac.2019.06.038]

[12] Picó Y. Pressurized liquid extraction of organic contaminants in environmental and food samples. Green Extraction Techniques. 2017; pp. 83-110.
[http://dx.doi.org/10.1016/bs.coac.2017.03.004]

[13] American Society for Testing and Materials (ASTM). ASTM WK19797. New test method for determing gel content in cross-linked ethylene plastics using pressurized liquid extraction.

[14] American Society for Testing and Materials (ASTM). ASTM D7567-09. Standard test method for determining gel content in crosslinked ethylene plastics using pressurized liquid extraction (Withdrawn 2015). 2015.

[15] Anastas P, Eghbali N. Green chemistry: Principles and practice. Chem Soc Rev 2010; 39(1): 301-12.
[http://dx.doi.org/10.1039/B918763B] [PMID: 20023854]

[16] Chemat F, Vian MA, Cravotto G. Green extraction of natural products: Concept and principles. Int J

Mol Sci 2012; 13(7): 8615-27.
[http://dx.doi.org/10.3390/ijms13078615] [PMID: 22942724]

[17] Mustafa A, Turner C. Pressurized liquid extraction as a green approach in food and herbal plants extraction: A review. Anal Chim Acta 2011; 703(1): 8-18.
[http://dx.doi.org/10.1016/j.aca.2011.07.018] [PMID: 21843670]

[18] Panja P. Green extraction methods of food polyphenols from vegetable materials. Curr Opin Food Sci 2018; 23: 173-82.
[http://dx.doi.org/10.1016/j.cofs.2017.11.012]

[19] Herrero M, Sánchez-Camargo AP, Cifuentes A, Ibáñez E. Plants, seaweeds, microalgae and food by-products as natural sources of functional ingredients obtained using pressurized liquid extraction and supercritical fluid extraction. Trends Analyt Chem 2015; 71: 26-38.
[http://dx.doi.org/10.1016/j.trac.2015.01.018]

[20] Herrero M, Castro-Puyana M, Mendiola JA, Ibañez E. Compressed fluids for the extraction of bioactive compounds. Trends Analyt Chem 2013; 43: 67-83.
[http://dx.doi.org/10.1016/j.trac.2012.12.008]

[21] Christen P, Kaufmann B. New trends in the extraction of natural products: Microwave-assisted extraction and pressurized liquid extraction. Encyclopedia of Analytical Chemistry. Chichester, UK: John Wiley & Sons, Ltd 2014; pp. 1-27.
[http://dx.doi.org/10.1002/9780470027318.a9904]

[22] Plaza M, Turner C. Pressurized hot water extraction of bioactives. In: Comprehensive Analytical Chemistry. Elsevier 2017; pp. 53-82.
[http://dx.doi.org/10.1016/bs.coac.2016.12.005]

[23] Lu J, Boughner EC, Liotta CL, Eckert CA. Nearcritical and supercritical ethanol as a benign solvent: Polarity and hydrogen-bonding. Fluid Phase Equilib 2002; 198(1): 37-49.
[http://dx.doi.org/10.1016/S0378-3812(01)00754-3]

[24] Luong D, Sephton MA, Watson JS. Subcritical water extraction of organic matter from sedimentary rocks. Anal Chim Acta 2015; 879: 48-57.
[http://dx.doi.org/10.1016/j.aca.2015.04.027] [PMID: 26002476]

[25] DDB. Dielectric Constant of Ethanol.

[26] Ballesteros-Vivas D, Ortega-Barbosa JP, Sánchez-Camargo A del P, Rodríguez-Varela LI, Parada-Alfonso F. Pressurized liquid extraction of bioactives. In: Cifuentes A (Ed.) Comprehensive Foodomics. 2021; 754-70.
[http://dx.doi.org/10.1016/B978-0-08-100596-5.22817-5]

[27] Camel V. Extraction methodologies in plants: General introduction. Encyclopedia of Analytical Chemistry. Chichester, UK: John Wiley & Sons, Ltd 2014; pp. 1-26.
[http://dx.doi.org/10.1002/9780470027318.a9902]

[28] Merlo TC, Molognoni L, Hoff RB, Daguer H, Patinho I, Contreras-Castillo CJ. Alternative pressurized liquid extraction using a hard cap espresso machine for determination of polycyclic aromatic hydrocarbons in smoked bacon. Food Control 2021; 120: 107565.
[http://dx.doi.org/10.1016/j.foodcont.2020.107565]

[29] Dessì C, Okoffo ED, O'Brien JW, *et al.* Plastics contamination of store-bought rice. J Hazard Mater 2021; 416: 125778.
[http://dx.doi.org/10.1016/j.jhazmat.2021.125778] [PMID: 33866293]

[30] Bichon E, Guiffard I, Vénisseau A, *et al.* Simultaneous analysis of historical, emerging and novel brominated flame retardants in food and feed using a common extraction and purification method. Chemosphere 2018; 205: 31-40.
[http://dx.doi.org/10.1016/j.chemosphere.2018.04.070] [PMID: 29679786]

[31] Trabalón L, Vilavert L, Domingo JL, Pocurull E, Borrull F, Nadal M. Human exposure to brominated

flame retardants through the consumption of fish and shellfish in Tarragona County (Catalonia, Spain). Food Chem Toxicol 2017; 104: 48-56.
[http://dx.doi.org/10.1016/j.fct.2016.11.022] [PMID: 27887975]

[32] Moret S, Sander M, Purcaro G, Scolaro M, Barp L, Conte LS. Optimization of pressurized liquid extraction (PLE) for rapid determination of mineral oil saturated (MOSH) and aromatic hydrocarbons (MOAH) in cardboard and paper intended for food contact. Talanta 2013; 115: 246-52.
[http://dx.doi.org/10.1016/j.talanta.2013.04.061] [PMID: 24054587]

[33] Khan Z, Kamble N, Bhongale A, Girme M, Bahadur Chauhan V, Banerjee K. Analysis of pesticide residues in tuber crops using pressurised liquid extraction and gas chromatography-tandem mass spectrometry. Food Chem 2018; 241: 250-7.
[http://dx.doi.org/10.1016/j.foodchem.2017.08.091] [PMID: 28958526]

[34] Chen H, Pan M, Liu X, Lu C. Evaluation of transfer rates of multiple pesticides from green tea into infusion using water as pressurized liquid extraction solvent and ultra-performance liquid chromatography tandem mass spectrometry. Food Chem 2017; 216: 1-9.
[http://dx.doi.org/10.1016/j.foodchem.2016.07.175] [PMID: 27596385]

[35] Rodrigues ET, Pardal MÂ, Salgueiro-González N, Muniategui-Lorenzo S, Alpendurada MF. A single-step pesticide extraction and clean-up multi-residue analytical method by selective pressurized liquid extraction followed by on-line solid phase extraction and ultra-high-performance liquid chromatography-tandem mass spectrometry for complex matrices. J Chromatogr A 2016; 1452: 10-7.
[http://dx.doi.org/10.1016/j.chroma.2016.05.036] [PMID: 27234845]

[36] Choi M, Lee IS, Jung RH. Rapid determination of organochlorine pesticides in fish using selective pressurized liquid extraction and gas chromatography–mass spectrometry. Food Chem 2016; 205: 1-8.
[http://dx.doi.org/10.1016/j.foodchem.2016.02.156] [PMID: 27006206]

[37] Fernández-Cruz T, Martínez-Carballo E, Simal-Gándara J. Optimization of selective pressurized liquid extraction of organic pollutants in placenta to evaluate prenatal exposure. J Chromatogr A 2017; 1495: 1-11.
[http://dx.doi.org/10.1016/j.chroma.2017.03.010] [PMID: 28320535]

[38] Salgueiro-González N, Turnes-Carou I, Viñas L, *et al.* Occurrence of alkylphenols and bisphenol A in wild mussel samples from the spanish atlantic coast and bay of biscay. Mar Pollut Bull 2016; 106(1-2): 360-5.
[http://dx.doi.org/10.1016/j.marpolbul.2016.03.003] [PMID: 27001713]

[39] Moras B, Rey S, Vilarem G, Pontalier PY. Pressurized water extraction of isoflavones by experimental design from soybean flour and soybean protein isolate. Food Chem 2017; 214: 9-15.
[http://dx.doi.org/10.1016/j.foodchem.2016.07.053] [PMID: 27507441]

[40] Setyaningsih W, Saputro IE, Palma M, Barroso CG. Pressurized liquid extraction of phenolic compounds from rice (*Oryza sativa*) grains. Food Chem 2016; 192: 452-9.
[http://dx.doi.org/10.1016/j.foodchem.2015.06.102] [PMID: 26304372]

[41] Montenegro-Landívar MF, Tapia-Quirós P, Vecino X, *et al.* Fruit and vegetable processing wastes as natural sources of antioxidant-rich extracts: Evaluation of advanced extraction technologies by surface response methodology. J Environ Chem Eng 2021; 9(4): 105330.
[http://dx.doi.org/10.1016/j.jece.2021.105330]

[42] Okiyama DCG, Soares ID, Toda TA, Oliveira AL, Rodrigues CEC. Effect of the temperature on the kinetics of cocoa bean shell fat extraction using pressurized ethanol and evaluation of the lipid fraction and defatted meal. Ind Crops Prod 2019; 130: 96-103.
[http://dx.doi.org/10.1016/j.indcrop.2018.12.063]

[43] Shang YF, Xu JL, Lee WJ, Um BH. Antioxidative polyphenolics obtained from spent coffee grounds by pressurized liquid extraction. S Afr J Bot 2017; 109: 75-80.
[http://dx.doi.org/10.1016/j.sajb.2016.12.011]

[44] Okiyama DCG, Soares ID, Cuevas MS, *et al.* Pressurized liquid extraction of flavanols and alkaloids

from cocoa bean shell using ethanol as solvent. Food Res Int 2018; 114: 20-9. [http://dx.doi.org/10.1016/j.foodres.2018.07.055] [PMID: 30361017]

[45] Dong W, Chen Q, Wei C, *et al.* Comparison of the effect of extraction methods on the quality of green coffee oil from Arabica coffee beans: Lipid yield, fatty acid composition, bioactive components, and antioxidant activity. Ultrason Sonochem 2021; 74: 105578. [http://dx.doi.org/10.1016/j.ultsonch.2021.105578] [PMID: 33965776]

[46] Tamkutė L, Liepuoniūtė R, Pukalskienė M, Venskutonis PR. Recovery of valuable lipophilic and polyphenolic fractions from cranberry pomace by consecutive supercritical CO_2 and pressurized liquid extraction. J Supercrit Fluids 2020; 159: 104755. [http://dx.doi.org/10.1016/j.supflu.2020.104755]

[47] Ballesteros-Vivas D, Alvarez-Rivera G, García Ocampo AF, *et al.* Supercritical antisolvent fractionation as a tool for enhancing antiproliferative activity of mango seed kernel extracts against colon cancer cells. J Supercrit Fluids 2019; 152: 104563. [http://dx.doi.org/10.1016/j.supflu.2019.104563]

[48] Ballesteros-Vivas D, Álvarez-Rivera G, Morantes SJ, *et al.* An integrated approach for the valorization of mango seed kernel: Efficient extraction solvent selection, phytochemical profiling and antiproliferative activity assessment. Food Res Int 2019; 126: 108616. [http://dx.doi.org/10.1016/j.foodres.2019.108616] [PMID: 31732074]

[49] Klejdus B, Plaza M, Šnóblová M, Lojková L. Development of new efficient method for isolation of phenolics from sea algae prior to their rapid resolution liquid chromatographic–tandem mass spectrometric determination. J Pharm Biomed Anal 2017; 135: 87-96. [http://dx.doi.org/10.1016/j.jpba.2016.12.015] [PMID: 28012309]

[50] Santos PH, Baggio Ribeiro DH, Micke GA, Vitali L, Hense H. Extraction of bioactive compounds from feijoa (*Acca sellowiana* (O. Berg) Burret) peel by low and high-pressure techniques. J Supercrit Fluids 2019; 145: 219-27. [http://dx.doi.org/10.1016/j.supflu.2018.12.016]

[51] He Y, Huang Z, Zhong C, Guo Z, Chen B. Pressurized liquid extraction with ethanol as a green and efficient technology to lipid extraction of Isochrysis biomass. Bioresour Technol 2019; 293: 122049. [http://dx.doi.org/10.1016/j.biortech.2019.122049] [PMID: 31484103]

[52] Conte R, Gullich LMD, Bilibio D, *et al.* Pressurized liquid extraction and chemical characterization of safflower oil: A comparison between methods. Food Chem 2016; 213: 425-30. [http://dx.doi.org/10.1016/j.foodchem.2016.06.111] [PMID: 27451200]

[53] Otero P, Gutierrez-Docio A, Navarro del Hierro J, Reglero G, Martin D. Extracts from the edible insects acheta domesticus and tenebrio molitor with improved fatty acid profile due to ultrasound assisted or pressurized liquid extraction. Food Chem 2020; 314: 126200. [http://dx.doi.org/10.1016/j.foodchem.2020.126200] [PMID: 31972408]

[54] Torres TMS, Álvarez-Rivera G, Mazzutti S, *et al.* Neuroprotective potential of extracts from leaves of ora-pro-nobis (*Pereskia aculeata*) recovered by clean compressed fluids. J Supercrit Fluids 2022; 179: 105390. [http://dx.doi.org/10.1016/j.supflu.2021.105390]

[55] Santana ÁL, Queirós LD, Martínez J, Macedo GA. Pressurized liquid- and supercritical fluid extraction of crude and waste seeds of guarana (Paullinia cupana): Obtaining of bioactive compounds and mathematical modeling. Food Bioprod Process 2019; 117: 194-202. [http://dx.doi.org/10.1016/j.fbp.2019.07.007]

[56] Gil-Ramírez A, Aldars-García L, Palanisamy M, *et al.* Sterol enriched fractions obtained from Agaricus bisporus fruiting bodies and by-products by compressed fluid technologies (PLE and SFE). Innov Food Sci Emerg Technol 2013; 18: 101-7. [http://dx.doi.org/10.1016/j.ifset.2013.01.007]

[57] Castejón N, Luna P, Señoráns FJ. Microencapsulation by spray drying of omega-3 lipids extracted

from oilseeds and microalgae: Effect on polyunsaturated fatty acid composition. Lebensm Wiss Technol 2021; 148: 111789.
[http://dx.doi.org/10.1016/j.lwt.2021.111789]

[58] Reyes-Giraldo AF, Gutierrez-Montero DJ, Rojano BA, Andrade-Mahecha MM, Martínez-Correa HA. Sequential extraction process of oil and antioxidant compounds from chontaduro epicarp. J Supercrit Fluids 2020; 166: 105022.
[http://dx.doi.org/10.1016/j.supflu.2020.105022]

[59] Santos PH, Kammers JC, Silva AP, Oliveira JV, Hense H. Antioxidant and antibacterial compounds from feijoa leaf extracts obtained by pressurized liquid extraction and supercritical fluid extraction. Food Chem 2021; 344: 128620.
[http://dx.doi.org/10.1016/j.foodchem.2020.128620] [PMID: 33223301]

[60] Tasfiyati AN, Antika LD, Dewi RT, Septama AW, Sabarudin A, Ernawati T. An experimental design approach for the optimization of scopoletin extraction from *Morinda citrifolia* L. using accelerated solvent extraction. Talanta 2022; 238(Pt 1): 123010.
[http://dx.doi.org/10.1016/j.talanta.2021.123010] [PMID: 34857344]

[61] Navarro del Hierro J, Gutiérrez-Docio A, Otero P, Reglero G, Martin D. Characterization, antioxidant activity, and inhibitory effect on pancreatic lipase of extracts from the edible insects *Acheta domesticus* and *Tenebrio molitor*. Food Chem 2020; 309: 125742.
[http://dx.doi.org/10.1016/j.foodchem.2019.125742] [PMID: 31704068]

[62] Suárez-Montenegro ZJ, Ballesteros-Vivas D, Gallego R, *et al.* Neuroprotective potential of tamarillo (*Cyphomandra betacea*) epicarp extracts obtained by sustainable extraction process. Front Nutr 2021; 8: 769617.
[http://dx.doi.org/10.3389/fnut.2021.769617] [PMID: 34869538]

[63] Hernández-Corroto E, Plaza M, Marina ML, García MC. Sustainable extraction of proteins and bioactive substances from pomegranate peel (*Punica granatum* L.) using pressurized liquids and deep eutectic solvents. Innov Food Sci Emerg Technol 2020; 60: 102314.
[http://dx.doi.org/10.1016/j.ifset.2020.102314]

[64] Stankiewicz AJ, Moulijn JA. Process intensification: Transforming chemical engineering. Chem Eng Prog 2000. 96(1): 22-33.

[65] Abraham MA. Encyclopedia of sustainable technologies. Front matter of volume 1. Encyclopedia of Sustainable Technologies. Elsevier 2017.
[http://dx.doi.org/10.1016/B978-0-12-804677-7.01001-X]

[66] Jiří KJ, Sabev VP, Rafidah WAS, Manan ZA, Van FY, Chin HH. 1. Process integration and intensification: An introduction. sustainable process integration and intensification. De Gruyter 2018; pp. 1-12.
[http://dx.doi.org/10.1515/9783110535365-003]

[67] Silva EK, Martelli-Tosi M, Vardanega R, Nogueira GC, Zabot GL, Meireles MAA. Technological characterization of biomass obtained from the turmeric and annatto processing by using green technologies. J Clean Prod 2018; 189: 231-9.
[http://dx.doi.org/10.1016/j.jclepro.2018.04.101]

[68] Tamires Vitor Pereira D, Vollet Marson G, Fernández Barbero G, *et al.* Concentration of bioactive compounds from grape marc using pressurized liquid extraction followed by integrated membrane processes. Separ Purif Tech 2020; 250: 117206.
[http://dx.doi.org/10.1016/j.seppur.2020.117206]

[69] Domínguez-Rodríguez G, García MC, Marina ML, Plaza M. Pressurized liquid extraction combined with enzymatic-assisted extraction to obtain bioactive non-extractable polyphenols from sweet cherry (*prunus avium* l.) pomace. Nutrients 2021; 13(9): 3242.
[http://dx.doi.org/10.3390/nu13093242] [PMID: 34579121]

[70] Viganó J, Assis BFP, Náthia-Neves G, *et al.* Extraction of bioactive compounds from defatted passion

fruit bagasse (*Passiflora edulis* sp.) applying pressurized liquids assisted by ultrasound. Ultrason Sonochem 2020; 64: 104999.
[http://dx.doi.org/10.1016/j.ultsonch.2020.104999] [PMID: 32092695]

[71] Viganó J, Sanches VL, de Souza Mesquita LM, *et al.* Comprehensive analysis of phenolic compounds from natural products: Integrating sample preparation and analysis. Anal Chim Acta 2021; 1178: 338845.
[http://dx.doi.org/10.1016/j.aca.2021.338845] [PMID: 34482871]

[72] Belandria V, Aparecida de Oliveira PM, Chartier A, Rabi JA, de Oliveira AL, Bostyn S. Pressurized-fluid extraction of cafestol and kahweol diterpenes from green coffee. Innov Food Sci Emerg Technol 2016; 37: 145-52.
[http://dx.doi.org/10.1016/j.ifset.2016.07.022]

[73] Setyaningsih W, Hidayah N, Saputro IE, Palma M, García Barroso C. Profile of phenolic compounds in Indonesian rice (*Oryza sativa*) varieties throughout post-harvest practices. J Food Compos Anal 2016; 54: 55-62.
[http://dx.doi.org/10.1016/j.jfca.2016.10.004]

[74] Mazzutti S, Rodrigues LGG, Mezzomo N, Venturi V, Ferreira SRS. Integrated green-based processes using supercritical CO_2 and pressurized ethanol applied to recover antioxidant compouds from cocoa (*Theobroma cacao*) bean hulls. J Supercrit Fluids 2018; 135: 52-9.
[http://dx.doi.org/10.1016/j.supflu.2017.12.039]

[75] Xu JL, Kim TJ, Kim JK, Choi Y. Simultaneous roasting and extraction of green coffee beans by pressurized liquid extraction. Food Chem 2019; 281: 261-8.
[http://dx.doi.org/10.1016/j.foodchem.2018.12.061] [PMID: 30658756]

[76] Colivet J, Oliveira AL, Carvalho RA. Influence of the bed height on the kinetics of watermelon seed oil extraction with pressurized ethanol. Separ Purif Tech 2016; 169: 187-95.
[http://dx.doi.org/10.1016/j.seppur.2016.06.020]

[77] Vásquez-Villanueva R, Plaza M, García MC, Marina ML. Recovery and determination of cholesterol-lowering compounds from *Olea europaea* seeds employing pressurized liquid extraction and gas chromatography-mass spectrometry. Microchem J 2020; 156: 104812.
[http://dx.doi.org/10.1016/j.microc.2020.104812]

[78] Oliveira NA, Cornelio-Santiago HP, Fukumasu H, Oliveira AL. Green coffee extracts rich in diterpenes – Process optimization of pressurized liquid extraction using ethanol as solvent. J Food Eng 2018; 224: 148-55.
[http://dx.doi.org/10.1016/j.jfoodeng.2017.12.021]

[79] He C, Du H, Tan C, *et al.* Semi-continuous pressurized hot water extraction of black tea. J Food Eng 2018; 227: 30-41.
[http://dx.doi.org/10.1016/j.jfoodeng.2018.02.001]

[80] Teixeira RF, Benvenutti L, Burin VM, Gomes TM, Ferreira SRS, Zielinski AAF. An eco-friendly pressure liquid extraction method to recover anthocyanins from broken black bean hulls. Innov Food Sci Emerg Technol 2021; 67: 102587.
[http://dx.doi.org/10.1016/j.ifset.2020.102587]

[81] Fernández-Ponce MT, Parjikolaei BR, Lari HN, Casas L, Mantell C, Martínez de la Ossa EJ. Pilot-plant scale extraction of phenolic compounds from mango leaves using different green techniques: Kinetic and scale up study. Chem Eng J 2016; 299: 420-30.
[http://dx.doi.org/10.1016/j.cej.2016.04.046]

[82] Martín-García B, Pimentel-Moral S, Gómez-Caravaca AM, Arráez-Román D, Segura-Carretero A. Box-Behnken experimental design for a green extraction method of phenolic compounds from olive leaves. Ind Crops Prod 2020; 154: 112741.
[http://dx.doi.org/10.1016/j.indcrop.2020.112741]

[83] Sánchez-Camargo AP, Mendiola JA, Valdés A, *et al.* Supercritical antisolvent fractionation of

rosemary extracts obtained by pressurized liquid extraction to enhance their antiproliferative activity. J Supercrit Fluids 2016; 107: 581-9.
[http://dx.doi.org/10.1016/j.supflu.2015.07.019]

[84] Manohar CM, Xue J, Murayyan A, Neethirajan S, Shi J. Antioxidant activity of polyphenols from ontario grown onion varieties using pressurized low polarity water technology. J Funct Foods 2017; 31: 52-62.
[http://dx.doi.org/10.1016/j.jff.2017.01.037]

[85] Zabot GL, Meireles MAA. On-line process for pressurized ethanol extraction of onion peels extract and particle formation using supercritical antisolvent. J Supercrit Fluids 2016; 110: 230-9.
[http://dx.doi.org/10.1016/j.supflu.2015.11.024]

[86] Horita CN, Farías-Campomanes AM, Barbosa TS, *et al.* The antimicrobial, antioxidant and sensory properties of garlic and its derivatives in Brazilian low-sodium frankfurters along shelf-life. Food Res Int 2016; 84: 1-8.
[http://dx.doi.org/10.1016/j.foodres.2016.02.006]

[87] Lama-Muñoz A, Contreras MM, Espínola F, Moya M, Romero I, Castro E. Content of phenolic compounds and mannitol in olive leaves extracts from six Spanish cultivars: Extraction with the Soxhlet method and pressurized liquids. Food Chem 2020; 320: 126626.
[http://dx.doi.org/10.1016/j.foodchem.2020.126626] [PMID: 32222659]

[88] Calvo MV, Villanueva-Bermejo D, Castro-Gómez P, Fornari T, Fontecha J. Appraisal of the suitability of two-stage extraction process by combining compressed fluid technologies of polar lipid fractions from chia seed. Food Res Int 2020; 131: 109007.
[http://dx.doi.org/10.1016/j.foodres.2020.109007] [PMID: 32247499]

[89] Castejón N, Luna P, Señoráns FJ. Alternative oil extraction methods from *Echium plantagineum* L. seeds using advanced techniques and green solvents. Food Chem 2018; 244: 75-82.
[http://dx.doi.org/10.1016/j.foodchem.2017.10.014] [PMID: 29120808]

[90] Castro-Puyana M, Pérez-Sánchez A, Valdés A, *et al.* Pressurized liquid extraction of *Neochloris oleoabundans* for the recovery of bioactive carotenoids with anti-proliferative activity against human colon cancer cells. Food Res Int 2017; 99(Pt 3): 1048-55.
[http://dx.doi.org/10.1016/j.foodres.2016.05.021] [PMID: 28865616]

[91] Gilbert-López B, Mendiola JA, van den Broek LAM, *et al.* Green compressed fluid technologies for downstream processing of *Scenedesmus obliquus* in a biorefinery approach. Algal Res 2017; 24: 111-21.
[http://dx.doi.org/10.1016/j.algal.2017.03.011]

[92] Gilbert-López B, Barranco A, Herrero M, Cifuentes A, Ibáñez E. Development of new green processes for the recovery of bioactives from *Phaeodactylum tricornutum.* Food Res Int 2017; 99(Pt 3): 1056-65.
[http://dx.doi.org/10.1016/j.foodres.2016.04.022] [PMID: 28865617]

[93] Ballesteros-Vivas D, Álvarez-Rivera G, del Pilar SCA, Ibáñez E, Parada-Alfonso F, Cifuentes A. A multi-analytical platform based on pressurized-liquid extraction, *in vitro* assays and liquid chromatography/gas chromatography coupled to high resolution mass spectrometry for food by-products valorisation. Part 1: Withanolides-rich fractions from golde. J Chromatogr A 2019; 1584: 155-64.
[http://dx.doi.org/10.1016/j.chroma.2018.11.055] [PMID: 30553502]

[94] Kitrytė V, Bagdonaitė D, Rimantas Venskutonis P. Biorefining of industrial hemp (*Cannabis sativa* L.) threshing residues into cannabinoid and antioxidant fractions by supercritical carbon dioxide, pressurized liquid and enzyme-assisted extractions. Food Chem 2018; 267: 420-9.
[http://dx.doi.org/10.1016/j.foodchem.2017.09.080] [PMID: 29934186]

[95] Basegmez HIO, Povilaitis D, Kitrytė V, *et al.* Biorefining of blackcurrant pomace into high value functional ingredients using supercritical CO 2, pressurized liquid and enzyme assisted extractions. J Supercrit Fluids 2017; 124: 10-9.

[http://dx.doi.org/10.1016/j.supflu.2017.01.003]

[96] Kraujalis P, Kraujalienė V, Kazernavičiūtė R, Venskutonis PR. Supercritical carbon dioxide and pressurized liquid extraction of valuable ingredients from *Viburnum opulus* pomace and berries and evaluation of product characteristics. J Supercrit Fluids 2017; 122: 99-108. [http://dx.doi.org/10.1016/j.supflu.2016.12.008]

[97] Sánchez-Camargo AP, Montero L, Stiger-Pouvreau V, *et al.* Considerations on the use of enzyme-assisted extraction in combination with pressurized liquids to recover bioactive compounds from algae. Food Chem 2016; 192: 67-74. [http://dx.doi.org/10.1016/j.foodchem.2015.06.098] [PMID: 26304321]

[98] Šulniūtė V, Pukalskas A, Venskutonis PR. Phytochemical composition of fractions isolated from ten Salvia species by supercritical carbon dioxide and pressurized liquid extraction methods. Food Chem 2017; 224: 37-47. [http://dx.doi.org/10.1016/j.foodchem.2016.12.047] [PMID: 28159282]

[99] Corazza GO, Bilibio D, Zanella O, *et al.* Pressurized liquid extraction of polyphenols from Goldenberry: Influence on antioxidant activity and chemical composition. Food Bioprod Process 2018; 112: 63-8. [http://dx.doi.org/10.1016/j.fbp.2018.09.001]

[100] Rebelatto EA, Rodrigues LGG, Rudke AR, Andrade KS, Ferreira SRS. Sequential green-based extraction processes applied to recover antioxidant extracts from pink pepper fruits. J Supercrit Fluids 2020; 166: 105034. [http://dx.doi.org/10.1016/j.supflu.2020.105034]

[101] Santos MP, Souza MC, Sumere BR, *et al.* Extraction of bioactive compounds from pomegranate peel (*Punica granatum* L.) with pressurized liquids assisted by ultrasound combined with an expansion gas. Ultrason Sonochem 2019; 54: 11-7. [http://dx.doi.org/10.1016/j.ultsonch.2019.02.021] [PMID: 30827900]

[102] Sumere BR, de Souza MC, dos Santos MP, *et al.* Combining pressurized liquids with ultrasound to improve the extraction of phenolic compounds from pomegranate peel (*Punica granatum* L.). Ultrason Sonochem 2018; 48: 151-62. [http://dx.doi.org/10.1016/j.ultsonch.2018.05.028] [PMID: 30080537]

[103] Kitrytė V, Narkevičiūtė A, Tamkutė L, Syrpas M, Pukalskienė M, Venskutonis PR. Consecutive high-pressure and enzyme assisted fractionation of blackberry (*Rubus fruticosus* L.) pomace into functional ingredients: Process optimization and product characterization. Food Chem 2020; 312: 126072. [http://dx.doi.org/10.1016/j.foodchem.2019.126072] [PMID: 31893552]

[104] Machado APDF, Pereira ALD, Barbero GF, Martínez J. Recovery of anthocyanins from residues of *Rubus fruticosus*, *Vaccinium myrtillus* and *Eugenia brasiliensis* by ultrasound assisted extraction, pressurized liquid extraction and their combination. Food Chem 2017; 231: 1-10. [http://dx.doi.org/10.1016/j.foodchem.2017.03.060] [PMID: 28449984]

[105] Grunovaitė L, Pukalskienė M, Pukalskas A, Venskutonis PR. Fractionation of black chokeberry pomace into functional ingredients using high pressure extraction methods and evaluation of their antioxidant capacity and chemical composition. J Funct Foods 2016; 24: 85-96. [http://dx.doi.org/10.1016/j.jff.2016.03.018]

[106] Brazdauskas T, Montero L, Venskutonis PR, Ibañez E, Herrero M. Downstream valorization and comprehensive two-dimensional liquid chromatography-based chemical characterization of bioactives from black chokeberries (*Aronia melanocarpa*) pomace. J Chromatogr A 2016; 1468: 126-35. [http://dx.doi.org/10.1016/j.chroma.2016.09.033] [PMID: 27658379]

[107] Domínguez-Rodríguez G, García MC, Plaza M, Marina ML. Revalorization of passiflora species peels as a sustainable source of antioxidant phenolic compounds. Sci Total Environ 2019; 696: 134030. [http://dx.doi.org/10.1016/j.scitotenv.2019.134030]

[108] Ballesteros-Vivas D, Alvarez-Rivera G, Ibánez E, Parada-Alfonso F, Cifuentes A. Integrated strategy

for the extraction and profiling of bioactive metabolites from *Passiflora mollissima* seeds combining pressurized-liquid extraction and gas/liquid chromatography–high resolution mass spectrometry. J Chromatogr A 2019; 1595: 144-57.
[http://dx.doi.org/10.1016/j.chroma.2019.02.031] [PMID: 30846312]

[109] Viganó J, Brumer IZ, Braga PAC, *et al.* Pressurized liquids extraction as an alternative process to readily obtain bioactive compounds from passion fruit rinds. Food Bioprod Process 2016; 100: 382-90.
[http://dx.doi.org/10.1016/j.fbp.2016.08.011]

[110] Viganó J, Aguiar AC, Moraes DR, *et al.* Sequential high pressure extractions applied to recover piceatannol and scirpusin B from passion fruit bagasse. Food Res Int 2016; 85: 51-8.
[http://dx.doi.org/10.1016/j.foodres.2016.04.015] [PMID: 29544852]

[111] Garcia-Mendoza MP, Espinosa-Pardo FA, Baseggio AM, *et al.* Extraction of phenolic compounds and anthocyanins from juçara (*Euterpe edulis* Mart.) residues using pressurized liquids and supercritical fluids. J Supercrit Fluids 2017; 119: 9-16.
[http://dx.doi.org/10.1016/j.supflu.2016.08.014]

[112] Náthia-Neves G, Tarone AG, Tosi MM, Maróstica Júnior MR, Meireles MAA. Extraction of bioactive compounds from genipap (*Genipa americana* L.) by pressurized ethanol: Iridoids, phenolic content and antioxidant activity. Food Res Int 2017; 102: 595-604.
[http://dx.doi.org/10.1016/j.foodres.2017.09.041] [PMID: 29195990]

[113] Hamany Djande CY, Piater LA, Steenkamp PA, Madala NE, Dubery IA. Differential extraction of phytochemicals from the multipurpose tree, *moringa oleifera*, using green extraction solvents. S Afr J Bot 2018; 115: 81-9.
[http://dx.doi.org/10.1016/j.sajb.2018.01.009]

[114] Rodríguez-Pérez C, Mendiola JA, Quirantes-Piné R, Ibáñez E, Segura-Carretero A. Green downstream processing using supercritical carbon dioxide, CO_2-expanded ethanol and pressurized hot water extractions for recovering bioactive compounds from *Moringa oleifera* leaves. J Supercrit Fluids 2016; 116: 90-100.
[http://dx.doi.org/10.1016/j.supflu.2016.05.009]

[115] Kamali H, Khodaverdi E, Hadizadeh F, Ghaziaskar SH. Optimization of phenolic and flavonoid content and antioxidants capacity of pressurized liquid extraction from *Dracocephalum kotschyi via* circumscribed central composite. J Supercrit Fluids 2016; 107: 307-14.
[http://dx.doi.org/10.1016/j.supflu.2015.09.028]

[116] Ferro DM, Mayer DA, Oliveira Müller CM, Ferreira SRS. Scale-up simulation of PLE process applied to recover bio-based materials from *Sida rhombifolia* leaves. J Supercrit Fluids 2020; 166: 105033.
[http://dx.doi.org/10.1016/j.supflu.2020.105033]

[117] Liau BC, Ponnusamy VK, Lee MR, Jong TT, Chen JH. Development of pressurized hot water extraction for five flavonoid glycosides from defatted *Camellia oleifera* seeds (byproducts). Ind Crops Prod 2017; 95: 296-304.
[http://dx.doi.org/10.1016/j.indcrop.2016.10.034]

[118] Mackėla I, Andriekus T, Venskutonis PR. Biorefining of buckwheat (*Fagopyrum esculentum*) hulls by using supercritical fluid, Soxhlet, pressurized liquid and enzyme-assisted extraction methods. J Food Eng 2017; 213: 38-46.
[http://dx.doi.org/10.1016/j.jfoodeng.2017.04.029]

[119] Smiderle FR, Morales D, Gil-Ramírez A, *et al.* Evaluation of microwave-assisted and pressurized liquid extractions to obtain β-d-glucans from mushrooms. Carbohydr Polym 2017; 156: 165-74.
[http://dx.doi.org/10.1016/j.carbpol.2016.09.029] [PMID: 27842810]

[120] Rodríguez-Sánchez S, Martín-Ortiz A, Carrero-Carralero C, Ramos S, Sanz ML, Soria AC. Pressurized liquid extraction of Aglaonema sp. iminosugars: Chemical composition, bioactivity, cell viability and thermal stability. Food Chem 2016; 204: 62-9.
[http://dx.doi.org/10.1016/j.foodchem.2016.02.091] [PMID: 26988476]

[121] Fraga S, Gonçalves D, Nasário F, *et al.* Sequential high-pressure extraction of caffeine and bioactive compounds from caferana seeds (*Bunchosia glandulifera*). J Supercrit Fluids 2020; 165: 104958. [http://dx.doi.org/10.1016/j.supflu.2020.104958]

[122] Kryževičiūtė N, Kraujalis P, Venskutonis PR. Optimization of high pressure extraction processes for the separation of raspberry pomace into lipophilic and hydrophilic fractions. J Supercrit Fluids 2016; 108: 61-8. [http://dx.doi.org/10.1016/j.supflu.2015.10.025]

[123] Rudke AR, Mazzutti S, Andrade KS, Vitali L, Ferreira SRS. Optimization of green PLE method applied for the recovery of antioxidant compounds from buriti (*Mauritia flexuosa* L.) shell. Food Chem 2019; 298: 125061. [http://dx.doi.org/10.1016/j.foodchem.2019.125061] [PMID: 31260983]

[124] Scopel E, Santos LC, Bofinger MR, Martínez J, Rezende CA. Green extractions to obtain value-added elephant grass co-products in an ethanol biorefinery. J Clean Prod 2020; 274: 122769. [http://dx.doi.org/10.1016/j.jclepro.2020.122769]

[125] de Mello BTF, Iwassa IJ, Cuco RP, dos Santos Garcia VA, da Silva C. Methyl acetate as solvent in pressurized liquid extraction of crambe seed oil. J Supercrit Fluids 2019; 145: 66-73. [http://dx.doi.org/10.1016/j.supflu.2018.11.021]

[126] Kitrytė V, Povilaitis D, Kraujalienė V, Šulniūtė V, Pukalskas A, Venskutonis PR. Fractionation of sea buckthorn pomace and seeds into valuable components by using high pressure and enzyme-assisted extraction methods. Lebensm Wiss Technol 2017; 85: 534-8. [http://dx.doi.org/10.1016/j.lwt.2017.02.041]

[127] Ballesteros-Vivas D, Álvarez-Rivera G, Ibáñez E, Parada-Alfonso F, Cifuentes A. A multi-analytical platform based on pressurized-liquid extraction, *in vitro* assays and liquid chromatography/gas chromatography coupled to high resolution mass spectrometry for food by-products valorisation. Part 2: Characterization of bioactive compounds from goldenberry (*Physalis peruviana* L.) calyx extracts using hyphenated techniques. J Chromatogr A 2019; 1584: 144-54. [http://dx.doi.org/10.1016/j.chroma.2018.11.054] [PMID: 30579639]

[128] Giombelli C, Iwassa IJ, da Silva C, Bolanho Barros BC. Valorization of peach palm by-product through subcritical water extraction of soluble sugars and phenolic compounds. J Supercrit Fluids 2020; 165: 104985. [http://dx.doi.org/10.1016/j.supflu.2020.104985]

[129] de Aguiar AC, Osorio-Tobón JF, Viganó J, Martínez J. Economic evaluation of supercritical fluid and pressurized liquid extraction to obtain phytonutrients from biquinho pepper: Analysis of single and sequential-stage processes. J Supercrit Fluids 2020; 165: 104935. [http://dx.doi.org/10.1016/j.supflu.2020.104935]

[130] Viganó J, Zabot GL, Martínez J. Supercritical fluid and pressurized liquid extractions of phytonutrients from passion fruit by-products: Economic evaluation of sequential multi-stage and single-stage processes. J Supercrit Fluids 2017; 122: 88-98. [http://dx.doi.org/10.1016/j.supflu.2016.12.006]

CHAPTER 10

Microwave-Assisted Extraction (MAE)

Milena Álvarez Viñas[1], **Vanesa Sanz**[1], **Paula Rodríguez-Seoane**[1], **Lucía López Hortas**[1], **Noelia Flórez Fernández**[1], **María Dolores Torres**[1] and **Herminia Domínguez**[1,*]

[1] *CINBIO, Centro de Investigaciones Biomédicas, Departamento de Enxeñería Química, Facultade de Ciencias, Universidade de Vigo, Campus Ourense, Edificio Politécnico, As Lagoas, 32004 Ourense, Spain*

Abstract: Microwave assistance is an optimum strategy to shorten time, solvent, and energy consumption during the extraction of target solutes from different sources. This intensification strategy has been successfully applied to laboratory methods to enhance the extraction performance of a number of bioactive compounds of interest for food, cosmetic and pharmaceutical applications. This chapter presents an overview of the fundamentals, equipment configurations, combinations with other techniques, and some representative applications for the extraction of compounds from food products and byproducts.

Keywords: Bioactive compounds, By-products, Biomass, Green extraction, Microwave treatment.

INTRODUCTION

Plant bioactives are an excellent resource that is gaining commercial interest and in order to provide standardized products, reliable extraction methods both for analysis and for industrial production are needed. Understanding the cellular and subcellular localization of plant metabolites is essential to study their extraction. Usually, the target solutes are not readily accessible to the extracting solvent and a previous pretreatment aimed at breaking the solid matrix structure is needed. Efficient solvent extraction techniques aim to reduce mass transfer limitations to enhance rates and to increase the yield. The use of intensification strategies can speed up the process without undesirable effects on the quality of the products. Their combination could also provide interesting and synergistic effects [1].

The benefits of microwaves in extraction processes are in relation to selective hea-

* **Corresponding Author Herminia Domínguez:** CINBIO, Centro de Investigaciones Biomédicas, Departamento de Enxeñería Química, Facultade de Ciencias, Universidade de Vigo, Campus Ourense, Edificio Politécnico, As Lagoas, 32004 Ourense, Spain; E-mail: herminia@uvigo.es

Merichel Plaza & Maria Luisa Marina (Eds.)

ting and thermal effects on the physicochemical and transport properties of solvent and solutes, as well as thermal-induced structural damage of the matrix. In analytical applications, microwave extraction offers a number of advantages over conventional methods, particularly in relation to the shortened extraction time, lower solvent consumption, improved efficiency and selectivity, as well as reproducibility and precision for analytes recovery [2, 3]. Furthermore, the costs and pollution are lower compared to conventional techniques and as a result, this technique is commonly found in laboratories and has been included in some standardized extraction protocols.

This chapter presents a short survey on the fundamentals, major types of extractor configuration, combination with other intensification strategies as well as representative examples of microwave-assisted extraction of bioactives from vegetal biomass and also from other sources.

GREEN EXTRACTION TECHNOLOGIES

Bioactive compounds from foods can be obtained using different extraction strategies which can be classified as conventional or non-conventional methodologies. The main advantage of the conventional extraction process is simplicity, but among the several disadvantages can be mentioned low selectivity, low yield, high temperature, long time, high energy consumption, and sometimes the use of organic solvents increases the cost. Besides, a possible thermal degradation of the compounds could happen. The common traditional techniques are: soxhlet extraction, maceration, and hydrodistillation [4]. Nowadays, innovative extraction technologies, focused on the reduction of energy consumption with the purpose of reducing the environmental impact by fostering eco-friendly technologies have been studied and developed. The environmental impact can be lower because these eco-friendly methodologies use a limited amount of toxic chemicals, which could be replaced using green solvents (*e.g.* ethanol solutions, ionic liquids, deep eutectic, and supramolecular solvents or water), therefore their resultant hazardous wastes are significantly reduced [5, 6].

In this context, the recent tendency of society towards consuming products obtained through environmentally sustainable processes is growing [7]. This trend fits into the "green chemistry" concept, which can be defined as a model aimed at designing products and processes that are less harmful to the environment and people [8]. On the other hand, a definition related to green extraction techniques has been proposed by Chemat *et al.* [8] as "*processes in which energy consumption and the use of harmful solvents will be reduced, in order to obtain a high quality and sustainable extract/product*".

A general overview of the conventional and innovative extraction technologies and their possible combinations is summarized in Fig. (**1**), the novelty of these green extraction techniques is presented in this section. Some examples of green extraction processes, defined by the extraction technique and the greener character of the solvents, have been summarized in Table **1**. The operation time can be significantly shortened compared to those of the conventional process and these innovative techniques can also be valid for the extraction of compounds from other biomass, not only from plant sources.

Table 1. Examples of green extraction of edible materials.

Extraction Technique	Solvent	Raw material	Compound extracted	Operation conditions T(°C), P(MPa), T(min), P(W), F(kHz)	References
EAE	Water	Fish scale	Collagen	135 °C; 360 rpm	[31]
EAE*	Water	Kinnow peel	Naringin	90 °C; 10 min	[32]
	Water	Lentils and white beans	Protein	50 °C; 180 min	[33]
	Water/EtOH	Vanilla green poods	Glucovanillin	8-70 °C	[34]
HHP	EtOH	Seafood (boone)	Astaxanthin	0.1-600 MPa; 0-20 min	[35]
	Water	Black garlic	Melanoidins	25 °C; 200 - 500 MPa; 5, 15, 25 min	[36]
PEF	Water	Tomato	Carotenoids	60 °C; 0.5 kV/cm; 1 kJ/kg	[37]
	EtOH	Purple-fleshed potato	Anthocyanins	40 °C; 3.4 kV/cm	[38]
	EtOH	Grape seeds	Polyphenols	5 kV/cm	[39]
SFE	EtOH/CO_2	Strawberry	Phenolic compounds	30 °C; 5 MPa; 60 min	[40]
	CO_2	Wine grapes (seeds)	Flavonoids and phenolic acids	39.85 °C; 35 MPa; 240 min	[41]
	EtOH/CO_2	Mushrooms	Antioxidants	35-55 °C; 10-30 MPa; 30 min	[42]
	EtOH/CO_2	Grape bagasse (stems, skins and seeds)	Anthocyanins, catechins, glucosides of flavonols	40 °C; 20 and 35 MPa; 300 min	[43]
	EtOH	Coffee (grounds and husk)	Phenolic compounds	20-20 MPa; 50-60 °C	[44]

(Table 1) cont.....

PLE	EtOH	Shrimp head and carapace	Oil	46 - 114 °C; 4.3-7.7 MPa; 7 - 24 min	[45]
SWE	Water	Mushrooms	Phenolic compounds	50 and 250 °C; 7.5 min	[46]
	Water	Mushrooms	Phenolic compounds	100 - 210 °C; 16 min	[42]
	Water	Citrus (peel)		160 °C; 10.13 MPa; 10 min	[47]
	Water	Oregano	Antioxidants	50 °C; 10.3 MPa; 20 min	[48]
	EtOH/Water	Lemon balm	Phenolic compounds	150 °C; 10.3 MPa; 10 min	[49]
	Water	Grape pomace	Phenolic compounds	140 °C; 10.0 MPa	[50]
	Water	Potato peel	Phenolic compounds	180 °C; 6.0 MPa; 30 min	[17]
	Water	Marigold	Phenolic compounds	220 °C; 6.0 MPa; 45 min	[51]
	Water	Hop samples	Prenylflavonoids	150 °C; 10.3 MPa; 30 min	[52]
	Water	Grape pomace	Flavonoids	124 °C; 10.3 MPa; 1 min	[53]
	Water	Citrus pomace	Sinensetin, nobiletin and tangeretin	200 °C; 14.0 MPa; 60 min	[54]
	Water	*Camelia sinensis* leaves	Flavonoids	140-220 °C	[55]
	Water	Ginger root	6-Gingerol	130 and 170 °C; 3.5 MPa; 60 min	[56]
	Water/EtOH/ formic acid	Red cabbage	Anthocyanins	99 °C; 5.0 MPa; 7 min	[51, 57]
	Water/EtOH	Red onion	Anthocyanins	110 °C; 1.5 MPa; 60 min	[58]
	Water	Sunflower seed	Oil	130 °C; 3.0 MPa; 30 min	[59]

(Table 1) cont.....

UAE	Water/EtOH	Grapes	Amino acids	10-70 °C, 200W; 24 kHz	[60]
	Water	Chinese star anise	Skimic acid	28 kHz; 240-600W; 10-40 min	[61]
	Water	Grape pomace	Phenolic compounds	35 and 50 °C; 55 kHz, 435 W	[62]
	EtOH	Fruits (blueberry, nectarine, raspeberry, watermelon)	Oligosaccharides	25-50 °C; 40 kHz; 5-15 min	[63]
	EtOH/Water	Chinese medicine root	Baicalin and baicalein	40 kHz; 185 W; 20 min	[64]
	EtOH	Grape marc		20-50 °C; 25 kHz	[65]
	Water/EtOH	Grapes	Flavonoids	0-75 °C; 24 kHz; 6 min	[66]
	EtOH/Water	Orange peel	Flavonoids	40 °C; 150 W; 30 min	[67]
	Citric acid	Grape pomace	Pectins	35-75 °C; 37 kHz; 140 W; 20-60 min	[68]
	-	Mulberry	Phytochemical compounds	35 °C; 36 kHz; 60 W; 18 min	[69]
	EtOH	Fruits (blueberry, nectarine, raspberry, watermelon)	Oligosaccharides	25-50 °C; 40 kHz; 5-15 min	[70]
	EtOH	Vegetables (garlic, Jerusalem artichoke, leek, scallion, spring garlic and white onion	Oligosaccharides	25-50 °C; 40 kHz; 5-15 min	[70]
	Water	Traditional Chinese medicines	Geniposide	40 kHz; 500 W; 30 min	[71]
	Sunflower oil	Carrot	Carotenoids	40 °C; 20 kHz; 9.5, 16 and 22.5 Wcm^{-2}; 20 min	[72]
	EtOH	Wheat bran	Flavonoids	60 °C; 40 kHz; 250W; 25 min	[73]
		Seafood (fish scales)	Gelatine	60, 70 and 80 °C; 60 min	[74]

(*Table 1*) *cont.....*

MAE	EtOH	Soybean	Isoflavone	50 °C; 500W; 20 min	[75]
	-	Sea-buckthorn (fruit, leaves and seeds)	Flavonoids	400W; 15 min	[76]
	EtOH	Apple (pomace)		MP 650.4 W; 53.7 s	[77]
MHG	-	Strawberry	Aromatic compounds	1000 W; 30 min	[78]
	-	Broccoli	Phenolic compounds; antioxidants	20 °C; 70 and 500 W	[79]
	-	Strawberry	Aromatic compounds	1000 W; 30 min	[78]
	-	Mushroom	Phenolic compounds; antioxidants	0.101 MPa; 1W	[79]
	-	Lettuce onions	Polyphenols	300-900 W; 5-70 min	[80]
	-	Mushroom	Phenolic compounds; antioxidants	100-900 W	[42]
SFME	No solvent or water	Basil, garden mint, thyme	Essential oil	100 °C; 500 W; 30 min	[81]
	-	Orange peel	Essential oil	100 °C; 1000 W; 30 min	[82]
	-	Orange	Essential oil	50 min	[83]
	-	Lettuce onions	Polyphenols	0.101 MPa; 1W; 15-50 min	[80]
	-	*Salvia officinalis*	Essential oil	650 W; 35 min	[80]
	-	*Ocimum basilicum L.*	Essential oil	100 °C; 0.101 MPa; 600 W; 30 min	[84]
	-	Lettuce, onions	Polyphenols	1 W/g; 15 - 50 min	[81]

Extrusion-Assisted Extraction (EAE) has been widely used in the food industry mainly to produce baby foods and ready-to-eat snacks [7, 9]. It is based on the application of high temperature and pressure, the main function of these variables is to help the macromolecules present in the foodstuff solubilize with the aim to make the bioactive compounds more accessible. The major advantages of the EAE process are the reduction of extraction time, energy consumption, and solvents used, increasing the extraction performance. However, the high temperature required could affect thermosensitive compounds [7].

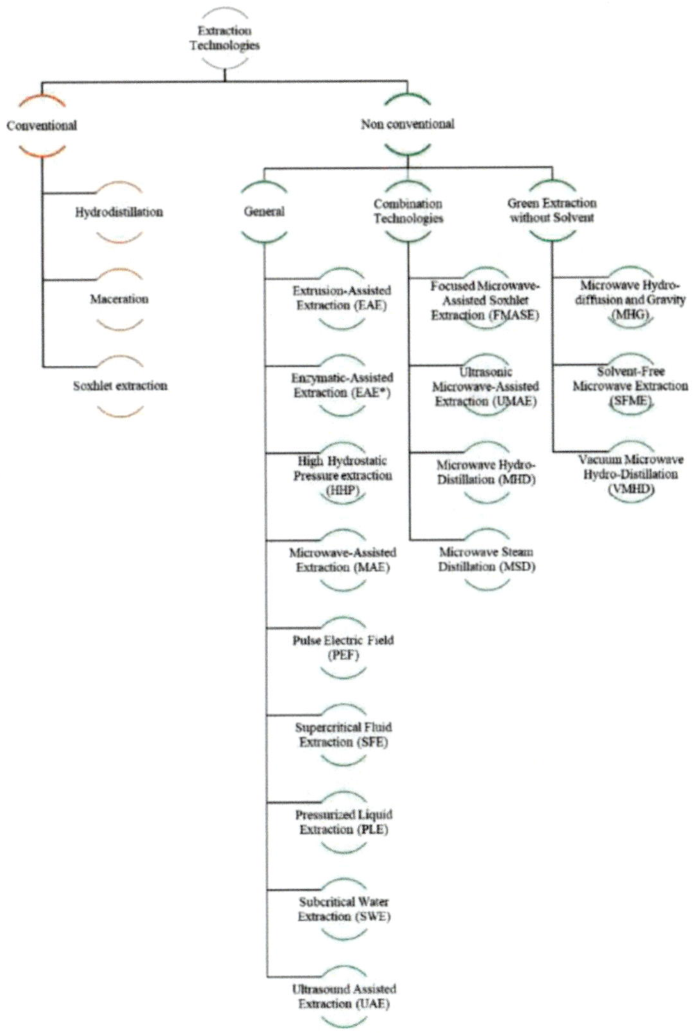

Fig. (1). Conventional and novel extraction technologies used independently or combined.

High Hydrostatic Pressure extraction (HHP) is a residue-free environmentally friendly technology mainly applied in the food industry for the inactivation of pathogens and enzymes [10]. It consists of pressurizing the material, causing deprotonation of charged groups and the breaking of salt bridges, in addition, breakage of weak bonds in cell membranes of foodstuffs can increase the cell permeability improving the recovery of biomolecules [7, 11]. In comparison with conventional extraction methodologies, HHP is more suitable for heat-sensitive molecules (such as flavoring agents, colorants, and bioactive compounds) because the covalent bonds are not affected by pressure [7].

Microwave-Assisted Extraction (MAE) is considered a green extraction technique because it saves time, solvent, and energy consumption [1, 12]. This non-ionizing radiation causes molecular motion, thus transforming the electromagnetic radiation into thermal energy while preserving the molecular structure of the compounds [12]. The process takes advantage of a synergistic combination of mass and energy transfer in the same direction [13]. The increase in temperature during the microwave process leads to the evaporation of the moisture contained in the cells, improving the porosity of the biomaterial, therefore, promoting mass transfer [14, 15]. Application in food, pharmacological and cosmetic industries, among others, is currently favorable considering their economic viability due to the reduction of time, lower energy requirements, improved yields, and maintained thermal stability of target compounds in comparison with conventional extraction. Different variables can affect the efficiency of the extraction process (power, frequency, extraction cycles, processing time, pressure, sample size and moisture content, solvent type, and solvent to sample ratio) [7]. Besides, it is possible to work in a variable range of operation scales [16]. The microwave extraction technology can be used alone or in combination with other traditional (*e.g.* steam distillation and Soxhlet extraction) or emerging extraction methods (*e.g.* ultrasonic, supercritical carbon dioxide, and enzymatic incubation).

Enzymatic-Assisted Extraction (EAE*) is a specific hydrolytic enzyme treatment where cellulases, pectinases, proteases or α-amylases, can be used in the extraction process to improve the recovery of the target compounds [17, 18]. Different strategies during enzyme-assisted extraction or as a pretreatment have been assessed to obtain different high-value compounds, such as polyphenols from plants [19] or for the recovery oils from seeds [20]. In this context, the key factors are the activity and concentration of enzymes, and on the other hand, the particle size of the biomaterials, the solid-to-water ratio, and also the pH, temperature, and hydrolysis time.

Pulsed Electric Field (PEF) is a technology that can be used in the food industry as a new non-thermal processing technology to increase the porosity of the biomaterials helping release intracellular biomolecules, a phenomenon known as electroporation or electropermeabilization [21]. The main advantage of this non-thermal process lies in the possibility to extract heat-sensitive compounds and has been applied to the extraction of different components, ranging from polysaccharides from edible mushrooms [22] to phenolics from industrial wastes [23].

Pressurized Liquid Extraction (PLE) is also referred to as accelerated solvent extraction, pressurized fluid extraction, pressurized hot solvent extraction, high-pressure solvent extraction, high-pressure high-temperature solvent extraction, and subcritical solvent extraction [24]. The main factors to be considered for an efficient extraction are the matrix characteristics, choice of solvent, extraction time, and temperature [24]. Operation at high pressure maintains the solvent in the liquid state, helping to achieve higher diffusion and mass transfer rates [25, 26]. Compared to the conventional, PLE offers advantages derived from reducing the extraction time and the amount of solvent used, improving the yields. A wide range of solvents can be used, and also their mixtures [26]. Since greener solvents are preferred, the use of water is promoted, for its non-toxic, non-flammable, and eco-friendly character [26]. This particular case is named subcritical water extraction, pressurized hot water extraction, pressurized low polarity water extraction, superheated water extraction, or autohydrolysis [27]. This process operates at temperatures between 100 and 374 ˚C and pressures from 10 to 60 bar in order to keep water in the liquid state, this technology is suitable to extract both polar and non-polar compounds [7, 28, 29]. Several compounds, such as polysaccharides, polyphenols, and antioxidants, among others have been extracted from different raw materials [22, 29, 30].

Supercritical Fluid Extraction (SFE) is an extraction technology using solvents over its critical point. Under such conditions of pressure and temperature, the properties are intermediate between those of gases and liquids, with higher diffusion coefficients and lower viscosity, respectively. These features favor solvent penetration in the matrix, enhancing mass transfer. Furthermore, with the adequate selection of operating conditions, the solvent density can be finely tuned to selectively extract the desired compounds from different raw materials [85, 86]. The low critical conditions of supercritical CO_2 (31.1 °C; 73.8 MPa), the absence of toxicity, chemical inertness, low cost, and easy recycling are some of the advantages of this solvent, widely used for food applications. In order to favor the extraction of polar compounds, a combination with a polar solvent or modifier is recommended. The major variables associated with the efficiency of the supercritical CO_2 extraction process are pressure, temperature, time, solvent flow rate, and the presence and concentration of polar modifiers [7, 87 - 89].

Ultrasound Assisted-Extraction (UAE) has emerged as an efficient, fast, clean, reproducible, and non-thermal alternative to conventional methodologies for the extraction of high-value compounds [90 - 92]. It is based on the phenomenon of acoustic cavitation and the process performance is influenced by ultrasonic device, power, and frequency as well as the shape and size of the extraction vessel, the type of solvent and the solvent to solid ratio, time, and temperature [7, 93].

Types of Technologies Used in Microwave-assisted Extraction Processes

The microwave energy may be used in different operational configurations, including pressurized operation, under controlled temperature and pressure operating with the solvent heated above its boiling point, or operation in open systems at atmospheric pressure. Microwave equipment can also operate either in mono mode, with radiation focused on a restricted zone, or in multimode, with randomly dispersed radiation. The main advantages of using focused-microwave radiation are safety, versatility, control of microwave energy released to the sample, and the possibility of a programmed addition of solutions [94]. Microwave-assisted extraction can be combined with other conventional or emerging physicochemical and biotechnological technologies to enhance extraction yields and solute selectivity. Some of these strategies to improve and increase the recovery of high-value compounds from biomaterials are discussed.

Ultrasonic Microwave-Assisted Extraction (UMAE) is a technology where the mass transfer can be improved by the ultrasonic waves in the equipment where the combined ultrasonic and microwaves facilitate the disruption of the raw material, releasing the bioactive compounds to the extracting solvent. UMAE has been proposed to extract a variety of active compounds with high value, such as carotenoids, vegetal oils, and polysaccharides from different plants [1, 12].

Vacuum Microwave-Assisted Extraction (VMAE) in this case, the combination of MAE in a vacuum system changes the boiling point of the solvents compared to the value at ambient pressure and favours the extraction of thermosensitive bioactive. Besides, the lower air pressure in this system reduces the oxidation of the sample being a suitable option for oxygen-sensitive compounds. The pressure difference between the inside and outside of the cell wall improves extraction efficiency. In addition to the variables influencing the classical MAE, the vacuum degree has to be controlled [12].

Extraction without added solvent: This technology has one distinctive feature its possibility to extract natural compounds from fresh or moisturized raw materials without needing to add solvents or water achieving solvent-free extraction using only the constitutive water of the plant matrix as the extraction solvent. An interesting application of this green process has been to assist in the extraction of essential oils and different water-soluble compounds [12].

Microwave Hydro-diffusion and Gravity (MHG) was designed for the extraction of essential oils from different matrices using microwave radiation at atmospheric pressure [95]. The sample is placed on a glass vessel inside the oven and the microwaves induce an increase in the temperature of the water inside the cell matrix, allowing the destruction of the cell walls and releasing the bioactive compounds. In this process, the internal water solubles of the matrix, are released and transferred from inside to the exterior, a phenomenon known as hydro-diffusion. A cooling system placed outside the oven allows the condensation of the liquid fraction drained by gravity. This extraction method, with the benefits of reduced cost and time, and energy savings [96], has been mainly applied to aromatic plants and citrus.

Solvent-Free Microwave Extraction (SFME) is a novel combination of microwave heating and dry distillation to obtain mainly essential oils, without solvent and water at atmospheric pressure. The fresh matrix is placed on a vessel in the microwave oven, and the water from the raw material is heated and breaks down the cell wall that contains the solutes, which are then propelled by the steam produced from the matrix water. A cooling system placed outside the microwave allows the continuous condensation of the distillate, usually composed of water and essential oil. The excess water returns to the interior of the vessel, to maintain the adequate humidity of the matrix [12, 97]. Several compounds from fresh or dry food matrixes have been extracted by this methodology, such as spices, aromatic herbs, and citrus fruits [12, 84, 98 - 100].

Vacuum Microwave Hydro-Distillation (VMHD) effectively evaporates water contained in the material and divides the cell structure of the plant sample, facilitating the release of natural compounds. During the application of microwave radiation, reduced pressure is intermittently applied to further cause the demolition of the cellular structure in the plant matrix. It also causes hydrodistillation by transporting natural compounds to the steam phase [12].

FUNDAMENTALS OF MICROWAVE-ASSISTED EXTRACTION

Electromagnetic (EM) radiation describes waves with a sinusoidal profile since their electric and magnetic field components oscillate and propagate in directions and phases perpendicular to each other [101]. Fig. (**2**) shows this behavior corresponding to two complete cycles. EM spectrum presents different regions depending on the operation frequency, wavelength, or photon energy regimes, namely γ-rays, X-rays, ultraviolet-visible, infrared, microwave, and radio [102]. Microwave is a non-ionizing EM energy that can be classified into three bands according to their signal, electromagnetic (EM), extremely high frequency (EHF), super high frequency (SHF), and ultra-high frequency (UHF) (Fig. **3**). Microwave electro-technology is successfully employed in telecommunications, scientific and medical applications [3]. Food processing is also an area in which microwaves are widely utilized, and the frequencies at $9.15 \cdot 10^8$ and $2.45 \cdot 10^9$ Hz are the most common for industrial-scale and domestic uses, respectively [103]. In particular, the first microwave oven was developed in the 1940s. Since then the continuous improvements in their design allowed the reduction in the equipment dimensions, weight, production cost, and energy consumption at the same time that increased their power, efficiency, and reliability of these systems [104].

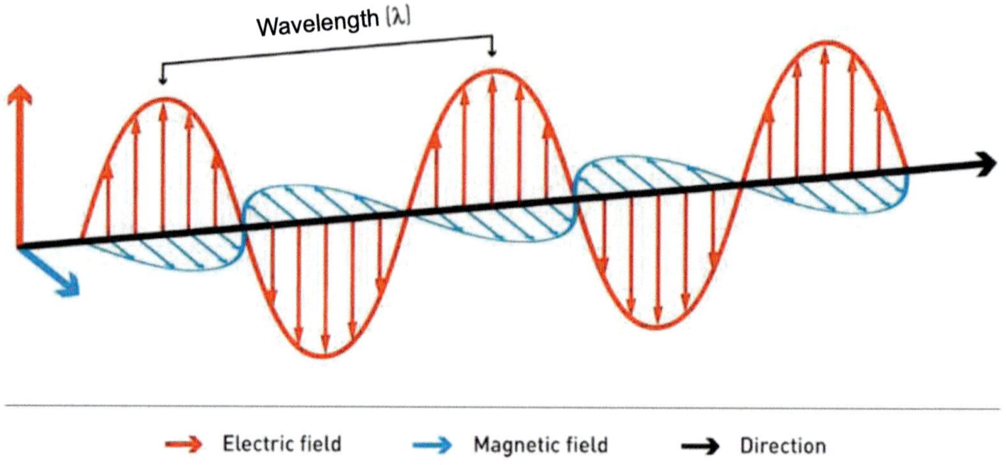

Fig. (2). Representation of oscillation and propagation of EM waves (adapted from [105]).

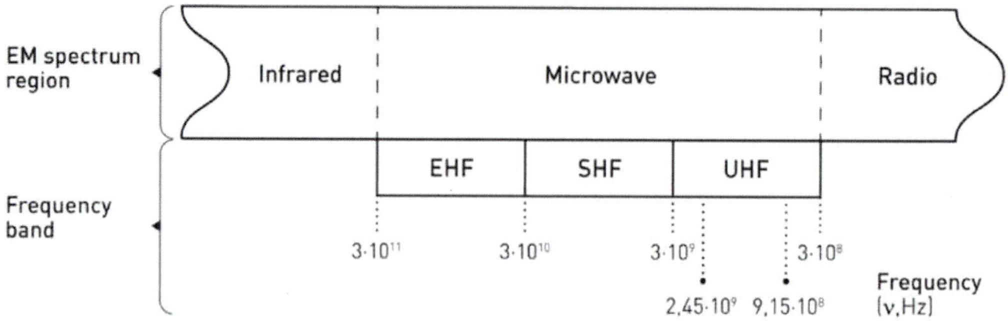

Fig. (3). Microwave spectrum (adapted from [103, 106]). EM: Electromagnetic; EHF: Extremely high frequency; SHF: Super high frequency; UHF: Ultra high frequency.

The application of EM microwave energy to a dielectric media causes a polarization effect that leads to a dielectric heating process. Specifically, ions migrate and dipoles rotate and perform translations trying to align themselves in opposition to the continuous alternating EM field producing molecular motion (Fig. **4**). These simultaneous mechanisms of intermolecular interaction and rotation produce a rapid energy transfer increasing the temperature of the treated samples. However, when the solid substance is a food product, the nutritional and sensorial quality characteristics are maintained at the same time with the molecular structure [103, 107, 108]. The physical (volume, density, and shape) and chemical (moisture content in free and bound state and organic matter) features of the solid matter determine the dielectric properties of the material and have a relevant role in the interaction between foodstuff and microwave radiation [107]. For example, food materials rich in hydrophobic carbohydrates, lipid, and protein molecules are relatively inert with EM radiation due to the low free water content in samples. On the other hand, the presence of hydroxyl groups in carbohydrate molecules and ionizable carboxyl groups of fatty acids, hydrated proteins, polypeptides, and free amino acid fractions promotes the reactive behavior of food matter in a microwave environment [103].

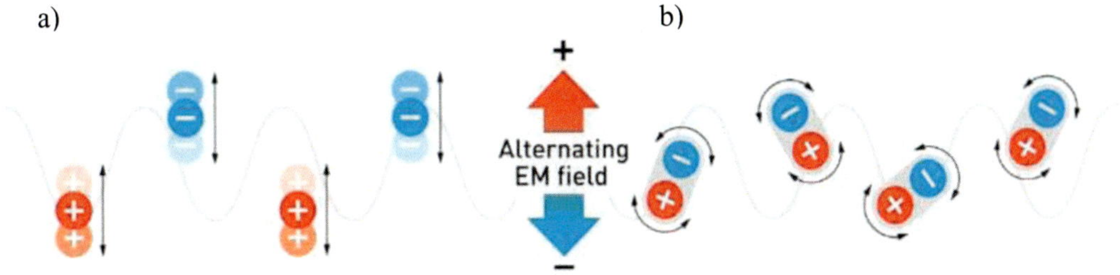

Fig. (4). Sketch of ionic conduction (a) and dipole orientation (b) in the presence of microwave energy (adapted from [14]). EM: Electromagnetic.

Molecular interaction with EM microwave generates direct heating and mass transfer of the internal structure of the samples, because microwave heating hardly produces energy dissipation to the environment. However, in conventional thermal processing (based on conduction, convention, and radiation thermal gradient phenomena), a portion of the heat transmitted is lost to the environment instead of being transferred from the medium to the inner part of the samples (Fig. 5) [6, 106, 109]. In comparison with conventional heating procedures, microwave systems provide a rapid, selective, and uniform heating method with suitable diffusion rates and thermal gradient allowing lower treatment temperatures and shorter time. For these reasons, this type of technology induces limited heat-affected areas as well as energy consumption and consequent detrimental effect on the environment [106, 107].

Food industries had integrated microwave systems in their transformation and preservation processes, as a pre-, post-treatment step or main heating operation, in order to take advantage of their positive effects, especially the reduction in the operation time. The use of microwave as a tool to assist other traditional or emerging procedures is a recent practice in this sector in such a way that the quality of end products is improved at the same time that the shortcomings of the process are amended [101]. EM heating methods, such as microwave technology can be used to ensure the safety of food products and increase their shelf life since this mechanism inactivates microorganisms by pasteurization and sterilization processes in several types of foodstuffs [14]. The microwave blanching process also allows the food resources to decrease enzymatic activity by heat deactivation. The thawing process using microwave equipment requires short treatment time so the chemical degradation and microbial contamination and proliferation of food materials are reduced [111]. Also, dehydration assisted by microwave of several food matrices can be carried out. Lowering the moisture content through solvent removal is favored due to the capillary force by the gas pressure gradient under microwave heating [107]. The microwave procedure operates efficiently for the extraction of high-value bioactive compounds from food materials at a laboratory scale as well as at pilot plant and industrial scale.

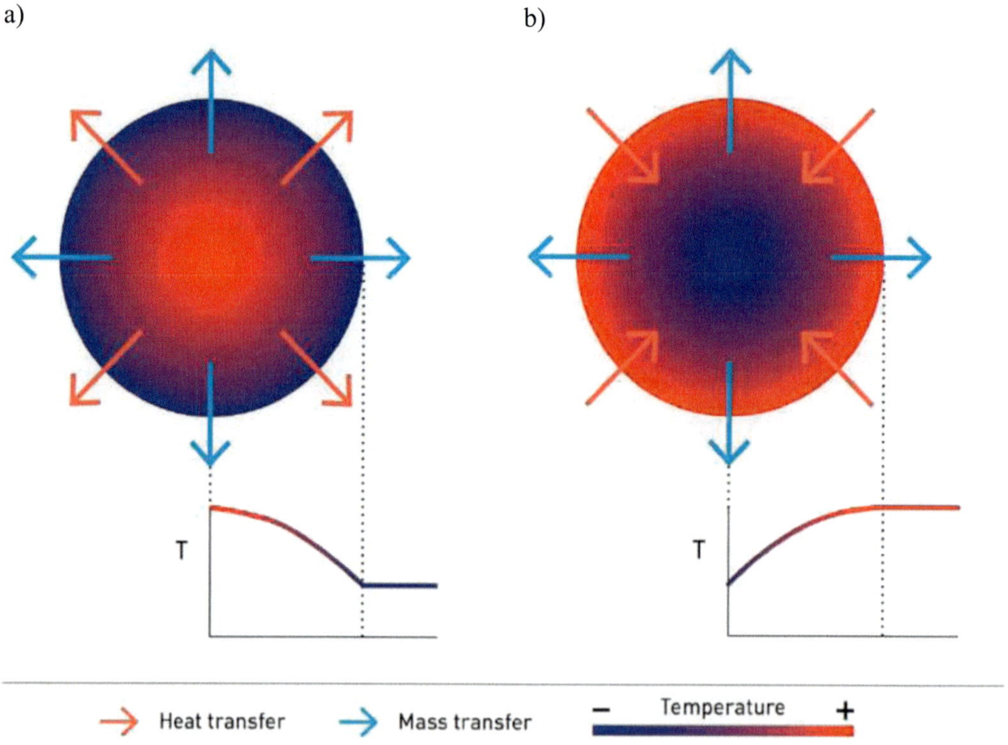

Fig. (5). Comparison of heat and mass transfer involved during microwave (a) and conventional thermal procedures (b) (adapted from [109, 110]).

APPLICATIONS OF MICROWAVE-ASSISTED EXTRACTION

Growing number of studies are using microwave-assisted extraction for enhancing the efficiency of the extraction of a number of compounds from different matrices, but initially, it was mainly applied for analytical purposes and also in order to perform online coupling of various extraction techniques with liquid and gas chromatography [112]. This technique has also been used for reactions during sample preparation, *i.e.* for saponification before the extraction of vitamins A and E in beverages prior to HPLC analysis [113], or for combined digestion and extraction procedures [94]. Some representative examples are summarized in Table **2**.

Table 2. Overview of the possibilities of application of microwave assistance during the extraction of different compounds from biomass.

Foodstuff	Target Compound	Microwave Technique	Microwave Treatment (P; ET; EA; SLR)	Remarks	References
Leatherleaf mahonia	Berberine	MAE	-, 2 min, E or M, 1:60	Destruction of starch particles	[127]
Garden angelica	Furanocoumarins	MAE, MO	40% (1 min), 60% (30 min), M (80%), -	Shorter times than Soxhlet	[115]
Licorice	Glycyrrhizic acid	MAE	60 ºC, 60 min, -, 1:6, 5 stages	Higher yield and lower time, energy and solvent consumption than conventional	[116]
Rubber tree	Benzoic and cinnamic acids	FMAE	50% power, 30 s, M:W:AcH (2:8:03), 1:10	Satisfactory repeatability	[114]
American gingseng	Ginsenosides	MAE	-; 3 min; E (50%), 1:40	Lower time and solvent consumption than Soxhlet and UAE; reliable for sample preparation	[128]
Ginseng root	Ginsenosides	VMW	2000 W; 13 min; -; -	Improves the efficiency of Rb1 and Rd ginsenosides	[118]
Caraway seeds	Terpenes	MAE	1200 W; 60 min; H; 1:20	Cell walls appeared thicker with uniform shapes. Better carvone yield	[117]
Peanuts	Arome-active	MAE	50 W; 11 min; -; -	Increase the formation of guaiacol and phenylacetaldehyde	[120]
Rosemary	Essential oils	MAD	400 W; 30 min; -; -	Saves energy, time (6-fold) or oxygenated fraction, improving yields	[119]
Meat	Fats	MAE	100 W; 5 min; H; 3:10	Saves time (4-fold) and solvent (3-fold), increasing hydrocarbons	[121]

(Table 2) cont.....

Foodstuff	Target Compound	Microwave Technique	Microwave Treatment (P; ET; EA; SLR)	Remarks	References
Propolis	Phenolics, flavonoids	MAE	800 W; 20 s; E (70%); 1:20	High extraction yield, in short timeframes	[122]
Cake	Maltol	MAD	300 W; 30 min; H:A (1:1); 5:20	Saves time (4-fold) and solvent (85%), increases yield (85%)	[123]
Olive oil	Pesticides	MAD	250 W (2 min) + 700 W (8 min); AN:D (90:10); 1:1	Allows recovering of pesticides > 73%	[124]
Cumin	Essential oils	MAM	440 W; <15 min; IL; -	Simplicity and saves time compared to HD, conventional MAE or SFME	[125]
Shiitake	Antioxidants	MAE	800 W; 4 min; -; -	Preserves antioxidants after freeze-drying	[126]
Apple pomace	Polyphenols	MAE	650 W; 54 s; E (62%); 1:23	Higher yield and lower solvent than RE & UAE	[77]
Purple corn	Anthocyanins	MAE	555 W; 19 min; HCl (1.5 M): E (95%) (15:85); 1:20	Higher efficiency and rates than CSE	[129]
Cacao	Caffeine (CF) and theobromine (TB)	FMAE	210 W; 5 min; DW; 0.5:90	Efficiency up to 150% (seeds) for CF and for TB up to 72% (peel)	[130]
Grapes	Pesticides	MAEGCD	600 W; 10 min; H:A (1:1); 2:10	Allows enhanced and simultaneous recoveries of eight pesticides	[131]
Sweet cherries	Anthocyanin	SFME	1000 W; 45 s (4 cycles); -; -	Obtains concentrated extracts only using the water present in fruit as solvent	[132]
Rice	Melatonin	MAE	1000 W; 20 min; M; 1:10	Recoveries between 16-264 mg/kg	[133]
Star anise	Oil	MAE	500 W; 16 min; E; 1:17	Higher yield in shorter time than distillation	[134]
Bread	Ochratoxin A	MAE	-; 10 min; AN; 1:25	Almost 90% recovery for fortification levels, 0.033-3.0 ng/g	[135]
Dragon fruit	Pectin	MAE	400 W; 20 min; -;1:24	Improves pectin yields	[136]

(Table 2) cont.....

Foodstuff	Target Compound	Microwave Technique	Microwave Treatment (P; ET; EA; SLR)	Remarks	References
Cabbage	Bioactive compounds	MAE	100 W; 5 min; E (99.9%); 5:50	Higher sulforaphane yield	[137]
Meat	Lipids	MAE	400 W; 16 min; A:M (2:1); 0.3:5	Reduces extraction time and solvent consumption compared to conventional	[138]
Mango	Antioxidants and pectin	MAE	800 W; 20 min; DW; 2.5:45	Combined recovery of pectin and phenolic compounds	[139]
Basil	Essential oils	SFME	600 W; 30 min; DW; 1:5	Saves time (4-fold) compared to HD	[140]
Green algae	Sulfated polysaccharides	MAHE	-; 10 min; DW; 1:20	Improves polysaccharide yields at reduced extraction time	[141]
Tomato	Sugars, proteins or phenolics	MAE	-; 20 min; E (47.4%); 1:45	High extraction yield and reduced solvent consumption	[142]
Fish	Fluoroquinolones	MAE	150 W; 8 min; M; 1:10	Using methanol as an alternative to reduce extractant and time	[143]
Broccoli	Dehydration and polyphenols	MHG	500 W; 43 min; -; 550 g	Previous dehydration increases extraction yield	[144]
Jackfruit	Pectin	UMAE	-; 29 min; OA; 1:48	Increases yield and reduces time	[145]
Banana	Phenolic compounds	HM	960 W; 6 min; DW (pH=1); 2:100	Improves the efficiency of the process	[146]
Black soybean	Anthocyanins	MAE	569.46 W; 262.54 s; E (59.99%); 1:40	Saves time and improves extraction yield	[147]
Brown macroalgae	Total soluble carbohydrates	MAE	250-600-1000 W; 2-5 min; -; -	Improved extraction compared to UAE	[148]
Legume	Inositol	MAE	1200 W; 3-30 min; E (0-100%); 1:5	Higher extraction yields than conventional	[149]
Artichoke	Polyphenols and flavonoid	HM	180 W; 50 s; E (67.81%); 1:41.78	Higher yield and reduction extraction time	[150]
Green algae	Pigments	MAE	300 W; 1 min; OS: 0.01	Enhanced pigment separation efficiency	[151]

(Table 2) cont.....

Foodstuff	Target Compound	Microwave Technique	Microwave Treatment (P; ET; EA; SLR)	Remarks	References
Red dragon fruit	Pectin	MAE	183 W, 12 min, pH 2.9, 1:56	Higher yield than conventional	[152]
Passion fruit	Pectin	MAE	218 W, 12 min, pH 2.9, 1:57	Higher yield than conventional	[152]

A: Acetone (content, %); AC: acetate; AC: antioxidant capacity; AcH: acetic acid; AN: acetonitrile (content, %); CF: caffeine; MAE: conventional microwave; CSE: conventional solvent extraction; D: dichloromethane (content, %); DMSO: dimethyl sulfoxide; DW: distilled water; E: ethanol (content, %); EA: extraction agent; ET: extraction time; FMAE: focused microwaves-assisted; FQs: fluoroquinolones antibiotics; H: hexane (content, %); HD: hydro-distillation; HD: hydrodistillation; HM: home microwave; IL: ionic liquid; M: Methanol; MAD: microwave assisted distillation; MAE: microwave assisted extraction; MAEGDC: microwave assisted-extraction gas chromatography detection; MAHE: microwave-assisted hydrothermal extraction; MAM: microwave absorption medium; MHG: microwave hydrodiffusion and gravity; MO: microwave oven; P: microwave power; OA: organic acidic; OS: organic solvents; RE: reflux extraction; SFME: solvent-free microwave extraction; SLR: solid-liquid ratio; TAC: Total anthocyanin content; TB: theobromine; TPC: total phenolic content; UAE: ultrasound assisted extraction; UMAE: ultrasonic-microwave-assisted extraction, VMW: vacuum microwave.

Li *et al.* [114] developed a method for the extraction of gallic acid, protocatechuic acid, chlorogenic acid, and caffeic acid from *Eucommia ulmodies* in an open MAE system using methanol:water:glacial acetic acid (2:8:0.3) (v/v/v) and this solvent could be directly used as the mobile phase in HPLC. Waksmundzka-Hajnos *et al.* [115] reported that the microwave-assisted extraction in an open system with methanol provided a yield of furanocoumarins from *Archangelica Officinalis* fruits comparable to the values obtained with ultrasonication at 60 °C and to those obtained in Soxhlet. However, the yields were lower than those provided by accelerated solvent extraction at 100-130 °C for 10 min. Wang *et al.* [116] developed a multi-stage countercurrent extraction of glycyrrhizic acid from *Glycyrrhiza uralensis* in a process offering higher extraction yield and considerable savings in time, energy, and solvent consumption.

Chemat *et al.* [117] studied the operation conditions for the extraction of carvone and limonene from caraway seeds (*Carum carvi L.*). The microwave-treated samples appeared to hit and cell walls were thicker with uniform shapes. Moreover, this green extraction resulted in better carvone and limonene yield, and the obtained extracts were more similar to the true composition of the raw material. Popovich *et al.* [118] evaluated the ginsenosides or triterpene saponins present in ginseng (*Panax quinquefolius*) pre-treated under several drying conditions. They found that vacuum microwave drying methods of ginseng improved both extraction efficiency and retention of individual ginsenoside.

Tigrine-Kordjani *et al.* [119] assessed microwave accelerated distillation in a single stage for the sustainable extraction of edible essential oils used in aromatherapy and in the food industry. The recovery of these compounds from rosemary with MAD was better when compared with hydrodistillation in terms of energy-saving, extraction time (30 min *vs* 3 h), oxygenated fraction (59% *vs* 46%), product yield (0.6% *vs* 0.4%), and product quality. Schirack *et al.* [120] explored the microwave blanching of peanuts as an attractive alternative to conventional oven methods based on the rapid operation, energy savings, and efficiency of process control. It should be indicated that these authors found the occurrence of stale/floral and ashy off-flavors at high process temperatures.

Kwon *et al.* [121] studied the efficiency of microwave-assisted and conventional methods in the extraction of radiation-induced hydrocarbons from commercial hams and sausages. Both methods were acceptable in extracting hydrocarbons from the meat samples, but microwave treatment needed less solvent content (3-fold) and extraction time (4-fold). Trusheva *et al.* [122] applied microwave-assisted extraction for the recovery of biologically active compounds of polar type from propolis. They found that the total phenolics and flavonoid content were adequately extracted with this procedure; although a large amount of non-phenolic and non-flavonoid material was also separated.

Barroso *et al.* [123] proposed a microwave-assisted extraction procedure to assist sample pre-treating steps of an electrochemical method in the determination of maltol in food. These authors found a notable reduction in processing time (from 2 h to 30 min) and extraction solvent (about 85%), as well as an increase of the extracted maltol (from 30 to 85%) when compared to conventional sample pre-treating procedures. Fuentes *et al.* [124] conducted a systematic study of the microwave-assisted extraction coupled with solid-phase extraction of several organophosphorus pesticides from olive oil. The optimized procedure allowed the recovery of compounds using acetonitrile-dichloromethane mixtures for partition ranging from 62 to 99%. The authors stated that under optimized conditions, recoveries of pesticides from olive oil were around 73%.

Zhai *et al.* [125] applied a microwave absorption medium based on ionic liquids for the extraction of essential oils from dried fruits of *Cuminum cyminum L.* and *Illicium verum Hook. f.*, saving time (less than 15 min) when compared to conventional hydrodistillation, microwave-assisted extraction, or solvent-free microwave extraction. Other authors [126] showed that freeze-drying, followed by conventional microwave treatment was more adequate than other drying methods in preserving the antioxidant features (total phenolic content and antioxidant capacity) of Shiitake.

Bai *et al.* [77] optimized a simple and efficient ethanol-based microwave-assisted extraction of phenolic compounds (*i.e.* chlorogenic acid, caffeic acid, syrigin, procyanidin B2, (-)-epicatechin, cinnamic acid, coumaric acid, phlorizin, or quercetin) from industrial apple pomace using response surface methodology. Under optimal extraction conditions, such as reflux or ultrasonic-assisted extraction methods, higher yields than conventional ones, were obtained. Yang and Zhai [129] also used a Box-Behnken design to determine the optimal conditions of microwave-assisted extraction for the recovery of anthocyanins (*i.e.* cyanidin-3- glucoside, pelargonidin-3-glucoside, peonidin-3-glucoside) present in Chinese purple corn cob. The optimal conditions provided higher total anthocyanins content in a highly efficient and rapid way when compared to the conventional solvent extraction.

González-Núñez and Cañizares-Macías [130] proposed the focused microwave-assisted extraction (FMAE) to improve the extraction efficiency of obromine and caffeine from cacao seeds, peels, and powder. Optimal conditions, determined by a full factorial design, provided higher amounts of these compounds than a conventional stirring method. Furthermore, theobromine extraction efficiency improved by 15% in seeds and by 72% in peels and caffeine extraction efficiency increased by 36% in peels and 153% in seeds.

Lagunas-Allué *et al.* [131] developed a microwave-assisted extraction gas chromatography detection (MAEGD) method for the simultaneous extraction of eight pesticides in grapes. The different MAE parameters were evaluated and optimized by applying an experimental design. Operation at 105 °C in 10 min and 600 W can provide good recoveries (82-107%) using a small volume of hexane/acetone (1:1, *v/v*) mixture used as a solvent.

Setyaningsih *et al.* [133] reported for the first time the use of MAE to optimize the extraction of melatonin from rice grains using Response Surface Methodology (RSM). Under optimal conditions (195 °C, 1000 W, 20 min, 100% methanol) the highest concentration of melatonin was obtained from exotic rice. This method showed satisfactory linearity and precision. Grigoras *et al.* [132] evaluated solvent-free microwave-assisted extraction (SFMAE) to extract anthocyanins contained in sweet cherries with only 4 irradiation cycles of 45 s each at 1000 W. The major compounds, cyanidin-3-O-glucoside and cyanidin-3-*O*-rutinoside, were further purified.

Cai *et al.* [134] applied microwave-assisted extraction with ethanol to extract star anise oil from *Illicium verum* Hook. f. obtaining higher yields than by steam distillation (SD). Operating under MAE conditions optimized by RSM, at 500 W and 16 min, the major components were oxygenated organic compounds.

Transanethole yields were improved over those with SD than with MAE, but the oils extracted by both technologies have strong antioxidant activities. Paíga *et al.* [135] described an analytical method using MAE and liquid chromatography with fluorescence detection (LC-FD) for the determination of ochratoxin A in bread samples. Operation of MAE at 80 °C and 10 minutes, optimized using an orthogonal composite design coupled with RSM, was accurate and precise, with recoveries in the range 79-87%, for fortification levels between 0.033 and 3.0 ng/g.

Thirugnanasambandham *et al.* [136] optimized the MAE of pectin from dragon fruit peel by using Box-Behnken design coupled with RSM. Under optimal conditions (400 W, 45 °C, 20 min and 24 g/mL solid-liquid ratio), 7.5% of pectin was extracted. Chaisamlitpol *et al.* [137] evaluated the effects of hot air drying and low-pressure superheated steam drying on the subsequent MAE of bioactives from cabbage outer leaves. The phenolic content and the antioxidant activity of the extracts from cabbages dried by both methods were significantly lower than those from the fresh sample, but the sulforaphane contents of the extracts from the dried cabbages were significantly higher.

Medina *et al.* [138] developed and validated a new method for MAE of meat lipids using a non-halogenated solvent and low temperature to minimize the risk of chemical changes during the process. Lipid extraction using MAE was optimized by an experimental design at 54 °C, 400 W, and 16 min using ethyl acetate:methanol (2:1, v/v) as a solvent. Rojas *et al.* [139] evaluated the microwave-assisted extraction of bioactives from mango (*Mangifera indica*) peels. Results demonstrated that operating under the best conditions (115 °C, 20 min), mango peels are an important source of pectin and polyphenols.

Studies from Chenni *et al.* [140] compared hydrodistillation and solvent-free microwave extraction (SFME) for the extraction of the essential oil from Egyptian Basil Leaves (*Ocimum basilicum* L.). The use of SFME saved time (50%) compared to hydrodistillation and provided essential oils with yields similar to conventional techniques and with adequate properties for applications in the food industry, pharmacy, and cosmetology. Tsubaki *et al.* [141] compared microwave-assisted hydrothermal extraction and conventional hot water to extract sulfated polysaccharides from dried *Ulva sp.* The microwave heated method efficiently improved polysaccharide yields, reduced extraction time and the obtained sulfated polysaccharides are potential raw materials for food and not food applications.

Pinela *et al.* [142] proposed MAE technology using 47% ethanol as solvent for the extraction of sugars, proteins, and phenolics from tomato by-products. The results confirmed that the technique saved time, increased the extraction yield

(75%), and reduced solvent consumption for the production of food ingredients with high levels of nutrients and antioxidants properties. Aufartová *et al.* [143] used a MAE coupled with solid-phase extraction using methanol as extractant for the determination of fluoroquinolones in fish samples from aquaculture farms. The technique was a viable alternative saving time and solvent and lowering costs.

Ferreira *et al.* [144] used a MHG process without the addition of any solvent/water for dehydrated broccoli wastes. The hydrodiffusion produced a dried material with 12% moisture in 43 min, preserving polysaccharides and proteins. The diffused water contained phenolic compounds, free sugars, 9 mg/mL amino acids, and 356 µg/mL glucosinolates. Studies from Xu *et al.* [145] showed the optimum conditions to obtain pectins from jackfruit peels by UMAE (ultrasonic-microwave-assisted extraction) using a nontoxic organic acidic solvent. The pectin yield by UMAE (21.5% in 30 min) was higher than by conventional extraction (17.2% in 2 h) and the products could be used as natural antioxidants in the food industry.

With a household microwave operating at 960 W, 6 min, Vu *et al.* [146] recovered phenolics from the dried banana peel, mostly recovered during the heating period (2 min) till the boiling point. Kumar *et al.* [147] used an experimental design to optimize the extraction of anthocyanins from the black soybean seed coat. Results showed a higher yield using a MAE technology (1.2 *vs* 5.1 g/L) compared to conventional methods. MAE seems to be much efficient and the phenolic extracted would be used as a functional ingredient in foods.

Garcia-Vaquero *et al.* [148] extracted sulfated fucose-containing polysaccharides (fucoidans) from brown macroalgae using MAE (250-1000 W, 2-5 min). This process provided higher carbohydrate and phenolic yields and antioxidant properties in the extracts than an ultrasound-assisted method. Studies from Zuluaga *et al.* [149] were pioneers obtaining bioactive inositol from legume pods and seeds using microwave-assisted extraction with aqueous ethanol as solvent. The solvent composition greatly influenced the inositol recovery and the sugar concentration in the extracts.

Rudić *et al.* [150] used RSM to develop a fast MAE procedure for the extraction of high yields of natural antioxidants from artichoke leaves dust. Their procedure yielded extracts with higher total phenolic and flavonoid content and antioxidant activities than conventional ones, and reduce the extraction time. The use of an aqueous ethanolic solution as a green solvent favored the food and pharmaceutical applications, due to the healthier character of the products compared to synthetic products. Martins *et al.* [151] showed an efficient process to separate pigments from green macroalgae, applicable to other raw materials. They developed a sequential process consisting of i) ultrasound-assisted solid-liquid extraction of pigments using ethanol, ii) separation of chlorophylls and xanthophylls by applying liquid-liquid extraction with a 50:30:20 ethanol:hexane:water mixture, and ending with the iii) pigments polishing and solvents recycle and reuse. Thu Dao *et al.* [152] found that microwave-assisted heating was helpful in achieving maximum pectin yield extracted from peels of red-flesh and white-flesh dragon fruit and from passion fruit. This flexible technique allowed modulating the esterification degree and to obtain a product similar to commercial ones. Macedo *et al.* [153] extracted antioxidant phenolics from olive pomace using microwave heating simultaneous with enzymatic treatment with cellulase, pectinase, and tannase. Microwave-assisted extraction allowed shorter times for the extraction of phenolics than conventional solvent extraction with water as a solvent and the use of enzymes during microwave processing led to increased extractability of phenolics at a higher temperature in a faster heating strategy. Buratto *et al.* [154] tried the valorization of Açaí berry pulp and seed wastes of the industrial processing by microwave-assisted techniques (ambient-pressure microwave extraction and pressurized microwave extraction at 1.5-3 bar). The pulp and seeds showed potential as antioxidant agents. Microwave pre-treatments increased the polyphenol contents and the antioxidant activities of the extracts obtained from the residues, yielding products of interest for the cosmetic or food industry.

CONCLUSION

The increasing demand for natural and safe ingredients for the food and cosmetic industry has promoted the utilization of greener solvents and more efficient technologies based on intensification strategies, to enhance the extraction performance both for analytical and industrial extraction. Microwave assistance offers advantages over other techniques, mainly derived from the dramatic reduction in processing times and the possibility of combining it with other strategies.

LIST OF ABBREVIATIONS

EAE	Extrusion-Assisted Extraction
HHP	High Hydrostatic Pressure Extraction
MAE	Microwave-Assisted Extraction
EAE*	Enzymatic-Assisted Extraction
PEF	Pulsed Electric Field
PLE	Pressurized Liquid Extraction
PLE	Pressurized Liquid Extraction

SFE	Supercritical Fluid Extraction
UAE	Ultrasound Assisted-Extraction
UMAE	Ultrasonic Microwave-Assisted Extraction
VMAE	Vacuum Microwave-Assisted Extraction
MHG	Microwave Hydro-diffusion and Gravity
SFME	Solvent-Free Microwave Extraction
VMHD	Vacuum Microwave Hydro-Distillation
EM	Electromagnetic
EHF	Extremely high frequency
SHF	Super high frequency
UHF	Ultra-high frequency
RSM	Response surface methodology
HPLC	High-performance liquid chromatography
LC-FD	Liquid chromatography with fluorescence detection

CONSENT FOR PUBLICATION

Not applicable.

CONFLICT OF INTEREST

The authors declare no conflict of interest, financial or otherwise.

ACKNOWLEDGEMENTS

M.A.-V. thanks the Ministry of Science, Innovation, and Universities of Spain for her predoctoral FPI grant (PRE2019-090567). N.F.F. thanks the Xunta de Galicia for her postdoctoral grant (ED481B 2018/071 and ED481D-2022-018). M.D.T. acknowledges the Ministry of Science, Innovation, and Universities of Spain for her postdoctoral grants (RYC2018-024454-I) and to the Consellería de Cultura, Educación e Universidade da Xunta de Galicia (ED431F 2020/01).

REFERENCES

[1] Wen L, Zhang Z, Sun DW, Sivagnanam SP, Tiwari BK. Combination of emerging technologies for the extraction of bioactive compounds. Crit Rev Food Sci Nutr 2020; 60(11): 1826-41. [http://dx.doi.org/10.1080/10408398.2019.1602823] [PMID: 30990060]

[2] Flórez N, Conde E, Domínguez H. Microwave assisted water extraction of plant compounds. J Chem Technol Biotechnol 2015; 90(4): 590-607. [http://dx.doi.org/10.1002/jctb.4519]

[3] Luque de Castro, M. D.; Castillo-Peinado, L. S. (2016). In K. Knoerzer, P. Juliano & G. Smithers (Eds.), Innovative food processing technologies. Extraction, separation, component modification and process intensification (pp. 57-110). Woodhead Publishing: Duxford.

[4] Azmir J, Zaidul ISM, Rahman MM, *et al.* Techniques for extraction of bioactive compounds from plant materials: A review. J Food Eng 2013; 117(4): 426-36.
[http://dx.doi.org/10.1016/j.jfoodeng.2013.01.014]

[5] de la Calle I, Costas-Rodríguez M. The application of green solvents in separation processes. Amsterdam: Elsevier 2017; pp. 253-300.
[http://dx.doi.org/10.1016/B978-0-12-805297-6.00009-7]

[6] Gomez L, Tiwari B, Garcia-Vaquero M. Sustainable seaweed technologies Cultivation, biorefinery, and applications. Amsterdam: Elsevier 2020; pp. 207-24.
[http://dx.doi.org/10.1016/B978-0-12-817943-7.00008-1]

[7] Bruno SF, Ekorong FJAA, Karkal SS, Cathrine MSB, Kudre TG. Green and innovative techniques for recovery of valuable compounds from seafood by-products and discards: A review. Trends Food Sci Technol 2019; 85: 10-22.
[http://dx.doi.org/10.1016/j.tifs.2018.12.004]

[8] Chemat F, Vian MA, Cravotto G. Green extraction of natural products: concept and principles. Int J Mol Sci 2012; 13(7): 8615-27.
[http://dx.doi.org/10.3390/ijms13078615] [PMID: 22942724]

[9] Riaz MN, Asif M, Ali R. Stability of vitamins during extrusion. Crit Rev Food Sci Nutr 2009; 49(4): 361-8.
[http://dx.doi.org/10.1080/10408390802067290] [PMID: 19234945]

[10] Galanakis CM. Recovery of high added-value components from food wastes: Conventional, emerging technologies and commercialized applications. Trends Food Sci Technol 2012; 26(2): 68-87.
[http://dx.doi.org/10.1016/j.tifs.2012.03.003]

[11] Shouqin Z, Junjie Z, Changzhen W. Novel high pressure extraction technology. Int J Pharm 2004; 278(2): 471-4.
[http://dx.doi.org/10.1016/j.ijpharm.2004.02.029] [PMID: 15196650]

[12] Destandau, E.; Michel, T.; Elfakir, C. (2013). In M. A. Rostagno & J. M. Prado (Eds.), Natural product extraction: Principles and applications (pp. 113-156). RSC Green Chemistry No. 21: Dorchester.

[13] Chemat, F.; Cravotto, G. (2013a). In F. Ghemat & G. Cravotto (Eds.), Microwave-assisted extraction for bioactive compounds: Theory and practice (pp. 15-52). Springer: London.
[http://dx.doi.org/10.1007/978-1-4614-4830-3]

[14] Martins CPC, Cavalcanti RN, Couto SM, *et al.* Microwave processing: Current background and effects on the physicochemical and microbiological aspects of dairy products. Compr Rev Food Sci Food Saf 2019; 18(1): 67-83.
[http://dx.doi.org/10.1111/1541-4337.12409] [PMID: 33337017]

[15] Araújo RG, Rodriguez-Jasso RM, Ruiz HA, Govea-Salas M, Pintado ME, Aguilar CN. Process optimization of microwave-assisted extraction of bioactive molecules from avocado seeds. Ind Crops Prod 2020; 154: 112623.
[http://dx.doi.org/10.1016/j.indcrop.2020.112623]

[16] Gallo M, Ferracane R, Graziani G, Ritieni A, Fogliano V. Microwave assisted extraction of phenolic compounds from four different spices. Molecules 2010; 15(9): 6365-74.
[http://dx.doi.org/10.3390/molecules15096365] [PMID: 20877228]

[17] Singh PP, Saldaña MDA. Subcritical water extraction of phenolic compounds from potato peel. Food Res Int 2011; 44(8): 2452-8.
[http://dx.doi.org/10.1016/j.foodres.2011.02.006]

[18] Fang J, Wang Z, Wang P, Wang M. Extraction, structure and bioactivities of the polysaccharides from *Ginkgo biloba*: A review. Int J Biol Macromol 2020; 162: 1897-905.
[http://dx.doi.org/10.1016/j.ijbiomac.2020.08.141] [PMID: 32827622]

[19] Gligor O, Mocan A, Moldovan C, Locatelli M, Crişan G, Ferreira ICFR. Enzyme-assisted extractions of polyphenols – A comprehensive review. Trends Food Sci Technol 2019; 88: 302-15. [http://dx.doi.org/10.1016/j.tifs.2019.03.029]

[20] Kitrytė V, Povilaitis D, Kraujalienė V, Šulniūtė V, Pukalskas A, Venskutonis PR. Fractionation of sea buckthorn pomace and seeds into valuable components by using high pressure and enzyme-assisted extraction methods. Lebensm Wiss Technol 2017; 85: 534-8. [http://dx.doi.org/10.1016/j.lwt.2017.02.041]

[21] Vorobiev E, Lebovka N. Processing of foods and biomass feedstocks by pulsed electric energy Processing of foods and biomass feedstocks by pulsed electric energy. Cham: Springer 2020; pp. 1-399. [http://dx.doi.org/10.1007/978-3-030-40917-3]

[22] Leong YK, Yang FC, Chang JS. Extraction of polysaccharides from edible mushrooms: Emerging technologies and recent advances. Carbohydr Polym 2021; 251: 117006. [http://dx.doi.org/10.1016/j.carbpol.2020.117006] [PMID: 33142573]

[23] Andreou V, Psarianos M, Dimopoulos G, Tsimogiannis D, Taoukis P. Effect of pulsed electric fields and high pressure on improved recovery of high-added-value compounds from olive pomace. J Food Sci 2020; 85(5): 1500-12. [http://dx.doi.org/10.1111/1750-3841.15122] [PMID: 32267966]

[24] Álvarez-Rivera G, Bueno M, Ballesteros-Vivas D, Mendiola JA, Ibañez E. Liquid-phase extraction (Handbooks in separation science). Amsterdam: Elsevier 2020; pp. 375-98. [http://dx.doi.org/10.1016/B978-0-12-816911-7.00013-X]

[25] Plaza M, Abrahamsson V, Turner C. Extraction and neoformation of antioxidant compounds by pressurized hot water extraction from apple byproducts. J Agric Food Chem 2013; 61(23): 5500-10. [http://dx.doi.org/10.1021/jf400584f] [PMID: 23675866]

[26] Ameer K, Shahbaz HM, Kwon JH. Green extraction methods for polyphenols from plant matrices and their byproducts: A review. Compr Rev Food Sci Food Saf 2017; 16(2): 295-315. [http://dx.doi.org/10.1111/1541-4337.12253] [PMID: 33371540]

[27] Herrero M, Castro-Puyana M, Mendiola JA, Ibañez E. Compressed fluids for the extraction of bioactive compounds. Trends Analyt Chem 2013; 43: 67-83. [http://dx.doi.org/10.1016/j.trac.2012.12.008]

[28] Álvarez-Viñas M, Rodríguez-Seoane P, Flórez-Fernández N, *et al.* Subcritical water for the extraction and hydrolysis of protein and other fractions in biorefineries from agro-food wastes and algae: a review. Food Bioprocess Technol 2021; 14(3): 373-87. [http://dx.doi.org/10.1007/s11947-020-02536-4]

[29] Ulug SK, Jahandideh F, Wu J. Novel technologies for the production of bioactive peptides. Trends Food Sci Technol 2021; 108: 27-39. [http://dx.doi.org/10.1016/j.tifs.2020.12.002]

[30] Cernadas H, Flórez-Fernández N, González-Muñoz MJ, Domínguez H, Torres MD. Retrieving of high-value biomolecules from edible *Himanthalia elongata* brown seaweed using hydrothermal processing. Food Bioprod Process 2019; 117: 275-86. [http://dx.doi.org/10.1016/j.fbp.2019.07.015]

[31] Huang CY, Kuo JM, Wu SJ, Tsai HT. Isolation and characterization of fish scale collagen from tilapia (*Oreochromis* sp.) by a novel extrusion–hydro-extraction process. Food Chem 2016; 190: 997-1006. [http://dx.doi.org/10.1016/j.foodchem.2015.06.066] [PMID: 26213067]

[32] Puri M, Kaur A, Schwarz WH, Singh S, Kennedy JF. Molecular characterization and enzymatic hydrolysis of naringin extracted from kinnow peel waste. Int J Biol Macromol 2011; 48(1): 58-62. [http://dx.doi.org/10.1016/j.ijbiomac.2010.09.012] [PMID: 20920523]

[33] Bildstein M, Lohmann M, Hennigs C, Krause A, Hilz H. An enzyme-based extraction process for the

purification and enrichment of vegetable proteins to be applied in bakery products. Eur Food Res Technol 2008; 228(2): 177-86.
[http://dx.doi.org/10.1007/s00217-008-0921-z]

[34] Ruiz-Terán F, Perez-Amador I, López-Munguia A. Enzymatic extraction and transformation of glucovanillin to vanillin from vanilla green pods. J Agric Food Chem 2001; 49(11): 5207-9.
[http://dx.doi.org/10.1021/jf010723h] [PMID: 11714304]

[35] Li J, Sun W, Ramaswamy HS, *et al.* High pressure extraction of astaxanthin from shrimp waste (*Penaeus vannamei* boone): Effect on yield and antioxidant activity. J Food Process Eng 2017; 40(2): e12353.
[http://dx.doi.org/10.1111/jfpe.12353]

[36] Zhao Y, Ding Y, Wang D, Deng Y, Zhao Y. Effect of high hydrostatic pressure conditions on the composition, morphology, rheology, thermal behavior, color, and stability of black garlic melanoidins. Food Chem 2021; 337: 127790.
[http://dx.doi.org/10.1016/j.foodchem.2020.127790] [PMID: 32799165]

[37] Pataro G, Carullo D, Bakar Siddique MA, Falcone M, Donsì F, Ferrari G. Improved extractability of carotenoids from tomato peels as side benefits of PEF treatment of tomato fruit for more energy-efficient steam-assisted peeling. J Food Eng 2018; 233: 65-73.
[http://dx.doi.org/10.1016/j.jfoodeng.2018.03.029]

[38] Puértolas E, Cregenzán O, Luengo E, Álvarez I, Raso J. Pulsed-electric-field-assisted extraction of anthocyanins from purple-fleshed potato. Food Chem 2013; 136(3-4): 1330-6.
[http://dx.doi.org/10.1016/j.foodchem.2012.09.080] [PMID: 23194531]

[39] Boussetta N, Vorobiev E, Le LH, Cordin-Falcimaigne A, Lanoisellé JL. Application of electrical treatments in alcoholic solvent for polyphenols extraction from grape seeds. Lebensm Wiss Technol 2012; 46(1): 127-34.
[http://dx.doi.org/10.1016/j.lwt.2011.10.016]

[40] Akay S, Alpak I, Yesil-Celiktas O. Effects of process parameters on supercritical CO_2 extraction of total phenols from strawberry (*Arbutus unedo* L.) fruits: An optimization study. J Sep Sci 2011; 34(15): 1925-31.
[http://dx.doi.org/10.1002/jssc.201100361] [PMID: 21695688]

[41] Prado JM, Dalmolin I, Carareto NDD, *et al.* Supercritical fluid extraction of grape seed: Process scale-up, extract chemical composition and economic evaluation. J Food Eng 2012; 109(2): 249-57.
[http://dx.doi.org/10.1016/j.jfoodeng.2011.10.007]

[42] Rodríguez-Seoane P, Díaz-Reinoso B, González-Muñoz MJ, Fernández de Ana Portela C, Domínguez H. Innovative technologies for the extraction of saccharidic and phenolic fractions from *Pleurotus eryngii.* Lebensm Wiss Technol 2019; 101: 774-82.
[http://dx.doi.org/10.1016/j.lwt.2018.11.062]

[43] Farías-Campomanes AM, Rostagno MA, Meireles MAA. Production of polyphenol extracts from grape bagasse using supercritical fluids: Yield, extract composition and economic evaluation. J Supercrit Fluids 2013; 77: 70-8.
[http://dx.doi.org/10.1016/j.supflu.2013.02.006]

[44] Andrade KS, Gonçalvez RT, Maraschin M, Ribeiro-do-Valle RM, Martínez J, Ferreira SRS. Supercritical fluid extraction from spent coffee grounds and coffee husks: Antioxidant activity and effect of operational variables on extract composition. Talanta 2012; 88: 544-52.
[http://dx.doi.org/10.1016/j.talanta.2011.11.031] [PMID: 22265539]

[45] Quan C, Turner C. Extraction of astaxanthin from shrimp waste using pressurized hot ethanol. Chromatographia 2009; 70(1-2): 247-51.
[http://dx.doi.org/10.1365/s10337-009-1113-0]

[46] Huamán-Leandro LR, González-Muñoz MJ, Fernández-de-Ana C, Rodríguez-Blanco A, Torres MD, Domínguez H. Autohydrolysis of *Lentinus edodes* for obtaining extracts with antiradical properties.

Foods 2020; 9(1): 74.
[http://dx.doi.org/10.3390/foods9010074] [PMID: 31936523]

[47] Cheigh CI, Chung EY, Chung MS. Enhanced extraction of flavanones hesperidin and narirutin from Citrus unshiu peel using subcritical water. J Food Eng 2012; 110(3): 472-7.
[http://dx.doi.org/10.1016/j.jfoodeng.2011.12.019]

[48] Miron TL, Plaza M, Bahrim G, Ibáñez E, Herrero M. Chemical composition of bioactive pressurized extracts of Romanian aromatic plants. J Chromatogr A 2011; 1218(30): 4918-27.
[http://dx.doi.org/10.1016/j.chroma.2010.11.055] [PMID: 21163488]

[49] Miron TL, Herrero M, Ibáñez E. Enrichment of antioxidant compounds from lemon balm (*Melissa officinalis*) by pressurized liquid extraction and enzyme-assisted extraction. J Chromatogr A 2013; 1288: 1-9.
[http://dx.doi.org/10.1016/j.chroma.2013.02.075] [PMID: 23528869]

[50] Rajha H, Ziegler W, Louka N, *et al.* Effect of the drying process on the intensification of phenolic compounds recovery from grape pomace using accelerated solvent extraction. Int J Mol Sci 2014; 15(10): 18640-58.
[http://dx.doi.org/10.3390/ijms151018640] [PMID: 25322155]

[51] Xu H, Wang W, Jiang J, Yuan F, Gao Y. Subcritical water extraction and antioxidant activity evaluation with on-line HPLC-ABTS(·+) assay of phenolic compounds from marigold (*Tagetes erecta* L.) flower residues. J Food Sci Technol 2015; 52(6): 3803-11.
[PMID: 26028765]

[52] Gil-Ramírez A, Mendiola JA, Arranz E, *et al.* Highly isoxanthohumol enriched hop extract obtained by pressurized hot water extraction (PHWE). Chemical and functional characterization. Innov Food Sci Emerg Technol 2012; 16: 54-60.
[http://dx.doi.org/10.1016/j.ifset.2012.04.006]

[53] Srinivas K, King JW, Monrad JK, Howard LR, Zhang D. Pressurized solvent extraction of flavonoids from grape pomace utilizing organic acid additives. Ital J Food Saf 2011; 23: 90-105.

[54] Kim JW, Nagaoka T, Ishida Y, Hasegawa T, Kitagawa K, Lee SC. Subcritical water extraction of nutraceutical compounds from citrus pomaces. Sep Sci Technol 2009; 44(11): 2598-608.
[http://dx.doi.org/10.1080/01496390903014375]

[55] Sanz V, Flórez-Fernández N, Domínguez H, Torres MD. Clean technologies applied to the recovery of bioactive extracts from *Camellia sinensis* leaves agricultural wastes. Food Bioprod Process 2020; 122: 214-21.
[http://dx.doi.org/10.1016/j.fbp.2020.05.007]

[56] Anisa NI, Azian N, Sharizan M, Iwai Y. Temperature effects on diffusion coefficient for 6-gingerol and 6-shogaol in subcritical water extraction. J Phys Conf Ser 2014; 495: 012009.
[http://dx.doi.org/10.1088/1742-6596/495/1/012009]

[57] Arapitsas P, Turner C. Pressurized solvent extraction and monolithic column-HPLC/DAD analysis of anthocyanins in red cabbage. Talanta 2008; 74(5): 1218-23.
[http://dx.doi.org/10.1016/j.talanta.2007.08.029] [PMID: 18371772]

[58] Liu J, Sandahl M, Sjöberg PJR, Turner C. Pressurised hot water extraction in continuous flow mode for thermolabile compounds: extraction of polyphenols in red onions. Anal Bioanal Chem 2014; 406(2): 441-5.
[http://dx.doi.org/10.1007/s00216-013-7370-7] [PMID: 24091739]

[59] Ravber M, Knez Ž, Škerget M. Simultaneous extraction of oil- and water-soluble phase from sunflower seeds with subcritical water. Food Chem 2015; 166: 316-23.
[http://dx.doi.org/10.1016/j.foodchem.2014.06.025] [PMID: 25053062]

[60] Carrera C, Ruiz-Rodríguez A, Palma M, Barroso CG. Ultrasound-assisted extraction of amino acids from grapes. Ultrason Sonochem 2015; 22: 499-505.

[http://dx.doi.org/10.1016/j.ultsonch.2014.05.021] [PMID: 24927904]

[61] Cai M, Luo Y, Chen J, Liang H, Sun P. Optimization and comparison of ultrasound-assisted extraction and microwave-assisted extraction of shikimic acid from Chinese star anise. Separ Purif Tech 2014; 133: 375-9.
[http://dx.doi.org/10.1016/j.seppur.2014.06.064]

[62] Samaram S, Mirhosseini H, Tan CP, Ghazali HM. Ultrasound-assisted extraction and solvent extraction of papaya seed oil: Crystallization and thermal behavior, saturation degree, color and oxidative stability. Ind Crops Prod 2014; 52: 702-8.
[http://dx.doi.org/10.1016/j.indcrop.2013.11.047]

[63] Yang YC, Wei MC, Hong SJ, Huang TC, Lee SZ. Development/optimization of a green procedure with ultrasound-assisted improved supercritical carbon dioxide to produce extracts enriched in oleanolic acid and ursolic acid from *Scutellaria barbata* D. Don. Ind Crops Prod 2013; 49: 542-53. a
[http://dx.doi.org/10.1016/j.indcrop.2013.05.013]

[64] Yang YC, Wei MC, Huang TC, Lee SZ, Lin SS. Comparison of modified ultrasound-assisted and traditional extraction methods for the extraction of baicalin and baicalein from Radix Scutellariae. Ind Crops Prod 2013; 45: 182-90. b
[http://dx.doi.org/10.1016/j.indcrop.2012.11.041]

[65] Bajerová P, Adam M, Bajer T, Ventura K. Comparison of various techniques for the extraction and determination of antioxidants in plants. J Sep Sci 2014; 37(7): 835-44.
[http://dx.doi.org/10.1002/jssc.201301139] [PMID: 24470390]

[66] Carrera C, Ruiz-Rodríguez A, Palma M, Barroso CG. Ultrasound assisted extraction of phenolic compounds from grapes. Anal Chim Acta 2012; 732: 100-4.
[http://dx.doi.org/10.1016/j.aca.2011.11.032] [PMID: 22688040]

[67] Khan MK, Abert-Vian M, Fabiano-Tixier AS, Dangles O, Chemat F. Ultrasound-assisted extraction of polyphenols (flavanone glycosides) from orange (*Citrus sinensis L.*) peel. Food Chem 2010; 119(2): 851-8.
[http://dx.doi.org/10.1016/j.foodchem.2009.08.046]

[68] Minjares-Fuentes R, Femenia A, Garau MC, Meza-Velázquez JA, Simal S, Rosselló C. Ultrasound-assisted extraction of pectins from grape pomace using citric acid: A response surface methodology approach. Carbohydr Polym 2014; 106: 179-89.
[http://dx.doi.org/10.1016/j.carbpol.2014.02.013] [PMID: 24721067]

[69] Abramov V, Abramov O, Bulgakov V, Sommer F. Solidification of aluminium alloys under ultrasonic irradiation using water-cooled resonator. Mater Lett 1998; 37(1-2): 27-34.
[http://dx.doi.org/10.1016/S0167-577X(98)00064-0]

[70] Jovanovic-Malinovska R, Kuzmanova S, Winkelhausen E. Application of ultrasound for enhanced extraction of prebiotic oligosaccharides from selected fruits and vegetables. Ultrason Sonochem 2015; 22: 446-53.
[http://dx.doi.org/10.1016/j.ultsonch.2014.07.016] [PMID: 25116595]

[71] Wang XS, Wu YF, Dai SL, Chen R, Shao Y. Ultrasound-assisted extraction of geniposide from *Gardenia jasminoides*. Ultrason Sonochem 2012; 19(6): 1155-9.
[http://dx.doi.org/10.1016/j.ultsonch.2012.03.012] [PMID: 22579801]

[72] Li Y, Fabiano-Tixier AS, Tomao V, Cravotto G, Chemat F. Green ultrasound-assisted extraction of carotenoids based on the bio-refinery concept using sunflower oil as an alternative solvent. Ultrason Sonochem 2013; 20(1): 12-8. b
[http://dx.doi.org/10.1016/j.ultsonch.2012.07.005] [PMID: 22884112]

[73] Wang J, Sun B, Cao Y, Tian Y, Li X. Optimisation of ultrasound-assisted extraction of phenolic compounds from wheat bran. Food Chem 2008; 106(2): 804-10.
[http://dx.doi.org/10.1016/j.foodchem.2007.06.062]

[74] Huang T, Tu Z, Xinchen-Shangguan, Wang H, Zhang L, Sha X. Rheological and structural properties of fish scales gelatin: Effects of conventional and ultrasound-assisted extraction. Int J Food Prop 2017; 20(2): 1210-20.

[75] Rostagno MA, Palma M, Barroso CG. Microwave assisted extraction of soy isoflavones. Anal Chim Acta 2007; 588(2): 274-82.
[http://dx.doi.org/10.1016/j.aca.2007.02.010] [PMID: 17386821]

[76] Périno-Issartier S, Zill-e-Huma , Abert-Vian M, Chemat F. Solvent free microwave-assisted extraction of antioxidants from sea buckthorn (*Hippophae rhamnoides*) food by products. Food Bioprocess Technol 2011; 4(6): 1020-8.
[http://dx.doi.org/10.1007/s11947-010-0438-x]

[77] Bai XL, Yue TL, Yuan YH, Zhang HW. Optimization of microwave-assisted extraction of polyphenols from apple pomace using response surface methodology and HPLC analysis. J Sep Sci 2010; 33(23-24): 3751-8.
[http://dx.doi.org/10.1002/jssc.201000430] [PMID: 21077127]

[78] Turk M, Perino S, Cendres A, Petitcolas E, Soubrat T, Chemat F. Alternative process for strawberry juice processing: Microwave hydrodiffusion and gravity. Lebensm Wiss Technol 2017; 84: 626-33.
[http://dx.doi.org/10.1016/j.lwt.2017.06.030]

[79] López-Hortas L, Falqué E, Domínguez H, Torres MD. Microwave Hydrodiffusion and Gravity (MHG) Extraction from different raw materials with cosmetic applications. Molecules 2019; 25(1): 92.
[http://dx.doi.org/10.3390/molecules25010092] [PMID: 31881788]

[80] Chemat F, Abert Vian M, Ravi HK, *et al.* Review of alternative solvents for green extraction of food and natural products: Panorama, principles, applications and prospects. Molecules 2019; 24(16): 3007.
[http://dx.doi.org/10.3390/molecules24163007] [PMID: 31430982]

[81] Lucchesi ME, Chemat F, Smadja J. Solvent-free microwave extraction of essential oil from aromatic herbs: comparison with conventional hydro-distillation. J Chromatogr A 2004; 1043(2): 323-7.
[http://dx.doi.org/10.1016/j.chroma.2004.05.083] [PMID: 15330107]

[82] Aboudaou M, Ferhat MA, Hazzit M, Ariño A, Djenane D. Solvent free-microwave green extraction of essential oil from orange peel (*Citrus sinensis* L.): effects on shelf life of flavored liquid whole eggs during storage under commercial retail conditions. J Food Meas Charact 2019; 13(4): 3162-72.
[http://dx.doi.org/10.1007/s11694-019-00239-9]

[83] Mohagheghniapour A, Saharkhiz MJ, Golmakani MT, Niakousari M. Variations in chemical compositions of essential oil from sour orange (*Citrus aurantium L.*) blossoms by different isolation methods. Sustain Chem Pharm 2018; 10: 118-24.
[http://dx.doi.org/10.1016/j.scp.2018.10.008]

[84] Chenni M, El Abed D, Neggaz S, Rakotomanomana N, Fernandez X, Chemat F. Solvent free microwave extraction followed by encapsulation of *O. basilicum* L. essential oil for insecticide purpose. J Stored Prod Res 2020; 86: 101575.
[http://dx.doi.org/10.1016/j.jspr.2020.101575]

[85] Pereira CG, Meireles MAA. Supercritical fluid extraction of bioactive compounds: fundamentals, applications and economic perspectives. Food Bioprocess Technol 2010; 3(3): 340-72.
[http://dx.doi.org/10.1007/s11947-009-0263-2]

[86] de Melo MMR, Silvestre AJD, Silva CM. Supercritical fluid extraction of vegetable matrices: Applications, trends and future perspectives of a convincing green technology. J Supercrit Fluids 2014; 92: 115-76.
[http://dx.doi.org/10.1016/j.supflu.2014.04.007]

[87] Sahena F, Zaidul ISM, Jinap S, Jahurul MHA, Khatib A, Norulaini NAN. Extraction of fish oil from the skin of Indian mackerel using supercritical fluids. J Food Eng 2010; 99(1): 63-9.
[http://dx.doi.org/10.1016/j.jfoodeng.2010.01.038]

[88] Sánchez-Camargo AP, Meireles MÂA, Ferreira ALK, Saito E, Cabral FA. Extraction of ω-3 fatty acids and astaxanthin from Brazilian redspotted shrimp waste using supercritical CO_2^+ethanol mixtures. J Supercrit Fluids 2012; 61: 71-7.
[http://dx.doi.org/10.1016/j.supflu.2011.09.017]

[89] Mezzomo N, Martínez J, Maraschin M, Ferreira SRS. Pink shrimp (*P. brasiliensis* and *P. paulensis*) residue: Supercritical fluid extraction of carotenoid fraction. J Supercrit Fluids 2013; 74: 22-33.
[http://dx.doi.org/10.1016/j.supflu.2012.11.020]

[90] Flórez-Fernández N, López-García M, González-Muñoz MJ, Vilariño JML, Domínguez H. Ultrasound-assisted extraction of fucoidan from *Sargassum muticum.* J Appl Phycol 2017; 29(3): 1553-61.
[http://dx.doi.org/10.1007/s10811-016-1043-9]

[91] Anticona M, Blesa J, Frigola A, Esteve MJ. High Biological Value Compounds Extraction from Citrus Waste with Non-Conventional Methods. Foods 2020; 9(6): 811.
[http://dx.doi.org/10.3390/foods9060811] [PMID: 32575685]

[92] Ojha KS, Aznar R, O'Donnell C, Tiwari BK. Ultrasound technology for the extraction of biologically active molecules from plant, animal and marine sources. Trends Analyt Chem 2020; 122: 115663.
[http://dx.doi.org/10.1016/j.trac.2019.115663]

[93] Flórez-Fernández N, González Muñoz MJ. Ultrasound-Assisted Extraction of Bioactive Carbohydrates. In: Domínguez H, González MJ (Eds.) Water Extraction of Bioactive Compounds: From Plants to Drug Development. NY: Elsevier 2017; pp. 317-31.
[http://dx.doi.org/10.1016/B978-0-12-809380-1.00012-7]

[94] Nóbrega JA, Trevizan LC, Araújo GCL, Nogueira ARA. Focused-microwave-assisted strategies for sample preparation. Spectrochim Acta B At Spectrosc 2002; 57(12): 1855-76.
[http://dx.doi.org/10.1016/S0584-8547(02)00172-6]

[95] Chemat F, Tomao V, Virot M. Handbook of food analysis instruments. Oxford: CRC Press 2008; pp. 85-103.

[96] Vian MA, Fernandez X, Visinoni F, Chemat F. Microwave hydrodiffusion and gravity, a new technique for extraction of essential oils. J Chromatogr A 2008; 1190(1-2): 14-7.
[http://dx.doi.org/10.1016/j.chroma.2008.02.086] [PMID: 18343393]

[97] Li Y, Fabiano-Tixier AS, Vian MA, Chemat F. Solvent-free microwave extraction of bioactive compounds provides a tool for green analytical chemistry. Trends Analyt Chem 2013; 47: 1-11. a
[http://dx.doi.org/10.1016/j.trac.2013.02.007]

[98] Chemat F, Cravotto G. Microwave-assisted extraction for bioactive compounds: Theory and practice. London: Springer 2013; pp. 53-68. b
[http://dx.doi.org/10.1007/978-1-4614-4830-3]

[99] Ciğeroğlu Z, Bayramoğlu M, Kırbaşlar Şİ, Şahin S. Comparison of microwave-assisted techniques for the extraction of antioxidants from *Citrus paradisi* Macf. biowastes. J Food Sci Technol 2021; 58(3): 1190-8.
[http://dx.doi.org/10.1007/s13197-020-04632-x] [PMID: 33678900]

[100] Teigiserova DA, Tiruta-Barna L, Ahmadi A, Hamelin L, Thomsen M. A step closer to circular bioeconomy for citrus peel waste: A review of yields and technologies for sustainable management of essential oils. J Environ Manage 2021; 280: 111832.
[http://dx.doi.org/10.1016/j.jenvman.2020.111832] [PMID: 33360259]

[101] Angoy A, Brianceau S, Chabrier F, *et al.* Green food processing techniques Preservation, transformation and extraction. London: Academic Press 2019; pp. 455-98.

[102] Sridharan K. Spectral methods in transition metal complexes. Amsterdam: Elsevier 2016; pp. 1-12.

[103] Dev SRS, Birla SL, Raghavan GSV, Subbiah J. Microbial decontamination in the food industry Novel

methods and applications. Cambridge: Woodhead Publishing Limited 2012; pp. 274-99. [http://dx.doi.org/10.1533/9780857095756.2.274]

[104] Cooper, N. S. (2020). In U. Erle, P. Pesheck, & M. Lorence (Eds.), Development of packaging and products for use in microwave ovens (pp. 177-199). Elsevier: Duxford.

[105] Braeuer A. Supercritical fluid science and technology. Amsterdam: Elsevier 2015; pp. 41-192.

[106] Singh S, Gupta D, Jain V, Sharma AK. Microwave processing of materials and applications in manufacturing industries: A review. Mater Manuf Process 2015; 30(1): 1-29. [http://dx.doi.org/10.1080/10426914.2014.952028]

[107] Li H, Zhao Z, Xiouras C, Stefanidis GD, Li X, Gao X. Fundamentals and applications of microwave heating to chemicals separation processes. Renew Sustain Energy Rev 2019; 114: 109316. [http://dx.doi.org/10.1016/j.rser.2019.109316]

[108] Ojeda CB, Rojas FS. Encyclopedia of analytical science. Amsterdam: Elsevier 2005; pp. 153-63. [http://dx.doi.org/10.1016/B0-12-369397-7/00540-9]

[109] Rodríguez Seoane P, Flórez-Fernández N, Conde Piñeiro E, Domínguez González H. Water extraction of bioactive compounds: From plants to drug development. Amsterdam: Elsevier 2017; pp. 163-98. [http://dx.doi.org/10.1016/B978-0-12-809380-1.00006-1]

[110] Zhang Y, Chen P, Liu S, *et al.* Pyrolysis. London: Headquarters Intech Open Limited 2017; pp. 129-66.

[111] Dar AH, Shams R, Rizvi QUEH, Majid I. Fresh-cut fruits and vegetables Technologies and mechanisms for safety control. London: Academic Press 2020; pp. 295-337. [http://dx.doi.org/10.1016/B978 0 12 816184-5.00013-6]

[112] Hyötyläinen T, Riekkola M-L. Approaches for on-line coupling of extraction and chromatography. Anal Bioanal Chem 2004; 378(8): 1962-81. [http://dx.doi.org/10.1007/s00216-004-2545-x] [PMID: 14997264]

[113] Höller U, Wolter D, Hofmann P, Spitzer V. Microwave-assisted rapid determination of vitamins a and e in beverages. J Agric Food Chem 2003; 51(6): 1539-42. [http://dx.doi.org/10.1021/jf0260038] [PMID: 12617580]

[114] Li H, Chen B, Nie L, Yao S. Solvent effects on focused microwave assisted extraction of polyphenolic acids from*Eucommia ulmodies.* Phytochem Anal 2004; 15(5): 306-12. [http://dx.doi.org/10.1002/pca.783] [PMID: 15508835]

[115] Waksmundzka-Hajnos M, Petruczynik A, Dragan A, Wianowska D, Dawidowicz AL. Effect of extraction method on the yield of furanocoumarins from fruits of *Archangelica officinalis* Hoffm. Phytochem Anal 2004; 15(5): 313-9. [http://dx.doi.org/10.1002/pca.784] [PMID: 15508836]

[116] Wang Q, Ma S, Fu B, Lee FSC, Wang X. Development of multi-stage countercurrent extraction technology for the extraction of glycyrrhizic acid (GA) from licorice (*Glycyrrhiza uralensis* Fisch). Biochem Eng J 2004; 21(3): 285-92. [http://dx.doi.org/10.1016/j.bej.2004.06.002]

[117] Chemat S, Aït-Amar H, Lagha A, Esveld DC. Microwave-assisted extraction kinetics of terpenes from caraway seeds. Chem Eng Process 2005; 44(12): 1320-6. [http://dx.doi.org/10.1016/j.cep.2005.03.011]

[118] Popovich DG, Hu C, Durance TD, Kitts DD. Retention of ginsenosides in dried ginseng root: Comparison of drying methods. J Food Sci 2005; 70(6): s355-8. [http://dx.doi.org/10.1111/j.1365-2621.2005.tb11455.x]

[119] Tigrinekordjani N, Meklati B, Chemat F. Microwave 'dry' distillation as an useful tool for extraction of edible essential oils. International Journal of Aromatherapy 2006; 16(3-4): 141-7. [http://dx.doi.org/10.1016/j.ijat.2006.09.007]

[120] Schirack AV, Drake MA, Sanders TH, Sandeep KP. Characterization of aroma-active compounds in microwave blanched peanuts. J Food Sci 2006; 71(9): C513-20.
[http://dx.doi.org/10.1111/j.1750-3841.2006.00173.x]

[121] Kwon J-H, Kausar T, Lee J, Kim H-K, Ahn DU. The microwave-assisted extraction of fats from irradiated meat products for the detection of radiation-induced hydrocarbons. Food Sci Biotechnol 2007; 16(1): 150-3.

[122] Trusheva B, Trunkova D, Bankova V. Different extraction methods of biologically active components from propolis: a preliminary study. Chem Cent J 2007; 1(1): 13.
[http://dx.doi.org/10.1186/1752-153X-1-13] [PMID: 17880743]

[123] Barroso MF, Sales MGF, Almeida SAA, Vaz MCVF, Delerue-Matos C. Maltol determination in food by microwave assisted extraction and electrochemical detection. Yao Wu Shi Pin Fen Xi 2008; 16(4): 30-6.

[124] Fuentes E, Báez ME, Quiñones A. Suitability of microwave-assisted extraction coupled with solid-phase extraction for organophosphorus pesticide determination in olive oil. J Chromatogr A 2008; 1207(1-2): 38-45.
[http://dx.doi.org/10.1016/j.chroma.2008.08.051] [PMID: 18774576]

[125] Zhai Y, Sun S, Wang Z, *et al.* Microwave extraction of essential oils from dried fruits of *Illicium verum* Hook. f. and *Cuminum cyminum* L. using ionic liquid as the microwave absorption medium. J Sep Sci 2009; 32(20): 3544-9.
[http://dx.doi.org/10.1002/jssc.200910204] [PMID: 19764052]

[126] Zhang Z, Lv G, Pan H, Wu Y, Fan L. Effects of different drying methods and extraction condition on antioxidant properties of Shiitake (*Lentinus edodes*). Food Sci Technol Res 2009; 15(5): 547-52.
[http://dx.doi.org/10.3136/fstr.15.547]

[127] Gao S, Han W, Deng X. Study of the mechanism of microwave-assisted extraction of *Mahonia bealei* (Fort.) leaves and *Chrysanthemum morifolium* (Ramat.) petals. Flavour Fragrance J 2004; 19(3): 244-50.
[http://dx.doi.org/10.1002/ffj.1296]

[128] Yang Y, Chen L, Zhang XX, Guo Z. Microwave assisted extraction of major active ingredients in *Panax quinquefolium* L. J Liq Chromatogr Relat Technol 2004; 27(20): 3203-11.
[http://dx.doi.org/10.1081/JLC-200034881]

[129] Yang Z, Zhai W. Optimization of microwave-assisted extraction of anthocyanins from purple corn (*Zea mays* L.) cob and identification with HPLC–MS. Innov Food Sci Emerg Technol 2010; 11(3): 470-6.
[http://dx.doi.org/10.1016/j.ifset.2010.03.003]

[130] González-Nuñez LN, Cañizares-Macías MP. Focused microwaves-assisted extraction of theobromine and caffeine from cacao. Food Chem 2011; 129(4): 1819-24.
[http://dx.doi.org/10.1016/j.foodchem.2011.05.118]

[131] Lagunas-Allué L, Sanz-Asensio J, Martínez-Soria MT. Validation of a microwave-assisted extraction gas chromatography detection method for the determination of fungicides in grapes. Anal Methods 2011; 3(12): 2881-92.
[http://dx.doi.org/10.1039/c1ay05406f]

[132] Grigoras CG, Destandau E, Zubrzycki S, Elfakir C. Sweet cherries anthocyanins: An environmental friendly extraction and purification method. Separ Purif Tech 2012; 100: 51-8.
[http://dx.doi.org/10.1016/j.seppur.2012.08.032]

[133] Setyaningsih W, Palma M, Barroso CG. A new microwave-assisted extraction method for melatonin determination in rice grains. J Cereal Sci 2012; 56(2): 340-6.
[http://dx.doi.org/10.1016/j.jcs.2012.02.012]

[134] Cai M, Guo X, Liang H, Sun P. Microwave-assisted extraction and antioxidant activity of star anise oil

from *Illicium verum* Hook. f. Int J Food Sci Technol 2013; 48(11): 2324-30.

[135] Paíga P, Morais S, Oliva-Teles T, *et al.* Determination of ochratoxin a in bread: evaluation of microwave-assisted extraction using an orthogonal composite design coupled with response surface methodology. Food Bioprocess Technol 2013; 6(9): 2466-77.
[http://dx.doi.org/10.1007/s11947-012-0876-8]

[136] Thirugnanasambandham K, Sivakumar V, Prakash Maran J. Process optimization and analysis of microwave assisted extraction of pectin from dragon fruit peel. Carbohydr Polym 2014; 112: 622-6.
[http://dx.doi.org/10.1016/j.carbpol.2014.06.044] [PMID: 25129791]

[137] Chaisamlitpol S, Hiranvarachat B, Srichumpoung J, Devahastin S, Chiewchan N. Bioactive compositions of extracts from cabbage outer leaves as affected by drying pretreatment prior to microwave-assisted extraction. Separ Purif Tech 2014; 136: 177-83.
[http://dx.doi.org/10.1016/j.seppur.2014.09.002]

[138] Medina AL, da Silva MAO, de Sousa Barbosa H, Arruda MAZ, Marsaioli A Jr, Bragagnolo N. Rapid microwave assisted extraction of meat lipids. Food Res Int 2015; 78: 124-30.
[http://dx.doi.org/10.1016/j.foodres.2015.10.028] [PMID: 28433273]

[139] Rojas R, Contreras-Esquivel JC, Orozco-Esquivel MT, Muñoz C, Aguirre-Joya JA, Aguilar CN. Mango peel as source of antioxidants and pectin: microwave assisted extraction. Waste Biomass Valoriz 2015; 6(6): 1095-102.
[http://dx.doi.org/10.1007/s12649-015-9401-4]

[140] Chenni M, El Abed D, Rakotomanomana N, Fernandez X, Chemat F. Comparative study of essential oils extracted from Egyptian basil leaves (*Ocimum basilicum* L.) using hydro-distillation and solvent-free microwave extraction. Molecules 2016; 21(1): 113.
[http://dx.doi.org/10.3390/molecules21010113] [PMID: 26797599]

[141] Tsubaki S, Oono K, Hiraoka M, Onda A, Mitani T. Microwave-assisted hydrothermal extraction of sulfated polysaccharides from *Ulva* spp. and *Monostroma latissimum.* Food Chem 2016; 210: 311-6.
[http://dx.doi.org/10.1016/j.foodchem.2016.04.121] [PMID: 27211652]

[142] Pinela J, Prieto MA, Barreiro MF, *et al.* Valorisation of tomato wastes for development of nutrient-rich antioxidant ingredients: A sustainable approach towards the needs of the today's society. Innov Food Sci Emerg Technol 2017; 41: 160-71.
[http://dx.doi.org/10.1016/j.ifset.2017.02.004]

[143] Aufartová J, Brabcová I, Torres-Padrón ME, Solich P, Sosa-Ferrera Z, Santana-Rodríguez JJ. Determination of fluoroquinolones in fishes using microwave-assisted extraction combined with ultra-high performance liquid chromatography and fluorescence detection. J Food Compos Anal 2017; 56: 140-6.
[http://dx.doi.org/10.1016/j.jfca.2016.12.012]

[144] Ferreira SS, Passos CP, Cardoso SM, Wessel DF, Coimbra MA. Microwave assisted dehydration of broccoli by-products and simultaneous extraction of bioactive compounds. Food Chem 2018; 246: 386-93.
[http://dx.doi.org/10.1016/j.foodchem.2017.11.053] [PMID: 29291863]

[145] Xu SY, Liu JP, Huang X, *et al.* Ultrasonic-microwave assisted extraction, characterization and biological activity of pectin from jackfruit peel. Lebensm Wiss Technol 2018; 90: 577-82.
[http://dx.doi.org/10.1016/j.lwt.2018.01.007]

[146] Vu HT, Scarlett CJ, Vuong QV. Maximising recovery of phenolic compounds and antioxidant properties from banana peel using microwave assisted extraction and water. J Food Sci Technol 2019; 56(3): 1360-70.
[http://dx.doi.org/10.1007/s13197-019-03610-2] [PMID: 30956315]

[147] Kumar M, Dahuja A, Sachdev A, *et al.* Evaluation of enzyme and microwave-assisted conditions on extraction of anthocyanins and total phenolics from black soybean (*Glycine max L.*) seed coat. Int J Biol Macromol 2019; 135: 1070-81.

[http://dx.doi.org/10.1016/j.ijbiomac.2019.06.034] [PMID: 31176863]

[148] Garcia-Vaquero M, Ummat V, Tiwari B, Rajauria G. Exploring ultrasound, microwave and ultrasound - Microwave assisted extraction technologies to increase the extraction of bioactive compounds and antioxidants from brown macroalgae. Mar Drugs 2020; 18(3): 172. [http://dx.doi.org/10.3390/md18030172] [PMID: 32244865]

[149] Zuluaga AM, Mena-García A, Soria Monzón AC, *et al.* Microwave assisted extraction of inositols for the valorization of legume by-products. Lebensm Wiss Technol 2020; 133: 109971. [http://dx.doi.org/10.1016/j.lwt.2020.109971]

[150] Rudić S, Dimitrijević-Branković S, Dimitrijević S, Milić M. Valorization of unexploited artichoke leaves dust for obtaining of extracts rich in natural antioxidants. Separ Purif Tech 2021; 256: 117714. [http://dx.doi.org/10.1016/j.seppur.2020.117714]

[151] Martins M, Oliveira R, Coutinho JAP, *et al.* Recovery of pigments from *Ulva rigida.* Separ Purif Tech 2021; 255: 117723. [http://dx.doi.org/10.1016/j.seppur.2020.117723]

[152] Thu Dao TA, Webb HK, Malherbe F. Optimization of pectin extraction from fruit peels by response surface method: Conventional *versus* microwave-assisted heating. Food Hydrocoll 2021; 113: 106475. [http://dx.doi.org/10.1016/j.foodhyd.2020.106475]

[153] Macedo GA, Santana ÁL, Crawford LM, Wang SC, Dias FFG, de Moura Bell JMLN. Integrated microwave- and enzyme-assisted extraction of phenolic compounds from olive pomace. Lebensm Wiss Technol 2021; 138: 110621. [http://dx.doi.org/10.1016/j.lwt.2020.110621]

[154] Buratto RT, Cocero MJ, Martín Á. Characterization of industrial açaí pulp residues and valorization by microwave-assisted extraction. Chem Eng Process 2021; 160: 108269. [http://dx.doi.org/10.1016/j.cep.2020.108269]

CHAPTER 11

Application of Enzyme-Assisted Extraction for Food Analysis and Release of Natural Products

María del Prado García Aparicio[1,2,*] and **María Luisa Marina**[1,2]

[1] *Departamento de Química Analítica, Universidad de Alcalá, Química Física e Ingeniería Química, Ctra. Madrid-Barcelona Km. 33.600, 28871 Alcalá de Henares (Madrid), Spain*

[2] *Universidad de Alcalá, Instituto de Investigación Química "Andrés M. del Río", Ctra. Madrid-Barcelona Km. 33.600, 28871 Alcalá de Henares (Madrid), Spain*

Abstract: The transition to a circular bioeconomic model that incorporates sustainable extraction processes such as enzyme-assisted extraction (EAE) is motivated by climate change, population growth, and changing diets to address food security and safety, and preserve natural resources (land, and water) and biodiversity. EAE can be applied to extract nutrients and bioactive molecules for food analysis and profiling, and for industrial exploitation of bioactive compounds from novel feedstocks. Commercial extraction processes require high recovery of the targeted compounds and must guarantee the preservation of the biological activity of the products, which is difficult to achieve using conventional methods. EAE is a possible alternative to preserve the quality of final products while reducing the industrial footprint in the food sector at a larger scale. This chapter describes the parameters that impact the extraction yield obtained in the EAE process and provides recent examples of its successful application for the extraction of polymers and bioactive compounds of very diverse matrices (plant, animal, mushrooms, yeast, food waste, and insects), with emphasis on process conditions. This chapter also identifies the challenges and opportunities of EAE and the emerging areas of research to facilitate the economic feasibility of the enzymatic extraction of bioactive molecules. Costs related to enzyme production and its use are one of the main impediments to the industrial application of the EAE process. Recent research progress suggests that reduction of EAE costs can be achieved by a holistic approach considering all steps: enzyme production (by using cheap enzyme production media, in-house enzyme production), selection of feedstock (*i.e.*, food byproducts), enzyme recycling (enzyme immobilization, nano-biocatalysts), the search of novel enzymes (marine degrading polysaccharides), more robust enzymes (*i.e.*, extremozymes) and/or enzyme improvement (bioengineering), and EAE process optimization (minimum optimal enzyme dosage). EAE technology for food analysis and production of bioactive molecules keeps building momentum as it is sustainable, environmentally friendly, and innovative.

[*] **Corresponding Author María del Prado García Aparicio:** Departamento de Química Analítica, Universidad de Alcalá, Química Física e Ingeniería Química, Ctra. Madrid-Barcelona Km. 33.600, 28871 Alcalá de Henares (Madrid), Spain; Tel: +34 918856389; E-mail: mprado.garcia@uah.es

Merichel Plaza & Maria Luisa Marina (Eds.)

Keywords: Bioactive compounds, Biorefinery, Biocatalysis, Environmentally friendly enzyme technology, Enzyme-assisted extraction, Sustainable natural sources.

INTRODUCTION

Enzymes have been applied as tools for research and development in almost every industrial sector [1]. Enzyme technology is one of the keystones of biotechnology, particularly industrial or white biotechnology, with an expanding market projected to reach $10,519 million by 2024 [2]. It involves the application of enzymes as catalysts in diverse industrial processes for the manufacture of products and thus fulfilling human needs. The primary goal of enzyme technology is the development of more sustainable processes to generate novel products and/or conventional products from traditional or new raw materials. Enzyme technology has important applications in the food and beverage industry as bioprocessing agents and as tools for food analysis. Factors that are driving the global food enzymes market are also related to changes in consumer preferences. There is a growing demand for: natural preservatives, antioxidants, and colorants in place of synthetic ones from health concerns [3, 4]; novel and exotic flavors; nutritious foods; and functional foods (*i.e.*, food enrichments, fortifiers). These increased requirements could be met by the extraction of bioactive compounds from different sources, including food waste [5]. Enzyme-assisted extraction (EAE) is gaining popularity owing to more stringent regulations over the industrial extraction of bioactive compounds and its impact on the environment and health [6]. The increasing weight of biocatalysis in the food and beverage sector is reflected by the estimated expansion of its enzyme market from about USD 2.8 billion in 2020 to around USD 3.7 billion by 2026 [7].

Enzymes were discovered in the 1830s when scientists demonstrated the hydrolysis of starch using an enzyme present in germinated cereal grains, the enzyme that we know these days as amylase. The production of rennet constitutes another example of biocatalysis where enzymes (aspartic protease) from calves' stomachs were used to coagulate the proteins in the milk for cheese making. The food industry has capitalized on enzyme applications not only for cheese products but also in other applications such as the clarification of fruit juices (pectinases). The microbial enzymes started to be applied in the food industry by 1960, with major applications in the starch industry. The starch industry is a good example of the adoption of a greener technology based on enzymes, α-amylases, and β-glucoamylases, in substitution of the acid hydrolysis process for the conversion of starch into glucose. The innovation of enzyme technology increased abruptly with the development of DNA technology that relies on the progress of other enzyme applications. Simultaneously, the discovery of more and more enzymes through metagenomics and the continuous progress in enzyme engineering is favoring the transition from conventional chemical processes to greener enzyme-based conversion technologies [1].

Enzymes are specialized proteins able to increase the speed of a chemical reaction (reduction of activation energy) up to more than 1010 times, promoting the transformation of different compounds into specific products of interest. There is an enzyme-catalyzed reaction corresponding to almost every type of known organic reaction. The nomenclature committee of the International Union of Biochemistry and Molecular Biology (IUBMB) classified the enzymes into six classes according to the reaction they catalyze: oxidoreductases, transferases, hydrolases, lyases, isomerases, and ligases. The IUBMB assigned a code of four digits to the enzymes, designated as Enzyme Commission number (EC number). This EC number also considers substrate specificity and the functional groups involved. The first digit refers to the enzyme class, the second digit refers to the sub-class and its substrate specificity, the third number corresponds to the functional group donor and the fourth one refers to the acceptor molecule. EC numbers do not identify specific enzymes, but enzyme-catalyzed reactions, meaning that different enzymes that catalyze the same reaction will share the same EC number. Databases dealing with enzyme nomenclature are ExplorEnz [8] and IntEnz [9]. Although the EC classification has been broadly used, it presents some limitations since it does not consider enzyme structure, its amino acid sequence, or mechanisms of action, all of which are significant parameters for elucidating the enzymatic reactions. Thus, enzyme databases have been created to cover a wide range of aspects including enzyme function and properties, distribution, kinetics of catalyzed reactions, structure (three-dimensional folds) and amino acid sequence, or metabolic function. BRENDA [10] supplies vast information of all classified enzymes whereas KEGG [11], EzCatDB [12], MEROPS [13], MetaCyc [14], REBASE [15], CAZy [16], ESTHER [17], RedoxiBase [18], and KinBase [19] in either certain aspects of enzyme function or specific enzyme classes, organisms, or metabolic pathways. There are other interesting databases on enzyme application that can be used to predict the cleavage sites of proteins of known sequence and structure when subjected to proteases, as well as to predict the potential biological activity of the products of the reaction [20].

The different enzyme classes and some examples used in the food industry are given below:

1. Oxidoreductases (EC 1) catalyze oxidation-reduction reactions where there is a transfer of electrons from an electron donor (reductant) to an electron acceptor molecule (oxidant). This group of enzymes usually utilizes nicotinamide adenine

dinucleotide (NADP) or NAD+ as cofactors. Although they are widespread in nature, only a few of them are produced commercially such as glucose oxidase (EC 1.1.3.4).

2. Transferases (EC 2) catalyze reactions where functional groups (*i.e.* acetyl, glycosyl, methyl, phosphate) are transferred from one molecule (donor) to another (acceptor). Transglutaminase (EC 2.3.2.13), cyclodextrin-glycosyltransferase (EC 2.4.1.19), and amylomaltase (EC 2.4.1.25) constitute common examples of transferases used in the food sector.

3. Hydrolases (EC 3) catalyze hydrolysis reactions where a molecule is cleaved into two or more smaller molecules by the addition of water. Examples of hydrolases used in the food sector are (phospho)-galactolipase (EC 3.1.1.26), triacylglycerol lipase (EC 3.1.1.3), lactase (EC 3.2.1.108), amylase (EC 3.2.1.1), endo-xylanase (EC 3.2.1.8), cellulase (EC 3.2.1.4), pectinase (EC 3.2.1.15), protease (EC 3.4.21.X) and phytase (EC 3.1.3.X).

4. Lyases (EC 4) are enzymes that only need one substrate to catalyze the split of chemical bonds such as C-C, C-O, C-S, and C-N by other methods different from hydrolysis and oxidation, normally creating a new double bond or a new ring structure. They also catalyze the inverse reaction for what they need both substrates. Pectin lyases (EC 4.2.2.10) find numerous applications in the food and beverage sector.

5. Isomerases (EC 5) catalyze the conversion of one molecule into a different isomer using atomic rearrangements. These re-arrangements can be facilitated either through the cleavage of bonds and formation of new ones, or through conformational changes. The most common use of isomerases in industrial applications is in sugar manufacturing (glucose isomerase, EC 5.3.1.18).

6. Ligases (EC 6) catalyze the reaction which links two molecules by forming a new chemical bond.

Except for ligases, all the enzyme classes have an established or emerging application in the food and beverage industry; with an impact on taste, texture, appearance, nutritional value, shelf-life, or process tolerance. Most of the enzymes used in EAE processes belong to the hydrolases group (for example, carbohydrases, proteases, lipases, and phospholipases) and lyases (pectin lyases) [21]. There are a few examples where laccases have been applied for the extraction of polysaccharides [22], lignans, and triterpenoids [23]. Most often a combination of these enzymes is required for the optimum extraction of bioactive molecules [21]. Most of the enzymes used in EAE are derived from microorganisms, but there are also some from plants and animal origin. For example, an enzyme mixture called snailase (consisting of cellulase, hemicellulase, pectinase, and ß-glucuronidase), obtained from the digestive tract of snails, has been applied as a digestion aid because of its proven hydrolysis ability [24]. In the case of emerging technologies based on nanomaterials with intrinsic enzyme-like characteristics, nanozymes, the reactions are mainly oxidoreductase-like [25].

This chapter will provide an overall overview of EAE processes for the extraction of specific compounds, including bioactive compounds, from diverse food matrices. The EAE process and its parameters, and the advantages and disadvantages compared to chemical-based methods, will be described. Recent examples will be given on the applications of EAE for food analysis, extraction of polymers, and bioactive compounds, with emphasis on process conditions, as a stand-alone process or combined with other extraction technologies. Challenges and opportunities of EAE will be identified, and the emerging areas of research to improve the economic feasibility of the enzymatic extraction of bioactive molecules will be described.

ENZYME-ASSISTED EXTRACTION PROCESS: PARAMETERS, OPERATIONAL CONDITIONS, ADVANTAGES AND DISADVANTAGES

Conventional extraction techniques may not be effective in the release of target molecules that are bound to structural compounds. The application of enzymes prior to conventional methods has been used as a treatment of the material, increasing the yield of extraction, reducing the extraction time and the amount of solvent required, while preserving the antioxidant capacity of the extracts [26]. Thus, the principle behind the EAE of food matrices relies on the hydrolysis of structural linkages to enhance the release of the component of interest. Several enzymes are generally required to disrupt the structural integrity of the cell walls, increasing their permeability, and thereby boosting the extraction of target biomolecules. Compared to physical methods, cell permeabilization by appropriate enzymes allows for an efficient disruption (that maximizes the extraction yield of targeted molecules) while reducing the release of unwanted products (for high purity) from the cells. Fig. (**1**) illustrates the different steps involved in an EAE of diverse food matrices and factors that influence the yield. EAE consists of the next main steps:

1. Preparation of the feedstocks that can be of plant, animal, algae, or fungi origin. This preparation aims to homogenize the material and reduce its particle size and can be conducted under dry or non-dry conditions, depending on the extraction process selected [27]. Addition of the buffer (pH) and adjustment of temperature.

3. Addition of enzyme or combination of enzymes and incubation process with agitation.

4. Inactivation of the enzyme.

5. Downstream processing: centrifugation/filtration/precipitation, pasteurization (juice), concentration by evaporation (concentrate) or spray drying (powder), purification.

6. Analysis of final product and further application.

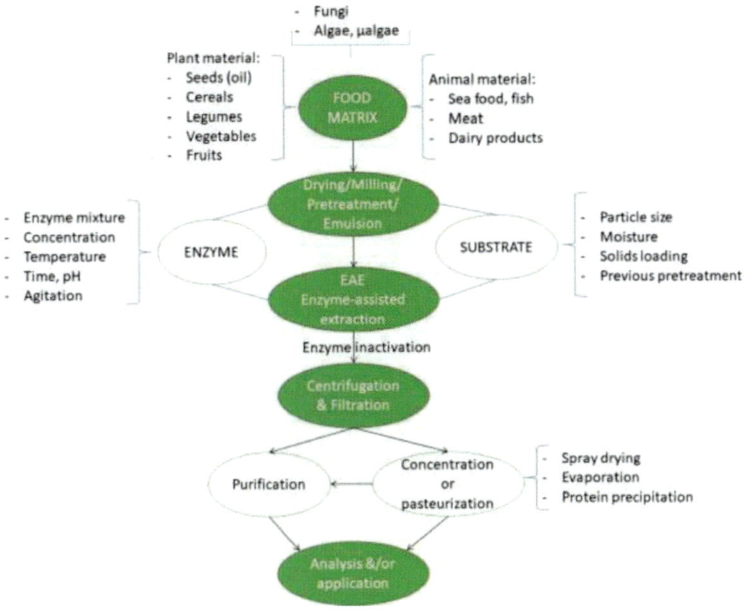

Fig. (1). Flow chart of the steps and operational parameters influencing enzyme-assisted extraction from different food matrices.

Parameters and Operating Conditions

The EAE process needs to be tailored to the food matrix and target molecule to be extracted. Examples of important value-added components that can be extracted by EAE from different food matrices are oil [28, 29], carbohydrates (dietary fibers, cellulose, hemicellulose, pectin, chitin, inulin, ulvan, and other heterogeneous polysaccharides such as β-glucans; oligosaccharides) [30, 31], phenolic compounds (polyphenols such as tannins and flavonoids; and simple phenols such as coumarins and phenolic acids; lignans) [6], terpenoids (*i.e.*, carotenoids such lycopene) [32], proteins [27] and flavors [33]. EAE has also been proven to be effective for the analysis of vitamins [34], metals [35], and mycotoxins [36, 37] in foods. Due to the vast range of structures of the compounds to be extracted and the food matrix itself, the enzymatic methods for their extraction need to be determined case-by-case. The structure of the food matrix, the location of the compound of interest, and its interaction with other components are factors that need to be evaluated in the selection of the enzyme(s) and process conditions. Besides, the EAE process depends significantly on operational parameters related to the enzyme itself (enzyme origin, enzyme mixture composition, and final enzyme concentration, temperature, pH and strength of ions, agitation), the substrate (particle size, moisture content, drying treatment applied, solids concentration) [38, 39] (Fig. **1**).

The concentration of the enzyme required to achieve the maximum extraction yield depends on the specificity of the substrate. Enzyme dosages used for EAE vary between 0.76% (w/w) up to 10% (w/w) (Table **1**). Increasing enzyme concentration for a given substrate concentration will accelerate the rate of the reaction, until reaching the saturation point. However, enzyme dosage should be kept to a minimum that provides satisfactory yields, since the costs associated with the enzymes are a bottleneck in the commercial application of EAE. The application of a synergistic enzyme mixture can improve the extraction yield while reducing the total enzyme dosage. Synergism is defined as an improvement of the yield that is superior to the sum of the yields obtained when using the enzymes individually. For example, enzymes such as cellulase, hemicellulase, pectinase, and proteases are generally applied to disrupt the plant cell wall of plant materials. The combination of these enzymes improves the release of more recalcitrant bioactive molecules that are linked covalently to carbohydrates through O-glucose or C-glucose bonds (flavonoids) and/or proteins (condensed tannins) [21]. These same enzymes have been proven to improve the yield and properties of extracted oil and proteins [28]. Phenolic acids can also be bound through ester bonds to carbohydrates and proteins, or through ether bonds to lignin [40]. Ferulic acid esterase has been applied successfully for the release of ferulic acid from corn cob. The combination of ferulic acid esterase with xylanase improved further the yield of the extraction of ferulic acid [41]. Bacterial β-etherases in combination with glutathione reductase liberate units of guaiacyl, syringyl, and tricin from lignin [42]. These enzymes hold the potential to release phenolic acids that are ether-linked to the lignin fraction of plant-based materials.

Enzyme-catalyzed reaction speed increases with the temperature, but the three-dimensional structure of the enzymes may not be stable if kept at high temperatures for long periods; normally required in an industrial setting. The most common temperatures applied during EAE studies varied from 45 °C to 60 °C (Table **1**), although some enzymes are quite thermostable, which allows their application combined with other green extraction technologies such as pressurized hot water extraction (PHWE) [43]. The pH has also an impact on the native 3D structure of the enzyme, and it is responsible for the state of protonation of amino acids and cofactors involved in a reaction. The pH during EAE should preserve enzyme function but should not be in the range of the isoelectric point of the protein [21]. Except for the extraction of proteins, most EAE are carried out at acidic pH conditions (Table **1**) that enable the disruption of the hydrogen bonds between some bioactive molecules with structural components. The enzyme reaction itself can produce pH changes and as a result a continuous decrease in activity. The buffering capacity of a buffer is influenced by its ionic strength, concentration, and composition. At the same time, the dissociation of compounds, including buffers, is temperature-dependent. The selection of the buffer and its concentration is thus important to keep the suitable range of pH for optimal enzyme activity during the process. Examples of buffers that have been employed in EAE studies are phosphate buffer (0.1-0.6 M), acetate (0.1 M), or citrate buffer (0.05 M) (Table **1**).

In terms of substrate related factors, one key parameter for any extraction process is particle size. A reduced particle size generally favors extraction yield because it disrupts structural components, increasing the surface accessible area to enzymes which in turn facilitates mass transfer. The selection of appropriate particle size (50–25 μm) of shrimp shells resulted in a reduction of reaction time from days to 3 h of the deproteination step for chitin extraction [51]. Nevertheless, if the particle size is reduced in excess, it could lead to the degradation of target molecules during cell wall disruption [52]. The reduction of size is normally carried out by milling or grinding the material. The initial moisture content will determine if the milling can be achieved through a wet or dry method. Generally, a dehydration process is applied to remove the water and preserve the substrate. The drying process and conditions should be chosen so that they preserve the bioactive molecules [45]. Freeze-dry treatment preserves both the antioxidant capacity and the original composition. Drying under vacuum is another method that maintains the antioxidant capacity of several feedstocks when the temperature applied does not exceed 60 °C. Other methods of drying that can find application in the industry are solar, oven, and hot-air drying [38]. Fewer studies have been conducted on the impact of the drying process on the properties of the fibers and the porosity of fibrous matrices. The application of cycles of drying (prior to the sample) and rewetting (during the EAE process) can induce changes in the structure of fibrous material, a phenomenon known as hornification [53, 54]. These changes include the formation of new hydrogen bonds, which in turn can collapse irreversibly the pores of cell walls between contiguous surfaces. Polysaccharides such as cellulose get packed more tightly during drying. These effects reduce the accessibility of enzymes to the internal surface area of lignocellulosic biomass [53].

Table 1. Optimal conditions for enzyme-assisted extraction of bioactive molecules from diverse food matrices.

Food matrix	Target compound	Conventional Extraction Method	Commercial Enzyme Preparations		Optimal Conditions					
			Company	Enzyme name	Solid loading	Enzyme dosage	Temperature, pH and buffer, time	Yield	Statistical design	Refs
Grounded tomato seeds	Oil	Hexane	Novozymes	Alcalase 2.4L (protease)	Not specified	2.25 mL	60°C, pH 7.4, 4 hours, 30 mL 0.6 M phosphate buffer	9.66%	Box-Behnken Design (BBD)	[29]
Veronicastrum axillare	Polysaccharides	Water extraction, alcohol precipitation, acid or alkali extraction	Shanghai Xilong Chemical Co., Ltd. (Shanghai, China).	Cellulase (from *Trichoderma reesei*)	13.2% (dw/v) solid/liquid ratio 1: 7.6	2% cellulase	47 °C for 2.5 h, then, 95°C for 2.5 h, Phosphate buffer pH 4.1	4.7%	BBD	[44]
Radish	Carotenoids and polyphenols (flavonoids).	Aqueous methanol	Sigma Aldrich	β-xylosidase, xyloglucanase, pectinases, xylanase, α-amylases, β-glucosidase, and protease	25% (fw/v)	3.53% (w/w)	46.08 °C, pH 6.10, 0.2 M phosphate buffer, 66.08 minutes.	45.9 mg GAE/g FW	Small Rotatable Central Composite Design (SRCD)	[45]
Dry grounded Bilberry pomace	Polyphenols (anthocyanins)	Conventional maceration	Novozymes	Viscozyme L (pectinases, cellulases, hemicellulases, and arabinases)	10% (dw/v)	2 active units (AU) /g DW 1.56% (w/w)	Citrate buffer pH: 4.5, 46 °C, 1 h	Extract yield: 56.15 g/100 g DW 12.15 mg GAE/g DW	Central Composite Design (CCD)	[46]
Sweet cherry pomace	Non-extractable polyphenols	Acid or alkali extraction	Biocatalyst	Promod 439L (protease)	38% (dw/v)	140 μL/g DW (5,32% w/w)	0.1 M sodium phosphate buffer, 70°C, pH 10, 40 min	2.8 mg GAE/g DW	BBD	[47]
				Depol 740L (pectinase)		90 μL/g DW (3,42% w/w)		1.1 mg GAE/g DW		
				Pectinase 62L		2 μL/g DW (0,76% w/w)	70°C, pH 10, 18.4 min	1.1 mg GAE/g DW		
Seaweeds	Protein	Aqueous, acidic and alkaline methods (yields from 24 to 59%)	Diestro, Girona SA Novozymes	Alpha-amylase Cellic CETC3 Pectinex Ultra	2% (dw/v)	10% (v/w)	50 °C, 0.1 M acetate buffer, pH 4.5, 12 h (red) to 18 h (brown)	36.1-74.6%	CCD	[48]
Mushrooms	Umami taste aminoacids	Dilute hydrochloric acid (0.1 M)	Novozymes	β-glucanase and Flavourzyme	2% (dw/v)	5% v/w of each enzyme	Water (initial pH = 7, 50 °C, 1 h	Up to 26.75 ± 1.07 mg/g DW of umami taste amino acids	CCD	[49]

(Table 1) cont.....

Food matrix	Target compound	Conventional Extraction Method	Commercial Enzyme Preparations		Optimal Conditions					
			Company	Enzyme name	Solid loading	Enzyme dosage	Temperature, pH and buffer, time	Yield	Statistical design	Refs
Adenanthera pavonina L. seeds	Phenolic compounds	Methanol-water	Novozymes	Neutrase® 0.8L (protease) Optimase® 56L (cellulase) Termamyl®2 X (α-amylase)	~11% (dw/v)	125 μL protease, 125 μL cellulase, 0 μL α-amylase	100mM, Phosphate buffer, pH 7	1183.20 ± 2.30 mg GAE/g DW (TPC) 366.98 ± 1.66 mg CE/g DW (flavonoid) 220.09 ± 7.07 mg CE/mg DW (tannin)	Mixture design	[50]

GAE: gallic acid equivalents; CE: catechin equivalents; FW: fresh weight; TPC: total phenolic content

The concentration of substrate is an important factor for the efficiency of the process industrial commercialization of the EAE process for the extraction of bioactive molecules. Like the particle size, the selection of solids loading (solids to liquid ratio) has an impact on mass transfer and thus the yield during the extraction. EAE at high solids' consistency could lead to an increased concentration of the biomolecules of interest, which could reduce the cost of product recovery. Increasing solid loading during EAE favors the economic feasibility as it would require a small reactor, and consequently lower energy requirements. However, at high solids content, several process problems can occur, such as those related to enzyme effectiveness (end-product inhibition) and insufficient substrate mixing. The agitation applied during the process facilitates the uniform mixing of constituents and thus mass transfer, and it reduces the time to achieve the maximum extraction yield.

Given the impact on the EAE process of all the above-mentioned parameters and their influence on each other, numerous studies have evaluated their effect and their interactions through multivariate analysis to maximize the yield of biologically active compounds from different food matrices [29, 44 - 50]. Response surface methodology (RSM) is the most common approach for multifactorial optimization of EAE, but there are other statistical designs that are gaining recognition in the food processing sector such as differential evolution technique (DET) and artificial neural networks (ANN) [39, 55, 56]. RSM is based on sequential experimental techniques to develop models through empirical equations predicting values of the dependent variables (the responses evaluated, for example, extraction yield and antioxidant capacity of the extracts) from the combination of independent variables (factors such as temperature, pH, enzyme concentration, solids to liquid ratio, *etc.*). Once a model is developed, the optimal conditions must be supported by the experimental value obtained under those conditions (process validation). The type of RSM models generally applied to determine optimal conditions of EAE are Central Composite Design (CCD) and Box-Behnken Design (BBD). A method less common in EAE studies is the mixture experimental design, which examines mixture-related elements (in this case enzymes). The relative proportion of the enzymes may depend not only on the substrate and target compound but also on the final protein loading in the extraction reaction. This type of design is useful to determine the minimum enzyme cocktail that is efficient for the EAE process while reducing the cost associated with enzyme dosage [50]. Some examples of optimal EAE conditions for the extraction of oil, polysaccharides, polyphenols, carotenoids, protein, and flavors (amino acids) from different food matrices are listed in Table **1**.

Advantages and Disadvantages

Enzyme catalysts present distinctive characteristics over traditional catalysts. The application of enzymes in industrial processes is generally favorable for the environment in almost every application, so EAE is no exception. EAE generally requires an aqueous solution and mild processing conditions, replacing chemicals or processes that present safety or environmental concerns. Advantages of industrial enzymes include:

- Reduction of production cost, easier separation, and improved quality of the product. For example, extracts from the EAE process could result in an improvement of the bio-accessibility of their bioactive molecules, a desired trait for their application as dietary supplements.
- Reduction of the number of steps and installations in an industrial process.
- Reduction of waste owing to the selectivity of the reaction, which minimizes side reactions, and the biodegradable nature of the enzymes.
- Replacement of petrochemicals and chemical usage.
- Energy savings since enzymatic reactions require milder conditions than their chemical counterparts.
- The process can be optimized by control of parameters such as enzyme dose, temperature, and time.
- High catalytic efficiency: relatively low concentrations are generally required for the reaction.
- Safer working conditions avoid the use of flammable or volatile solvents.

However, the enzyme use, and application also present some drawbacks such as possible allergic reactions when inhaled or ingested, limitations in the operating conditions (extreme pH and high temperatures result in protein denaturalization), substrate or product inhibition, and that they normally require water as solvent (high boiling point and heat of evaporation). Moreover, some of the commercial enzymes are still very expensive, some require expensive cofactors, some are sensitive to metals and strong solvents, and are susceptible to hydrolysis by peptidases. Pure enzyme preparations of other enzymes with potential for EAE are provided in small amounts by different suppliers at prohibitive costs for industrial application. In most of the cases, the enzymes do not possess the attributes required to conduct the reaction in industrial conditions. Nonetheless, nearly all the biocatalyst properties can be tuned with protein engineering methods to adjust to the desired process conditions. Enzyme immobilization can also increase the stability of the enzymes as well as favor the economics as it allows for several cycles of reuse [57, 58].

Besides the sustainability of a process, the industry is interested in reducing production costs, and finding the balance between product quality and solvent consumption. The EAE process may disrupt the wall or liberate the bioactive molecules from organelles, but it may occur that these do not solubilize adequately to allow their extraction [59]. In these cases, enzymes can be combined with the conventional process or greener extraction methods, *i.e.*, ultrasounds [60], microwaves [28], supercritical fluids extraction [61], deep eutectic solvents [62] or pressurized liquid extraction [43, 63, 64]. Some examples of EAE in combination with these techniques will be shown in the next section.

APPLICATIONS OF ENZYME ASSISTED EXTRACTION

EAE process can be used as a step prior to the analysis of food nutrients and toxins, as well as a method for the commercial extraction of the so-called bioactive molecules: carbohydrates, polyphenols, terpenes and terpenoids (*i.e.*, carotenoids), vitamins, alkaloids, oils, proteins, peptides, and others (Fig. **2**). These compounds find application not only in the food and beverage industry (as a nutraceutical or added in functional foods) but also in other industrial sectors such as chemical, pharmaceutical, biomedical, and cosmeceutical. The properties of the feedstock and the physicochemical characteristics of the target molecules influence the extraction yield of EAE, which needs to be designed accordingly.

Extraction of Nutrients and Bioactive Molecules by EAE for Food Analysis and Profiling

Analytical procedures for the analysis of nutrients such as vitamins in food samples are complicated because of their poor stability, low concentrations, variability (vitamers), and complexity of food matrix where there is interaction with other compounds such as carbohydrates or proteins. An increasing number of technologies are thus being developed to improve conventional analysis approaches in terms of time, reliability, and costs. The introduction of EAE alone or in combination with other extraction technologies represents an alternative to traditional extraction methods for the detection of important nutrients in foodstuffs: organic acids [65], fat-soluble vitamins (vitamins A, E; K, and D), water-soluble vitamins (vitamin B_1 active compounds, pyridoxine (B_6), folates (B_9), cobalamins (B_{12}), riboflavin (B_2), nicotinamide (B_3) and nicotinic acid, vitamin C), cholesterol [34, 66] (Table **2**), organic acids. EAE was already applied in the analysis of total dietary fiber (TDF), but its optimization has led to a more accurate determination [67].

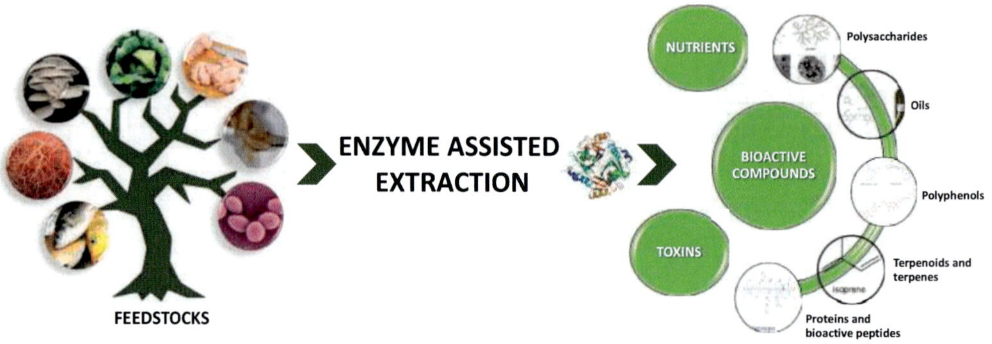

Fig. (2). Schematic representation of possible applications of enzyme assisted extraction prior to the analysis of nutrients, toxins, and bioactive profile as well as for the commercial exploitation of bioactive molecules (polysaccharides, oils, polyphenols, terpenoids and terpenes, proteins, and bioactive peptides) from diverse feedstocks.

Dietary fiber (DF) is an important element of a healthy diet as its consumption has been proven to prevent cardiovascular diseases, obesity, diabetes, and certain cancers. Foods that are rich in DF are vegetables, fruits, and cereals, but they are also present in legumes and nuts. DF is composed of carbohydrate polymers (cellulose, hemicellulose, pectin, gums), oligosaccharides, waxes, and lignin that cannot be digested by the endogenous enzymes of the human small intestine [68]. These DF polysaccharides can be intrinsic to food, extracted from different sources, or synthetic, all with proven evidence of their beneficial effects on health. The beneficial effects of DF differ according to its composition, structure, and association with other compounds. For instance, DF is normally associated with bioactive compounds (*i.e.*, polyphenols), which potentiate the health benefits of their consumption [68]. The procedure for the determination of TDF is based on enzymatic and gravimetric methods but it needed some modifications to adapt to improve food composition data for dietary fiber [69]. A modified method for the determination of TDF is called Rapid Integrated Total Dietary Fiber (RITDF) (AOAC 2017.16) where the EAE steps using pancreatic α-amylase (PAA) plus amylo-glucosidase (AMG) followed by a protease treatment have been optimized (concentration and time) [34, 67]. The concentration of PAA and AMG was increased to ensure the hydrolysis of non-resistant starch. The incubation time with the PAA+AMG treatment was carried out under conditions that resemble physiological digestion conditions, reducing analysis time from 16 to 4 h. RIRDF simplifies sample preparation, minimizes the use of hazardous chemicals, and improves the accuracy of fiber analysis, which also improves the determination of available carbohydrates.

The analysis of other nutrients such as fat-soluble vitamins, water-soluble vitamins, and cholesterol has also benefitted from incorporating an EAE process [34]. EAE assists in the release of these nutrients from the complex food matrix. In the case of vitamins, there are some additional difficulties such as their low concentration and poor stability, and a mild treatment such as EAE could improve their preservation [34]. EAE adoption can also help in the automatization of the analysis because it facilitates sample throughput.

Table 2. EAE conditions in recent methods for the analysis of total dietary fiber (TDF), fat-soluble vitamins (FSV), water-soluble vitamins (WSV), cholesterol and polyphenols in several matrices.

Food matrix	Compound	EAE Conditions	Recovery (RSD)	Analyte(s) Detection	Refs.
Total dietary fiber					
Fibersym, kidney beans, bran cereal, defatted cookies with FOS or polydextrose, oat bran, dark rye crispbread, whole meal bread.	Dietary fiber	Solids loading: ~3% (dw/v) (0.5 mm sieved material) PAA (4 KU/g) + AMG (1.7 KU/g) Ph 6.0 (50 Mm maleate buffer or MES buffer), 37°C, 4 h, 150 rpm > Inactivation and adjustment of pH to 8.2> Protease (30 U tyrosine/g), 60°C, 30 min	(2.14-10.62)	Fructo-oligosaccharides (FOS) in soluble dietary fiber are analyzed by HPLC with TSKgel®G2500PWXL gel permeation columns with in-line deionization	[67]
Fat-soluble vitamins					
Infant formula powder, cereals	Fat-soluble vitamins (vitamins A, E, K, and D)	Digestion: Solids loading: ~11% (dw/v) α-amylase (100 U/g), 45°C, 30 min. Extraction: 2-5 grams of digested substrate, 0.1 g papain, 45°C, 30 min, intermittent shaking.	>90% (0.7-5.1) Vit K from cereals (>80%)	SFC-APCI(+)-MS/MS	[70]

(Table 2) cont.....

Food matrix	Compound	EAE Conditions	Recovery (RSD)	Analyte(s) Detection	Refs.
Water-soluble vitamins					
Milk	Vit B1 active compounds: thiamine, thiamine monophosphate an thiamine diphosphate.	Substrate: Hydrolysed milked (10%v/v) ~0.6 Mm sodium acetate buffer, 25°C or 45°C, constant shaking, 9 h Enzymes (alone or combined): Acid phosphatase (150 U) Papain (3000 U) α-amylase (100 U) Claradiastase (7500 U)	>100%	UPLC, with fluorescence (FLR) detector, and an Acquity UPLC column	[71]
Anchovies	Riboflavin, nicotinamide, and nicotinic acid	Solids loading: 6% (fw/v) 5% Takadistase (α-amylase) or 5% claradiastase (α-amylase, cellulase, invertase, peptidase, phosphatase, and sulfatase) 45°C-2.5 h	93.2-101.9% (6.2-10.3)	LC-MS/MS	[72]
Grain Products	B6 Vitamers	Substrate: 4% (w/v) Acid hydrolysis followed by enzyme treatment: ~0.6 Mm sodium acetate buffer, pH 4.5, 37°C, 18 h Enzymes: Papain (40 mg/g), Diastase (100 mg/g)	99% (1.88)	HPLC-ED Reversed phase HPLC with coulometric electrochemical detection.	[73]
Infant formula and adult nutritionals	Folates (vitamin B9)	Solids loading: 1-20% (w/v) (estimated 20-500 ng folate) 10% (v/v) protease-37°C-60 rpm-3 h> 10% (v/v) α-amylase and 3%(v/v) rat plasma conjugase 37°C-60 rpm-16 h	Not provided	LC-MS/MS Quantitation	[74]
Water soluble Vitamins					

(Table 2) cont.....

Food matrix	Compound	EAE Conditions	Recovery (RSD)	Analyte(s) Detection	Refs.
Vegetables and fruits, vitamin tables, juice	Cobalamins (Vitamin B_{12})	Improved homogenization procedure Solids loading: ~1-16% (w/v) (depending on estimates of B12 concentration) 50 mM acetate buffer pH 4 1.6% (w/v) pepsin, 0.3 mL α-amylase 37°C-agitation-1.5 h	80-100% (0.7-6.73)	RP-HPLC with UV detection (DAD), after prior matrix isolation by immunoaffinity chromatography (IAC).	[75]
Tea	Vitamin C	Tea powder (20 mesh) Solids loading: ~10-25% (w/v) Acetic acid or sodium acetate buffer solution pH 5-7, microwave power 80-280 W-for 5-15 min/Ultrasound 100-250 W for 8-20 min/ 1-5% (w/v) cellulase,10-30 minutes	> 90%	-	[66]
Cholesterol					
Bovine and human milk	Cholesterol	Saponification followed by enzymatic treatment Cholesterol esterase (>200 U) horseradish peroxidase (500 U), cholesterol oxidase (>300 U).	95% (CV of 25.19)	Spectrophotometer ʎ: 502 nm	[76]
Polyphenols					
Milk-tea beverages	Catechins	Substrate loading: 25% (v/v) Pepsin: 120 mg/mL tea 37°C, 15 min, gentle mixing	89-102%	RP-HPLC-DAD (200-500 nm)	[77]

(Table 2) cont.....

Food matrix	Compound	EAE Conditions	Recovery (RSD)	Analyte(s) Detection	Refs.
Expresso coffee	Lignans	Substrate loading: 80% (v/v) Clara-Diastase: 10% (w/v) 37°C, 3 h	93-98%	HPLC-MS/MS triple quadrupole	[78]
Commercial tea	NEPP	Substrate loading: 2% (dw/v) Cellulase: 200 U/g Pectinase: 20 U/g sample) 0.2 M hydrochloric acid buffer solution Room temperature, 4 hr	-	UPLC-DAD	-
Residues of apple, peach, and nectarine	NEPA	- Pectinase 690L (74 µL/g) Sodium acetate buffer pH 5, 37°C, 24 h - Depol 740L (pectinase) (120 µL/g) Fosfate buffer pH 7, 55°C, 5 h - Promod 439L (protease) (20 µL/g) Tris maleate buffer pH 9, 55°C, 5 h - Mixture of the three enzymes (214 µL/g): Phosphate buffer pH 7, 55°C, 5 h	-	HPLC normal-phase separation – fluorescence detection	[80]

APCI – Atmospheric pressure chemical ionization; CV– coefficient of variance; ED – Electrochemical detection; FLR– fluorescence; LC-MS/MS – Liquid chromatography–tandem mass spectrometry; IAC – immunoaffinity chromatography; SFC – Supercritical fluid chromatography; UPLC – Ultra High-Performance Liquid Chromatography;

Fat-soluble compounds are commonly extracted by applying an alkaline saponification step before liquid-liquid extraction with organic solvents. These methods are cumbersome, time-consuming, require a large amount of solvents and provide low recovery of the compounds of interest. EAE has been applied successfully as a digestion pretreatment and extraction method for the simultaneous determination of FSV vitamers in selected food products, reducing cost and enhancing safety [70]. This method was demonstrated to be reliable for the analysis of FSV in powdered milk and more complex matrices such as cereals (Table **2**).

Vitamin B1 active compounds (free thiamine: thia; thiamine monophosphate: TMP; and thiamine diphosphate: TDP) are normally analyzed by a combination of acid hydrolysis and enzymatic digestion, to ensure the release of these compounds from the sample matrix as well as to complete the hydrolysis of the ester bonds of TMP and TDP. Recent work has proven that enzymatic treatments on their own resulted in higher values of total vitamin B1 from milk samples [71]. Among the enzymes evaluated, papain and acid phosphatase provided better yields than the optimized acidic treatment [71]. Conventional analytical methods for other WSVs (*i.e.*, Vit B_6, Vit B_9, Vit B_{12}) in food samples are based on standard microbiological assays that are laborious and time-consuming. Furthermore, these methods may not be appropriate for complex food matrices (*i.e.*, cereal products or fruits) with interfering compounds (*i.e.*, starch, vitamin C, and minerals such as copper). Dilute acid hydrolysis followed by enzyme treatment with papain and diastase improved sensitivity and selectivity for the simultaneous analysis of vitamin B_6 vitamers in several cereal products [73]. Folates have also been effectively extracted from infant formula and adult Nutritionals by a tri enzyme extraction consisting of sequential treatments with protease, α-amylase, and rat plasma conjugase [74] (Table **2**). The enzyme treatment with pepsin and α-amylase is effective for the analysis of Vit B12 in several vegetable matrices [75]. However, more complex matrices such as berries required an improved homogenization pretreatment (consisting of mill, mortar and pestle, homogenizer, and ultrasonic bath) to minimize interferences and increase vitamin B12 concentration [75]. Enzyme treatment of different food matrices (flour, pork, peas, milk, and yeast) provided similar yields of riboflavin than when it was combined with acidic pretreatment [72]. EAE treatment with Takadistase (α-amylase) or claradiastase (α-amylase, cellulase, invertase, peptidase, phosphatase, and sulfatase) has also proven to be an accurate method for the simultaneous extraction of riboflavin, nicotinamide, and nicotinic acid in anchovies [72] (Table **2**). EAE combined with other green extraction technologies (microwaves and ultrasounds) has been applied for the analysis of vitamin C in tea [66].

A simplified spectrophotometric method proposed by Kamelksa *et al.* [76] consisting of direct saponification followed by enzymatic treatment (cholesterol esterase, horseradish esterase, and cholesterol oxidase) has been demonstrated suitable for the rapid analysis of cholesterol in milk samples. In this specific case, the cholesterol esterase is the enzyme assisting in the extraction of the cholesterol from the matrix, while the other two enzymes are used to generate O_2 that generates quinone imine, the dye compound that can be measured spectrophotometrically.

Compared to other methods based on acid or methanol deproteinization, the application of EAE increased the recovery of polyphenols, for example, catechins from milk-tea beverages [77] and lignans in expresso coffee [78]. A significant fraction of polyphenols in foodstuff is not extracted by conventional methods based on aqueous-organic solvents. These non-extractable polyphenols (NEPP) can be mainly divided into two types: (i) the hydrolyzable polyphenols and tannins, consisting of low molecular weight phenolics strongly linked to structural compounds such as polysaccharides of dietary fiber or proteins, and (ii) non-extractable proanthocyanidins (NEPA) of high molecular weight [81]. The procedure for the detection and quantification of these recalcitrant polyphenols in foodstuff must include a hydrolysis step that is generally carried out by acid or alkali methods [52, 79, 82]. EAE has also been applied for the analysis of the NEPP in commercial teas at different fermentation degrees [79] and of NEPA in fruits such as apple, nectarine, and peach [80], but they are not as effective as the alkaline or acid hydrolysis step.

Extraction of Trace Elements and Pharmaceuticals, and Mycotoxin Monitoring by EAE

Humans and animals are exposed to trace elements (*i.e.*, arsenic (As)) and emerging pollutants (*i.e.*, pharmaceuticals) primarily through consumption of contaminated food and water. Similarly, mycotoxins (*i.e.*, aflatoxin 1 (AFM_1) and ochratoxin A (OTA)) can contaminate cereal crops by fungal infection, thereby ending up in the food chain (dairy products, blood-derived meat products). Mycotoxins have been also frequently found in spices, cocoa, dried fruits, and beverages (beer, wine, grape juice) [83]. Moreover, these toxins resist sterilization and pasteurization processes, complicating their elimination. Prolonged exposure to these compounds, even at low concentrations, is correlated with several adverse health and/or environmental impacts. Consequently, the determination of their presence and concentration in water and food is necessary to carry out health and environmental risk assessments.

The detection of trace elements and emerging pollutants in solid samples (for example, fish, chicken meat, rice) involves appropriate extraction, coupled with an efficient separation system and sensitive detection [84]. The extraction conditions should be chosen so these minor elements integrity is maintained. At the same time, they should guarantee their efficient release from the food matrix, as they could be potentially attached to proteins, lipids, or carbohydrates. In this context, EAE mild conditions of pH and temperature would ensure that these species are not altered, while the incorporation of enzymes would improve extraction efficiency [84 - 86]. EAE with pepsin and trypsin has been used to develop a rapid, simple, sensitive, and reliable method for the detection of salicylic acids in meat products [87]. Examples of the analysis of arsenic, pharmaceuticals, and mycotoxins based on EAE are given in Table **3**. EAE using proteases was effective (88 to 106% of As recovery) to release up to 11 arsenic species from chicken meat [88] and fish [35] in combination with ultrasound reaching detection levels of up to 0.11 µg/kg. The inclusion of proteinase K for the extraction of pharmaceuticals from crayfish resulted in improvements compared with microwave treatment with no enzyme [89]. The combined EAE-microwave treatment had different recoveries that ranged from 69% (trimethoprim and danofloxacin) to 106% (for salicylic acid). EAE has also been applied for defatting (lipase) and disrupting possible protein–toxin links (protease) for the analysis of AFM_1 in dairy products that are commercially available [37]. EAE method showed better yields of mycotoxins AFM_1 [36] and OTA [90] compared to the conventional method based on chloroform extraction. EAE using pepsin proved to be simpler and more effective for the detection of AFM_1 in cheese, with added benefits such as improved accuracy and higher recoveries [36]. Comparably, EAE using pancreatin demonstrated significantly higher values of OTA on dry-cured ham samples [90].

Table 3. Performance parameters (detection limit, recovery, and relative standard deviation -RSD) for the methods applied to the analysis of arsenic, pharmaceuticals, and mycotoxins, after applying EAE on diverse food matrices.

Food matrix	Compound(s)	Pretreatment EAE Conditions	Analytical Technique (detection limit)	Refs.
		Trace elements		
Chicken meat	Arsenic species*: AsB, AsIII, DMA, MMA, AsV, 3-nitro- Roxarsone, and NAHAA.	Blending and freeze-drying. Solids loading: 6% (dw/v) 30 mg papain, ultrapure water. Ultrasound: 20% amplitude (20 KHz) for 2 min at 60°C > 1 min pause> 2 min sonication> incubation for 6 h, 60°C	HPLC coupled with ICP-MS, and ESI-MS (1.0-1.8 µg/kg) Recovery: 88%	[88]
Fish	Arsenic species*: AsB, DMA, As (III), AsC, MMA, ASA, As (V), CA, NIT, NAPP, ROX.	Blending Solids loading: 10% (fw/v) 40 mg protease in 0.075 HCl> Sonication for 15 minutes> overnight 37°C	HPLC coupled with ICP-MS (0.11–0.59 µg/kg) Recovery: 94-106% (RSD:1.1-4.4)	[35]
		Pharmaceuticals		

(Table 3) cont.....

Food matrix	Compound(s)	Pretreatment EAE Conditions	Analytical Technique (detection limit)	Refs.
Crayfish	Pharmaceuticals**: Fluroquinolons- CPR, DNF, ENR, FMQ, GTF, GPF, MRB, NRF Tetracyclines- CLT, OXT Sulphonamides- SMX, SDZ, SMT, SMR Penicillins- AMX Anfenicols- CLF, TIF, FLF Non-steroideal steroidal anti-inflammatory drugs (IBU, SAC) Diaminopyrimidines (TMP)	Freeze-drying and grinding Solids loading: 10% (dw/v) 50 µL proteinase K in acetonitrile: acidified water (0.2% formic acid)> Microwaved at 50 W for 5 minutes	UPLC-MS/MS (0.6–12 ng/g) Recovery: 69-106% (RSD: 1.5-17.2%)	[89]
Mycotoxins				
Cheese, sour cream	Aflatoxin 1 (AFM_1)	Homogenization at 8500 rpm to obtain a paste> 50 grams of cheese/250 grams sour cream + 20 mL of 10 mM sodium acetate buffer (pH 7.5) + 1 mg of porcine pancreas Type II lipase (cheese) or 1 mg porcine pancreas Type II lipase (sour cream)> Incubation at 35°C for 24 h.	HPLC-FLD (22-23 ng/Kg) Recovery: 80.4-92.3% RSD: 2.2-4.7%	[37]
Pork and dry-cured ham	Ochratoxin A (OTA)	Homogenization with a blender> Solids loading: 5% (fw/v) 0.2M phosphate buffer pH 7.5, 1% pancreatin> Incubation at 37°C for 3 h	HPLC-FLD (0.06 µg/Kg) Recovery: 90% RSD: 5.5%	[90]

ESI -FLD - Fluorescence detection; ICP - Inductively coupled plasma mass spectrometry (*) Inorganic arsenic species: As (III) and As (V). Organic arsenic species: arsenobetaine (AsB), dimethylarsinic acid (DMA), arsenocholine (AsC), monomethylarsonic acid (MMA), 4-nitrophenylarsonic acid (NIT), 4-hydrox--3-nitrobenzenearsonic acid (ROX), 4-aminobenzenearsonic acid (ASA), carbarsone (CA) and p-hydroxyphenylarsonic acid (NAPP), 3-nitro-4-hydroxyphenylarsonic acid (Roxarsone), N-acetyl-4-hydr-xy-m-arsanilic acid (NAHAA). (**) Pharmaceuticals as emerging pollutants: ciprofloxacin (CPR), danofloxacin (DNF), enrofloxacin (ENR), flumequine (FMQ), gatifloxacin (GTF), grepafloxacin (GPF), mar-bofloxacin (MRB), norfloxacin (NRF), chlortetracycline (CLT), oxytetracycline (OXT), sulfamethoxazole (SMX), sulfadiazine (SDZ), sulfamethazine (SMT), sulfamerazine (SMR), trimethoprim (TMP), ibuprofen (IBU), salicylic acid (SAC), amoxicillin (AMX), chloramphenicol (CLF), thiamphenicol (TIF) and florfenicol (FLF).

Extraction of Bioproducts and Bioactive Molecules by EAE

Carbohydrates

Bioactive carbohydrates are found in terrestrial plants, macroalgae, microalgae, bacteria, the cell wall of yeasts, mushrooms, animals, and fish mucus coating [21, 91 - 93]. Carbohydrates are a relatively unexploited source of bioactive compounds that can be used in the food and pharmaceutical sectors. For example, water-soluble polysaccharides of high molecular weight have been used to alter food rheological properties such as thickness, texture, or emulsion stability [21]. Polysaccharides have also bioactive properties as immunostimulants of a broad-spectrum and a high degree of biocompatibility [93]. Other bioactive properties ascribed to carbohydrates are the ability to regulate intestinal microflora by their prebiotic and antiadherent properties (*i.e.*, oligosaccharides, inulin, chitin, chitosan, fucoidan, alginate, ulvan); cardiovascular protective, hepatoprotective, anticoagulant, anti-inflammatory, antioxidant, and antitumoral properties [91, 92, 94, 95]. These specific bioactive activities are associated with structural characteristics (molecular weight, type of bonds, functional groups, degree of polymerization). Some functional polysaccharides such as mucilage (arabinoxylans and rhamnogalacturonan) can also be used as an ingredient in new edible films and used to purify water from heavy metals [96].

Most functional polysaccharides are polar molecules highly soluble in water and can be extracted using water extraction methods coupled with physical methodologies. The process temperature, time, and pH should be controlled to not cause loss of functional properties through depolymerization and degradation, as well as problems related to the viscosity of some polysaccharides such as pectin. Moreover, polysaccharides of high molecular weight present limited solubility, which makes the extraction process difficult. Carbohydrates are generally found bound to other molecules (proteins, lipids, polynucleotides, and some inorganic mineral compounds) within the food matrix. In this context, EAE can assist in the liberation of polysaccharides. Several carbohydrases and proteases are already commercially available for polysaccharide extraction. The use of commercially available enzymes is practical and cost-effective for industrial applications, although the availability of specific hydrolytic enzymes, such as mannanases and those for algae polysaccharides (*i.e.*, fucoidanases and alginases) is limited [97, 98]. EAE has been applied to extract yeast mannan using diverse enzyme cocktails consisting of proteases combined with glucanases, amylases, and/or lipases followed by glucoamylases [99]. EAE adds selectivity and specificity, allowing a more efficient extraction and/or purification of the target carbohydrates. EAE can also be tuned to hydrolyze partially the polysaccharides into small fragments to facilitate their extraction, and/or to modify some of the properties (*i.e.*, gelation) [95, 100, 101].

EAE is superior to conventional methods in terms of extraction yield and rate, energy, and solvent consumption. EAE process has been proven superior to hot water extraction (HWE) of polysaccharides from edible mushrooms (increments of the yield ranging from 34.4 to 75.8%) [102] and ulvan from green algae (increment of 43%) [95]. EAE also provided better yields of pectin than acid-based extraction in diverse agricultural byproducts such as kiwifruit pomace [103] and apple pomace [104]. Moreover, the milder conditions of EAE do not generally alter the original structure of the polysaccharides, which assist in keeping its bioactivity [91 - 94, 105]. For example, EAE was superior to HWE and ultrasound-assisted extraction for polysaccharide extraction from pomegranate peel in terms of yields and antioxidant properties [106]. Enzymes such as cellulase, pectinase, α-amylase, and protease are commonly used to extract polysaccharides from plant sources [105]. For example, β-glucan of sclerotium of Cooke has been extracted by a sequential process consisting of hot water extraction, followed by treatments with α-amylase and protease [92] (Table **4**). EAE of lotus leaves using pectinase provided pectic-like polysaccharides with superior immunostimulatory properties when compared to extracts from EAE using α-amylase, cellulase, or protease [107].

There are several insoluble polysaccharides such as chitin from animal origin that require harsher conditions of temperature and /or pressure for their extraction. For example, the production of chitin from shell waste involves three steps: demineralization, deproteinization, and decolorization. The first two steps are carried out through alkali and acidic treatments, respectively. The use of enzymes combined with mild acidic and basic treatments has been proposed as a greener alternative that could yield a chitin of higher quality [51, 108]. Enzymatic deproteinization of white shrimp waste before mild alkali treatment was proven to be effective (92% recovery), yielding a chitin with higher crystallinity structure and higher degree of acetylation than the commercial chitin produced through chemical treatment [108].

Polyphenols

The main antioxidants in the human diet are polyphenols, naturally present in many plants, especially fruits, vegetables, cereals, and pulses. The antioxidant capacity is provided by their ability to act as donors of electron/hydrogen, reducing agents, or metal chelators. Numerous studies suggest that the consumption of polyphenols has a main role in the prevention of diseases correlated to oxidative stress such as cancer, cardiovascular disease, and neurogenerative disorders.

Table 4. EAE conditions applied for extraction of bioactive polysaccharides from diverse matrices.

Source	Polysaccharide	Pretreatment	EAE conditions	Refs.
Plant material				

(Table 4) cont.....

Source	Polysaccharide	Pretreatment	EAE conditions	Refs.
Sclerotium of *Lignosus rhinocerotis* (Cooke)	β-glucan	Freeze-drying> milling (0.2 mm) >100°C for 30 min	Solid to liquid: 33% (w/v) α-amylase treatment (1% v/w, pH 4.5, 90°C for 24 h) > protease treatment (pH 7, 37°C for 3 h)	[92]
Lotus leaf	Peptic-like polysaccharide	Drying, milling.	Solid to liquid: 5% (w/v) Enzyme treatment (1% w/w), 50°C for 48 h. (a) Diazyme® (amylase): pH 4.5-5 (b) Cellulase KN (cellulase and hemicellulase), pH 4.5-5 (c) Plantase MAX® pH (polygalacturonase and pectinesterase), pH 4.5-5 (d) Protex 7L (protease) pH 7	[107]
Edible fronds of *Asplenium australasicum* (J. Sm.) Hook	Mucilage	Freeze-drying	Enzymes: endo-1,4-ß-xylanase (50 IU/g) and ß-glucanase (50 IU/g) Solids loading: 2% (dw/v) 120 rpm, pH 4.5, 50°C for 4/5 h.	[109]
		Algae		
Ulva pertusa Kjellm seaweed	Ulvan	Ethanol extraction treatment> milling 60/80 mesh sieve.	Solid to liquid: 20% (w/v) cellulase treatment (5% w/w, 2500 units/g ulvan), pH 4.5, 50°C for 2 h (with and without ultrasound treatment)	[95]
		Mushrooms		
Medicinal fungus *Shiraia bambusicola*	β-D-galactofuranose containing polysaccharides	Drying and milling into powder>defatting 12h	Solid to liquid: 5% (dw/v), distilled water, 45 °C, neutral proteinase (0.8%, w/v), and extraction time of 2 h	[110]
Pleurotus djamor	Enzyme extractable polysaccharides	Pulverizing, sieving (100 mesh screen), and vacuum drying (50°C)	Solid to liquid: ~0.6% (dw/v), pH 7.36, 44.77 min, 35.36°C Enzyme snailase	[93]
		Fish byproducts		

(Table 4) cont.....

Source	Polysaccharide	Pretreatment	EAE conditions	Refs.
Shell waste from white shrimp	Chitin	Washing> drying at 60°C for 2h> milling> demineralization (lactic acid) and washing	Solid to liquid: ~33% (dw/v) Phosphate buffer with sodium metabisulphite (0.02% w/w) Room temperature, continuous stirring, pH controlled, time (1-3 h) Enzyme dosages: 0.03-0.1% (w/w) -Trypsin pH 7.7 -Ficin pH 7.5	[108]

Polyphenols are classified into phenolics, stilbenes, flavonoids, tannins, and lignins. The structure of phenolic compounds is based on an aromatic ring containing one or more hydroxyls, ranging from simple molecules (benzoquinones, hydroxybenzoic acids) to highly polymerized polyphenols (condensed tannins or NEPA). Extractable phenolic compounds have been one of the most studied hydrophilic bioactive molecules. However, there is an increased interest in the role of non-extractable or bound phenolic compounds. Insoluble phenolics are covalently linked to structural components of the plant cell such as cellulose, hemicellulose, pectin, lignin, and structural proteins. The most common methods to release the bound phenolics are alkaline or acid hydrolysis, effective for the hydrolysis of glycosidic and ester bonds, respectively. Although these chemical methods provide good yields and are useful to study the profile of bound phenolics, they require harsh acid/alkali conditions and temperature (in the case of acid hydrolysis) that are not environmentally friendly. EAE is a milder method for the release of bound phenolics by the action of carbohydrase enzymes (cellulases, pectinases, glucanases, feruloyl esterase) and proteases. EAE has been applied for the extraction of all types of polyphenols. A recent and comprehensive review on the extraction of polyphenols by EAE or EAE combined with other green extraction techniques can be consulted [6]. Some examples of EAE extraction conditions of different types of polyphenols are given in Table **5**: phenolic compounds of watermelon rinds [111], flavonoids of pectin waste [112], tannins of acorn [113], stilbenes of peanut seeds [114], and lignans from medicinal plants [115].

The characterization of the polyphenols in the extracts is generally carried out by Reversed-phase high-performance liquid chromatography (RP-HPLC) with Diode-Array Detection (DAD) (Table **5**). HPLC-DAD can be coupled to mass spectrometry (MS) detector [82]. Electrospray ionization mass spectrometry (ESI-MS) can be used for the identification of NEPP owing to its efficiency in the analysis of multicharged ions, with several high-resolution mass analyzers, for example, Orbitrap, quadrupole-time of flight (Q-TOF-MS), and triple quadrupole-TOF (TQ-TOF-MS) [116]. Fig. (**3**) illustrates the signal of NEPP [M-H]- ion mass spectrum of quercetin of enzymatic extract from fruit peel (*Passiflora edulis*) after treatment with a commercial enzyme preparation containing ferulic acid esterase (Depol 740L, Biocatalyst).

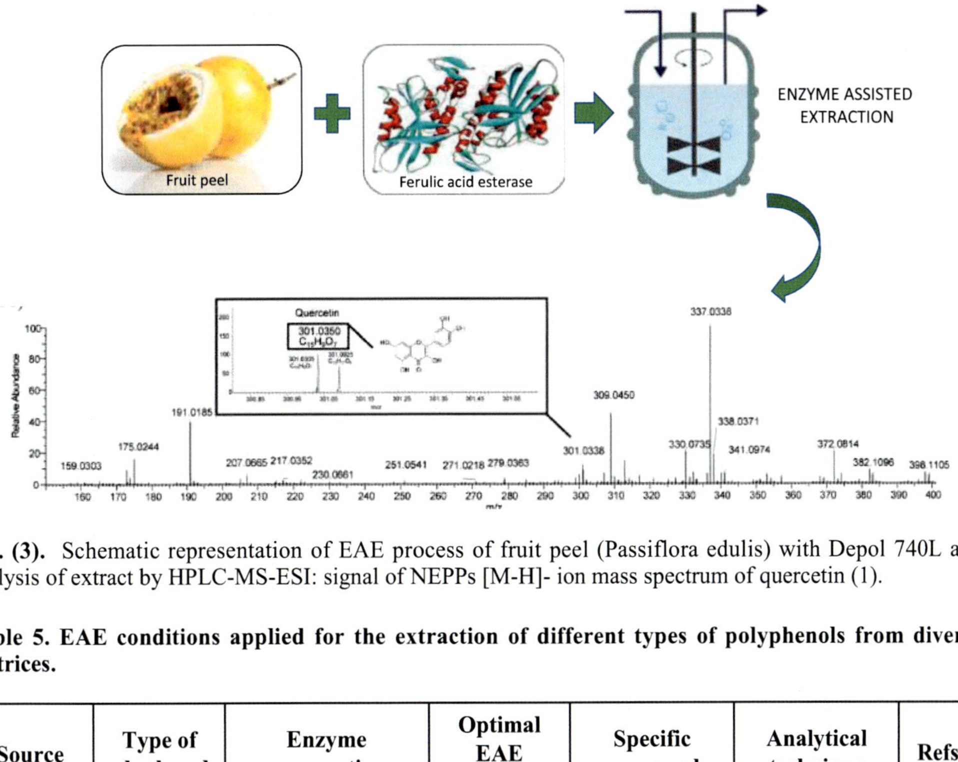

Fig. (3). Schematic representation of EAE process of fruit peel (Passiflora edulis) with Depol 740L and analysis of extract by HPLC-MS-ESI: signal of NEPPs [M-H]- ion mass spectrum of quercetin (1).

Table 5. EAE conditions applied for the extraction of different types of polyphenols from diverse matrices.

Source	**Type of polyphenols**	**Enzyme preparation**	**Optimal EAE Conditions**	**Specific compounds**	**Analytical technique**	**Refs.**
Phenolic acids						
Watermelon rind	Phenolic compounds	“Kemzyme”: pectinase, endo-1,3 (4)-β-glucanase, α-amylase, endo-1,4-β-xylanase and bacillolysine (protease)	Enzyme: 2.24% Phosphate buffer pH 6.58, 51.8°C, 30 min	Chlorogenic acid, Vanillic acid, and Sinapic acid.	RP-HPLC-DAD	[111]
Flavonoids						

(Table 5) cont.....

Source	Type of polyphenols	Enzyme preparation	Optimal EAE Conditions	Specific compounds	Analytical technique	Refs.
Pectin waste	Flavanone aglycones	β-glucosidase Cellulase Tannase and combinations thereof.	Solids loading: 8% (dw/v), Enzyme: 20 IU β-glucosidase 20 mM Acetate buffer pH 5.0, 40°C, 120 rpm, 24 h	Naringenin (77.63 mg/ 100 g) and hesperetin (766.44 mg/ 100 g)	RP-HPLC-DAD (280 nm)	[112]
Tannins						
Acorn	Tannin	Cellulase (Sigma)	Ultrasonic time: 2.51 h. Ultrasonic power: 97.92 W. Cellulase: 0.344% (dw/v) 38°C, pH 5	Tannins (63.16% yield)	Gravimetric method	[113]
Stilbenes						
Peanut seeds	Resveratrol	Cellulase (Beijing Suo Lai Bao Technology)	Solids loading: 60% (dw/v), Enzyme: 0.38% (dw/v), 57.06 min, pH 5.88, and 51.05 °C	Total trans-resveratrol (free or combined with glucose)	RP-HPLC-DAD (360 nm)	[114]
Lignans						
Phyllanthus niruri	Lignans	Cellulase, mixtures of cellulase and protease or cellulase and hemicellulase	Solids loading: 5% (w/v) Cellulase treatment (9 U/g for 12 h) followed by protease (4 U/g up to 72 h)	Total lignans (85.87 mg/g extract), phyllanthin (25.9 mg/g extract) Extraction yield of 13.92 (% w/w)	RP-HPLC-DAD (230, 280, and 320 nm) UPLC-ESI-MS	[115]

Alkaloids

Alkaloids are the major group of secondary metabolites, produced by a myriad of organisms including bacteria, fungi, plants, and animals. Alkaloids exert important biological properties such as anti-endotoxin, anti-inflammatory, antioxidant, and antiproliferative. The interest in the extraction of alkaloids is due to their preventive effect towards several diseases such as diabetes, Parkinson's, cancer, cardiovascular disease, and depression.

There is no standardized method for the extraction of bioactive alkaloids, but it is generally carried out by conventional organic solvents (*e.g.*, methanol, alcohol, ethyl acetate, acetone, and chloroform). However, these methods are not environmentally friendly and present drawbacks such as toxicity, low extraction yields, long extraction times, and the potential existence of trace organic solvent in final products which reduces their quality. Thus, EAE has been proposed as an alternative method for the extraction of several alkaloids. The effect of enzyme treatment on the release of eritadenine, a purine alkaloid with cholesterol-reducing properties, was evaluated in shitake mushrooms [117]. The carbohydrase treatment (5% solids loading (dw/v), 0.02% of commercial enzyme preparation NS 33075, Novozymes, in 0.1 M acetate buffer pH 4.8, at 50°C for 3 hours) before methanol extraction, did not significantly increase the yield (3.82 ± 0.30 mg/g) compared with no enzyme treatment (3.50 ± 0.26 mg/g). However, this study did not optimize the EAE conditions or study the differences in the bioactivities among the different extracts. For example, pilocarpine, an imidazole alkaloid found in plants of the genus *Pilocarpus*, has been extracted enzymatically with yields similar to conventional methods [118]. The EAE optimum conditions were found to be: 0.25% solids loading (dw/v), 10% of Viscozyme enzyme in 50 mM acetate buffer pH 4, at 45°C for 30 h [118]. EAE has also been used to extract camptothecin, a modified pentacyclic alkaloid produced by some plants (angiosperms) and endophytic fungus [119]. This study showed that cellulase improved the extraction of camptothecin from *Aspergillus* with increased anticancer activity against the MCF7 cell line [119].

Terpenes and Terpenoids

Terpenoids, also known as isoprenoids, are the most abundant and structurally diverse type of natural compounds present mainly in many plants and fungi, although they have been detected in bacteria too. Terpenes are simple hydrocarbons, while terpenoids are a modified class of terpenes with different functional groups and oxidized methyl groups. Terpenoids can be classified according to the number of isoprene units (hemiterpene (1) < monoterpenes (2) < sesquiterpenes (3) < diterpenes (4) < sesterpenes (5) < triterpenes (6) <

tetraterpene (8)) or the sort and number of cyclic structures in the molecule.

Terpenoids have been exploited for many applications in different sectors [120]. Terpenoids have microbicidal and insecticidal properties, making them a good food preservative, and an ingredient in disinfectants, mouthwashes, cough syrups, and insect repellents. Terpenoids have also medicinal applications such as hormonal treatment, treatment of heart diseases, malaria prophylaxis, and cancer. Among all applications, they are mostly used in the food sector for the improvement of organoleptic properties (color, flavor, and aroma) and antioxidant capacity. Carotenoids (*i.e.*, lycopene and β-carotene) are the main group of natural pigments responsible for the color of fruits, vegetables, and algae. They have been used as food colorants due to their desirable properties such as no toxicity and high versatility. Carotenoids are generally extracted by organic solvents: non-polar (hexane, petroleum ether, or tetrahydrofuran) for non-polar carotenes (ß-carotene and lycopene) or esterified xanthophylls; and polar (acetone, ethanol, and ethyl acetate) for polar carotenoids (*i.e.*, lutein and zeaxanthin). The simultaneous extraction of carotenoids is however difficult due to their diverse range of polarity and their oxidative nature, limiting the extraction time and exposure to high temperatures, light, and acids. In this regard, milder conditions of EAE could prevent the degradation of these compounds, while assisting in the reduction of solvent required for their extraction. For example, the enzyme treatment with cellulase and pectinase prior to ethyl acetate extraction of tomato waste yielded up to 10 times more carotenoids compared with no enzyme treatment [121]. Optimization of EAE conditions (30 °C, 3.5 h for EAE, enzyme dosage = 0.17%, substrate loading=7%) before hexane extraction has led up to a 19-fold improvement of lycopene yield from tomato waste [32]. The improvement of lycopene yield with enzyme treatments has not been observed when other solvents (*i.e.*, acetone) were used for the extraction, highlighting the importance of the selection of enzyme mixture-solvent combination [122]. The extraction of lycopene from tomato peels through EAE combined with other green technologies such as supercritical fluids increased the yield 3 times compared with the control [61]. Similarly, the addition of a commercial enzyme preparation (β-1,3-glucanase and protease) improved the yields of extraction of xanthophyll from microalgae during ultrasound extraction [123].

Oils

Oils are obtained mainly from oilseeds, but also from oleaginous yeast, algae, and animal sources (mainly fish and shellfish). Oils and lipids are used as ingredients of foodstuff and as additives for other nonedible products such as cosmetics, soaps, adhesives, waxes and synthetic resins, varnishes, lubricants, and paints, to name a few [28]. Some oils are recognized as functional food owing to the presence of some bioactive components (phenolics, terpenoids, and tocopherols) with antioxidant properties, and essential fatty acids (omegas 3 and 6, omega 9, lauric acid). The consumption of functional food oil is associated with health benefits such as reduction of inflammation, regulation of cholesterol levels, cancer prevention, and mitigation of hypertension, diabetes, and Alzheimer's. Another application of oil within a biorefinery system is the production of biodiesel. In this regard, insect fat is being considered as alternative feedstock for biodiesel production [124].

The industrial extraction of vegetal oil is generally carried out by mechanical pressing or solvent extraction [28]. Physical extraction achieves recoveries of 80%, so alternative methods are required to extract the remaining 20%. Solvent extraction with organic solvents such as hexane provides the highest yields (96%). Fish oil extraction also relies on the use of organic solvents and wet rendering, that can cause degradation of heat-labile compounds [125]. The use of organic solvents adds regulatory problems associated with their use and disposal, and increases toxicity and environmental footprint, besides influencing the cost and oil quality. Mechanical high-pressure homogenization methods are used industrially for microbial oil extraction, which can cause the thermal degradation of the lipids [126]. EAE is an alternative to these methods and has been applied as a stand-alone or combined with other techniques. The selection of enzymes for oil extraction depends on the composition of the material. The enzymes normally used for the EAE of oils in plant-based materials are cellulase, hemicellulase, pectinase, protease, and phospholipase. In the case of fungal cell walls, the enzyme mixtures consist of glucanases, chitinases, N-acetyl--d-glucosaminidases, and proteases [126]. Proteases have been used for the extraction of oil from fish [125] and insect larvae [124].

Examples of EAE conditions for the oil extraction of seeds [127 - 130], algae [131], fish [125, 132], yeast [126], and insects [124] are compiled in Table **6**. The application of EAE prior to hexane extraction of fat from fly larvae provided up to 2.3 times more fat yield than when no protease was added [124]. The combined extraction process proved economic viability for its industrial application [124]. EAE provided a good yield (92%) in the extraction of an oil-protein emulsion of coconut [133] and extraction of oil from spirulina (89-100%) [131], with an improved composition of the fatty acids [131] compared to the solvent extraction process (Table **6**). However, the combination of EAE with other green extraction technologies, such as microwave or ultrasound combined with microwave, was required to improve the extraction yield from pumpkin seeds [134], cherry seeds [128], and tiger nuts [129] (Table **6**). Similar composition of the extracted oils was obtained by EAE and the conventional process based on organic solvents, but the bioactivity of the enzyme extracts was improved, as well as their organoleptic properties. Added advantages of EAE in aqueous media are that it allows the separation of components without altering their properties, and it can extract other components such as proteins simultaneously [133]. Further, the aqueous media facilitates the simultaneous removal of phospholipids, circumventing the degumming process in the extracted oil and thus reducing production costs [28]. However, the amphipathic nature of the lipid molecules causes the formation of an emulsion of the extract in aqueous media, which calls for an additional de-emulsification step. This can be achieved by other enzymes (phospholipases and proteases) or by changes in temperature. This additional step reduced the oil recovery hampering the application of EAE at the industrial level.

Table 6. EAE conditions applied for the extraction of functional food oil from different matrices.

Source	Compounds analyzed	Pretreatment	EAE Conditions (optimal)	Yield	Analytical technique	Refs.
			Oilseeds and nuts			
Cherry seed	Total phenolics, α-tocopherol, β-carotene, phospholipids and phytosterols	Ultrasonic power: 560 W Microwave power: 323 W Extraction time: 38 min, Temperature: f 40°C (prior to EAE)	Particle size < 0.425 mm Solids loading: ~8% (dw/v) Enzyme (cellulase, hemi-cellulase and pectinase in 1:1:1 ratio) dosage: 2.7% 40°C, pH 3.5, 4 h	83.9%	Spectrophotometric HPLC-DAD AOAC standard methods 992. 06 (II), Ca 12–55 and Ch 6–91	[128]
Pumpkin seeds	Fatty acid, total phenolics, total tocopherols	Irradiation power: 419 W (simultaneous)	Particle size: 80-mesh Solids loading: ~17% (dw/v) Enzyme (cellulase, pectinase, protease in 1:1:1 ratio) concentration: 1.4%, w/w. 44 °C, pH 5, 66 min, and 200 rpm.	64.17%	Gas chromatography (GC) Spectrophotometric	[134]

(Table 6) cont.....

Source	Compounds analyzed	Pretreatment	EAE Conditions (optimal)	Yield	Analytical technique	Refs.
Tiger nut	Fatty acids, triglycerides,	Microwave power: 300 W Radiation temperature: 40 °C. Ultrasonic power: 460 W, Time 30 min	Particle size: 600 µm Solids loading: 10% (dw/v) Enzyme (cellulase, pectinase and hemicellulase in 1:1:1 ratio) loading: 2%. 45°C, pH 4.9, 3 h	85.23%	GC HPLC-APCI-MS/MS	[129]
Coconut oil	Oil and protein emulsion	-	Solids loading: 17% (dw/v) Viscozyme: 0.6% (w/w) Neutrase: 0.3% (w/w) 30 minutes, pH 7.	92%	Gravimetric	[133]
Fish						
Baltic herring and filleting byproducts (head, fins, tail)	Fatty acids	-	Substrate loading: 50% (fw/v) 55°C, 35 minutes (whole) or 70 minutes (byproducts) Enzyme dosage: 0.4% (w/w) Enzymes: Alcalase, Neutrase or Protamex.	43-69%	Gas chromatography	[125]

(Table 6) cont.....

Source	Compounds analyzed	Pretreatment	EAE Conditions (optimal)	Yield	Analytical technique	Refs.
Rahu (head)	Oil (DHA and EPA)	Ultrasounds: 20 KHz, 40% amplitude, 5-10-15 minutes	Substrate loading: 50% (fw/v) 55°C, pH 8, 2 h, 150 rpm Enzyme dosage: 1% (w/w) Enzymes Protamex.	58.74–68.08%	GC-FID	[132]
Algae						
Spirulina	Fatty acids	-	Substrate loading: 20% (dw/v) Vinoflow: 1% (w/v) 40°C, pH 6.5, 25 minutes	89–100% of total lipids recovery	HPLC-ELSD GC-FID	[131]
Yeast						
Schwanniomyces occidentalis	Fatty acids	1 M NaOH, 24 h at room temperature in a rotary wheel at 20 rpm	Yeast concentration: 7% (v/v) Enzyme dosage-time: 187 mg/ L-4 h 46.8 mg/L-24 h pH5, 40 °C	100%	GC-FID	[126]
Insect						
Black soldier fly larvae	Fatty acids	-	Substrate loading: 23% (dw/v) Protamex: 3.85% (w/v) 38.1°C, pH 6.5, 4.27 h	36.09%	GC-FID	[124]

APCI – Atmospheric pressure chemical ionization; DHA – Docosahexaenoic acid; ELSD – Evaporative light scattering detector; EPA – Eicosapentaenoic acid; FID –Flame ionization detection; GC- Gas Chromatography

Proteins and Bioactive Peptides

Proteins are not only important in nutrition but are also extensively employed in the food industry. The recovered proteins can have different applications such as food fortification or dietary supplements. Food proteins have long been known as a source of bioactive peptides, with documented antioxidative capacity, angiotensin I converting enzyme (ACE) inhibitory properties, hepatoprotective activity, and antibacterial activity [135, 136]. Besides their nutritional value and health benefits, proteins are used in the food industry as stabilizing agents and flavor enhancers, and modifiers of food properties (emulsifying, gelling, foaming, and water binding properties). They are also used to encapsulate other bioactive compounds and flavors (such as fish oil and orange oil). Algal phycobiliproteins can substitute synthetic dyes and find other applications such as pharmaceutical and medical [137]. Further, proteins can be used for biomaterials production, or as a building block for further conversion processes within a biorefinery [138].

The expanding food industry and consumer awareness are boosting the protein market and the search for alternative protein ingredients. The protein demand has increased recently with an overall protein ingredient market valued at USD 38 billion in 2019 with a projected growth rate of 9.1% until 2027 [139]. Proteins can be obtained from animals, plants, and microorganisms. The increased demand for non-animal protein can be met by the use of proteins from novel sources such as fungi (19-37% protein) [140, 141], algae (50-60% protein), food waste (up to 38% protein in sesame cakes [142]) and insects (13–77%) [142 - 144]. The extraction of proteins from different sources is generally based on alkali or acid conditions. The proteins in the extract are then precipitated by means of salts (*i.e.*, ammonium sulfate) or by dewatering solvents (*i.e.*, ethanol). Depending on the downstream processing, the resulting products have different protein content and are categorized as protein flours (up to 65%), protein concentrates (65–90%), and isolates (> 90%).

Chemical-based protein extraction can degrade the proteins being extracted, with the subsequent loss of yield and bioactivity [145]. Alkaline conditions improve the extractability of protein, but very harsh conditions can lead to formation of salt, racemization of amino acids, reduced protein bioavailability, and degradation of some amino acids such as lysine and cysteine. EAE can be used to respect the 3D structure of the proteins, which favors their bioavailability. EAE also presents a lower environmental impact, and the products are more suitable for human consumption. Enzymatic modification could significantly improve the emulsifying and foaming properties of plant-derived proteins. Another added benefit of EAE of proteins is that enzymes can minimize complex formation among the extracted proteins and other molecules (carbohydrates, phytates). Despite the benefits, the yields of extraction are generally lower than those obtained by chemical processes. For example, the solvent-based chemical method provided higher protein extraction yields (84%) compared to the enzymatic-assisted process (67%) from black soldier fly prepupae [146]. The efficiency of protein extraction from sugar beet leaves, however, increased by 43.27% with an EAE process [147]. The EAE of proteins can be improved by removing the lipids from the substrate [148]. The incorporation of an EAE step prior to alkaline extraction of defatted moringa leaves increased protein yields by more than double compared to alkaline extraction [149]. The protein extract from moringa contained a balanced amino acid profile enriched in the essential amino acids and with an improved digestibility (99.9%) [149]. The combination of an optimized EAE process (proteases and carbohydrases mixtures) followed by alkaline extraction has been successful for the extraction of proteins from red algae [150]. Similarly, protein yields were significantly improved by sequential mild alkali extraction followed by EAE (proteases) of rapeseed meal and microalgae [148]. The optimal conditions of EAE of protein from defatted food waste cake (*Sacha inchi* cake meal) resulted also in 1.46-fold more yield than the optimal conditions determined for alkaline extraction [151]. The combination of EAE with other processes such as ultrasounds and microwave have been applied for extraction of protein from chicken bone [60] and sesame bran [152], and the production of bioactive peptides from protein extract of ginkgo biloba nuts [153], respectively. Examples of EAE conditions for protein/peptide extraction from diverse feedstocks are listed in Table 7. Most of thc processes are based on protease extraction/digestion, but some materials benefit from the inclusion of carbohydrases for protein extraction [149, 150, 154]. Proteases used for protein hydrolysis are generally derived from plants (bromelain and papain), animals (pepsin and trypsin), and microorganisms (for example Alcalase®, flavourzyme®, neutrase®, collagenase, and proteinase K) [20, 155].

Table 7. EAE conditions for protein extraction and bioactive peptide production from diverse matrices.

Source	Compound	Pretreatment	EAE Conditions	Analytical Technique	Refs.
Animal					
Egg white	Bioactive peptides from albumin	Freeze-dry	Solids loading: 1% albumin Pepsine: 50%, 0.1 M glycine-HCl buffer pH 2, 37 °C for 4 h	UPLC-ESI-MS	[156]
Plant and Algae					
Moringa leaves	Protein, amino acid composition of extracted protein	Defatting	Particle size: 500 µm Enzyme (Viscozyme):60 FBG), pH 5, 30 °C for 30 min	HPLC-FLD Yield: 14%	[149]

(Table 7) cont.....

Source	Compound	Pretreatment	EAE Conditions	Analytical Technique	Refs.
Rapeseed meal Soybean meal Microalgae meal	Protein	Milled defatting	Solids loading: 4% protein. Enzymes: Protex 40XL, Protex P, Protex 5L, Protex 50FP, and Protex 26L. Enzyme dosage: 1 or 5% (w/w) pH 2.5-11.5 (optimal of each enzyme) time: 2, 3, and 24 h	DUMAS analysis (total nitrogen) Spectrophotometric (degree of hydrolysis)	[148]
Ginkgo biloba nut	Protein, peptides fragments	Microwaved assisted protein digestion 300-600 W	Immobilized bromelain (4500 U/g ginkgo) 55°C with magnetic stirring	Tricine SDS PAGE LC–MS^n	[153]
Red seaweed *Palmaria palmata*	Protein, amino acid composition of extracted protein	Sun-dried algae and cut-milling	0.5–1 cm particle size Solids loading: 5% (dw/v) Enzymes loading: Celluclast® + Alcalase® (0.2% w/w of each, pH 4.5) or Shearzyme® + Alcalase® (0.2% w/w of each, pH 4.5)	DUMAS analysis (total nitrogen) HPLC-MSD Trap 86-90% efficiency	[150]
Porphyra	Phycobiliproteins (phycoerythrin and phycocyanin)	Grounding, milling	Particle size: 0.38 mm. Enzyme: 15% of crude extract Marine bacterial crude enzyme (agarase, carragenase, α-amylase, cellulase, xylanase, and mannanase)	HPLC-UV-Visible detector SDS-PAGE	[154]
Food waste					

(Table 7) cont.....

Source	Compound	Pretreatment	EAE Conditions	Analytical Technique	Refs.
Chicken bone	Protein	Ultrasounds Frequency: 40 kHz; Power: 400 W; Time: 120 min; Temperature: 45 °C.	Solids loading: 3.1% protein Papain: 4019 U/g, pH 6.9, 42 °C for 3 h	Soluble protein Total nitrogen (Kjeldahl) Yield~ 57%	[60]
Tuna dark muscle	Protein, amino acid composition	Homogenization	Solids loading: ~33% (w/w) Enzyme (Alcalase PROLYVE 1000) loading: 1%, pH 8.5, 55 °C, 1 h	HPLC using the ion exchange column Yied~80%	[157]
Sacha inchi cake meal	Protein	Defatting	Solids loading: 2% (w/v) Enzyme (Alcalase 2.4L) concentration: 5.6% 50°C, 40.4 min , pH 9.0, 300 rpm	Soluble protein Total nitrogen (Kjeldahl) (Yield 44.7% protein and hydrolysis degree of 7.8%)	[151]
Sugar beet leaves	Protein	-	Enzyme (Pectinex Ultra SP-L) loading 8%, 45 °C for 30 min	Soluble protein Yield 79%	[147]
Sesame bran	Protein	ns	836 W Ultrasound power Solids loading: 10% (w/v) Enzyme (Alcalase 2L) concentration: 1.248% 43°C, 98 min , pH 9.8	-	[152]
Insects					
Black Soldier Fly Prepupae	**Protein-** Lipids_Chitin	Homogeneization	Enzyme (protease): 1% (w/w) Solids loading: 11% (dw/v) 10 mM phosphate buffer pH 6.5, 60 °C for 14 h	HPLC-FLD (Yield 67%)	[146]
Yeast					

(Table 7) cont.....

Source	Compound	Pretreatment	EAE Conditions	Analytical Technique	Refs.
Brewer's spent yeast (BSY)	Protein	ns	Solids loading: 50% (w/v) Enzyme loading: 1%, pH 6, 55°C, 5 h (Protamex® and Flavourzyme® enzymes)	HPLC-FLD ~30%	[158]

FLD fluorescence detector; MSD mass selective detector; ns: not specified; SDS-PAGE – Sodium dodecyl sulfate-polyacrylamide gel electrophoresis

OPPORTUNITIES AND CHALLENGES FOR EAE COMMERCIALIZATION

A biorefinery is a facility that incorporates several processes for the integral use of biomass to produce energy, biofuels, and added-value coproducts such as bioactive compounds following sustainability criteria. EAE follows several sustainable principles in line with the biorefinery: innovation (selection of novel and renewable feedstocks, novel extraction technique); use of alternative solvents (preferably water or agro-solvents); reduction of energy consumption; production of co-products (feedstock for other conversion processes); reduction of unit operations; prioritize controlled process based on safety and robustness; production of biodegradable products of high quality (non-denatured, without contaminants) [159]. The incorporation of new process technologies such as EAE within a biorefinery faces numerous barriers that hinder their implementation at a commercial scale. To promote the use of these green technologies, it is necessary to identify the barriers and opportunities of such systems within a specific context. Biomass deployment consists of several steps: biomass production, harvesting, distribution and transformation by multiple conversion technologies to provide the specific service. This complex chain involves the collaboration of numerous different sectors (energetic, transport, agricultural and forestry industries, food and beverage industries, research, and development, *etc.*) whose cooperation is crucial to implement the EAE process at a commercial scale. To facilitate communication, collaboration, and innovation among the different stakeholders, several food networks and organizations have been launched: EIT Food (established by the European Institute of Innovation & Technology (EIT)) [160], EAT forum [161], Food Science Research Network [162], Agri-food Research Network [163], Sustainable Food Systems Programme from OnePlanet Network [164].

Although many of these barriers are common for any process using biomass as feedstock, the application of EAE has its own peculiarities. This is because the enzymes are considered additives since they are in small quantities and inactivated in the final product. Moreover, the enzymes applied to obtained products for human consumption must have a generally regarded as safe (GRAS) status [165, 166]. Moreover, when the enzyme has been genetically modified, the final products cannot be certified as organic. There are other barriers specific to each feedstock. For example, the extraction of components from lignocellulosic biomass is difficult owing to its recalcitrant and heterogeneous nature. In the case of food byproducts, seasonality and storage are also factors to be considered. Developing storing and pre-processing strategies of this waste to minimize spoilage and improve homogeneity will be required prior to the evaluation of conversion processes. Novel feedstocks such as algae, mushrooms, and insects are proposed to limit the use of natural resources (land, water). These feedstocks call for a different set of enzymes whose mechanism of action is less known compared to plant degrading enzymes. As a result, there has been an increased interest in the study of novel sources of chitinases [167] and enzymes degrading marine polysaccharides (such as carrageenan, β-agarase, β-porphyranase, laminerase, exo-alginate lyase, β-glucuronoyl hydrolase, *etc.*) [97]. New enzyme sources from shrimp have been applied for assisting in extracting astaxanthin from green algae [168].

The main obstacles to the EAE process are the high cost of enzymes produced in GMP (good manufacturing practices) and the enzyme inactivation during EAE [169]. Among the different areas of research and aspects that are directly involved in the cost and efficiency of the EAE process, the following should be pointed out:

- *Reduction of enzyme production* cost by near-site or *in situ* enzyme production, an improvement on expression, and the use of cheaper culture media (for example, by using agricultural byproducts) [170]. On-site enzyme production has been proposed to cut costs related to enzyme production within a lignocellulose biorefinery [171]. On-site production reduces dependence on third parties, and it avoids formulation and transportation costs. Enzyme production and extraction can be carried out simultaneously reaching maximum process integration to reduce costs, as it has been tested for the extraction of polysaccharides [172] and flavones [173].
- Improvement of EAE by *optimization of parameters* to prevent reduction of enzyme activity during EAE process [40].
- Reutilization of enzymes by *enzyme immobilization*. Immobilized EAE using an immobilized enzyme column has been applied for the extraction of soybean oil completing the extraction process in 35 min [174]. The column can work in continuous form for over 24 hours, with the enzyme retaining 80% of its initial

activity [174]. Dendronized silica supports have shown great potential for the immobilization of enzymes such as thermolysin to obtain reusable biocatalysts [57]. Enzymes can be immobilized in nanoparticles (nanobiocatalyst) to enhance the enzyme capacity, stability, and reusability during the extraction. Magnetic nanoparticles are the most common nanomaterials used to immobilize natural enzymes. The immobilization of cellulase and α-amylase on amino-functionalized magnetic nanoparticles allowed for up to 6 cycles of EAE of anthocyanin from black rice [175]. Another trend is the use of nanomaterials with enzyme-like characteristics, nanozymes, that hold potential in food analysis and detection [25].

- Design of hybrid techniques that enhance the extraction ability (*i.e.*, ultrasound-assisted enzymatic extraction (UAEE), microwaved-assisted enzymatic extraction (MWAE) [59].
- *Reduction of enzyme dosage.* Several approaches have been suggested to decrease the costs associated with the use of enzymes during extraction (*i.e.*, high dosage and enzyme deactivation). These strategies can be classified in terms of *improvement of the enzyme systems* and *process development.* The first class comprises the development of enzymes with improved specific activity by protein engineering of known enzymes, and/or by search for novel enzymes (bio-prospecting). The extremozymes hold great potential for the industrial development of EAE within biorefineries [176]. For example, extremozymes such as cold-adapted enzymes are suited for the extraction of thermosensitive bioactive compounds [177]. Other processes would benefit from extremozymes that are stable at high temperatures, such as PLE [43], or enzymes stable in the presence of other solvents such as ionic liquids [178, 179]. Extremozymes can also be used as a prototype for the design of new enzymes with suitable properties for the extraction process.

Concurrently, there is a trend to build tailor-made enzyme cocktails adapted to specific feedstocks and a targeted biomolecule, to improve the extraction efficiency with minimum enzyme dosage. The boost effect of enzymes in the appropriate ratio can lead to the reduction of the final protein concentration needed for maximum extraction [50]. The supplementation with non-catalytic additives such as surfactants can also lead to improvements in extraction yield at lower enzyme dosages [180 - 182].

Despite these challenges, there has been continuous progress in research and development, with a growing trend of ongoing projects on the biobased circular economy [183, 184] and an increase of launched bioactive ingredients by important chemical companies such as Cargill (US), BASF SE (Germany), Dupont (US), Arla Foods (Denmark), Kerry Group (Ireland), Ajinomoto (Japan), Chr. Hansen Holding A/S (Denmark), BENEO (Germany), Royal Cosun

(Netherlands), and Roquette Frères (France), to name a few. Similarly, the progress in EAE processes is highlighted by the increasing number of published articles about EAE (Scopus search of "enzyme assisted extraction" showed 15 articles in 2012 and 205 in 2020) and patents on EAE processes for the extraction of bioactive molecules (Table **8**). The proportion of scientific articles and patents on EAE of each main group of bioactive compounds is represented in Fig. (**4**). EAE patented processes for the extraction of polyphenols and polysaccharides are the most numerous while there are more scientific publications about EAE of proteins than any other bioactive compound.

Overall, the economic feasibility of enzymatic extraction must achieve a good extraction yield at high solids loadings, with reduced enzyme requirements over short periods. The commercial suitability of any process is determined by the results of its upscaling, where analytical criteria and factors are evaluated: the ratio between mass and solvent, solvent concentration, use of continuous or batch mode production, instrumentation design, energy consumption, and economic efficiency, mass transfer, and kinetics. There are few studies of the scale-up of the EAE process on soybeans (pilot plant, 75 Kg) [185], and grape skin (5 Kg) [186]. The current scale-up status of EAE technology is in its infancy, with a capacity of a few tons/day in batch mode [169]. The development of hyphenated or hybrid techniques combining EAE with other green techniques is recommended to favor the scale-up process [169]. Once a process is scaled up, the techno-economic feasibility and environmental impact should be evaluated by environmental assessment tools (such as life cycle analysis or LCA) and economic studies that are compared to conventional processes [159].

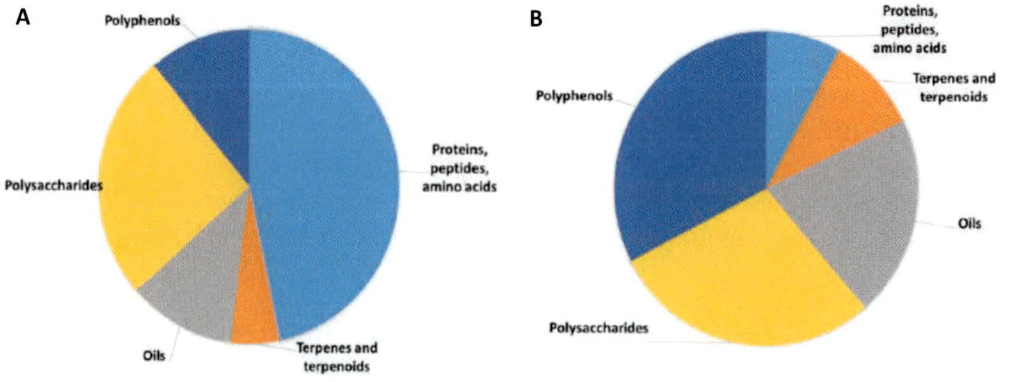

Fig. (4). Pie chart representing the proportion of (A) scientific articles (Scopus search "enzyme assisted extraction" AND "specific bioactive molecule") and (B) patents (Espacenet search "enzyme assisted extraction") on EAE process for polyphenols, polysaccharides, oils, terpenes and terpenoids, and protein and bioactive peptides.

Table 8. Selected search results of patents about enzyme-assisted extraction, Espacenet patent database.

Patent Publication Number	Source	Compound	Hyphenated	Claims	Earliest Priority
Polysaccharides					
CN108157982A	Ginseng	Dietary fiber	-	Economic benefits of the process, the safety of process, product rich in nutrients, and safe to consume.	2018
CN104004108A	*Cynomorium songaricum*	Polysaccharide	-	Simple in process, low in extraction cost, and suitable for industrial mass production	2016
CN111423524A	Lotus leaf	Polysaccharide	-	Concise, efficient, high-activity lotus leaf polysaccharide, large-scale production	2020
CN107778375A	*Dioscorea opposita* Thunb slime	Polysaccharides and resistant starch	high-pressure pulsed electric field	Processing method for agricultural products	2017
CN111607012A	*Paeonia rockii*	Polysaccharide	-	Polysaccharide extract high in biological activity and purity suitable for health-care foods, high extract yield.	2020
US2019032099A1	Fiber feedstock	bio-based fiber gum	-	Increased efficiency in the production of bio-based fiber gum from bio-based fiber feedstocks.	2017
CN108752495A	*Enteromorpha*	Alkali-soluble polysaccharide	Ultrasound wave; thermal reflux	No waste discharge, increased extraction rate, suitable for large-scale production	2018
CN111793142A	*Polygonatum sibiricum*	Polysaccharide	Ultrasounds	Extraction rate improved	2020
CN107141372A	-	Heparin	-	High yield, and the quality of the obtained crude heparin sodium.	2017
CN107098990A	-	Heparin sodium	-	Improved extraction rate and yield of crude heparin sodium.	2017
CN108003251A	Pineapple pomace	Polysaccharide	Ultrasonic wave-assisted extraction	High extraction rate, high purity, and good antioxidant activity.	2017

(Table 8) cont.....

Patent Publication Number	Source	Compound	Hyphenated	Claims	Earliest Priority
CN104911229A	-	Agar	-	The amount of alkali and acid is reduced, and thus the discharge of sewage is reduced; extracting rate is increased, the transparency, the viscosity, and the strength of agar are improved.	2015
CN112481134A	Mulberry leaf	Polysaccharide	Hot water ultrasonic water extraction	Increments in 35% on extraction yield	2020
CN104558232A	Herbaceous plant	Polysaccharide	-	High extraction efficiency, maximum activity of polysaccharide, production is economic, environment-friendly and safe.	2015
US2010119651A1	Jerusalem artichoke	Inulin	Hot water	High extraction rate and short water bath extraction time.	2008
CN108913334A	Soybean	Polysaccharide	-	Product concentration and solvent recovery technology are simple and environmentally friendly, efficient process.	2018
CN102964461A	Dendrobium	Bioactive polysaccharide	-	Higher dissolution rate, improved product quality, and yield. The process is energy-saving and environment-friendly.	2012
CN107312107A	*Dendrobium officinale*	Polysaccharide	-	High efficiency and safety, the process is suitable for industrial production	2017
CN106282260A	*Dendrobium officinale*	Polysaccharide	-	Extraction rate increased to 65.36%, mild, convenient to operate, economical.	2016
CN108503719A	*Dendrobium officinale*	Polysaccharide	-	Lower impurities, reduced loss of solvent, extracting rate increased by 10% or above, and the content of polysaccharide is increased by 10% or above	2018

(Table 8) cont.....

Patent Publication Number	Source	Compound	Hyphenated	Claims	Earliest Priority
CN109053926A CN109053926B	*Dendrobium officinale*	Polysaccharide	-	Extracted polysaccharide is small in molecular structure, high in biological activity, and beneficial to be absorbed and utilized by a human body	2018
Terpenes and terpenoids					
WO2018130682A1	Plant materials	Phyto-cannabinoids and terpenes	-	High cannabinoid content, enhanced phyto-cannabinoids and terpenes/terpenoids stability	2017
AU2013252503A1	Cashew pseudofruit wastes	Carotenoid	Pressing operations in successive aqueous-extraction cycles	Colouring in foods and beverages (high soluble in water).	2012
CN107669721A	*Panax notoginseng*	Saponins	Water extraction and multistage countercurrent alcohol extraction sequentially	Potential for industrial exploitation. Simple process, high in extraction efficiency short time, yield of saponins of 80% or higher, content of saponins can reach 98% or higher.	2017
CN107384907A	*Chlorella* powder	Astaxanthin	Optimized high-pressure homogeneous wall breaking	Innovative process	2017
CN105032282A	*Gleditsia sinensis*	Saponins (natural surfactant)	Ultrasonic-assisted extraction	Avoidance of toxic solvent, improvement of extraction rate and the purity of the extract; the content of total saponins in extractive reaches 50-80%, and the aglycone purity reaches 80-95%.	2015
US10463065B2, US2018153196A1	*Stevia rebaudiana Bertoni* (leaves)	Steviol glycosides	Pressurized aqueous extraction	High purity steviol glycosides (Stevioside and Rebaudioside-A) with improved organoleptic properties	2016

(Table 8) cont.....

Patent Publication Number	Source	Compound	Hyphenated	Claims	Earliest Priority
CN102443037A, CN102443037B	*Camellia oleifera*	Saponin	-	Purity of saponin in tea of more than 90%, yielding rate of more than 85%. Simple process that can be easily incorporated using traditional equipment, suitable for mass production.	2010
CN102424700A, CN102424700B	*Dioscorea zingiberensis* C.H. Wright	Diosgenin	Microwave acidolysis	Reduce considerably amount of acid used in the extraction process; recovery of other fractions (starch, cellulose, *etc.*)	2011
CN112056352A	*Cordyceps militaris*	Carotenoid	-	Process operability is high, the product yield is high, suitable for industrial production.	2020
IN909DE2013A	Plant matrix such as capsicum	Carotene, phenolic and ascorbic acid	-	Solvent free and highly effective	2013
		Polyphenols			
CN112370474A	*Acanthopanax senticosus* (stem bark)	Flavonoids	Ultrasound- assisted extraction	Process easy and convenient to operate, the flavone extraction rate can be increased by 10% compared to conventional process.	2020
CN102702163A	Leftovers of tea leaves	Epigallocatechin gallate	-	Chloroform usage is avoided, integration of extraction and purification process, and the quality and yield of the product are high.	2012
CN105777822A	apple tree branches	Phloridzin	Supercritical CO_2 assisted extraction	Safe and simplified process with high extraction rate, high yield and purity of phloridzin, shorter separation, and purification process.	2016
CN111440218A	Plant	Tannin, flavanol, flavonoid aglycone, phenolic acid, and other glycosyl-free polyphenol components	Ultrasonic or heating extraction	High in purity, light in color, and good in stability	2020
CN104435062A	Blueberry wine mud	Anthocyanins and functional oligosaccharides	Ultrasonic-assisted hot water extraction	Easy to operate, economic and high efficiency	2014

(Table 8) cont.....

Patent Publication Number	Source	Compound	Hyphenated	Claims	Earliest Priority
CN107759647A	*Rosa davurica*	Hyperoside and luteolin	Ethyl alcohol aqueous solution with the assistance of an ionic liquid	Innovative process with high content of hyperoside (> 33 μg/g) and luteolin (16 μg/g) in the extract	2017
CN111440735A	*Scutellaria baicalensis Georgi*	Baicalin	Simultaneous enzyme production by endophytic bacteria and extraction	Increment of yield by 71.34%.	2019
CN105777697A	purple sweet potatoes	Anthocyanin	Aqueous weak acid ethanol solution	Production cost is reduced, the quality of the extract is increased, the anthocyanin content is > 80%; no pesticide and heavy metal ions are detected.	2016
CN105175380A	Yunnan pine bark	Procyanidine	High-voltage pulsed electric field processing	Simple technology, mild conditions, low production cost, the extraction rate (>95%) and the yield are raised, the purity is at least 75%, and the extraction time is shortened. Organic solvents are avoided.	2015
CN111265446A	*Houttuynia cordata*	Flavonoids	Alcohol extraction	Extraction rate of 95%; total flavonoids up to 86%; excellent antibacterial properties; excellent antioxidant capacities (ABTS scavenging rate up to 96%), the elastin's inhibiting rate can reach 92%.	2020
CN109010626A	*Rhizoma Gastrodiae*	Gastrodin	Ultrasound assisted extraction	Extract is safe and effective and has no toxic and side effect; extraction process is simple and suitable for industrial production. Gastrodin content is increased by 25-37%	2018
CN111358821A	*Artemisiae argyi* (folium)	total flavone	Ultrasound-assisted extraction	Optimization of EAE improved the total flavone content to at least 10.99 mg/g, and the ultrasonic extraction time is 20min.	2020

(Table 8) cont.....

Patent Publication Number	Source	Compound	Hyphenated	Claims	Earliest Priority
CN107987046A	Peanut shells	Luteolin	Organic solvent extraction	Accurate extraction technology, simplicity and feasibility in operation, moderate conditions and easiness for realizing industrialization.	2017
CN102461642A	*Orange peels*	Polymethoxylated flavonoids extract	Ethanol extraction	The extract is used to prepare an antistaling agent that can be used for keeping refrigerated fresh meat fresh.	2010
CN106806160A	Green Tea	polyphenol	Ultrasonic-microwave assisted extraction	High in extraction efficiency, polyphenol extracted can be used as ingredient of shower gel and have skin protective effect.	2017
CN112574890A	Mulberry leaves	Flavones	Simultaneous enzyme production by *Mucor racemosus* SY5-47 and enzyme extraction followed by ultrasonic ethanol extraction	Extraction yield can be increased by nearly 50%	2020
CN106108019A	Orange peel	Total flavone	Microwave assisting complex enzyme	Process is easy to operate, high in extraction rate and good in product quality, the total flavones extracted reaches 39.5 mg/g.	2016
CN107468719A, CN107468719B	Algae	Polyphenols	Ultrasound assisted extraction	Increase of the content of the total phenols preserving their anti-oxidation activity, easy absorption, and good bioactivity of the polyphenols.	2017
CN102948758A CN102948758B	Buckwheat bran	Flavone	Ultrasound assisted extraction	Process, operation and its implementation are simple; environmentally friendly; suitable for industrial production; purity of product is higher	2012
CN102977164A, CN102977164B	*Acanthopanax senticosus* Harms	Hyperine	Alcohol extraction	Environmental protection, no pollution, convenient operation, and production cost reduction.	2012

(Table 8) cont.....

Patent Publication Number	Source	Compound	Hyphenated	Claims	Earliest Priority
CN109260300A	*Schisandra chinensis* vine	Lignans and triterpenoids	-	Process low in cost, fast, and high in efficiency and is suitable for large-scale industrialized production	2018
CN106619744A	Algae	Polyphenols	Combined ultrasonic-microwave extraction	The extraction rate is increased, higher yield, reduced cost	2017
CN111743930A	Olive	Polyphenol	-	High extraction efficiency, good preservation.	2019
CN102020868A, CN102020868B	Purple sweet potato	Anthocyanin	-	Extract of excellent quality, low impurity content, stable, and long shelf life. The number of recycling times of the macroporous absorbing resin is increased, and the production efficiency is improved.	2010
CN104448915A	Purple sweet potato	Anthocyanin pigment	-	-	2013
CN108727327A	*Lycium ruthenicum*	Anthocyanin	Ultrasound-assisted extraction	The extracting method can increase the anthocyanin extracting rate under an economic condition.	2017
CN107982296A, CN107982296B	*Radix echinopsis*	Flavones	Ultrasound-assisted extraction	The extraction rate is increased. Compounds are stable in acid environments	2017
CN108815210A	*Ginkgo biloba* L. (dry leaves)	Ginkgetin	-	High extraction rate, mild conditions, simple process, sustainable, and suitable for industrial production	2018
CN101775417A	Blueberry pomace	Anthocyanin	Cavitation suspension	Simple and practicable, suitable for the industry production	2010
CN102489038A, CN102489038B	Chinese herbal medicine	Flavones	Ultrasound-assisted extraction, surfactant	Extraction time shorter, the solvent can be saved, the yield is high, and the stability is good.	2011

(Table 8) cont.....

Patent Publication Number	Source	Compound	Hyphenated	Claims	Earliest Priority
CN102040519A, CN102040519B	*Eucommia ulmoides* leaves	Chlorogenic acid	Warm water extraction	The extraction rate can reach 95%, the water extraction temperature is lower, and the extraction time is shortened, purity > 60% and the extraction ratio is more than 1.4%	2009
CN101830881A, CN101830881B	Larch processing leftover	Toxifolin	Solid cavitation under negative pressure cavitation	Simple and feasible operation, which is environment-friendly and suitable for industrial application. Increase the toxifolin content of by 60.65%	2010
CN106491697A	Peanut shell	Polyphenol	Ethyl alcohol microwave-assisted extraction	Polyphenol yield is high; process is simple and convenient; the time is relatively short; the energy consumption is low; extraction rate and the production efficiency are improved; industrial production is	2016
Oils					
WO2017182665A1	Palm	Oil	-	Milling line with a continuous sterilizer and unit to add enzyme.	2016
WO2019038418A1	Palm	Oil	-	EAE process by means of a GH10 xylanase	2017
WO2019068850A1	Palm	Oil	-	High solids EAE (40% w/w)	2017
CN109022118A	By-product of *thunnus obesus*	Oil, EPA	-	Total content of EPA and DHA is at least 80%. High fish oil yield, good product quality, high added value of products.	2018
WO2018135937A1	Palm fruitlet	Oil	-	An extraction device consisting of threshing unit, an enzyme dosing unit, a maceration unit, a digester unit, and a press unit.	2017

(Table 8) cont.....

Patent Publication Number	Source	Compound	Hyphenated	Claims	Earliest Priority
CN112126504A	*Cinnamomum longepaniculatum*	Oil	Microbial and enzyme extraction followed by distillation	Improved extraction yield and purity of product, reduction of the time required by subsequent distillation, residues of an organic solvent can be avoided	2020
CN109111981A	Peanut	Oil	-	Simple and short process, high extraction efficiency, high purity, mild technological, suitable for industrial application	2018
WO2017182665A1	Palm	Oil	-	Milling line with a continuous sterilizer and unit to add enzyme.	2016
WO2019038418A1	Palm	Oil	-	EAE process by means of a GH10 xylanase	2017
WO2019068850A1	Palm	Oil	-	High solids EAE (40% w/w)	2017
CN109022118A	By-product of *thunnus obesus*	Oil, EPA	-	Total content of EPA and DHA is at least 80%. High fish oil yield, good product quality, high added value of products.	2018
WO2018135937A1	Palm fruitlet	Oil	A extraction device consisting of threshing unit, an enzyme dosing unit, a maceration unit, a digester unit, and a press unit.	-	2017
CN112126504A	*Cinnamomum longepaniculatum*	Oil	Microbial and enzyme extraction followed by distillation	Improved extraction yield and purity of product, reduction of the time required by subsequent distillation, residues of an organic solvent can be avoided	2020
CN109111981A	Peanut	Oil	-	Simple and short process, high extraction efficiency, high purity, mild technological, suitable for industrial application	2018
CN106753802A	Jasmine flower	essential oil crystal	-	Increased yield, low impurity, better quality, stability, and final composition of essential oil.	2017

(Table 8) cont.....

Patent Publication Number	Source	Compound	Hyphenated	Claims	Earliest Priority
CN103981031A	Flos Magnoliae	Oil	Ultrasonic-assisted enzymolysis	Extraction at room temperature, short extraction process, high extraction capability.	2014
CN101940240A	Fish pomace	oil ethyl ester microcapsule	Supercritical CO_2 extraction	Microcapsule can hide the fishy smell of the fish oil ethyl ester, improve the mouth feel, and be beneficial to packaging and storage.	2010
CN105255577A	Fresh olives	Oil	-	Extraction of high-quality olive oil efficiently and rapidly. Extraction rate up to 84.31%. Mild operation conditions, short time, high efficiency, and good quality and is suitable for industrialization.	2015
CN108760910A	Krill	Oil (phospholipids)	-	Higher extraction rate, product with more kinds of fatty acids and higher content of unsaturated fatty acids, nutritional quality of the phospholipids is better.	2018
CN102816639A	Soybean	Oil	-	Extraction time of only 35min, low residual oil (0.5%), immobilized enzyme up to 24 hours keeping 80% activity.	2012
CN107653059A CN107653059B	semen coicis bran	-	Radio-frequency and ethanol assistance	High oil extraction rate, low cost, good quality	2017
		Proteins, peptides and amino acids			
WO2019229288A1	*Arthrospira platensi*	Amino acids, low molecular weight peptides, and polyphenols, amongst other antioxidants	-	Extract can be used to prevent and treat hypertension or hyperlipidemia.	2018
WO2015004255A1	Brown algae	Glycoproteins	-	Novel process for the isolation of unique anti-oxidative glycoproteins	2013
WO2019077145A1	Potato rasp	SAME (S-adenosyl-L-methionine)	Ethanol extraction	Antioxidant extract from potato	2017
CN107446978A	Dendrobium	small molecule peptide	-	Process time is shortened, the production efficiency is increased	2017

(Table 8) cont.....

Patent Publication Number	Source	Compound	Hyphenated	Claims	Earliest Priority
CN104404094A	Clams	taurine	-	Simple operation, safety, low cost, green environmental protection, continuous production, high taurine yield, and the like	2014
CN104982643A	Rice bran	Protein	-	Extraction rate reaches 76.19%, the purity > 70%. Extraction efficiency is high; the process is simple, the stability is good.	2015
CN110279038A	-	Phycocyanin	Microwave treatment on the raw materials	Phycocyanin-containing pig feed is easily obtained, poison-less and harmless, and rich in nutrients, and can improve the immunity of pigs and help the pigs grow quickly	2019
WO2018195657A1	Raw biomasses such as oil seeds and grains.	Soluble protein/peptide and purified dietary fiber	Alkaline pretreatment	Increased product yield and solubility	2017

DHA – Docosahexaenoic acid; EPA – Eicosapentaenoic acid

CONCLUDING REMARKS

There is a worldwide challenge to address food security while limiting the usage of natural resources (land, water) due to global warming, growing population, and change in diets. Global environmental awareness has prompted an increased interest in sustainable and biodiverse food systems. Consequently, food industries are interested in new products that contain natural ingredients from sustainable crops that are processed by green technologies such as EAE. The EAE process, however, faces several challenges for its application at a commercial scale. Food matrices are generally complex, and heterogeneous and some are recalcitrant. The combination of EAE with other green extraction methods such as ultrasound and microwave has been proven to improve yields. There is also an interest in the development of more efficient enzymes by manipulating desirable properties for the process such as thermostability, interaction with the substrate, and compatibility with solvents. The enzyme cocktail applied during EAE should be optimized considering the substrate and the molecule targeted.

Cost related to enzyme production and its use is one of the main obstacles that hamper the industrial expansion of the EAE process. A better understanding of the structure of the feedstocks and the mechanism of action of the enzyme mixtures is crucial to improving the extraction process. This will require an integrated approach to the process considering all the steps (feedstock preparation, structure, and composition; enzyme production; enzyme mixture and properties; downstream processing) and requires the expertise of different fields such as genetics, biochemistry, biotechnology, and process engineering. The development of specific models and simulations to evaluate the EAE process is necessary. This assessment will determine the economic feasibility and current bottlenecks that need to be addressed to improve the economics and sustainability of the process.

LIST OF ABBREVIATIONS

ACE	Angiotensin I converting enzyme
AFM_1	Aflatoxin 1
AMG	Amylo-glucosidase
ANN	Artificial neural networks
AOAC	Association of official agricultural chemists
BBD	Box-behnken design
CCD	Central composite design
DAD	Diode-array detection
DET	Differential evolution technique
DF	Dietary fiber
DW	Dry weight
EAE	Enzyme-assisted extraction
ESI	Electrospray ionization
FOS	Fructo-oligosaccharides
GMP	Good manufacturing practices
HWE	Hot water extraction
IUBMB	International union of biochemistry and molecular biology
LCA	Life cycle analysis
NADP	Nicotinamide adenine dinucleotide
NEPA	Non-extractable proanthocyanidins
NEPP	Non-extractable polyphenols
PAA	Pancreatic α-amylase
PHWE	Pressurised hot water extraction
Q-TOF	Quadrupole time of flight
RITDF	Rapid integrated total dietary fiber
RP-HPLC	Reversed-phase high-performance liquid chromatography
RSM	Response surface methodology

TDF	Total dietary fiber
TQ-TOF-MS	Triple quadrupole time of flight mass spectrometry
UAE	Ultrasound-assisted extraction

CONSENT FOR PUBLICATION

Not applicable.

CONFLICT OF INTEREST

The authors declare that they have no known conflict of interests, financial or personal relationships that could have influenced the work reported in this chapter.

ACKNOWLEDGEMENTS

The authors would like to acknowledge the following funders: European Union's Horizon 2020 Research and Innovation program under Grant Agreement No 754382, the Comunidad of Madrid and European funding from FSE and FEDER programs (S2018/BAA-4393, AVANSECAL-II-CM) and the University of Alcalá (Creación y consolidación de grupos UAH CCG20/CC-055).

REFERENCES

[1] Li S, Yang X, Yang S, Zhu M, Wang X. Technology prospecting on enzymes: Application, marketing and engineering. Comput Struct Biotechnol J 2012; 2(3): e201209017. [http://dx.doi.org/10.5936/csbj.201209017] [PMID: 24688658]

[2] Hernández BE, De Jesús PLE, Zartha SJW. Recovery of biomolecules from agroindustry by solid-liquid enzyme-assisted extraction: A review. Food Anal Methods 2021; 14(8): 1744-77. [http://dx.doi.org/10.1007/s12161-021-01974-w]

[3] Pandal N. Global markets for flavors and fragances. BBC Research 2014; p. 7215.

[4] Sowbhagya HB, Chitra VN. Enzyme-assisted extraction of flavorings and colorants from plant materials. Crit Rev Food Sci Nutr 2010; 50(2): 146-61. [http://dx.doi.org/10.1080/10408390802248775] [PMID: 20112157]

[5] Dahiya S, Kumar AN, Shanthi Sravan J, Chatterjee S, Sarkar O, Mohan SV. Food waste biorefinery: Sustainable strategy for circular bioeconomy. Bioresour Technol 2018; 248(Pt A): 2-12. [http://dx.doi.org/10.1016/j.biortech.2017.07.176] [PMID: 28823499]

[6] Gligor O, Mocan A, Moldovan C, Locatelli M, Crişan G, Ferreira ICFR. Enzyme-assisted extractions of polyphenols – A comprehensive review. Trends Food Sci Technol 2019; 88: 302-15. [http://dx.doi.org/10.1016/j.tifs.2019.03.029]

[7] Food Enzymes Market Research Report by Type, by Formulation, by Source, by Region - Global Forecast to 2026 - Cumulative Impact of COVID-19 2021, Research and Markets.. 2021. Available from: https://www.researchandmarkets.com/reports/4807813/food-enzymes-market-research-re-ort-by-type-by [accessed 21/06/2021].

[8] ExplorEnz: Search the Enzyme List. Available from: https://www.enzyme-database.org/ [accessed 21/06/2021].

[9] IntEnz home. Available from: https://www.ebi.ac.uk/intenz/ [accessed 21/06/2021].

[10] Enzyme Database - BRENDA. Available from: https://www.brenda-enzymes.org/ [accessed 21/06/2021].

[11] KEGG ENZYME Database. Available from: https://www.genome.jp/kegg/annotation/enzyme.html [accessed 21/06/2021].

[12] EzCatDB. Available from: http://ezcatdb.cbrc.jp/EzCatDB/ [accessed 21/06/2021].

[13] MEROPS - the Peptidase Database. Available from: https://www.ebi.ac.uk/merops/ [accessed 21/06/2021].

[14] MetaCyc: Metabolic Pathways From all Domains of Life.. Available from: https://metacyc.org/ [accessed 21/06/2021]/

[15] Official REBASE Homepage | The Restriction Enzyme Database | NEB. Available from: http://rebase.neb.com/rebase/rebase.html [accessed 21/06/2021].

[16] CAZy - Home. Available from: http://www.cazy.org/ [accessed 21/06/2021].

[17] Esther Home Page. Available from: http://bioweb.supagro.inra.fr/ESTHER/general?what=index [accessed 21/06/2021].

[18] RedOxiBase. Available from: http://peroxibase.toulouse.inra.fr/ [accessed 21/06/2021].

[19] KinBase. Available from: http://kinase.com/kinbase/ [accessed 21/06/2021].

[20] Rani S, Pooja K, Pal GK. Exploration of rice protein hydrolysates and peptides with special reference to antioxidant potential: Computational derived approaches for bio-activity determination. Trends Food Sci Technol 2018; 80: 61-70. [http://dx.doi.org/10.1016/j.tifs.2018.07.013]

[21] Nadar SS, Rao P, Rathod VK. Enzyme assisted extraction of biomolecules as an approach to novel extraction technology: A review. Food Res Int 2018; 108: 309-30. [http://dx.doi.org/10.1016/j.foodres.2018.03.006] [PMID: 29735063]

[22] Xianhuan X. Process for extracting dendrobium officinale polysaccharide by enzyme hydrolysis method CN107312107A. Patent 2013.

[23] LI W, Li Y, Liu W, *et al.* A method for extracting effective ingredient from stem of *Schisandra chinensis* vine by enzyme assisted extraction. CN109260300A Patent 2018. .

[24] Liang K, Zhang Q, Cong W. Enzyme-assisted aqueous extraction of lipid from microalgae. J Agric Food Chem 2012; 60(47): 11771-6. [http://dx.doi.org/10.1021/jf302836v] [PMID: 23072503]

[25] Zhang Y, Rui X, Simpson BK. Trends in nanozymes development *versus* traditional enzymes in food science. Curr Opin Food Sci 2021; 37: 10-6. [http://dx.doi.org/10.1016/j.cofs.2020.08.001]

[26] Gil-Chávez JG, Villa JA, Ayala-Zavala FJ, *et al.* Technologies for extraction and production of bioactive compounds to be used as nutraceuticals and food ingredients: An overview. Compr Rev Food Sci Food Saf 2013; 12(1): 5-23. [http://dx.doi.org/10.1111/1541-4337.12005]

[27] Contreras MM, Lama-Muñoz A, Gutiérrez-Pérez MJ, Espínola F, Moya M, Castro E. Protein extraction from agri-food residues for integration in biorefinery: Potential techniques and current status. Bioresour Technol 2019; 280: 459-77. [http://dx.doi.org/10.1016/j.biortech.2019.02.040] [PMID: 30777702]

[28] Mwaurah PW, Kumar S, Kumar N, *et al.* Novel oil extraction technologies: Process conditions, quality parameters, and optimization. Compr Rev Food Sci Food Saf 2020; 19(1): 3-20. [http://dx.doi.org/10.1111/1541-4337.12507] [PMID: 33319516]

[29] Ozyurt VH, Çakaloğlu B, Otles S. Optimization of cold press and enzymatic-assisted aqueous oil extraction from tomato seed by response surface methodology: Effect on quality characteristics. J Food Process Preserv 2021; 45(5): 1-16.
[http://dx.doi.org/10.1111/jfpp.15471]

[30] Ma M, Mu T, Sun H, Zhang M, Chen J, Yan Z. Optimization of extraction efficiency by shear emulsifying assisted enzymatic hydrolysis and functional properties of dietary fiber from deoiled cumin (*Cuminum cyminum* L.). Food Chem 2015; 179: 270-7.
[http://dx.doi.org/10.1016/j.foodchem.2015.01.136] [PMID: 25722165]

[31] Huang G, Chen F, Yang W, Huang H. Preparation, deproteinization and comparison of bioactive polysaccharides. Trends Food Sci Technol 2021; 109: 564-8.
[http://dx.doi.org/10.1016/j.tifs.2021.01.038]

[32] Zuorro A, Lavecchia R. Optimization of enzyme-assisted lycopene extraction from tomato processing waste. Adv Mat Res 2013; 800: 173-6.
[http://dx.doi.org/10.4028/www.scientific.net/AMR.800.173]

[33] Poojary MM, Orlien V, Passamonti P, Olsen K. Enzyme-assisted extraction enhancing the umami taste amino acids recovery from several cultivated mushrooms. Food Chem 2017; 234: 236-44.
[http://dx.doi.org/10.1016/j.foodchem.2017.04.157] [PMID: 28551231]

[34] Md Noh MF, Gunasegavan RDN, Mustafa Khalid N, Balasubramaniam V, Mustar S, Abd Rashed A. Recent techniques in nutrient analysis for food composition database. Molecules 2020; 25(19): 4567.
[http://dx.doi.org/10.3390/molecules25194567] [PMID: 33036314]

[35] Zhao F, Liu Y, Zhang X, *et al.* Enzyme-assisted extraction and liquid chromatography-inductively coupled plasma mass spectrometry for the determination of arsenic species in fish. J Chromatogr A 2018; 1573: 48-58.
[http://dx.doi.org/10.1016/j.chroma.2018.08.068] [PMID: 30195857]

[36] Pietri A, Fortunati P, Mulazzi A, Bertuzzi T. Enzyme-assisted extraction for the HPLC determination of aflatoxin M1 in cheese. Food Chem 2016; 192: 235-41.
[http://dx.doi.org/10.1016/j.foodchem.2015.07.006] [PMID: 26304342]

[37] Chavarría G, Granados-Chinchilla F, Alfaro-Cascante M, Molina A. Detection of aflatoxin M_1 in milk, cheese and sour cream samples from Costa Rica using enzyme-assisted extraction and HPLC. Food Addit Contam Part B Surveill 2015; 8(2): 128-35.
[http://dx.doi.org/10.1080/19393210.2015.1015176] [PMID: 26111266]

[38] Çoklar H, Akbulut M. Effect of sun, oven and freeze-drying on anthocyanins, phenolic compounds and antioxidant activity of black grape (Ekşikara) (*Vitis vinifera* L.) S Afr J Enol Vitic 2017; 38(2): 264-72.

[39] Belwal T, Ezzat SM, Rastrelli L, *et al.* A critical analysis of extraction techniques used for botanicals: Trends, priorities, industrial uses and optimization strategies. Trends Analyt Chem 2018; 100: 82-102.
[http://dx.doi.org/10.1016/j.trac.2017.12.018]

[40] De Luna RSL, Ramírez-Garza RE, Serna SSO. Environmentally friendly methods for flavonoid extraction from plant material: Impact of their operating conditions on yield and antioxidant properties. ScientificWorldJournal 2020; 2020: 1-38.
[http://dx.doi.org/10.1155/2020/6792069] [PMID: 32908461]

[41] Lau T, Harbourne N, Oruña-Concha MJ. Optimization of enzyme-assisted extraction of ferulic acid from sweet corn cob by response surface methodology. J Sci Food Agric 2020; 100(4): 1479-85.
[http://dx.doi.org/10.1002/jsfa.10155] [PMID: 31756272]

[42] Gall DL, Kontur WS, Lan W, *et al. In vitro* enzymatic depolymerization of lignin with release of syringyl, guaiacyl, and tricin units. Appl Environ Microbiol 2018; 84(3): e02076-17.
[http://dx.doi.org/10.1128/AEM.02076-17] [PMID: 29180366]

[43] Ekman A, Campos M, Lindahl S, *et al.* Bioresource utilisation by sustainable technologies in new

value-added biorefinery concepts – two case studies from food and forest industry. J Clean Prod 2013; 57: 46-58.
[http://dx.doi.org/10.1016/j.jclepro.2013.06.003]

[44] Hongkulsup C, Khutoryanskiy VV, Niranjan K. Enzyme assisted extraction of chitin from shrimp shells (*Litopenaeus vannamei)*. J Chem Technol Biotechnol 2016; 91(5): 1250-6.
[http://dx.doi.org/10.1002/jctb.4714]

[45] Rani A, Arfat Y, Aziz RS, *et al.* Enzymatically assisted extraction of antioxidant and anti-mutagenic compounds from radish (*Raphanus sativus*). Environmental Technology & Innovation 2021; 23: 101620.
[http://dx.doi.org/10.1016/j.eti.2021.101620]

[46] Luo X, Zhu JY. Effects of drying-induced fiber hornification on enzymatic saccharification of lignocelluloses. Enzyme Microb Technol 2011; 48(1): 92-9.
[http://dx.doi.org/10.1016/j.enzmictec.2010.09.014] [PMID: 22112776]

[47] Domínguez-Rodríguez G, Marina ML, Plaza M. Enzyme-assisted extraction of bioactive non-extractable polyphenols from sweet cherry (*Prunus avium* L.) pomace. Food Chem 2021; 339: 128086.
[http://dx.doi.org/10.1016/j.foodchem.2020.128086] [PMID: 33152877]

[48] Dragoi EN, Curteanu S. The use of differential evolution algorithm for solving chemical engineering problems. Rev Chem Eng 2016; 32(2): 149-80.
[http://dx.doi.org/10.1515/revce-2015-0042]

[49] Poojary MM, Orlien V, Passamonti P, Olsen K. Enzyme-assisted extraction enhancing the umami taste amino acids recovery from several cultivated mushrooms. Food Chem 2017; 234: 236-44.
[http://dx.doi.org/10.1016/j.foodchem.2017.04.157] [PMID: 28551231]

[50] Peixoto Araujo NM, Pereira GA, Arruda HS, *et al.* Enzymatic treatment improves the antioxidant and antiproliferative activities of Adenanthera pavonina L. seeds. Biocatal Agric Biotechnol 2019; 18: 101002.
[http://dx.doi.org/10.1016/j.bcab.2019.01.040]

[51] Qi Y, Ren W, Zhang H, *et al.* Optimization of extraction and purification of polysaccharides from *Veronicastrum axillare*, and evaluation of their biological activities. Chem Biodivers 2021; 18(3): e2000864.
[http://dx.doi.org/10.1002/cbdv.202000864] [PMID: 33533083]

[52] Pérez-Jiménez J, Arranz S, Tabernero M, *et al.* Updated methodology to determine antioxidant capacity in plant foods, oils and beverages: Extraction, measurement and expression of results. Food Res Int 2008; 41(3): 274-85.
[http://dx.doi.org/10.1016/j.foodres.2007.12.004]

[53] Syrpas M, Valanciene E, Augustiniene E, Malys N. Valorization of bilberry (*Vaccinium myrtillus* l.) pomace by enzyme-assisted extraction: Process optimization and comparison with conventional solid-liquid extraction. Antioxidants 2021; 10(5): 773.
[http://dx.doi.org/10.3390/antiox10050773] [PMID: 34068178]

[54] Giacomozzi DE, Joutsimo O. Drying temperature and hornification of industrial never-dried *Pinus radiata* pulps. 1. strength, optical, and water holding properties. BioResources 2015; 10(3): 5791-808.
[http://dx.doi.org/10.15376/biores.10.3.5791-5808]

[55] Vásquez V, Martínez R, Bernal C. Enzyme-assisted extraction of proteins from the seaweeds *Macrocystis pyrifera* and *Chondracanthus chamissoi*: Characterization of the extracts and their bioactive potential. J Appl Phycol 2019; 31(3): 1999-2010.
[http://dx.doi.org/10.1007/s10811-018-1712-y]

[56] Sonawane A, Sudhir P, Sumit, Pradhan RC. Optimization of a process for the enzymatic extraction of nutrient enriched bael fruit juice using artificial neural network (ANN) and response surface methodology (RSM). Int J Fruit Sci 2020; 20(sup3): S1845-61.

[57] Hernández-Corroto E, Sánchez-Milla M, Sánchez-Nieves J, de la Mata FJ, Marina ML, García MC. Immobilization of thermolysin enzyme on dendronized silica supports. Evaluation of its feasibility on multiple protein hydrolysis cycles. Int J Biol Macromol 2020; 165(Pt B): 2338-48. [http://dx.doi.org/10.1016/j.ijbiomac.2020.10.138] [PMID: 33132126]

[58] Andler SM, Goddard JM. Transforming food waste: How immobilized enzymes can valorize waste streams into revenue streams. NPJ Sci Food 2018; 2(1): 19. [http://dx.doi.org/10.1038/s41538-018-0028-2] [PMID: 31304269]

[59] Wen L, Zhang Z, Sun DW, Sivagnanam SP, Tiwari BK. Combination of emerging technologies for the extraction of bioactive compounds. Crit Rev Food Sci Nutr 2020; 60(11): 1826-41. [http://dx.doi.org/10.1080/10408398.2019.1602823] [PMID: 30990060]

[60] Dong ZY, Li MY, Tian G, Zhang TH, Ren H, Quek SY. Effects of ultrasonic pretreatment on the structure and functionality of chicken bone protein prepared by enzymatic method. Food Chem 2019; 299: 125103. [http://dx.doi.org/10.1016/j.foodchem.2019.125103] [PMID: 31284247]

[61] Patil PD, Patil SP, Kelkar RK, Patil NP, Pise PV, Nadar SS. Enzyme-assisted supercritical fluid extraction: An integral approach to extract bioactive compounds. Trends Food Sci Technol 2021; 116: 357-69. [http://dx.doi.org/10.1016/j.tifs.2021.07.032]

[62] Makkliang F, Siriwarin B, Yusakul G, *et al.* Biocompatible natural deep eutectic solvent-based extraction and cellulolytic enzyme-mediated transformation of *Pueraria mirifica* isoflavones: A sustainable approach for increasing health-bioactive constituents. Bioresour Bioprocess 2021; 8(1): 76. [http://dx.doi.org/10.1186/s40643-021-00428-9]

[63] Costa JR, Tonon RV, Cabral L, Gottschalk L, Pastrana L, Pintado ME. Valorization of agricultural lignocellulosic plant byproducts through enzymatic and enzyme-assisted extraction of high-value-added compounds: A review. ACS Sustain Chem& Eng 2020; 8(35): 13112-25. [http://dx.doi.org/10.1021/acssuschemeng.0c02087]

[64] Domínguez-Rodríguez G, García MC, Marina ML, Plaza M. Pressurized liquid extraction combined with enzymatic-assisted extraction to obtain bioactive non-extractable polyphenols from sweet cherry (*Prunus avium* L.) pomace. Nutrients 2021; 13(9): 3242. [http://dx.doi.org/10.3390/nu13093242] [PMID: 34579121]

[65] Juan W. Method for extracting organic acids from fruits. CN106588637A Patent 2017.

[66] Zhenxing H, Wanliang M, Yang Y, Qin Y, Fenfen Z. Method for extracting vitamin C from sweet tea, CN104292196A. CN104292196B Patent 2015.

[67] McCleary BV, Ames N, Cox J, *et al.* Total dietary fiber (codex definition) in foods and food ingredients by a rapid enzymatic-gravimetric method and liquid chromatography: Collaborative study, first Action 2017.16. J AOAC Int 2019; 102(1): 196-207. [http://dx.doi.org/10.5740/jaoacint.18-0180] [PMID: 30107867]

[68] Macagnan FT, da Silva LP, Hecktheuer LH. Dietary fibre: The scientific search for an ideal definition and methodology of analysis, and its physiological importance as a carrier of bioactive compounds. Food Res Int 2016; 85: 144-54. [http://dx.doi.org/10.1016/j.foodres.2016.04.032] [PMID: 29544829]

[69] Phillips KM, Haytowitz DB, Pehrsson PR. Implications of two different methods for analyzing total dietary fiber in foods for food composition databases. J Food Compos Anal 2019; 84: 103253. [http://dx.doi.org/10.1016/j.jfca.2019.103253]

[70] Oberson JM, Campos-Giménez E, Rivière J, Martin F. Application of supercritical fluid chromatography coupled to mass spectrometry to the determination of fat-soluble vitamins in selected food products. J Chromatogr B Analyt Technol Biomed Life Sci 2018; 1086: 118-29. [http://dx.doi.org/10.1016/j.jchromb.2018.04.017] [PMID: 29665470]

[71] Schmidt A, Pratsch H, Schreiner MG, Mayer HK. Determination of the native forms of vitamin B_1 in bovine milk using a fast and simplified UHPLC method. Food Chem 2017; 229: 452-7. [http://dx.doi.org/10.1016/j.foodchem.2017.02.092] [PMID: 28372200]

[72] Lebiedzińska A, Marszałł ML, Grembecka M, *et al.* Detection of B6 vitamers in grain products: Experimental and computational studies. Food Anal Methods 2018; 11(3): 725-32. [http://dx.doi.org/10.1007/s12161-017-1045-5]

[73] Bhandari SD, Gao M, Szpylka J, *et al.* Total folate in infant formula and adult nutritionals by trienzyme extraction and LC-MS/MS quantitation: A multilaboratory testing study, final action 2011.06. J AOAC Int 2018; 101(6): 1881-94. [http://dx.doi.org/10.5740/jaoacint.18-0114] [PMID: 30049295]

[74] Nakos M, Pepelanova I, Beutel S, Krings U, Berger RG, Scheper T. Isolation and analysis of vitamin B12 from plant samples. Food Chem 2017; 216: 301-8. [http://dx.doi.org/10.1016/j.foodchem.2016.08.037] [PMID: 27596424]

[75] Caprioli G, Sagratini G, Vittori S, Torregiani E. Optimization of an extraction procedure for the simultaneous quantification of riboflavin, nicotinamide and nicotinic acid in anchovies (*Engraulis enrasicolus*) by high-performance liquid chromatography–tandem mass spectrometry. J Food Compos Anal 2018; 66: 23-9. [http://dx.doi.org/10.1016/j.jfca.2017.11.004]

[76] Kamelska AM, Jarmołowska B, Bryl K. A simplified enzymatic method for total cholesterol determination in milk. Int Dairy J 2015; 50: 50-7. [http://dx.doi.org/10.1016/j.idairyj.2015.06.004]

[77] Ferruzzi MG, Green RJ. Analysis of catechins from milk–tea beverages by enzyme assisted extraction followed by high performance liquid chromatography. Food Chem 2006; 99(3): 484-91. [http://dx.doi.org/10.1016/j.foodchem.2005.08.010]

[78] Angeloni S, Navarini L, Sagratini G, Torregiani E, Vittori S, Caprioli G. Development of an extraction method for the quantification of lignans in espresso coffee by using HPLC-MS/MS triple quadrupole. J Mass Spectrom 2018; 53(9): 842-8. [http://dx.doi.org/10.1002/jms.4251] [PMID: 29925121]

[79] Yan S, Zhou Z, Wang K, Song S, Shao H, Yang X. Chemical profile and antioxidant potential of extractable and non-extractable polyphenols in commercial teas at different fermentation degrees. J Food Process Preserv 2020; 44(6): 1-13. [http://dx.doi.org/10.1111/jfpp.14487]

[80] Pérez-Jiménez J, Arranz S, Saura-Calixto F. Proanthocyanidin content in foods is largely underestimated in the literature data: An approach to quantification of the missing proanthocyanidins. Food Res Int 2009; 42(10): 1381-8. [http://dx.doi.org/10.1016/j.foodres.2009.07.002]

[81] Pérez-Jiménez J, Saura-Calixto F. Macromolecular antioxidants or non-extractable polyphenols in fruit and vegetables: Intake in four European countries. Food Res Int 2015; 74: 315-23. [http://dx.doi.org/10.1016/j.foodres.2015.05.007] [PMID: 28411997]

[82] Domínguez-Rodríguez G, Marina ML, Plaza M. Strategies for the extraction and analysis of non-extractable polyphenols from plants. J Chromatogr A 2017; 1514: 1-15. [http://dx.doi.org/10.1016/j.chroma.2017.07.066] [PMID: 28778531]

[83] Visconti A, De Girolamo A. Fitness for purpose – Ochratoxin A analytical developments. Food Addit Contam 2005; 22(sup1) (Suppl. 1): 37-44. [http://dx.doi.org/10.1080/02652030500410307] [PMID: 16332620]

[84] Bermejo-Barrera P, Moreda-Piñeiro A, Bermejo-Barrera A. Sample pre-treatment methods for the trace elements determination in seafood products by atomic absorption spectrometry. Talanta 2002; 57(5): 969-84.

[http://dx.doi.org/10.1016/S0039-9140(02)00139-X] [PMID: 18968702]

[85] Mattusch J, Cimpean M, Wennrich R. Enzyme-assisted extraction of arsenic species from plant material. Int J Environ Anal Chem 2006; 86(9): 629-40. [http://dx.doi.org/10.1080/03067310600557489]

[86] Komorowicz I, Hanć A, Lorenc W, Barałkiewicz D, Falandysz J, Wang Y. Arsenic speciation in mushrooms using dimensional chromatography coupled to ICP-MS detector. Chemosphere 2019; 233: 223-33. [http://dx.doi.org/10.1016/j.chemosphere.2019.05.130] [PMID: 31176123]

[87] Tingquan S. Method for detecting salicylic acids in meat balls, CN104634777A. CN104634777B Patent 2015.

[88] Liu Q, Peng H, Lu X, Le XC. Enzyme-assisted extraction and liquid chromatography mass spectrometry for the determination of arsenic species in chicken meat. Anal Chim Acta 2015; 888: 1-9. [http://dx.doi.org/10.1016/j.aca.2015.05.001] [PMID: 26320952]

[89] Kazakova J, Fernández-Torres R, Ramos-Payán M, Bello-López MÁ. Multiresidue determination of 21 pharmaceuticals in crayfish (*Procambarus clarkii*) using enzymatic microwave-assisted liquid extraction and ultrahigh-performance liquid chromatography-triple quadrupole mass spectrometry analysis. J Pharm Biomed Anal 2018; 160: 144-51. [http://dx.doi.org/10.1016/j.jpba.2018.07.057] [PMID: 30086507]

[90] Pietri A, Gualla A, Rastelli S, Bertuzzi T. Enzyme-assisted extraction for the HPLC determination of ochratoxin A in pork and dry-cured ham. Food Addit Contam Part A Chem Anal Control Expo Risk Assess 2011; 28(12): 1-7. [http://dx.doi.org/10.1080/19440049.2011.609490] [PMID: 22007860]

[91] Hayes M, Tiwari B. Bioactive carbohydrates and peptides in foods: An overview of sources, downstream processing steps and associated bioactivities. Int J Mol Sci 2015; 16(9): 22485-508. [http://dx.doi.org/10.3390/ijms160922485] [PMID: 26393573]

[92] Veeraperumal S, Qiu HM, Tan CS, *et al.* Restitution of epithelial cells during intestinal mucosal wound healing: The effect of a polysaccharide from the sclerotium of *Lignosus rhinocerotis* (Cooke) Ryvarden. J Ethnopharmacol 2021; 274: 114024. [http://dx.doi.org/10.1016/j.jep.2021.114024] [PMID: 33727110]

[93] Jiao F, Wang X, Song X, *et al.* Processing optimization and anti-oxidative activity of enzymatic extractable polysaccharides from *Pleurotus djamor.* Int J Biol Macromol 2017; 98: 469-78. [http://dx.doi.org/10.1016/j.ijbiomac.2017.01.126] [PMID: 28167111]

[94] Liu J, Willför S, Xu C. A review of bioactive plant polysaccharides: Biological activities, functionalization, and biomedical applications. Bioactive Carbohydrates and Dietary Fibre 2015; 5(1): 31-61. [http://dx.doi.org/10.1016/j.bcdf.2014.12.001]

[95] Chen J, Zeng W, Gan J, *et al.* Physicochemical properties and anti-oxidation activities of ulvan from *Ulva pertusa* Kjellm. Algal Res 2021; 55: 102269. [http://dx.doi.org/10.1016/j.algal.2021.102269]

[96] Kamel R, Afifi SM, Kassem IAA, Elkasabgy NA, Farag MA. Arabinoxylan and rhamnogalacturonan mucilage: Outgoing and potential trends of pharmaceutical, environmental, and medicinal merits. Int J Biol Macromol 2020; 165(Pt B): 2550-64. [http://dx.doi.org/10.1016/j.ijbiomac.2020.10.175] [PMID: 33115647]

[97] Hehemann JH, Boraston AB, Czjzek M. A sweet new wave: Structures and mechanisms of enzymes that digest polysaccharides from marine algae. Curr Opin Struct Biol 2014; 28(1): 77-86. [http://dx.doi.org/10.1016/j.sbi.2014.07.009] [PMID: 25136767]

[98] Dobrinčić A, Balbino S, Zorić Z, *et al.* Advanced technologies for the extraction of marine brown algal polysaccharides. Mar Drugs 2020; 18(3): 168.

[http://dx.doi.org/10.3390/md18030168] [PMID: 32197494]

[99] Sedmak J J. Production of beta-glucans and mannans. US 8,753,668 B2 Patent 2014.

[100] Zeng WW, Lai LS. Characterization of the mucilage extracted from the edible fronds of bird's nest fern (*Asplenium australasicum*) with enzymatic modifications. Food Hydrocoll 2016; 53: 84-92. [http://dx.doi.org/10.1016/j.foodhyd.2015.03.026]

[101] Rivas MÁ, Casquete R, Martín A, Córdoba MG, Aranda E, Benito MJ. Strategies to increase the biological and biotechnological value of polysaccharides from agricultural waste for application in healthy nutrition. Int J Environ Res Public Health 2021; 18(11): 5937. [http://dx.doi.org/10.3390/ijerph18115937] [PMID: 34205897]

[102] Leong YK, Yang FC, Chang JS. Extraction of polysaccharides from edible mushrooms: Emerging technologies and recent advances. Carbohydr Polym 2021; 251: 117006. [http://dx.doi.org/10.1016/j.carbpol.2020.117006] [PMID: 33142573]

[103] Yuliarti O, Goh KKT, Matia-Merino L, Mawson J, Brennan C. Extraction and characterisation of pomace pectin from gold kiwifruit (*Actinidia chinensis*). Food Chem 2015; 187: 290-6. [http://dx.doi.org/10.1016/j.foodchem.2015.03.148] [PMID: 25977029]

[104] Wikiera A, Mika M, Starzyńska-Janiszewska A, Stodolak B. Application of Celluclast 1.5L in apple pectin extraction. Carbohydr Polym 2015; 134: 251-7. [http://dx.doi.org/10.1016/j.carbpol.2015.07.051] [PMID: 26428122]

[105] Cheng L, Zhang X, Hong Y, Li Z, Li C, Gu Z. Characterisation of physicochemical and functional properties of soluble dietary fibre from potato pulp obtained by enzyme-assisted extraction. Int J Biol Macromol 2017; 101: 1004-11. [http://dx.doi.org/10.1016/j.ijbiomac.2017.03.156] [PMID: 28373048]

[106] Zhai X, Zhu C, Li Y, Zhang Y, Duan Z, Yang X. Optimization for pectinase-assisted extraction of polysaccharides from pomegranate peel with chemical composition and antioxidant activity. Int J Biol Macromol 2018; 109: 244-53. [http://dx.doi.org/10.1016/j.ijbiomac.2017.12.064] [PMID: 29242125]

[107] Song YR, Han AR, Park SG, Cho CW, Rhee YK, Hong HD. Effect of enzyme-assisted extraction on the physicochemical properties and bioactive potential of lotus leaf polysaccharides. Int J Biol Macromol 2020; 153: 169-79. [http://dx.doi.org/10.1016/j.ijbiomac.2020.02.252] [PMID: 32105695]

[108] Chiang CF, Lai LS. Effect of enzyme-assisted extraction on the physicochemical properties of mucilage from the fronds of *Asplenium australasicum* (J. Sm.) Hook. Int J Biol Macromol 2019; 124: 346-53. [http://dx.doi.org/10.1016/j.ijbiomac.2018.11.181] [PMID: 30465842]

[109] Zhou D, Li P, Dong Z, *et al.* Structure and immunoregulatory activity of β-d-galactofurano-e-containing polysaccharides from the medicinal fungus *Shiraia bambusicola.* Int J Biol Macromol 2019; 129: 530-7. [http://dx.doi.org/10.1016/j.ijbiomac.2019.01.179] [PMID: 30710588]

[110] Marzieh MN, Zahra F, Tahereh E, Sara KN. Comparison of the physicochemical and structural characteristics of enzymatic produced chitin and commercial chitin. Int J Biol Macromol 2019; 139: 270-6. [http://dx.doi.org/10.1016/j.ijbiomac.2019.07.217] [PMID: 31376451]

[111] Mushtaq M, Sultana B, Bhatti HN, Asghar M. RSM based optimized enzyme-assisted extraction of antioxidant phenolics from underutilized watermelon (*Citrullus lanatus* Thunb.) rind. J Food Sci Technol 2015; 52(8): 5048-56. [http://dx.doi.org/10.1007/s13197-014-1562-9] [PMID: 26243925]

[112] Barbosa PPM, Ruviaro AR, Macedo GA. Conditions of enzyme-assisted extraction to increase the recovery of flavanone aglycones from pectin waste. J Food Sci Technol 2021; 58(11): 4303-12.

[http://dx.doi.org/10.1007/s13197-020-04906-4] [PMID: 34538913]

[113] Luo X, Bai R, Zhen D, *et al.* Response surface optimization of the enzyme-based ultrasound-assisted extraction of acorn tannins and their corrosion inhibition properties. Ind Crops Prod 2019; 129: 405-13.
[http://dx.doi.org/10.1016/j.indcrop.2018.12.029]

[114] Xu S, Luo H, Chen H, *et al.* Optimization of extraction of total trans-resveratrol from peanut seeds and its determination by HPLC. J Sep Sci 2020; 43(6): 1024-31.
[http://dx.doi.org/10.1002/jssc.201900915] [PMID: 31916409]

[115] Meselhy MR, Abdel-Sattar OE, El-Mekkawy S, *et al.* Preparation of lignan-rich extract from the aerial parts of *Phyllanthus niruri* using non-conventional methods. Molecules 2020; 25(5): 1179.
[http://dx.doi.org/10.3390/molecules25051179] [PMID: 32151037]

[116] Domínguez-Rodríguez G, Marina ML, Plaza M. Rapid fingerprinting of extractable and non-extractable polyphenols from tropical fruit peels using direct analysis in real time coupled to orbitrap mass spectrometry. Food Chem 2022; 371: 131191.
[http://dx.doi.org/10.1016/j.foodchem.2021.131191] [PMID: 34600365]

[117] Enman J, Rova U, Berglund KA. Quantification of the bioactive compound eritadenine in selected strains of shiitake mushroom (*Lentinus edodes*). J Agric Food Chem 2007; 55(4): 1177-80.
[http://dx.doi.org/10.1021/jf062559+] [PMID: 17256958]

[118] Jun-Ho J-H, Bhattarai S, Oh TJ, Jang JH. Enzymatic extraction of pilocarpine from *Pilocarpus jaborandi.* Korean J Microbiol Biotechnol 2013; 41(2): 236-41.
[http://dx.doi.org/10.4014/kjmb.1303.03005]

[119] Natarajan S, Alagar YS. Enzyme-enhanced extraction of camptothecin and its anticancer activity on breast cancer cell line. Asian J Pharm Clin Res 2018; 11(6): 468-72.
[http://dx.doi.org/10.22159/ajpcr.2018.v11i6.26333]

[120] Pichersky E, Raguso RA. Why do plants produce so many terpenoid compounds? New Phytol 2018; 220(3): 692-702.
[http://dx.doi.org/10.1111/nph.14178] [PMID: 27604856]

[121] Strati IF, Gogou E, Oreopoulou V. Enzyme and high pressure assisted extraction of carotenoids from tomato waste. Food Bioprod Process 2015; 94: 668-74.
[http://dx.doi.org/10.1016/j.fbp.2014.09.012]

[122] Deng Y, Zhao S, Yang X, *et al.* Evaluation of extraction technologies of lycopene: Hindrance of extraction, effects on isomerization and comparative analysis - A review. Trends Food Sci Technol 2021; 115: 285-96.
[http://dx.doi.org/10.1016/j.tifs.2021.06.051]

[123] Machado FRS Jr, Trevisol TC, Boschetto DL, *et al.* Technological process for cell disruption, extraction and encapsulation of astaxanthin from *haematococcus pluvialis*. J Biotechnol 2016; 218: 108-14.
[http://dx.doi.org/10.1016/j.jbiotec.2015.12.004] [PMID: 26685712]

[124] Su CH, Nguyen HC, Bui TL, Huang DL. Enzyme-assisted extraction of insect fat for biodiesel production. J Clean Prod 2019; 223: 436-44.
[http://dx.doi.org/10.1016/j.jclepro.2019.03.150]

[125] Aitta E, Marsol-Vall A, Damerau A, Yang B. Enzyme-assisted extraction of fish oil from whole fish and by-products of baltic herring (*Clupea harengus* membras). Foods 2021; 10(8): 1811.
[http://dx.doi.org/10.3390/foods10081811] [PMID: 34441588]

[126] Heshof R, Visscher B, van de Zilver E, *et al.* Production of tailor-made enzymes to facilitate lipid extraction from the oleaginous yeast *Schwanniomyces occidentalis.* AMB Express 2020; 10(1): 41.
[http://dx.doi.org/10.1186/s13568-020-00974-z] [PMID: 32112299]

[127] Wei G, Zhang Z, Fu D, *et al.* Enzyme-assisted solvent extraction of high-yield *paeonia suffruticosa*

andr. seed oil and fatty acid composition and anti-alzheimer's disease activity. J Oleo Sci 2021; 70(8): 1133-46.
[http://dx.doi.org/10.5650/jos.ess21040] [PMID: 34248097]

[128] Hu B, Wang H, He L, *et al.* A method for extracting oil from cherry seed by ultrasonic-microwave assisted aqueous enzymatic process and evaluation of its quality. J Chromatogr A 2019; 1587: 50-60.
[http://dx.doi.org/10.1016/j.chroma.2018.12.027] [PMID: 30578025]

[129] Hu B, Li Y, Song J, *et al.* Oil extraction from tiger nut (*Cyperus esculentus* L.) using the combination of microwave-ultrasonic assisted aqueous enzymatic method - design, optimization and quality evaluation. J Chromatogr A 2020; 1627: 461380.
[http://dx.doi.org/10.1016/j.chroma.2020.461380] [PMID: 32823093]

[130] Rohman A, Irnawati , Erwanto Y, *et al.* Virgin coconut oil: Extraction, physicochemical properties, biological activities and its authentication analysis. Food Rev Int 2021; 37(1): 46-66.
[http://dx.doi.org/10.1080/87559129.2019.1687515]

[131] Verdasco-Martín CM, Díaz-Lozano A, Otero C. Advantageous enzyme selective extraction process of essential spirulina oil. Catal Today 2020; 346: 121-31.
[http://dx.doi.org/10.1016/j.cattod.2019.02.066]

[132] Bruno SF, Kudre TG, Bhaskar N. Impact of pretreatment-assisted enzymatic extraction on recovery, physicochemical and rheological properties of oil from *Labeo rohita* head. J Food Process Eng 2019; 42(3): e12990.
[http://dx.doi.org/10.1111/jfpe.12990]

[133] Sant'Anna BPM, Freitas SP, Coelho MAZ. Enzymatic aqueous technology for simultaneous coconut protein and oil extraction. Grasas Aceites 2003; 54(1): 77-80.
[http://dx.doi.org/10.3989/gya.2003.v54.i1.281]

[134] Jiao J, Li ZG, Gai QY, *et al.* Microwave-assisted aqueous enzymatic extraction of oil from pumpkin seeds and evaluation of its physicochemical properties, fatty acid compositions and antioxidant activities. Food Chem 2014; 147: 17-24.
[http://dx.doi.org/10.1016/j.foodchem.2013.09.079] [PMID: 24206680]

[135] Hernández-Corroto E, Marina ML, García MC. Multiple protective effect of peptides released from *Olea europaea* and *Prunus persica* seeds against oxidative damage and cancer cell proliferation. Food Res Int 2018; 106: 458-67.
[http://dx.doi.org/10.1016/j.foodres.2018.01.015] [PMID: 29579948]

[136] Priya S. Therapeutic perspectives of food bioactive peptides: A mini review. Protein Pept Lett 2019; 26(9): 664-75.
[http://dx.doi.org/10.2174/0929866526666190617092140] [PMID: 31215368]

[137] Bilal M, Iqbal HMN. Biologically active macromolecules: Extraction strategies, therapeutic potential and biomedical perspective. Int J Biol Macromol 2020; 151: 1-18.
[http://dx.doi.org/10.1016/j.ijbiomac.2020.02.037] [PMID: 32035954]

[138] Kumar MBA, Gao Y, Shen W, He L. Valorisation of protein waste: An enzymatic approach to make commodity chemicals. Front Chem Sci Eng 2015; 9(3): 295-307.
[http://dx.doi.org/10.1007/s11705-015-1532-4]

[139] Ismail BP, Senaratne-Lenagala L, Stube A, Brackenridge A. Protein demand: Review of plant and animal proteins used in alternative protein product development and production. Anim Front 2020; 10(4): 53-63.
[http://dx.doi.org/10.1093/af/vfaa040] [PMID: 33391860]

[140] González A, Cruz M, Losoya C, *et al.* Edible mushrooms as a novel protein source for functional foods. Food Funct 2020; 11(9): 7400-14.
[http://dx.doi.org/10.1039/D0FO01746A] [PMID: 32896845]

[141] Xu X, Yan H, Chen J, Zhang X. Bioactive proteins from mushrooms. Biotechnol Adv 2011; 29(6):

667-74.
[http://dx.doi.org/10.1016/j.biotechadv.2011.05.003] [PMID: 21605654]

[142] Kamal H, Le CF, Salter AM, Ali A. Extraction of protein from food waste: An overview of current status and opportunities. Compr Rev Food Sci Food Saf 2021; 20(3): 2455-75.
[http://dx.doi.org/10.1111/1541-4337.12739] [PMID: 33819382]

[143] Pojić M, Mišan A, Tiwari B. Eco-innovative technologies for extraction of proteins for human consumption from renewable protein sources of plant origin. Trends Food Sci Technol 2018; 75: 93-104.
[http://dx.doi.org/10.1016/j.tifs.2018.03.010]

[144] Roselló-Soto E, Parniakov O, Deng Q, *et al.* Application of non-conventional extraction methods: Toward a sustainable and green production of valuable compounds from mushrooms. Food Eng Rev 2016; 8(2): 214-34.
[http://dx.doi.org/10.1007/s12393-015-9131-1]

[145] Baker PW, Charlton A. A comparison in protein extraction from four major crop residues in Europe using chemical and enzymatic processes-a review. Innov Food Sci Emerg Technol 2020; 59: 102239.
[http://dx.doi.org/10.1016/j.ifset.2019.102239]

[146] Rosa R, Spinelli R, Neri P, *et al.* Life cycle assessment of chemical *vs* enzymatic-assisted extraction of proteins from black soldier fly prepupae for the preparation of biomaterials for potential agricultural use. ACS Sustain Chem& Eng 2020; 8(39): 14752-64.
[http://dx.doi.org/10.1021/acssuschemeng.0c03795]

[147] Akyüz A, Ersus S. Optimization of enzyme assisted extraction of protein from the sugar beet (*Beta vulgaris* L.) leaves for alternative plant protein concentrate production. Food Chem 2021; 335: 127673.
[http://dx.doi.org/10.1016/j.foodchem.2020.127673] [PMID: 32745844]

[148] Sari YW, Bruins ME, Sanders JPM. Enzyme assisted protein extraction from rapeseed, soybean, and microalgae meals. Ind Crops Prod 2013; 43(1): 78-83.
[http://dx.doi.org/10.1016/j.indcrop.2012.07.014]

[149] Benhammouche T, Melo A, Martins Z, *et al.* Nutritional quality of protein concentrates from *Moringa Oleifera* leaves and *in vitro* digestibility. Food Chem 2021; 348: 128858.
[http://dx.doi.org/10.1016/j.foodchem.2020.128858] [PMID: 33508602]

[150] Naseri A, Marinho GS, Holdt SL, Bartela JM, Jacobsen C. Enzyme-assisted extraction and characterization of protein from red seaweed *Palmaria palmata.* Algal Res 2020; 47: 101849.
[http://dx.doi.org/10.1016/j.algal.2020.101849]

[151] Chirinos R, Aquino M, Pedreschi R, Campos D. Optimized methodology for alkaline and enzyme-assisted extraction of protein from sacha inchi (*Plukenetia volubilis*) kernel cake. J Food Process Eng 2017; 40(2): e12412.
[http://dx.doi.org/10.1111/jfpe.12412]

[152] Görgüç A, Bircan C, Yılmaz FM. Sesame bran as an unexploited by-product: Effect of enzyme and ultrasound-assisted extraction on the recovery of protein and antioxidant compounds. Food Chem 2019; 283: 637-45.
[http://dx.doi.org/10.1016/j.foodchem.2019.01.077] [PMID: 30722922]

[153] Chen Z, Li Y, Lin S, Wei M, Du F, Ruan G. Development of continuous microwave-assisted protein digestion with immobilized enzyme. Biochem Biophys Res Commun 2014; 445(2): 491-6.
[http://dx.doi.org/10.1016/j.bbrc.2014.02.025] [PMID: 24530398]

[154] Huang CH, Chen WC, Gao YH, Chen GW, Lin HTV, Pan CL. Enzyme-assisted method for phycobiliproteins extraction from *Porphyra* and evaluation of their bioactivity. Processes 2021; 9(3): 560.
[http://dx.doi.org/10.3390/pr9030560]

[155] Olivares-Galván S, Marina ML, García MC. Extraction and characterization of antioxidant peptides from fruit residues. Foods 2020; 9(8): 1018.
[http://dx.doi.org/10.3390/foods9081018] [PMID: 32751284]

[156] Al-Mohammadi AR, Osman A, Enan G, *et al.* Powerful antibacterial peptides from egg albumin hydrolysates. Antibiotics 2020; 9(12): 901.
[http://dx.doi.org/10.3390/antibiotics9120901] [PMID: 33322196]

[157] Saidi S, Deratani A, Belleville M P, Amar R B. Production and fractionation of tuna by-product protein hydrolysate by ultrafiltration and nanofiltration: Impact on interesting peptides fractions and nutritional properties. Food Res Int 2014; 65(PC): 453-61.

[158] San Martin D, Ibarruri J, Iñarra B, *et al.* Valorisation of brewer's spent yeasts' hydrolysates as high-value bioactive molecules. Sustainability 2021; 13(12): 6520.
[http://dx.doi.org/10.3390/su13126520]

[159] Herrero M, Ibañez E. Green extraction processes, biorefineries and sustainability: Recovery of high added-value products from natural sources. J Supercrit Fluids 2018; 134: 252-9.
[http://dx.doi.org/10.1016/j.supflu.2017.12.002]

[160] EIT Food is Europe's leading food innovation initiative, working to make the food system more sustainable, healthy and trusted | EIT Food.. Available from: https://www.eitfood.eu/ [accessed 21/06/2021].

[161] About EAT. Available from: https://eatforum.org/about/ [accessed 21/06/2021].

[162] FoodSciRN: SSRN. Available from: https://www.ssrn.com/index.cfm/en/foodscirn/ [accessed 21/06/2021].

[163] Agri-Food Research Network. Available from: https://afrn.co/[accessed 21/06/2021].

[164] Sustainable Food Systems Programme | One Planet Network.. Available from:https://www.oneplanetnetwork.org/sustainable-food-system [accessed 21/06/2021].

[165] Food enzymes | European Food Safety Authority. Available from:https://www.efsa.europa.eu/en/topics/topic/food-enzymes [accessed 21/06/2021].

[166] Enzyme Preparations Used in Food (Partial List) | FDA. Available from:https://www.fda.gov/food/generally-recognized-safe-gras/enz-me-preparations-used-food-partial-list [accessed 21/06/2021].

[167] Dhole NP, Dar MA, Pandit RS. Recent advances in the bioprospection and applications of chitinolytic bacteria for valorization of waste chitin. Arch Microbiol 2021; 203(5): 1953-69.
[http://dx.doi.org/10.1007/s00203-021-02234-5] [PMID: 33710379]

[168] Han C, Jiaxiang H, Shanfeng L, Shishui L, Yongjian S. Preparation method for selenium-rich astaxanthin chlorella powder. CN107384907A Patent 2017.

[169] Belwal T, Chemat F, Venskutonis PR, *et al.* Recent advances in scaling-up of non-conventional extraction techniques: Learning from successes and failures. Trends Analyt Chem 2020; 127: 115895.
[http://dx.doi.org/10.1016/j.trac.2020.115895]

[170] Teles ASC, Chávez DWH, Santiago MCPA, Gottschalk LMF, Tonon RV. Composition of different media for enzyme production and its effect on the recovery of phenolic compounds from grape pomace. Biocatal Agric Biotechnol 2021; 35: 102067.
[http://dx.doi.org/10.1016/j.bcab.2021.102067]

[171] Kachrimanidou V, Kopsahelis N, Alexandri M, *et al.* Integrated sunflower-based biorefinery for the production of antioxidants, protein isolate and poly(3-hydroxybutyrate). Ind Crops Prod 2015; 71: 106-13.
[http://dx.doi.org/10.1016/j.indcrop.2015.03.003]

[172] Junqing J, Jianfeng M, Xudong W, Yu Y, Guoqing Y, Shihang Z. Method for extracting mulberry leaf

polysaccharide by utilizing microbial fermentation method. CN112481134A Patent 2021.

[173] Xiaodi M, Xinguo Z. Cellulase-producing endophytic bacteria of *Scutellaria baicalensis* Georgi and application of the enzyme produced in extracting baicalin from *Scutellaria baicalensis* Georgi. CN111440735A Patent 2020.

[174] Lianzhou J, Wanzhen L, Xin L, Min S, Peng S, Yunhua S, *et al.* Immobilized enzyme-assisted extraction method of soybean oil. CN102816639A Patent 2012.

[175] Yi J, Qiu M, Zhu Z, Dong X, Andrew Decker E, McClements DJ. Robust and recyclable magnetic nanobiocatalysts for extraction of anthocyanin from black rice. Food Chem 2021; 364: 130447. [http://dx.doi.org/10.1016/j.foodchem.2021.130447] [PMID: 34214946]

[176] Chettri D, Verma AK, Sarkar L, Verma AK. Role of extremophiles and their extremozymes in biorefinery process of lignocellulose degradation. Extremophiles 2021; 25(3): 203-19. [http://dx.doi.org/10.1007/s00792-021-01225-0] [PMID: 33768388]

[177] de Melo RR, de Lima EA, Persinoti GF, *et al.* Identification of a cold-adapted and metal-stimulated β-1,4-glucanase with potential use in the extraction of bioactive compounds from plants. Int J Biol Macromol 2021; 166: 190-9. [http://dx.doi.org/10.1016/j.ijbiomac.2020.10.137] [PMID: 33164774]

[178] Pedersen JN, Pérez B, Guo Z. Stability of cellulase in ionic liquids: Correlations between enzyme activity and COSMO-RS descriptors. Sci Rep 2019; 9(1): 17479. [http://dx.doi.org/10.1038/s41598-019-53523-5] [PMID: 31767916]

[179] Hebal H, Boucherba N, Binay B, Turunen O. Activity and stability of hyperthermostable cellulases and xylanases in ionic liquids. Biocatal Biotransform 2021; 39(4): 242-59. [http://dx.doi.org/10.1080/10242422.2021.1882430]

[180] Tavanandi HA, Vanjari P, Raghavarao KSMS. Synergistic method for extraction of high purity Allophycocyanin from dry biomass of *Arthrospira platensis* and utilization of spent biomass for recovery of carotenoids. Separ Purif Tech 2019; 225: 97-111. [http://dx.doi.org/10.1016/j.seppur.2019.05.064]

[181] Gümüş Yılmaz G, Gómez Pinchetti JL, Cifuentes A, Herrero M, Ibáñez E. Comparison of extraction techniques and surfactants for the isolation of total polyphenols and phlorotannins from the brown algae *lobophora variegata*. Anal Lett 2019; 52(17): 2724-40. [http://dx.doi.org/10.1080/00032719.2019.1597878]

[182] Han W, Ma W, Liu X, Luo K, Ban Y, Chen H. Method for extracting flavones in Chinese herbal medicine by surfactant cooperated supersonic wave-enzymatic hydrolysis. CN102178739B Patent 2011.

[183] Bio-based Industries Joint Undertaking (BBI JU) Lauching the circular bio-based economy. Available from: https://www.bbi.europa.eu/projects [accessed 21/06/2021].

[184] Bio-based Industries Joint Undertaking (BBI JU) Lauching the circular bio-based economy. Available from: https://www.bbi.europa.eu/projects?field_project_year_tid=All&field_project_category_tid=All&field_project_classification_tid=All&combine=enzyme [accessed 21/06/2021].

[185] Towa LT, Kapchie VN, Hauck C, Wang H, Murphy PA. Pilot plant recovery of soybean oleosome fractions by an enzyme-assisted aqueous process. J Am Oil Chem Soc 2011; 88(5): 733-41. [http://dx.doi.org/10.1007/s11746-010-1716-5]

[186] Maier T, Göppert A, Kammerer DR, Schieber A, Carle R. Optimization of a process for enzyme-assisted pigment extraction from grape (*Vitis vinifera* L.) pomace. Eur Food Res Technol 2008; 227(1): 267-75. [http://dx.doi.org/10.1007/s00217-007-0720-y]

CHAPTER 12

Pulsed Electric Field

Ester Hernández-Corroto[1], **Nadia Boussetta**[2], **María Luisa Marina**[1,3], **María Concepción García**[1,3] and **Eugène Vorobiev**[2,*]

[1] *Departamento de Química Analítica, Universidad de Alcalá, Química Física e Ingeniería Química, Ctra. Madrid-Barcelona Km. 33.600, 28871 Alcalá de Henares (Madrid), Spain*

[2] *Sorbonne Universités, Université de Technologie de Compiègne, Laboratoire Transformations Intégrées de la Matière Renouvelable (UTC/ESCOM, EA 4297 TIMR), Centre de Recherches de Royallieu, CS 60 319, 60 203 Compiègne Cedex, France*

[3] *Universidad de Alcalá, Instituto de Investigación Química "Andrés M. del Río", Ctra. Madrid-Barcelona Km. 33.600, 28871 Alcalá de Henares (Madrid), Spain*

Abstract: This chapter reviews the fundamentals of the Pulsed Electric Field (PEF) and its applications to the extraction of high-added value substances from food matrices. The electroporation process on the cell membrane is explained and the most recent works dealing with the use of PEF for extracting essential molecules for the human body such as lipids, phenolic compounds, carotenoids, proteins, carbohydrates, and vitamins, from food and plant matrices, and food waste, are described in detail. The combination of PEF with other extraction techniques is a common practice and improves the extractability of specific compounds to increase the recovery yields.

Keywords: Electroporation, Extraction, Food matrices, Food waste, High-added value substances, Pulsed electric field.

INTRODUCTION

Pulsed electric field (PEF) technology consists of an electric treatment of repetitive short pulses with high voltage to a sample placed between two electrodes. The application of an electric field leads to the formation of pores on the cell membrane due to the electroporation phenomenon. This process improves cell permeability, increasing mass transfer by the diffusion of solutes located inside the cells toward the solvent. In this way, PEF enhances the extraction yield of intracellular compounds. Moreover, the application of electric pulses at short time periods allows for decreasing the processing time avoiding the elevation of

* **Corresponding Author Eugène Vorobiev:** Sorbonne Universités, Université de Technologie de Compiègne, Laboratoire Transformations Intégrées de la Matière Renouvelable (UTC/ESCOM, EA 4297 TIMR), Centre de Recherches de Royallieu, CS 60 319, 60 203 Compiègne Cedex, France; Tel. +33 3 44 23 44 23 E-mail: eugene.vorobiev@utc.fr

Merichel Plaza & Maria Luisa Marina (Eds.)

temperature, so this non-thermal technique reduces the degradation of heat-sensitive compounds. In addition, this technology facilitates obtaining purified extracts due to the release of selective molecules and easy separation between the solid particles and the extracted compounds in the solvent. In addition, it reduces the economic and energetic impact and the extraction times, so it is considered an environmentally friendly technique [1, 2].

PEF technology presents numerous applications in food processing. The main applications are the extraction of intracellular compounds and the food preservation, especially juices. PEF enhances the diffusion extraction, osmotic treatment, pressing extraction, drying, and freezing [2, 3]. PEF has also been applied to pasteurization and sterilization processes, inactivation of pathogenic and spoilage microorganisms, and enzyme deactivation, among others [2, 4 - 6].

The extraction of compounds by PEF takes place in a treatment chamber in batch or continuous flow systems [7]. Special attention must be paid to the chamber to avoid fouling and corrosion of the electrodes [8]. The PEF technology is used for treating liquid, solid, and semisolid foods. Food waste has also been treated by PEF. The valorisation of food residues supposes a big challenge from the environmental, economic, and social point of views. Food residues are cheap sources of valuable compounds whose recovery contributes to reach the target of a circular economy (zero residues). In order to increase the recovery yields of specific compounds, many researchers have combined a PEF pre-treatment with other extraction techniques such as solid-liquid extraction, ultrasounds, and high hydrostatic pressure, among others [9 - 11].

This chapter reviews the action mechanism of a pulsed electric field based on the electroporation phenomenon and focuses on the application of PEF to extract essential compounds for the human body from different food matrices.

MECHANISM OF PULSED ELECTRIC FIELD

PEF is based on the electroporation phenomenon. This phenomenon is related to the formation of pores inside the cell membrane under the influence of an external electrical field. The electroporation process increases the electric conductivity and improves the cell permeability and the extraction of intracellular compounds [12]. PEF is an environmentally friendly technique that enhances mass transfer in food products.

Generator of Electrical Pulse and Pulse Waveforms

A PEF equipment is composed of a pulsed power generator and a treatment chamber. PEF treatments are based on the application of electric pulses for very short periods of time (μs) [13]. Different types of pulses can be produced by a PEF system. The exponential decay and square wave pulses are the most used, but each one is produced by a different electrical circuit (Fig. **1**) [14]. Fig. (**1a**) shows an electrical circuit formed by a high-voltage power supply, a capacitor connected in series with a charging resistor, a high-voltage switch, and a treatment chamber. When the switch is opened, the high-voltage power supply charges the capacitor of energy. When the trigger signal is applied, the switch is closed and the electrical energy stored in the capacitor flows toward the treatment chamber. In that moment, the exponential decay pulses are produced due to the voltage is unidirectional and it rises quickly to a maximum value. Then, it decays slowly to zero [14, 15]. These pulses offer short peaks of high electric field intensity and long tails of low electric field strength [16]. On the other hand, square wave pulses are more energetic than the exponential decay pulses. Square waveforms offer width peaks with fast rising and fast falling. This electrical circuit is formed by an array of capacitors, inductors, and switching devices (Fig. **1b**) [14, 15]. In both cases, exponential decay and square wave pulses, the duration of the pulse is determined by the control signal of the switch, because the pulse starts by closing the switch and it ends by opening it [13].

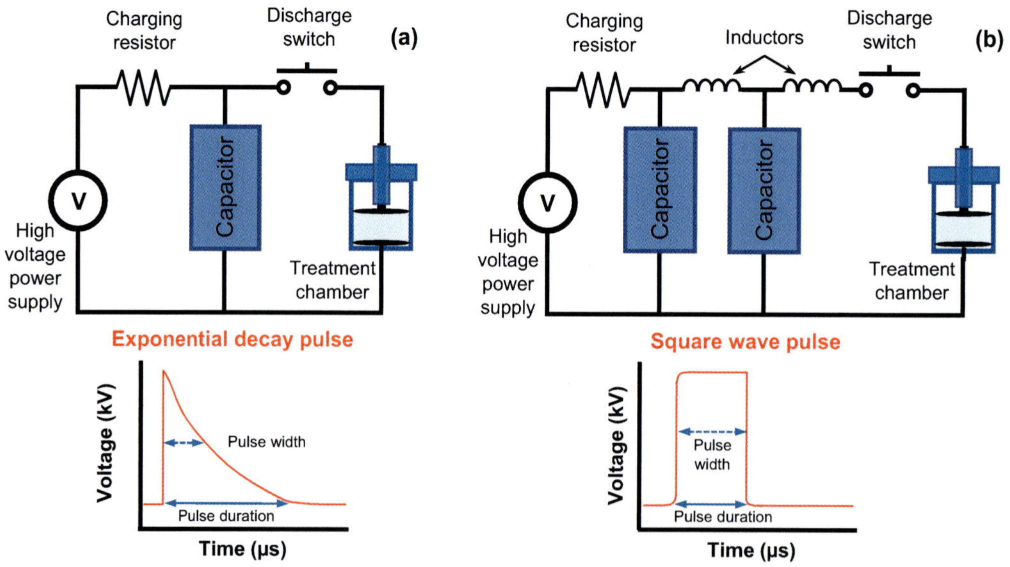

Fig. (1). Schemes of electrical circuits employed to generate exponential decay pulse (a) and square wave pulse (b).

The application of a pulsed electric field can take place in batch or static systems and in dynamic or continuous flow systems [17]. Batch systems are the most employed at the lab and pilot scale. The treatment chamber has two parallel plate geometry electrodes (Fig. **2a**) for a homogeneous electric field distribution. Both electrodes are separated by a certain distance from where the sample and solvent are located. One of the electrodes is connected to a high-voltage generator that produces short electric pulses. Electrical current is transmitted from the positive electrode to the opposite electrode, resulting in the generation of an electric field. The strength of this electric field depends on the applied voltage and the inter-electrode distance.

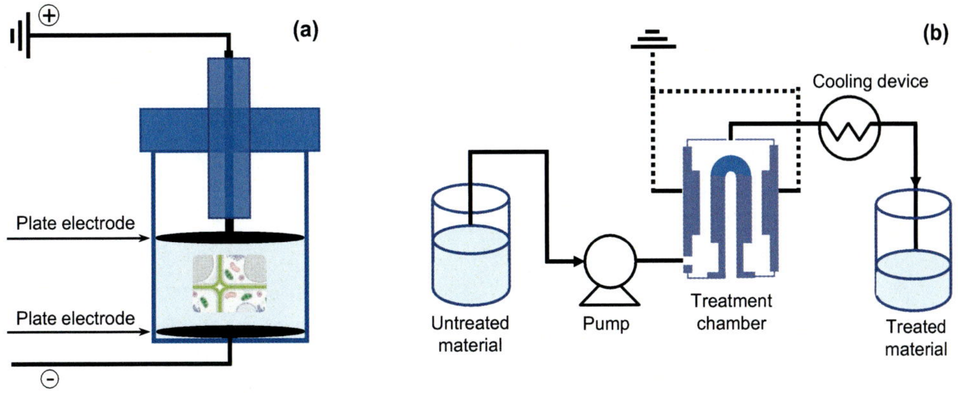

Fig. (2). Scheme of a treatment chamber in a batch PEF extraction system (a) and a continuous PEF extraction system (b).

Continuous flow systems (Fig. **2b**) are more required at the industrial scale. The treatment chamber is composed of electrode configurations with parallel plates, coaxial cylinders, or colinear cylinders [18]. In addition, this system needs a pump to drive the material and solvent into the treatment chamber and a cooling device to control the temperature [7].

Cell Membrane Electroporation

Electroporation process takes place when the applied electric field intensity exceeds a critical value and the potential difference across the cell membrane (transmembrane potential) increases to a threshold value (from 0.7 to 2.2 V in plant and animal cells) [19]. Fig. (**3**) shows a scheme of the electroporation process. When the cells are exposed to an electric field (Fig. **3a**), the ions migrate towards the cell plasmatic membrane creating free charges with opposite signs and producing the polarization of the cell. The accumulation of free charges of different polarities on each side of the membrane induces the electro-compression of the membrane. As a consequence, the thickness of the membrane is reduced by the forces of attraction between the oppositely charged ions. Moreover, the accumulation of charges generates a transmembrane potential that overlaps with the potential induced by the external electric field until reaching a critical value. When the external electric field intensity is high enough (Fig. **3b** and **3c**), the formation of pores in the cell membrane takes place. The membrane is directly perforated and the barrier function is lost. Nevertheless, this mechanism can be reversible depending on the electric field intensity or the exposure of the cell to the electric field. In the case of low electric field intensity (< 200 V/cm) and very short expositions (1-100 μs), the membrane could return to its initial state (Fig. **3b**). In contrast, intense electric fields or longer expositions can cause irreversible electroporation, that is, a permanent damage on the cell membrane and cell lysis (Fig. **3c**). In this case, a moderate PEF treatment with electric field intensities ranging from 0.5 to 3 kV/cm would produce irreversible permeabilization in plant and animal tissues, while a PEF treatment with higher intensities (from 15 to 40 kV/cm) could produce irreversible permeabilization in microbial cells [20 - 24]. Despite the formation of pores in the cell membrane due to the electroporation process, cell walls remain intact [25, 26], except under very intense PEF conditions. There is no fragmentation of the particle. Therefore, this technique can be used for the extraction of specific compounds. In addition, the solid particles and the extracted compounds in the solvent can be easily separated [27]. In the food industry applications, irreversible electroporation is the most usual [12, 28].

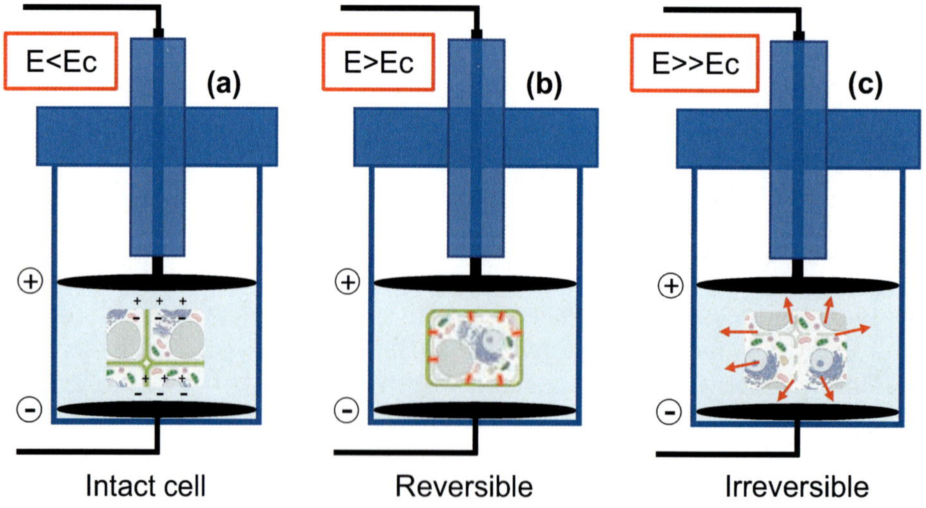

Fig. (3). Scheme of the electroporation process by the application of an electric field (E). Cell polarization and compression of membrane due to the electric charges attraction (a), and reversible (b) and irreversible (c) membrane perforation when E exceeds a critical value (Ec).

APPLICATIONS

The application of PEF on food matrices leads to the formation of pores in the cell membrane enhancing the cell permeability. Therefore, the high-added value compounds contained in the matrix can be easily released. It is a non-thermal technique, so it is possible to extract thermolabile compounds since they are not exposed to high temperatures. Moreover, this technique does not affect the nutritional and quality characteristics of the released compounds [29]. On the other hand, PEF is not a destructive technique at low/moderate electric field intensities (< 20 kV/cm) [30], that is, the cell membrane can be perforated, but the cell walls are intact, so the solid particles and the extracted compounds in the solvent can be easily separated. Therefore, this technique allows extracting specific compounds. It is more selective than the high-voltage electrical discharges (HVED) technique and it is a great alternative to conventional extraction methodologies.

The efficiency of PEF for the intracellular compounds extraction depends on the characteristics of the PEF process (electric field strength, pulse-specific energy, number of pulses, pulse shape, pulse width, pulse duration, and frequency), extraction variables (temperature, product size, solvent nature, and concentration), and cell properties (size, shape, orientation, and electric conductivity) [1, 13, 18, 21].

Many researchers applied PEF to the extraction of lipids, phenolic compounds, carotenoids, proteins, carbohydrates, and vitamins, among others, from different food matrices or, even, food waste. In some cases, PEF was used as a sample pre-treatment step, followed by an extraction step using other techniques. This enables to increase the amount of extracted compounds. Some applications have been previously reviewed [2, 30 - 32]. This chapter just reviews the most recent and relevant applications to food extraction, which are described below.

Extraction of Lipids

Lipids are hydrophobic compounds soluble in non-polar organic solvents. These molecules are formed mainly by carbon, hydrogen, and oxygen atoms, although many structures also contain phosphorus, nitrogen, and sulfur. Lipids are essential molecules required for the suitable performance of many human body functions [33]. They exist in the cell cytoplasm from which they can be extracted by applying PEF. The action mechanism of PEF for the extraction of lipids is shown in Fig. (**4**). PEF technique enhances the cell permeability due to the perforation of the cell membrane by the electroporation process. Thus, the intracellular compounds, which are more soluble in water, are released from the food matrix.

Nevertheless, lipids and other hydrophobic compounds would need another solid-liquid extraction step using an organic solvent for their extraction.

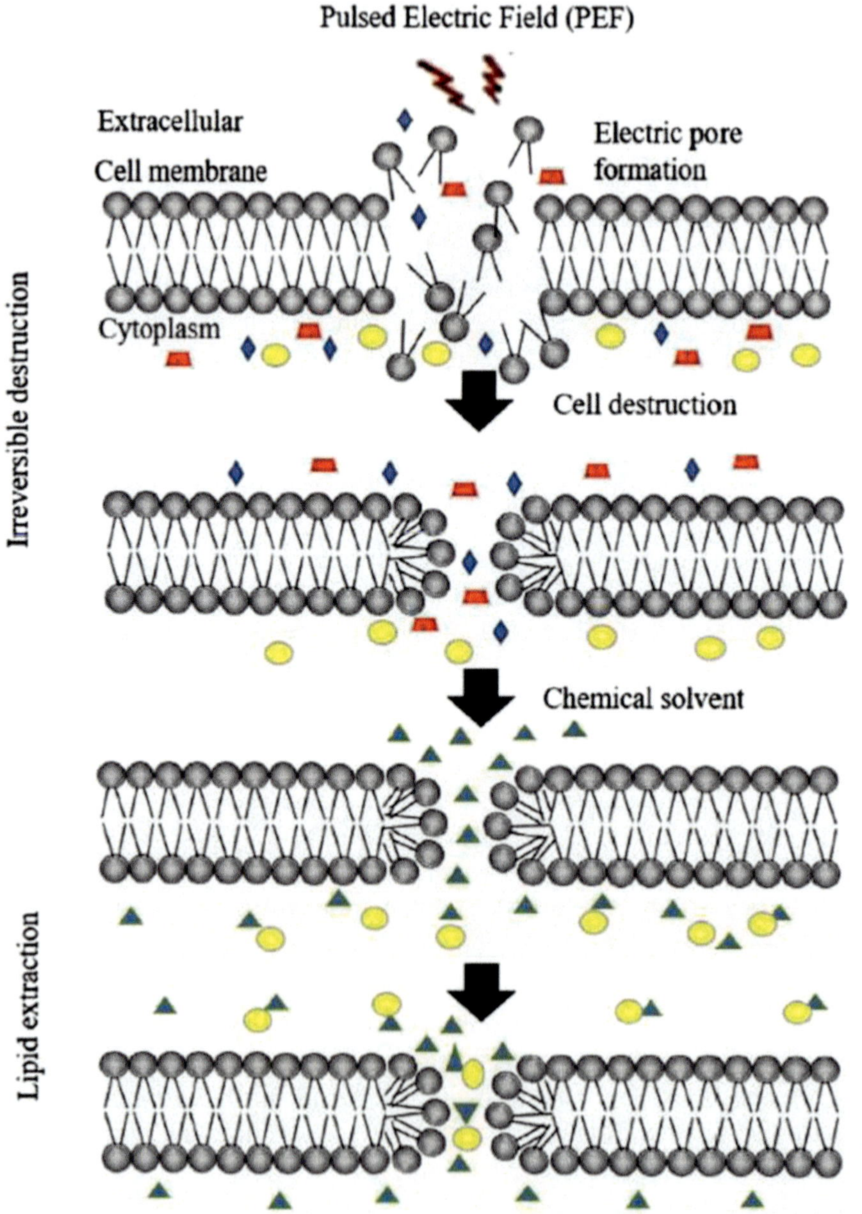

Fig. (4). Action mechanism of pulsed electrical field for lipid extraction. Symbols , , and represent proteins, carbohydrates, lipids, and organic solvents, respectively (Reproduced with permission from Han *et al*. [34].

Table **1** shows the most recent examples related to lipid extraction from shrimp, Niger seeds, and different seaweeds. Most works were based on the use of PEF like a sample pre-treatment to improve the extractability of the compounds and, a subsequent solid-liquid extraction using 2-propanol, hexane, ethanol, chloroform, methanol, or mixtures of them, to promote lipids solubilization and, consequently, their extraction. In some cases, a continuous system using flow rates from 6 to 90 mL/min improves the extraction of lipids, being not necessary a second extraction step [35]. In the PEF process for the extraction of lipids, the inter-electrode distance is found between 0.4 and 2.7 cm, and the electric field strength ranges from 0.31 to 40 kV/cm. Therefore, the extraction yield of lipids varies from 0.72% to 35%.

Table 1. Application of PEF technique for the extraction of lipids from food matrices.

Matrix	Extracting solvent	S-L ratio[a] (w/v)	Experimental Conditions	Extracted Compounds	Extraction Yield	Refs.
			Food			
Pacific white shrimp cephalothorax	Tap water	1:5	1st step: PEF (interelectrode distance, 1.5 cm; electric field strength, 16 kV/cm; 240 pulses). 2nd step: solid-liquid extraction (2-propanol:hexane (1:1); 1:5 (w/v); 9500 rpm; 4 °C; 2 min). 3rd step: UAE (amplitude, 80%; 25 min).	Lipids	31.5%	[36]
				PUFA	41.0 ± 0.3 g/100 g lipids	
			Food waste			
Niger seed	-	-	1st step: MW (900 W; 156.34 s). 2nd step: PEF (electric field strength, 1.18 kV/cm; 30 pulses; pulse width, 20 μs). 3rd step: screw press (20 rpm).	Oil	35%	[37]
			Seaweeds			
Scenedesmus almeriensis	Water	1:5	1st step: PEF (Interelectrode distance, 0.4 cm; electric field strength, 40 kV/cm; energy input, 1500 kJ/kg; pulse duration, 1 μs; frequency, 3 Hz, flow rate of the solution, 6 mL/min). 2nd step: solid-liquid extraction (EtOH/hexane (1:0.41, v/v); 1:81 (w/v); 24 h).	Lipids	16.3% DW	[38]

(Table 1) cont.....

Matrix	Extracting solvent	S-L ratio[a] (w/v)	Experimental Conditions	Extracted Compounds	Extraction Yield	Refs.
Chlorella	Water	-	Generator (35 kV); interelectrode distance, 0.4 cm; electric field strength, 35 kV/cm; square wave pulses; pulse width, 10 µs; frequency, 50 kHz, flow rate of the solution, 50 mL/min.	Lipids	25.5%	[35]
Chlorella pyrenoidosa	-	-	1st step: PEF (generator (20 kV); electric field strength, 20 kV/cm; pulse width, 6 µs; frequency, 130 Hz; flow rate of the solution, 90 mL/min; 3 cycles). 2nd step: solid-liquid extraction (chloroform/MeOH (2:1, v/v); 150 rpm, 30 h).	FAME	12.8% DW	[34]
Cystoseira sauvageauana	-	-	1st step: PEF (interelectrode distance, 2.7 cm; electric field strength, 1.04 kV/cm; exponential decay pulses; total treatment time, 5.56 ms). 2nd step: solid-liquid extraction (chloroform/MeOH (2:1, v/v); 30 min).	Lipids	0.72%	[39]
Laurencia pinnatifida	-	-	1st step: PEF (interelectrode distance, 2.7 cm; electric field strength, 0.31 kV/cm; exponential decay pulses; total treatment time, 3.12 ms). 2nd step: solid-liquid extraction (chloroform/MeOH (2:1, v/v); 30 min).	Lipids	0.76%	[39]

PEF: pulsed electric field; UAE: ultrasound-assisted extraction; PUFA: poly-unsaturated fatty acids; MW: microwave; FAME: fatty acid methyl ester; DW: dry weight; EtOH: ethanol; MeOH: methanol. [a] solid-liquid ratio

Mohseni *et al.* [37] employed a double pre-treatment using microwave in combination with PEF followed by screw press to extract oil from Niger seeds. They reported that a high microwave irradiation time (200 s) and PEF intensity (5 kV/cm) decreased the oil extraction efficiency due to destruction of cell internal structures and blocking of oils extraction pores. Thus, the best conditions for the oil extraction were 156.34 s of microwave time and 1.18 kV/cm of PEF intensity. Under these conditions, it was possible to obtain 35% of oil from Niger seeds.

Gulzar and Benjakul [36] reported the application of a PEF pre-treatment (16 kV/cm) followed by two consecutive extraction processes (solid-liquid extraction and ultrasounds). The combination of three processes enhanced the extraction yield of lipids from the cephalothorax of Pacific white shrimp up to 31.5%. Moreover, Papachristou *et al.* [38] studied the application of PEF pre-treatment (40 kV/cm) followed by three solid-liquid extraction protocols from *Scenedesmus almeriensis* microalgae. They showed that the extraction of lipids with ethanol and hexane for 24 h offered the highest lipid yield (16.3%). Kord *et al.* [39] obtained a 0.72% lipid extraction yield from *Cystoseira sauvageauana* and a 0.76% from *Laurencia pinnatifida*, in both cases by the application of a PEF pre-treatment followed by an extraction with chloroform and methanol. The PEF pre-treatment increased the lipid yield in 0.24% and 0.25%, respectively.

In addition to the extraction of lipids for food applications, the extraction of these compounds can also be employed for biodiesel production. Indeed, seaweeds are a cheap source of lipids that have been employed for this purpose. Zhang *et al.* [35] reported that the application of PEF increased the lipid extraction yield from *Chlorella* up to 385.7%, compared to non-PEF treatment. Besides, Han *et al.* [34] concluded that PEF was a more effective technique for cell destruction and for the release of lipid components than ultrasounds.

Extraction of Phenolic Compounds

Phenolic compounds are a type of secondary metabolites present in plant tissues. Some of them are inside cell vacuoles. Electric field intensities ranging from 1 to 20 kV/cm and long expositions of the cells (from 2 μs to 100 ms) produce irreversible electroporation and allow the release of phenolic compounds. Since polyphenols might be sensitive to high temperatures, the application of this non-thermal technique is highly recommended. Table **2** presents the experimental conditions employed in the last two years for the extraction of phenolic compounds using PEF from different food matrices. In all cases, the application from 5 to 700 pulses with energy input from 0.36 to 100 kJ/kg and a frequency from 0.1 to 1000 Hz increased the release of phenolic compounds.

Some researchers used PEF in combination with other extraction techniques, such as solid-liquid extraction and ultrasounds, to increase the extractability and the solubilization of phenolic compounds. Siddeeg *et al.* [40] reported that the application of PEF in date palm fruit followed by ethanolic extraction increased the content of phenolic compounds and flavonoids when rising the electric field strength (up to 3 kV/cm). Similarly, Manzoor *et al.* [41] observed an increase in the extraction of total phenolic compounds, flavonoids, tannins, and anthocyanins from almond seeds by the combination of PEF and ultrasounds. Nevertheless,

López-Gámez *et al.* [42] associated the increase in total phenolic acids with the stress induced by PEF and not with the extractability caused by electroporation. For that, they studied the effect of PEF in carrots after its application reaching the maximum content at 24 h.

Furthermore, PEF has also been applied for the extraction of phenolic compounds from food waste observing significant yield increases. Indeed, Andreou *et al.* [43] reported that treating tomato waste with PEF doubled the recovery of phenolic compounds as compared to untreated samples while Wang *et al.* [44] extracted almost three times more phenolic compounds from apple peels. Other examples were: Andreou *et al.* [45] showed an increase up to 28.2% from olive pomace; Frontuto *et al.* [46] obtained an increase of a 10% from potato peels; and Ozkan *et al.* [47] observed an increase between a 10% and 14% from Cranberrybush purée. Lončarić *et al.* [48] reported the highest extraction of phenolic compounds (10.52 mg GAE/ g DW) by PEF treatment from blueberry pomace compared to HVED (6.58 mg GAE/ g DW) or ultrasounds (5.46 mg GAE/ g DW) extractions, under optimized conditions for every treatment. However, Rajha *et al.* [49] showed lower amounts of phenolic compounds from pomegranate peel using PEF (39 ± 2 mg GAE/ g DW) than HVED (46.0 ± 0.5 mg GAE/ g DW), although both amounts were higher than the obtained with ultrasound extraction (14.5 ± 0.8 mg GAE/ g DW).

Table 2. Application of PEF technique for the extraction of phenolic compounds from food matrices.

Matrix	Extracting Solvent	S-L ratio[a] (w/v)	Experimental Conditions	Extracted Compound	Extraction Yield	Refs.
			Food			
Date palm fruit	-	100 g	1st step: PEF (two parallel electrodes; electric field strength, 3 kV/cm; 30 pulses; pulse duration, 100 μs; frequency, 10 Hz). 2nd step: solid-liquid extraction (EtOH (80%); 1:3 (w/v); 6 h).	Phenolic compounds	0.6735 ± 0.0007 mg GAE/g	[40]
				Flavonoids	0.068 ± 0.006 mg CE/g	
Carrot	Aqueous solution	1:3	Two parallel electrodes; interelectrode distance, 5 cm; electric field strength, 3.5 kV/cm; energy input, 0.58 kJ/kg; exponential decay pulses; 5 pulses; pulse duration, 4 μs; frequency, 0.1 Hz.	Phenolic compounds	0.97 mg/g	[42]

(Table 2) cont.....

Matrix	Extracting Solvent	S-L ratio[a] (w/v)	Experimental Conditions	Extracted Compound	Extraction Yield	Refs.
Almond seeds	MeOH	1:17	1st step: incubation in the water bath (1 h). 2nd step: PEF (250 mL extract; electric field strength, 18 kV/cm; pulse duration, 500 μs; frequency, 1 Hz; flow rate of the solution, 40 mL/min). 3rd step: UAE (200 W; 40 Hz; 35 °C; 20 min).	Phenolic compounds	21.20 ± 0.06 mg GAE/g DE	[41]
				Flavonoids	9.72 ± 0.05 mg CE/g DE	
				Tannin	1.48 ± 0.01 mg CE/g DE	
				Anthocyanins	1.28 ± 0.09 mg/L	
Food waste						
Tomato waste	-	-	Interelectrode distance, 4 cm; electric field strength, 2 kV/cm; energy input, 22.8 kJ/kg; 700 pulses; pulse width, 15 μs, frequency, 20 Hz.	Phenolic compounds	0.056 mg GAE/g	[43]
Blueberry Pomace	EtOH (50%) and HCl (1%)	1:50	Generator (25 kV); two parallel plate electrodes; interelectrode distance, 1 cm; electric field strength, 20 kV/cm; energy input, 41.03 kJ/kg; exponential decay pulses; 100 pulses; pulse duration, 2 μs.	Phenolic compounds	10.52 mg GAE/g DW	[48]
Olive pomace	EtOH (50%)	1:10	1st step: PEF (interelectrode distance, 4 cm; electric field strength, 3 kV/cm; energy input, 10.9 kJ/kg, 3000 pulses; pulse width, 15 μs; frequency, 20 Hz). 2nd step: solid-liquid extraction (EtOH (50%); 1:10 (w/v); 25 °C; 0-60 min).	Phenolic compounds	3.07 mg GAE/mL	[45]

(Table 2) cont.....

Matrix	Extracting Solvent	S-L ratio[a] (w/v)	Experimental Conditions	Extracted Compound	Extraction Yield	Refs.
Potato peel	-	5 g	1st step: PEF (generator (25 kV, 500 A); two parallel plate cylindrical electrodes; interelectrode distance up to 5 cm; electric field strength, 1 kV/cm; energy input, 5 kJ/kg; square wave pulses; pulse width, 3-25 µs; frequency, 1-450 Hz). 2nd step: solid-liquid extraction (EtOH (52%); 1:20 (w/v);160 rpm; 230 min; 50 ºC).	Phenolic compounds	1.3 mg GAE/g FW	[46]
Pomegranate peels	Hot water (50 ºC)	1:10	Generator (40 kV); two parallel plate electrodes; interelectrode distance, 4 cm; electric field strength, 10 kV/cm; exponential decay pulses.	Phenolic compounds	39 ± 2 mg GAE/g DW	[49]
Apple peels	Distilled water	1:1	1st step: PEF (generator (1.5 kV, 20 A); two parallel plate cylindrical electrodes; electric field strength, 1.2 kV/cm; energy consumption, 100 kJ/kg; square wave pulses; pulse width, 100 µs; 50 pulses; frequency, 500 Hz). 2nd step: solid-liquid extraction (water; 1:10 (w/v);150 rpm; 36 min).	Phenolic compounds	1.75 mg GAE/g DW	[44]
Cranberrybush purée	-	2 kg	Generator (30 kV); interelectrode distance, 2 cm; electric field strength, 3 kV/cm; energy input, 15 kJ/kg; exponential decay pulses; pulse width, 20 µs; pulse duration, 40 ms; frequency, 2 Hz.	Phenolic compounds	4.5 mg GAE/g FW	[47]
				Flavonoids	13 mg RE/g FW	
				Anthocyanins	0.5 mg Cy---Glc/g FW	
			Juice			

(Table 2) cont.....

Matrix	Extracting Solvent	S-L ratio[a] (w/v)	Experimental Conditions	Extracted Compound	Extraction Yield	Refs.
Apricot juice	-	-	Electric field strength, 14 kV/cm; energy input, 41.03 kJ/kg; 100 pulses; pulse duration, 500 μs; frequency, 1000 Hz; flow rate of the juice, 40 mL/min.	Phenolic compounds	63.14 ± 0.09 μg/g GAE	[50]
				Flavonoids	564.5 ± 0.2 μg/g	
Grape juice	-	-	1st step: PEF (generator (3 kV, 30 A); two circular electrodes; interelectrode distance, 4 cm; energy input, 23.8 kJ/kg; square waveform pulses; pulse duration, 10 μs; frequency, 1500 Hz). 2nd step: dynamic maceration (mechanical stirrer, 2 h, 156 rpm). 3rd step: static maceration (12 h).	Phenolic compounds	0.44 mg GAE/mL	[51]
Seaweeds						
Tetraselmis chuii	Tap water	1:100	1st step: PEF (two parallel plate cylindrical electrodes; interelectrode distance, 1 cm; electric field strength, 3 kV/cm; energy input, 100 kJ/kg; 45 pulses; pulse duration, 100 ms; frequency, 2 Hz). 2nd step: solid-liquid extraction (water; 1:1 (v/v); 400 rpm; 4 h).	Phenolic compounds	6.42 mg GAE/g DW	[52]
			1st step: PEF (two parallel plate cylindrical electrodes; interelectrode distance, 1 cm; electric field strength, 3 kV/cm; energy input, 100 kJ/kg; 45 pulses; pulse duration, 100 ms; frequency, 2 Hz). 2nd step: solid-liquid extraction (DMSO (50%); 1:1 (v/v); 400 rpm; 4 h).	Phenolic compounds	6.70 mg GAE/g DW	

(Table 2) cont.....

Matrix	Extracting Solvent	S-L ratio[a] (w/v)	Experimental Conditions	Extracted Compound	Extraction Yield	Refs.
Phaeodactylum tricornutum	Tap water	1:100	1st step: PEF (two parallel plate cylindrical electrodes; interelectrode distance, 1 cm; electric field strength, 1-3 kV/cm; energy input, 100 kJ/kg; 400-45 pulses; pulse duration, 100 ms; frequency, 2 Hz). 2nd step: solid-liquid extraction (DMSO (50%); 1:1 (v/v); 400 rpm; 4-24 h).	Phenolic compounds	8 mg GAE/g DW	[52]
Laminaria digitata	Water	1:116	Generator (18 kV); two plate electrodes; electric field strength, 7.5 kV/cm; energy input, 45.3 kJ/kg; 140 pulses; exponential decay pulses; pulse width, 0.96 μs; frequency, 1.2 Hz.	Phenolic compounds	3 ± 1 mg GAE/ g DW	[53]
Other plants						
Linum usitalissmum L. seeds	MeOH (80%)	1:50	Interelectrode distance, 4 cm; electric field strength, 6 kV/cm; 30 pulses.	Phenolic compounds	10.4 ± 0.8 mg GAE/g extract	[54]
Salvia leriifolia seeds	MeOH (80%)	1:50	Interelectrode distance, 4 cm; electric field strength, 6 kV/cm; 30 pulses.	Phenolic compounds	64 ± 1 mg GAE/g extract	
Barberry integerrima stem	EtOH	1:11	1st step: PEF (two electrodes; interelectrode distance, 4 cm; electric field strength, 1.25 kV/cm; 50 pulses; frequency, 1 Hz). 2nd step: solid-liquid extraction (EtOH (68%); 1:11 (w/v); stirring; 24 h).	Phenolic compounds	14.6 mg GAE/g DW	[55]
Panax ginseng roots	Tap water	1:12	Generator (5 kV); parallel electrodes; interelectrode distance, 8 cm; electric field strength, 1.5 kV/cm; 500 pulses; pulse width, 25 μs; frequency, 50 Hz.	Phenolic compounds	8.9 ± 0.3 mg TAE/g	[56]

(Table 2) cont.....

Matrix	Extracting Solvent	S-L ratio[a] (w/v)	Experimental Conditions	Extracted Compound	Extraction Yield	Refs.
Rosemary by-products	NaCl (0.1%)	1:1.4	1st step: PEF (generator (5 kV); two parallel plate electrodes; electric field strength, 1.1 kV/cm; energy input, 0.36 kJ/kg; square wave pulses; 167 pulses; pulse duration, 30 µs; frequency, 10 Hz). 2nd step: UAE (EtOH (55.19%); 1:20 (w/v); 200 W; amplitude, 50%; non-continuous mode (0.5 s on, 0.5 s off); energy, 409.31 kJ/kg; 12.48 min, 40 °C).	Phenolic compounds	2.97 mg GAE/ g FW	[57]
Thyme by-products	NaCl (0.1%)	1:1.5	1st step: PEF (generator (5 kV); two parallel plate electrodes; electric field strength, 1.1 kV/cm; energy input, 0.46 kJ/kg; square wave pulses; 167 pulses; pulse duration, 30 µs; frequency, 10 Hz). 2nd step: UAE (EtOH (55.19%); 1:20 (w/v); 200 W; amplitude, 50%; non-continuous mode (0.5 s on, 0.5 s off); energy, 409.31 kJ/kg; 12.48 min, 40 °C).	Phenolic compounds	4.6 mg GAE/g FW	

GAE: gallic acid equivalent; DE: dried extract; DW: dry weight; FW: fresh weight; RE: rutin equivalent; CE: catechin equivalent; Cy-3-Glc: cyanidin-3-O-glucoside equivalents; TAE: tannic acid equivalent; PEF: pulsed electric field; UAE: ultrasound-assisted extraction; EtOH: ethanol; MeOH: methanol; DMSO: dimethyl sulfoxide. [a] solid-liquid ratio

This technique was also applied to liquid samples like juices. Rahaman *et al.* [50] and Ricci *et al.* [51] reported the extraction of phenolic compounds under continuous flow conditions from apricot and grape juices, respectively. While a high electric field intensity (14 kV/cm) extracted up to 63.14 ± 0.09 µg GAE/g from apricot juice [50], moderate electric field strengths (0.9 – 3 kV/cm) were required when working with grape juices, obtaining 0.44 mg GAE/mL [51].

The electropermeabilization in seaweeds addressed by PEF has also been studied in different works. Kokkali *et al.* [52] showed the extraction of phenolic compounds from two different microalgae. *Tetraselmis chuii* was influenced by the electric field strength and pulses treatment. However, no significant differences were observed on *Phaeodactylum tricornutum*. Einarsdóttir *et al.* [53] studied the effect of PEF in *Laminaria digitata* macroalgae. They observed that the number of pulses during PEF treatment and the initial temperature did not have a significant effect on the total polyphenol extraction. However, a lower concentration of biomass (0.15%) raised the specific energy (up to 45 kJ/kg), which increased the polyphenolic recovery. Furthermore, the extraction yield from macroalgae (3 mg GAE/g) was lower than the obtained from microalgae (from 6 to 8 mg GAE/g). Probably, a second extraction step could increase this yield.

Shirazi *et al.* [54] studied phenolic compounds extraction from other plant matrices, such as *Linum usitalissmum* and *Salvia leriifolia* seeds. They stated that the highest amount of phenolic compounds was obtained with the maximal electric field intensity (6 kV/cm) due to the electrical degradation of cells by PEF, which improved cell permeabilization and intracellular extraction. In the same way, Sarraf *et al.* [55] corroborated that the phenolic compounds extraction increased up to 31% after applying 100 pulses of 1.25 kV/cm in *Barberry integerrima* stem. Kim *et al.* [56] observed that PEF treatment in ginseng led to biological cell membrane disruption at field strengths >1.5 kV/cm. Thus, this technology increased the phenolic recovery up to 11% using less energy in a shorter time. Finally, Tzima *et al.* [57] stated that the application of PEF did not significantly affect the phenolic compound extraction. However, the use of PEF before ultrasounds increased the phenolic recovery in 1.51-fold from rosemary. In the case of thyme, PEF significantly increased the phenolic compound extraction up to 2.4-fold.

Extraction of Carotenoids

Carotenoids are a group of unsaturated molecules with a large conjugate double-bonds structure, which makes them highly susceptible to chemical reactions such as oxidation and isomerization during processing and storage [30]. These compounds are located in chloroplasts and chromoplasts of plant matrices [29]. The application of PEF can affect the membranes of plant cells and, thus, promote the release of carotenoid compounds [42]. Table **3** shows the most recent works dealing with the extraction of these compounds from food matrices, tomato waste, and microalgae. The application at moderate electric field intensities (between 1 and 5 kV/cm) from 5 to 600 pulses, with energy inputs ranging from 0.61 to 100 kJ/kg, and cell expositions from 10 μs to 100 ms increased the carotenoid yield up to 14.3 mg/g.

López-Gámez *et al.* [58] studied the feasibility of PEF treatment for the extraction of carotenoids from carrots. The total carotenoid content in extracts increased up to 83.8% compared to untreated carrots, after applying 5 pulses of 3.5 kV/cm. Siddeeg *et al.* [40] also reported that the application of PEF in date palm fruit increased the carotenoids extractability 2.14-fold, compared to untreated extract. Thus, the application of 30 pulses of 5 kV/cm raised the carotenoid content up to 6.1 ± 0.1 μg/mL. Rybak *et al.* [59] applied two different energy inputs (1 and 3 kJ/kg) for the extraction of carotenoids from red bell pepper. The juice pressed from bell pepper was composed of small fragments of tissue and cells that contained cellular organelles and bioactive compounds, among others. The highest carotenoid content was obtained at 1 kJ/kg (14.3 ± 0.2 mg β-carotene/g DW).

Lycopene is the most abundant carotenoid in tomato processing waste. This compound is thermosensitive, so its extraction is difficult. Thus, the application of a non-thermal technique, such as PEF, constitutes an interesting possibility. Therefore, Andreou *et al.* [43] reported the extraction of 0.178 mg/g of carotenoids from tomato waste after applying 500 pulses of 1 kV/cm. They stated that more intense PEF conditions (from 1 to 5 kV/cm) improved the extractability of carotenoid compounds from 35.9% to 56.4% compared to the untreated sample. In addition, under optimized conditions, they extracted 0.14 mg/g of lycopene, which corresponded to 81% of carotenoids. Pataro *et al.* [60] extracted the highest amount of lycopene (13.9 ± 0.6 mg/g DW) from tomato peels. In this study, a PEF pre-treatment of 10 pulses at 5 kV/cm was followed by a subsequent solid-liquid extraction with acetone for 240 min.

Table 3. Application of PEF technique for the extraction of carotenoids from food matrices.

Matrix	Extracting Solvent	S-L ratio[a] (w/v)	Experimental Conditions	Extracted Compound	Extraction Yield	Refs.
			Food			
Carrots	NaCl solution	1:3	Two parallel electrodes; interelectrode distance, 5 cm; electric field strength, 3.5 kV/cm; energy input, 0.61 kJ/kg; 5 pulses; exponential wave pulses; pulse width, 4 μs; frequency, 0.1 Hz.	Carotenoids	1.34 mg/g DW	[58]

(Table 3) cont.....

Matrix	Extracting Solvent	S-L ratio[a] (w/v)	Experimental Conditions	Extracted Compound	Extraction Yield	Refs.
Date palm fruit	-	100 g	1st step: PEF (two parallel electrodes; electric field strength, 3 kV/cm; 30 pulses; pulse duration, 100 µs; frequency, 10 Hz). 2nd step: solid-liquid extraction (EtOH (80%); 1:3 (w/v); 6 h).	Carotenoids	6.1 ± 0.1 µg/mL	[40]
Red bell pepper	Tap water	1:24	1st step: PEF (generator (30 kV), interelectrode distance, 28 cm; electric field strength, 1.07 kV/cm; energy input, 1 kJ/kg; exponential decay pulses; pulse width, 40 ms; frequency, 2 Hz. 2nd step: screw press.	Carotenoids (β-carotene)	14.3 ± 0.2 mg/g DW	[59]
Food waste						
Tomato peels	-	10 g	1st step: PEF (generator (25 kV, 500 A); two parallel plate cylindrical electrodes; interelectrode distance, 1.4 cm; electric field strength, 5 kV/cm; energy input, 5 kJ/kg; 10 pulses; square wave pulses; pulse width, 20 µs; frequency, 10 Hz). 2nd step: solid-liquid extraction (acetone; 1:40 (w/v); 160 rpm; 240 min).	Lycopene	13.9 ± 0.6 mg/g DW	[60]
Tomato waste	-	-	Interelectrode distance, 4 cm; electric field strength, 1 kV/cm; energy input, 28.5 kJ/kg; 500 pulses; pulse width, 15 µs; frequency, 20 Hz.	Lycopene	0.14 mg/g	[43]
				Carotenoids	0.178 ± 0.004 mg/g	
Microalgae						

(Table 3) cont.....

Matrix	Extracting Solvent	S-L ratio[a] (w/v)	Experimental Conditions	Extracted Compound	Extraction Yield	Refs.
Tetraselmis chuii	Tap water	1:100	1st step: PEF (interelectrode distance, 1 cm; electric field strength, 1 kV/cm; energy input, 100 kJ/kg; 400 pulses; pulse duration, 100 ms; frequency, 2 Hz). 2nd step: solid-liquid extraction (DMSO (50%); 1:1 (v/v); 400 rpm; 24 h).	Carotenoids	0.48 mg/g DW	[52]
			1st step: PEF (interelectrode distance, 1 cm; electric field strength, 1-3 kV/cm; energy input, 100 kJ/kg; 400-45 pulses; pulse duration, 100 ms; frequency, 2 Hz). 2nd step: solid-liquid extraction (DMSO (50%); 1:1 (v/v); 400 rpm; 4-24 h).	Chlorophyll-a	0.7 mg/g DW	
			1st step: PEF (interelectrode distance, 1 cm; electric field strength, 3 kV/cm; energy input, 100 kJ/kg; 45 pulses; pulse duration, 100 ms; frequency, 2 Hz). 2nd step: solid-liquid extraction (water; 1:1 (v/v); 400 rpm; 24 h).	Chlorophyll-b	0.79 mg/g DW	

(Table 3) cont.....

Matrix	Extracting Solvent	S-L ratio[a] (w/v)	Experimental Conditions	Extracted Compound	Extraction Yield	Refs.
Phaeodactylum tricornutum	Tap water	1:100	1st step: PEF (interelectrode distance 1 cm, electric field strength 3 kV/cm, energy input 100 kJ/kg, 45 pulses, pulse duration 100 ms, frequency 2 Hz). 2nd step: solid-liquid extraction (DMSO (50%), 1:1 (v/v), 400 rpm, 4 h).	Carotenoids	1.5 mg/g DW	[52]
			1st step: PEF (interelectrode distance, 1 cm; electric field strength, 3 kV/cm; energy input, 100 kJ/kg; 45 pulses; pulse duration, 100 ms; frequency, 2 Hz). 2nd step: solid-liquid extraction (DMSO (50%); 1:1 (v/v); 400 rpm; 4 h).	Chlorophyll-a	4.42 mg/g DW	
	Water	1:100	1st step: PEF (generator (40 kV, 10 kA); two parallel plate electrodes; interelectrode distance, 2 cm; electric field strength, 20 kV/cm; 600 pulses; exponential wave pulses; pulse duration, 10 μs; frequency, 0.5 Hz). 2nd step: solid-liquid extraction (EtOH (95%); 1:20 (w/v); stirring; 24 h; 150 rpm).	Chlorophyll-a	0.14 mg/g	[61]
Parachlorella kessleri	Water	1:100	1st step: PEF (generator (40 kV, 10 kA); two parallel plate electrodes; interelectrode distance, 2 cm; electric field strength, 20 kV/cm; 600 pulses; exponential wave pulses; pulse duration, 10 μs; frequency, 0.5 Hz). 2nd step: solid-liquid extraction (EtOH (95%); 1:20 (w/v); stirring; 24 h; 150 rpm).	Chlorophyll-a	1 mg/g	[61]

(Table 3) cont.....

Matrix	Extracting Solvent	S-L ratio[a] (w/v)	Experimental Conditions	Extracted Compound	Extraction Yield	Refs.
Nannochloropsis sp.	Water	1:100	1st step: PEF (generator (40 kV, 10 kA); two parallel plate electrodes; interelectrode distance, 2 cm; electric field strength, 20 kV/cm; 600 pulses; exponential wave pulses; pulse duration, 10 μs; frequency, 0.5 Hz). 2nd step: solid-liquid extraction (EtOH (95%); 1:20 (w/v); stirring; 24 h; 150 rpm).	Chlorophyll-a	3.5 mg/g	[61]

PEF: pulsed electric field; EtOH: ethanol; DMSO: dimethyl sulfoxide; DW: dry weight; FW: fresh weight. [a] solid-liquid ratio

PEF was also applied for the extraction of carotenoids from microalgae. These compounds were obtained after a PEF pre-treatment followed by a solid-liquid extraction with DMSO, water, or ethanol. Kokkali *et al.* [52] extracted carotenoids from *Tetraselmis chuii* and *Phaeodactylum tricornutum* microalgae. They stated that the shape and size of the cell and its orientation in the medium may influence the electroporation. Total carotenoids and chlorophyll-a were more efficiently extracted using organic solvents like DMSO, while chlorophyll-b was better extracted using aqueous solvents. The highest recovery of chlorophyll-a (4.42 mg/g DW) was found after applying 45 pulses of 3 kV/cm followed by DMSO (50%) for 4 h from *Phaeodactylum tricornutum*. Nevertheless, more intense PEF conditions (600 pulses and 20 kV/cm) did not improve carotenoid extraction, as observed by Zhang *et al.* [61]. In fact, they obtained 0.14 mg/g of chlorophyll-a from *Phaeodactylum tricornutum* using ethanol (95%) after PEF pre-treatment. They also carried out the extraction of chlorophyll-a from *Parachlorella kessleri* and *Nannochloropsis sp.* microalgae, obtaining 1 mg/g and 3.5 mg/g, respectively.

Extraction of Proteins

Proteins are very important ingredients in the food that provide nutritional, functional, and textural properties. PEF technology has been applied to facilitate protein extraction, change the catalytic activity of enzymes, and modify the physiological and nutritional properties of bioactive proteins [31]. The application of PEF enhances the diffusion of proteins located inside the cytoplasm, chloroplasts, and cell walls [61]. Table **4** shows the most recent PEF applications for the extraction of proteins from food waste and microalgae. Moderate electric field intensities from 0.6 to 20 kV/cm, the application of up to 3000 pulses, energy inputs from 5.45 to 300 kJ/kg, and frequencies between 0.5 and 20 Hz were employed.

Andreou *et al.* [43] studied the release of proteins by PEF from tomato waste. They reported that the amount of proteins increased with the electric field strength and time. For instance, they could extract 1.45 mg/g after applying 25 pulses at 5 kV/cm for 1.5 ms. This result showed an increase of 63% in the recovery of proteins, compared to untreated samples. Andreou *et al.* [45] also employed a second solid-liquid extraction step to improve the protein recovery from olive pomace. They also observed that more intense PEF conditions resulted in a higher recovery of proteins. Thus, they applied 3000 pulses at 3 kV/cm to obtain 5.4 mg/mL of proteins, which was 1.5 times higher protein content than with untreated. On the other hand, Kumari *et al.* [62] studied the extraction of proteins from different brewer's spent grains samples. They extracted 23.8 ± 0.2% proteins by applying a PEF pre-treatment (3000 pulses at 2.8 kV/cm) followed by aqueous extraction. Nevertheless, this protein recovery could be overestimated since PEF could extract some non-proteinaceous nitrogen components such as nucleotides caused by cell disruption. PEF was also applied for the extraction of proteins from animal waste. Wang *et al.* [63] reported the protein extraction from sole skin using a PEF pre-treatment (3 kV/cm and 300 kJ/kg) followed by magnetic stirring for 24 h. PEF pre-treatment improved the protein extraction efficiency by almost 2-fold. Ghosh *et al.* [64] combined high voltage and short pulses with low voltage and long pulses for protein extraction from waste chicken breast muscle. In this way, protein concentration in the crude extract was 78 ± 8 mg/mL.

Table 4. Application of PEF technique for the extraction of proteins from food matrices.

Matrix	Extracting Solvent	S-L ratio[a] (w/v)	Experimental Conditions	Extraction Yield	Refs.
Food waste					
Tomato waste	-	-	Interelectrode distance, 4 cm; electric field strength, 5 kV/cm; energy input, 7.1 kJ/kg; 25 pulses; pulse width, 15 μs; frequency, 20 Hz.	1.45 mg/g	[43]
Olive pomace	EtOH (50%)	1:10	1st step: PEF (interelectrode distance, 4 cm; electric field strength, 3 kV/cm; energy input, 10.9 kJ/kg, 3000 pulses; pulse width, 15 μs; frequency, 20 Hz). 2nd step: solid-liquid extraction (EtOH (50%); 1:10 (w/v); 25 °C; 0-60 min).	5.4 mg/mL	[45]

(Table 4) cont.....

Matrix	Extracting Solvent	S-L ratio[a] (w/v)	Experimental Conditions	Extraction Yield	Refs.
Brewer's spent grains	Water	1:06	1st step: PEF (generator (25 kV); two parallel electrodes; interelectrode distance, 4 cm; electric field strength, 2.8 kV/cm; 3000 pulses; square waveform pulses; pulse width, 20 µs; frequency, 10 Hz). 2nd step: solid-liquid extraction (water, 55 °C; 220 rpm; 16 h).	23.8 ± 0.2% DW extracts	[61]
Sole skin	Tap water	-	1st step: PEF (electric field strength, 3 kV/cm; energy input, 300 kJ/kg). 2nd step: magnetic stirring (24 h).	70%	[63]
Waste chicken breast muscle	-	5.14 g	1st step: PEF (interelectrode distance, 1.7 cm; electric field strength, 0.604 kV/cm; energy input, 5.45 kJ/kg; 100 pulses; square wave pulses; pulse duration, 50 µs; frequency, 3 Hz). 2nd step: PEF (interelectrode distance, 1.6 cm; electric field strength, 0.309 kV/cm; energy input, 15.9 kJ/kg; 50 pulses; square wave pulses; pulse duration, 1 ms; frequency, 3 Hz). 3rd step: PEF (interelectrode distance, 1.3 cm; electric field strength, 0.157 kV/cm; energy input, 16.2 kJ/kg; 50 pulses; square wave pulses; pulse duration, 5 ms; frequency, 3 Hz).	78 ± 8 mg/mL	[64]
Microalgae					
Parachlorella kessleri	Water	1:200	Generator (40 kV, 10 kA); two parallel plate electrodes; interelectrode distance, 2 cm; electric field strength, 20 kV/cm; 800 pulses; exponential wave pulses; pulse duration, 10 µs; frequency, 0.5 Hz.	4 mg/g	[65]
			1st step: PEF (generator (40 kV, 10 kA); two parallel plate electrodes; interelectrode distance, 2 cm; electric field strength, 20 kV/cm; 800 pulses; exponential wave pulses; pulse duration, 10 µs; frequency, 0.5 Hz). 2nd step: HPH (80 MPa; 4 passes; 204.8 kJ/g).	68 mg/g	
	Water	1:100	Generator (40 kV, 10 kA); two parallel plate electrodes; interelectrode distance, 2 cm; electric field strength, 20 kV/cm; 600 pulses; exponential wave pulses; pulse duration, 10 µs; frequency, 0.5 Hz.	1.8%	[61]

(Table 4) cont.....

Matrix	Extracting Solvent	S-L ratio[a] (w/v)	Experimental Conditions	Extraction Yield	Refs.
Phaeodactylum tricornutum	Water	1:100	Generator (40 kV, 10 kA); two parallel plate electrodes; interelectrode distance, 2 cm; electric field strength, 20 kV/cm; 600 pulses; exponential wave pulses; pulse duration, 10 μs; frequency, 0.5 Hz.	4%	[61]
Nannochloropsis sp.	Water	1:100	Generator (40 kV, 10 kA); two parallel plate electrodes; interelectrode distance, 2 cm; electric field strength, 20 kV/cm; 600 pulses; exponential wave pulses; pulse duration, 10 μs; frequency, 0.5 Hz.	3%	[61]
Chlorella vulgaris	PBS (151.7 mM, pH 7)	1:167	1st step: PEF (two parallel plate electrodes; interelectrode distance, 0.2 cm; electric field strength, 20 kV/cm; energy input, 7.76 kJ/kg; 2 pulses; exponential decay pulses; pulse width, 50 μs). 2nd step: Incubation (24 h).	13.5%	[66]

PEF: pulsed electric field; HPH: high pressure homogenization; EtOH: ethanol; DMSO: dimcthyl sulfoxide; DW: dry weight; PBS: phosphate buffered saline. [a] solid-liquid ratio

Microalgae also contain proteins, which are located in the cell wall, cytoplasm, chloroplast, and organelles. They can be extracted by the application of pulsed electric energy. Zhang *et al.* [65] reported that PEF treatment opened the pores on cell membranes, allowing the release of small-sized cytoplasmic proteins, but most proteins were large and were bound to the cell structure, so their extraction was difficult by electroporation. Nevertheless, the combination of PEF with high pressure homogenization led to the intensive generation of cell debris that released large intracellular molecules, improving protein extraction efficiency. Therefore, this double treatment increased the protein extraction up to 68 mg/g from *Parachlorella kessleri* microalgae. Zhang *et al.* [61] also evaluated the protein extraction from three microalgal species (*Parachlorella kessleri*, *Phaeodactylum tricornutum*, and *Nannochloropsis sp.*) by PEF treatment. The highest protein recovery was observed in the *Phaeodactylum tricornutum* (4%) followed by *Nannochloropsis sp.* (3%) and, finally, *Parachlorella kessleri* (1.8%). They demonstrated that the electroporation mechanism was not as effective for the extraction of large intracellular molecules as other more destructive treatments, for example, HVED and ultrasounds. Buchmann *et al.* [66] reported a continuous extraction of proteins from *Chlorella vulgaris* microalgae using PEF. They observed that the protein extraction increased with the electric field strength, obtaining a 13.5% of proteins by the application of 20 kV/cm.

Extraction of Carbohydrates

Carbohydrates are primary macronutrients that humans consume in large quantities in their daily diet. These molecules provide energy to support various human metabolic processes. Carbohydrates are found in a wide variety of foods, mostly from plant tissues [67]. Carbohydrates are water soluble compounds located inside the cells. Table **5** shows the experimental conditions employed for the extraction of carbohydrates from carrots, brewer's spent grains, and seaweeds using PEF treatment. In these works, the inter-electrode distance was between 2 and 4 cm, the electric field strength from 0.75 to 20 kV/cm, and frequencies ranged from 0.5 to 10 Hz.

Table 5. Application of PEF technique for the extraction of carbohydrates from food matrices.

Matrix	Extracting Solvent	S-L ratio[a] (w/v)	Experimental Conditions	Extracted Compound	Extraction Yield	Refs.
			Food			
Carrots	Water	1:2	1st step: PEF (generator (1.75 kV, 20 A); interelectrode distance, 4 cm; electric field strength, 0.75 kV/cm; 10 pulses; exponential decay pulses; frequency, 1 Hz). 2nd step: Stirring (250 rpm; 45 °C).	Sugar	64%	[68]
			Food waste			
Brewer's spent grains	Water	1:06	1st step: PEF (generator (25 kV); two parallel electrodes; interelectrode distance, 4 cm; electric field strength, 2.8 kV/cm; 3000 pulses; square waveform pulses; pulse width, 20 μs; frequency, 10 Hz). 2nd step: solid-liquid extraction (water, 55 °C; 220 rpm; 16 h).	Carbohydrates	75.3 ± 0.6% DW extracts	[62]
			Seaweeds			

(Table 5) cont.....

Matrix	Extracting Solvent	S-L ratio[a] (w/v)	Experimental Conditions	Extracted Compound	Extraction Yield	Refs.
Parachlorella kessleri	Water	1:200	Generator (40 kV, 10 kA); two parallel plate electrodes; interelectrode distance, 2 cm; electric field strength, 20 kV/cm; 800 pulses; exponential wave pulses; pulse duration, 10 μs; frequency, 0.5 Hz.	Carbohydrates	40.4 mg/g	[65]
			1st step: PEF (generator (40 kV, 10 kA); two parallel plate electrodes; interelectrode distance, 2 cm; electric field strength, 20 kV/cm; 400 pulses; exponential wave pulses; pulse duration, 10 μs; frequency, 0.5 Hz). 2nd step: HPH (80 MPa; 4 passes; 134.4 kJ/g).	Carbohydrates	104.8 mg/g	
	Water	1:100	Generator (40 kV, 10 kA); two parallel plate electrodes; interelectrode distance, 2 cm; electric field strength, 20 kV/cm; 600 pulses; exponential wave pulses; pulse duration, 10 μs; frequency, 0.5 Hz.	Carbohydrates	10%	[61]
Phaeodactylum tricornutum	Water	1:100	Generator (40 kV, 10 kA); two parallel plate electrodes; interelectrode distance, 2 cm; electric field strength, 20 kV/cm; 600 pulses; exponential wave pulses; pulse duration, 10 μs; frequency, 0.5 Hz.	Carbohydrates	18.2%	[61]
Nannochloropsis sp.	Water	1:100	Generator (40 kV, 10 kA); two parallel plate electrodes; interelectrode distance, 2 cm; electric field strength, 20 kV/cm; 600 pulses; exponential wave pulses; pulse duration, 10 μs; frequency, 0.5 Hz.	Carbohydrates	17.4%	[61]

(Table 5) cont.....

Matrix	Extracting Solvent	S-L ratio[a] (w/v)	Experimental Conditions	Extracted Compound	Extraction Yield	Refs.
Laminaria digitata	Water	1:116	Generator (18 kV); two plate electrodes; electric field strength, 7.5 kV/cm; energy input, 11.9 kJ/kg; 210 pulses; exponential decay pulses; pulse width, 0.96 μs; frequency, 1.2 Hz.	Carbohydrates	13.0 ± 0.1 mg GE/ g DW	[53]

PEF: pulsed electric field; HPH: high pressure homogenization; DW: dry weight; GE: glucose equivalents. [a] solid-liquid ratio

Dastangoo *et al.* [68] reported the sugar extraction from carrots by PEF treatment. They evaluated the extraction kinetics using different electric field intensities and temperatures. The extraction yield increased with the electric field intensity, due to the degradation of cell membranes; however, high temperatures (70 ºC) decreased the extraction yield, since some compounds were degraded. Under optimized conditions, the sugar extraction yield was 64%. However, Kumari *et al.* [62] observed a much lower increase (5.42%) in carbohydrate recovery from brewer's spent grains pre-treated by PEF, compared to untreated samples.

The feasibility of PEF treatment for the extraction of carbohydrates from microalgae depends on cell shape, size, cell wall structure, composition, and orientation. Zhang *et al.* [65] reported that the electroporation phenomenon was not very effective for the extraction of carbohydrates from *Parachlorella kessleri* microalgae. Nevertheless, the combination of PEF with high pressure homogenization increased the carbohydrate content up to 104.8 mg/g. In another work, Zhang *et al.* [61] studied carbohydrate extraction by PEF treatment from three microalgal species. *Parachlorella kessleri* and *Nannochloropsis sp.* had spherical forms and strong and rigid cell walls mainly composed of cellulose and hemicelluloses, while *Phaeodactylum tricornutum* had fusiform shape and its cell wall was poor in silica and composed of sulfated glucomannan. Thus, the highest extraction yield of carbohydrates was from *Phaeodactylum tricornutum* since the resistance of cell walls was lower than that of other species. Einarsdóttir *et al.* [53] reported the carbohydrate extraction from *Laminaria digitata* macroalgae. They observed that the lowest biomass concentrations (0.25-0.86%) originated 20 times higher carbohydrate yield at 20ºC. They concluded that the number of treatment pulses, temperature, and biomass concentration could affect the selectivity of the extraction.

Extraction of Vitamins

Vitamins are vital organic molecules that have an essential role in human metabolism. Some of them have the high antioxidant capacity, so their extraction as well as their retention in processed foods are of high importance [30]. Vitamin C is one of the most studied vitamins, although it is highly sensitive to processing and storage. Many researchers have applied PEF to preserve this vitamin in liquid foods [30], since it can be easily degraded under the influence of pH, temperature, and the presence of metal ions, oxygen, light and oxidative enzymes [47]. Indeed, PEF (3 kV/cm for 40 ms) was employed to preserve vitamin C in cranberry bush purée [47] and in apple juice (30 kV/cm for 400 pulses) for 72 h under refrigeration [69]. Moreover, PEF has also been applied to promote the extraction of vitamin C from red bell peppers. The application of PEF at 1.07 kV/cm for 40 ms improved the extractability of vitamin C by 18–25% as compared to untreated bell peppers [59].

CONCLUDING REMARKS

The application of pulsed electric fields on food processing has been very studied in the last few years. The use of high voltage and short electric pulses leads to the formation of pores on the cell membrane due to the electroporation phenomenon. This strategy has been applied to the extraction of intracellular compounds located in cell walls, cytoplasm, chloroplasts, and vacuoles, among others, from different food matrices, food waste, seaweeds, or other plants. The extraction of compounds is favoured by increasing the electric field intensity, the treatment time, and the number of pulses.

PEF is a sustainable technique since it requires low energy and solvent consumption and reduces extraction time. It is a non-thermal and non-destructive technique, so thermolabile compounds can be extracted. The PEF technique has been demonstrated to be suitable for the extraction of essential phytochemicals for the human body such as lipids, phenolic compounds, carotenoids, proteins, carbohydrates, and vitamins.

PEF can be combined with other extraction techniques for enhancing the diffusion of solutes located inside the cells and thus increasing their recovery. The use of organic solvents in a second extraction step improves the extraction of hydrophobic compounds such as lipids and some carotenoids. The use of emerging technologies such as ultrasounds and high-pressure homogenization, increases the recovery yields of the extracted compounds. Additionally, PEF is also employed in food preservation. PEF guarantees the nutritional value of extracted molecules and prologues their shelf life. This technique is a great opportunity to obtain safe, healthy, and nutritious food.

LIST OF ABBREVIATIONS

CE	Catechin equivalent
Cy-3-Glc	Cyanidin-3-O-glucoside equivalents
DE	Dried extract
DMSO	Dimethyl sulfoxide
DW	Dry weight
E	Electric field intensity
Ec	Critical electric field intensity
EtOH	Ethanol
FAME	Fatty acid methyl ester
FW	Fresh weight
GAE	Gallic acid equivalent
GE	Glucose equivalents
HPH	High pressure homogenization
MeOH	Methanol
MW	Microwave
PBS	Phosphate buffered saline
PEF	Pulsed electric field
PUFA	Poly-unsaturated fatty acids
RE	Rutin equivalent
S-L ratio	Solid-liquid ratio
TAE	Tannic acid equivalent
UAE	Ultrasound assisted extraction

CONSENT FOR PUBLICATION

Not applicable.

CONFLICT OF INTEREST

The authors declare no conflict of interest, financial or otherwise.

ACKNOWLEDGEMENTS

Authors thank the Spanish Ministry of Science and Innovation (PID2020-114891RB-I00) and the Comunidad of Madrid (Spain) and European funding from FSE and FEDER programs (S2018/BAA-4393, AVANSECAL-II-CM).

REFERENCES

[1] Barba FJ, Puértolas E, Brnčić M, et al. Emerging extraction. In: Galanakis CM (Ed.) Food Waste Recovery. Academic Press. 2015; pp. 249-72.

[2] Barba FJ, Parniakov O, Pereira SA, *et al.* Current applications and new opportunities for the use of pulsed electric fields in food science and industry. Food Res Int 2015; 77: 773-98. [http://dx.doi.org/10.1016/j.foodres.2015.09.015]

[3] Vorobiev E, Lebovka N. Pulsed electric field in green processing and preservation of food products. In: Green FPT, Chemat F, Vorobiev E, Eds. Academic Press. 2019; pp. 403-30. [http://dx.doi.org/10.1016/B978-0-12-815353-6.00015-X]

[4] Maroun RG, Rajha HN, El Darra N, *et al.* Emerging technologies for the extraction of polyphenols from natural sources. In: Galanakis CM, Ed. Polyphenols: Properties, Recovery, and Applications. Woodhead Publishing 2018; pp. 265-93.

[5] Barba FJ, Ahrné L, Xanthakis E, Landerslev MG, Orlien V. Innovative technologies for food preservation. In: Barba FJ, Sant'Ana AS, Orlien V, Koubaa M, Eds. Innovative Technologies for Food Preservation: Inactivation of Spoilage and Pathogenic Microorganisms. Academic Press 2018; pp. 25-51. [http://dx.doi.org/10.1016/B978-0-12-811031-7.00002-9]

[6] Toepfl S, Heinz V, Knorr D. Applications of pulsed electric fields technology for the food industry. In: Raso J, Heinz V, Eds. Pulsed Electric Fields Technology for the Food Industry. Food Engineering SeriesBoston, MA: Springer 2006; pp. 197-221. [http://dx.doi.org/10.1007/978-0-387-31122-7_7]

[7] Xi J, Li Z, Fan Y. Recent advances in continuous extraction of bioactive ingredients from food-processing wastes by pulsed electric fields. Crit Rev Food Sci Nutr 2021; 61(10): 1738-50. [http://dx.doi.org/10.1080/10408398.2020.1765308] [PMID: 32406247]

[8] Arshad RN, Abdul-Malek Z, Roobab U, *et al.* Pulsed electric field: A potential alternative towards a sustainable food processing. Trends Food Sci Technol 2021; 111: 43-54. [http://dx.doi.org/10.1016/j.tifs.2021.02.041]

[9] Kumari B, Tiwari BK, Hossain MB, Brunton NP, Rai DK. Recent advances on application of ultrasound and pulsed electric field technologies in the extraction of bioactives from agro-industrial by-products. Food Bioprocess Technol 2018; 11(2): 223-41. [http://dx.doi.org/10.1007/s11947-017-1961-9]

[10] Khadhraoui B, Ummat V, Tiwari BK, Fabiano-Tixier AS, Chemat F. Review of ultrasound combinations with hybrid and innovative techniques for extraction and processing of food and natural products. Ultrason Sonochem 2021; 76: 105625-53. [http://dx.doi.org/10.1016/j.ultsonch.2021.105625] [PMID: 34147916]

[11] Khan MK, Ahmad K, Hassan S, Imran M, Ahmad N, Xu C. Effect of novel technologies on polyphenols during food processing. Innov Food Sci Emerg Technol 2018; 45: 361-81. [http://dx.doi.org/10.1016/j.ifset.2017.12.006]

[12] Lebovka NI, Bazhal MI, Vorobiev E. Simulation and experimental investigation of food material breakage using pulsed electric field treatment. J Food Eng 2000; 44(4): 213-23. [http://dx.doi.org/10.1016/S0260-8774(00)00029-7]

[13] Raso J, Condón S, Álvarez I. Pulsed electric field. In: Batt C, Ed. Encyclopedia of Food Microbiology. Academic Press 2014; Vol. 2: pp. 966-73. [http://dx.doi.org/10.1016/B978-0-12-384730-0.00397-9]

[14] Barbosa-Canovas GV, Pierson MD, Zhang QH, Schaffner DW. Pulsed electric fields. J Food Sci 2000; 65: 65-79. [http://dx.doi.org/10.1111/j.1750-3841.2000.tb00619.x]

[15] Guerrero-Beltrán JA, Welti-Chanes J. Pulsed electric field. In: Caballero B, Finglas PM, Toldrá F, Eds. Encyclopedia of Food and Health. Academic Press 2016; pp. 561-5.
[http://dx.doi.org/10.1016/B978-0-12-384947-2.00579-1]

[16] Arshad RN, Buntat Z, Sidik MAB, Alamgir A, Nawawi Z, Hafizi M. Proficiency of double-exponential pulse waveform in food treatment through pulsed electric field. 2nd International Conference on High Voltage Engineering and Power Systems (ICHVEPS). 218-22.
[http://dx.doi.org/10.1109/ICHVEPS47643.2019.9011148]

[17] Puértolas E, Barba FJ. Electrotechnologies applied to valorization of by-products from food industry: Main findings, energy and economic cost of their industrialization. Food Bioprod Process 2016; 100: 172-84.
[http://dx.doi.org/10.1016/j.fbp.2016.06.020]

[18] Vorobiev E, Lebovka N. Effects of pulsed electric energy in food and agricultural products: A review of recent research advances. In: Knoerzer K, Muthukumarappan K, Eds. Innovative Food Processing Technologies: A Comprehensive Review. Elsevier 2021; Vol. 1: pp. 173-98.
[http://dx.doi.org/10.1016/B978-0-12-815781-7.00019-6]

[19] Vorobiev E, Chemat F. Principles of physically assisted extractions and applications in the food, beverage and nutraceutical industries. In: Rizvi SSH, Ed. Separation, Extraction and Concentration Processes in the Food, Beverage and Nutraceutical Industries. Woodhead Publishing 2010; pp. 71-108.

[20] Lebovka NI, Bazhal MI, Vorobiev E. Pulsed electric field breakage of cellular tissues: visualisation of percolative properties. Innov Food Sci Emerg Technol 2001; 2(2): 113-25.
[http://dx.doi.org/10.1016/S1466-8564(01)00024-8]

[21] Jaeger H, Meneses N, Knorr D. Pulsed electric field technology. In: Motarjemi Y, Ed. Encyclopedia of Food Safety. Academic Press 2014; Vol. 3: pp. 561-5.
[http://dx.doi.org/10.1016/B978-0-12-378612-8.00260-2]

[22] Saldaña G, Luengo E, Puértolas E, Álvarez I, Raso J. Pulsed electric fields in wineries: Potential applications. In: Miklavčič D, Ed. Handbook of Electroporation. Switzerland: Springer 2017; pp. 2825-42.
[http://dx.doi.org/10.1007/978-3-319-32886-7_155]

[23] Tappi S, Tylewicz U, Dalla-Rosa M. Effect of nonthermal technologies on functional food compounds. In: Betoret N, Betoret E, Eds. Sustainability of the Food System. Academic Press 2020; pp. 147-65.
[http://dx.doi.org/10.1016/B978-0-12-818293-2.00008-2]

[24] Pagan R, Manas P. Fundamental aspects of microbial membrane electroporation. In: Raso J, Heinz V, Eds. Pulsed Electric Fields Technology for the Food Industry. Boston, MA: Springer 2006; pp. 73-94.
[http://dx.doi.org/10.1007/978-0-387-31122-7_3]

[25] Maroun RG, Rajha HN, Vorobiev E, Louka N. Emerging technologies for the recovery of valuable compounds from grape processing by-products. In: Galanakis CM, Ed. Handbook of Grape Processing By-Products. Academic Press 2017; pp. 155-81.
[http://dx.doi.org/10.1016/B978-0-12-809870-7.00007-7]

[26] Vorobiev E, Lebovka N. Enhancing extraction from solid foods and biosuspensions by electrical pulsed energy (pulsed electric field, ohmic heating, and high-voltage electrical discharge). In: Yanniotis S, Taoukis P, Stoforos NG, Karathanos VT, Eds. Advances in Food Process Engineering Research and Applications. Boston, MA: Springer 2013; pp. 415-28.
[http://dx.doi.org/10.1007/978-1-4614-7906-2_19]

[27] Sarkis JR, Boussetta N, Vorobiev E. Application of pulsed electric energies for oil and polyphenol extraction from sesame cake and sesame seeds. In: Miklavčič D, Ed. Handbook of Electroporation. Switzerland: Springer 2017; pp. 2699-712.
[http://dx.doi.org/10.1007/978-3-319-32886-7_119]

[28] Dragomir MCB, Zeca ED, Ivan AS, Stoica M. Pulsed electric field and high voltage electrical discharge-innovative food electrotechnologies. A review. J Agroaliment Process Technol 2020; 26: 34-9.

[29] Barba FJ, Koubaa M, Grimi N, Vorobiev E. Selective extraction of biocompounds from stevia rebaudiana bertoni leaves using electrotechnologies. In: Miklavčič D, Ed. Handbook of Electroporation. Switzerland: Springer 2017; pp. 2750-61. [http://dx.doi.org/10.1007/978-3-319-32886-7_123]

[30] Morales-de la Peña M, Rábago-Panduro LM, Soliva-Fortuny R, Martín-Belloso O, Welti-Chanes J. Pulsed electric fields technology for healthy food products. Food Eng Rev 2021; 13(3): 509-23. [http://dx.doi.org/10.1007/s12393-020-09277-2]

[31] Zhang S, Sun L, Ju H, Bao Z, Zeng X, Lin S. Research advances and application of pulsed electric field on proteins and peptides in food. Food Res Int 2021; 139: 109914-26. [http://dx.doi.org/10.1016/j.foodres.2020.109914] [PMID: 33509481]

[32] Salehi F. Physico-chemical properties of fruit and vegetable juices as affected by pulsed electric field: a review. Int J Food Prop 2020; 23(1): 1036-50. [http://dx.doi.org/10.1080/10942912.2020.1775250]

[33] Pateiro M, Domínguez R, Munekata PES, Barba FJ, Lorenzo JM. Lipids and fatty acids. In: Barba FJ, Jorge MAS, Cravotto G, Lorenzo JM, Eds. Innovative Thermal and Non-Thermal Processing, Bioaccessibility and Bioavailability of Nutrients and Bioactive Compounds. Woodhead Publishing 2019; pp. 107-37. [http://dx.doi.org/10.1016/B978-0-12-814174-8.00004-4]

[34] Han SF, Jin W, Yang Q, *et al.* Application of pulse electric field pretreatment for enhancing lipid extraction from *Chlorella pyrenoidosa* grown in wastewater. Renew Energy 2019; 133: 233-9. [http://dx.doi.org/10.1016/j.renene.2018.10.034]

[35] Zhang R, Gu X, Xu G, Fu X. Improving the lipid extraction yield from Chlorella based on the controllable electroporation of cell membrane by pulsed electric field. Bioresour Technol 2021; 330: 124933-41. [http://dx.doi.org/10.1016/j.biortech.2021.124933] [PMID: 33721737]

[36] Gulzar S, Benjakul S. Impact of pulsed electric field pretreatment on yield and quality of lipid extracted from cephalothorax of Pacific white shrimp (*Litopenaeus vannamei*) by ultrasound□assisted process. Int J Food Sci Technol 2020; 55(2): 619-30. [http://dx.doi.org/10.1111/ijfs.14316]

[37] Mohseni NM, Mirzaei H, Moghimi M. Optimized extraction and quality evaluation of Niger seed oil via microwave□pulsed electric field pretreatments. Food Sci Nutr 2020; 8(3): 1383-93. [http://dx.doi.org/10.1002/fsn3.1396] [PMID: 32180948]

[38] Papachristou I, Akaberi S, Silve A, *et al.* Analysis of the lipid extraction performance in a cascade process for *Scenedesmus almeriensis* biorefinery. Biotechnol Biofuels 2021; 14(1): 20. [http://dx.doi.org/10.1186/s13068-020-01870-1] [PMID: 33446259]

[39] Kord A, Cherif Y, Amiali M, *et al.* Fatty acids composition of *Cystoseira sauvageauana* and *Laurencia pinnatifida* collected from the Algerian coast. Acta Period Technol 2019; 50(50): 113-22. [http://dx.doi.org/10.2298/APT1950113K]

[40] Siddeeg A, Faisal Manzoor M, Haseeb Ahmad M, *et al.* Pulsed electric field-assisted ethanolic extraction of date palm fruits: Bioactive compounds, antioxidant activity and physicochemical properties. Processes (Basel) 2019; 7(9): 585-97. [http://dx.doi.org/10.3390/pr7090585]

[41] López-Gámez G, Elez-Martínez P, Martín-Belloso O, Soliva-Fortuny R. Pulsed electric fields affect endogenous enzyme activities, respiration and biosynthesis of phenolic compounds in carrots. Postharvest Biol Technol 2020; 168: 111284-91.

[http://dx.doi.org/10.1016/j.postharvbio.2020.111284]

[42] Manzoor MF, Zeng XA, Rahaman A, *et al.* Combined impact of pulsed electric field and ultrasound on bioactive compounds and FT-IR analysis of almond extract. J Food Sci Technol 2019; 56(5): 2355-64.
[http://dx.doi.org/10.1007/s13197-019-03627-7] [PMID: 31168118]

[43] Andreou V, Dimopoulos G, Dermesonlouoglou E, Taoukis P. Application of pulsed electric fields to improve product yield and waste valorization in industrial tomato processing. J Food Eng 2020; 270: 109778-90.
[http://dx.doi.org/10.1016/j.jfoodeng.2019.109778]

[44] Wang L, Boussetta N, Lebovka N, Vorobiev E. Cell disintegration of apple peels induced by pulsed electric field and efficiency of bio-compound extraction. Food Bioprod Process 2020; 122: 13-21.
[http://dx.doi.org/10.1016/j.fbp.2020.03.004]

[45] Andreou V, Psarianos M, Dimopoulos G, Tsimogiannis D, Taoukis P. Effect of pulsed electric fields and high pressure on improved recovery of high-added-value compounds from olive pomace. J Food Sci 2020; 85(5): 1500-12.
[http://dx.doi.org/10.1111/1750-3841.15122] [PMID: 32267966]

[46] Frontuto D, Carullo D, Harrison SM, *et al.* Optimization of pulsed electric fields-assisted extraction of polyphenols from potato peels using response surface methodology. Food Bioprocess Technol 2019; 12(10): 1708-20.
[http://dx.doi.org/10.1007/s11947-019-02320-z]

[47] Ozkan G, Stübler AS, Aganovic K, Dräger G, Esatbeyoglu T, Capanoglu E. Retention of polyphenols and vitamin C in cranberrybush purée (*Viburnum opulus*) by means of non-thermal treatments. Food Chem 2021; 360: 129918-27.
[http://dx.doi.org/10.1016/j.foodchem.2021.129918] [PMID: 34051454]

[48] Lončarić A, Celeiro M, Jozinović A, *et al.* Green extraction methods for extraction of polyphenolic compounds from blueberry pomace. Foods 2020; 9(11): 1521-42.
[http://dx.doi.org/10.3390/foods9111521] [PMID: 33113955]

[49] Rajha HN, Abi-Khattar AM, El Kantar S, *et al.* Comparison of aqueous extraction efficiency and biological activities of polyphenols from pomegranate peels assisted by infrared, ultrasound, pulsed electric fields and high-voltage electrical discharges. Innov Food Sci Emerg Technol 2019; 58: 102212-20.
[http://dx.doi.org/10.1016/j.ifset.2019.102212]

[50] Rahaman A, Zeng XA, Farooq MA, *et al.* Effect of pulsed electric fields processing on physiochemical properties and bioactive compounds of apricot juice. J Food Process Eng 2020; 43(8): e13449-56.
[http://dx.doi.org/10.1111/jfpe.13449]

[51] Ricci A, Parpinello GP, Banfi BA, Olivi F, Versari A. Preliminary study of the effects of pulsed electric field (PEF) treatments in wines obtained from early-harvested sangiovese grapes. Beverages 2020; 6(2): 34-50.
[http://dx.doi.org/10.3390/beverages6020034]

[52] Kokkali M, Martí-Quijal FJ, Taroncher M, Ruiz MJ, Kousoulaki K, Barba FJ. Improved extraction efficiency of antioxidant bioactive compounds from *Tetraselmis chuii* and *Phaedoactylum tricornutum* using pulsed electric fields. Molecules 2020; 25(17): 3921-32.
[http://dx.doi.org/10.3390/molecules25173921] [PMID: 32867350]

[53] Einarsdóttir R, Þórarinsdóttir KA, Aðalbjörnsson BV, Guðmundsson M, Marteinsdóttir G, Kristbergsson K. The effect of pulsed electric field-assisted treatment parameters on crude aqueous extraction of *Laminaria digitata.* J Appl Phycol 2021; 33(5): 3287-96.
[http://dx.doi.org/10.1007/s10811-021-02500-5]

[54] Arab SS, Pedram NA, Saeidi AMR, Fariba N, Hamid T. Antioxidant activity of aqueous and alcohol

extracts of *Salvia leriifolia* L. and *Linum usitalissmum* L. subjected to a pulsed electric field. Foods Raw Mater 2020; 8: 186-95.

[55] Sarraf M, Beig-babaei A, Naji-Tabasi S. Optimizing extraction of berberine and antioxidant compounds from barberry by maceration and pulsed electric field-assisted methods. J Berry Res 2021; 11(1): 133-49.
[http://dx.doi.org/10.3233/JBR-200603]

[56] Kim YN, Kwon HJ, Lee DU. Effects of pulsed electric field (PEF) treatment on physicochemical properties of *Panax ginseng.* Innov Food Sci Emerg Technol 2019; 58: 102232-8.
[http://dx.doi.org/10.1016/j.ifset.2019.102232]

[57] Tzima K, Brunton NP, Lyng JG, Frontuto D, Rai DK. The effect of Pulsed Electric Field as a pre-treatment step in Ultrasound Assisted Extraction of phenolic compounds from fresh rosemary and thyme by-products. Innov Food Sci Emerg Technol 2021; 69: 102644-55.
[http://dx.doi.org/10.1016/j.ifset.2021.102644]

[58] López-Gámez G, Elez-Martínez P, Martín-Belloso O, Soliva-Fortuny R. Changes of carotenoid content in carrots after application of pulsed electric field treatments. Lebensm Wiss Technol 2021; 147: 111408-14.
[http://dx.doi.org/10.1016/j.lwt.2021.111408]

[59] Rybak K, Samborska K, Jedlinska A, *et al.* The impact of pulsed electric field pretreatment of bell pepper on the selected properties of spray dried juice. Innov Food Sci Emerg Technol 2020; 65: 102446-53.
[http://dx.doi.org/10.1016/j.ifset.2020.102446]

[60] Pataro G, Carullo D, Falcone M, Ferrari G. Recovery of lycopene from industrially derived tomato processing by-products by pulsed electric fields-assisted extraction. Innov Food Sci Emerg Technol 2020; 63: 102369-78.
[http://dx.doi.org/10.1016/j.ifset.2020.102369]

[61] Zhang R, Lebovka N, Marchal L, Vorobiev E, Grimi N. Pulsed electric energy and ultrasonication assisted green solvent extraction of bio-molecules from different microalgal species. Innov Food Sci Emerg Technol 2020; 62: 102358-64.
[http://dx.doi.org/10.1016/j.ifset.2020.102358]

[62] Kumari B, Tiwari BK, Walsh D, *et al.* Impact of pulsed electric field pre-treatment on nutritional and polyphenolic contents and bioactivities of light and dark brewer's spent grains. Innov Food Sci Emerg Technol 2019; 54: 200-10.
[http://dx.doi.org/10.1016/j.ifset.2019.04.012]

[63] Wang M, Zhou J, Collado MC, Barba FJ. Accelerated solvent extraction and pulsed electric fields for valorization of rainbow trout (*Oncorhynchus mykiss*) and sole (*Dover sole*) by-products: protein content, molecular weight distribution and antioxidant potential of the extracts. Mar Drugs 2021; 19(4): 207-21.
[http://dx.doi.org/10.3390/md19040207] [PMID: 33916965]

[64] Ghosh S, Gillis A, Sheviryov J, Levkov K, Golberg A. Towards waste meat biorefinery: Extraction of proteins from waste chicken meat with non-thermal pulsed electric fields and mechanical pressing. J Clean Prod 2019; 208: 220-31.
[http://dx.doi.org/10.1016/j.jclepro.2018.10.037]

[65] Zhang R, Marchal L, Vorobiev E, Grimi N. Effect of combined pulsed electric energy and high pressure homogenization on selective and energy efficient extraction of bio-molecules from microalga *Parachlorella kessleri*. Lebensm Wiss Technol 2021; 141: 110901-7.
[http://dx.doi.org/10.1016/j.lwt.2021.110901]

[66] Buchmann L, Brändle I, Haberkorn I, Hiestand M, Mathys A. Pulsed electric field based cyclic protein extraction of microalgae towards closed-loop biorefinery concepts. Bioresour Technol 2019; 291: 121870-7.

[http://dx.doi.org/10.1016/j.biortech.2019.121870] [PMID: 31382092]

[67] Leong SY, Duque SM, Abduh SBM, Oey I. Carbohydrates. In: Barba FJ, Jorge MAS, Cravotto G, Lorenzo JM, Eds. Innovative Thermal and Non-Thermal Processing, Bioaccessibility and Bioavailability of Nutrients and Bioactive Compounds. Woodhead Publishing 2019; pp. 171-206. [http://dx.doi.org/10.1016/B978-0-12-814174-8.00006-8]

[68] Dastangoo S, Hamed Mosavian MT, Yeganehzad S. Optimization of pulsed electric field conditions for sugar extraction from carrots. Food Sci Nutr 2020; 8(4): 2025-34. [http://dx.doi.org/10.1002/fsn3.1490] [PMID: 32328269]

[69] Dziadek K, Kopeć A, Dróżdż T, *et al.* Effect of pulsed electric field treatment on shelf life and nutritional value of apple juice. J Food Sci Technol 2019; 56(3): 1184-91. [http://dx.doi.org/10.1007/s13197-019-03581-4] [PMID: 30956298]

CHAPTER 13

High-Voltage Electrical Discharges

Ester Hernández-Corroto[1], **María Luisa Marina**[1,2] and **María Concepción García**[1,2,*]

[1] *Universidad de Alcalá, Departamento de Química Analítica, Química Física e Ingeniería Química, Ctra. Madrid-Barcelona Km. 33.600, 28871 Alcalá de Henares (Madrid), Spain*

[2] *Universidad de Alcalá, Instituto de Investigación Química "Andrés M. del Río", Ctra. Madrid-Barcelona Km. 33.600, 28871 Alcalá de Henares (Madrid), Spain*

Abstract: High-Voltage Electrical Discharges (HVED) are considered an emergent extraction technique based on the application of high-pulsed voltages. The aim of this chapter was to review its fundamentals for applications at laboratory and industrial scales. The configuration of devices and employed electrodes is described. Moreover, main steps required for using HVED and most important factors affecting this technique are also highlighted. Extraction of high added-value compounds from food waste and plant matrices using batch HVED has been the most usual application in last five years. In many cases, the low selectivity of the technique has made the use of a solid-liquid extraction step after HVED application necessary.

Keywords: Application, Electrodes, Extraction, Food, High-voltage electrical discharges, Treatment chamber.

INTRODUCTION

High-voltage electrical discharges (HVED) is a non-thermal technology based on the application of high-pulsed voltages within a liquid phase. Main applications of this technique are the preservation of foods and the extraction of compounds. The application of HVED to a liquid phase can lead to the fragmentation of cell membranes and the permeation of cell walls. This fact has been applied for increasing food shelf life by the deactivation of microorganisms, which constitutes an interesting alternative to other thermal methods that can result in sample degradation. Moreover, HVED technique is also an interesting tool for the extraction of compounds by promoting the availability of intracellular compounds [1]. Indeed, HVED enhances mass transfer and increases extraction yields.

* **Corresponding Author María Concepción García:** Universidad de Alcalá, Departamento de Química Analítica, Química Física e Ingeniería Química e Instituto de Investigación Química "Andrés M. del Río", Ctra. Madrid-Barcelona Km. 33.600, 28871 Alcalá de Henares (Madrid), Spain; Tel. +34 918854915 E-mail: concepcion.garcia@uah.es

Merichel Plaza & Maria Luisa Marina (Eds.)

Additionally, it is considered an environmentally friendly technique since it requires low solvent and energy consumptions and short treatment times [2]. Furthermore, HVED is suitable for extracting thermolabile compounds since it can use mild conditions and temperatures.

Nevertheless, this technique also shows some limitations. It can disintegrate food particles, making more difficult the separation of target compounds. Moreover, it is not a selective technique and can generate reactive oxidizing species [3, 4]. In order to overcome these drawbacks, some researchers have suggested the application of electrical discharges at low energy followed by an additional step to promote mass transfer and increase the extraction yield of compounds [5]. Thus, HVED has often been used as a pre-treatment followed by a conventional solid-liquid extraction [6 - 8], enzymatic hydrolysis [4] or other non-conventional techniques such as high-pressure homogenization [9].

This chapter reviews the main designs and theoretical aspects related to this technique and the application of HVED to the sustainable extraction of compounds.

DEVICE CONFIGURATION

HVED can be applied in three different regimes: batch, continuous, and circulating. The design of the treatment chamber, the geometry of electrodes, and the process scale limit the kind of regime. The batch HVED system is the most used when the aim is the extraction of high added-value compounds from food matrices, at the lab or pilot scale. In this case, electrical discharges are generated in a treatment chamber usually constituted by two needle-plate geometry electrodes (point-plane) working in an intermittent way [1]. The continuous HVED system is preferred for industrial applications. In this case, the material and solvent are continuously pumped through a cooling system into a treatment chamber where electrical discharges are applied. After the application of HVED, the treated material enters again into the cooling system. There are two continuous systems with different treatment chambers: the converged electric field system and the annular gap system. The converged electric field system is constituted by a pair of parallel plates electrodes, separated by a small gap where electrical discharges are produced. The annular gap system presents a treatment chamber that consists of two concentric electrodes of different diameters with an annular gap between them for hosting the material to treat [4]. Samples with high amounts of solids could provoke the clogging of the converged electric field system, being the annular gap system the best choice in this case [10]. Finally, the circulating HVED system contains a needle electrode located in the centre of a ring electrode. The material and solvent are introduced into the bore region between electrodes for their treatment. Afterward, the treated material is diffused into an extraction tank, located at the end of the treatment chamber. After extraction, a valve enables the separation of the supernatant from the residues, which can be driven into the treatment chamber again [4]. This system can enhance the extraction yield since the material can be exposed to electrical discharges several times [10].

THEORETICAL FUNDAMENTALS

HVED technology is based on the phenomenon of electrical breakdown in water. The application of high-voltage electrical discharges generates physical processes and chemical reactions that lead to the fragmentation of the cell membrane promoting the extraction of intracellular compounds. It is considered a green extraction technique since it is a non-thermal and low energy technique [11]. The batch extraction system has been the most applied at lab and pilot scale [4]. The mechanism, in this case, involves the generation of the electrical pulse, the streamer propagation and the electrical arc formation, the fragmentation of the raw material, and the extraction of compounds. A brief description of these steps is given below.

Generation of the Electrical Pulse

HVED are generated through a system connected to a pulsed power generator. This generator consists of a high voltage power supply, a high energy storage capacitor, and a high voltage switch (Fig. **1**) [12]. The power source supplies a high voltage and generates electrical energy that is stored in the capacitor. The maximum stored energy in the capacitor depends on its capacitance and can be calculated using the following equation [12]:

$$E_e = \frac{1}{2} C V^2 \quad \textbf{(1)}$$

where E_e is the stored electrical energy (Joules), C is the capacitance (Farads), and V is the charging voltage (V) in the capacitor.

When the capacitor reaches its maximum voltage, the stored energy is transferred towards the switch. The high voltage switch is constituted by two spherical electrodes separated a certain distance. When the voltage reaches a threshold value, an electric arc is formed in the air, and the switch is closed. Then, the electrical energy from the capacitor is transmitted to the chamber electrodes. When the electrical current arrives at the needle electrode, the voltage rises in few seconds (nanoseconds) and the streamer reaches the ground electrode, which leads to the formation of the electrical discharge. After the energy stored in the capacitors is transferred to the chamber, the switch is opened again [12]. This system generates an electrical pulse with every discharge. The repetition of this process increases the number of pulses and, consequently, the energy input. The energy input can be calculated by the following equation [6]:

$$W_{HVED} = \frac{W_i \times n}{m} \quad \textbf{(2)}$$

where W_{HVED} is the energy input (kJ/kg), W_i is the energy per pulse (kJ/pulse), n is the number of pulses, and m is the mass of the material (kg) located within the two electrodes.

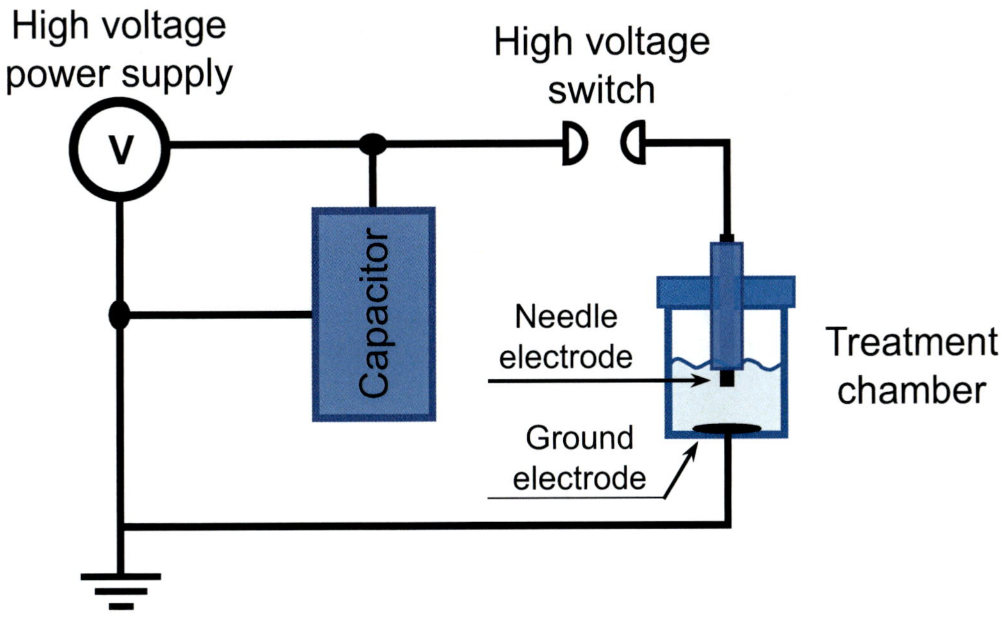

Fig. (1). Scheme of a high voltage pulsed power generator.

Streamer Propagation and Electrical arc Formation

The generation and propagation of the streamer (pre-breakdown phase) and the formation of an electrical arc (breakdown phase) are next steps. The streamer is initiated close to one of the electrodes due to the high local electric field. Then, it is propagated through a plasma channel until the opposite electrode is reached. The electrodes shape can affect the generation and the propagation of the streamer [12]. HVED electrodes can present different geometries: needles [13], needle-plate [6 - 8], needle-ring [14], and wires-plate [15], among others. Needle-plate electrodes are the most common. They have separated a certain distance in order to provide the electrical breakdown. The needle electrode is connected to a high-voltage generator, so the streamer initiates near the needle tip. The streamer is, then, propagated within the plasma channel through different pathways resulting in a heterogeneous field distribution. Finally, the streamer reaches the ground electrode, and an electric arc is formed [1].

The voltage and the electric field intensity of the streamer are limited by the interelectrode distance according to the following equation [16, 17]:

$$E = \frac{V}{d} \qquad \textbf{(3)}$$

where E (V/m) is the electric field intensity, V is the applied voltage (V), and d is the interelectrode distance (m). Based on this equation, if the interelectrode distance is small, the electric field between electrodes will be high and the electrical discharges between electrodes will be formed in a shorter time and will present a higher intensity [12]. However, if the interelectrode distance is large, the intensity of the resulting electric field will be lower. This is because formed plasma channels are longer and promote the dissipation of more energy. Therefore, the optimization of interelectrode distance is a requirement for a suitable breakdown [10].

Fig. (**2**) shows a scheme of the formation of the electrical breakdown in water. There are two different theories for explaining streamer propagation. The first hypothesis states that the streamer starts in a small part of the gap between the two electrodes, specifically in a gas phase, inside a bubble. This bubble can be already present in the water or can result from the boiling of water due to Joule heating and water conductivity [18]. Under an intense electric field, neutral gases, such as oxygen, are ionized creating conducting paths within the liquid medium. Ionization of the neutral gas is due to free electrons present in the gas that are accelerated by the electric force. If the kinetic energy of electrons is greater than the ionization potential of gas molecules, the collision of electrons against them will result in the formation of a positive ion and the release of an electron. If this process is repeated several times, the number of electrons will increase exponentially provoking an electron avalanche that will lead to the formation of a streamer (pre-breakdown phase) [12, 19]. The second hypothesis assumes that the ionization of neutral molecules is directly produced without the requirement of a gas phase [1]. In this case, the electron avalanche is generated due to the collision of free electrons with water molecules.

Once the electrical breakdown is formed, the electrical energy transported by the streamer through plasma channels is dissipated and this results in the formation of an electrical arc when the streamer reaches the ground electrode. As the arc resistance diminishes in a short time, all the electrical energy transferred through the system will be dissipated. Therefore, the electrical current will raise while the voltage will drop quickly. Consequently, the temperature will also increase due to the Joule effect. The change in temperature will produce different secondary phenomena: shock waves, cavitation bubbles, intense light emission, and the formation of radicals, among others [1, 12].

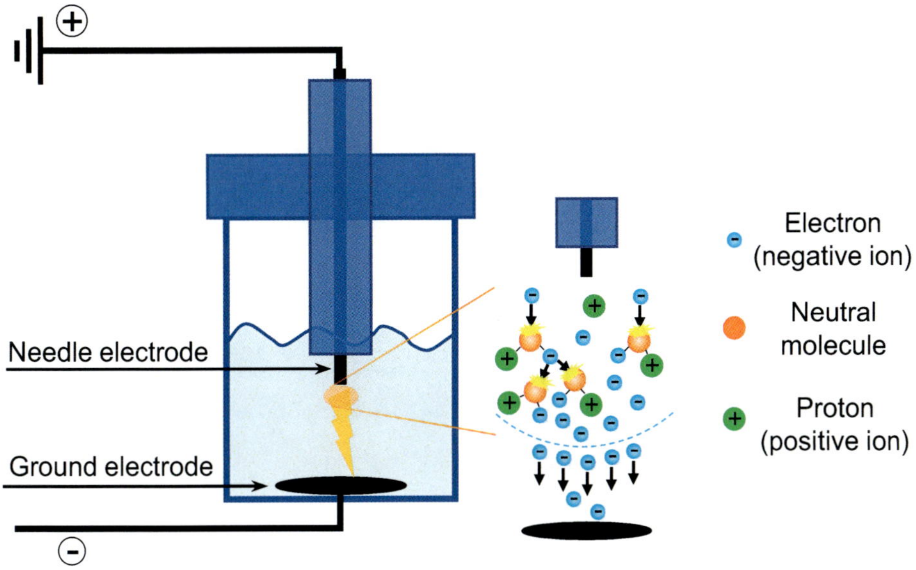

Fig. (2). Scheme of the electrical breakdown in water and the formation of the electron avalanche.

Fragmentation of Raw Material and Extraction of Compounds

The fragmentation of sample particles is produced because of water electrical breakdown. Fig. (**3**) shows the different phenomena generated by the water electrical breakdown that finally results in sample fragmentation.

The raw material is placed between the two electrodes of the chamber and these electrodes are submerged in a liquid medium, usually water. When the electrical energy transported by the streamer is dissipated resulting in the formation of an electrical arc, both temperature and pressure significantly increase in a short time producing powerful shock waves. The generation of shock waves is due to liquid incompressibility. The propagation of these waves contributes to the formation of small gaseous bubbles in the liquid medium due to changes in the liquid temperature and pressure that produce expansion and compression cycles known as cavitation. The expansion of the cavitation bubbles leads to their explosion causing damage of cell walls. Only a small part of the electrical energy released by the formation of the electrical arc (5%) is converted into the acoustic energy while the rest of the energy is turned into intense light. Moreover, these processes lead to strong turbulence in the liquid, which increases the rupture and the permeabilization of cell membranes. Therefore, the combination of shock waves, cavitation bubbles, and liquid turbulence promotes the fragmentation of the sample and the mass transfer from the cells to the solution [1, 12].

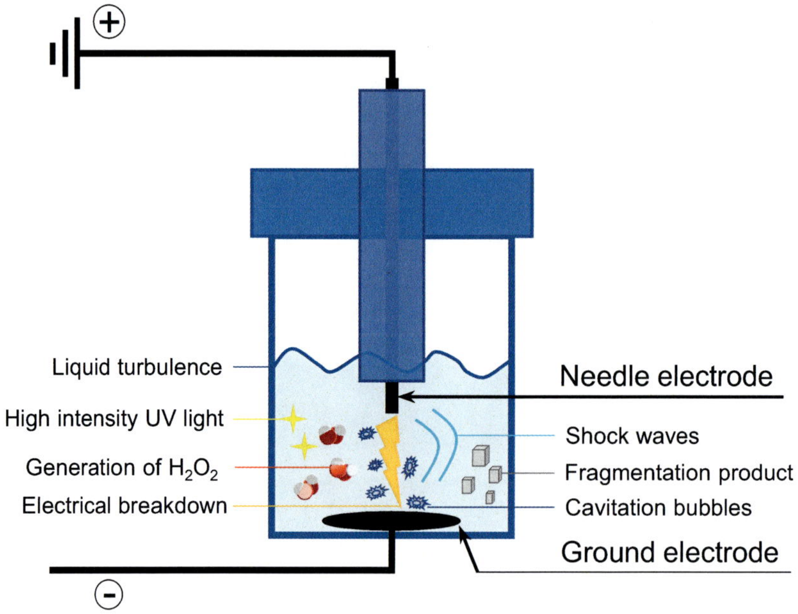

Fig. (3). Phenomena generated by high-voltage electrical discharges.

The release of intracellular compounds by HVED causes an increase in the electrical conductivity of the sample. The performance of HVED treatment can be evaluated from this electrical conductivity through the degree of disintegration index (Z). This index is calculated by the following equation [1, 20]:

$$Z = \frac{\sigma - \sigma_i}{\sigma_f - \sigma_i} \quad (4)$$

where σ is the value of the electrical conductivity, σ_i is the initial electrical conductivity (untreated sample), and σ_f is the final electrical conductivity (maximum cellular damage). A Z value equal to 0 means that the cell tissue is intact and there is no cell damage, while a Z value equal to 1 is expected when the cell tissue is completely damaged.

Different authors have shown that the Z value increases gradually with the energy input or the number of pulses [1, 20]. Nevertheless, a maximum disintegration (Z = 1) is not recommended due to the formation of small particles and the difficulty in the separation of compounds from the cell membrane [21]. In addition, the extraction yield usually decreases due to the high presence of oxidizing agents. Similarly, a degree of fragmentation close to zero is not adequate since the cell tissue is scarcely damaged and the compounds cannot be released. Therefore, an intermediate degree of fragmentation would be the best option to improve the mass transfer from cells to the solution.

Application of HVED can be accompanied by other physical processes and chemical reactions that can hinder the extraction of compounds. The electrical arc emits an intense electromagnetic radiation, which is absorbed by water and that can provoke many chemical reactions. The main products of these reactions are reactive oxidizing agents, especially free radicals, which can degrade some extracted compounds as well as damage cell structure [12].

APPLICATIONS

Despite most applications of HVED are focused on the extraction of compounds from food and plant matrices, another use of this technique is food preservation.

Extraction of Compounds

Extraction with HVED is considered a green alternative to conventional methods that require the use of polluting reagents and solvents and high-energy inputs. Moreover, an additional characteristic of HVED, as compared to other techniques, is its low heating effects that makes its application to the extraction of thermolabile compounds possible.

Plant cells present large vacuoles and thick cell walls that make difficult the extraction of compounds in comparison with animal cells. The application of HVED allows increasing the permeabilization of cell walls and improving mass transfer. This comes along with the formation of shock waves, cavitation bubbles, liquid turbulence, and other phenomena that lead to the fragmentation of matrix particles. All these promote the release of compounds from plant matrices such as proteins, phenolic compounds, pigments, and carbohydrates [4, 22, 23].

HVED has been explored as an extraction technique for different kinds of compounds in the last decade [11]. Different works have reviewed these applications [1, 2, 4, 5, 19, 24]. This chapter just reviews the most recent works reporting the application of this technique to the extraction of compounds. Table **1** collects the experimental conditions employed for this purpose in the last five years. Most works are devoted to the extraction of compounds, mainly bioactive ones, from food waste. Additionally, some works describing the extraction of different target compounds from microalgae and other plant matrices, including non-edible parts of plants, have been carried out .

Table 1. Application of HVED technique to the extraction of compounds from food waste and plant matrices in the last five years.

Material	Extracting Solvent	S-L ratio[a] (w/v)	Experimental Conditions	Extracted Compounds	Extraction Yield	Refs.
Extraction of compounds from food waste						
Pomegranate peels	Hot water	1:10	Generator, 40 kV; needle-plate electrodes (interelectrode distance, 40 mm); electric field strength, 10 kV/cm; energy input, 90-100 kJ/kg; extraction time, 7 min; temperature, 50 °C.	Phenolic compounds	46.0 ± 0.5 mg GAE/ g DM	[25]
	Water	1:35	Generator, 20 kV and 10 kA; plate-plate electrodes (interelectrode distance, 3.1 mm); electric field strength, 29 kV/cm; flow rate of the solution, 12 mL/min; frequency, 100 Hz; voltage, 9 kV; extraction time, 26 min.	Phenolic compounds	197 ± 6 mg/g	[26]
Blueberry pomace	MeOH (50%) and HCl (1%)	1:50	Generator, 30 kV; needle-plate electrodes (interelectrode distance, 5 mm); electric field strength, 60 kV/cm; energy input, 22.27 kJ/kg; frequency, 50 Hz; extraction time, 15 min.	Phenolic compounds	6.58 mg GAE/ g DM	[27]

(Table 1) cont.....

Material	Extracting Solvent	S-L ratio[a] (w/v)	Experimental Conditions	Extracted Compounds	Extraction Yield	Refs.
Grape pomace	EtOH (50%) and HCl (1%)	1:50	Generator, 30 kV; needle-plate electrodes (interelectrode distance, 5 mm); electric field strength, 60 kV/cm; energy input, 22.27 kJ/kg; frequency, 100 Hz; extraction time, 15 min.	Phenolic compounds	30.24 mg GAE/ g DM	[28]
	EtOH (30%)	1:40	Generator, 40 kV; needle-ring electrodes (interelectrode distance, 40 mm); flow rate of the solution, 50 mL/min; frequency, 3 Hz; voltage, 12 kV; extraction time, 6 min.	Phenolic compounds	6.8 ± 0.2%	[14]
Grapefruit peels	Water	1:9	1st step: HVED (generator, 40 kV and 10 kA; needle-plate electrodes (interelectrode distance, 5 mm); energy input, 72 kJ/kg; frequency, 0.5 Hz). 2nd step: solid-liquid diffusion (glycerol (30%), 1:10 (w/v), 160 rpm, 1 h, 50 °C).	Phenolic compounds	19.3 mg GAE/ g DM	[29]
Grape stems	Water	1:7.5	1st step: HVED (generator, 40 kV and 10 kA; needle-plate electrodes (interelectrode distance, 5 mm); energy input, 188 kJ/kg; frequency, 0.5 Hz; number of pulses, 400; treatment duration, 4 ms). 2nd step: solid-liquid diffusion (EtOH (50%), 1:15 (w/v), pH 2.5, 120 min, 20 °C, 160 rpm).	Phenolic compounds	66 ± 2 mg GAE/ g RM	[30]
				Flavan-3-ols*	7.9 ± 0.7 mg/ g RM	
				Flavonols*	0.56 ± 0.05 mg QUE/ g RM	
				Stilbenes*	0.24 ± 0.04 mg RES/ g RM	

(Table 1) cont.....

Material	Extracting Solvent	S-L ratio[a] (w/v)	Experimental Conditions	Extracted Compounds	Extraction Yield	Refs.
Orange peels	Water	1:8	1st step: HVED (generator, 40 kV and 10 kA; needle-plate electrodes (interelectrode distance, 5 mm); energy input, 222 kJ/kg; frequency, 0.5 Hz; temperature, 50 °C). 2nd step: enzymatic hydrolysis (Viscozyme L (12 FBGU/g), 50 °C, pH 4.5, 160 rpm, 180 min)	Phenolic compounds	9 mg/g DM	[31]
				Reducing sugar	500 mg/g DM	
Orange pomace	Water (pH 4.8) and pectinase enzyme (1.5%)	1:20	Generator, 60 kV; wires-plate electrodes; peak voltage, 18 kV; extraction time, 10 min.	Phenolic compounds	618 ± 6 mg/L,	[32]
Mango peel	Water	1:10	Generator, 40 kV and 10 kA; needle-plate electrodes (interelectrode distance, 10 mm); electric field strength, 40 kV/cm; energy input, 160 kJ/kg; number of pulses, 300.	Proteins	0.3 mg/g	[33]
				Phenolic compounds	1 mg GAE/g	
				Carbohydrates	150 mg/g	
Carrot pomace	EtOH	1:10	Generator, 60 kV; wires-plate electrodes (interelectrode distance, 50 mm); peak voltage, 20 kV; extraction time, 20 min.	β-carotene	433 ± 2 mg/L	[15]
				Phenolic compounds	155 ± 8 mg/L	
Sugar beet pulp	Water	1:20	1st step: HVED (generator, 40 kV and 10 kA; needle-plate electrodes (interelectrode distance, 5 mm); energy input, 76.2 kJ/kg; frequency, 0.5 Hz; number of pulses, 100; extraction time, 4 min). 2nd step: solid-liquid extraction (pH 2.0, 1 h, 90 °C).	Pectin	135 mg/g DM	[34]

(Table 1) cont.....

Material	Extracting Solvent	S-L ratio[a] (w/v)	Experimental Conditions	Extracted Compounds	Extraction Yield	Refs.
Mustard seeds	Water	1:5	Generator, 40 kV and 10 kA; needle-plate electrodes (interelectrode distance, 5 mm); energy input, 233 kJ/kg; frequency, 1 Hz; number of pulses, 350; electrical treatment time, 3.5 ms; extraction time, 5.8 min.	Sinigrin	97.2%	[35]
			Generator, 40 kV and 10 kA; needle-plate electrodes (interelectrode distance, 5 mm); frequency, 1 Hz; number of pulses, 500; electrical treatment time, 5 ms; extraction time, 8 min.	Gluconapin	74.8%	
			Generator, 40 kV and 10 kA; needle-plate electrodes (interelectrode distance, 5 mm); energy input, 233 kJ/kg; frequency, 1 Hz; number of pulses, 350; electrical treatment time, 3.5 ms; extraction time, 5.8 min.	Proteins	3.2%	
Cacao shells	Water	1:33	Generator, 30 kV; needle-plate electrodes (interelectrode distance, 20 mm); electric field strength, 15 kV/cm; frequency, 40 Hz; extraction time, 15 min.	Tannins*	17.07 mg TA/ g	[36]
			Generator, 30 kV; needle-plate electrodes (interelectrode distance, 20 mm); frequency, 80 Hz; extraction time, 15 min.	Theobromine*	3.28 ± 0.02 mg/g	[37]
Spent coffee grounds	EtOH (24%)	1:50	Generator, 40 kV; needle-ring electrodes; flow rate of the solution, 200 mL/min; energy input, 43.6 kJ/kg; frequency, 3 Hz; voltage, 11 kV; extraction time, 20 min.	Phenolic compounds	60 ± 2 mg/g	[38]

(Table 1) cont.....

Material	Extracting Solvent	S-L ratio[a] (w/v)	Experimental Conditions	Extracted Compounds	Extraction Yield	Refs.
Peanut shells	EtOH (25%)	1:30	Generator, 20 kV and 10 kA; annular plate electrodes (interelectrode distance, 5 mm); flow rate of the solution, 60 mL/min; frequency, 150 Hz; voltage, 13 kV; extraction time, 5 min.	Flavonoids*	0.95 ± 0.01%	[39]
	EtOH (20%)	1:40	Generator, 20 kV and 10 kA; plate-plate electrodes (interelectrode distance, 5 mm); flow rate of the solution, 10 mL/min; frequency, 150 Hz; voltage, 14 kV; extraction time, 40 min.	Flavonoids*	1.00 ± 0.01%	
Extraction of compounds from microalgae						
Parachlorealla kessleri	Water	1:200	Generator, 40 kV and 10 kA; needle-plate electrodes (interelectrode distance, 10 mm); electric field strength, 40 kV/cm; frequency, 0.5 Hz; number of pulses, 800; electrical treatment time, 8 ms.	Carbohydrates	83 mg/g	[40]
				Proteins	22 mg/g	

(Table 1) cont.....

Material	Extracting Solvent	S-L ratio[a] (w/v)	Experimental Conditions	Extracted Compounds	Extraction Yield	Refs.
Parachlorealla kessleri	Water	1:200	1st step: HVED (generator, 40 kV and 10 kA; needle-plate electrodes (interelectrode distance, 10 mm); electric field strength, 40 kV/cm; frequency, 0.5 Hz; number of pulses, 800; electrical treatment time, 8 ms). 2nd step: HPH (80 MPa, 4 passes, 204.8 kJ/g).	Carbohydrates	151.9 mg/g	[40]
			1st step: HVED (generator, 40 kV and 10 kA; needle-plate electrodes (interelectrode distance, 10 mm); electric field strength, 40 kV/cm; frequency, 0.5 Hz; number of pulses, 600; electrical treatment time, 6 ms). 2nd step: HPH (80 MPa, 4 passes, 169.6 kJ/g).	Proteins	74.1 mg/g	
		1:100	1st step: HVED (generator, 40 kV and 10 kA; needle-plate electrodes (interelectrode distance, 10 mm); electric field strength, 40 kV/cm; frequency, 0.5 Hz; number of pulses, 600; electrical treatment time, 6 ms).	Carbohydrates	22%	[41]
				Proteins	4%	
			2nd step: EtOH extraction (EtOH (95%), 1:20 (w/v), stirring, 24 h, 150 rpm).	Chlorophyll-a	1 mg/g DM	

(Table 1) cont.....

Material	Extracting Solvent	S-L ratio[a] (w/v)	Experimental Conditions	Extracted Compounds	Extraction Yield	Refs.
Phaeodactylum tricornutum	Water	1:100	1st step: HVED (generator, 40 kV and 10 kA; needle-plate electrodes (interelectrode distance, 10 mm); electric field strength, 40 kV/cm; frequency, 1 Hz; number of pulses, 800; electrical treatment time, 8 ms).	Carbohydrates	22 mg/g	[42]
				Proteins	61 mg/g	
			2nd step: aqueous extraction (stirring, 1 h, 150 rpm).	Carbohydrates	24 mg/g	
				Proteins	60 mg/g	
			3rd step: EtOH extraction (EtOH (95%), 1:20 (w/v), stirring, 500 s, 150 rpm).	Chlorophyll-a	1.3 mg/g DM	
			4th step: lipid extraction ($CHCl_3$:MeOH (2:1, v/v), stirring, 6 h, in dark).	Lipids	39 ± 1 mg/g DM	
			1st step: HVED (generator, 40 kV and 10 kA; needle-plate electrodes (interelectrode distance, 10 mm); electric field strength, 40 kV/cm; frequency, 0.5 Hz; number of pulses, 600; electrical treatment time, 6 ms).	Carbohydrates	37%	[41]
				Proteins	5.5%	
			2nd step: EtOH extraction (EtOH (95%), 1:20 (w/v), stirring, 24 h, 150 rpm).	Chlorophyll-a	0.15 mg/g DM	

(Table 1) cont.....

Material	Extracting Solvent	S-L ratio[a] (w/v)	Experimental Conditions	Extracted Compounds	Extraction Yield	Refs.
Nannochloropsis sp.	Water	1:20	1st step: HVED (generator, 40 kV and 10 kA; needle-plate electrodes (interelectrode distance, 10 mm); electric field strength, 40 kV/cm; frequency, 0.5 Hz; number of pulses, 400; electrical treatment time, 4 ms).	Carbohydrates	38 mg/g DM	[43]
				Proteins	33 mg/g DM	
			2nd step: EtOH extraction (EtOH (95%), 1:20 (w/v), stirring, 8 h, 150 rpm).	Chlorophyll-a	1 mg/g DM	
				Carotenoids	0.25 mg/g DM	
			3rd step: lipid extraction ($CHCl_3$:MeOH, 2:1 (v/v), stirring, 6 h, in dark).	Lipids	200 ± 2 mg/g DM	
		1:100	1st step: HVED (generator, 40 kV and 10 kA; needle-plate electrodes (interelectrode distance, 10 mm); electric field strength, 40 kV/cm; frequency, 0.5 Hz; number of pulses, 600; electrical treatment time, 6 ms).	Carbohydrates	38 ± 2%	[41]
				Proteins	3%	
			2nd step: EtOH extraction (EtOH (95%, 1:20, w/v), stirring, 24 h, 150 rpm).	Chlorophyll-a	4.2 mg/g DM	
Gracilaria Lemaneiformis	Water	1:50	Generator, 40 kV; needle-ring electrodes; flow rate of the solution, 250 mL/min; energy input, 343.64 kJ/kg; frequency, 3 Hz; voltage, 13 kV; extraction time, 24 min.	Polysaccharides	164 ± 5 mg/g	[44]
Extraction of compounds from other plant matrices						

(Table 1) cont.....

Material	Extracting Solvent	S-L ratio[a] (w/v)	Experimental Conditions	Extracted Compounds	Extraction Yield	Refs.
Wild thyme	EtOH (50%)	1:50	Generator, 40 kV and 10 kA; needle-needle electrodes (interelectrode distance, 15 mm); flow rate of nitrogen, 0.75 L/min: peak voltage, 20 kV; frequency, 100 Hz; pulse width, 400 ns; extraction time, 9 min.	Phenolic compounds	43 ± 3 mg GAE/g	[45]
Rosemary	EtOH (50%)	1:50	Generator, 25 kV and 30 mA; needle-needle electrodes (interelectrode distance, 15 mm); flow rate of nitrogen, 0.75 L/min; peak voltage, 25 kV; frequency, 100 Hz; pulse width, 400 ns; extraction time, 9 min.	Phenolic compounds	31.64 mg GAE/g	[13]
Dalmatian sage	EtOH (33.16%)	1:50	Generator, 25 kV and 30 mA; needle-needle electrodes (interelectrode distance, 15 mm); flow rate of nitrogen, 5 L/min; peak voltage, 25 kV; frequency, 100 Hz; pulse width, 400 ns; extraction time, 9 min.	Phenolic compounds	57.21 mg GAE/g	[46]
Oregano leaves	EtOH (25%)	1:50	Generator, 25 kV and 30 mA; needle-needle electrodes (interelectrode distance, 15 mm); electric field strength, 20-80 kV/cm; flow rate of nitrogen, 5 L/min; frequency, 100 Hz; pulse width, 400 ns; extraction time, 9 min.	Phenolic compounds	191 ± 7 mg GAE/ g	[47]
Olive Leaves	EtOH (50%)	1:50	Generator, 25 kV and 30 mA; needle-needle electrodes (interelectrode distance, 15 mm); electric field strength, 20-80 kV/cm; flow rate of argon, 5 L/min; frequency, 100 Hz; pulse width, 400 µs; extraction time, 9 min.	Phenolic compounds	65.99 ± 0.06 mg GAE/ g	[48]

(Table 1) cont.....

Material	Extracting Solvent	S-L ratio[a] (w/v)	Experimental Conditions	Extracted Compounds	Extraction Yield	Refs.
Stevia rebaudiana leaves	Water	1:30	Generator, 40 kV and 10 kA; needle-plate electrodes (interelectrode distance, 5 mm); energy input, 178 kJ/kg; frequency, 0.5 Hz; number of pulses, 200; electrical treatment time, 2 ms.	Phenolic compounds	26.2 ± 0.5 mg/g	[49]
	EtOH (50%)	1:30	Generator, 40 kV and 10 kA; needle-plate electrodes (interelectrode distance, 5 mm); energy input, 178 kJ/kg; frequency, 0.5 Hz; number of pulses, 20; electrical treatment time, 2 ms.	Flavonoids*	19.2 ± 0.1 mg/g	
				Carotenoids	5.5 ± 0.2 mg/g	
				Chlorophyll-a	13.4 ± 0.2 mg/g	
Spruce sawdust	NaOH (1M)	1:15	<u>1st step</u>: HVED (generator, 40 kV and 10 kA; needle-plate electrodes (interelectrode distance, 10 mm); electric field strength, 40 kV/cm; frequency, 0.5 Hz; number of pulses, 400; electrical treatment time, 4 ms). <u>2nd step</u>: solid-liquid extraction (stirring, 3 h, 50 °C).	Hemicelluloses	19.78 mg/g	[50]
Vine shoot	Water	1:20	<u>1st step</u>: HVED (generator, 40 kV and 10 kA; needle-plate electrodes (interelectrode distance, 5 mm); energy input, 101.6 kJ/kg; frequency, 0.5 Hz; number of pulses, 200; pulse width, 10 µs). <u>2nd step</u>: enzymatic hydrolysis (Viscozyme L (24 FBGU/g), pH 4.5, 4 h, 50 °C, 160 rpm). <u>3rd step</u>: alkaline hydrolysis (NaOH (1 M), 4 h, 50 °C, 160 rpm).	Phenolic compounds	20 mg GAE/g DM	[51]
				Reducing sugars	110 mg/g DM	
				Lignin	15%	

(Table 1) cont.....

Material	Extracting Solvent	S-L ratio[a] (w/v)	Experimental Conditions	Extracted Compounds	Extraction Yield	Refs.
Rapeseed straw	Water	1:30	1st step: HVED (generator, 40 kV and 10 kA; needle-plate electrodes (interelectrode distance, 5 mm); energy input, 800 kJ/kg; frequency, 0.5 Hz; extraction time, 39 min). 2nd step: alkaline extraction (NaOH (45%), 150 min, 80 °C, 750 rpm). 3rd step: enzymatic hydrolysis (cellulase (1 mg/mL), acetate buffer (50 mM, pH 4.8), 30 min, 40 °C, 175 rpm).	Lignin	29.2%	[52]
Rapeseed straw	Water	1:30		Polysaccharides	85%	[52]
Rapeseed straw	Water	1:30	1st step: HVED (generator, 40 kV and 10 kA; needle-plate electrodes (interelectrode distance, 5 mm); energy input, 800 kJ/kg; frequency, 0.5 Hz; extraction time, 39 min). 2nd step: EtOH extraction (EtOH (65%), H_2SO_4 (1%), 80 min, 200 °C, 750 rpm). 3rd step: enzymatic hydrolysis (cellulase (1 mg/mL), acetate buffer (50 mM, pH 4.8), 30 min, 40 °C, 175 rpm).	Lignin	25%	[52]
Rapeseed straw	Water	1:30		Polysaccharides	96%	[52]

DM: dry matter; RM: raw material; GAE: gallic acid equivalent; QUE: quercetin, RES: resveratrol; TA: tannic acid; HPH: high-pressure homogenization; MeOH: methanol; EtOH: ethanol; FBGU: fungal β-glucanase units. * Specific compound belongs to phenolic compounds group, and it was determined by a specific method, while "phenolic compounds" correspond to a total phenolic compound content determined by the Folin-Ciocalteu method. [a] solid-liquid ratio

Extraction of Compounds from Food Waste

Electrical discharges have been extensively applied to the extraction of intracellular compounds from fruit and vegetable wastes. Operating conditions depend on the extracting solvent, the liquid to solid ratio, the energy input, the extraction time, the geometry of the electrodes, and the interelectrode distance, among others [1, 4, 19]. The extracting solvent affects the electrical conductivity and optimal solvents should promote the production of electrical discharges. However, excessive electrical conductivity is not recommended since it can result in treatment chamber overpressure and, consequently, the complete disintegration of cell tissues. More preferred solvents in HVED treatments are water and ethanol. Water is the most common solvent when target compounds are polar while the addition of organic solvents is usual to enhance the solubility of more hydrophobic compounds. Regarding the liquid to solid ratio, high ratios improve compounds' diffusivity at higher energy consumption. A suitable balance between extractability and energy consumption is reached at liquid to solid ratios from 5 mL/g to 50 mL/g (energy between 22 and 240 kJ/kg, respectively). The extraction time is related to the number of pulses and the energy input. Long extraction times result in the complete destruction of cell tissues and promote the formation of reactive species. In this case, smaller particles are formed and the separation of compounds from the cell membrane is hindered. Additionally, the higher presence of reactive species can decrease the extraction yield due to the degradation of target compounds. Thus, extraction times ranging from 4 to 40 min are usual in HVED treatments.

On the other hand, the efficiency of the electrical discharges is also related to the geometry of electrodes and their distance. Needle-plate electrodes are usually the most common in HVED intermittent extraction systems (batch systems) since they provide a heterogeneous electric field distribution, although wires-plate electrodes have also been used. The optimum separation of electrodes is between 2 and 50 mm. This electrode distance enables the suitable formation of electrical discharges. The HVED technique is more selective for the extraction of small molecules. That is why there is a higher number of works related to the extraction of molecules like phenolic compounds, glucosinolates, and carotenoids. Indeed, as observed in Table **1**, the extraction yield of phenolic compounds varied from 1 to 46 mg/g and from 155 to 618 mg/L, while the extraction yield for carotenoids was 433 mg/mL (corresponding to the carrot pomace) [15] and for glucosinolates, it ranged from 75 to 97% (corresponding to gluconapin and sinigrin, respectively, from mustard seeds) [35]. Nevertheless, other molecules with higher molecular weight were also extracted from food waste using HVED batch systems. Parniakov *et al.* [33] obtained 0.3 mg/g of proteins and 150 mg/g of carbohydrates from mango peel, while Hebert *et al.* [35] extracted only 3.2% of proteins from mustard seeds.

Other systems with different electrodes' geometries, such as continuous and circulating systems, have also been employed for the application of HVED. For example, Xi *et al.* [26] used a "converged electric field type" continuous system. It consisted of a pair of parallel plate electrodes with a small hole between them where electrical discharges were produced. The separation between electrodes was 3.1 mm. This system was applied to the extraction of phenolic compounds from pomegranate peel obtaining an extraction yield of 197 mg/g. Yan *et al.* [39]

also employed a continuous HVED extraction system based on a "converged electric field type" and an "annular gap type" continuous system. In the last case, they used two concentric plates' electrodes with an annular gap between them where electrical discharges were produced. In both systems, the distance between electrodes was 5 mm. The "annular gap type" design resulted in a shorter extraction time and higher efficiency for the extraction of flavonoids from peanut shells than the "converged electric field type". Deng *et al.* [14] also employed a continuous HVED extraction system but, in this case, they used needle-ring type electrodes. The needle electrode was located in the centre of the ring electrode at a distance of 40 mm. In this work, 68 mg/g of phenolic compounds were extracted from grape pomace. The same authors in a previous publication [38], also used needle-ring electrodes but in a circulating regime. In this case, they extracted 60 mg/g of phenolic compounds from spent coffee grounds. In both continuous and circulating HVED extraction systems, the sample flow rate is an additional parameter to be studied since it affects the mass transfer. Flow rates from 10 to 60 mL/min were employed in the continuous HVED system, while values of 200 mL/min were used in the circulating HVED system. In addition, a higher yield in the extraction of phenolic compounds was obtained as compared to the batch system. These HVED modes are very interesting from industrial and economic points of view [10].

HVED technique has also been employed as a pre-treatment followed by a solid-liquid extraction or enzymatic hydrolysis. Indeed, HVED treatment is not selective and, in some cases, a second step is required to extract just target compounds. El Kantar *et al.* [29] reported that the solid-liquid extraction of polyphenols in glycerol after HVED treatment, reduced the HVED pre-treatment energy by 6 times and increased the solubility of polyphenols from grapefruit peels. Brianceau *et al.* [30] established a predictive model based on three levels to extract the highest amount of phenolic compounds from grape stems by using a HVED pre-treatment followed by a solid-liquid extraction. The suitable selection of the pH (2.5–8.5), ethanol concentration (39.8–50%), and pre-treatment duration (0–4 ms) enabled to tune the kind of phenolic compounds that could be extracted: total phenolic compounds, flavan-3-ols, flavonols or stilbenes. El Kantar *et al.* [31] also reported the combination of a HVED pre-treatment with enzymatic hydrolysis for the extraction of polyphenols and fermentable sugars from orange peels. They observed that HVED pre-treatment enhanced the accessibility of enzymes to the substrate, resulting in a higher release of biomolecules. Almohammed *et al.* [34] improved the extraction of pectin from sugar beet pulp by introducing a solid-liquid extraction step with acidified water (pH 2.0) and high temperature (90 °C) after HVED pre-treatment.

Extraction of Compounds from Microalgae

Like in the case of food waste, the most common solvents employed in the extraction of compounds from microalgae by HVED are water and ethanol. Zhang *et al.* [40 - 43] optimized the HVED conditions using a batch extraction system for the recovery of carbohydrates and proteins from three microalgae types (*Parachlorealla kessleri*, *Phaeodactylum tricornutum*, and *Nannochloropsis sp.*). They employed liquid to solid ratios ranging from 20 mL/g to 200 mL/g and applied between 400 and 800 pulses. The treatment chamber contained two electrodes with needle-plate geometry. The electric field was applied in the needle tip and provided a heterogeneous electric field distribution within the 10 mm gap between electrodes. In the case of the microalgae *Parachlorealla kessleri*, HVED treatment could extract 83 mg/g of carbohydrates and 22 mg/g of proteins. Moreover, the combination of HVED with HPH (high-pressure homogenization) increased the extraction efficiency of both compounds to 152 mg/g and 74 mg/g, respectively [40]. In the case of the microalgae *Phaeodactylum tricornutum*, a multi-step procedure was performed for the selective extraction of multiple compounds. More hydrophilic compounds, like carbohydrates and proteins, were extracted during the HVED pre-treatment (1st step) and in the aqueous extraction step (2nd step), while more hydrophobic biomolecules, like pigments and lipids, were extracted during the 3rd (ethanol extraction) and 4th (chloroform:methanol extraction) steps, respectively [42]. A similar procedure was applied in the extraction of biomolecules from *Nannochloropsis sp.* The HVED pre-treatment allowed extracting 38 mg/g of carbohydrates and 33 mg/g of proteins. Next, the ethanol extraction step promoted the extraction of pigments like chlorophyll and carotenoids (1 mg/g DM and 0.25 m/g DM, respectively). The last step (extraction using chloroform:methanol) recovered the lipid fraction from microalgae (200 ± 2 mg/g DM) [43]. Zhang *et al.* [41] also compared the effect of HVED, followed by an ethanol extraction step, on the extraction of different compounds from the three microalgae. The cell wall structure of *P. tricornutum* was more fragile than that of *P. kessleri* and *Nannochloropsis sp.* and, consequently, *P. tricornutum* resulted in larger yields in the extraction of carbohydrates and proteins. This fact seemed to be related to cell wall composition. In fact, *P. tricornutum* is rich in sulfated glucomannan, while *P. kessleri* and *Nannochloropsis sp.* present a higher concentration in cellulose and hemicellulose. Additionally, the concentration of chlorophyll extracted by HVED was also higher for *P. tricornutum*, although the final concentration was the lowest (0.15 mg/g DM).

Ju *et al.* [44] optimized the conditions of a circulating HVED system for the extraction of polysaccharides from *Gracilaria Lemaneiformis*. For that purpose, a water:solid ratio of 50 mL/g and a treatment chamber with two needle-ring geometry electrodes were employed. The circulating process at a sample flow rate of 250 mL/min enhanced the extraction of polysaccharides (164 mg/g) and reduced the extraction time (24 min) in comparison with a two-step batch HVED extraction.

Extraction of Compounds from other Plant Matrices

Nutrizio *et al.* [13, 45, 46] evaluated the extraction of phenolic compounds from three aromatic plants belonging to the *Lamiaceae* family (wild thyme, rosemary, and Dalmatian sage). For that purpose, they used a HVED treatment combined with computational simulation methods. Ethanol was chosen as an extracting solvent due to its high capacity to solubilize hydrophobic compounds and its electrical conductivity. The plant matrix and solvent were introduced in a 15 mm gap between two electrodes in a baker-shaped reactor at a liquid to solid ratio of 50 mL/g. Nitrogen gas was flowed at 0.75 and 5 mL/min through the bottom needle electrode to improve the ionization and the cold plasma formation. The duration of the treatment was 9 min. Under optimal HVED conditions, the highest amount of phenolic compounds was obtained from Dalmatian sage (57.21 mg GAE/g) [46] followed by wild thyme (43 mg GAE/g) [45], and rosemary (31.64 mg GAE/g) [13].

HVED technique has also been useful in the extraction of bioactive molecules from leaves and other non-edible plant parts. Some plant leaves are considered aromatic and medicinal, and others are treated as waste from food processing. Phenolic compounds from oregano or olive leaves were extracted by HVED using ethanol (ethanol:sample ratio of 50 mL/g) in a beaker-shaped reactor constituted by two electrodes with a 15 mm separation [47, 48]. A stream of nitrogen or argon at a flow rate of 5 mL/min was employed to improve compounds' diffusivity. It was possible to obtain extracts with 191 mg GAE/g, from oregano leaves, and 66 mg GAE/g, from olive leaves. On the other hand, *stevia rebaudiana* leaves were also treated by HVED using a chamber with needle-plate geometry electrodes with a 5 mm separation. A lower amount of phenolic compounds was extracted from *stevia rebaudiana* leaves (26 mg/g) than from oregano and olive leaves. Additionally, flavonoids, carotenoids, and chlorophyll-a were extracted by increasing the ethanol concentration [49].

Spruce sawdust contains a high amount of hemicellulose, which is mainly composed of acetylated galactoglucomannan. This polysaccharide presents interesting properties and has been employed for the preparation of biodegradable films. Extraction of hemicellulose was carried out by Chadni *et al.* [50] using HVED followed by solid-liquid extraction with NaOH. The HVED pre-treatment produced the fragmentation of the particles, and the solid-liquid extraction promoted the transference of hemicellulose chains from the wood matrix to the extracellular medium. Electrical discharges were applied between two needle-plate geometry electrodes with a separation of 10 mm, and, after 400 pulses, the extracted compounds were left to diffuse for 3 h. The use of NaOH enabled it to keep a basic environment and, thus, avoid the depolymerization of hemicellulose due to the decrease in pH produced by the release of carboxylic acids resulting in the application of electrical discharges.

Another plant matrix that has been treated with HVED is vine shoot. Rajha *et al.* [51] reported a multistep extraction process for the recovery of phenolic compounds, reducing sugars, and lignin from this material. The process consisted of a HVED pre-treatment followed by enzymatic hydrolysis and alkaline hydrolysis. HVED treatment was carried out using two needle-plate electrodes that applied 200 discharges on a water:vine shoot mixture at a 20 mL/g ratio. Next, the enzymatic hydrolysis promoted the extraction of reducing sugars while a final delignification step was carried out by alkaline hydrolysis. Results showed that HVED treatment favored the extraction of phenolic compounds, reducing sugars, and lignin as compared to the control experiment consisting of enzymatic hydrolysis. Moreover, HVED also improved the later delignification process carried out by alkaline hydrolysis.

Rapeseed straw is a residue mainly employed for bioethanol production. Nevertheless, processing of rapeseed straw is hindered by the presence of lignin-carbohydrate complexes. Brahim *et al.* [52] evaluated the effect of HVED and ultrasound treatments to increase sugars accessibility for bioethanol production. While ultrasounds promoted hemicelluloses extractability, HVED induced the partial degradation of cellulose.

Food Preservation

HVED technique has also been employed for the preservation of food and the decontamination of food products by microbial inactivation [53]. HVED can lead to the generation of hydrogen peroxide, free radicals, and other reactive species that can cause the inactivation of microorganisms [54]. For example, the application of electrical discharges (30 kV, 40 Hz, 25 ºC for 3 min) to a tangerine juice allowed the complete disintegration of *Escherichia coli* [55]. Moreover, the addition of a cupper solution could even accelerate the inactivation of these bacteria [56]. However, this does not have to be a permanent situation since reactive oxygen species can react with macromolecules like proteins with capability to repair oxidized compounds and, thus, promote the recovery of cells. Likewise, treated cells could mutate and adapt to stress conditions [54].

Other work demonstrated the reduction of the presence of *Salmonella enterica* in coconut water by a treatment with HVED. The reduction of *Salmonella* provoked a change in the pH due to the generation of reactive nitrogen species, which generated nitrous acids causing a pH decrease. Nevertheless, this pH reduction was less significant than the observed when adding citric acid for food preservation. Again, this is not a permanent situation since the presence of nutrients like phosphates and magnesium can provide a mechanism for repairing damaged cells [57].

On the other hand, reactive oxygen species generated by the treatment of water with HVED can oxidize organic compounds and remove organic impurities from water [58]. Moreover, HVED can also remove other polluting compounds like heavy metals, polycyclic aromatic hydrocarbons (PAHs), mycotoxins, and microorganisms present in the cocoa shell. The inactivation of these compounds is a key step for the revalorization of this cocoa residue rich in dietary fiber, proteins, polyphenols, *etc* [59]. Furthermore, acrylamide and 5-hydroxymethylfurfural are two undesirable products from Maillard's reaction that could be carcinogenic. Barišić *et al.* [60] reported the reduction of the content of both compounds from cocoa shells by using a HVED treatment.

However, the formation of free radicals can also deteriorate the quality of products through the oxidation of compounds. Jambrak *et al.* [61] showed that a HVED treatment at 90 Hz for 10 min with an argon flow of 2 L/min caused the highest amount of hydroxyl radicals in water. Moreover, the number of radicals increased when the samples were stored at room temperature for 24 h as compared to samples stored at 4 ºC.

The application of HVED can also lead to the modification of food physicochemical and organoleptic properties. Indeed, HVED can lead to the deterioration of phenolic compounds resulting in a change in the color and flavor of the sample [62]. Lukić *et al.* [62] evaluated the effect of HVED on red and white wines. They employed two needle electrodes and an argon flow of 4 L/min. They reported that some parameters related to the color slightly decreased with the duration and frequency of the treatment, but these changes were not perceived in any wine. On the other hand, phenolic compounds also slightly decreased with the treatment duration and frequency. This decrease could be related to the degradation caused by shock waves, cavitation, light emissions, and reactive species generated by HVED.

Buniowska *et al.* [63] studied the changes in bioactive compounds induced by the treatment of a fruit juice (mixture of papaya, mango, and stevia infusion) with HVED. They employed a treatment chamber with electrodes having a needle-plate geometry at two different energy inputs (32 kJ/kg and 256 kJ/kg). They did not observe any significant difference in carotenoids, phenolic compounds, and anthocyanins contents at an energy input of 32 kJ/kg. However, the use of a higher energy input (256 kJ/kg) resulted in a reduced carotenoid content and a slight increase in phenolic compound content. Moreover, they also observed that HVED improved the bioaccessibility of bioactive compounds.

CONCLUDING REMARKS

The use of HVED on food and plant matrices has been investigated in the last few years. The main application of HVED is the extraction of compounds from vegetal food waste and plant samples due to the difficulty in the extraction of intracellular compounds from these matrices. HVED promotes the breakage of vegetal cell membranes and makes easier the extraction of compounds from the inner part of cells. It is considered a sustainable technology since it uses green solvents and requires low energy and solvent consumption. HVED can be used in batch, continuous or circulating regimes. The configuration and geometry of electrodes and treatment chamber, along with the scale of the process, is established by the employed regime. The application of HVED is accompanied by different physical processes (formation of shock waves, cavitation bubbles, and turbulence) and chemical reactions (formation of reactive oxygen species). Physical processes promote cell permeability and induce the release of intracellular compounds. The formation of reactive oxygen species is considered detrimental and can result in compounds degradation. Therefore, the application of HVED requires a suitable optimization of parameters such as the solvent type, liquid to solid ratio, energy input, geometry of electrodes, and interelectrode distance. Since it is not a selective technique, it is usual the combination of HVED with a subsequent solid-liquid extraction. Despite small molecules, mainly polyphenols, vitamins, and pigments, have been the most extracted by HVED, the technique has also demonstrated its suitability for the extraction of larger molecules like proteins and carbohydrates. HVED has also been used as an alternative to thermal processes to reduce food-borne pathogens.

LIST OF ABBREVIATIONS

C	Capacitance
d	Interelectrode distance
DM	Dry matter
E	Electric field intensity
E_e	Electrical energy
EtOH	Ethanol
FBGU	Fungal β-glucanase units
GAE	Gallic acid equivalent
HPH	High-pressure homogenization

HVED	High-voltage electrical discharges
m	Mass of the suspension
MeOH	Methanol
n	Number of pulses
PAHs	Polycyclic aromatic hydrocarbons
QUE	Quercetin
RES	Resveratrol
RM	Raw material
S-L ratio	Solid-liquid ratio
TA	Tannic acid
V	Voltage
W_{HVED}	Energy input
W_i	Energy per pulse
Z	Degree of disintegration index
σ	Electrical conductivity
σ_f	Final electrical conductivity (maximum cellular damage)
σ_i	Initial electrical conductivity (untreated sample)

CONSENT FOR PUBLICATION

Not applicable.

CONFLICT OF INTEREST

The authors declare no conflict of interest, financial or otherwise.

ACKNOWLEDGEMENTS

Authors thank the Spanish Ministry of Science and Innovation (PID2020-114891RB-I00) and the Comunidad of Madrid (Spain) and European funding from FSE and FEDER programs (S2018/BAA-4393, AVANSECAL-II-CM).

REFERENCES

[1] Boussetta N, Vorobiev E. Extraction of valuable biocompounds assisted by high voltage electrical discharges: A review. C R Chim 2014; 17(3): 197-203. [http://dx.doi.org/10.1016/j.crci.2013.11.011]

[2] Barba FJ, Zhu Z, Koubaa M, Sant'Ana AS, Orlien V. Green alternative methods for the extraction of antioxidant bioactive compounds from winery wastes and by-products: A review. Trends Food Sci Technol 2016; 49: 96-109. [http://dx.doi.org/10.1016/j.tifs.2016.01.006]

[3] Dragomir MCB, Zeca ED, Ivan AS, Stoica M. Pulsed electric field and high voltage electrical

discharge-innovative food electrotechnologies. A review. J Agroaliment Proc Technol 2020; 26: 34-9.

[4] Li Z, Fan Y, Xi J. Recent advances in high voltage electric discharge extraction of bioactive ingredients from plant materials. Food Chem 2019; 277: 246-60.
[http://dx.doi.org/10.1016/j.foodchem.2018.10.119] [PMID: 30502141]

[5] Puértolas E, Barba FJ. Electrotechnologies applied to valorization of by-products from food industry: Main findings, energy and economic cost of their industrialization. Food Bioprod Process 2016; 100: 172-84.
[http://dx.doi.org/10.1016/j.fbp.2016.06.020]

[6] Sarkis JR, Boussetta N, Blouet C, Tessaro IC, Marczak LDF, Vorobiev E. Effect of pulsed electric fields and high voltage electrical discharges on polyphenol and protein extraction from sesame cake. Innov Food Sci Emerg Technol 2015; 29: 170-7.
[http://dx.doi.org/10.1016/j.ifset.2015.02.011]

[7] Roselló-Soto E, Barba FJ, Parniakov O, *et al.* High Voltage Electrical Discharges, Pulsed Electric Field, and Ultrasound Assisted Extraction of Protein and Phenolic Compounds from Olive Kernel. Food Bioprocess Technol 2015; 8(4): 885-94.
[http://dx.doi.org/10.1007/s11947-014-1456-x]

[8] Rajha HN, Boussetta N, Louka N, Maroun RG, Vorobiev E. A comparative study of physical pretreatments for the extraction of polyphenols and proteins from vine shoots. Food Res Int 2014; 65: 462-8.
[http://dx.doi.org/10.1016/j.foodres.2014.04.024]

[9] Shynkaryk MV, Lebovka NI, Lanoisellé JL, Nonus M, Bedel-Clotour C, Vorobiev E. Electrically-assisted extraction of bio-products using high pressure disruption of yeast cells (Saccharomyces cerevisiae). J Food Eng 2009; 92(2): 189-95.
[http://dx.doi.org/10.1016/j.jfoodeng.2008.10.041]

[10] Rodríguez De Luna SL, Ramírez-Garza RE, Serna Saldívar SO. Environmentally friendly methods for flavonoid extraction from plant material: Impact of their operating conditions on yield and antioxidant properties. ScientificWorldJournal 2020; 2020: 1-38.
[http://dx.doi.org/10.1155/2020/6792069] [PMID: 32908461]

[11] Sarkis JR, Boussetta N, Vorobiev E. Application of pulsed electric energies for oil and polyphenol extraction from sesame cake and sesame seeds.Handbook of Electroporation. Switzerland: Springer 2017; pp. 2699-712.
[http://dx.doi.org/10.1007/978-3-319-32886-7_119]

[12] Adda P, Boussetta N, Lesaint O, Vorobiev E. High-voltage electrical discharge-assisted extraction of phenolic compounds from grape seeds.Handbook of Electroporation. Switzerland: Springer 2017; pp. 2799-824.
[http://dx.doi.org/10.1007/978-3-319-32886-7_154]

[13] Nutrizio M, Gajdoš Kljusurić J, Marijanović Z, *et al.* The Potential of High Voltage Discharges for Green Solvent Extraction of Bioactive Compounds and Aromas from Rosemary (*Rosmarinus officinalis* L.)—Computational Simulation and Experimental Methods. Molecules 2020; 25(16): 3711-41.
[http://dx.doi.org/10.3390/molecules25163711] [PMID: 32823941]

[14] Deng Y, Ju T, Xi J. Optimization of Continuous Extraction of Polyphenols from Grape Pomace by a Pulsed Electrical Discharge System with a "Needle-Ring Type" Treatment Chamber. ACS Sustain Chem& Eng 2019; 7(10): 9342-51.
[http://dx.doi.org/10.1021/acssuschemeng.9b00480]

[15] Salehi L, Taghian Dinani S. Application of electrohydrodynamic-ultrasonic procedure for extraction of β-carotene from carrot pomace. J Food Meas Charact 2020; 14(6): 3031-9.
[http://dx.doi.org/10.1007/s11694-020-00542-w]

[16] Grémy-Gros C, Lanoisellé JL, Vorobiev E. Application of high-voltage electrical discharges for the

aqueous extraction from oilseeds and other plants.Electrotechnologies for extraction from food plants and biomaterials. New York: Springer 2009; pp. 217-35.
[http://dx.doi.org/10.1007/978-0-387-79374-0_8]

[17] Boussetta N, Lebovka N, Vorobiev E, Adenier H, Bedel-Cloutour C, Lanoisellé JL. Electrically assisted extraction of soluble matter from chardonnay grape skins for polyphenol recovery. J Agric Food Chem 2009; 57(4): 1491-7.
[http://dx.doi.org/10.1021/jf802579x] [PMID: 19173604]

[18] Vorobiev E, Lebovka N. Enhanced extraction from solid foods and biosuspensions by pulsed electrical energy. Food Eng Rev 2010; 2(2): 95-108.
[http://dx.doi.org/10.1007/s12393-010-9021-5]

[19] Dalvi-Isfahan M, Hamdami N, Le-Bail A, Xanthakis E. The principles of high voltage electric field and its application in food processing: A review. Food Res Int 2016; 89(1): 48-62.
[http://dx.doi.org/10.1016/j.foodres.2016.09.002] [PMID: 28460942]

[20] Rajha HN, Boussetta N, Louka N, Maroun RG, Vorobiev E. Pulsed electric fields and high-voltage electrical discharge-assisted extraction of biocompounds from vine shoots.Handbook of Electroporation. Switzerland: Springer 2017; pp. 2683-98.
[http://dx.doi.org/10.1007/978-3-319-32886-7_118]

[21] Barba FJ, Puértolas E, Brnčić M, *et al.* Emerging extraction.Food waste recovery. Academic Press 2015; pp. 249-72.
[http://dx.doi.org/10.1016/B978-0-12-800351-0.00011-0]

[22] Maroun RG, Rajha HN, El Darra N, *et al.* Emerging technologies for the extraction of polyphenols from natural sources.Polyphenols: Properties, recovery, and applications. Woodhead Publishing 2018; pp. 265-93.

[23] Maroun RG, Rajha HN, Vorobiev E, Louka N. Emerging technologies for the recovery of valuable compounds from grape processing by-products.Handbook of Grape Processing By-Products. Academic Press 2017; pp. 155-81.
[http://dx.doi.org/10.1016/B978-0-12-809870-7.00007-7]

[24] Rocha CMR, Genisheva Z, Ferreira-Santos P, *et al.* Electric field-based technologies for valorization of bioresources. Bioresour Technol 2018; 254: 325-39.
[http://dx.doi.org/10.1016/j.biortech.2018.01.068] [PMID: 29395742]

[25] Rajha HN, Abi-Khattar AM, El Kantar S, *et al.* Comparison of aqueous extraction efficiency and biological activities of polyphenols from pomegranate peels assisted by infrared, ultrasound, pulsed electric fields and high-voltage electrical discharges. Innov Food Sci Emerg Technol 2019; 58: 102212-20.
[http://dx.doi.org/10.1016/j.ifset.2019.102212]

[26] Xi J, He L, Yan L. Continuous extraction of phenolic compounds from pomegranate peel using high voltage electrical discharge. Food Chem 2017; 230: 354-61.
[http://dx.doi.org/10.1016/j.foodchem.2017.03.072] [PMID: 28407922]

[27] Lončarić A, Celeiro M, Jozinović A, *et al.* Green Extraction Methods for Extraction of Polyphenolic Compounds from Blueberry Pomace. Foods 2020; 9(11): 1521-42.
[http://dx.doi.org/10.3390/foods9111521] [PMID: 33113955]

[28] Lončarić A, Jozinović A, Kovač T, Kojić N, Babić J, Šubarić D. High voltage electrical discharges and ultrasound-assisted extraction of phenolics from indigenous fungus-resistant grape by-product. Pol J Food Nutr Sci 2020; 70: 101-11.
[http://dx.doi.org/10.31883/pjfns/117716]

[29] El Kantar S, Rajha HN, Boussetta N, Vorobiev E, Maroun RG, Louka N. Green extraction of polyphenols from grapefruit peels using high voltage electrical discharges, deep eutectic solvents and aqueous glycerol. Food Chem 2019; 295: 165-71.
[http://dx.doi.org/10.1016/j.foodchem.2019.05.111] [PMID: 31174746]

[30] Brianceau S, Turk M, Vitrac X, Vorobiev E. High voltage electric discharges assisted extraction of phenolic compounds from grape stems: Effect of processing parameters on flavan-3-ols, flavonols and stilbenes recovery. Innov Food Sci Emerg Technol 2016; 35: 67-74. [http://dx.doi.org/10.1016/j.ifset.2016.04.006]

[31] El Kantar S, Boussetta N, Rajha HN, Maroun RG, Louka N, Vorobiev E. High voltage electrical discharges combined with enzymatic hydrolysis for extraction of polyphenols and fermentable sugars from orange peels. Food Res Int 2018; 107: 755-62. [http://dx.doi.org/10.1016/j.foodres.2018.01.070] [PMID: 29580544]

[32] Shahram H, Taghian Dinani S. Influences of electrohydrodynamic time and voltage on extraction of phenolic compounds from orange pomace. Lebensm Wiss Technol 2019; 111: 23-30. [http://dx.doi.org/10.1016/j.lwt.2019.05.002]

[33] Parniakov O, Barba FJ, Grimi N, Lebovka N, Vorobiev E. Extraction assisted by pulsed electric energy as a potential tool for green and sustainable recovery of nutritionally valuable compounds from mango peels. Food Chem 2016; 192: 842-8. [http://dx.doi.org/10.1016/j.foodchem.2015.07.096] [PMID: 26304419]

[34] Almohammed F, Koubaa M, Khelfa A, Nakaya M, Mhemdi H, Vorobiev E. Pectin recovery from sugar beet pulp enhanced by high-voltage electrical discharges. Food Bioprod Process 2017; 103: 95-103. [http://dx.doi.org/10.1016/j.fbp.2017.03.005]

[35] Hebert M, Mhemdi H, Vorobiev E. Selective and eco-friendly recovery of glucosinolates from mustard seeds (Brassica juncea) using process optimization and innovative pretreatment (high voltage electrical discharges). Food Bioprod Process 2020; 124: 11-23. [http://dx.doi.org/10.1016/j.fbp.2020.04.009]

[36] Barišić V, Flanjak I, Kopjar M, *et al.* Does high voltage electrical discharge treatment induce changes in tannin and fiber properties of cocoa shell? Foods 2020; 9(6): 810-24. [http://dx.doi.org/10.3390/foods9060810] [PMID: 32575607]

[37] Barišić V, Flanjak I, Križić I, *et al.* Impact of high-voltage electric discharge treatment on cocoa shell phenolic components and methylxanthines. J Food Process Eng 2020; 43(1): e13057-66. [http://dx.doi.org/10.1111/jfpe.13057]

[38] Deng Y, Ju T, Xi J. Circulating polyphenols extraction system with high-voltage electrical discharge: Design and performance evaluation. ACS Sustain Chem& Eng 2018; 6(11): 15402-10. [http://dx.doi.org/10.1021/acssuschemeng.8b03827]

[39] Yan LG, Deng Y, Ju T, Wu K, Xi J. Continuous high voltage electrical discharge extraction of flavonoids from peanut shells based on "annular gap type" treatment chamber. Food Chem 2018; 256: 350-7. [http://dx.doi.org/10.1016/j.foodchem.2018.02.129] [PMID: 29606459]

[40] Zhang R, Marchal L, Vorobiev E, Grimi N. Effect of combined pulsed electric energy and high pressure homogenization on selective and energy efficient extraction of bio-molecules from microalga *Parachlorella kessleri*. Lebensm Wiss Technol 2021; 141: 110901-7. [http://dx.doi.org/10.1016/j.lwt.2021.110901]

[41] Zhang R, Lebovka N, Marchal L, Vorobiev E, Grimi N. Pulsed electric energy and ultrasonication assisted green solvent extraction of bio-molecules from different microalgal species. Innov Food Sci Emerg Technol 2020; 62: 102358-64. [http://dx.doi.org/10.1016/j.ifset.2020.102358]

[42] Zhang R, Lebovka N, Marchal L, Vorobiev E, Grimi N. Multistage aqueous and non-aqueous extraction of bio-molecules from microalga *Phaeodactylum tricornutum*. Innov Food Sci Emerg Technol 2020; 62: 102367-73. [http://dx.doi.org/10.1016/j.ifset.2020.102367]

[43] Zhang R, Marchal L, Lebovka N, Vorobiev E, Grimi N. Two-step procedure for selective recovery of bio-molecules from microalga *Nannochloropsis oculata* assisted by high voltage electrical discharges. Bioresour Technol 2020; 302: 122893-900.
[http://dx.doi.org/10.1016/j.biortech.2020.122893] [PMID: 32018087]

[44] Ju T, Deng Y, Xi J. Optimization of circulating extraction of polysaccharides from *Gracilaria lemaneiformis* using pulsed electrical discharge. ACS Sustain Chem& Eng 2019; 7(3): 3593-601.
[http://dx.doi.org/10.1021/acssuschemeng.8b06183]

[45] Nutrizio M, Pataro G, Carullo D, *et al.* High Voltage Electrical Discharges as an Alternative Extraction Process of Phenolic and Volatile Compounds from Wild Thyme (*Thymus serpyllum* L.): In Silico and Experimental Approaches for Solubility Assessment. Molecules 2020; 25(18): 4131-57.
[http://dx.doi.org/10.3390/molecules25184131] [PMID: 32927598]

[46] Nutrizio M, Gajdoš Kljusurić J, Badanjak Sabolović M, *et al.* Valorization of sage extracts (*Salvia officinalis* L.) obtained by high voltage electrical discharges: Process control and antioxidant properties. Innov Food Sci Emerg Technol 2020; 60: 102284-95.
[http://dx.doi.org/10.1016/j.ifset.2019.102284]

[47] Nutrizio M, Maltar-Strmečki N, Chemat F, Duić B, Jambrak AR. High-Voltage Electrical Discharges in Green Extractions of Bioactives from Oregano Leaves (*Origanum vulgare* L.) Using Water and Ethanol as Green Solvents Assessed by Theoretical and Experimental Procedures. Food Eng Rev 2021; 13(1): 161-74.
[http://dx.doi.org/10.1007/s12393-020-09231-2]

[48] Žuntar I, Putnik P, Bursać Kovačević D, *et al.* Phenolic and antioxidant analysis of olive leaves extracts (*Olea europaea* L.) obtained by high voltage electrical discharges (HVED). Foods 2019; 8(7): 248-76.
[http://dx.doi.org/10.3390/foods8070248] [PMID: 31288471]

[49] Carbonell-Capella JM, Šic Žlabur J, Rimac Brnčić S, *et al.* Electrotechnologies, microwaves, and ultrasounds combined with binary mixtures of ethanol and water to extract steviol glycosides and antioxidant compounds from *Stevia rebaudiana* leaves. J Food Process Preserv 2017; 41(5): e13179-86.
[http://dx.doi.org/10.1111/jfpp.13179]

[50] Chadni M, Grimi N, Ziegler-Devin I, Brosse N, Bals O. High voltage electric discharges treatment for high molecular weight hemicelluloses extraction from spruce. Carbohydr Polym 2019; 222: 115019-26.
[http://dx.doi.org/10.1016/j.carbpol.2019.115019] [PMID: 31320084]

[51] Rajha HN, El Kantar S, Afif C, *et al.* Selective multistage extraction process of biomolecules from vine shoots by a combination of biological, chemical, and physical treatments. C R Chim 2018; 21(6): 581-9.
[http://dx.doi.org/10.1016/j.crci.2018.02.013]

[52] Brahim M, Checa Fernandez BL, Regnier O, *et al.* Impact of ultrasounds and high voltage electrical discharges on physico-chemical properties of rapeseed straw's lignin and pulps. Bioresour Technol 2017; 237: 11-9.
[http://dx.doi.org/10.1016/j.biortech.2017.04.003] [PMID: 28411049]

[53] Zuckerman H, Krasik YE, Felsteiner J. Inactivation of microorganisms using pulsed high-current underwater discharges. Innov Food Sci Emerg Technol 2002; 3(4): 329-36.
[http://dx.doi.org/10.1016/S1466-8564(02)00052-8]

[54] Stulić V, Vukušić T, Jambrak AR, *et al.* Quantitative microbial assessment for *Escherichia coli* after treatment by high voltage gas phase plasma. Innov Food Sci Emerg Technol 2019; 53: 26-35.
[http://dx.doi.org/10.1016/j.ifset.2018.08.007]

[55] Yannam SK, Estifaee P, Rogers S, Thagard SM. Application of high voltage electrical discharge plasma for the inactivation of *Escherichia coli* ATCC 700891 in tangerine juice. Lebensm Wiss

Technol 2018; 90: 180-5.
[http://dx.doi.org/10.1016/j.lwt.2017.12.018]

[56] Su X, Feng M, Rogers S, Holsen TM, Thagard SM. The role of high voltage electrode material in the inactivation of *E. coli* by direct-in-liquid electrical discharge plasma. Plasma Chem Plasma Process 2019; 39(3): 577-96.
[http://dx.doi.org/10.1007/s11090-019-09980-x]

[57] Mahnot NK, Mahanta CL, Keener KM, Misra NN. Strategy to achieve a 5-log Salmonella inactivation in tender coconut water using high voltage atmospheric cold plasma (HVACP). Food Chem 2019; 284: 303-11.
[http://dx.doi.org/10.1016/j.foodchem.2019.01.084] [PMID: 30744862]

[58] Sugiarto AT, Sato M. Pulsed plasma processing of organic compounds in aqueous solution. Thin Solid Films 2001; 386(2): 295-9.
[http://dx.doi.org/10.1016/S0040-6090(00)01669-2]

[59] Barišić V, Jozinović A, Flanjak I, *et al.* Difficulties with use of cocoa bean shell in food production and high voltage electrical discharge as a possible solution. Sustainability (Basel) 2020; 12(10): 3981-91.
[http://dx.doi.org/10.3390/su12103981]

[60] Barišić V, Flanjak I, Tot A, *et al.* 5-Hydroxymethylfurfural and acrylamide content of cocoa shell treated with high voltage electrical discharge. Food Control 2020; 110: 107043-51.
[http://dx.doi.org/10.1016/j.foodcont.2019.107043]

[61] Režek Jambrak A, Ojha S, Šeremet D, *et al.* Free radical detection in water after processing by means of high voltage electrical discharges and high power ultrasound. J Food Process Preserv 2021; 45(2): e15176-90.
[http://dx.doi.org/10.1111/jfpp.15176]

[62] Lukić K, Vukušić T, Tomašević M, Ćurko N, Gracin L, Kovačević Ganić K. The impact of high voltage electrical discharge plasma on the chromatic characteristics and phenolic composition of red and white wines. Innov Food Sci Emerg Technol 2019; 53: 70-7.
[http://dx.doi.org/10.1016/j.ifset.2017.11.004]

[63] Buniowska M, Carbonell-Capella JM, Frigola A, Esteve MJ. Bioaccessibility of bioactive compounds after non-thermal processing of an exotic fruit juice blend sweetened with Stevia rebaudiana. Food Chem 2017; 221: 1834-42.
[http://dx.doi.org/10.1016/j.foodchem.2016.10.093] [PMID: 27979170]

CHAPTER 14

High Hydrostatic Pressure: A Green Extraction Technique for Food and Pharmaceutical Industries

Ergin Murat Altuner[1,*]

[1] *Department of Biology, Faculty of Science, Kastamonu University, Kastamonu, Turkey*

Abstract: High Hydrostatic Pressure (HHP) is a green extraction method, which finds several uses in different branches of science. HHP is a novel non-thermal technique mostly used in food processing. The "high pressure" in HHP states an ultra-high cold isostatic hydraulic pressure, which processes basically at low or mild process temperatures (<45 °C) ranging between 100 and 800 MPa. In some applications, this pressure can extend up to 1000 MPa. In food processing, there are several purposes for using HHP, such as sterilizing, coagulating, and gelatinizing food samples. Alternatively, HHP has many remarkable uses in some branches of science besides food processing. This chapter aims to present the capabilities of HHP as a green extraction technique in the food and pharmaceutical industries.

Keywords: Extraction, HHP, High hydrostatic pressure, UHP, Ultra-high pressure.

INTRODUCTION

High hydrostatic pressure (HHP) or alternatively known as ultra-high pressure (UHP), is one of the novel food processing techniques. It is a non-thermal food processing method that has become popular, especially in the last three decades, due to its wide application potential. As a non-thermal process, it has minor effects on both the quality and nutritional characteristics of processed food [1]. The "high pressure" in HHP states an ultra-high cold isostatic hydraulic pressure [2], which processes basically at low or mild process temperatures (<45 °C) ranging between 100 and 800 MPa. In some applications, this pressure can extend up to 1000 MPa [3, 4].

In HHP processing, water is generally the medium to transfer pressure. As the pressurizing process starts, pressure acts immediately on the whole sample being processed. Thus, HHP is a quicker process than thermal food processing techniq-

* **Corresponding Author Ergin Murat Altuner:** Department of Biology, Faculty of Science, Kastamonu University, Kastamonu, Turkey; Tel: + 90 366 280 1914, Fax: +90 366 215 4969; E-mail: ergin.murat.altuner@gmail.com

Merichel Plaza & Maria Luisa Marina (Eds.)

ues, which require a longer time to act on the whole food [5, 6]. In addition, using HHP in food processing does not require to use heat which can damage proteins, enzymes, etc [7]. On the other hand, HHP can cause cellular deformation, membrane damage, and protein denaturation [8 - 11].

In food processing, there are several purposes for using HHP, such as sterilizing, coagulating, and gelatinizing food samples [12]. In addition, some new potential uses of HHP in processing food are under research by several researchers. Alternatively, HHP has many remarkable uses in some branches of science besides food processing [7, 12 - 20].

This chapter aims to present the capabilities of HHP as a green extraction technique in the food and pharmaceutical industries.

The Definition of Pressure

The particles in a closed container hit the inner walls of this thermodynamic system and create a force on the walls of the container. The total net force on the container's wall is known as pressure [21].

In thermodynamics, pressure, which is quite important, is a characteristic element of a system. It is affected by volume, temperature, entropy, and the number of particles.

Pressure (*p* or *P*) is the amount of force that acts on a unit area (Fig. **1**).

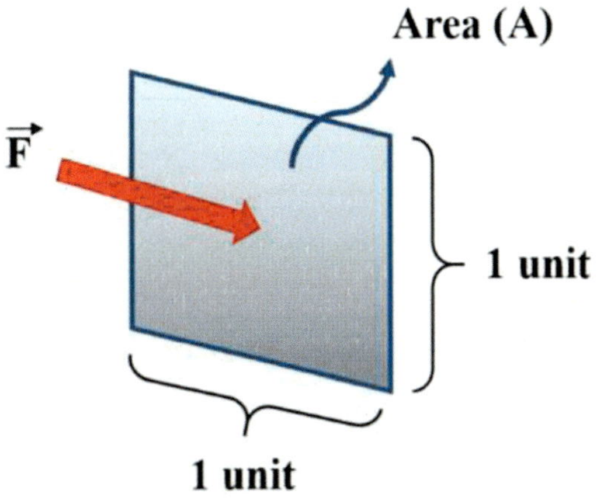

Fig. (1). Definition of pressure.

The following formula (1) shows how to express pressure mathematically.

$$p = \frac{\vec{F}}{A} \tag{1}$$

In the above equation, "*p*" stands for pressure, "*F*" for force, and "*A*" for surface area.

The official SI unit of pressure is the Pascal (Pa). It is possible to convert Pascal to other pressure units as follows [22].

1 Pa = 10^{-5} bar = 1 N/m^2

Since 1 atmosphere (atm) equals 1.01325 bar, the pressure unit Pascal represents a low pressure. For this reason, the megapascal (MPa) is the pressure unit in HHP studies.

1 MPa = 10^6 Pa

Table **1** shows the conversion of different pressure units.

Table 1. Conversion of different pressure units (adapted from Rivalain *et al.* [22]).

-	Atm	Bar	MPa	PSI*
Atm	1	0.987	9.901	0.068
Bar	1.013	1	10.000	0.069
MPa	0.101	0.100	1	0.00689
PSI	14.696	14.504	145.038	1

* PSI: pounds/inch^2

What is Hydrostatic Pressure?

Fluid systems are either static or dynamic. Thus, a fluid system can be either defined as hydrostatic or hydrodynamic. In a hydrostatic system, the fluid does not move. Therefore, it is called a static system.

In a static system, the pressure can be isostatic or non-isostatic. Isostatic pressure has the same value at any point in a system and acts in all directions. But a pressure gradient is present in the non-isostatic type. The structure of the pressure-generating equipment and the non-homogeneous compressibility of the medium are the main factors in generating pressure gradients [22].

Using HHP in food processing has several benefits. In HHP processing, the applied pressure can be conducted evenly and in all directions independent of the shape of the food (Fig. **2**) [4].

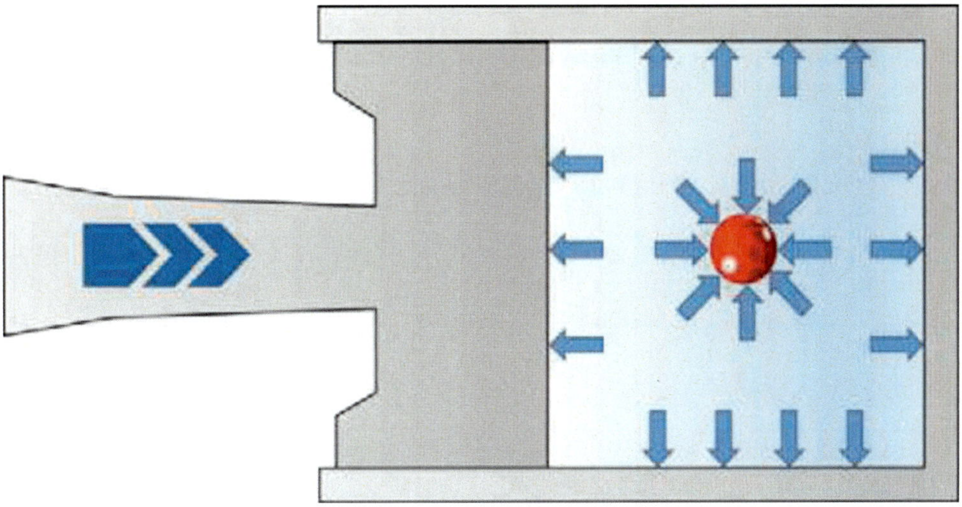

Fig. (2). Isostatic hydraulic pressure conducted evenly and in all directions.

A BRIEF HISTORY OF HHP

Regnard [23], Royer [24], and Hite [25] firstly introduced the idea of using pressure in biology and related sciences (Fig. **3**).

Especially between 1910 and 1955, research groups of Bridgman in the USA [26, 27] and Basset in France [28 - 30] concentrated on the uses of HHP. The studies of these groups encouraged the development of HHP applications.

Another discovery of HHP is synthesizing diamonds by high-pressure application [31, 32].

In 1965, Timson and Short [33] applied pressure on raw milk. They presented how high pressure affects, especially bacteria naturally found in milk.

In 1970, Gould and Sale [34] observed especially bacterial spore germination, during high-pressure application.

By applying pressure as a sterilization parameter, in 1974 Wilson successfully sterilized foods having low pH [35].

In 1976, Marquis showed a change in microbial physiology under HHP up to 100 MPa of pressure [36].

In 1977, Charm *et al.* [37] used pressure to keep foods refrigerated for a long term.

Marquis and Matsumara [38] presented the response of biological systems under the same pressure in 1978.

Later, several researchers discovered how to use and apply high pressure in geology, geochemistry, chemistry, and physics.

Between 1980 and 1990, scientists rediscovered the use of HHP in Japan and applied this technique mainly in food processing. In the last three decades, more researchers have been interested in HHP and researched the applications of HHP in biology, biochemistry, molecular biology, biotechnology, and bioengineering [22].

In 1981, Morild [39] revealed how high pressure may affect the enzymatic activity in foods. Right after this study, in 1982, Heremans [40] showed that proteins and some other biomolecules could be affected during high-pressure applications.

In 1989, Hoover *et al.* [41] proved that HHP processing affected microorganisms found in food.

In 1990, two separate studies applied high-pressure homogenization as food processing technology [42, 43].

Due to the increasing demand for introducing new food processing technologies, between 1989 and 1992, several researchers conducted experiments, especially in Japan [44 - 48]. These studies provided new data on the mechanisms involved in the inactivation of food microorganisms in novel food processing technologies such as HHP [49 - 56].

Progress in the application of HHP in different sciences has led to the establishment of an international association to enable scientists and engineers to share information.

The first industrial application of HHP was the production of NH_3 from N_2 and H_2 due to a volume decrease [57].

In 1993, the Japanese food industry released the first food product processed by HHP to the market in Japan [22].

Some research groups proposed HHP as an effective tool for extracting several pharmaceutical ingredients from plants [11, 58, 59]. They proved that high-pressure extraction (HPE) is a fast tool for extracting active ingredients from plants and is more effective than further extraction techniques [20, 60, 61].

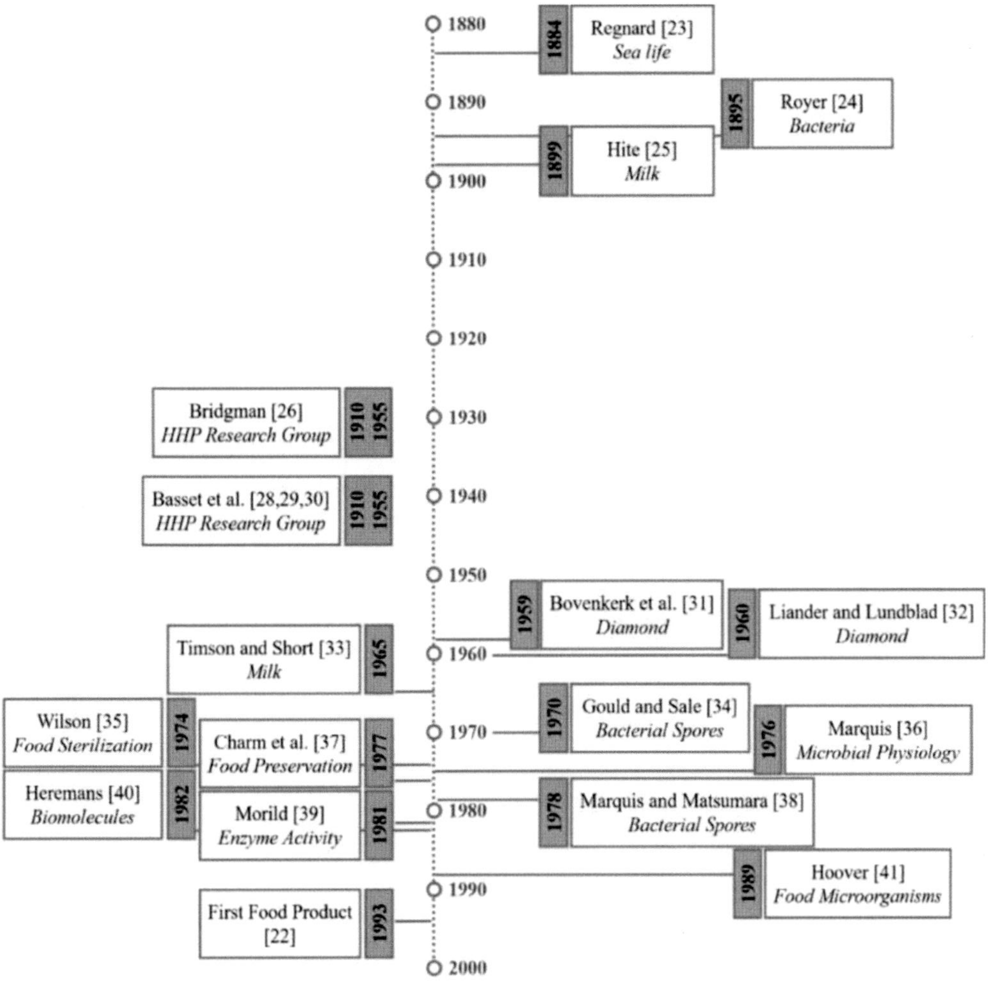

Fig. (3). A brief timeline of HHP history (1884 - 1993).

Previously, several studies have posted that HPE has advantages, such as reducing the duration of the processing and obtaining an improved extraction efficiency [11, 59, 60, 62 - 64]. As a result, they proposed HPE as an advanced extraction technique having some advantages in natural product extraction [65, 66]. Moreover, there are several studies in the literature showing that this method is promising in obtaining flavonoids from propolis [11, 63], anthocyanins from grape skin [63], and ginsenosides from *Panax ginseng* roots [67].

THE PRINCIPLES OF HHP PROCESSING

HHP is one of the food processing methodologies used for food products, for which heat pasteurization adversely affects their quality. But due to the compressibility differences between water and air, the food may not return to its original shape after HHP application. Because as pressure increases, it causes a decrease in volume.

The compression in HHP can be either direct or indirect. A basic HHP processing device consists of a high-pressure chamber, a pressure transmitting medium, a pressure generating device, a pressure intensifier, and a temperature controlling system.

In direct compression (Fig. **4**), a piston takes a role in pressurizing the pressure transmitting medium. This compression type is fast but has limitations. As a reason, it is mostly used in laboratory-scale systems [68].

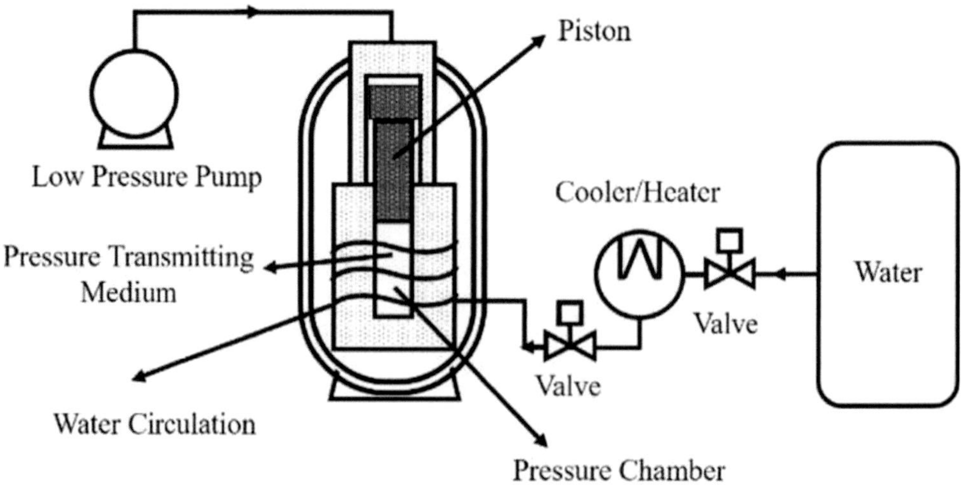

Fig. (4). Direct compression system.

In indirect compression (Fig. **5**), a high-pressure intensifier takes a role in pressurizing the pressure transmitting medium in a reservoir and transfers this pressure into a chamber [68].

For cold HHP processing, the compression system should contain a cooling device. The cooling system decreases the internal temperature both for cold processing and controlling the temperature rise caused by the pressure-temperature effect. While applying pressure, the volume of the food product decreases as a function of pressure but returns the initial value upon decompression.

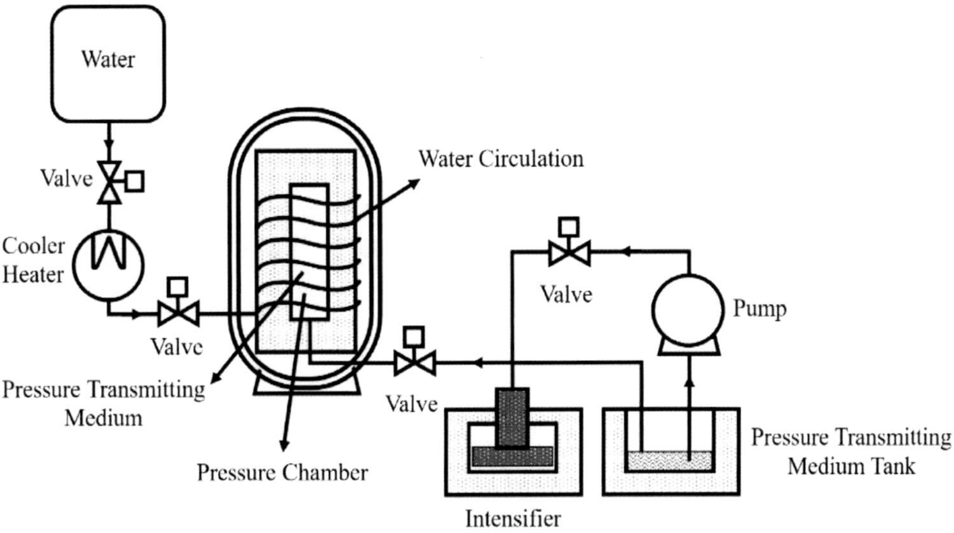

Fig. (5). Indirect compression system.

The alteration in volume leads to a modification in the temperature, which is called the pressure-temperature effect. Due to the pressure-temperature effect, for every 100 MPa pressure, the temperature of water increases by 3 °C [68 - 73].

THE USE OF HHP IN THE FOOD INDUSTRY

HHP preserves most foods with minimal effect on their nutritional value, texture, and taste [2]. In addition, it can inactivate microorganisms without the need for a heating process, but while inactivating microorganisms, it does not generate energy to break down the covalent bonds. Thus, HHP doesn't affect the quality of food while extending the shelf-life of the food without a need for excessive heating processes or chemical preservatives [68]. HHP also has the potential of inactivating food enzymes without changing the freshness and flavor of food [74].

HHP is suitable to process liquid or solid foods containing moisture. There are some pros and cons (Table **2**) of using HHP in food processing [75].

HHP has applications in the pasteurization or sterilization of vegetables, meats, fruits, salad dressings, yogurts, yogurt-based dressings, sauces, jams, fruit juices, seafood, milk, and many ready-to-eat foods [76]. HHP is suitable for the decontamination of food products containing heat-sensitive ingredients such as vitamins [75].

Table 2. Pros and cons of using HHP in food processing.

Pros	Cons
Low processing cost per batch	Relatively expensive investment
Inactivating microorganisms at low temperatures	Food should contain at least 40% moisture to inactivate microorganisms
Low processing time	Limited batch processing
Reduced need for chemical preservatives	Limited packaging options
Preserves food quality	Limited effect on food enzymes
No record of its toxicity	

HHP preserves both the sensory and nutritional characteristics of food, but the foremost reason for its use is related to food safety issues. Previous studies have shown that HHP can inactivate many types of microorganisms, which is one of the reasons for increasing food safety. But spores are resistant to pressure application. Thus, scientists recommend using HHP in combination with other food processing techniques to prevent spore germination [77].

Previous studies proved that several parameters contribute to the inactivation of microorganisms in HHP processed foods (Table **3**).

Table 3. Parameters important in the inactivation of microorganisms in HHP processing (adapted from San Martin *et al.* [77]).

Parameters	Comments
The intensity of pressure	Mostly between 100 and 800 MPa or even higher
Pressure application type	Constant, pulsed or multi-pulsed
Application temperature	Mostly between 5 and 30 °C
Application time	Mostly between 5 and 15 minutes
Microorganism in concern	Each species or strain reacts differently according to its shape and Gram type
Concentration of microorganism	May increase the rate of survival
Growth stage	Whether the microorganism in concern is at lag, log (exponential), stationary, or death phase

(Table 3) cont.....

Parameters	Comments
Age of culture	Closely related to the growth stage
Medium properties	Such as ionic strength, a_w (water activity), pH, and type of ions

Patterson and Kilpatrick [78] proved the effect of pressure intensity in the HHP processing. They applied 500 and 600 MPa of pressure on poultry meat containing *Staphylococcus aureus* and kept the time (15 minutes) and temperature (55°C) constant. As a result, they observed that the inactivation of *S. aureus* drastically increased from 2.0 log cycles to 6.0 log cycles. In addition, Yuste *et al.* [79] presented how pressure intensity affects the inactivation of *Listeria innocua* CECT 910 in HHP processing.

Palou *et al.* [80] showed how pressure application type could influence the inactivation in HHP processing. They used *Zygossaccharomyces bailii* and applied 276 MPa pressure for 15 minutes continuous and 15 minutes (5 minutes cycles) as pulsed pressure. According to the results obtained from their study, they observed higher inactivation in pulsed pressure applications.

Pandya *et al.* [81] tried the effect of temperature in the HHP processing. In HHP processing, they applied 253 MPa pressure for 10 minutes at 25°C and 45°C. Accordingly, they saw a change in inactivation from 3 log cycles to 7 log cycles for *Saccharomyces cerevisiae* in spaghetti sauce.

Benito *et al.* [82] proved that time is a valid parameter affecting the inactivation of *Escherichia coli* O157 in phosphate buffer saline (PBS). They applied 200 MPa of pressure and changed the application time to 2, 6, and 8 minutes. As a result, they observed that the *E. coli* O157 inactivation increased from 3.7 log cycles to 5.2 log cycles.

There are many studies in the literature showing how microorganism species and strains can affect HHP processing. For example, Patterson *et al.* [69] tested how microorganism species and strain affect the HHP processing. They selected *Salmonella typhimurium*, *S. aureus*, *L. monocytogenes* NCTC 2433, and *Listeria monocytogenes* NCTC 11994 and applied different pressures for 15 minutes at 20 °C. They observed that the difference in microorganisms was an important parameter.

Garcia-Graells *et al.* [83] showed how medium properties could affect the inactivation in HHP processing. They used *E. coli* MG1655 and applied 500 MPa pressure for 15 minutes at 20 °C. They only changed the medium and used milk, phosphate buffer, skim milk, and skim milk with nisin. As a result, they observed that the *E. coli* inactivation rate was different in a different medium. The lowest value of inactivation was in skim milk and the highest was in phosphate buffer. In addition, McClements *et al.* [84] also showed the effect of medium properties in HHP processing on *Listeria innocua* CECT 910.

On the other hand, Berlin *et al.* [85] showed how time and pressure together affect *Vibrio vulnificus* in artificial seawater. Consequently, they proposed that an increase in pressure (from 200 MPa to 250 MPa) and time (from 5 to 15 minutes) dramatically increased the inactivation of *V. vulnificus* from 1.0 to 6.9 log cycles.

In addition, Linton *et al.* [86] examined the combined effect of time and temperature on *E. coli* O157:H7 during the treatment of orange juice with HHP. They applied 550 MPa of pressure and changed the time (1, 2, 5 minutes) and temperature (20, 30 °C). As a result, they observed a change in the inactivation of *E. coli* ranging between 4 and 7 log cycles.

Gervilla *et al.* [87] tested the combined effect of pressure (400 and 500 MPa), time (15 min), and temperature (2, 10, and 25 °C) to inactivate *S. aureus* in sheep milk. They observed that the inactivation of *S. aureus* was 1.2 log cycles for 400 MPa, 15 min, and 2 °C. The inactivation increased to 3.4 log cycles for 500 MPa, 15 min, and 25 °C.

THE USE OF HHP IN EXTRACTION

Secondary Metabolites in Food and Pharmaceutical Industries

Food additives increase food quality by altering color, texture, and taste. They also improve the shelf-life of food [88]. Phytochemicals have several applications in the food industry to improve quality. Phytochemicals are either synthesized through primary or secondary metabolism. Primary metabolites are metabolites that take part in the physiological functions of plants, such as reproduction, respiration, development, and growth. Carbohydrates, proteins, lipids, nucleic acid, and vitamins are examples of primary metabolites. Eventually, these metabolites are identical in all organisms [89]. On the other hand, secondary metabolites take roles in the plants' adaptation to their environment [90]. Secondary metabolites have great importance in protecting plants from UV radiation and radicals, and they are also responsible for defending plants against herbivores, microorganisms, and viruses [88].

Secondary metabolites are also signal agents attracting pollinators and seed-dispersing animals. These agents provide symbiotic interactions between plants and microorganisms [91, 92]. Primary metabolites and primary metabolism are the basis for producing secondary metabolites.

Plant secondary metabolites have four major classes conforming to their chemical structures. These are alkaloids, phenolic compounds, sulfur-containing compounds, and terpenes [93].

Alkaloids are typically odorless and colorless crystals, however, occasionally they are yellowish liquids. Mostly, they have a bitter flavor [94]. Alkaloids contain a nitrogen atom in a heterocyclic ring [89]. Most of the alkaloids are toxic and protect plants from other organisms [95].

Phenolic compounds contain phenol rings and one or more OH groups, and they are critical for the quality of plant-based foods. Phenolic compounds are responsible for the flavor and color of red fruits, juices, and wines [96]. Flavonoids, phenolic acids, tannins, stilbenes, anthocyanins, xanthines, and lignans are the most abundant phenolic compounds. These compounds have several roles in plants, such as protecting against UV radiation, and defense against herbivores, parasites, and microorganisms, besides having pharmacological activities [97].

Sulfur-containing compounds are uncommon secondary metabolites. They are involved in pest-plant interactions [98]. Terpenes are simple hydrocarbons, and terpenoids are a modified class of terpenes. Terpenoids have sub-classes such as monoterpenes, sesquiterpenes, diterpenes, sesterpens, and triterpenes [99]. Most of the compounds used as flavors, fragrances, pharmaceuticals, and industrial compounds are terpenoids [100, 101].

Some of the secondary metabolites are plant pigments, some take a significant role in the taste and color of most plants, and some others have antimicrobial or antioxidant activities [89]. Plant secondary metabolites have several applications in the food industry owing to these properties.

Scientists and engineers extract secondary metabolites directly from plant samples to use in scientific research and industrial applications.

Extraction Methods

Extraction is a process used to separate a compound of interest from a matrix, such as plant tissues [102]. Extraction processes have applications such as sample preparations, scientific research, and qualitative analysis of bioactive compounds [103].

The extraction process progresses in four steps. Firstly, the extraction solvent enters the plant cells. Secondly, the solvent dissolves the compounds present in the cell. Thirdly, the compounds diffuse out of the cells, and lastly, these compounds are collected [104].

The chemical properties of the compounds determine the extraction method, especially the extraction solvent. Scientists advise using water for the extraction of water-soluble compounds and organic solvents for water-insoluble compounds [102].

There are three critical stages in all extraction methods:

1. Grinding or homogenizing the ruptured cells in the extraction solvent,

2. Allowing enough time for the extraction solvent to penetrate ruptured cells at a temperature, which does not cause a degradation in the active compounds, and

3. Filtrating the slurry to remove the raw plant materials [105].

Conventional Extraction Methods

Maceration

Maceration is one of the traditional extraction methods in which ground plant samples are placed into an extraction solvent in an air-tight container and left for extraction at room temperature. The maceration process takes at least three days, and the container is agitated frequently during this time. The next step in the extraction process is discarding the pressed plant material and filtering the extract. Maceration is simple, but the extraction process takes a long time, and the extraction efficiency is low [104].

Infusion

Infusion is another traditional extraction process used to extract compounds from plant samples by macerating them in a shorter time. In this extraction process, the extraction solvent can be water (cold or boiling), oil, or alcohol [102].

Percolation

Percolation is a process to prepare tinctures and fluid extracts. In this process, the extraction solvent is strained through plant samples by a filter [106].

Decoction

The decoction is a method of extracting compounds by boiling hard plant parts of plant samples such as barks and roots. After cooling the mixture, it is possible to filter the extract to remove plant samples. The decoction is suitable to extract water-soluble and heat-resistant compounds [107].

Heat Reflux Extraction

Heat reflux extraction (HRE) is a type of extraction where the solution is continuously evaporated and condensed at a constant temperature. This extraction technique is capable of extracting heat-resistant compounds [108, 109].

Soxhlet Extraction

Soxhlet extraction is a widely preferred continuous extraction technique used to extract compounds from plant samples [110]. This technique incorporates the benefits of percolation and reflux extraction. The extraction efficiency is high, and it requires less solvent and time than percolation and maceration. This extraction technique can also extract heat-resistant compounds [104].

Alternative Approaches in Extraction

Supercritical Fluid Extraction

Supercritical fluid extraction (SFE) is an effective method that is an alternative to traditional extraction techniques [111]. The extraction solvent in SFE is a supercritical fluid (SF). The advantage of SF is demonstrating similar diffusivity to gas and similar solubility to liquid. SF has the capability of dissolving a wide variety of natural compounds. The commonly used SF in SFE is supercritical carbon dioxide (S-CO_2). It is selective, inert, cheap, and non-toxic. The polarity of S-CO_2 is low, which is critical to extract non-polar compounds [104].

Ultrasound-Assisted Extraction

Ultrasound-assisted extraction (UAE) is an alternative method with an ultrasound frequency ranging between 50 and 2000 kHz [102]. Previous studies showed that UAE, which leads to cell disruption, increases extraction efficiency [112]. As a result of cell disruption, the extraction is faster and more reproducible. UAE is a non-thermal extraction method. Thus, it is suitable to extract heat-sensitive compounds [113].

Microwave-Assisted Extraction

Microwave-assisted extraction (MAE) combines microwaves with a traditional solvent extraction system [114]. MAE is preferred because it is rapid and efficient. In addition, it enhances the penetration of extraction solvents into plant samples as well [115]. Microwave produces heat by interacting with water in plant samples, accelerating extraction, and increasing extraction efficiency [104, 116, 117].

Pulsed Electric Field Extraction

The pulsed electric field (PEF) method accelerates the extraction efficiency and decreases the application time because it increases mass transfer by damaging membranes. Since PEF is a non-thermal extraction method, it is convenient to extract heat-sensitive compounds [104].

Pressure-Assisted Extraction

The pressurized liquid extraction (PLE) method uses high pressure to extract compounds. Many studies showed that PLE decreases the extraction time, temperature needs, and amount of solvent. This extraction method has preferable replicability in contrast to other extraction methods [104].

The Parameters Affecting Extraction Efficiency

The extraction solvent properties, the size of plant samples, the solvent-to-solid ratio, the temperature of extraction, process duration, and pressure are critical factors affecting extraction efficiency [104].

The properties of extraction solvent such as polarity, selective solubility, volatility, and toxicity affect extraction efficiency.

Nawaz *et al.* [118] showed that the polarity of extraction solvent affects the extraction yield. In addition to polarity, Lin and Giusti [119] proved that the acidity of extraction solvent changes the recovery of isoflavones from soybean.

Selective solubility is one of the properties of extraction solvents, which is closely related to polarity. Ramluckan *et al.* [120] revealed that polarities change the selective solubility of solvents and selective solubility directly affects the extraction efficiency. They explained the effect of solubility on extraction yield by a well-known phrase; "*like dissolves like*". The extraction solvent must dissolve the target compound free of impurities.

The volatility of the extraction solvent also affects the extraction yield. Volatility is the tendency of the solvent to evaporate efficiently at low temperatures. Holser and Abbott [121] showed that volatility is a critical parameter for an extraction solvent. Ncube *et al.* [122] and Gurjar *et al.* [123] presented that the extraction solvent should be non-toxic because the end product in extraction may contain traces of residual solvent.

The size of plant samples is another parameter affecting extraction efficiency. Zhang *et al.* [104] proposed that the size of the plant changes the surface area interacting with the extraction solvent. Therefore, the increased surface area increases the efficiency of extraction.

Zhang *et al.* [104] also suggested that the higher the solvent-to-solid ratio, the higher the extraction efficiency. However, if the solvent-to-solid ratio is high, it will need additional time to concentrate the extract.

According to Zhang *et al.* [104], the efficiency of extraction increases when process duration increases in a specific time range. Conversely, several studies in the literature show that the concentration of extracted secondary metabolites decreased in a longer process duration. Bae *et al.* [124] showed that increasing extraction time decreased the flavonoid concentration in the pepper. Perva-Uzunalic *et al.* [125] also proved that prolonged process time decreased the catechin concentration in green tea (*Camellia sinensis*).

Doran [126] explained that the effective diffusivity value depends on the pore structure of the solid and shows the penetration tendency of a solvent into a solid matrix. Cacace and Mazza [127] defined effective diffusivity by using the effective diffusion coefficient for the mass transfer process while extracting phenolic compounds from milled berries. Turker and Erdogdu [128] proposed that the effective diffusion coefficient changes in responding to both temperature and pressure. Therefore, HHP processing is an effective candidate that can be used in the extraction of secondary metabolites by controlling pressure which means changing the effective diffusivity.

HHP as a Tool in Extraction

Previous sections have explained how controlling pressure can improve extraction efficiency. Based on the pressure effect on extraction efficiency, scientists proposed to use HHP as an alternative tool in extraction. In addition, Food and Drug Administration (FDA) recognized HHP as an eco-friendly technique [129, 130].

Xi *et al.* [131] extracted and characterized polyphenols from green tea leaves by using HHP. They investigated the effect of several parameters, such as extraction solvents (acetone, methanol (MeOH), ethanol (EtOH), and water), process duration (1, 4, 7, and 10 minutes), ethanol concentration (between 0 and 100% v/v), solid-liquid ratio (between 1:10 and 1:25 g/mL), and pressure (100, 200, 300, 400, 500, and 600 MPa) on the extraction efficiency. They also compared the extraction efficiency to some other traditional extraction methods, such as extraction at room temperature [131], UAE [132], and HRE [132, 133]. As a result of the study, they observed that increasing pressure from 100 MPa to 600 MPa also increased the extraction yield from 15 ± 1.4 to 30 ± 1.3%. Based on these results, they proposed that increasing pressure could increase extraction efficiency. They also showed that the optimal conditions for HHP processing were 1:20 (g/mL) solid-liquid ratio, 50% (v/v) EtOH concentration, and 500 MPa HHP application for 1 minute. Under optimal conditions, the extraction yield for tea polyphenols was observed to be 30 ± 1.3%.

Sanchez-Moreno *et al.* [134] tested how high pressure affects the extraction of carotenoids from tomato puree (*Lycopersicum esculentum*). They applied pressure ranging between 50 and 400 MPa. As a result, they also proposed that increasing pressure accelerated the extraction efficiency of carotenoids.

De Ancos *et al.* [135] also used HHP in the extraction of β-cryptoxanthin, lutein, zeaxanthin, α-carotene, and β-carotene, which are carotenoids, from orange juice. They tested the effect of different pressures (between 50 and 350 MPa), temperatures (30 and 60°C), and application times (2.5, 5, and 15 minutes). They observed that 350 MPa of pressure application at 30 °C for 5 minutes led to a remarkable improvement in the carotenoid concentration. But increasing application time over 5 minutes and temperature higher than 30 °C did not enhance the carotenoid amount.

Furthermore, Plaza *et al.* [136] observed the effect of pressure (200 and 400 MPa), temperature (25 °C), and application time (1, 3, and 6 minutes). They proved that 200 MPa pressure application at 25 °C for 6 minutes increased the carotenoid and vitamin A concentration in astringent persimmon (*Diospyros kaki*). However, there was no meaningful alteration in carotenoid concentration for 400 MPa, which conversely caused a decrease in non-astringent compounds.

In another study, Plaza *et al.* [137] explored how HHP (400 MPa at 40 °C for 1 minute) and PEF affected the bioactive compounds of orange juice and compared the results with low pasteurization. They demonstrated that HHP caused a serious raise in flavanone and total carotenoid content, but they did not observe any changes in PEF and low pasteurization. As a result, they put forward that flavanones and carotenoids were extracted in higher concentrations in HHP processing than in PEF and low pasteurization.

Altuner *et al.* [20] extracted phenolic compounds from *Maclura pomifera* fruits. They used methanol and a solvent cocktail containing water, EtOH, MeOH, acetone, and dichloromethane with a ratio of 1:2.5:2.5:2:2 as the extraction solvent. They compared the results of HHP processing against two other extraction methods, namely shaking at room temperature for 2 hours and Soxhlet extraction for 14 hours. The HHP processing conditions were 250 MPa and 500 MPa for 10 minutes at room temperature. Therefore, they discovered the best

phenolic compound concentration in HHP extraction at 500 MPa by using the solvent cocktail.

Lee *et al.* [138] extracted ginsenosides from ginseng (*Panax ginseng*) by using HHP extraction. In their study, they used four different pressures (0.1, 30, 50, and 80 MPa) and compared the results to the heat extraction method. They observed the maximum ginsenoside amount in 80 MPa and proposed that increasing pressure accelerates the concentration of ginsenosides.

Zhang *et al.* [67] also applied HHP (500 MPa for 2 minutes) and observed an increase in the extraction efficiency of ginsenoside in ginseng EtOH extracts. In addition, they compared the extraction efficiency of HHP with conventional water cooking, heat reflux, and UAE. They proved that HHP showed much higher extraction efficiency.

Furthermore, Chen *et al.* [139] tested the effect of HHP on the extraction of ginsenoside. They reported that HHP (200 MPa, at 60 °C for 5 minutes) presented a better extraction rate than heat reflux (80 °C) and Soxhlet (90 °C for 4 hours) extractions. They also observed that HHP extracted higher concentrations of ginsenoside from ginseng than MAE and UAE.

Prasad *et al.* [140] reported that HHP extraction increased the extraction efficiency of flavonoids from litchi fruit pericarp (*Litchi chinensis*). They observed that HHP extraction increased flavonoid extraction efficiency in comparison with UAE and conventional solvent extraction. According to the results, 400 MPa pressure significantly increased the catechin level in the extract.

In another study, Prasad *et al.* [141] observed the antioxidant and anticancer activities of HHP (500 MPa) processed EtOH (50%) extract of longan (*Dimocarpus longan*) fruit pericarp. They extracted phenolic compounds from longan fruit pericarp by HHP and conventional extraction methods. As a result, they observed that HHP presents higher extraction efficiency with a shorter process time.

Again Prasad *et al.* [141] and Yang *et al.* [142] showed that for extracting polysaccharides, lignin, and corilagin from longan fruit pericarp, HHP is better than other extraction methods.

Bi *et al.* [143] applied HHP to extract salidroside from *Rhodiola sachalinensis* and compared results to UAE, reflux extraction, and Soxhlet extraction. They also tested the effect of different extraction solvents, solid-liquid ratios, temperatures, and application times. The results presented that the extraction efficiency for Soxhlet extraction was 100% for the following conditions: MeOH as the extraction solvent, 1:10 as the solid-liquid ratio, at 80 °C for 240 minutes. The extraction efficiency for reflux extraction was 62.27% for MeOH extract with a solid-liquid ratio of 1:4 at 80 °C for 120 minutes. The extraction efficiency for the HHP method was 82.68% for ethanol extract, with a solid-liquid ratio of 1:50, but only at 25 °C for 3 minutes. The results showed that HHP extraction needed a moderate concentration of EtOH at room temperature to obtain a high extraction efficiency of salidroside in a short time.

Moreira *et al.* [144] used HHP to extract phenolic compounds, phenolic acids, and flavonoids from stinging nettle. They carried out HHP extraction at room temperature, different extraction times, and different pressures with different extraction solvents. The extraction solvents used in this study were water, 35% EtOH, and 70% EtOH where the pressures were 200 MPa and 500 MPa. They obtained the highest amount of total phenolic compounds for water extracts by 200 MPa. In addition, the highest concentration of neochlorogenic acid, caftaric acid, 5-*p*-coumaroylquinic acid, chlorogenic acid, fertaric acid, 4-*p*-coumaroylquinic acid, 2-*O*-caffeoylmalic acid, isoferulic acid, and rutin was for 70% EtOH in 200 MPa.

In another study, Moreira *et al.* [145] observed the effect of HHP on the extraction of phenolic compounds from winter savory (*Satureja montana*). They tested different pressures (0.1, 348, and 500 MPa), extraction solvents (water, 35% EtOH, and 70% EtOH), temperatures (20 and 25°C), and application times (1 and 20 minutes). As a result, they observed that 348 MPa with 35% EtOH for 20 minutes was the best extraction condition.

Pinela *et al.* [146] applied HHP to extract phenolic compounds from the watercress (*Nasturtium officinale*). In their study, they showed the effect of time (between 1.5 and 33.5), pressure (between 0.1 and 600 MPa), and EtOH concentration (between 0 and 100%). The application temperature was 20 °C for all applications. As a result, they proposed the optimum condition as 100% EtOH pressurized at 600 MPa for 3.1 minutes and stressed HHP as a favorable method for cold extracting phenolic compounds from watercress. They also stated that with the HHP processing technology, it is possible to perform the extraction in shorter times by using green solvents.

Anthocyanins have several applications both in the food and cosmetic industry. Organic solvents such as acetone and acidified methanol are the common extraction solvents to extract anthocyanins. The use of these solvents causes risk both for the environment and the residual organic solvent in the final extract causing problems for consumers. Corrales *et al.* [63] applied HHP in extracting anthocyanins from the grape skin. They used 600 MPa pressure at 70 °C and compared the results with PEF, UAE, and hot water extraction. As a result, they proposed that HHP revealed better extraction than other extraction methods.

Briones-Labarca *et al.* [147] applied HHP to extract lycopene and flavonoids from tomato pulp and used three different pressures, namely 250, 350, and 450 MPa, at room temperature. They prepared three different polar and non-polar solvent mixtures by using hexane and EtOH-acetone mixture. As a result, they observed that the optimal extraction conditions were 450 MPa and hexane:EtOH:acetone (60:20:20). They compared the results with a conventional extraction method and concluded that HHP is a tool to improve the extraction of flavonoids and polyphenols.

Furthermore, Briones-Labarca *et al.* [148] extracted some antioxidant compounds, sulforaphane, and fatty acids from Chilean papaya (*Vasconcellea pubescens*) by 80% MeOH using HHP and UAE. The pressure used in the study was 500 MPa for 5, 10, and 15 min with 1-minute pulses at room temperature. According to the data obtained from the study, they put forward that HHP provided more effective extraction than UAE and conventional extraction.

Jamaludin *et al.* [149] used HHP to extract bioactive compounds by EtOH from noni (*Morinda citrifolia*) fruits. They investigated the effect of several parameters, such as EtOH concentration (30%, 50%, and 70%), process duration (5, 10, and 15 minutes), and pressure (400, 500, and 600 MPa) on the extraction efficiency for rutin, alizarin, and scopoletin. They predicted the optimal HHP conditions as 544 MPa, with 65% EtOH for 15 minutes.

Grassino *et al.* [150] tried HHP to extract polyphenols from tomato peel waste obtained from the canning industry. They observed the effect of application time (5 and 10 minutes), temperature (25, 35, 45, and 55 °C), different extraction solvents, and a constant 600 MPa of pressure. The extraction solvents used in the study were (with or without the addition of HCl) water, HCl, 50% EtOH, 70% EtOH, 50% MeOH, and 70% MeOH. The results presented a noteworthy difference in the concentration of phenolic compounds regarding the temperature and solvent used. In contrast, the time did not affect the polyphenol concentration. As a result, they put forward that the minimum application time in HHP application is possible to extract polyphenols from tomato peel efficiently.

Naghshineh *et al.* [151] aimed to extract pectin from dried lime peel by HHP application. Pectin is one of the polysaccharides present mainly in the plant cell wall. It has several functional properties for the food and pharmaceutical industries [152 - 154]. They applied 100 and 200 MPa pressure at room temperature, and the results showed that pressure enhanced pectin extraction.

Choi *et al.* [155] processed turmeric (*Curcuma longa*) by HHP in extracting bioactive compounds to enhance the antioxidant capacity. They observed the effect of pressure, processing time, and EtOH concentration but used a processing procedure different than other studies given here. They first kept processing time and EtOH concentration constant at 15 minutes and 70%, respectively but changed pressure (0.1, 100, 250, 400, and 550 MPa). As a result of the experiments, they determined that 400 MPa presented the best results. Then they kept pressure constant (400 MPa) and EtOH concentration (70%) constant and changed processing time (5, 10, 15, 20, and 30 minutes). In the end, they kept pressure (400 MPa) and processing time (20 minutes) constant and changed EtOH concentration (0, 20, 40, 70, and 95%). As a result, they observed that the best combination, which yields the highest antioxidant activity, was 400 MPa, by 70% EtOH for 20 minutes. In addition, they determined that the concentration of ferulic acid and vanillic acid also increased with increasing processing time.

Briones-Labarca *et al.* [156] observed the effect of HHP on the extraction of antioxidant compounds, vitamin C, and β-carotene from cape gooseberry (*Physalis peruviana*) and compare the results with conventional extraction methods [157]. They processed cape gooseberry samples at 500 MPa for three different processing times (30, 60, and 90 seconds) at room temperature to extract vitamin C. For other compounds, they changed the processing time (5, 10, and 15 minutes). And they applied pressure as pulses for every minute. As a result, they observed that HHP provided higher extraction efficiency with less processing time.

Teles *et al.* [158] tried HHP and enzyme-assisted extraction to recover phenolic compounds from grape pomace, which comprises the skin and the pulp, usually the seeds of the grape. They applied three different pressures (50, 100, and 200 MPa) and different application times (0-30 minutes). According to the results, they suggested that HHP is an efficient and cost-effective extraction method.

Altuner [159] conducted a predictive modeling study for using HHP in protein extraction and observed that increasing pressure increased the extraction efficiency for proteins.

Zhang *et al.* [160] used HHP to extract flavonoids from propolis. They tested several experimental conditions, such as ethanol as an extraction solvent (between 35 and 95% (v/v)), pressure (between 100 and 600 MPa), process duration (between 1 and 10 minutes), and solid-liquid ratio (between 1:5 and 1:45 g/cm^3) to optimize the process. They compared the HHP processing to extraction at room temperature (7 days) and reflux extraction (4 hours). As a result, they proposed that HHP processing was more effective than traditional methods according to extraction efficiency and time.

As a result of the studies, scientists proposed that both the food and pharmaceutical industries could benefit from HHP, which is safe, fast, and more eco-friendly than traditional extraction methods.

The Efficiency of HHP in Extraction

It is possible to explain the efficiency of HHP in extraction by mass transfer theory. The permeability of cells under pressure increases due to the mass transfer theory [59, 160 - 163].

High pressure changes the effective diffusivity and increases the rate of dissolution. The application of high pressure establishes a high differential pressure between the cell interior and exterior of the cell membrane. This differential pressure causes rapid permeation [160]. The rapid permeation leads to solvent penetration through membranes into cells or increased permeability, increasing the mass transfer rate [60].

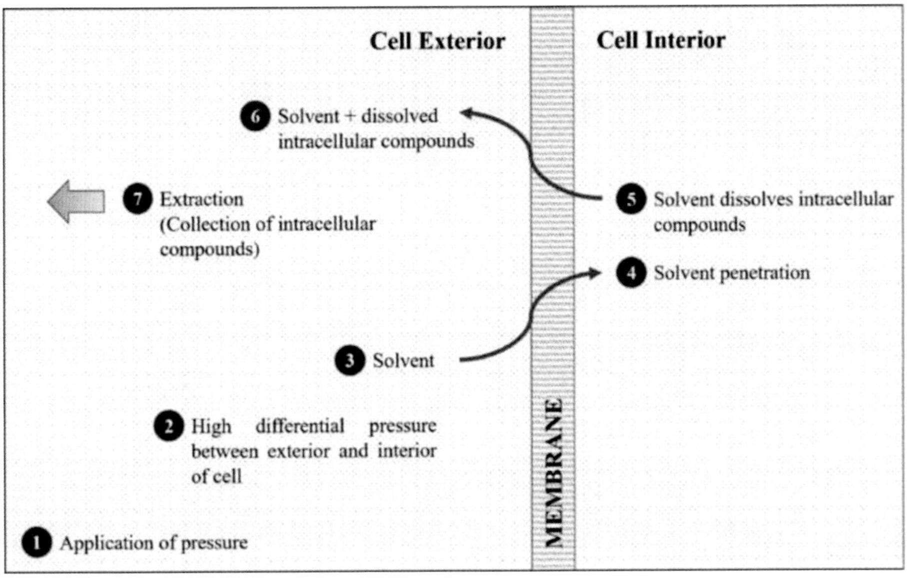

Fig. (6). Extraction procedure by HHP.

As a rule, the higher the hydrostatic pressure, the more solvent enters the cell. As more solvent enters a cell, more compounds can permeate out to the extraction solvent (Fig. **6**). Keeping pressure constant establishes an equilibrium between the extracellular and intracellular parts of the cell. However, when releasing pressure, the cell releases the solvent containing target compounds, and this causes a higher yield of extraction [131]. An improvement in the extraction yield makes HHP an effective tool in the extraction procedure.

CONCLUDING REMARKS

Green extraction methods (GEMs) eliminate or reduce the use and production of hazardous substances. In addition, GEMs reduce energy consumption, permit using alternative solvents and renewable natural products, and warrant high-quality and safe products [164].

Previous studies proved that using safe extraction solvents is possible in HHP. In addition, HHP needs low processing time and does not require high temperature, which means reducing energy consumption. As the transmission of pressure in HHP is instant, it does not need a long processing time. Furthermore, HHP yields safe products. For example, the possibility of processing at low temperatures prevents the Maillard reaction. Thus, there is a reduced risk of arising carcinogen acrylamide in samples containing amino acids and reducing sugars.

As a result, it is possible to propose HHP as a green extraction technique for the food and pharmaceutical industries.

LIST OF ABBREVIATIONS

CECT	Spanish-type culture collection
EtOH	Ethanol
FDA	Food and Drug Administration
GEM	Green extraction method
HHP	High hydrostatic pressure
HPE	High-pressure extraction
HRE	Heat reflux extraction
MAE	Microwave-assisted extraction
MeOH	Methanol
MPa	Megapascal
NCTC	National Collection of Type Cultures (UK)
PBS	Phosphate buffer saline
PEF	Pulsed electric field
PLE	Pressurized liquid extraction
PSI	Pounds/inch^2
S-CO_2	Supercritical carbon dioxide

SF	Supercritical fluid
SFE	Supercritical fluid extraction
UAE	Ultrasound-assisted extraction
UHP	Ultra high pressure
UV	Ultraviolet

CONSENT FOR PUBLICATION

Not applicable.

CONFLICT OF INTEREST

The authors declare no conflict of interest, financial or otherwise.

ACKNOWLEDGEMENT

Declared none.

REFERENCES

[1] Tiwari BK, O'Donnell CP, Cullen PJ. Effect of non thermal processing technologies on the anthocyanin content of fruit juices. Trends Food Sci Technol 2009; 20(3-4): 137-45. [http://dx.doi.org/10.1016/j.tifs.2009.01.058]

[2] Muntean MV, Marian O, Barbieru V, *et al.* High Pressure Processing in Food Industry - Characteristics and Applications. 5th International Conference - Agriculture for Life, Life for Agriculture. 377-83. [http://dx.doi.org/10.1016/j.aaspro.2016.09.077]

[3] Altuner EM, Çeter T, Alpas H. Effect of high hydrostatic pressure on the profile of proteins extracted from *Betula pendula* pollens. High Press Res 2014; 34(4): 470-81. [http://dx.doi.org/10.1080/08957959.2014.975698]

[4] Altuner EM, Tokuşoğlu Ö. The effect of high hydrostatic pressure processing on the extraction, retention and stability of anthocyanins and flavonols contents of berry fruits and berry juices. Int J Food Sci Technol 2013; 48(10): n/a. [http://dx.doi.org/10.1111/ijfs.12185]

[5] Oey I, Van der Plancken I, Van Loey A, Hendrickx M. Does high pressure processing influence nutritional aspects of plant based food systems? Trends Food Sci Technol 2008; 19(6): 300-8. [http://dx.doi.org/10.1016/j.tifs.2007.09.002]

[6] Rawson A, Patras A, Tiwari BK, Noci F, Koutchma T, Brunton N. Effect of thermal and non thermal processing technologies on the bioactive content of exotic fruits and their products: Review of recent advances. Food Res Int 2011; 44(7): 1875-87. [http://dx.doi.org/10.1016/j.foodres.2011.02.053]

[7] Altuner EM, Alpas H, Erdem YK, Bozoglu F. Effect of high hydrostatic pressure on physicochemical and biochemical properties of milk. Eur Food Res Technol 2006; 222(3-4): 392-6. [http://dx.doi.org/10.1007/s00217-005-0072-4]

[8] Le Noble WJ. Organic High Pressure Chemistry. Amsterdam: Elsevier 1988.

[9] Richard JS. High Pressure Phase Behaviour of Multicomponent Fluid Mixtures. Amsterdam: Elsevier

1992.

[10] Altuner EM. Effect of high hydrostatic pressure (HHP) on the surface hydrophobicity of milk proteins, MSc Thesis, Middle East Technical University. Ankara, Turkey 2002.

[11] Shouqin Z, Jun X, Changzheng W. Effect of high hydrostatic pressure on extraction of flavonoids in propolis. Food Sci Technol Int 2005; 11(3): 213-6. [http://dx.doi.org/10.1177/1082013205054420]

[12] Bertucco A, Vetter G. High pressure process technology: Fundamentals and applications. Amsterdam: Elsevier 2001.

[13] Alpas H, Özilgen M, Faruk Bozoğlu T, Katnas S. Mathematical modeling of survival and weight loss of Baker's yeast during drying. Enzyme Microb Technol 1996; 19(5): 348-51. [http://dx.doi.org/10.1016/S0141-0229(96)00027-0]

[14] Peter, B B; Ivan, D; Robert, E M, High Pressure Biology and Medicine. New York: University of Rochester Press 1998.

[15] Alpas H, Kalchayanand N, Bozoglu F, Ray B. Interactions of high hydrostatic pressure, pressurization temperature and pH on death and injury of pressure-resistant and pressure-sensitive strains of foodborne pathogens. Int J Food Microbiol 2000; 60(1): 33-42. [http://dx.doi.org/10.1016/S0168-1605(00)00324-X] [PMID: 11014520]

[16] Alpas H, Bozoglu F. Inactivation of Staphylococcus aureus and Listeria monocytogenes in milk and cream of chicken soup by high hydrostatic pressure and bacteriocins. High Press Res 2002; 22(3-4): 681-4. [http://dx.doi.org/10.1080/08957950212440]

[17] Bozoglu F, Alpas H, KaletunÃ GÃÃ. Injury recovery of foodborne pathogens in high hydrostatic pressure treated milk during storage. FEMS Immunol Med Microbiol 2004; 40(3): 243-7. [http://dx.doi.org/10.1016/S0928-8244(04)00002-1] [PMID: 15039101]

[18] Peñas E, Gómez R, Frías J, Vidal-Valverde C. Application of high-pressure treatment on alfalfa (*Medicago sativa*) and mung bean (*Vigna radiata*) seeds to enhance the microbiological safety of their sprouts. Food Control 2008; 19(7): 698-705. [http://dx.doi.org/10.1016/j.foodcont.2007.07.010]

[19] Peñas E, Gómez R, Frías J, Vidal-Valverde C. Effects of combined treatments of high pressure, temperature and antimicrobial products on germination of mung bean seeds and microbial quality of sprouts. Food Control 2010; 21(1): 82-8. [http://dx.doi.org/10.1016/j.foodcont.2009.04.008]

[20] Altuner EM, Çeter T, Alpas H. High hydrostatic pressure processing: a method having high success potential in pollen protein extraction. High Press Res 2012; 32(2): 1-8. [http://dx.doi.org/10.1080/08957959.2012.678341]

[21] Altuner EM. Potential of High Hydrostatic Pressure to Improve the Production of Plants Used as Food. In: Holban AM, Grumezescu AM, Eds. Advances in Biotechnology for Food Industry. London: Academic Press Ltd-Elsevier Science Ltd 2018; Vol. 14: pp. 213-41. [http://dx.doi.org/10.1016/B978-0-12-811443-8.00008-6]

[22] Rivalain N, Roquain J, Demazeau G. Development of high hydrostatic pressure in biosciences: Pressure effect on biological structures and potential applications in Biotechnologies. Biotechnol Adv 2010; 28(6): 659-72. [http://dx.doi.org/10.1016/j.biotechadv.2010.04.001] [PMID: 20398747]

[23] Regnard P. Note sur les conditions de la vie dans les profondeurs de la mer. C R Seances SocBiol 1884; 36: 164-8.

[24] Royer H. Action des hautes pressions sur quelques bactéries. Archives de Physiologie Normale et Pathologique 1895; 7: 12-7.

[25] Hite BH. The effect of pressure in the preservation of milk: a preliminary report. W Va Agr Exp St Bull 1899; 58: 1866-921.
[http://dx.doi.org/10.33915/agnic.58]

[26] Bridgman PW. The coagulation of albumen by pressure. J Biol Chem 1914; 19(4): 511-2.
[http://dx.doi.org/10.1016/S0021-9258(18)88287-4]

[27] Bridgman PW. The physics of high pressure. London: Bell and Sons 1949.

[28] Basset J. Appareil pour réaliser des expériences physiques ou chimiques à toutes températures variées sous des pressions liquides de 15000 kg/cm3. C R Acad Hebd Seances Acad Sci 1927; 185: 343-5.

[29] Basset J, Macheboeuf MA. Etude sur les effets biologiques des ultra-pressions: résistance des bactéries, des diastases et des toxines aux pressions très élevées. C R Acad Hebd Seances Acad Sci 1932; 195: 1431-3.

[30] Basset J, Lepine P, Chaumont L. [Effects of high pressures on the poliomyelitis virus (Lansing strain)]. Ann Inst Pasteur (Paris) 1956; 90(5): 575-93.
[PMID: 13327515]

[31] Bovenkerk HP, Bundy FP, Hall HT, Strong HM, Wentorf RH. Preparation of Diamond. Nature 1959; 184(4693): 1094-8.
[http://dx.doi.org/10.1038/1841094a0]

[32] Liander H, Lundblad E. Some observations on the synthesis of diamonds. Ark Kemi 1960; 1960(16): 139-49.

[33] Timson WJ, Short AJ. Resistance of microorganisms to hydrostatic pressure. Biotechnol Bioeng 1965; 7(1): 139-59.
[http://dx.doi.org/10.1002/bit.260070124]

[34] Gould GW, Sale AJH. Initiation of germination of bacterial spores by hydrostatic pressure. J Gen Microbiol 1970; 60(3): 335-46.
[http://dx.doi.org/10.1099/00221287-60-3-335] [PMID: 5487618]

[35] Wilson DC. High pressure sterilization 34th Annual Meeting of the Institute of Food Technologists. New Orleans. 1974.

[36] Marquis RE. High-pressure microbial physiology. Adv Microb Physiol 1976; 14(11): 159-241.
[http://dx.doi.org/10.1016/S0065-2911(08)60228-3] [PMID: 795275]

[37] Charm SE, Longmaid HE III, Carver J. A simple system for extending refrigerated, nonfrozen preservation of biological material using pressure. Cryobiology 1977; 14(5): 625-36.
[http://dx.doi.org/10.1016/0011-2240(77)90174-2] [PMID: 409589]

[38] Marquis RE, Matsumara P. Microbial life under pressure. In: Kushner DJ, Ed. Microbial life in extreme environments. New York: Academic Press 1978; pp. 105-57.

[39] Morild E. The theory of pressure effects on enzymes. In: Anfinsen C B, Edsall J T, Richards F M, Eds. Advances in Protein Chemistry Academic Press. 1981; 34: pp. 93-166.
[http://dx.doi.org/10.1016/S0065-3233(08)60519-7]

[40] Heremans K. High pressure effects on proteins and other biomolecules. Annu Rev Biophys Bioeng 1982; 11(1): 1-21.
[http://dx.doi.org/10.1146/annurev.bb.11.060182.000245] [PMID: 7049058]

[41] Hoover DG, Metrick C, Papineau AM, Farkas DF, Knorr D. Biological effects of high hydrostatic-pressure on food microorganisms. Food Technol 1989; 43(3): 99-107.

[42] Popper L, Knorr D. Applications of high-pressure homogenization for food preservation. Food Technol 1990; 44(7): 84.

[43] Farr D. High pressure technology in the food industry. Trends Food Sci Technol 1990; 1: 14-6.
[http://dx.doi.org/10.1016/0924-2244(90)90004-I]

[44] Hayashi R. Application of high pressure to food processing and preservation: philosophy and development. In: Spiess WEL, Schubert H, Eds. Engineering and food. London: Elsevier 1989; Vol. 2: pp. 815-26.

[45] Hayashi R. Pressure processed food: research and development. Kyoto, Japan: San-Ei Publishing 1990.

[46] Horie Y, Kimura K, Ida M, Yosida Y, Ohki K. Jam preparation by pressurization. Nippon Nogeikagaku Kaishi 1991; 65(6): 975-80. [http://dx.doi.org/10.1271/nogeikagaku1924.65.975]

[47] Ogawa H, Fukuhisa K, Kubo Y, Fukumoto H. Pressure inactivation of yeasts, molds, and pectinesterase in satsuma mandarin juice - effects of juice concentration, ph, and organic-acids, and comparison with heat sanitation. Agric Biol Chem 1990; 54(5): 1219-25.

[48] Tanaka T, Hatanaka K. Application of hydrostatic-pressure to yogurt to prevent its after-acidification. Nippon Shokuhin Kagaku Kogaku Kaishi J Jpn Soc Food Sci 1992; 39(2): 173-7.

[49] Balny C, Hayashi R, Heremans K, Masson P. High pressure and biotechnology In: Balny C, Hayashi R, Heremans K, Masson P, Eds. First European seminar on high pressure and biotechnology joint meeting with the fifth symposium on high pressure and food science. La Grande Motte (France): John Libbey Eurotext 1992; pp. 329-31.

[50] Cheftel JC. Review : High-pressure, microbial inactivation and food preservation. Food Sci Technol Int 1995; 1(2-3): 75-90. [http://dx.doi.org/10.1177/108201329500100203]

[51] Cheftel JC, Culioli J. Effects of high pressure on meat: A review. Meat Sci 1997; 46(3): 211-36. [http://dx.doi.org/10.1016/S0309-1740(97)00017-X] [PMID: 22062123]

[52] Demazeau G. Les hautes pressions: dévelopements potentiels. In: Agro-alimentaire O-NP, Ed. L'avenir des hautes pressions dans l'industrie agroalimentaire. Nantes, France 1993; pp. 91-109.

[53] Hoover DG. Pressure effects on biological-systems. Food Technol 1993; 47(6): 150-5.

[54] Knorr D. Effects of high-hydrostatic-pressure processes on food safety and quality. Food Technol., 1993, 47, 6, 156.

[55] Tauscher B. Pasteurization of food by hydrostatic high pressure: chemical aspects. Z Lebensm Unters Forsch 1995; 200(1): 3-13. [http://dx.doi.org/10.1007/BF01192901] [PMID: 7732731]

[56] Tonello C, Largeteau A, Jolibert F, Deschamps A, Demazeau G. Destruction par hautes pressions de micro-organismes contaminant des produits laitiers. L'électricité dans l'agro-alimentaire, 1993, 43-48.

[57] Travis AS. High pressure industrial chemistry: the first steps, 1909-1913, and the impact. In: Travis AS, Ed. Determinants in the evolution of the European chemical industry, 1900-1939: new technologies, political frameworks, markets and companies. London: Kluwer 1998; pp. 3-21. [http://dx.doi.org/10.1007/978-94-017-1233-0_1]

[58] Tokuşoğlu Ö, Swanso BG. Improving Food Quality with Novel Food Processing: Chemistry, Functionality & Rheology. Florida: CRC Press, Taylor & Francis Group 2011.

[59] Altuner EM, Çeter T, Alpas H. High hydrostatic pressure extraction of phenolic compounds from Maclura pomifera fruits. Afr J Biotechnol 2012; 11(4): 930-7.

[60] Shouqin Z, Junjie Z, Changzhen W. Novel high pressure extraction technology. Int J Pharm 2004; 278(2): 471-4. [http://dx.doi.org/10.1016/j.ijpharm.2004.02.029] [PMID: 15196650]

[61] US Food and Drug Administration Center for Food Safety and Applied Nutrition. "Kinetics of microbial inactivation for alternative food processing technologies - high pressure processing." Available: http://www.fda.gov/Food/FoodScienceResearch/SafePracticesforFoodProcesses/ucm101

456.htm

[62] Corrales M, Toepfl S, Butz P, Knorr D, Tauscher B. Extraction of anthocyanins from grape by-products assisted by ultrasonics, high hydrostatic pressure or pulsed electric fields: A comparison. Innov Food Sci Emerg Technol 2008; 9(1): 85-91. [http://dx.doi.org/10.1016/j.ifset.2007.06.002]

[63] Corrales M, García AF, Butz P, Tauscher B. Extraction of anthocyanins from grape skins assisted by high hydrostatic pressure. J Food Eng 2009; 90(4): 415-21. [http://dx.doi.org/10.1016/j.jfoodeng.2008.07.003]

[64] Tokuşoğlu Ö, Stoner G. Phytochemical Bioactives in Berry Fruits. In: Tokuşoğlu Ö, Hall C, Eds. Fruit and Cereal Bioactives: Sources, Chemistry & Applications. Florida: CRC Press, Taylor & Francis Group 2011.

[65] Tokuşoğlu Ö, Hall C. Fruit and Cereal Bioactives: Sources, Chemistry & Applications. Florida: CRC Press, Taylor & Francis Group 2011. [http://dx.doi.org/10.1201/b10786]

[66] Tokuşoğlu Ö, Doona CJ. High Pressure Processing Technology on Bioactives in Fruits & Cereals. In: Tokuşoğlu Ö, Hall C, Eds. Fruit and Cereal Bioactives: Sources, Chemistry & Applications. Florida: CRC Press, Taylor & Francis Group 2011. [http://dx.doi.org/10.1201/b10786-25]

[67] Shouqin Z, Ruizhan C, Changzheng W. Experiment study on ultrahigh pressure extraction of Ginsenosides. J Food Eng 2007; 79(1): 1-5. [http://dx.doi.org/10.1016/j.jfoodeng.2005.12.048]

[68] Yordanov DG, Angelova GV. High pressure processing for foods preserving. Biotechnol Biotechnol Equip 2010; 24(3): 1940-5. [http://dx.doi.org/10.2478/V10133-010-0057-8]

[69] Patterson MF, Quinn M, Simpson R, Gilmour A. Sensitivity of vegetative pathogens to high hydrostatic-pressure treatment in phosphate-buffered saline and foods. J Food Prot 1995; 58(5): 524-9. [http://dx.doi.org/10.4315/0362-028X-58.5.524] [PMID: 31137263]

[70] Patterson MF, Quinn M, Simpson R, Gilmour A. Studies on the inactivation by pressure of microorganisms. In: Ledward DA, Johnston DE, Earnshaw RG, Hasting APM, Eds. High Pressure Processing of Foods. Nottingham: Nottingham University Press 1995; pp. 47-63.

[71] Perrier-Cornet JM, Maréchal PA, Gervais P. A new design intended to relate high pressure treatment to yeast cell mass transfer. J Biotechnol 1995; 41(1): 49-58. [http://dx.doi.org/10.1016/0168-1656(95)00052-R] [PMID: 7640002]

[72] Rasanayagam V, Balasubramaniam VM, Ting E, Sizer CE, Bush C, Anderson C. Compression heating of selected fatty food materials during high-pressure processing. J Food Sci 2003; 68(1): 254-9. [http://dx.doi.org/10.1111/j.1365-2621.2003.tb14148.x]

[73] Patazca E, Koutchma T, Balasubramaniam VM. Quasi-adiabatic temperature increase during high pressure processing of selected foods. J Food Eng 2007; 80(1): 199-205. [http://dx.doi.org/10.1016/j.jfoodeng.2006.05.014]

[74] Polydera AC, Stoforos NG, Taoukis PS. Comparative shelf life study and vitamin C loss kinetics in pasteurised and high pressure processed reconstituted orange juice. J Food Eng 2003; 60(1): 21-9. [http://dx.doi.org/10.1016/S0260-8774(03)00006-2]

[75] Penchalaraju M, Shireesha B. Preservation of foods by high-pressure processing - a review. Indian J Sci Technol 2013; 1: 30-8.

[76] Bello E, Martínez G, Ceberio B, Rodrigo D, López A. High Pressure Treatment in Foods. Foods 2014; 3(3): 476-90. [http://dx.doi.org/10.3390/foods3030476] [PMID: 28234332]

[77] San Martín MF, Barbosa-Cánovas GV, Swanson BG. Food processing by high hydrostatic pressure. Crit Rev Food Sci Nutr 2002; 42(6): 627-45.
[http://dx.doi.org/10.1080/20024091054274] [PMID: 12487422]

[78] Patterson MF, Kilpatrick DJ. The combined effect of high hydrostatic pressure and mild heat on inactivation of pathogens in milk and poultry. J Food Prot 1998; 61(4): 432-6.
[http://dx.doi.org/10.4315/0362-028X-61.4.432] [PMID: 9709206]

[79] Yuste J, Mor-Mur M, Capellas M, Pla R. Listeria innocua and aerobic mesophiles during chill storage of inoculated mechanically recovered poultry meat treated with high hydrostatic pressure. Meat Sci 1999; 53(4): 251-7.
[http://dx.doi.org/10.1016/S0309-1740(99)00064-9] [PMID: 22063467]

[80] Palou E, López-Malo A, Barbosa-Cánovas G, Welti-Chanes J, Swanson BG. Oscillatory high hydrostatic pressure inactivation of Zygosaccharomyces bailii. J Food Prot 1998; 61(9): 1213-5. [http://dx.doi.org/10.4315/0362-028X-61.9.1213] [PMID: 9766081]

[81] Pandya Y, Jewett FF Jr, Hoover DG. Concurrent effects of high hydrostatic-pressure, acidity and heat on the destruction and injury of yeasts. J Food Prot 1995; 58(3): 301-4.
[http://dx.doi.org/10.4315/0362-028X-58.3.301] [PMID: 31137294]

[82] Benito A, Ventoura G, Casadei M, Robinson T, Mackey B. Variation in resistance of natural isolates of Escherichia coli O157 to high hydrostatic pressure, mild heat, and other stresses. Appl Environ Microbiol 1999; 65(4): 1564-9.
[http://dx.doi.org/10.1128/AEM.65.4.1564-1569.1999] [PMID: 10103251]

[83] García-Graells C, Masschalck B, Michiels CW. Inactivation of Escherichia coli in milk by high-hydrostatic-pressure treatment in combination with antimicrobial peptides. J Food Prot 1999; 62(11): 1248-54.
[http://dx.doi.org/10.4315/0362-028X-62.11.1248] [PMID: 10571312]

[84] McClements JMJ, Patterson MF, Linton M. The effect of growth stage and growth temperature on high hydrostatic pressure inactivation of some psychrotrophic bacteria in milk. J Food Prot 2001; 64(4): 514-22.
[http://dx.doi.org/10.4315/0362-028X-64.4.514] [PMID: 11307889]

[85] Berlin DL, Herson DS, Hicks DT, Hoover DG. Response of pathogenic Vibrio species to high hydrostatic pressure. Appl Environ Microbiol 1999; 65(6): 2776-80.
[http://dx.doi.org/10.1128/AEM.65.6.2776-2780.1999] [PMID: 10347079]

[86] Linton M, McClements JMJ, Patterson MF. Inactivation of Escherichia coli O157:H7 in orange juice using a combination of high pressure and mild heat. J Food Prot 1999; 62(3): 277-9.
[http://dx.doi.org/10.4315/0362-028X-62.3.277] [PMID: 10090248]

[87] Gervilla R, Sendra E, Ferragut V, Guamis B. Sensitivity of *Staphylococcus aureus* and *Lactobacillus helveticus* in ovine milk subjected to high hydrostatic pressure. J Dairy Sci 1999; 82(6): 1099-107.
[http://dx.doi.org/10.3168/jds.S0022-0302(99)75332-4] [PMID: 10386295]

[88] Kallscheuer N, Classen T, Drepper T, Marienhagen J. Production of plant metabolites with applications in the food industry using engineered microorganisms. Curr Opin Biotechnol 2019; 56: 7-17.
[http://dx.doi.org/10.1016/j.copbio.2018.07.008] [PMID: 30114548]

[89] Hussein RA, El-Anssary AA. Plants Secondary Metabolites: The Key Drivers of the Pharmacological Actions of Medicinal Plants. In: Builders PF, Ed. Herbal Medicine. IntechOpen 2018; pp. 25-30.

[90] Bourgaud F, Gravot A, Milesi S, Gontier E. Production of plant secondary metabolites: a historical perspective. Plant Sci 2001; 161(5): 839-51.
[http://dx.doi.org/10.1016/S0168-9452(01)00490-3]

[91] Singer AC, Crowley DE, Thompson IP. Secondary plant metabolites in phytoremediation and biotransformation. Trends Biotechnol 2003; 21(3): 123-30.

[http://dx.doi.org/10.1016/S0167-7799(02)00041-0] [PMID: 12628369]

[92] Wink M. Plant secondary metabolism: Diversity, function and its evolution. Nat Prod Commun 2008; 3(8): 1934578X0800300.
[http://dx.doi.org/10.1177/1934578X0800300801]

[93] Guerriero G, Berni R, Muñoz-Sanchez J, *et al.* Production of Plant Secondary Metabolites: Examples, Tips and Suggestions for Biotechnologists. Genes (Basel) 2018; 9(6): 309.
[http://dx.doi.org/10.3390/genes9060309] [PMID: 29925808]

[94] Kurek J. Introductory Chapter: Alkaloids - Their Importance in Nature and for Human Life. In: Kurek J, Ed. Alkaloids. IntechOpen 2019.
[http://dx.doi.org/10.5772/intechopen.85400]

[95] Debnath B, Singh WS, Das M, *et al.* Role of plant alkaloids on human health: A review of biological activities. Mater Today Chem 2018; 9: 56-72.
[http://dx.doi.org/10.1016/j.mtchem.2018.05.001]

[96] Cheynier V. Phenolic compounds: from plants to foods. Phytochem Rev 2012; 11(2-3): 153-77.
[http://dx.doi.org/10.1007/s11101-012-9242-8]

[97] Rupasinghe HPV, Nair SVG, Robinson RA. Chemopreventive Properties of Fruit Phenolic Compounds and Their Possible Mode of Actions. In: Atta-ur-Rahman (Ed.) Studies in Natural Products Chemistry. Elsevier 2014, Vol. 42: pp. 229-66.
[http://dx.doi.org/10.1016/B978-0-444-63281-4.00008-2]

[98] Burow M, Wittstock U, Gershenzon J. Sulfur-Containing Secondary Metabolites and Their Role in Plant Defense. In: Hell R, Dahl C, Knaff D, Leustek T, Eds. Sulfur Metabolism in Phototrophic Organisms. Dordrecht: Springer 2008; pp. 201-22.
[http://dx.doi.org/10.1007/978-1-4020-6863-8_11]

[99] Perveen S. Introductory Chapter: Terpenes and Terpenoids. In: Perveen S, Al-Taweel A, Eds. Terpenes and Terpenoids. IntechOpen 2018.
[http://dx.doi.org/10.5772/intechopen.79683]

[100] Tholl D. Biosynthesis and Biological Functions of Terpenoids in Plants. In: Schrader J, Bohlmann J, Eds. Biotechnology of Isoprenoids Advances in Biochemical Engineering-Biotechnology. Cham: Springer International Publishing Ag 2015; 148: pp. 63-106.
[http://dx.doi.org/10.1007/10_2014_295]

[101] Sadeek AMM, Abdallah EM. Phytochemical Compounds as Antibacterial Agents: A Mini Review. Glo J Pharm Pharmac Sci 2019; 7(4): 555720.

[102] Handa SS. An overview of extraction techniques for medicinal and aromatic plants. In: Handa SS, Khanuja SPS, Longo G, Rakesh DD, Eds. Extraction technologies for medicinal and aromatic plants. Trieste: International Centre for Science and HighTechnology 2008; pp. 21-54.

[103] Huang HW, Hsu CP, Yang BB, Wang CY. Advances in the extraction of natural ingredients by high pressure extraction technology. Trends Food Sci Technol 2013; 33(1): 54-62.
[http://dx.doi.org/10.1016/j.tifs.2013.07.001]

[104] Zhang QW, Lin LG, Ye WC. Techniques for extraction and isolation of natural products: a comprehensive review. Chin Med 2018; 13(1): 20.
[http://dx.doi.org/10.1186/s13020-018-0177-x] [PMID: 29692864]

[105] Cseke LJ, Setzer WN, Vogler B, Kirakosyan A, Kaufman PB. Traditional, analytical and preparative separations of natural products. In: Cseke LJ, Kirakosyan A, Kaufman PB, Warber SL, Duke JA, Brielmann HL, Eds. Natural products from plants. Boca Raton: CRC Press, Taylor & Francis Group 2006; pp. 263-317.

[106] Dembicki H. Basin Modeling. Practical Petroleum Geochemistry for Exploration and Production. Amsterdam: Elsevier Science Bv 2017; pp. 273-308.
[http://dx.doi.org/10.1016/B978-0-12-803350-0.00008-8]

[107] Suna S, Tamer CE, Ozcan-Sinir G. Trends and possibilities of the usage of medicinal herbal extracts in beverage production. In: Grumezescu AM, Holban AM (Eds.) Natural Beverages, Volume 13: The Science of Beverages. Elsevier 2019: pp. 361-98. [http://dx.doi.org/10.1016/B978-0-12-816689-5.00013-4]

[108] Wang DG, Liu WY, Chen GT. A simple method for the isolation and purification of resveratrol from *Polygonum cuspidatum.* J Pharm Anal 2013; 3(4): 241-7. [http://dx.doi.org/10.1016/j.jpha.2012.12.001] [PMID: 29403824]

[109] Chua LS, Latiff NA, Mohamad M. Reflux extraction and cleanup process by column chromatography for high yield of andrographolide enriched extract. J Appl Res Med Aromat Plants 2016; 3(2): 64-70. [http://dx.doi.org/10.1016/j.jarmap.2016.01.004]

[110] Weggler BA, Gruber B, Teehan P, Jaramillo R, Dorman FL. Chapter 5 - Inlets and sampling. In: Snow N H, Ed. Separation Science and Technology. Academic Press 2020; 12: pp. 141-203.

[111] Ibáñez E, Mendiola JA, Castro-Puyana M. Supercritical Fluid Extraction. In: Caballero B, Finglas PM, Toldrá F, Eds. Encyclopedia of Food and Health. Academic Press 2016; pp. 227-33. [http://dx.doi.org/10.1016/B978-0-12-384947-2.00675-9]

[112] Costa Freitas AM, Gomes da Silva MDR, Cabrita MJ. Sampling Techniques for the Determination of Volatile Components in Grape Juice, Wine and Alcoholic Beverages. In: Pawliszyn J, Ed. Comprehensive Sampling and Sample Preparation. Academic Press 2012; pp. 27-41. [http://dx.doi.org/10.1016/B978-0-12-381373-2.00126-5]

[113] Gachanja AN. Polycyclic Aromatic Hydrocarbons, Environmental Applications. In: Worsfold P, Townshend A, Poole C, Eds. Encyclopedia of Analytical Science. 2nd ed. Elsevier 2005; pp. 234-42. [http://dx.doi.org/10.1016/B0-12-369397-7/00474-X]

[114] Mukherjee PK, Harwansh RK, Bahadur S, Chanda J, Biswas S, Banerjee S. Enzyme inhibition assay for metabolic disorders-exploring leads from medicinal plants. In: Verma AS, Singh A, Eds. Animal Biotechnology. Academic Press 2020; pp. 631-53. [http://dx.doi.org/10.1016/B978-0-12-811710-1.00033-1]

[115] Pingret D, Fabiano-Tixier AS, Chemat F. Accelerated Methods for Sample Preparation in Food. In: Pawliszyn J, Ed. Comprehensive Sampling and Sample Preparation. Academic Press 2012; pp. 441-55. [http://dx.doi.org/10.1016/B978-0-12-381373-2.00143-5]

[116] Chemat F, Cravotto G. Microwave-assisted extraction for bioactive compounds. Boston: Springer 2013. [http://dx.doi.org/10.1007/978-1-4614-4830-3]

[117] Vinatoru M, Mason TJ, Calinescu I. Ultrasonically assisted extraction (UAE) and microwave assisted extraction (MAE) of functional compounds from plant materials. *Trac-Trend.* Anal Chem 2017; 97: 159-78.

[118] Nawaz H, Shad MA, Rehman N, Andaleeb H, Ullah N. Effect of solvent polarity on extraction yield and antioxidant properties of phytochemicals from bean (*Phaseolus vulgaris*) seeds. Braz J Pharm Sci 2020; 56: e17129. [http://dx.doi.org/10.1590/s2175-97902019000417129]

[119] Lin F, Giusti MM. Effects of solvent polarity and acidity on the extraction efficiency of isoflavones from soybeans (*Glycine max*). J Agric Food Chem 2005; 53(10): 3795-800. [http://dx.doi.org/10.1021/jf050030r] [PMID: 15884799]

[120] Ramluckan K, Moodley KG, Bux F. An evaluation of the efficacy of using selected solvents for the extraction of lipids from algal biomass by the soxhlet extraction method. Fuel 2014; 116: 103-8. [http://dx.doi.org/10.1016/j.fuel.2013.07.118]

[121] Holser RA, Abbott TP. Extraction of simmondsins from defatted jojoba meal using aqueous ethanol. Ind Crops Prod 1999; 10(1): 41-6.

[http://dx.doi.org/10.1016/S0926-6690(99)00008-4]

[122] Ncube NS, Afolayan AJ, Okoh AI. Assessment techniques of antimicrobial properties of natural compounds of plant origin: current methods and future trends. Afr J Biotechnol 2008; 7(12): 1797-806.
[http://dx.doi.org/10.5897/AJB07.613]

[123] Gurjar MS, Ali S, Akhtar M, Singh KS. Efficacy of plant extracts in plant disease management. Agric Sci 2012; 3(3): 425-33.
[http://dx.doi.org/10.4236/as.2012.33050]

[124] Bae H, Jayaprakasha GK, Jifon J, Patil BS. Extraction efficiency and validation of an HPLC method for flavonoid analysis in peppers. Food Chem 2012; 130(3): 751-8.
[http://dx.doi.org/10.1016/j.foodchem.2011.07.041]

[125] Perva-Uzunalić A, Škerget M, Knez Ž, Weinreich B, Otto F, Grüner S. Extraction of active ingredients from green tea (*Camellia sinensis*): Extraction efficiency of major catechins and caffeine. Food Chem 2006; 96(4): 597-605.
[http://dx.doi.org/10.1016/j.foodchem.2005.03.015]

[126] Doran PM. Heterogeneous Reactions. Bioprocess Engineering Principles. 2nd ed. Amsterdam: Elsevier Science Bv 2013; pp. 705-59.
[http://dx.doi.org/10.1016/B978-0-12-220851-5.00013-7]

[127] Cacace JE, Mazza G. Mass transfer process during extraction of phenolic compounds from milled berries. J Food Eng 2003; 59(4): 379-89.
[http://dx.doi.org/10.1016/S0260-8774(02)00497-1]

[128] Türker N, Erdogdu F. Effects of pH and temperature of extraction medium on effective diffusion coefficient of anthocynanin pigments of black carrot (*Daucus carota* var. L.). J Food Eng 2006; 76(4): 579-83.
[http://dx.doi.org/10.1016/j.jfoodeng.2005.06.005]

[129] Jun X. High-pressure processing as emergent technology for the extraction of bioactive ingredients from plant materials. Crit Rev Food Sci Nutr 2013; 53(8): 837-52.
[http://dx.doi.org/10.1080/10408398.2011.561380] [PMID: 23768146]

[130] Scepankova H, Martins M, Estevinho L, Delgadillo I, Saraiva JA. Enhancement of Bioactivity of Natural Extracts by Non-Thermal High Hydrostatic Pressure Extraction. Plant Foods Hum Nutr 2018; 73(4): 253-67.
[http://dx.doi.org/10.1007/s11130-018-0687-9] [PMID: 30269189]

[131] Xi J, Shen D, Zhao S, Lu B, Li Y, Zhang R. Characterization of polyphenols from green tea leaves using a high hydrostatic pressure extraction. Int J Pharm 2009; 382(1-2): 139-43.
[http://dx.doi.org/10.1016/j.ijpharm.2009.08.023] [PMID: 19715745]

[132] Hu QH, Jiang M, Zhu JC. Research on technology of extraction of tea caffeine and polyphenols. Nat Prod Res Dev Chin 1997; 9: 63-6.

[133] Ge YZ, Jin H. New methods for extraction of tea polyphenols. Chin Herb Med 1994; 25: 124-5.

[134] Sánchez-Moreno C, Plaza L, de Ancos B, Cano MP. Effect of combined treatments of high-pressure and natural additives on carotenoid extractability and antioxidant activity of tomato puree (*Lycopersicum esculentum* Mill.). Eur Food Res Technol 2004; 219(2): 151-60.
[http://dx.doi.org/10.1007/s00217-004-0926-1]

[135] de Ancos B, Sgroppo S, Plaza L, Cano MP. Possible nutritional and health-related value promotion in orange juice preserved by high-pressure treatment. J Sci Food Agric 2002; 82(8): 790-6.
[http://dx.doi.org/10.1002/jsfa.1093]

[136] Plaza L, Colina C, Ancos B, Sánchez-Moreno C, Pilar Cano M. Influence of ripening and astringency on carotenoid content of high-pressure treated persimmon fruit (*Diospyros kaki* L.). Food Chem 2012; 130(3): 591-7.

[http://dx.doi.org/10.1016/j.foodchem.2011.07.080]

[137] Plaza L, Sánchez-Moreno C, De Ancos B, Elez-Martínez P, Martín-Belloso O, Cano MP. Carotenoid and flavanone content during refrigerated storage of orange juice processed by high-pressure, pulsed electric fields and low pasteurization. Lebensm Wiss Technol 2011; 44(4): 834-9. [http://dx.doi.org/10.1016/j.lwt.2010.12.013]

[138] Lee D, Ghafoor K, Moon S, *et al.* Phenolic compounds and antioxidant properties of high hydrostatic pressure and conventionally treated ginseng (*Panax ginseng*) products. Qual Assur Saf Crops Foods 2015; 7(4): 493-500. [http://dx.doi.org/10.3920/QAS2014.0416]

[139] Chen R, Meng F, Zhang S, Liu Z. Effects of ultrahigh pressure extraction conditions on yields and antioxidant activity of ginsenoside from ginseng. Separ Purif Tech 2009; 66(2): 340-6. [http://dx.doi.org/10.1016/j.seppur.2008.12.026]

[140] Prasad KN, Yang B, Zhao M, Ruenroengklin N, Jiang Y. Application of ultrasonication or high-pressure extraction of flavonoids from litchi fruit pericarp. J Food Process Eng 2009; 32(6): 828-43. [http://dx.doi.org/10.1111/j.1745-4530.2008.00247.x]

[141] Prasad KN, Hao J, Shi J, *et al.* Antioxidant and anticancer activities of high pressure-assisted extract of longan (*Dimocarpus longan* Lour.) fruit pericarp. Innov Food Sci Emerg Technol 2009; 10(4): 413-9. [http://dx.doi.org/10.1016/j.ifset.2009.04.003]

[142] Yang B, Jiang Y, Wang R, Zhao M, Sun J. Ultra-high pressure treatment effects on polysaccharides and lignins of longan fruit pericarp. Food Chem 2009; 112(2): 428-31. [http://dx.doi.org/10.1016/j.foodchem.2008.05.097]

[143] Bi HM, Zhang SQ, Liu CJ, Wang CZ. High hydrostatic pressure extraction of salidroside from *Rhodiola sachalinensis*. J Food Process Eng 2009; 32(1): 53-63. [http://dx.doi.org/10.1111/j.1745-4530.2007.00202.x]

[144] Moreira SA, Silva S, Costa EM, Saraiva JA, Pintado M. Effect of high hydrostatic pressure extraction on biological activities of stinging nettle extracts. Food Funct 2020; 11(1): 921-31. [http://dx.doi.org/10.1039/C9FO02442E] [PMID: 31950959]

[145] Moreira SA, Silva S, Costa E, *et al.* Effect of High Hydrostatic Pressure Extraction on Biological Activities and Phenolics Composition of Winter Savory Leaf Extracts. Antioxidants 2020; 9(9): 841. [http://dx.doi.org/10.3390/antiox9090841] [PMID: 32911721]

[146] Pinela J, Prieto MA, Barros L, *et al.* Cold extraction of phenolic compounds from watercress by high hydrostatic pressure: Process modelling and optimization. Separ Purif Tech 2018; 192: 501-12. [http://dx.doi.org/10.1016/j.seppur.2017.10.007]

[147] Briones-Labarca V, Giovagnoli-Vicuña C, Cañas-Sarazúa R. Optimization of extraction yield, flavonoids and lycopene from tomato pulp by high hydrostatic pressure-assisted extraction. Food Chem 2019; 278: 751-9. [http://dx.doi.org/10.1016/j.foodchem.2018.11.106] [PMID: 30583438]

[148] Briones-Labarca V, Plaza-Morales M, Giovagnoli-Vicuña C, Jamett F. High hydrostatic pressure and ultrasound extractions of antioxidant compounds, sulforaphane and fatty acids from Chilean papaya (*Vasconcellea pubescens*) seeds: Effects of extraction conditions and methods. Lebensm Wiss Technol 2015; 60(1): 525-34. [http://dx.doi.org/10.1016/j.lwt.2014.07.057]

[149] Jamaludin R, Kim DS, Md Salleh L, Lim SB. Optimization of high hydrostatic pressure extraction of bioactive compounds from noni fruits. J Food Meas Charact 2020; 14(5): 2810-8. [http://dx.doi.org/10.1007/s11694-020-00526-w]

[150] Grassino AN, Pedisić S, Dragović-Uzelac V, Karlović S, Ježek D, Bosiljkov T. Insight into High-Hydrostatic Pressure Extraction of Polyphenols from Tomato Peel Waste. Plant Foods Hum Nutr 2020; 75(3): 427-33.

[http://dx.doi.org/10.1007/s11130-020-00831-1] [PMID: 32572675]

[151] Naghshineh M, Olsen K, Georgiou CA. Sustainable production of pectin from lime peel by high hydrostatic pressure treatment. Food Chem 2013; 136(2): 472-8. [http://dx.doi.org/10.1016/j.foodchem.2012.08.036] [PMID: 23122086]

[152] Fraeye I, Duvetter T, Doungla E, Van Loey A, Hendrickx M. Fine-tuning the properties of pectin–calcium gels by control of pectin fine structure, gel composition and environmental conditions. Trends Food Sci Technol 2010; 21(5): 219-28. [http://dx.doi.org/10.1016/j.tifs.2010.02.001]

[153] Willats WGT, Knox JP, Mikkelsen JD. Pectin: new insights into an old polymer are starting to gel. Trends Food Sci Technol 2006; 17(3): 97-104. [http://dx.doi.org/10.1016/j.tifs.2005.10.008]

[154] Yapo BM, Robert C, Etienne I, Wathelet B, Paquot M. Effect of extraction conditions on the yield, purity and surface properties of sugar beet pulp pectin extracts. Food Chem 2007; 100(4): 1356-64. [http://dx.doi.org/10.1016/j.foodchem.2005.12.012]

[155] Choi Y, Kim W, Lee JS, Youn SJ, Lee H, Baik MY. Enhanced Antioxidant Capacity of Puffed Turmeric (*Curcuma longa* L.) by High Hydrostatic Pressure Extraction (HHPE) of Bioactive Compounds. Foods 2020; 9(11): 1690. [http://dx.doi.org/10.3390/foods9111690] [PMID: 33218182]

[156] Briones-Labarca V, Giovagnoli-Vicuña C, Figueroa-Alvarez P, Quispe-Fuentes I, Pérez-Won M. Extraction of β-carotene, vitamin C and antioxidant compounds from *Physalis peruviana* (Cape Gooseberry) assisted by high hydrostatic pressure. Food Nutr Sci 2013; 4(8): 109-18. [http://dx.doi.org/10.4236/fns.2013.48A014]

[157] Campos FM, Ribeiro SMR, Della Lucia CM, Pinheiro-Sant'Ana HM, Stringheta PC. Optimization of methodology to analyze ascorbic and dehydroascorbic acid in vegetables. Quim Nova 2009; 32(1): 87-91. [http://dx.doi.org/10.1590/S0100-40422009000100017]

[158] Cascaes Teles AS, Hidalgo Chávez DW, Zarur Coelho MA, Rosenthal A, Fortes Gottschalk LM, Tonon RV. Combination of enzyme-assisted extraction and high hydrostatic pressure for phenolic compounds recovery from grape pomace. J Food Eng 2021; 288: 110128. [http://dx.doi.org/10.1016/j.jfoodeng.2020.110128]

[159] Altuner EM. A predictive modelling study for using high hydrostatic pressure, a food processing technology, for protein extraction. 9th International Conference on Predictive Modelling in Food 2016; 7: 121-4. [http://dx.doi.org/10.1016/j.profoo.2016.02.103]

[160] Shouqin Z, Jun X, Changzheng W. High hydrostatic pressure extraction of flavonoids from propolis. J Chem Technol Biotechnol 2005; 80(1): 50-4. [http://dx.doi.org/10.1002/jctb.1153]

[161] Dörnenburg H, Knorr D. Cellular permeabilization of cultured plant tissues by high electric field pulses or ultra high pressure for the recovery of secondary metabolites. Food Biotechnol 1993; 7(1): 35-48. [http://dx.doi.org/10.1080/08905439309549844]

[162] Yan H. Separation Engineering. Beijing: China Petrochemical Press 2002.

[163] Ahmed J, Ramaswamy HS. High pressure processing of fruits and vegetables. Stewart Postharvest Rev 2006; 2(1): 1-10. [http://dx.doi.org/10.2212/spr.2006.1.8]

[164] Chemat F, Vian MA, Cravotto G. Green extraction of natural products: concept and principles. Int J Mol Sci 2012; 13(7): 8615-27. [http://dx.doi.org/10.3390/ijms13078615] [PMID: 22942724]

SUBJECT INDEX

A

B

Merichel Plaza & Maria Luisa Marina (Eds.)

C

D

E

F

M

N

S

T

U

V

W

Made in United States
Orlando, FL
05 November 2026

90200479R10371